Confinement, Duality, and Nonperturbative Aspects of QCD

NATO ASI Series

Advanced Science Institutes Series

A series presenting the results of activities sponsored by the NATO Science Committee, which aims at the dissemination of advanced scientific and technological knowledge, with a view to strengthening links between scientific communities.

The series is published by an international board of publishers in conjunction with the NATO Scientific Affairs Division

A Life Sciences **B Physics**	Plenum Publishing Corporation New York and London
C Mathematical and Physical Sciences **D Behavioral and Social Sciences** **E Applied Sciences**	Kluwer Academic Publishers Dordrecht, Boston, and London
F Computer and Systems Sciences **G Ecological Sciences** **H Cell Biology** **I Global Environmental Change**	Springer-Verlag Berlin, Heidelberg, New York, London, Paris, Tokyo, Hong Kong, and Barcelona

PARTNERSHIP SUB-SERIES

1. Disarmament Technologies	Kluwer Academic Publishers
2. Environment	Springer-Verlag
3. High Technology	Kluwer Academic Publishers
4. Science and Technology Policy	Kluwer Academic Publishers
5. Computer Networking	Kluwer Academic Publishers

The Partnership Sub-Series Incorporates activities undertaken in collaboration with NATO's Cooperation Partners, the countries of the CIS and Central and Eastern Europe, in Priority Areas of concern to those countries.

Recent Volumes in this Series:

Volume 365 —Techniques and Concepts of High-Energy Physics IX
edited by Thomas Ferbel

Volume 366 — New Developments in Quantum Field Theory
edited by Poul Henrik Damgaard and Jerzy Jurkiewicz

Volume 367— Electron Kinetics and Applications of Glow Discharges
edited by U. Kortshagen and L. D. Tsendin

Volume 368 — Confinement, Duality, and Nonperturbative Aspects of QCD
edited by Pierre van Baal

Series B: Physics

Confinement, Duality, and Nonperturbative Aspects of QCD

Edited by

Pierre van Baal

Institute Lorentz
University of Leiden
Leiden, The Netherlands

KLUWER ACADEMIC PUBLISHERS
NEW YORK, BOSTON, DORDRECHT, LONDON, MOSCOW

eBook ISBN: 0-306-47056-X
Print ISBN: 0-306-45826-8

Created in the United States of America

PREFACE

Before you lies the proceedings of the NATO Advanced Study Institute/Newton Institute Workshop "Confinement, duality and non-perturbative aspects of QCD". The school covered the most important techniques to study Quantum Chromodynamics (QCD) and confinement, from lattice gauge theory, through Wilson's renormalisation group, to electromagnetic duality. The organising committee existed of: Ian Drummond (DAMTP, Cambridge), Mikhail Shifman (Minneapolis), Peter West (King's, London), and Pierre van Baal (Leiden), who acted as director of the school.

This summer school was the concluding activity of a six-month programme on "Non-perturbative Aspects of Quantum Field Theory" taking place at the Isaac Newton Institute for Mathematical Sciences in Cambridge, UK, which started in January 1997, organised by David Olive, Pierre van Baal, and Peter West. A large number of the lecturers also participated in the programme and a few programme participants were asked to present a seminar at the school. Not contained in these proceedings are the seminars by Peter Landshoff (DAMTP, Cambridge) on "The Pomeron" and Ludwig Faddeev (Steklov Math. Inst., St. Petersburg) on "Knot-like solitons in 3+1 dimensional field theory". In addition to the lectures and seminars there were two poster sessions at which participants presented their work. Authors and titles of these posters are listed on a separate page.

These proceedings address the longstanding question of understanding how quarks are confined within subnuclear particles. It covers aspects of hadron spectroscopy, the running of the strong coupling constant, and the Wilson renormalisation group, setting out the framework in which non-perturbative issues in QCD should be discussed. The renormalisation group is discussed in the context of the lattice, momentum-cutoff and light-front approaches. Instantons feature prominently in the context of a successful low-energy description for QCD. Some of these properties relevant to chiral symmetry breaking are universal and can be captured in random matrix models. Another important theme that is represented in these proceedings is the conjecture that monopoles condensate, such that QCD behaves as a dual superconductor. These proceedings address how the so-called abelian projection introduced fifteen years ago by 't Hooft, might provide an explicit scenario to test this dual superconductor picture. The recent Seiberg-Witten duality results add a new dimension to this conjecture. Supersymmetry has developed into an important tool to learn more about non-perturbative aspects of field theories. In the presence of supersymmetry, both instantons and monopoles will contribute in very special ways, revealing deep results relevant for the dynamics of the theory. It is hoped that some of these lessons are relevant to QCD.

This school was unique in bringing so many different approaches together, each expected to carry part of the solution towards the confinement problem. Any student that wishes to make progress on the confinement problem should be aware of these approaches. The lectures are ordered as much as possible so as to assist the reader in acquiring the necessary background, assumed known in some of the lectures.

This school would not have taken place without the generous financial support of NATO. I also thank the Newton Institute for financial and administrative support, in particular its conference secretary, Heather Dawson who has been of much help. Also I would like to thank DAMTP and in particular David Harris to provide computer access for the participants during the school. I am grateful to all the lecturers for their efforts to make this school and the proceedings a successful one. Last, but not least, I thank all students for their enthusiastic participation in this summer school.

Pierre van Baal

Lecturers

Adriano Di Giacomo - INFN, Pisa, Italy
Ludwig Faddeev - St.Petersburg, Russia
Peter Hasenfratz - Bern, Switzerland
Gerard 't Hooft - Utrecht, The Netherlands
Richard Kenway - Edinburgh, UK
Peter Landshoff - DAMTP, Cambridge, UK
Peter Lepage - Cornell Univ., US
Chris Michael - Liverpool, UK
Robert Perry - Ohio State Univ., US
Mikhail Polikarpov - ITEP, Moscow, Russia
Adam Schwimmer - Weizmann Institute, Israel
Mikhail Shifman - Minneapolis, US
Edward Shuryak - Stony Brook, US
Tsuneo Suzuki - Kanazawa, Japan
Michael Teper - Oxford, UK
Pierre van Baal - Leiden, The Netherlands
Jac Verbaarschot - Stony Brook, US
Peter Weisz - MPI, Munich, Germany
Peter West - King's, London, UK
Christof Wetterich - Heidelberg, Germany
Daniel Zwanziger - NYU, New York, US

Students

Gert Aarts - Utrecht, The Netherlands
Tom Albert - Bonn, Germany
Avetis Avakyan - Yerevan, Armenia
Zoltan Bajnok - Budapest, Hungary
Massimiliano Baldicchi - Milano, Italy
Silas Beane - Maryland, US
Roberto Begliuomini - Trento, Italy
Andree Blotz - Los Alamos, US
Martina Brisudova - Los Alamos, US
Stephane Bronoff - Marseille, France
Boris Chibisov - Minneapolis, US
Attilio Cucchieri - Bielefeld, Germany
Aldo Deandrea - Marseille, France
Luigi Del Debbio - Marseille, France
Massimo Di Pierro - Southampton, UK
Frank Ferrari - LPTENS, Paris, France
Cesar Fosco - Bariloche, Argentina
Martyn Foster - Liverpool, UK
Amit Ghosh - Calcutta, India
Harald Griesshamer - Erlangen, Germany
Elena Gubankova - Heidelberg, Germany
Miklos Adam Halasz - Stony Brook, US
Anthony Hams - Groningen, The Netherlands
Alistair Hart - Louisiana, US
Ronald Horgan - Cambridge, UK
Edmund Iancu - Saclay, France
Alfonso Jaramillo - Valencia, Spain
Dirk Jungnickel - Heidelberg, Germany
Seikou Kato - Kanazawa, Japan
Arjan Keurentjes - Leiden, The Netherlands
Elyakum Klepfish - King's, London, UK
Stefano Kovacs - Tor Vergata, Rome, Italy
Markus Leibundgut - Bern, Switzerland
Maxim Libanov - INR, Moscow, Russia
David Lin - Edinburgh, UK
Daniel Litim - Imperial, UK
Carlos Lozano - Santiago de Compostela, Spain
Jani Lukkarinen - Helsinki, Finland
Thomas Manke - Cambridge, UK
Fotini Markopoulou - Imperial, UK
Manu Mathur - Pisa, Italy
Christopher Maynard - Edinburgh, UK
Shiraz Minwalla - Princeton, US
Vapharsh Mkhitaryan - Yerevan, Armenia
Leszek Motyka - Krakow, Poland
Guido Mueller - Bonn, Germany
Avijit Mukherjee - Brandeis, US
Shinsuke Nishigaki - NBI, Copenhagen, Denmark
Thomas Pause - Regensburg, Germany
Mike Peardon - Kentucky, US
Petrus Pennanen - Helsinki, Finland
Adam Ritz - Imperial, UK
Joao Rodrigues - Lisbon, Portugal
Sinead Ryan - Fermilab, US
Ricardo Schiappa - MIT, Boston, US
Frederik Scholtz - Stellenbosch, South Africa
Myckola Schwetz - Yale, US
Konstantin Selivanov - ITEP, Moscow, Russia
Melih Sener - Stony Brook, US
Sergei Shabanov - FU Berlin, Germany
Peter Skala - Vienna, Austria
Matthew Slater - Durham, UK
Corneliu Sochichiu - JINR, Dubna, Russia
Mikhail Stephanov - Urbana-Champaign, US
Kazunori Takenaga - Kobe, Japan
Sergey Troitsky - INR, Moscow, Russia
Tanmay Vachaspati - Case Western, US
Federica Vian - Parma, Italy
Thomas Waindzoch - Darmstadt, Germany
Axel Weber - Mexico, Mexico
Pawel Wegrzyn - Krakow, Poland
Maxim Zabzine - St.Petersburg, Russia
Martin Zach - Vienna, Austria

List of Posters

- *Attilio Cucchieri* - Infrared behavior and Gribov noise for gluon and ghost propagators in minimal Landau gauge
- *Cesar Fosco* - On bosonization in higher dimensions
- *Amit Ghosh* - Understanding the area proposal for extremal black hole entropy
- *Elena Gubankova* - Modified similarity renormalization
- *Adam Halasz* - Higher order level statistics in lattice QCD spectra
- *Alistair Hart* - Ehrenfest theorems for field strength and electric current in Abelian projected SU(2) gauge theory
- *Ronald Horgan* - The nature of the continuum limit in 2D-RP(n) models
- *Alfonso Jaramillo* - Confinement through a local vacuum wave functional
- *Seikou Kato* - Various representations of infrared effective lattice QCD
- *Jani Lukkarinen* - Lattice simulations of the microcanonical ensemble
- *Manu Mathur* - Magnetic monopoles, gauge invariant dynamical variables and the Georgi-Glashow model
- *Myckola Schwetz* - Softly Broken SQCD
- *Peter Skala* - Confinement and colour magnetic currents in QCD
- *Matthew Slater*- One-instanton tests of the exact results in N=2 supersymmetric QCD
- *Corneliu Sochichiu* - On the connection between symplectic form and commutator anomaly for chiral SU(N) Yang-Milss models
- *Tanmay Vachaspati* - The dual standard model
- *Federica Vian* - How to compute the chiral anomaly without breaking global chiral symmetry
- *Thomas Waindzoch* - Collective coordinate description of soliton dynamics
- *Axel Weber* - The heavy quark potential from Wilson's exact renormalization group
- *Maxim Zabzine* - Zamolodchikov's C-theorem and phase transitions
- *Martin Zach* - Flux tubes in dually transformed U(1) lattice gauge theory - Do they attract or repel each other?

CONTENTS

HADRONIC PHYSICS FROM THE LATTICE

Chris Michael

Theoretical Physics Division, Department of Mathematical Sciences
University of Liverpool, L69 3BX, UK

INTRODUCTION

Topics covered in this review arc the lattice gauge theory approach to the evaluation of non-perturbative hadronic interactions from first principles, particularly applications to glueballs, inter-quark potentials, the running coupling constant and hybrid mesons. Also discussed are the limitations of the quenched approximation.

The theory of the strong interactions is accurately provided by Quantum Chromodynamics (QCD). The theory is defined in terms of elementary components: quarks and gluons. The only free parameters are the quark mass values (apart from an overall energy scale imposed by the need to regulate the theory). This formulation is essentially the unique candidate for the theory of the strong interactions. The only feasible way to describe a departure from QCD would be in terms of quark and gluon substructure. At least at the energy scales up to which it is tested, there is no evidence for such substructure.

QCD provides a big challenge to theoretical physicists. It is defined in terms of quarks and gluons but the physical particles are composites: the mesons and baryons. Any complete description must then yield these bound states: this requires a non-perturbative approach. One can see the limitations of a perturbative approach by considering the vacuum: this will be approximated in perturbation theory as basically empty with rare quark or gluon loop fluctuations. Such a description will allow quarks and gluons to propagate essentially freely which is not the case experimentally. The true (non-perturbative) vacuum can be better thought of as a disordered medium with whirlpools of colour on different scales. Such a non-perturbative treatment then has the possibility to explain why quarks and gluons do not propagate (i.e. quark confinement). Furthermore it will be able to explore other situations than experiment — different quark masses, etc. My notation is that there are N_f flavours of light quark with N colours.

Any mathematically precise description of QCD must introduce some regularisation of the ultra-violet divergences. Any such regularisation may spoil the symmetries, for example 4-dimensional Lorentz invariance is broken by dimensional regularisation. Introducing a space time lattice likewise breaks Lorentz invariance. Using a space time lattice with exact gauge invariance retained proves to be a very successful regularisation.

Confinement, Duality, and Nonperturbative Aspects of QCD
Edited by Pierre van Baal, Plenum Press, New York, 1998

In the continuum limit, as the lattice spacing is reduced to zero, Lorentz invariance will be found to be restored.

For a general review of lattice gauge theory, see standard textbooks[1]. Here I give a brief introduction to the salient points. The simplest discretisation of space-time is that introduced by Wilson[2]. Space-time is replaced by a discrete grid (the lattice) but gauge invariance is retained exactly. Then gluonic colour fields which relate the colour coordinate system at different space time points will be represented as links on this lattice. Thus a link of the lattice from x to $x + a$ in the μ-direction (with unit vector e_μ) is represented by a $N \times N$ colour matrix which can be thought of as the path-ordered exponential of the continuum colour fields $A_\mu(x)$,

$$U_\mu(x) \approx e^{igaA_\mu(x+\frac{1}{2}ae_\mu)}. \qquad (1)$$

From these link matrices, the simplest non-trivial gauge invariant is the trace of the path-ordered product of links around a unit square: the plaquette.

$$S_\Box = \frac{1}{N}\mathrm{ReTr}(\prod U). \qquad (2)$$

Note that, with this definition, a cold lattice with $A \approx 0$ will have all link matrices $U \approx 1$ and hence have $S_\Box \approx 1$, while a hot lattice with random U will have $S_\Box \approx 0$ This gauge invariant $S_\Box$, when summed over all space-time, is the simplest candidate for the gluonic component of the action: the Wilson gauge action. This follows since naively (i.e. neglecting quantum fluctuations) as $a \to 0$

$$\beta \sum_{\text{hypercube}} S_\Box - 1 \to -\frac{1}{4}\int F_{\mu\nu}F_{\mu\nu}\, d^4x \qquad (3)$$

where $\beta = 2N/g^2$

Using periodic boundary conditions in space and time, the system will have a finite number of degrees of freedom: the gluon and quark fields at the lattice sites. This finite number of degrees of freedom implies that the theory is a quantum many-body problem rather than a field theory. The step which makes this many-body problem tractable is to consider Euclidean time. With the Euclidean time approach, the formulation of QCD (consider, for example, the functional integral over the gauge fields) is converted into a multiple integral which is well defined mathematically and has a positive definite integrand:

$$Z = \int DU e^{\beta \sum S_\Box}, \qquad (4)$$

where the integral is over the group manifold of $SU(N)$ of colour for each link matrix U. Because of gauge invariance not all links are independent, but integrating over the dependent links only introduces a finite constant which is irrelevant. For a lattice of L^4 sites with a colour gauge group of SU(3), this would be a $8 \times 4 \times L^4$ dimensional integral (8 from the colour group manifold, $4L^4$ from the gauge fields on each link). For any reasonable value of L, this is a very high dimension indeed. Simpson's rule is not the way forward! Because the integrand is positive definite, the standard approach is to use a Monte Carlo approximation to the integrand. This is implemented in an 'importance sampling' version so that a stochastic estimate of the integral is made from a finite number of samples (called configurations) of equal weight. The construction of efficient algorithms to achieve this is a topic in itself. Here I will concentrate on the analysis of the outcome, assuming that such configurations have been generated.

So what is at our disposal is a set of samples of the vacuum. It is then straightforward to evaluate the average of various products of fields over these samples — this gives the Green functions by definition. The Green function can then be continued from Euclidean to Minkowski time (in most cases this is trivial) and compared to experiment. Thus masses and matrix elements can be evaluated readily. Scattering, hadronic decays, real-time processes etc will not be directly accessible.

A simple example of a lattice measurement is that of a Wilson loop. This is the expectation value in the vacuum of the path ordered product of links around a closed loop. For example, a rectangular loop of size $R \times T$ can be used to extract the potential between heavy quarks and thus to explore confinement.

So far I have concentrated on the gluonic degrees of freedom of QCD, where an elegant formulation was available. In contrast, the inclusion of quarks in a lattice approach is very inelegant and computationally challenging. The quarks will be represented by Grassmannian variables ψ and the fermionic action for each species of quark will be bilinear $\bar{\psi}Q(m,U)\psi$ where Q is the fermionic matrix for a quark of mass m in the gauge fields $U(x)$, namely $\not{D} + m$ in the continuum. Various discretisations of this have been proposed. They all have the feature that each fermionic species needs to be at least doubled in a local lattice formulation. The Wilson fermionic approach kills the doublers at the expense of breaking chiral symmetry while the staggered (Kogut-Susskind) approach retains chiral symmetry but mixes space, flavour and colour degrees of freedom. Both discretisations are expected to yield the same continuum result as $a \to 0$.

The treatment of these discretised fermions is still difficult. The fermionic contribution to the action is not positive definite so straightforward Monte Carlo methods are excluded. The usual approach is to exploit the fact that the action is bilinear to explicitly integrate out the N_f Grassmannian fields leaving an effective action in terms of the gauge fields:

$$S = \sum S_{\Box} + N_f \mathrm{Tr} \log[Q(m,U)] \tag{5}$$

Algorithms to deal with the non-local trace-log term exist but they are very computationally intensive. It is frustrating that adding quark degrees of freedom results in a computational increase by a factor of 1000 or more. This is the reason that the so-called *quenched* approximation is so popular. Here the limit $N_f = 0$ is taken in constructing vacuum samples — i.e. just pure gluouic QCD. Then quarks can be propagated in this gluonic vacuum by solving the lattice Dirac equation $(Q(m,U)\psi = 0)$. This approach is not unitary — the quarks do not feel any back reaction from quark pairs in the vacuum. However, it appears to be a rather good approximation for many purposes.

Going beyond the quenched approximation, most approaches use 2 flavours of equal mass quarks to give a satisfactory algorithm and then vary the quark mass. The limit of large quark mass is of course just the quenched approximation since heavy quark loops are suppressed.

The validation of the lattice approach calls for a series of checks that everything is under control.

- the lattice spacing should be small enough (discretisation errors)
- the lattice must be big enough in space and time (finite size errors)
- the statistical errors must be under control
- Green functions must be extracted with no contamination (eg a ground state mass could be contaminated with a piece coming from an excited state)
- both the quark contribution to the vacuum (sea quarks) and the quark constituents of hadrons (valence quarks) are usually treated by using larger mass

values than the experimental ones and then extrapolating. This extrapolation must be treated accurately.

The most subtle of these is the discretisation error. In order to extract the continuum limit of the lattice, one must show that the physical results will not change if the lattice spacing is decreased further. This is subtle because the lattice spacing is not known directly — in effect, it is measured. I first discuss what is expected on general grounds to be the appropriate way to achieve small a. The lattice simulation is undertaken at a value of a parameter conventionally called β — see eq(4). In the limit of small coupling g^2, where perturbation theory applies, $\beta = 2N/g^2$. Thus large β corresponds to small g^2. Now, perturbatively for $N = 3$ colours, the coupling g^2 corresponds to the lattice spacing a as

$$g^2 \approx \frac{1}{b_0 \log a^{-2}} \quad \text{where} \quad b_0 = \frac{11 - \frac{2}{3}N_f}{16\pi^2} \tag{6}$$

for N_f flavours of quarks. For the case of interest, $N_f \leq 3$, this corresponds to small values of g^2 at small distance scale a — as expected from asymptotic freedom.

The perturbative argument is appropriate to the study of results at large β (small g^2). The lattice simulation of QCD uses values of $\beta \approx 6$ and hence bare couplings corresponding to $\alpha = g^2/(4\pi) \approx 0.08$. In the pioneering years of lattice work, this was thought to be a sufficiently small number that the perturbation series would converge rapidly. One of the major advances, in recent years, has been the realisation that the bare lattice coupling (our g^2 above) is a very poor expansion parameter and the perturbation series in the bare coupling does not converge well at the β values of interest. The theoretical explanation for this poor convergence is that the lattice action differs from the continuum action and allows extra interactions. These include tadpole diagrams[3] which have the property that they sum up to give a contribution that involves high order terms in the perturbation series. The way to avoid this problem with tadpole terms is to use a perturbation series in terms of a renormalised coupling — rather than the bare lattice coupling. I return to this topic when discussing the lattice determination of the QCD coupling α later.

This change of attitude to the method of determining a from β has had considerable implications for lattice predictions, as I now explain, since the aim is to work in a region of lattice spacing a where perturbation theory in the bare coupling is not precise. Because of this, in practice, a is determined from the non-perturbative lattice results themselves. Thus if the energy of some particle is measured on the lattice, it will be available as the dimensionless combination $\hat{E}$. From a value for E in physical units, then a can be determined since, on dimensional grounds, $\hat{E} = Ea$. Furthermore, by increasing β, the change in the observable $\hat{E}$ gives information about the change in a since E is fixed — assuming it is the physical mass. This should allow a calibration of β in terms of a to be established.

This procedure is overly optimistic, however. The discretised lattice theory is different from the continuum theory on scales of the order of the lattice spacing a. For the Wilson action formulation of gauge theory, this implies that the continuum energy E_c is related to the lattice observable $\hat{E}(\beta) = Ea$ as

$$\frac{\hat{E}(\beta)}{a} = E_c + O(a^2) \tag{7}$$

A direct consequence is that the ratio of two energies (of different particles, for example) will have discretisation errors of order a^2.

Thus, to cope with discretisation errors, the procedure required is to evaluate dimensionless ratios of quantities of physical interest at a range of values of the lattice spacing a and then extrapolate the ratio to the continuum limit $(a \to 0)$.

Note that when the fermionic terms are included, the discretisation error is of order a for the Wilson fermionic action. By adding further terms in the fermionic action, the error can be reduced — to order αa for the SW-clover fermion formulation.

GLUEBALL MASSES

I choose to illustrate the workings of the lattice method by describing the determination of the glueball spectrum. Of course, glueballs are only defined unambiguously in the quenched approximation — where quark loops in the vacuum are ignored. In this approximation, glueballs are stable and do not mix with quark-antiquark mesons. This approximation is very easy to implement in lattice studies: the full gluonic action is used but no quark terms are included. This corresponds to a full non-perturbative treatment of the gluonic degrees of freedom in the vacuum. Such a treatment goes much further than models such as the bag model.

The glueball mass can be measured on a lattice through evaluating the correlation $C(t)$ of two closed colour loops (called Wilson loops) at separation t lattice spacings. Formally

$$C(t) = \langle 0|G(0)G^{\dagger}(t)|0\rangle = \sum_{i=0} c_i \langle g_i(0)|g_i(t)\rangle c_i^{*} = \sum_{i=0} |c_i|^2 e^{-\hat{m}_i t} \tag{8}$$

where G represents the closed colour loop which can be thought of as creating a glueball state g_i from the vacuum. Summing over a complete set of such glueball states (strictly these are eigenstates of the lattice transfer matrix where $\exp -\hat{m}_i$ is the lattice eigenvalue corresponding to a step of one lattice spacing in time) then yields the above expression. As $t \to \infty$, the lightest glueball mass will dominate. This can be expressed as

$$\hat{m}_0 = \lim_{t\to\infty} \hat{m}_{\text{eff}}(t) \quad \text{where} \quad \hat{m}_{\text{eff}}(t) = \log\left(\frac{C(t-1)}{C(t)}\right) \tag{9}$$

Note that since for the excited states $\hat{m}_i > \hat{m}_0$, then $\hat{m}_{\text{eff}}(t) > \hat{m}_{\text{eff}}(t+1) > \hat{m}_0$ This implies that the effective mass, defined above, is an upper bound on the ground state mass. In practice, sophisticated methods are used to choose loops G such that the correlation $C(t)$ is dominated by the ground state glueball (i.e. to ensure $|c_0| >> |c_i|$). By using several different loops, a variational method can be used to achieve this effectively. These techniques are needed to obtain accurate estimates of $\hat{m}_0$ from modest values of t since the signal to noise decreases as t is increased. Even so, it is worth keeping in mind that upper limits on the ground state mass are obtained in principle.

The method also needs to be tuned to take account of the many glueballs: with different J^{PC} and different momenta. On the lattice the Lorentz symmetry is reduced to that of a hypercube. Non-zero momentum sates can be created (momentum is discrete in units of $2\pi/L$ where L is the lattice spatial size). The usual relationship between energy and momentum is found for sufficiently small lattice spacing. Here I shall concentrate on the simplest case of zero momentum (obtained by summing the correlations over the whole spatial volume). Charge conjugation C is a good quantum number in lattice studies of glueballs: C interchanges the direction of all links.

For a state at rest, the rotational symmetry becomes a cubic symmetry. The lattice states (the g_i above) will transform under irreducible representations of this cubic symmetry group (called O_h). These irreducible representations can be linked to the representations of the full rotation group SU(2). Thus, for example, the five spin components of a $J^{PC} = 2^{++}$ state should be appear as the two-dimensional $\mathbf{E}^{++}$ and thethree-dimensional T_2^{++} representations on the lattice, with degenerate masses. This degeneracy requirement then provides a test for the restoration of rotational invariance — which is expected to occur at sufficiently small lattice spacing.

The results of lattice measurements[4, 5, 6, 7] of the $0^{++}$and 2^{++} states are shown in fig. 1. The restoration of rotational invariance is shown by the degeneracy of the two O_h representations that make up the 2^{++} state. Fig. 1 shows the dimensionless combination of the lattice glueball mass $\hat{m}_0$ to a lattice quantity $\hat{r}_0$. I will return to describe the lattice determination of $\hat{r}_0$ in more detail — here it suffices to accept it as a well measured quantity on the lattice that can be used to calibrate the lattice spacing and so explore the continuum limit. The quantity plotted, $\hat{m}_0\hat{r}_0$, is expected to be equal to the product of continuum quantities $m_0 r_0$ up to corrections of order a^2. This is indeed seen to be the case. The extrapolation to the continuum limit $(a \to 0)$ can then be made with confidence. This is an important result: continuum quenched QCD does have massive states and their properties are determined.

The only other candidate for a relatively light glueball is the pseudoscalar. Values quoted of $\hat{r}_0 m(0^{-+}) = 5.6(6)$, 7.1(1.1) and 5.3(6) from refs(5, 6) suggest an average of 6.0(1.0), not appreciably lighter than the tensor glueball. This is confirmed by preliminary results from the group of ref(10) that the pseudoscalar is heavier than the tensor glueball.

The value of r_0 in physical units is about 0.5 fm and I will adopt a scale equivalent to $r_0^{-1} = 0.372$ GeV with a 10% systematic error on this scale since different physical observables differ from the quenched approximation values by this amount. Then the masses of lightest glueballs in the continuum limit are $m(0^{++}) = 1611(30)(160)$ MeV; $m(2^{++}) = 2232(220)(220)$ MeV and $m(0^{-+}) = 2232(370)(220)$ MeV, where the second error is the overall scale error.

Recently a lattice approach using a large spatial lattice spacing with an improved action and a small time spacing has been used to study glueball masses. The results [10] in the continuum limit are that $r_0 m(0^{++}) = 3.98(15)$, $r_0 m(2^{++}) = 5.85(2)$, $r_0 m(1^{+-}) = 7.21(2)$ and $r_0 m'(2^{++}) = 8.11(4)$. There remains a small discrepancy with the result for the 0^{++} glueball obtained above $(r_0 m(0^{++}) = 4.33(5))$ from lattice spacings much closer to the continuum limit. When this is fully understood, the new method looks to be very promising for access to excited glueball masses.

The predictions for the other J^{PC} states are that they lie higher in mass and the present state of knowledge is summarised in fig. 2. Remember that the lattice results are strictly upper limits. For the J^{PC} values not shown, these upper limits are too weak to be of use.

Since quark-antiquark mesons can only have certain J^{PC} values, it is of special interest to look for glueballs with J^{PC} values not allowed for such mesons: 0^{--}, 0^{+-}, 1^{-+}, 2^{+-}, etc. Such spin-exotic states, often called 'oddballs', would not mix directly with quark-antiquark mesons. This would make them a very clear experimental signal of the underlying glue dynamics. Various glueball models (bag models, flux tube models, QCD sum-rule inspired models,..) gave different predictions for the presence of such oddballs (eg. 1^{-+}) at relatively low masses. The lattice mass spectra clarify these uncertainties but, unfortunately for experimentalists, do not indicate any low-lying oddball candidates. The lightest candidate is from the T_2^{+-} spin combination. Such a state could correspond to an 2^{+-} oddball. Another interpretation is also possible, however, namely that anon-exotic 3^{+-} state is responsible (this choice of interpretation can be resolved in principle by finding the degenerate 5 or 7 states of a $J = 2$ or 3

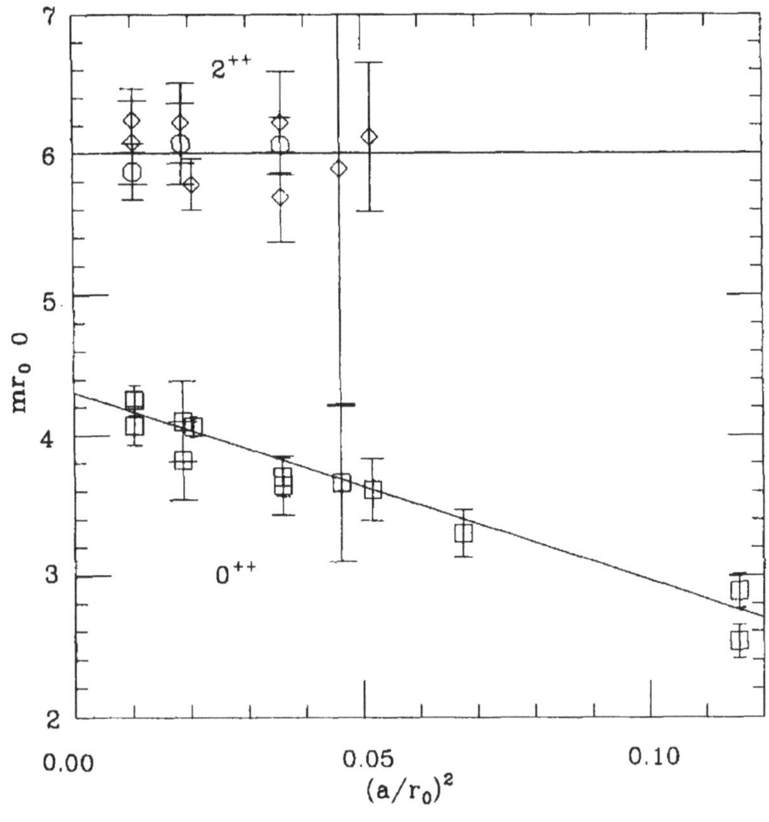

Figure 1. The value of mass of the $J^{PC} = 0^{++}$ and 2^{++} glueball states from refs(4, 5, 6, 7) in units of r_0. The T_2 and E representations are shown by octagons and diamonds respectively. The straight line shows a fit describing the approach to the continuum limit as $a \to 0$.

meson). The overall conclusion at present is that there is no evidence for any oddballs of mass less than 3 GeV.

Returning briefly to the independence of the results on the volume of the lattice, in the early days of glueball mass determination, it was expected that a spatial size L should satisfy $\hat{m}(0^{++})L > 1$ and, hence, that values of $\hat{m}(0^{++})L$ of 1 to 4 would suffice. A careful lattice study[11] showed that $\hat{m}(0^{++})L > 8$ was required to obtain rotational invariance and a result independent of L. The results collected in fig. 1 all satisfy this latter inequality so can be regarded as the infinite volume determination.

The small volume lattice results illustrate several points. One important feature is that closed loops of colour flux acting on the vacuum create glueballs but, if the loop encircles the periodic spatial boundary of length L, a torelon state can be created with energy given approximately by KL where K is the string tension. Thus at small L, the torelon state will be lighter than a glueball. In the quenched approximation, there is a $Z(N)$ symmetry of the lattice action which puts glueballs and torelons in different representations so they are unmixed. The fermion term in the action, however, breaks this symmetry so that for $N_f \neq 0$, the lightest glueball will be very heavily contaminated by torelon-like contributions. In a semi-analytic study[8] this was explored and what is clear is that in a small volume the limit $N_f \to 0$ of full QCD is *not* equivalent to the quenched approximation with $N_f = 0$.

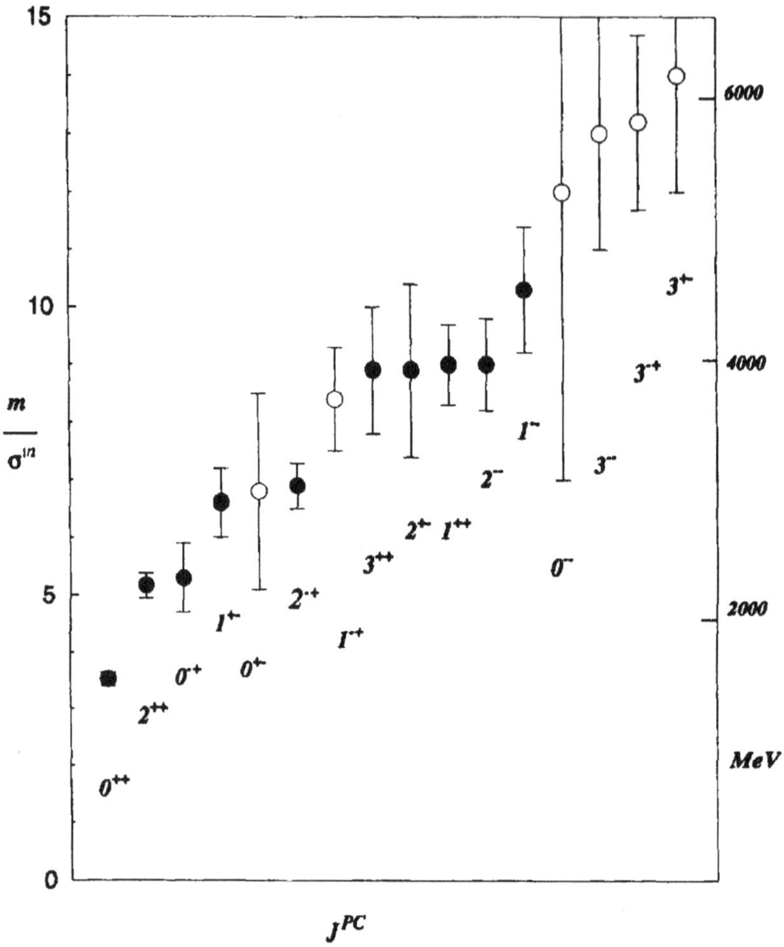

Figure 2. The mass of the glueball states with quantum numbers J^{PC} from ref(6). The scale is set by $\sqrt{\sigma} \approx 0.44$ GeV which yields the right hand scale in GeV. The solid points represent mass determinations whereas the open points are upper limits.

Glueballs are defined in the quenched approximation and, hence, they do not decay into mesons since that would require quark-antiquark creation. It is, nevertheless, still possible to estimate the strength of the matrix element between a glueball and a pair of mesons within the quenched approximation. For the glueball to be a relatively narrow state, this matrix element must be small. A very preliminary attempt has been made to estimate the size of the coupling of the 0^{++} glueball to two pseudoscalar mesons[9]. A relatively small value of order 100 MeV is found. Further work needs to be done to investigate this in more detail, in particular to study the mixing between the glueball and 0^{++} mesons since this mixing may be an important factor in the decay process.

Another lattice study will become feasible soon. This is to study the glueball spectrum in full QCD vacua with sea quarks of mass m_{sea}. For large m_{sea}, the result is just the quenched result described above. For m_{sea} equal to the experimental light quark masses, the results should just reproduce the experimental meson spectrum — with the resultant uncertainty between glueball interpretations and other interpretations. The lattice enables these uncertainties to be resolved in principle: one obtains the spectrum for a range of values of m_{sea} between these limiting cases, so mapping glueball states at large m_{sea} to the experimental spectrum at light m_{sea}. Studies conducted so far show no significant change of the glueball spectrum as dynamical quark effects are added — but the sea quark masses used are still rather large[12].

From the point of view of comparing quenched lattice results with models, a very useful system to study is the *gluelump.* This is the hydrogen atom of gluonic QCD: a system with one very heavy gluon which is treated as a static adjoint colour source and the surrounding gluonic field that makes a colour singlet hadron. In terms of experiment: this is the gluinoball formed from a gluino-gluon bound state which will be observable should massive gluinos exist and be sufficiently stable. The spectrum and spatial distribution of these states have been explored[13] for $SU(2)$ of colour. Because one colour source is fixed, the spatial size is easier to measure than for the glueballs themselves: the distributions were found to extend out to a radius of 0.5 fm. The ground state and first excited state were found to have quantum numbers consistent with being bound states of a magnetic gluon and an electric gluon respectively. For $SU(3)$ of colour the spectrum in the continuum limit has been determined[14] and the mass splitting between the two lightest of these states is found to be 350 MeV.

POTENTIALS BETWEEN QUARKS

A very straightforward quantity to determine from lattice simulation is the interquark potential in the limit of very heavy quarks (static limit). This potential is of direct physical interest because solving the Schrödinger equation in such a potential provides a good approximation to the Υ spectrum. It is also relevant to exploring both confinement and asymptotic freedom on a lattice.

The basic route to the static potential is to evaluate the average $W(\hat{R},\hat{T})$ in the vacuum samples of a rectangular closed loop of colour flux (a Wilson loop of size $\hat{R}\times\hat{T}$). This can be visualised as involving static sources R apart with potential energy $V(R)$ for time T so that $W \approx ce^{-V(R)T}$. More precisely, the lattice quantities $\hat{R}$ and $\hat{T}$ are related to the physical distances R and T by $\hat{R} = R/a$, etc where a is the lattice spacing which is not known explicitly. Then it can be shown that the required static potential in lattice units is given by

$$\hat{V}(\hat{R}) = \lim_{\hat{T}\to\infty} \log \frac{W(\hat{R},\hat{T}-1)}{W(\hat{R},\hat{T})} \tag{10}$$

The limit of large T is needed to separate the required potential from excited potentials. This limit can be made tractable in practice by using more complicated loops than the simple rectangular loop described above. The motivation for this is to generalise the straight paths of length R at $t = 0$ and T by considering sums over paths that reflect more fully the colour flux between static quark and antiquark at separation R. Typically a smearing or fuzzing algorithm is used to create suitable wandering paths, then several such paths are combined in a variational approach to find the linear combination that best describes the ground state of the system: the potential $V(R)$.

A summary of results[15, 16] for the potential at large R is shown in figs 3, 4. The result that the force dV/dR tends to a constant at large R (and thus $V(R)$ continues to rise as R increases) is a manifestation of the confinement of heavy quarks (in the quenched approximation). The force appears to approach a constant at large R. A simple parametrisation is traditional in this field:

$$V(R) = V_0 - \frac{e}{R} + KR \tag{11}$$

where K (sometimes written σ) is the string tension. The term e/R is referred to as the Coulombic part in analogy to the electromagnetic case. The equivalent relationship in terms of quantities defined on a lattice is

$$\hat{V}(\hat{R}) = \hat{V}_0 - \frac{e}{\hat{R}} + \hat{K}\hat{R} \tag{12}$$

Since the string tension is given by the slope of $V(R)$ against R as $R \to \infty$, this implies that some error will arise in determining K coming from the extrapolation of lattice data at finite R. A practical resolution[17] is to define a value of R where the potential takes a certain form. The convention is to use r_0 where

$$\hat{R}^2 \frac{d\hat{V}(\hat{R})}{d\hat{R}}\bigg|_{\hat{r}_0} = 1.65 \tag{13}$$

Thus $\hat{r}_0$ can be determined by interpolation in $\hat{R}$ rather than extrapolation. In practice, this means that $\hat{r}_0$ is very accurately determined by lattice measurements and so is a useful quantity to use to set the scale since $\hat{r}_0 = r_0/a$. With the simple parametrisation above, $r_0^2 = (1.65 - e)/K$ so r_0 is closely related to the string tension since $e \approx 0.25$. The string tension is usually taken from experiment as $\sqrt{K} = 0.44\text{GeV}$ where the value comes from $c\bar{c}$ and $b\bar{b}$ spectroscopy and from the light meson spectrum interpreted as

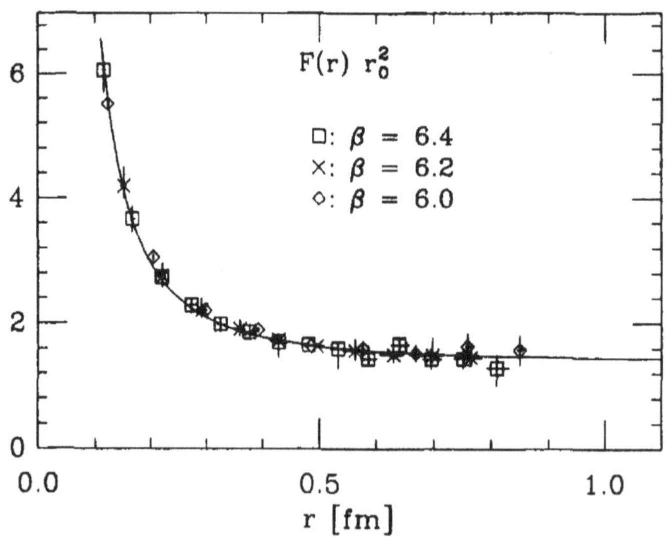

Figure 3. The force between static quarks at separation r as measured on lattices with different lattice spacing a (with the β-values shown).

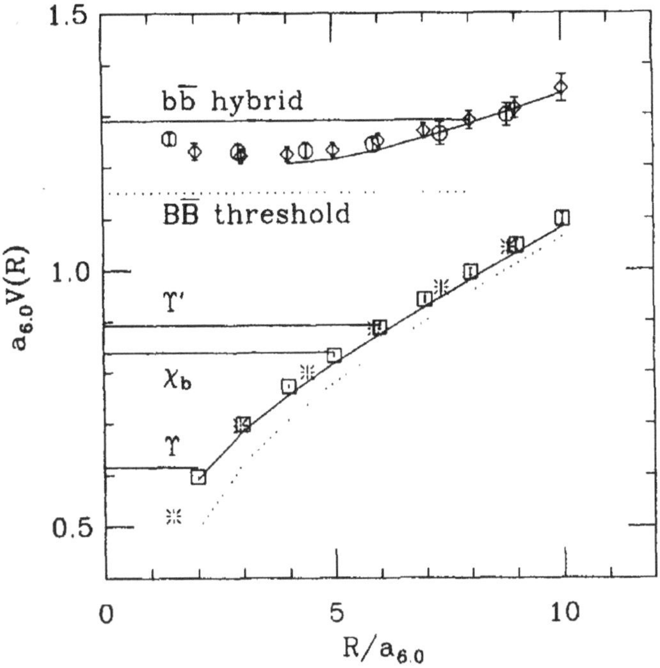

Figure 4. Potentials $V(R)$ between static quarks at separation R for the ground state (□ and *) and for the E_u symmetry which corresponds to the first excited state of the gluonic flux (octagons and diamonds). Results in lattice units ($a_{6.0}^{-1} = 2.02$ GeV) from the quenched calculations of ref(16) are shown by symbols corresponding to different lattice spacings. For the ground state potential the continuous curve is an interpolation of the lattice data while the dotted curve with enhanced Coulomb term fits the spectrum and yields the masses shown. The lightest hybrid level in the excited gluonic potential is also shown.

excitations of a relativistic string. Similar analyses also imply that $r_0 \approx 0.5$ fm. Here I use $r_0^{-1} = 0.372$ GeV to be specific. Since I shall be describing quenched lattice results, the energy scale set from different physical quantities will not necessarily agree (since *experiment* has full QCD not the quenched vacuum) and so a systematic error of order 10% must be applied to any such choice of scale. This was discussed when taking glueball mass values from quenched lattice calculations.

The lattice potential $V(R)$ can be used to determine the spectrum of $b\bar{b}$ mesons by solving Schrödinger's equation since the motion is reasonably approximated as non-relativistic. The lattice result is similar to the experimental Υ spectrum. The main difference is that the Coulombic part (e) is effectively too small (0.28 rather than 0.4). This produces[16] a ratio of mass differences $(1P - 1S)/(2S - 1S)$ of 0.71 to be compared with the experimental ratio of 0.78. This difference is understandable as a consequence of the Coulombic force at short distances which would be increased by $33/(33-2N_f)$ in perturbation theory in full QCD compared to quenched QCD. I will return to discuss this in the context of lattice dynamical fermion calculations[18].

Running coupling constant

At small R, the static potential can be used, in principle, to study the running coupling constant. Small R corresponds to large momentum and the coupling should decrease at small R. Thus the Coulombic coefficient e introduced above should actually decrease logarithmically as R decreases. Perturbation theory can be used to determine this behaviour of the potential at small R.

In the continuum the potential between static quarks is known perturbatively to two loops in terms of the scale $\Lambda_{\overline{MS}}$. For $SU(3)$ colour, the continuum force is given for $N_f = 0$ by[19]

$$\frac{dV}{dR} = \frac{4}{3}\frac{\alpha(R)}{R^2} \tag{14}$$

with the effective coupling $\alpha(R)$ given by

$$\frac{1}{4\pi[b_0 \log(R\Lambda_F)^{-2} + (b_1/b_0)\log\log(R\Lambda_F)^{-2}]} \tag{15}$$

where $b_0 = 11/16\pi^2$ and $b_1 = 102\ b_0^2/121$ are the usual coefficients in the perturbative expression for the β-function, neglecting quark loops in the vacuum. Here $\Lambda_F = 1.048\Lambda_{\overline{MS}}$.

This perturbative result can be used to *define* a running coupling constant non-perturbatively. Thus a coupling in the 'force' scheme can be defined by

$$\alpha_F(R) \equiv \frac{3}{4}R^2\frac{dV}{dR} \tag{16}$$

Different non-perturbative definitions of α can be related using perturbation theory, for example the one loop comparison of α_F and $\alpha_{\overline{MS}}$ gives the relationship between the respective Λ values quoted above.

On a lattice the force can be estimated by a finite difference and one can extract the running coupling constant by using[20]

$$\alpha_F(\frac{\hat{R}_1 + \hat{R}_2}{2}) = \frac{3}{4}\hat{R}_1\hat{R}_2\frac{\hat{V}(\hat{R}_1) - \hat{V}(\hat{R}_2)}{\hat{R}_1 - \hat{R}_2} \tag{17}$$

where the error in using a finite difference is negligible in practice.

This is plotted in fig. 5 versus $RK^{1/2}$: this combination is dimensionless and so can be determined from lattice results since $RK^{1/2} = \hat{R}\hat{K}^{1/2}$, where $\hat{K}$ is taken from the fit to $\hat{V}(\hat{R})$. The interpretation of α_F as defined above as an effective running coupling constant is only justified at small R where the perturbative expression dominates. Also shown are the two-loop perturbative results for $\alpha(R)$ for different values of Λ_F.

Fig. 5 clearly shows a *running* coupling constant. Moreover the result is consistent with the expected perturbative dependence on R at small R. There are systematic errors, however. At larger R, the perturbative two-loop expression will not be an accurate estimate of the measured potentials, while at smaller R, the lattice artefact corrections (which arise because $R/a \approx 1$) are relatively big. Setting the scale using $\sqrt{K} = 0.44$ GeV implies $1/a(\beta = 6.5) = 4.13$ GeV, so $R < 4a(6.5)$ corresponds to values of $1/R > 1$ GeV. This R-region is expected to be adequately described by perturbation theory. Other methods to extract a running coupling from lattice results have been used and some have smaller systematic errors. For a comparison see ref(22).

This determination from the interquark force of the coupling α allows us to compare the result with the bare lattice coupling determined from $\beta = 6/g^2$. At $\beta = 6.5$, $\alpha_{bare} = g^2/4\pi = 0.073$. The values of α shown in fig. 5 are much larger. The effective coupling constant is thus almost twice the bare coupling. This is quite acceptable in a renormalisable field theory. The message is that the bare coupling should be disregarded — it is not a good expansion parameter. The measured α_F, however, proves to be a reasonable expansion parameter in the sense that the first few terms of the

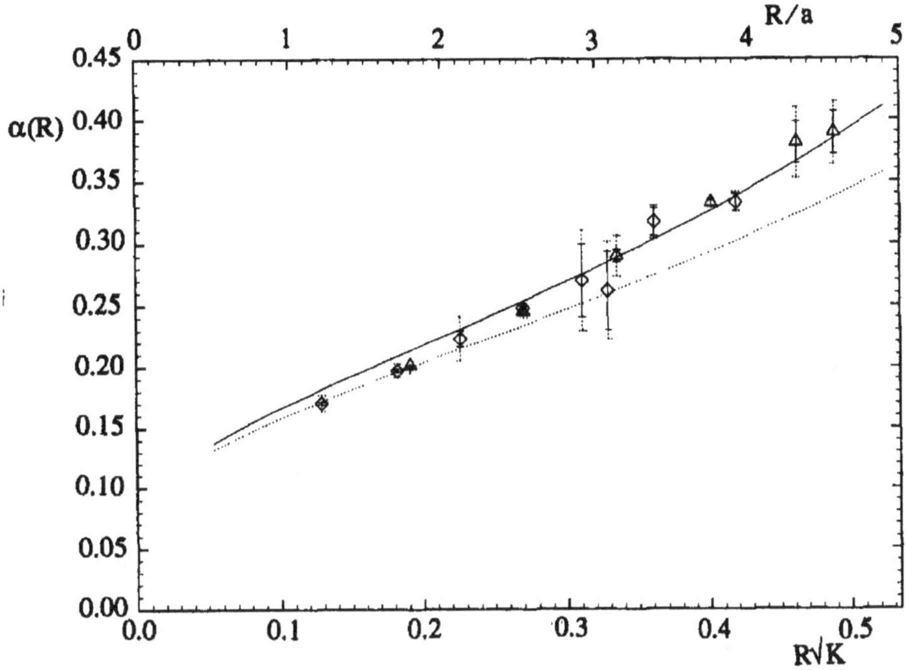

Figure 5. The effective running coupling constant $\alpha_F(R)$ obtained from the force between static quarks at separation R from ref(21). The scale is set by the string tension K. Data are at $\beta = 6.5$ (diamonds) and at $\beta = 6.2$ (triangles). The dotted error bars represent an estimate of the systematic error due to lattice artefact corrections. The curves are the two-loop perturbative expression with $a(6.5)\Lambda_F = 0.060$ (dotted) and 0.070 (continuous).

perturbation series converge. This successful calibration of perturbation theory on a lattice is important in practice. For instance, when matrix elements are measured on a lattice they have finite correction factors (usually called Z) to relate them to continuum matrix elements. These Z factors are evaluated perturbatively — so an accurate continuum prediction needs trustworthy perturbative calculations.

Excited gluonic modes

The situation of a static quark and antiquark is a very clear case in which to discuss hybrid mesons which have excited gluonic contributions. A discussion of the colour representation of the quark and antiquark is not useful since they are at different space positions and the combined colour is not gauge invariant. A better criterion is to focus on the spatial symmetry of the gluonic flux. As well as the symmetric ground state of the colour flux between two static quarks, there will be excited states with different symmetries. These were studied on a lattice[23] and the conclusion was that the E_u symmetry (corresponding to flux states from an operator which is the difference of U-shaped paths from quark to antiquark of the form ⊓ − ⊔) was the lowest lying gluonic excitation. Results for this potential are shown in fig. 4.

This gluonic excitation corresponds to a component of angular momentum of one unit along the quark antiquark axis. Then one can solve for the spectrum of hybrid mesons using the Schrödinger equation in the adiabatic approximation. The spatial wave function necessarily has non zero angular momentum and corresponds to $L^{PC} = 1^{+-}$ and 1^{-+}. Combining with the quark and antiquark spins then yields[23] a set of 8 degenerate hybrid states with $J^{PC} = 1^{--}$, 0^{-+}, 1^{-+}, 2^{-+} and 1^{++}, 0^{+-}, 1^{+-}, 2^{+-} respectively. These contain the spin-exotic states with $J^{PC} = 1^{-+}$, 0^{+-} and 2^{+-} which will be of special interest.

Since the lattice calculation of the ground state and hybrid masses allows a direct prediction for their difference, the result for this 8-fold degenerate hybrid level is illustrated in fig. 4 and corresponds[16] to masses of 10.81(25) GeV for $b\bar{b}$ and 4.19(15) GeV for $c\bar{c}$. Here the errors take into account the uncertainty in setting the ground state mass using the quenched potential as discussed above. Recently a different lattice technique[24] has been used to explore the excited gluonic levels in the quenched approximation. The results above are confirmed and preliminary values quoted for the lightest hybrid mesons are 10.83 and 4.25 GeV respectively for $b\bar{b}$ and $c\bar{c}$ with no error estimates given.

The quenched lattice results show that the lightest hybrid mesons lie above the open $B\bar{B}$ threshold and are likely to be relatively wide resonances. This could also be checked by comparing with quenched masses for the B meson itself[26], but at present there are quite large uncertainties on that mass determination. The very flat potential implies a very extended wavefunction: this has the implication that the wavefunction at the origin will be small, so hybrid vector states will be weakly produced from e^+e^-.

It would be useful to explore the splitting among the 8 degenerate J^{PC} values obtained. This could come from different excitation energies in the $L^{PC} = 1^{+-}$ (magnetic) and 1^{-+} (pseudo-electric) gluonic excitations, spin-orbit terms, as well as mixing between hybrid states and $Q\bar{Q}$ mesons with non-exotic spin. One way to study this on a lattice is to use the NRQCD formulation which describes non-static heavy quarks which propagate non-relativistically. Preliminary results for hybrid excitations from several groups[25] give at present similar results to those with the static approximation as described above, but future results may be more precise and able to measure splittings among different states.

As well as comparing excited gluonic states from the lattice with experimental spectra for $\bar{b}b$ systems, it is also worthwhile to compare with phenomenological models. One such model is the hadronic string. This has the simple prediction that gluonic excitations are at multiples of π/R in energy higher at large R. A detailed comparison[16] for SU(2) of colour shows qualitative agreement of the lattice excited potentials with the string model provided an appropriate expression is used for excited level j:

$$V_j(R) = (K^2R^2 - \frac{\pi K}{6} + 2\pi jK)^{\frac{1}{2}}. \tag{18}$$

This expression also shows that even the ground state string mode ($j = 0$) will have a contribution from a string fluctuation, namely[27] $V(R) \approx KR - \pi/(12R)$ as $R \to \infty$. In practice this $1/R$ string fluctuation term is very hard to disentangle from the Coulomb term e/R. One way to get round this in lattice studies is to consider a hadronic string that encircles the periodic spatial boundaries of length L. Then there are no sources and hence no Coulomb component. The appropriate string fluctuation term in the energy of this state, called the torelon, is then given by $E(L) = KL + \pi/(3L)$ as $L \to \infty$. Lattice studies[28] have confirmed the presence of this string fluctuation term with the correct coefficient *as* given to a precision of 3%. This is impressive evidence that the hadronic string is a good model of the energy of the colour flux tube at large distances.

Confinement

The simplest manifestation of confinement is that the potential $V(R)$ between static colour sources in the fundamental representation continues to rise with increasing R in quenched QCD. This raises the question of the nature of the colour fields between the sources. Lattice studies have been undertaken[29, 30] to probe the energy momentum tensor of these colour fields. The probe used is a plaquette so the momentum scale of the probe increases as $a \to 0$. Lattice sum rules [30] can be used to normalise these distributions and relate them to the β-function.For separation $R > 0.7$ fm, a string-like spatial distribution is found where the transverse width of the flux tube increases slowly at most with increasing R. For these R values, the averages of the squared components of the colour fields (these are gauge invariant quantities) are found to be roughly equal (i.e. $E_i^2 \approx B_j^2$, for all $i,\ j = 1,\ 2,\ 3$). This implies that the energy density is much smaller than the action density. As well as components of the energy-momentum tensor which are gauge invariant, it is possible to study the colour field tensors directly by choosing a suitable gauge[31].

Another detail concerning the confining force is its spin dependence. A lattice study of the spin-spin and spin-orbit potentials between static quarks allows this to be explored. The basic conclusion[32] is that the only long-range force is a spin-orbit force of the type usually called scalar. More recent studies[33] confirm this. This lattice result is confirmed by phenomenological studies of the observed splitting between P-wave $\bar{b}b$ and $\bar{c}c$ states.

Another route to explore confinement is to measure on a lattice the potential energy between two static sources in the adjoint representation of the colour group. At large R the adjoint potential $V_A(R)$ must become a constant because each adjoint colour source can be screened by a gluonic field. Indeed at large R, $V_A \to 2m_{\text{gluelump}}$ where the gluelump is the ground state hadron with a gluon field around a static adjoint: the gluinoball. Of interest to model builders is the adjoint potential at smaller R values. The most precise data are for $SU(2)$ of colour[34] and they do show a region of linear rise, although with a slope less steep than that given by the Casimir ratio (namely $\frac{8}{3}V(R)$ where $V(R)$ is the fundamental colour source potential discussed previously). Attempts have also been made[35] to explore the colour field distributions in this case.

LIGHT QUARKS — HYBRIDS

Unlike very heavy quarks, light quark propagation in the gluonic vacuum sample is very computationally intensive — involving inversion ofhuge $(10^7 \times 10^7)$ sparse matrices. Current computer power is sufficient to study light quark physics thoroughly in the quenched approximation. The state of the art[36] is the Japanese CPPACS Collaboration who are able to study a range of large lattices (up to about 64^4) with a range of light quark masses. Qualitatively the meson and baryon spectrum of states made of light and strange quarks is reproduced with discrepancies of order 10% in the quenched approximation.

Here I will focus on hybrid mesons made from light quarks. In the quenched approximation, there will be no mixing involving spin-exotic hybrid mesons and so these are of special interest. The first study of this area was by the UKQCD Collaboration[37] who used operators motivated by the $Q\bar{Q}$ studies referred to above. Using non-local operators, they studied all 8 J^{PC} values coming from $L^{PC} = 1^{+-}$ and 1^{-+} excitations. The resulting mass spectrum is shown in fig. 6 where the $J^{PC} = 1^{-+}$ state is seen to be the lightest spin-exotic state with a statistical significance of 1 standard deviation.

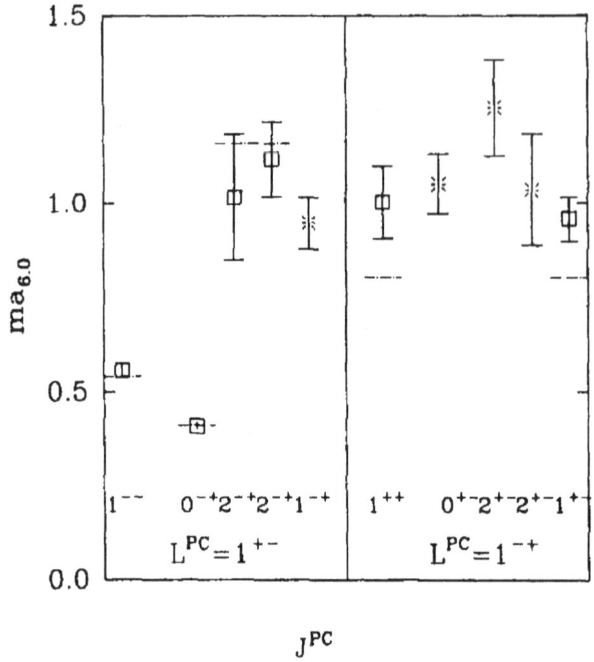

Figure 6. The masses in lattice units (with $a_{6.0}^{-1} \approx 2$ GeV) of states of J^{PC} built from hybrid operators with strange quarks, spin-exotic (*) and non-exotic (□). The dot-dashed lines are the mass values found for $s\bar{s}$ operators. Results from ref(37).

The statistical error on the mass of this lightest spin-exotic meson is 7% but to take account of systematic errors from the lattice determination a mass of 2000(200) MeV is quoted for the $s\bar{s}$ meson. Although not directly measured, the corresponding light quark meson would be expected to be around 120 MeV lighter. In view of the small statistical error, it seems unlikely that the 1^{-+} meson in the quenched approximation could lie as light as 1.4 GeV where there are experimental indications for such a state[38]. Beyond the quenched approximation, there will be mixing between such a hybrid meson and $q\bar{q}q\bar{q}$ states such as $\eta\pi$ and this may be significant.

One feature clearly seen in fig. 6 is that non spin-exotic mesons created by hybrid meson operators have masses which are very similar to those found when the states are created by $q\bar{q}$ operators. This suggests that there is quite strong coupling between hybrid and $q\bar{q}$ mesons even in the quenched approximation. This would imply that the $\pi(1800)$ is unlikely to be a pure hybrid, for example.

A second lattice group has also evaluated hybrid meson spectra from light quarks. They obtain[39] masses with statistical and various systematic errors for the 1^{-+} state of 1970(90)(300) MeV, 2170(80)(100)(100) MeV and 4390(80)(200) MeV for $n\bar{n}$, $s\bar{s}$ and $c\bar{c}$ quarks respectively. For the 0^{+-} spin-exotic they have a noisier signal but evidence that it is heavier. They also explore mixing matrix elements between spin-exotic hybrid states and 4 quark operators.

FULL QCD

So far I have discussed the glueball spectrum, interquark potentials and α_S in the quenched approximation. This corresponds to treating the sea quarks as of infinite mass (so they don't contribute to the vacuum). To make direct comparison with experiment, it is necessary to estimate the corrections from these dynamical quark loops in the vacuum.

The strategy is to use a finite sea-quark mass but still a value larger than the empirical light quark mass. The reason is computational: the algorithms become very inefficient as the sea-quark mass is reduced. The target is to study the effects as the sea-quark mass is reduced and then extrapolate to the physical value. The present situation, in broad terms, is that there is no significant change as the sea-quark mass is reduced. This could be because there are no corrections to the quenched approximation. Alternatively, the corrections may only turn on at a much lower quark mass than has been explored so far.

Let us try to make this argument a little more quantitative. For heavy sea-quarks of mass m, their contribution will be approximately proportional to $e^{-2m/E}$ where E is a typical hadronic energy scale (a few hundred MeV). Thus the quark loop contributions will be negligible for $m >> E$, which corresponds to the quenched approximation. As $m \approx E$, the effects will turn on in a non-linear way.

The computational overhead of full QCD on a lattice is so large because the quark loops effectively introduce a long range interaction. The quark interaction in the action is quadratic and so can be integrated out analytically — see eq(5). This leaves an effective action for the gluonic fields which couples together the fields at all sites. This implies that, in a Monte Carlo method, a change in gluon field at one site involves the evaluation of the interaction with all other sites. In practice, one makes small changes at all sites in parallel, but this still amounts to inverting a large sparse matrix for each update. This is computationally slow.

As the sea-quark mass becomes small, one would expect to need a larger lattice size to hold the quarks. For heavy quarks, the effective range of the quark loops in the vacuum will be of order $1/m$. Thus the quenched approximation corresponds to $m \to \infty$ and a local interaction. For light quarks of a few MeV mass, the range will not be $1/m$, because quarks are confined. The lattice studies that have been made suggest that spatial sizes of order twice those adequate for the quenched approximation are needed for full QCD. This also implies considerable computational commitment.

A popular indication of how close a full QCD study is to experiment is to ask whether the ρ meson can decay to two pions. Since the decay is P-wave, it needs non-zero momentum. On a lattice spatial momentum is quantised in units of $2\pi/L$. Thus $m_V > 2(m_P{}^2 + 4\pi^2/L^2)^{1/2}$ for the decay channel to be allowed energetically. At present this criterion is rarely satisfied in quenched studies, let alone in the more computationally demanding case of full QCD.

The conclusion of current full QCD lattice calculations is that the expected sea-quark effects are not yet fully present. The main effect observed in full QCD calculations is that the lattice parameter β which multiplies the gluonic interaction term in the action is shifted. Apart from this renormalisation of β, there is little sign of any other statistically significant non-perturbative effect.

Consider the changes to be expected for the inter-quark potential when the full QCD vacuum is used:

- At small separation R, the quark loops will increase the size of the effective coupling α compared to the pure gluonic case. This effect can be estimated in perturbation theory and the change at lowest order will be from 1/33 to $1/(33\text{-}2N_f)$.

- At large separation R, the potential energy will saturate at a value corresponding to two 'heavy-quark mesons'. In other words, the flux tube between the static quarks will break by the creation of a $q\bar{q}$ pair from the vacuum.

Current lattice simulation[18] shows some evidence for the former effect but no statistcally significant signal for the latter.

OUTLOOK

Lattice QCD is good for asking 'what' not 'why'. Lattice results for masses and matrix elements are obtained from first principles without approximations (except in many cases the quenched approximation is still needed to keep the computational resource manageable). No model is used but no understanding of the underlying physics is obtained. By varying the quark masses, boundary conditions etc, it is possible to explore a much wider range of circumstances than is available directly from experiment. This is a very valuable tool for validating models and learning 'why'.

Lattice techniques can extract reliable continuum properties from QCD. At present, the computational power available combined with the best algorithms suffices to give accurate results for many quantities in the quenched approximation. The future is to establish accurate values for more subtle quantities in the quenched approximation (eg. weak matrix elements of strange particles) and to establish the corrections to the quenched approximation by full QCD calculations.

I hope that soon we reach the stage where an experimentalist saying 'as calculated in QCD' is assumed to be speaking of non-perturbative lattice calculations rather than perturbative estimates only.

REFERENCES

1. H. Rothe, *Lattice Gauge Theories*, World Scientific, 1992; I. Montvay and G. Münster, *Quantum Fields on a Lattice*, CUP, 1994.
2. K. G. Wilson, *Phys. Rev.* D10:2445 (1974).
3. G. P. Lepage and P. B. Mackenzie, *Phys. Rev.* D48:2250 (1993).
4. P. De Forcrand, et al., *Phys. Lett.* B152:107 (1985).
5. C. Michael and M. Teper, *Nucl. Phys.* B314:347 (1989).
6. UKQCD collaboration, G. Bali, K. Schilling, A. Hulsebos, A. C. Irving, C. Michael and P. Stephenson, *Phys. Lett.* B309:378 (1993).
7. H. Chen, J. Sexton, A. Vaccarino, and D. Weingarten, *Nucl. Phys. B (Proc. Suppl.)* 34:357 (1994).
8. J. Kripfganz and C. Michael, *Nucl. Phys.* B314:25 (1989).
9. J. Sexton, A. Vaccarino, and D. Weingarten, *Phys. Rev. Lett.* 75:4563 (1995).
10. C. Morningstar, and M. Peardon, hep-lat/9704011.
11. C. Michael, G. A. Tickle and M. Teper, *Phys. Lett.* B207:313 (1988).
12. SESAM Collaboration, G. Bali et al., *Nucl. Phys.* B *(Proc. Suppl.)* 53:239 (1997).
13. I. Jorysz and C. Michael, *Nucl. Phys.* B302:448 (1988).
14. M. Foster and C. Michael, *Nucl. Phys. B (Proc. Suppl.)* (LAT97 in press), hep-lat/9709051.
15. G.S. Bali and K. Schilling, *Phys. Rev.* D47:661 (1993); H. Wittig (UKQCD collaboration), *Nucl. Phys. B (Proc. Suppl)* 42:288 (1995).
16. S. Perantonis and C. Michael, *Nucl. Phys.* B347:854 (1990).
17. R. Sommer, *Nucl. Phys.* B411:839 (1994).
18. SESAM Collaboration, U. Glässner, et al., *Phys. Lett.* B383:98 (1966); S. Güsken , *Nucl. Phys. B (Proc. Suppl.)* (LAT97 in press).
19. A. Billoire, *Phys. Lett.* B104:472 (1981).
20. C. Michael, *Phys. Lett.* B283:103 (1992)
21. UKQCD collaboration, A. Hulsebos et al., *Phys. Lett.* B294:385 (1992).
22. C. Michael, *Nucl. Phys. B (Proc. Suppl.)* 42:147 (1995).

23. L. Griffiths, C. Michael, and P. Rakow, *Phys. Lett.* B129:351 (1983).
24. K. Juge, J. Kuti and C. Morningstar, *Nucl. Phys.* B *(Proc. Suppl.)* (in press), heplat/9709131.
25. A. Alikhan, *Nucl. Phys. B (Proc. Suppl.)* (LAT97 in press).
26. R. Sommer, *Phys. Rep.* 275:1 (1996).
27. M. Lüscher, *Nucl. Phys.* B180[FS2]:317 (1981).
28. C. Michael and P. Stephenson, *Phys. Rev.* D50:4634 (1994).
29. G. Bali, K. Schilling and C. Schlichter, *Phys. Rev.* D51:5165 (1995).
30. A. M. Green, C. Michael and P. Spencer, *Phys. Rev.* D55:1216 (1997).
31. A. Di Giacomo et al., *Nucl Phys.* B347:441 (1990).
32. C. Michael, *Phys. Rev. Lett.* 56:1219 (1986).
33. G. Bali, A. Wachter and K. Schilling, *Phys. Rev.* D55:5309 (1997); D56:2566 (1997).
34. C. Michael, *Nucl. Phys.* B *(Proc. Suppl.)* 26:417 (1992).
35. H. Trottier, *Nucl. Phys.* B. *(Proc. Suppl.)* 47:286 (1996).
36. T. Yoshie, *Nucl. Phys.* B *(Proc. Suppl.)* (LAT97 in press).
37. UKQCD Collaboration, P. Lacock, C. Michael, P. Boyle, and P. Rowland, *Phys. Rev.* D54:6997 (1996); *Phys. Lett.* B401:308 (1997); *Nucl. Phys.* B *(Proc. Suppl.)* (LAT97 in press), hep-lat/9708013.
38. A. Ostrovidov, *Proc. Hadron97, BNL.*
39. C. Bernard, et al., *Nucl. Phys.* B *(Proc. Suppl.)* 53:228 (1996); hep-lat/9707008.

MONTE CARLO RESULTS FOR THE HADRON SPECTRUM

Richard Kenway

Department of Physics & Astronomy
The University of Edinburgh
The King's Buildings
Edinburgh EH9 3JZ, Scotland

INTRODUCTION

In these lectures, I will try to explain the important part played by hadron spectrum calculations in developing a reliable and quantitative numerical calculation scheme from first principles for QCD.

Why Calculate What is Well-Known?

The successful calculation of the spectrum of light hadrons, composed of u, d and s quarks, would, first of all, demonstrate that QCD correctly describes quark confinement at low energies. It would validate Monte Carlo simulation of QCD as a quantitative phenomenological tool, and this is vital, because many Standard Model parameters are obscured by non-perturbative QCD effects. Finally, computing the hadron spectrum would provide the most direct determination of the light quark masses, which are some of the most poorly-known Standard Model parameters.

Simply because we have to fix the quark masses before we can do anything else in QCD, spectrum calculations are the starting point for every simulation. Any error in the hadron mass estimates propagates through the calculation, to affect all predictions to some extent. So accuracy is crucial.

A Brief History

The first calculation was reported in 1981 by Hamber and Parisi[1], using a 6^3 spatial lattice and a computer with a speed of about 1 Mflops. They employed the quenched approximation, which neglects virtual quark-antiquark pairs, and Wilson quarks at $\beta = 6.0$ (see later), and they obtained reasonable agreement with experiment:

$$M_\rho = 800 \pm 100 \text{ MeV} \tag{1}$$

$$M_N = 950 \pm 100 \text{ MeV}. \tag{2}$$

However, their estimate of the scale (inverse lattice spacing), $a^{-1} = 1120$ MeV, is wrong by a factor two, most likely because of huge finite-size effects on such a small lattice. So the agreement with experiment was fortuitous! Nevertheless, their work showed that the low-energy properties of QCD were amenable to numerical simulation.

Subsequent progress may appear to have been slow. This is because it has largely entailed the controlling of systematic approximations, rather than any dramatic discoveries – after all, Hamber and Parisi *did agree* with experiment – and this has depended on significant progress in algorithms and machines. Nevertheless, I believe that no other area of computational science can match the rigour and sophistication that has been brought to bear on the calculation of the hadron masses.

In 1993, this topic hit the headlines with the announcement by the GF11 Collaboration that they had computed the hadron spectrum and found results within 6%, and 1.6 standard deviations, of the experimental values[2], using a purpose-built computer five thousand times faster than the one Hamber and Parisi had used. Remarkably, this claim was based on quenched simulations, not full QCD.

Today, following further progress in controlling lattice-spacing errors, the calculations are becoming accurate enough to show that the quenched results do disagree with experiment[3,4] as, of course, they must!

HADRONS FROM LATTICE QCD

The spectrum, as well as many useful matrix elements, may be obtained from hadron correlation functions in Euclidean spacetime. The corresponding path integral is defined on a hypercubic lattice, in the limit of zero lattice spacing, $a \to 0$ in physical units. This limit occurs at a critical point of the lattice theory and the methods of statistical physics, Monte Carlo simulation and the renormalisation group, provide a tractable non-perturbative scheme for precise calculations[5].

The Lattice Formulation of QCD

Wilson Lattice Action. At every site, x, of the 4-dimensional lattice is a quark and antiquark field, $q(x)$ and $\bar{q}(x)$, which carry Dirac, colour and flavour indices. Associated with every link, (x,μ), of the lattice, from x to $x + a\hat{\mu}$, $\mu = 0,1,2,3,$ is a gauge field $U_\mu(x) \in SU(3)$. The Wilson lattice action[6] is

$$S[U,\bar{q},q] = S_G[U] + S_F[U,\bar{q},q]. \tag{3}$$

The gauge action is

$$S_G[U] = \frac{6}{g^2}\sum_{\Box}\left[1 - \frac{1}{3}\mathrm{Re\ Tr}\ U_\Box\right] = \frac{1}{4}\int d^4x\ F_{\mu\nu}^2 + O(a^2), \tag{4}$$

where $U_\Box$ is the product of the gauge fields around a plaquette. The quark action is

$$S_F[U,\bar{q},q] = a^4\sum_x \bar{q}(x)(D+m)q(x) = \int d^4x\ \bar{q}(\partial\!\!\!/ + igA\!\!\!/ + m)q + O(a) \tag{5}$$

where the lattice derivative is

$$\begin{aligned} Dq(x) = \frac{1}{2a}\sum_\mu \Big\{ & \gamma_\mu\left[U_\mu(x)q(x+a\hat{\mu}) - U_\mu^\dagger(x-a\hat{\mu})q(x-a\hat{\mu})\right] \\ & -\left[U_\mu(x)q(x+a\hat{\mu}) + U_\mu^\dagger(x-a\hat{\mu})q(x-a\hat{\mu}) - 2q(x)\right]\Big\}. \end{aligned} \tag{6}$$

This choice gives the unwanted fermionic modes, which are artefacts of the lattice theory[7], masses of order $1/a$, so these should decouple in the continuum limit. It does this by introducing a chiral-symmetry-breaking term of $O(a)$ (from the second term in square brackets in Eq (6)), which is the dominant source of discretisation errors. Since the breaking is proportional to a, chiral symmetry should be restored in the continuum limit. But, the absence of chiral symmetry at non-zero lattice spacing allows additive renormalisation of the quark mass, so that a renormalisation condition must be imposed to define zero quark mass - usually that the pseudoscalar meson mass should vanish at zero quark mass, corresponding to a Goldstone pion. The explicit breaking of chiral symmetry also creates technical problems for the calculation of matrix elements, because operators of different chirality mix.

Note that, although I have explicitly included the lattice spacing, a, simulations are performed in terms of dimensionless variables, in which the quark fields are rescaled, and the quark action is

$$S_F[U,\bar{q},q] = \sum_x \bar{q}(x)\Big\{q(x) - \kappa\sum_\mu\big[(1-\gamma_\mu)U_\mu(x)q(x+\hat{\mu}) + (1+\gamma_\mu)U_\mu^\dagger(x-\hat{\mu})q(x-\hat{\mu})\big]\Big\}. \tag{7}$$

The bare parameters are the gauge coupling, $g^2 = 6/\beta$, and the masses of the quark flavours, generically denoted by $m = (1-8\kappa)/2\kappa$. These parameters must be tuned to achieve continuum results corresponding to the physical quark masses.

Staggered Fermions. The alternative staggered-fermion action[8, 9] retains a continuous chiral symmetry and incurs discretisation errors only of $O(a^2)$, just like the gauge action. Unfortunately, it describes degenerate opposite-chirality pairs of quark flavours and so cannot be realistic[5, 7]. Also, it has a complicated flavour structure and large operator renormalisations, which have to be controlled non-perturbatively, making matrix element calculations more difficult than for Wilson quarks.

Hadron Correlators. The spectrum may be computed from hadron correlators of the form

$$\langle \mathcal{O}^\dagger(x)\mathcal{O}(0)\rangle = \frac{1}{Z}\int \mathcal{D}U\,\mathcal{D}\bar{q}\mathcal{D}q\;\mathcal{O}^\dagger(x)\mathcal{O}(0)\;e^{-S_G[U]-S_F[U,\bar{q},q]}, \tag{8}$$

where $\mathcal{O}$ is a gauge-invariant product of quark and gluon fields with the quantum numbers of a specific hadron. For a meson, $\mathcal{O}(x)$ is a bilinear in the quark fields. The simplest possibility is the local product $\bar{q}(x)\Gamma q(x)$. A pseudoscalar meson is generated by using $\Gamma = \gamma_5$ and a vector meson by using $\Gamma = \gamma_i$. However, as the lattice spacing decreases, such local composite fields have decreasing overlap with the meson states, which are of non-zero spatial extent. A bigger overlap with the lightest meson can be achieved using smeared quark fields,

$$\mathcal{O}(t,\vec{x}) = \bar{q}(t,\vec{x})\Gamma\sum_{\vec{y}}\Psi(\vec{x}-\vec{y})q(t,\vec{y}), \tag{9}$$

where Ψ is an approximation to the groundstate wavefunction (which should be gauge covariant, unless the computation is done in a fixed gauge). The nucleon correlator has $\mathcal{O} = \epsilon_{abc}(u^aC\gamma_5d^b)u^c$, or its smeared version, and the Δ has $\mathcal{O} = \epsilon_{abc}(u^aC\gamma_\mu u^b)u^c$.

The meson correlator is expressed in terms of quark propagators by performing the path integral over the Grassmann variables in Eq (8):

$$\langle\bar{q}(x)\Gamma_1q(x)\bar{q}(0)\Gamma_2q(0)\rangle = \frac{1}{Z}\int\mathcal{D}U\;\det(D+m)^{N_f}\,e^{-S_G[U]} \times\mathrm{Tr}\,\Gamma_1(D+m)^{-1}_{x0}\Gamma_2(D+m)^{-1}_{0x}. \tag{10}$$

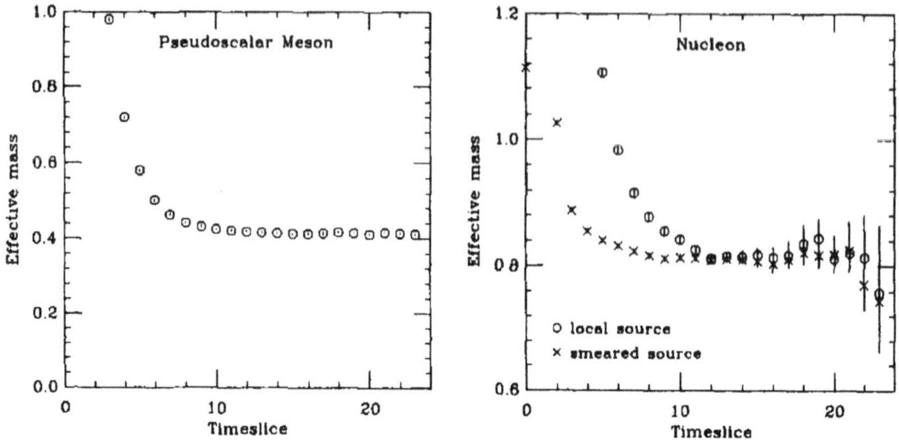

Figure 1. The pseudoscalar meson (left) and nucleon (right) effective masses from a quenched QCD simulation with Wilson quarks at $\beta = 6.0$.

The integration over the gauge fields is performed by Monte Carlo.

The timeslice correlator projects onto zero-momentum eigenstates, $|h_n\rangle$, of the Hamiltonian, $\hat{H}$:

$$\begin{aligned}\sum_{\vec{x}}\langle \mathcal{O}^\dagger(t,\vec{x})\mathcal{O}(0,\vec{0})\rangle &= \sum_{\vec{x}}\langle 0|\hat{\mathcal{O}}(\vec{x})^\dagger e^{-\hat{H}t}\hat{\mathcal{O}}(\vec{0})|0\rangle \\ &= \sum_n \langle 0|\hat{\mathcal{O}}^\dagger|h_n\rangle \frac{e^{-M_n t}}{2M_n}\langle h_n|\hat{\mathcal{O}}|0\rangle \\ &\stackrel{t\to\infty}{\longrightarrow} |\langle 0|\hat{\mathcal{O}}|h_0\rangle|^2\,\frac{e^{-M_0 t}}{2M_0} \end{aligned} \tag{11}$$

where $|h_0\rangle$ is the lightest hadron state which has a non-zero overlap with $\hat{\mathcal{O}}$, with mass M_0. By fitting the exponential fall-off of the timeslice correlator at large Euclidean times, we can extract the lightest hadron mass, M_0, and the decay-constant matrix element, $\langle 0|\hat{\mathcal{O}}|h_0\rangle$. (Actually, on a periodic lattice, the fit is to a cosh or sinh for mesons and to an exponential for baryons.)

The approach of a timeslice correlator, $C(t)$, to saturation by the lightest hadronic state can be displayed by plotting the effective mass,

$$M_{\rm eff}(t) = \log\frac{C(t)}{C(t+1)} \tag{12}$$

versus timeslice, t, examples of which are shown in Fig 1. On a finite lattice, a systematic error is introduced in the hadron mass estimates, because of the contamination of the timeslice correlator, Eq (11), by excited states. This can be reduced by choosing the smearing function, Ψ, to maximise the overlap with the state of interest, as can be seen for the nucleon in Fig 1, and by using multi-exponential fits to absorb some of the excited state contributions. Unfortunately, the statistical error grows at large times for any hadron other than the pion, typically,

$$\frac{\text{noise}}{\text{signal}} \sim e^{(M_{\rm meson}-M_\pi)t}, \quad \text{or} \quad \sim e^{(M_{\rm baryon}-3M_\pi/2)t} \tag{13}$$

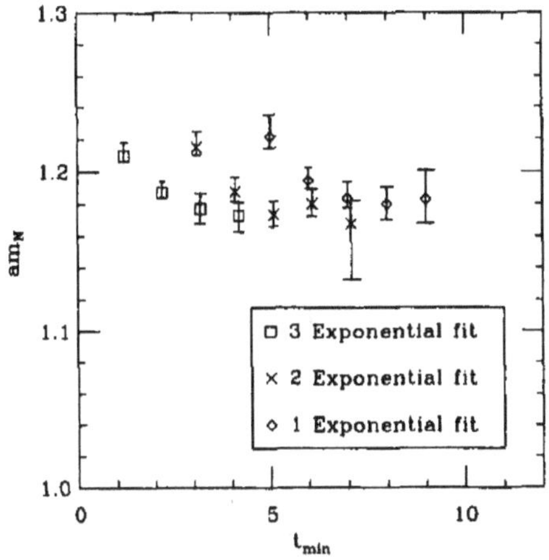

Figure 2. Nucleon mass from 1-, 2- and 3-exponential fits to the timeslice correlator as the minimum time, $t_{\min}$, in the range of timeslices included in the fit, is increased[10].

and so the range of timeslices available is limited. The number of exponentials and the range of timeslices are chosen to minimise the contamination, while maintaining a good signal, which can be judged to occur when the result of the fit becomes independent of the minimum timeslice included, as shown in Fig 2.

Continuum Limit. The hadron masses extracted from the fitted data are dimensionless functions of g (at fixed quark mass, which I take to be $m = 0$ for simplicity). The lattice spacing in physical units is then,

$$a(g) = \frac{M_h(g)}{M_h^{\text{expt}}} \tag{14}$$

and $a = 0$ corresponds to the critical point $M_h(g^*) = 0$, for all hadrons, h.

It is not necessary to reach $a = 0$ directly. Instead, a needs only to be small enough that the masses scale:

$$\frac{M_h(g)}{M_{h'}(g)} = \text{constant} = \frac{M_h^{\text{expt}}}{M_{h'}^{\text{expt}}}. \tag{15}$$

This implies a universal dependence on the gauge coupling,

$$M_h(g) = c_h\, f(g) \tag{16}$$

and the existence of a cut-off independent mass,

$$\Lambda_{\text{QCD}} = \frac{1}{a} f(g). \tag{17}$$

Then, from Eq (14), the requirement that

$$\frac{d}{da} M^{\text{expt}} = 0 \tag{18}$$

implies that

$$f(g) = \exp - \int^{g} \frac{dg'}{\beta(g')}, \quad \beta(g) = -a\frac{dg}{da}. \tag{19}$$

Asymptotic freedom tells us that $a = 0$ at $g = 0$. So $\beta(g)$ may be reliably computed in perturbation theory, with the result that

$$\beta(g) = -\beta_0 g^3 - \beta_1 g^5 - \ldots, \qquad \beta_0 = \frac{33 - 2N_f}{3(16\pi^2)}. \tag{20}$$

Combining Eqs (17), (19) and (20), we know the behaviour of masses as g *is* tuned towards zero. In practice, mass ratios from the current best data still have a lattice-spacing dependence and are not yet scaling, i.e.,

$$\frac{M_h(g)}{M_{h'}(g)} = \text{constant} + O(a^p). \tag{21}$$

But we are close enough for the a^p dependence to be extrapolated away, with sufficient confidence, using the theoretically-expected exponent, p.

The Monte Carlo Method for Quenched QCD

The Algorithm. The idea is to approximate the path integral over the gauge fields in Eq (10) by an average over a finite ensemble of gauge configurations, $\{U^{(n)} | n = 1, \ldots, N_{\text{config}}\}$, drawn from the probability distribution

$$\mathcal{P}[U] = \frac{1}{Z} \det(D[U] + m)^{N_f} e^{-S_G[U]}. \tag{22}$$

Unfortunately, the inclusion of the determinant greatly increases the computational cost, typically by a factor of a thousand at current parameter values, and so it is usually ignored in what is called the quenched approximation.

An expectation value, such as the meson correlator, is approximated by

$$\langle \bar{q}(x)\Gamma_1 q(x)\bar{q}(0)\Gamma_2 q(0)\rangle \simeq \frac{1}{N_{\text{config}}} \sum_{n=1}^{N_{\text{config}}} \text{Tr}\,\Gamma_1 (D[U^{(n)}] + m)^{-1}_{x0} \Gamma_2 (D[U^{(n)}] + m)^{-1}_{0x}. \tag{23}$$

Provided the gauge configurations are statistically independent, the error in the mean decreases like $1/\sqrt{N_{\text{config}}}$. There is some subtlety here: because the ensemble is generated in a Markov process, which converges to the distribution in Eq (22) after some number of steps (which must be discarded), successive configurations are correlated. So the ensemble must be built up from sufficiently widely separated configurations. These days, the statistical errors are largely under control and a sophisticated statistical analysis is employed to account for correlations between different quantities computed using the same ensemble of configurations.

Quark Propagators. The quark propagator,

$$G_{x0}[U^{(n)}] = (D[U^{(n)}] + m)^{-1}_{x0} \tag{24}$$

for a given gaugeconfiguration, $U^{(n)}$, is obtained by solving the large sparse system of linear equations,

$$(D[U^{(n)}] + m)_{xy} G_{y0} = \delta_{x0}. \tag{25}$$

Because of the high computational cost, considerable effort has gone into finding efficient algorithms for this[11]. These are iterative in nature and, at fixed quark mass, the cost scales with the number of lattice sites.

Quenched Approximation. Replacing $\det(D\!\!\!/+m)$ by a constant, its mean value, becomes valid in the limit of large quark mass. In Feynman-diagram language, it corresponds to omitting virtual quark loops. There are two ways of viewing Quenched QCD (QQCD):

- as an effective theory for QCD, in which its bare parameters at fixed lattice spacing are chosen empirically to best match experiment (and by assumption QCD);
- as a peculiar type of quantum field theory in its own right, which almost certainly is pathological in the limit $m \to 0$

When doing phenomenology the first approach is taken. Some dynamical quark effects are incorporated in the choice of parameters. This is why QQCD has provided such a successful basis for QCD phenomenology, and why it has taken so long to reach the point where quenching errors dominate for most physical quantities. The second approach can provide insight as to when to expect this effective theory to fail.

Mass Ratios and the Edinburgh Plot

QCD should have a well-defined continuum limit for any specific choice of the quark masses. Simulations are usually carried out for a range of quark masses and the data extrapolated/interpolated to the physical quark mass values. For simplicity, if we consider a single quark flavour, then we can fix the quark mass by fixing M_{PS}/M_V, say. Then continuum physics corresponds to all other mass ratios being independent of g for g small enough.

The spectra obtained at different lattice spacings may be compared using the Edinburgh Plot[12] in which one mass ratio, usually M_N/M_V is plotted versus the ratio used to fix the quark mass, usually M_{PS}/M_V. Scaling corresponds to these data falling on a universal curve. Typically, the data are compared with a phenomenological quark model[13],

$$M_{\rm baryon} = M_b + \sum_{i=1}^{3} m_i + \xi_b \sum_{i>j} \frac{\vec{s}_i \cdot \vec{s}_j}{m_i m_j} \tag{26}$$

$$M_{\rm meson} = M_m + \sum_{i=\bar{q},q} m_i + \xi_m \frac{\vec{s}_{\bar{q}} \cdot \vec{s}_q}{m_{\bar{q}} m_q} \tag{27}$$

where the parameters, M, m and ξ are taken from experiment. This is expected to be a good model for large quark mass, but does not have the correct chiral behaviour.

The original Edinburgh Plot[12] is shown in Fig 3. Evidently, as was concluded at the time, the data does not scale down to $\beta = 5.7$. The corresponding plot today, also shown in Fig 3, includes the same lattice spacings, but an improved lattice formulation with reduced discretisation error, lattices twice the linear size and much higher statistics. The error bars are much smaller and now the data scale at fixed quark mass. Notice that this progress has been achieved without significantly reducing the lattice spacing.

THE CHOICE OF PARAMETERS

The Computational Scaffolding

Finite volume. Numerical simulations are done in a finite volume, so that the number of degrees of freedom is finite and they can be stored in the finite memory of a

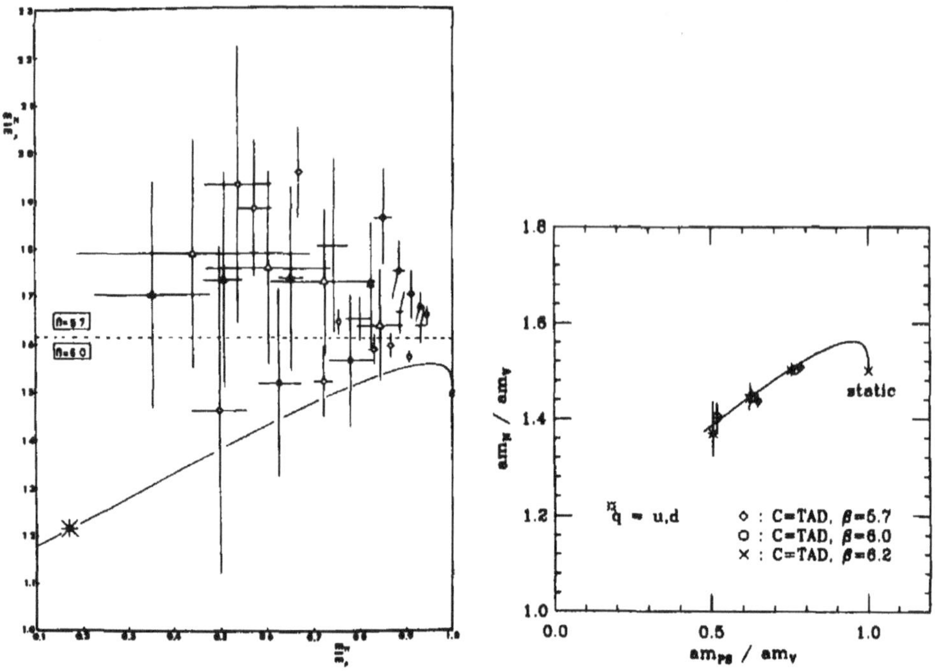

Figure 3. The original 1985 Edinburgh plot[12] (left), showing data from quenched simulations on 8^3, 10^3 and 16^3 spatial lattices, at two lattice spacings, and the 1997 Edinburgh plot (right) showing scaling at fixed m_{PS}/m_V using the tadpole-improved action.

computer! Ideally, with enough computer power, we could obtain all physical quantities in a range of physical volumes and extrapolate the results to the infinite-volume limit. Unfortunately, we are far from this situation, and do not really understand the dependence on lattice size for the existing range of data. Finite-size effects are expected to fall exponentially with lattice size[14], so it is reasonable, and the best we can hope for at present, to increase the lattice size until they are obscured by the statistical errors. For QQCD, a linear dimension of around 2.5 fm seems to be large enough[15] for the nucleon. In full QCD, finite-size effects are expected to be larger, so larger volumes will probably be needed, but existing data is too sparse to conclude anything.

Non-zero lattice spacing. Considerable progress has been made recently in systematically exploring the discretisation errors due to non-zero lattice spacing in QQCD. Today's simulations span a range of lattice spacings corresponding to momentum cutoffs extending from $a^{-1} \simeq 1.1$ GeV at $\beta = 5.7$ to 3.5 GeV at $\beta = 6.4$. These numbers are only a rough guide, of course, because estimates from Eq (14) using different hadrons, do not have to agree in the quenched theory. There is evidence from matrix-element calculations that discretisation errors, due to the chiral-symmetry-breaking term in the Wilson action may be large in this range, eg as much as 30% in the normalisation of the vector current, and over 50% when the PCAC relation,

$$\partial_\mu A^a_\mu = 2mP^a \tag{28}$$

is sandwiched between different external states[16]. So reducing these errors has been the major goal in recent years. Significantly reducing the lattice spacing itself is not a practical option!

Computational cost. In QQCD, for an N^4 lattice, the computational cost grows like N^6 at fixed physical length scales ($aM_\pi N$ = **constant**) and fixed statistical error (it comprises two factors, the lattice volume and $1/(aM_\pi)^2$, where the latter comes from either the random walk by which the Monte Carlo algorithm explores the physical length scale, or the condition number of the fermion matrix). So, to reduce the lattice spacing by a factor 2 costs around 64 times more computer power. This is a compelling argument for trying to improve the lattice formulation by introducing irrelevant terms which cancel the leading discretisation errors, even though these terms have coefficients which have to be determined non-perturbatively. For dynamical quarks the computational cost grows like $N^{10.5}$, and improvement is the only hope, short of an algorithmic miracle!

Improvement and Irrelevant Parameters

Local effective theory. The effects of the leading discretisation errors in Eqs (4) and (5) on on-shell quantities (particle energies, matrix elements of local composite operators, etc) can be systematically eliminated by means of the Symanzik improvement programme[17], in which higher-dimension operators are added to the lattice action and to the composite fields in correlation functions[18].

Close to the continuum limit, the lattice action, Eq (3), may be expanded around the continuum action, $S_{\rm ctm}$,

$$S = S_{\rm ctm} + a\int d^4x\, \mathcal{L}_1(x) + O(a^2) \tag{29}$$

where $\mathcal{L}_1$ is a linear combination of dimension 5 composite fields:

$$\mathcal{O}_1 = \bar{q}\sigma_{\mu\nu}F_{\mu\nu}q, \tag{30}$$

$$\mathcal{O}_2 = \bar{q}D_\mu D_\nu q + \bar{q}\,\overleftarrow{D}_\mu\overleftarrow{D}_\nu\, q \tag{31}$$

$$\mathcal{O}_3 = m\mathrm{Tr}\, F_{\mu\nu}F_{\mu\nu} \tag{32}$$

$$\mathcal{O}_4 = m\{\bar{q}\gamma_\mu D_\mu q - \bar{q}\,\overleftarrow{D}_\mu\, \gamma_\mu q\} \tag{33}$$

$$\mathcal{O}_5 = m^2\bar{q}q. \tag{34}$$

Similarly, any renormalised composite field, $Z_\phi\phi(x)$, built from the quark and gluon fields, can be expanded in terms of local composite fields,

$$Z_\phi\phi(x) = \phi_0(x) + a\phi_1(x) + O(a^2). \tag{35}$$

The classical field equations can be used to find linear combinations of the fields in Eqs (30)–(34) which vanish, and a linearly-independent basis is $\mathcal{O}_1$, $\mathcal{O}_3$ and $\mathcal{O}_5$.

Improved lattice action. A lattice action in which $\mathcal{L}_1$ vanishes can be obtained by adding a counterterm of the form

$$a^5\sum_x\left\{c_1\hat{\mathcal{O}}_1(x) + c_3\hat{\mathcal{O}}_3(x) + c_5\hat{\mathcal{O}}_5(x)\right\} \tag{36}$$

where $\hat{\mathcal{O}}_k$ is some lattice representation of $\mathcal{O}_k$[19]. The $O(a)$ counterterms in $\hat{\mathcal{O}}_3$ and $\hat{\mathcal{O}}_5$ renormalise the bare coupling and mass. Hence, the $O(a)$-improved action involves one additional term, the so-called 'clover term',

$$S_{\rm SW} = S[U,\bar{q},q] + a^5\sum_x \bar{q}(x)\frac{ic\kappa}{2}\sigma_{\mu\nu}F_{\mu\nu}(x)q(x) \tag{37}$$

where the name derives from the four plaquettes used to approximate $F_{\mu\nu}$, which was first written down by Sheikholeslami and Wohlert[20, 21].

Improved axial current. Composite fields such as the isovector axial current and the pseudoscalar density,

$$A_\mu^a(x) = \bar{q}(x)\gamma_\mu\gamma_5\frac{\tau^a}{2}q(x) \tag{38}$$

$$P^a(x) = \bar{q}(x)\gamma_5\frac{\tau^a}{2}q(x), \tag{39}$$

where τ^a is a Pauli matrix acting on flavour, must also be improved if all $O(a)$ effects are to be removed from their matrix elements. Again, the classical field equations may be used to eliminate some of the field combinations, and we find that for the axial current only one additional term is needed[19]:

$$A_\mu^{a,\mathrm{imp}}(x) = A_\mu^a(x) + \frac{c_A}{2}\left[P^a(x+a\hat{\mu}) - P^a(x-a\hat{\mu})\right]. \tag{40}$$

A_μ^a must be renormalised if it is to obey the correct current algebra[22]. If a mass-independent renormalisation scheme is used to compute the matching factor $Z_A(g^2)$, then the renormalised improved axial current receives an additional term of $O(ma)$[18]:

$$A_\mu^{a,\mathrm{ren}}(x) = Z_A(g^2)\left(1 + b_A(g^2)am\right)A_\mu^{a,\mathrm{imp}}(x). \tag{41}$$

The coefficients c, c_A, b_A, etc, are functions of g^2 and are determined by imposing improvement conditions.

Choice of improvement coefficients. At tree-level in perturbation theory,

$$c = 1 \tag{42}$$

$$c_A = 0 \quad \text{(tree level)} \tag{43}$$

$$b_A = 1 \tag{44}$$

and it can be shown that the leading discretisation errors are reduced to $O(g^2a)$, provided the quark fields are 'rotated' by an amount of $O(am)$, corresponding to the b_A term in Eq (41)[21].

Perturbative expansions in terms of the bare coupling are useless for most quantities, because of large contributions from tadpole graphs. If these tadpole contributions are treated non-perturbatively, eg by writing $U_\mu = u_0[1 + igaA_\mu + \ldots]$ with $u_0 = \langle\frac{1}{3}(\mathrm{Tr}\,U_\Box\rangle^{1/4}$ from the simulations, and the perturbative expressions are adjusted to remove the one-loop contribution from u_0, then the perturbative estimates are much more reliable[23]. The mean-field, or tadpole, prescription for the action is to replace

$$U_\mu(x) \to \frac{U_\mu(x)}{u_0} \tag{45}$$

everywhere, and then to use perturbation theory (excluding the perturbative contribution from u_0) for the coefficients, ie,

$$g^2 = \tilde{g}^2u_0^4, \quad \kappa = \frac{\tilde{\kappa}}{u_0}, \quad c = \frac{\tilde{c}}{u_0^3} \tag{46}$$

where $\tilde{\kappa}_c$ and $\tilde{c}$ should be close to their perturbative values of 1/8 and 1, respectively. In other words,

$$c = \frac{\tilde{c}}{u_0^3}, \quad \tilde{c} = 1 \tag{47}$$

$$c_A = 0 \qquad \text{(tadpole)} \tag{48}$$

$$b_A = \frac{\tilde{b}_A}{u_0}, \quad \tilde{b}_A = 1 \tag{49}$$

and the tadpole-improved axial vector current is[24, 25, 26],

$$A_\mu^{a,\mathrm{ren}}(x) = Z_A\, 2\kappa u_0\left(1+\frac{am}{u_0}\right)A_\mu^a(x) \quad \text{(tadpole)} \tag{50}$$

$$Z_A = 1 + \left(\frac{1}{12} - 0.116458\right)\frac{g^2}{u_0^4}. \tag{51}$$

The Alpha Collaboration have determined c and c_A non-perturbatively[27] by requiring that the PCAC relation, Eq (28), holds irrespective of the states between which it is sandwiched[19]. The resulting best fit to the numerical data is

$$c = \frac{1-0.656g^2-0.152g^4-0.054g^6}{1-0.922g^2} \quad \text{(non-perturbative, } g^2 \le 1) \tag{52}$$

$$c_A = -0.00756g^2\,\frac{1-0.748g^2}{1-0.977g^2} \tag{53}$$

$$b_A = 1 + 0.1522\, g^2 \quad \text{(1-loop perturbation theory}^{28}) \tag{54}$$

The non-perturbative values should remove all $O(a)$ discretisation errors, whereas this is not guaranteed using the tadpole prescription. In both cases, the hope is that the remaining a dependence is weak enough to be extrapolated away with confidence.

Relevant Parameters

QCD has $N_f + 1$ free parameters, the quark masses and the gauge coupling. So N_f+1 dimensionful quantities, usually hadron masses, must be taken from experiment. One of these determines the lattice spacing in physical units, or, equivalently, the QCD scale, Λ_{QCD}, via the usual dimensional transmutation. The others determine the renormalised quark masses.

Chiral behaviour. Present day simulations are carried out for quark masses in the vicinity of m_s or higher. There are three reasons for this:

- The linear equation solvers for computing quark propagators typically have a condition number $\propto 1/(aM_{PS})^2 \propto 1/am$, so that the number of iterations grows linearly as the quark mass decreases and the computations become too expensive.
- Finite-size effects are likely to be large when the pseudoscalar meson mass is small and these would have to be modelled.
- Quenched QCD is pathological in the limit $m \to 0$, so the simulations cannot be regarded as a good approximation to QCD.

Consequently, the results for physical quantities have to be extrapolated in m from $m \simeq m_s$, where they are computed, to $m = m_{u,d}$.

The form of this extrapolation is given by leading-order chiral perturbation theory[29]. This assumes that the shift in mass away from the $m = 0$ limit is given by the matrix element of the quark mass term in the action:

$$\delta M = \frac{\langle|m\bar{q}q|\rangle}{\langle|\rangle} = \frac{m}{2M}\langle\bar{q}q\rangle, \tag{55}$$

using the covariant normalisation of states, ie,

$$\delta M^2 = m\langle\bar{q}q\rangle. \tag{56}$$

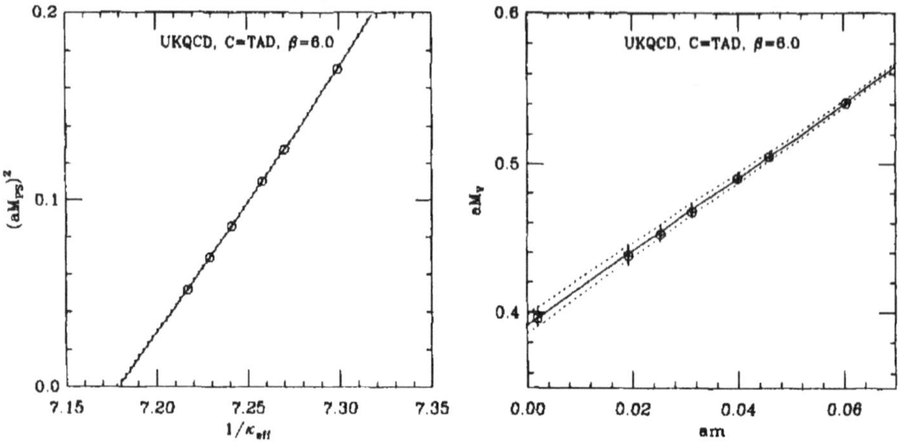

Figure 4. The squared pseudoscalar meson mass in lattice units versus the sum of the quark masses, $1/\kappa_{\text{eff}} = 1/2\kappa_1 + 1/2\kappa_2$, (left) and the vector meson mass versus quark mass, am, (right) from a quenched simulation with tadpole-improved Wilson quarks at $\beta = 6.0$.

Assuming the pseudoscalar meson's mass comes solely from this mechanism,

$$(aM_{PS})^2 = C_{PS}\, am \tag{57}$$

whereas, for hadrons whose masses do not vanish at $m = 0$,

$$aM_h = A_h + B_h\, am. \tag{58}$$

For hadrons made of non-degenerate quarks, m is the average quark mass. Chiral perturbation theory is believed to be reliable for quark masses up to that of the strange quark. Eq (57) and (58) should be used also for QQCD if we are using it as an effective theory.

The numerical simulation data for Wilson quarks is consistent with this picture, although chiral symmetry is broken explicitly at non-zero lattice spacing. As a consequence, there is no symmetry to prevent additive renormalisation of the quark mass. We find that the square of the pseudoscalar meson mass vanishes at $m = m_c$ (equivalently, $\kappa = \kappa_c$), as shown in Fig 4. For this set of data, slight curvature is observed over this range of quark masses, which extends to just above the strange quark mass, and the best fit is to,

$$(aM_{PS})^2 = C_{PS}\, am + D_{PS}(am)^2 \tag{59}$$

where

$$am = \frac{1}{2\kappa_{\text{eff}}} - \frac{1}{2\kappa_c}. \tag{60}$$

Also shown in Fig 4 is the fit of Eq (58) to data for the vector meson mass. The vanishing of the pseudoscalar meson mass at non-zero lattice spacing indicates a critical point of the Wilson theory[30]. The associated critical exponents determine the rate at which the pseudoscalar meson mass vanishes with quark mass, and the square-root behaviour of Eq (57) corresponds to mean-field exponents.

The values chosen for the quark masses in Fig 4 are too large to expose quenching effects, but it is interesting to compute the modification to the chiral behaviour due to QQCD.

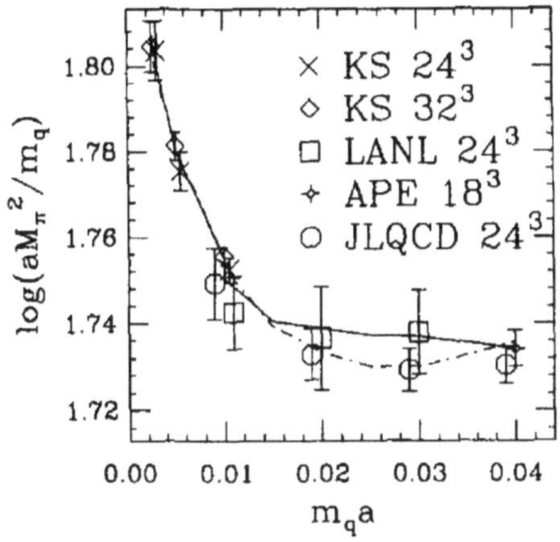

Figure 5. The observed chiral behaviour of the pseudoscalar meson mass in QQCD[32].

Quenched chiral perturbation theory. There are two distinct quenching effects[31].

1. Quenching removes loops which, at the underlying quark level, involve internal quark loops. This changes the values of the coefficients in the chiral expansion and may remove some terms completely.

2. Because of the absence of an anomaly, the η' remains light in QQCD, so there are new contributions coming from η' loops. Some of these are singular in the chiral limit. If these terms are large, then one can be sure that quenching errors are large. Consequently, it is necessary to carry out simulations at large enough quark masses for these terms to be negligible.

The η' loops modify the behaviour of the pseudoscalar meson mass in the chiral limit,

$$M_{PS}^2 = C'_{PS} m\left[1 + \delta \ln\left(\frac{\Lambda^2}{M_{PS}^2}\right)\right] \tag{61}$$

where $\delta \simeq 0.2$, so that

$$\frac{M_{PS}^2}{m} \propto m^{-\frac{\delta}{1+\delta}} \tag{62}$$

and this divergence has been observed in simulations, as shown in Fig 5[32].

Renormalised quark masses. The light quark masses are amongst the most poorly known parameters of the Standard Model. The Particle Data Book[33] gives $2 \leq m_u \leq 8$, $5 \leq m_d \leq 15$, and $100 \leq m_s \leq 300$ MeV (evaluated in the $\overline{\text{MS}}$ scheme at $\mu = 1$ GeV). Chiral perturbation theory[29, 34] relates their masses to those of the pseudoscalar mesons, but only predicts dimensionless ratios, eg, $2m_s/(m_u + m_d) = 25$ at lowest order, and 31 at next order.

Lattice QCD can provide a precise determination, because it directly relates input quark masses to output hadron masses. However, since electromagnetic effects are not included in the simulations, we can only calculate the isospin symmetric mass

$\bar{m} = (m_u + m_d)/2$. The lattice scale is usually taken from $M_\rho = 770$ MeV, then $\bar{m}$ is fixed by the isospin-averaged value $M_\pi = 137$ MeV, and m_s is fixed by one of $M_K = 495$ MeV, $M_{K^*} = 894$ MeV and $M_\phi = 1020$ MeV.

The $\overline{\text{MS}}$ mass at scale μ is related to the lattice mass by[35]

$$\begin{aligned} m_{\overline{\text{MS}}}(\mu) &= Z_m(\mu a)m(a) \\ &= \Big\{1 - \alpha_s\big[\gamma_0 \ln(\mu a) - C_m\big]\Big\}m(a) \\ &= \Big\{1 - \tilde{\alpha}_s\big[\gamma_0 \ln(\mu a) - \tilde{C}_m\big]\Big\}\frac{m(a)}{u_0}. \end{aligned} \tag{63}$$

Here $Z_m = Z_S^{-1}$ is the mass renormalisation, at one-loop order, relating the lattice and continuum regularisation schemes. It is computed by demanding that on-shell Green functions calculated with both regulators be equal[36,37]. $\gamma_0 = 2/\pi$ is the leading quark mass anomalous dimension. In the last part of Eq (63), the tadpole-improved mass is used, Eq (46), and the perturbative matching adjusted to remove the one-loop contribution to u_0. The lattice mass in Eq (63) is taken to be

$$m(a) = \frac{1}{a}\Big(\frac{1}{2\kappa} - \frac{1}{2\kappa_c}\Big) \tag{64}$$

for Wilson quarks, and is the bare mass in the lattice action for staggered quarks. There is some freedom in the precise choice of scale, $1/a \le \mu \le \pi/a$, in Eq (63) at which to match the continuum and lattice theories, and in the corresponding value of α_s (or $\tilde{\alpha}_s$). The usual choice follows the approach of Lepage and Mackenzie, with $\mu = q^*$, the dominant momentum scale in the loop integration, and $\tilde{\alpha}_s = \alpha_{\overline{\text{MS}}}(q^*)^{23}$. Once $m_{\overline{\text{MS}}}(\mu)$ has been computed, its value at any other scale Q is given by the two-loop running[38].

It turns out that the resulting quark masses are not very sensitive to tadpole effects (because the operators involved only contain one link variable), or to the choice of q^*. For Wilson quarks the one-loop contribution to Z_m is small, so the perturbative estimate of the renormalisation factor is likely to be reliable. But for staggered quarks the one-loop term is large, regardless of whether tadpole improvement is used, and the renormalisation factor is the major source of uncertainty. However, extrapolation of the lattice data to the continuum limit should remove any systematic error in the renormalisation prescription.

The u and d quark masses are determined by solving Eq (57) and Eq (58) with $m_h = m_V$, for $m = \bar{m}$, given $M_{PS} = M_\pi$ and $M_V = M_\rho$. The data from a range of simulations are shown in Fig 6. The expected leading term in the discretisation error is $O(a)$ for Wilson, $O(g^2 a)$ for tree-level improved (clover), and $O(a^2)$ for staggered quarks. This leading behaviour is used to extrapolate to $a = 0$, giving a consistent continuum limit of[35]

$$\bar{m}_{\overline{\text{MS}}}(2\ \text{GeV}) = 3.4 \pm 0.4 \pm 0.3\ \text{MeV}, \tag{65}$$

where the first error is the largest of the extrapolation errors and the second is due to the uncertainty in the scale. The data for the s quark mass in QQCD, obtained by fixing $M_V(m_s) = M_\phi$, are also shown in Fig 6. The Wilson data again show large $O(a)$ effects, although a consistent extrapolation to $a = 0$ is possible for both the Wilson and staggered data, giving[35]

$$m_{s,\overline{\text{MS}}}(2\ \text{GeV}, M_\phi) = 100 \pm 21 \pm 10\ \text{MeV}. \tag{66}$$

This is at the lower end of the expected range of values, but the ratio $m_s/\bar{m}$ is in good agreement with second-order chiral perturbation theory.

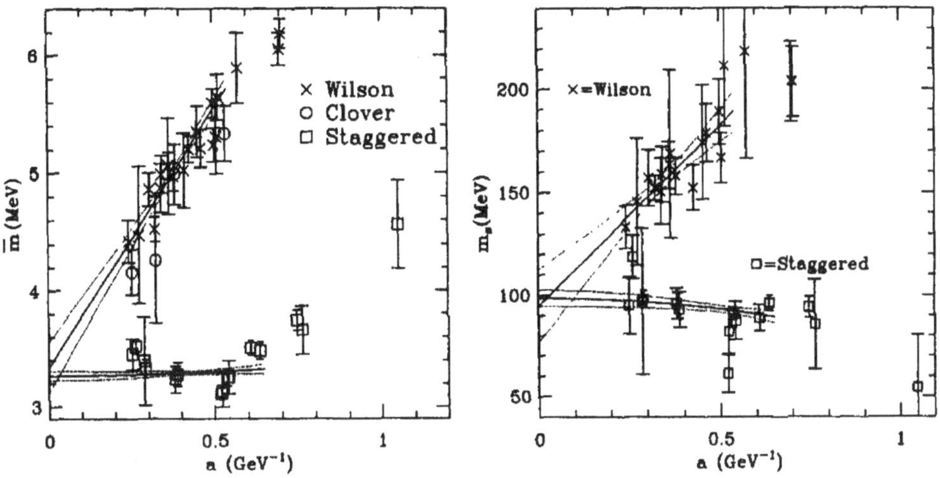

Figure 6. The behaviour with lattice spacing of $\bar{m}_{\overline{MS}}(2\ \text{GeV})$ obtained from M_π (left) and $m_{s,\overline{MS}}(2\ \text{GeV})$ obtained from M_ϕ (right) in QQCD with the scale set by M_ρ[35].

LIGHT HADRONS IN QUENCHED QCD

Early Results with Improved Actions

As an example of the present state of the art, I now present UKQCD's latest results for the spectrum of light hadrons and pseudoscalar decay constants in QQCD[3], obtained from simulations at $\beta = 5.7$, 6.0 and 6.2, which enable us to compare extrapolations to zero lattice spacing using the tadpole and the non-perturbative prescriptions for the clover coefficient. We find good agreement between the continuum estimates, which suggests that the extrapolation is under control. Although our data is at a fixed, and unfortunately rather small, box size with linear dimension of only about 1.7 fm, we also have data for a larger volume on the coarsest lattice for each improvement prescription and this gives us a check of finite-size effects.

At both volumes, our ensembles at $\beta = 6.0$ for the non-perturbatively-improved action contain a small number of exceptional configurations, in which the fermion matrix has a (near) zero eigenvalue[39]. For these configurations, which have been excluded from our analysis, the quark propagators at our lightest κ value fail to converge using several different solvers, although we have successfully converged some of them with a QMR algorithm[11]. The resulting pseudoscalar-meson timeslice correlator does not have a simple exponential decay from $t = 0$, as in Eq (11), and would distort the Monte Carlo average. Such zero modes are an artefact of the quenched approximation, because they appear with zero probability in the ensemble for full QCD (the determinant in Eq (22) would vanish), and so it is legitimate to omit them from the QQCD ensemble.

The Edinburgh plot for the light-hadron spectrum at our chosen quark mass values and using tadpole improvement is shown in Fig 3. The results using non-perturbative improvement are similar. In both cases, the mass ratios scale at fixed quark mass, defined by the value of M_{PS}/M_V. As shown in Fig 4, we observe slight curvature in our data for M_{PS}^2 versus quark mass and so we extrapolate to the chiral limit using Eq (59). For all other masses, our data are consistent with linear behaviour and we use Eq (58).

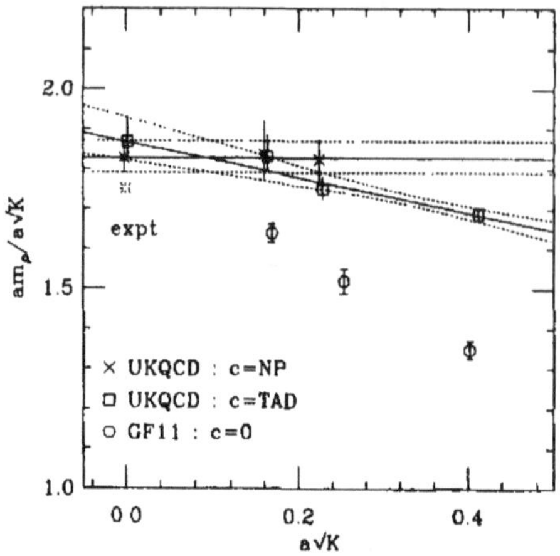

Figure 7. M_ρ **versus lattice spacing, with the scale set by the string tension (points have been slightly displaced horizontally for clarity).**

The tadpole-improvement prescription[23] is expected to reduce the leading discretisation error in physical quantities, but not to eliminate it completely, so some $O(a)$ dependence should remain in this data. In contrast, the leading lattice-spacing dependence for the non-perturbative improvement prescription[18] should be $O(a^2)$. In comparing the continuum extrapolations for our two data sets, we should work to a consistent order in a, ie, either we ignore $O(a^2)$ corrections in both sets of data, or we include them in both. Unfortunately, we only have three different lattice spacings for the tadpole-improved data and two for the non-perturbatively improved data, so only the first option is available to us. Hence, in the following, I will compare the results from linearly extrapolating the tadpole-improved data to $a = 0$, with the results from fitting the non-perturbatively improved data to a constant. This latter choice certainly underestimates the error in the continuum limit, but I shall be concerned only with showing consistency between the two continuum estimates, and this turns out to be remarkably good even within the unrealistically small errors. I note that, as a consequence of this agreement, we could fit all five data consistently to $O(a^2)$, with a single continuum value, to obtain a more reliable error estimate for the continuum limit.

Meson Masses

The lattice-spacing dependence of the ρ mass in units of the square root of the string tension, is shown in Fig 7. As expected, the tadpole-improved data are consistent with linear behaviour, whereas the dependence of the non-perturbatively improved data is weaker and consistent with a constant. The continuum extrapolations agree. This limit agrees with a linear extrapolation of GF11's data for the Wilson fermion action[2], which show a much stronger lattice-spacing dependence.

The strange quark mass can be fixed using any one of M_K, M_{K^*} and M_ϕ. These prescriptions are inconsistent in quenched QCD at fixed lattice spacing[24, 35]. In other words, the strange meson spectrum, at non-zero lattice spacing, is inconsistent with experiment. However, we find consistent results in the continuum limit. For example,

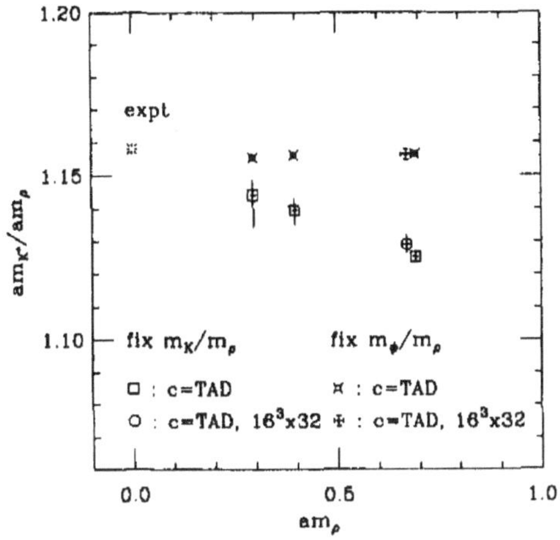

Figure 8. K^* mass in units of M_ρ, obtained from two different definitions of the strange quark mass.

Fig 8 shows our tadpole-improved estimates for the K^* mass, obtained using the K and the ϕ to define m_s. Although there are big differences between the results at non-zero lattice spacing, the continuum extrapolations are consistent with each other and with experiment. Fig 8 also shows results in a bigger volume at the coarsest lattice spacing, which indicate that finite-size effects for mesons are small.

Quark Masses

The light-quark masses at 2 GeV obtained with tadpole improvement are shown in Fig 9. Our results using the non-perturbative value of c are essentially identical to these, because the c values are almost the same and we do not yet have a non-

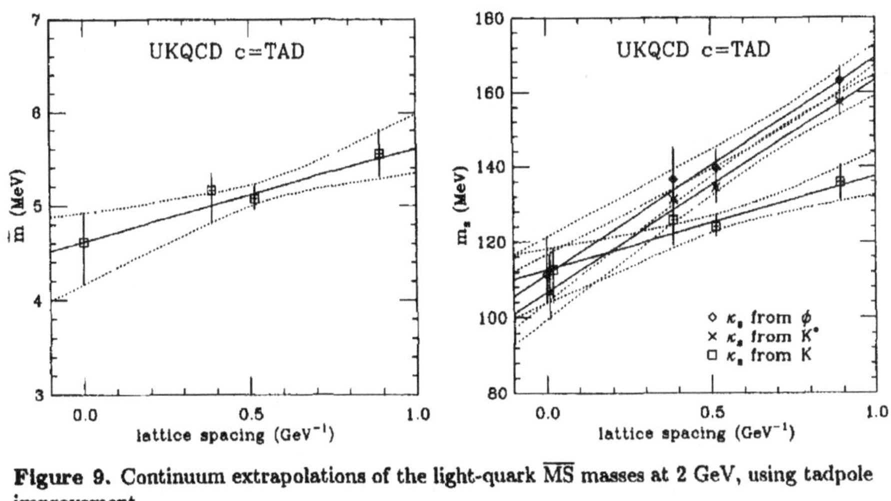

Figure 9. Continuum extrapolations of the light-quark $\overline{\text{MS}}$ masses at 2 GeV, using tadpole improvement.

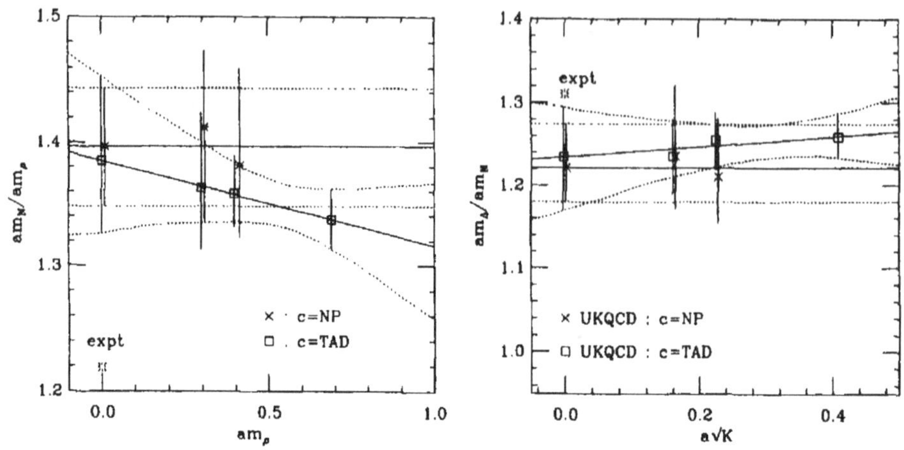

Figure 10. Continuum extrapolations of the nucleon mass in units of M_ρ (left) and the ratio of the Δ mass to the nucleon mass (right).

perturbative determination of Z_m, so we have to use the tadpole-improved values. Our u and d masses are 2σ above the estimate, Eq (65), based on world data, but this is not significant given the spread of data in Fig 6. Our result for m_s does agree with world data (Eq (66)), and confirms our observation from the strange meson spectrum that differences in the definitions of m_s at fixed lattice spacing, go away in the continuum limit.

Baryon Masses

Continuing this analysis for the baryons gives a similar picture of consistent continuum extrapolations for the two improvement schemes. This can be seen for the nucleon mass in Fig 10, although the continuum value appears to disagree with experiment (a bigger discrepancy is obtained if we use the scale from the string tension, rather than from M_ρ).

The ratio of the Δ mass to the nucleon mass, also shown in Fig 10, is only weakly dependent on lattice spacing (in both improvement schemes) and the continuum estimate is, in this case, in good agreement with experiment.

It is necessary to confirm that the baryons are not affected by the small volume of the lattice. Our results on a lattice of twice the linear size show that there is no discernable effect, given the size of our statistical errors.

Pseudoscalar Meson Decay Constants

The pseudoscalar meson decay constant, f_{PS}, is defined by the matrix element of the axial vector current,

$$\langle 0|A_0^{\text{ren}}|PS\rangle = i f_{PS} M_{PS}, \tag{67}$$

which is computed from the ratio of the axial-pseudoscalar and the pseudoscalar-pseudoscalar correlation functions, cf Eq (11). The tadpole-improved axial vector current is given in Eq (50). In the non-perturbatively improved scheme, the lattice axial vector and pseudoscalar currents are mixed (Eq (40)). This effect appears to be significant at our lattice spacings, but unfortunately, at the time of writing, the

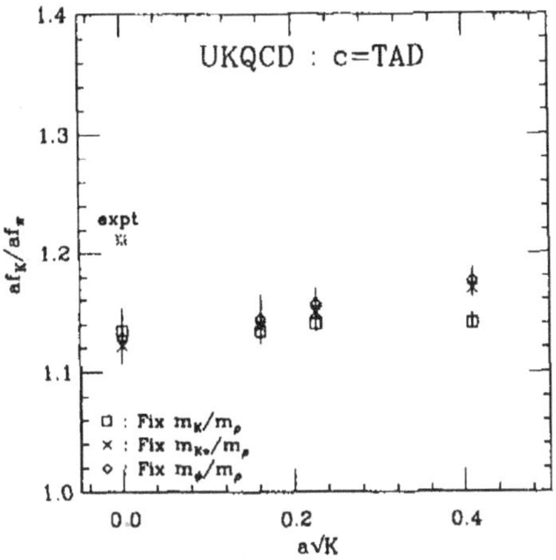

Figure 11. **Lattice-spacing dependence of f_K/f_π, obtained using different definitions of the strange quark mass and tadpole improvement.**

results show a puzzlingly-strong a dependence, which requires further investigation. So I cannot compare continuum extrapolations in the two improvement schemes.

Current normalisations cancel in the ratio f_K/f_π and the tadpole-improved data alone can be extrapolated reliably to the continuum limit. As shown in Fig 11, we find a significant disagreement between the quenched QCD estimate and experiment, and that, within our statistics, the continuum limit for this quantity is independent of the definition of the strange quark mass.

CONCLUSION

There is a striking level of agreement between the continuum extrapolations of the light-hadron spectrum obtained from non-perturbative $O(a)$ improvement and from tadpole improvement. Our continuum estimates have been obtained at fixed physical volume, corresponding to a linear size of about 1.7 fm. The smallness of this volume is a cause for concern, but our checks of finite-size effects, particularly for baryons, which are expected to be the worst affected, indicate that there is no significant effect.

Although there are differences at fixed lattice spacing, we obtain compatible continuum estimates for f_K, M_{K^*} and/or M_K, using different definitions of the strange quark mass, ie, from M_K, M_{K^*} and M_ϕ. Equivalently, we obtain consistent continuum estimates for the strange quark mass, in agreement with world data and at the low end of the range expected phenomenologically. It will be interesting to see if this low value turns out to be a quenching effect.

It appears, therefore, that QQCD works better as an effective theory in the continuum limit than it does at a fixed, non-zero lattice spacing. However, for some quantities in the light-hadron sector, such as M_N/M_ρ and f_K/f_π, the evidence is growing that the continuum limit of QQCD disagrees significantly with experiment. Thus, despite considerable success as an effective theory, numerical simulations are now becoming sufficiently precise to expose the limitations of the quenched approximation.

This progress has been due to a 10^5-fold increase in computer speed, to the development of more efficient and robust linear equation solvers, and to the implementation of the Symanzik improvement scheme to $O(a)$. Since 1981, lattice volumes have increased by 5^4, but the lattice spacing is almost unchanged! This last fact is strong encouragement for applying the improvement programme in simulations of QCD with dynamical quarks, where the computational cost is a much higher power of the number of lattice sites and finite-size effects are probably bigger.

Acknowledgements

This work supported by PPARC Grants GR/J98202, GR/K54601 and GR/K55745, by EPSRC Grant GR/K41663, and by EU Network Grant CHRX-CT92-0051. Figures 3 (left) and 5 are reprinted from references 12 and 32, respectively, with kind permission of Elsevier Science - NL, Sara Burgerhartstraat 25, 1055 KV Amsterdam, The Netherlands.

REFERENCES

1. H. Hamber and G. Parisi, *Phys. Rev. Lett.* 47 (1981) 1792.
2. F. Butler, H. Chen, A. Vaccarino and D. Weingarten, *Phys. Rev. Lett.* 70 (1993) 2849; *Nucl. Phys.* B421 (1994) 217; B430 (1994) 179.
3. R.D. Kenway (UKQCD Collaboration), Improvement and the continuum limit for quenched light hadrons, in Proceedings of the International Workshop "Lattice QCD on Parallel Computers", ed. Y. Iwasaki and A. Ukawa, *Nucl. Phys. B (Proc. Suppl.)* (1997) in press.
4. T. Yoshié (CP-PACS Collaboration), CP-PACS results with the Wilson action, in Proceedings of the International Workshop "Lattice QCD on Parallel Computers", ed. Y. Iwasaki and A. Ukawa, *Nucl. Phys. B (Proc. Suppl.)* (1997) in press.
5. I. Montvay and G. Münster, "Quantum Fields on a Lattice", Cambridge University Press, Cambridge (1994).
6. K.G. Wilson, *Phys. Rev.* D10 (1974) 2445; *in* "New Phenomena in Subnuclear Physics", ed. A. Zichichi, Plenum Press, New York (1975).
7. H.B. Nielsen and M. Ninomiya, *Nucl. Phys.* B185 (1981) 20; *Nucl. Phys.* B193 (1981) 173.
8. T. Banks et al., *Phys. Rev.* D15 (1976) 1111.
9. L. Susskind, *Phys. Rev.* D16 (1977) 3031.
10. H.P. Shanahan et al. (UKQCD Collaboration), *Phys. Rev.* D55 (1997) 1548.
11. A. Frommer, *Nucl. Phys. B (Proc. Suppl.)* 53 (1997) 120.
12. K.C. Bowler, D.L. Chalmers, A. Kenway, R.D. Kenway, G.S. Pawley and D.J. Wallace, *Phys. Lett.* 162B (1985) 354.
13. S. Ono, *Phys. Rev.* D17 (1978) 888.
14. M. Lüscher, *Commun. Math. Phys.* 104 (1986) 117; 105 (1986) 153.
15. S. Gottlieb, *Nucl. Phys. B (Proc. Suppl.)* 53 (1997) 155.
16. K. Jansen, C. Liu, M. Lüscher, H. Simma, S. Sint, R. Sommer, P. Weisz and U. Wolff, *Phys. Lett.* B372 (1996) 275.
17. K. Symanzik, in "Mathematical Problems in Theoretical Physics", eds R. Schrader et al., *Lecture Notes* in *Physics*, Vol. 153 (Springer, New York, 1982); *Nucl. Phys.* B226 (1983) 187 and 205.
18. M. Lüscher et al. (Alpha Collaboration), *Nucl. Phys.* B478 (1996) 365.
19. P. Weisz, *these Proceedings.*
20. B. Sheikholeslami and R. Wohlert, *Nucl. Phys.* B259 (1985) 572.
21. G. Heatlie et al., *Nucl. Phys.* B352 (1991) 266.
22. M. Bochicchio, L. Maiani, G. Martinelli, G. Rossi and M. Testa *Nucl. Phys.* B262 (1985) 331.
23. G.P. Lepage and P.B. Mackenzie, *Phys. Rev.* D48 (1993) 2250.
24. R.D. Kenway (UKQCD Collaboration), *Nucl. Phys.* B (Proc. Suppl.) 53 (1997) 206.
25. P.A. Rowland, *Nucl. Phys. B (Proc. Suppl,)* 53 (1997) 308.
26. M. Lüscher et al. (Alpha Collaboration), DESY preprint 96-222, hep-lat/9611015.
27. M. Lüscher et al. (Alpha Collaboration), *Nucl. Phys.* B491 (1997) 323.
28. S. Sint and P. Weisz, FSU preprint FSU-SCRI-97-24, hep-lat/9704001.
29. J. Gasser and H. Leutwyler, *Phys. Rep.* C87 (1982) 108.

30. S. Aoki and A. Gocksch, *Phys. Rev.* D45 (1992) 3845.
31. C. Bernard and M. Golterman, *Phys. Rev.* D46 (1992) 853.
32. S.R. Sharpe, *Nucl. Phys.* B (Proc. Suppl.) 53 (1997) 181.
33. R.M. Barnett et al. (Particle Data Group), *Phys. Rev.* D54 (19%) 1.
34. J. Donoghue, B. Holstein and D. Wyler, *Phys. Rev. Lett.* 69 (1992) 3444.
35. R. Gupta and T. Bhattacharya, Los Alamos preprint LAUR-96-1840, hep-lat/9605039.
36. G. Martinelli and Y. Zhang, *Phys. Lett.* B123 (1983) 433.
37. M. Golterman and J. Smit, *Phys. Lett.* B140 (1984) 392.
38. C.R. Allton et al. (APE Collaboration), *Nucl. Phys.* B431 (1994) 667.
39. W. Bardeen et al., Fermilab preprint FERMILAB-PUB-97/121-T, hep-lat/9705008.

PHYSICS FROM THE LATTICE: GLUEBALLS IN QCD; TOPOLOGY; SU(N) FOR ALL N

Michael Teper

Theoretical Physics, University of Oxford
1 Keble Road, Oxford, OX1 3NP, U.K.

INTRODUCTION

In these lectures I will show, through three examples, how current lattice calculations are able to tell us interesting things about the continuum physics of non-Abelian gauge theories.

My first topic concerns the glueball spectrum. The physics question here is: where, in the experimentally determined hadron spectrum, are the glueballs hiding? I will first summarise what lattice calculations tell us about the continuum glueball spectrum of the $SU(3)$ gauge theory. I will then discuss what this tells us about the masses of the corresponding 'bare' glueballs in QCD. I will then turn to the experimental spectrum with some discussion of the interpretation of the observed states in the quark model. Finally I will pinpoint the experimental states most likely to have large glueball components.

The second topic concerns topological fluctuations in the $SU(3)$ gauge theory. Here the simplest physics question is: are these fluctuations large enough to be consistent with the observed large η' mass? Thanks to Witten and Veneziano this is a question that can be posed in the pure gauge theory. We shall see that the fluctuations of the topological charge do indeed have the required magnitude. Along the way I will discuss the problems with topology on a lattice. I then move onto the much less straightforward question concerning the structure of these vacuum fluctuations: e.g. what is the size distribution of instantons? I will discuss some preliminary calculations that show the mean size to be about $\bar{\rho} \sim 0.5$ fm. I will also point to some intriguing evidence for a long distance polarisation of the topological fluctuations.

My third topic concerns the physics of $SU(N_c)$ gauge theories as a function of the number of colours, N_c. This will be mainly in $2+1$ dimensions, since that is where we have good calculations. I will describe calculations of the mass spectrum for $N_c \leq 4$ which explicitly show that for $N_c \geq 2$ mass ratios are independent of N_c up to a modest $\sim 1/N_c^2$ correction. This is a very elegant result: it tells us that all the apparently different $SU(N_c)$ theories are actually one single theory, $SU(\infty)$, to a reasonable first approximation. There is some very preliminary evidence, as I will show, that the same is true for $D = 3 + 1$.

GLUEBALLS IN QCD

Ideally I should be telling you what happens when you simulate QCD with realistically light quarks. But it is going to be a few years yet before I can do that. What current lattice Monte Carlo calculations are able to provide is predictions for the low-lying mass spectrum of the continuum $SU(3)$ gauge theory without quarks. These states are glueballs - there being nothing other than gluons in the theory. If you want hadrons with quarks then you can propagate quarks in this gluonic vacuum and then tie such propagators together so that the object propagating has the appropriate hadronic quantum numbers. That is to say, you calculate hadron masses in the relativistic valence quark approximation. (This is usually referred to as the 'quenched approximation' to QCD.) The spectrum one obtains this way is a remarkably good approximation to the observed hadron spectrum. This is not too surprising: one reason we were able to learn of the existence of quarks in the first place is because the low-lying hadrons are in fact well described by a simple valence quark picture.

Suppose that we begin with our SU(3) gauge theory and then couple to it 3 flavours of very heavy quarks. Initially the spectrum will contain the usual light glueball spectrum, supplemented by a spectrum of very heavy quarkonia that can be well accounted for in terms of a valence quark potential model. Let us now gradually reduce the quark masses towards their physical values. In principle the glueball and quarkonia states might entirely change their character once their masses become comparable. However, as we remarked above, the experimental light quark spectrum still seems to retain the essential features of valence quark physics. If the quarkonia are not qualitatively altered, it seems reasonable to think that neither will the glueballs be. Of course if a quarkonium state and a glueball are close enough in mass they will mix. However there is reason to believe that this mixing is weak. The reason is the Zweig (OZI) rule: hadron decays where the initial quarks all have to annihilate are strongly suppressed. The classic example is the ϕ meson. Such a decay may be thought of as $quarks \rightarrow glue \rightarrow quarks$. Glueball mixing with quarks should therefore be $\sqrt{OZI}$ suppressed. As should glueball decays into hadrons composed of quarks. The existence of such a suppression is supported by a recent lattice calculation[1].

The picture we have in mind is therefore as follows. The glueballs will only be mildly affected by the presence of light quarks. They will decay into, say, pions but their decay width will be relatively small; and there will be a correspondingly small mass shift. Only if there happens to be a flavour singlet quarkonium state close by in mass will things be very different, because of the mixing of these nearly degenerate states. In this context we expect 'close by' to mean within ~ 100 MeV. So we view the glueballs in the pure SU(3) gauge theory as being the 'bare' glueballs of QCD which may mix with nearby quarkonia to produce the hadrons that are observed in experiments. All this is an assumption of course, albeit a reasonable one. If true it tells us that the glueballs, whether mixed with quarkonia or not, should lie close to the masses they have in the gauge theory. So we now turn to the calculation of those masses.

I shall begin by reviewing the available lattice calculations, placing a particular emphasis on exposing the sources of systematic error in arriving at a final mass prediction in MeV units. I do this in some detail, so that you are able to judge for yourselves the credibility of lattice mass estimates.

As we shall soon see, it is only for three states that the lattice calculations are reliable enough that we can extract continuum predictions: the lightest scalar, 0^{++}, tensor, 2^{++}, and pseudoscalar, 0^{-+}, glueballs. For other states we do have calculations for one or two values of the lattice spacing a, but that is not enough to extrapolate to $a = 0$. Nonetheless the lattice results strongly suggest that glueballs with other J^{PC} are heavier[2].

The lightest glueball is the 0^{++}, and it is for this state that we have the most accurate lattice predictions. Although there have been recent estimates for the mass that appear to differ, e.g.[2] 1.55 ± 0.05 GeV and[3,1], 1.74 ± 0.07 GeV we shall see that this difference is illusory. In fact the apparent difference reflects different ways of extrapolating to the continuum limit and different ways of introducing physical MeV units into the pure gauge theory. That is to say, it reflects particular systematic errors which we need to estimate and this we try to do. Since we find that the various lattice calculations are consistent, we are able to carry out a global analysis that provides the best available glueball mass estimate. We find $m_{0^{++}} = 1606 \pm 73 \pm 130$ MeV where the first error is statistical and the second is systematic. Performing a similar analysis for the lightest tensor and pseudoscalar glueballs we find $m_{2^{++}} = 2200 \pm 73 \pm 130$ MeV and $m_{0^{-+}} = 2100 \pm 73 \pm 130$ MeV.

From the experimental and phenomenological point of view the scalar sector is complex, and I will review the current state of play. As we shall see, there appear to be too many scalar states to be explained as quarkonia, and the strongest candidates for states with large gluonic components are the $f_0(1500)$ and the $f_{(J=0?)}(1710)$. This possibility is strongly reinforced by the fact that these are the only experimental states that are compatible with the lattice mass estimate in the previous paragraph.

In order to assess which states have the largest gluonic components, it is important to understand the mixing between nearby quarkonia and glueballs. I shall briefly discuss what happens in the case of mixing between a glueball and the lightest singlet and octet scalar quarkonium multiplets.

For this review I have drawn very heavily on Ref.[4].

Calculating Glueball Masses

I begin by briefly reminding you of some general aspects of lattice calculations. (For more detail see the lectures by Chris Michael at this School and the books by Creutz[5] and Montvay and Münster[6].) In a lattice calculation (Euclidean) space-time is discretised, usually onto a hypercubic lattice. What we calculate is the mass spectrum of the discretised theory, but what we actually want is the corresponding spectrum of the continuum theory. Because the theory is renormalisable, the effects of the lattice spacing, a, on physical length scales will vanish as $a \to 0$. Since QCD has effectively one length scale, ~ 1 fm, one expects the effects of the discretisation to become negligible once $a << 1$ fm. The same is true of the gauge theory without quarks. (By '1 fm' I really mean some characteristic physical length scale. How one introduces actual fermi units into the pure gauge theory is something we shall return to below.) The lattice spacing is varied by changing the value of the bare (inverse) coupling which appears in the lattice action: $\beta = 6/g^2$. Since the theory is asymptotically free, we know that in order to approach the continuum limit, $a \to 0$, we need to take $g^2 \to 0$ and so $\beta \to \infty$. Indeed for sufficiently small g^2 one can determine the relationship between a and g^2 in low-order perturbation theory. (Although in practice the latter does not work well for the range of couplings currently accessible.) Reducing a makes the calculation numerically more intensive for various reasons. An obvious one is that if we wish to maintain a constant volume, the number of lattice sites grows $\propto 1/a^4$. It is only in recent years that calculations for very small values of a have become practical. As we shall see, the mass calculations I shall use here have been performed over a range of lattice spacings $0.18 \text{ fm} \geq a \geq 0.06$ fm, corresponding to couplings $5.7 \leq \beta \leq 6.4$.

We now outline the main steps in a lattice Monte Carlo calculation of a glueball mass with a view to exposing the main different sources of systematic error. (So we ignore various unilluminating technicalities and skate over various qualifications that are irrelevant for this purpose.)

• *Extracting lattice masses.* The first step is to calculate the glueball spectrum on a lattice, for a given space-time volume V and for a particular lattice spacing a. The Euclidean time translation operator is e^{-Ht}, where H is the Hamiltonian of the theory. Thus the correlation function of an operator with some particular J^{PC} quantum numbers will, for large enough values of t, vary as $e^{-m_0 t}$ where m_0 is the lightest mass with those quantum numbers. We calculate such propagators numerically for t large enough that we see this asymptotic exponential decay. From the exponent we extract the mass. Since t is given in lattice units ($t = an$ where n is the number of lattice spacings) the exponent is $m_0 t = m_0 an$ and so what we actually obtain is the mass in lattice units, i.e. am_0. The statistical errors are straightforward to estimate. However there is also a systematic error that has to do with determining the range of t where the asymptotic exponential dominates. We do not attempt to quantify this error but simply note that it will become increasingly important for the heavier glueball states (such as the tensor and pseudoscalar) where the exponential decrease with t of the 'signal', and hence its immersion into the statistical noise, is more rapid.

• Finite *V corrections.* The second step is to determine the corrections due to the fact that the volume is finite. For most quantities the functional forms of the leading large-V corrections are known theoretically; typically they will be of the form $\delta m \propto \exp\{-camL\}$ where am and L are the mass-gap and lattice size in lattice units[7]. By doing calculations for a variety of volumes at some chosen value of a, any unknown constants in these expressions can be fitted and the resulting formulae can be used to apply corrections at other values of a. Since the calculations we use here have been performed on periodic volumes of typical sizes 1.5 to 2.0 fm, these corrections are smaller than our typical statistical errors; but it is hard to determine them more accurately than that. This provides another source of systematic error.

• Finite a corrections. The calculated glueball mass will depend on a; and it will do so in two ways. Firstly, it is obtained in lattice units, am. This trivial dependence is removed when we take the ratio of two masses: $am_1/am_2 \equiv m_1/m_2$ The non-trivial a dependence is due to the distortion of the dynamics by the discretisation. For the lattice action we use and for quantities such as glueball mass ratios it is known[8] that the leading small-a correction is $O(a^2)$. So for small enough a we can extrapolate our calculated mass ratios to $a = 0$ using

$$\frac{m_1(a)}{m_2(a)} = \frac{m_1(a=0)}{m_2(a=0)} + c(am)^2 \tag{1}$$

where m may be chosen to be m_1 or m_2 or some other physical mass: the difference between these choices is clearly higher order in a^2. Such neglected higher order terms in the extrapolation are another source of systematic error.

• Introducing the *MeV* scale. Having obtained the continuum mass spectrum in the form of mass ratios, we want to express the masses in usual MeV units. This can be done if at least one of the masses corresponds to a quantity whose value is known in MeV. For example the potential between heavy quarks is linear for large separations: $V(r) \simeq \sigma r$ where σ is called the string tension. In simple string pictures for high J hadrons σ is related to the slope, α', of Regge trajectories by $\alpha' = 1/2\pi\sigma$ and this provides the conventional estimate[9] $\sqrt{\sigma} \simeq 400 - 450$ MeV. Knowing the continuum value of $m_G/\sqrt{\sigma}$ we can now express m_G in MeV units. Of course one may distrust this particular argument for the value of σ. An alternative is to calculate quark propagators in the pure gauge theory and from these to form hadron propagators. From the asymptotic exponential decay of the latter we obtain quarkonium masses in the (relativistic) valence quark approximation. For example, we can obtain the continuum limit of the mass ratio $m_\rho/\sqrt{\sigma}$. Setting $m_\rho = 770$ MeV we obtain a value for $\sqrt{\sigma}$ and hence for the glueball mass m_G. Or we might set the scale using the ϕ-meson or the nucleon instead of the ρ. While the mass spectrum one obtains in the quenched approximation is remarkably close to that which is experimentally observed, it is not exactly the same and so these different ways of setting the physical scale will lead to slightly different glueball masses. This is a source of systematic error.

The first three types of systematic error can be made arbitrarily small by sufficiently improving the numerical calculations (in obvious ways). The error in setting the MeV scale is qualitatively different. It is intrinsic to working within the quenched approximation. How do we estimate it? There are some quantities which we know are going to be sensitive to the absence of vacuum $q\bar{q}$ fluctuations, e.g. the η' mass or the topological susceptibility. These should obviously not be used to set the MeV scale in the pure gauge theory. There are other quantities which we expect to be no more sensitive to the absence of vacuum $q\bar{q}$ fluctuations than the glueball masses themselves and which are therefore suitable quantities with which to attempt to set the MeV scale. These include the masses of typical quarkonia such as the ρ meson, the ϕ meson, the nucleon, certain matrix elements, etc. Of course the ratios of these quantities cannot be exactly the same in full and quenched QCD and so the scale we extract will vary according to which of these quantities we choose to use. The extent of this variation can be used as an estimate of the systematic error.

The Lightest Glueballs

The values of the glueball masses that we shall use are from Refs.[11, 10, 2, 3] and those of the string tension are taken from Refs.[10, 2, 12]. As we have mentioned already, only the 0^{++}, 2^{++} and 0^{-+} glueballs are determined accurately enough that a continuum extrapolation is possible. In Ref.[2] you can find mass estimates for glueballs of widely varying J^{PC}, obtained for a very small lattice spacing. These do suggest that the three masses we shall obtain are in fact the lightest ones. There is clearly an urgent need for a new generation of lattice glueball calculations which will provide information on a much larger part of the continuum mass spectrum. For a preview of what these are likely to look like, see Ref.[13].

The first step is to take ratios of masses so that the scale, a, in which they are expressed cancels. We choose to take ratios of the glueball masses, am_G, to $a\sqrt{\sigma}$ since this latter quantity has been very accurately calculated. Now we remarked above that for small enough a the leading discretisation effects in such mass ratios are $O(a^2)$. So for small enough a we expect

$$\frac{m_G(a)}{\sqrt{\sigma(a)}} = \frac{m_G(a=0)}{\sqrt{\sigma(a=0)}} + ca^2\sigma \tag{2}$$

We take all the available mass values and try to fit them using eq. (2). If a good fit is not possible we assume that this is because the largest values of a used is too large for the $O(a^2)$ correction to be adequate. So we drop the mass corresponding to the largest value of a and try again. We keep doing this until we get a good fit. We find that the 0^{++} glueball can be well fitted in this way over a range of lattice spacings $0.18\ \text{fm} \geq a \geq 0.06\ \text{fm}$, corresponding to couplings $5.7 \leq \beta \leq 6.4$. (Note that where

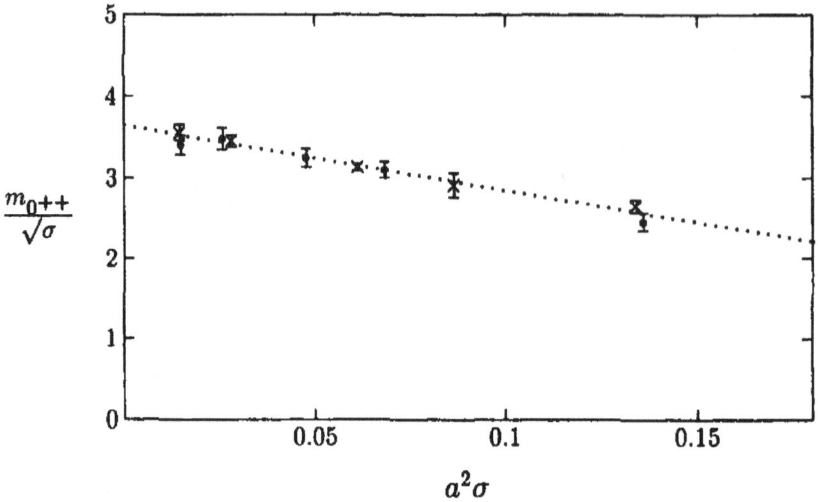

Figure 1. The scalar glueball mass: the GF11 values (×) and the rest (•). The best linear extrapolation to the continuum limit is shown.

we employ fermi units, these have been introduced using $\sqrt{\sigma} = 440$ **MeV**; a value that will be made plausible later on.) Such lattice spacings are small enough that we are not surprised that higher order terms in a^2 should be small. We show the mass ratios in Fig. 1 together with the best fit of the form in eq. (2). We obtain the *continuum* mass ratio:

$$\frac{m_{0^{++}}}{\sqrt{\sigma}} = 3.65 \pm 0.11 \tag{3}$$

The fit is obviously a very good one (with a confidence level of 85%) and it is clear that the calculations of the different groups are entirely consistent with each other.

In Fig. 2 and Fig. 3 I show corresponding plots for the 2^{++} and 0^{-+} glueballs. The

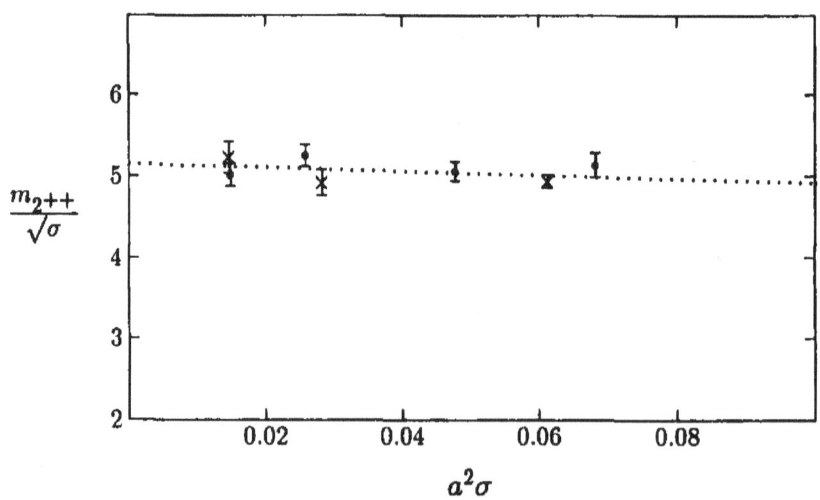

Figure 2. The tensor glueball mass: the GF11 values (×) and the rest (•). The best linear extrapolation to the continuum limit is shown.

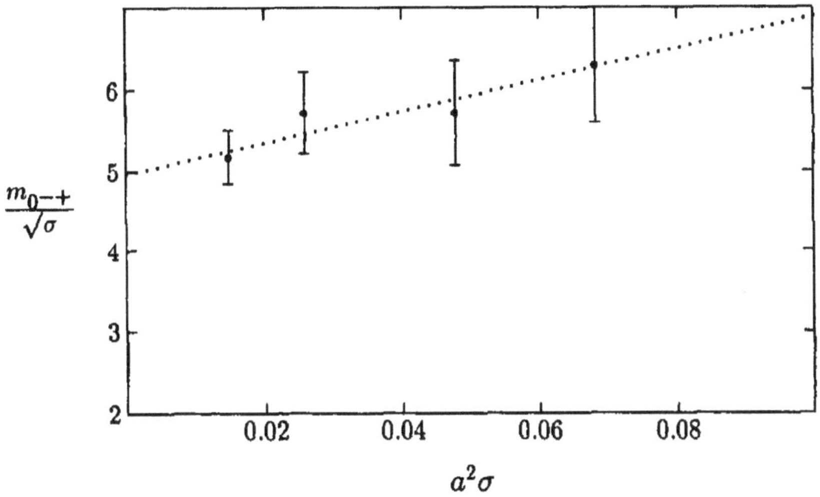

Figure 3. **The pseudoscalar glueball mass. The best linear extrapolation to the continuum limit is shown.**

former is well determined, and we obtain the continuum ratio

$$\frac{m_{2^{++}}}{\sqrt{\sigma}} = 5.15 \pm 0.21. \tag{4}$$

It is however clear that our control over the 0^{-+} is marginal. This translates into a very large error when we perform the continuum extrapolation:

$$\frac{m_{0^{-+}}}{\sqrt{\sigma}} = 4.97 \pm 0.58. \tag{5}$$

We now wish to transform the above continuum glueball masses to physical MeV units. There are several reasonable ways to do this and how they differ will give us an estimate of the systematic error intrinsic to introducing physical units into a theory that is not quite physical.

As we remarked earlier, one way is to infer from the observed Regge slopes that $\sqrt{\sigma} \simeq 400 - 450$ MeV. This estimate does however suffer from being somewhat model dependent. An alternative is to take lattice calculations of the mass of the ρ and extrapolate the ratio $m_\rho/\sqrt{\sigma}$ to the continuum limit, just as we did for the glueball. The only difference with eq. (2) is that the leading correction will be $O(a)$ rather than $O(a^2)$. We do this for two recent 'state-of-the-art' calculations: the GF11 collaboration[14] and the UKQCD collaboration[15] who use quite different discretisations for the quark action. UKQCD uses an improved action which should have smaller discretisation errors. We plot the two sets of ratios in Fig. 4 with their corresponding continuum extrapolations. These give us

$$\frac{m_\rho}{\sqrt{\sigma}} = 1.72 \pm 0.08 \qquad GF11 \tag{6}$$

and

$$\frac{m_\rho}{\sqrt{\sigma}} = 1.78 \pm 0.08 \qquad UKQCD \tag{7}$$

respectively. It is reassuring that these two calculations are entirely consistent in the continuum limit, despite the fact that they have very different lattice discretisation

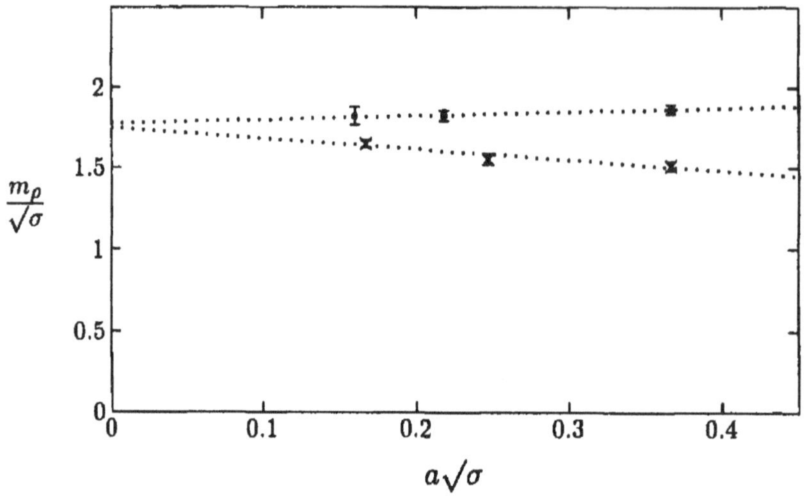

Figure 4. The ρ mass: GF11 (×) and UKQCD (•) values. The best linear extrapolations to the respective continuum limits are shown.

corrections. (Indeed one can argue that the dominant correction in the UKQCD calculation will be $0(a^2)$ rather than $0(a)$.) To extract a value for σ in MeV units we average the above results, set $m_\rho = 770$ MeV and so obtain $\sqrt{\sigma} = 440 \pm 15$ MeV. We observe that this is entirely consistent with the scale we inferred from Regge slopes; but the argument is much cleaner here.

The error on σ is largely statistical. We now need to estimate the systematic errors as well. These are discussed in detail in Ref.[4]. There we consider errors due to finite volume corrections, to extrapolations in the quark mass, to uncertainties in our estimates of the lattice values of σ and to using the K^* or nucleon to set the scale rather than the ρ. We estimate a ±8% systematic error in total. This leads to our final estimate for the value of the string tension as being:

$$\sqrt{\sigma} = 440 \pm 15 \pm 35 \text{ MeV} \tag{8}$$

where the first error is statistical and the second is systematic.

We can now use this value in eqs. (3–5) to express our glueball masses in MeV units. We obtain

$$m_{0^{++}} = 1.61 \pm 0.07 \pm 0.13 \text{ GeV} \tag{9}$$

$$m_{2^{++}} = 2.26 \pm 0.12 \pm 0.18 \text{ GeV} \tag{10}$$

and

$$m_{0^{-+}} = 2.19 \pm 0.26 \pm 0.18 \text{ GeV}. \tag{11}$$

The first error combines the statistical errors on the glueball and ρ continuum extrapolations, and the second is our estimate of the systematic error. This, then, is our best lattice prediction for the lightest glueballs prior to any mixing with nearby quarkonium states.

Where are the quarkonia? We are interested in flavour-singlet states because those are the ones that can mix with glueballs. It would clearly be very useful if lattice calculations were to provide estimates for the masses of the $u\bar{u} + d\bar{d}$ and $s\bar{s}$ mesons prior to their mixing with glue. One could then introduce the mixing with the scalar glueball using the (standard) formalism described below and compare the resulting states with the experimental spectrum. There are some indications from recent lattice calculations[16, 17] that there is a scalar $s\bar{s}$ state close to the scalar glueball mass. However these quenched calculations do not try to incorporate the essential quark annihilation contributions and so must be regarded as indicative at best. We will therefore turn now to experiment and phenomenology.

Experiment and Phenomenology

An illuminating (if arbitrary) starting point is provided by first considering the lightest 2^{++} mesons. In the quark model[18] these are very similar to the scalars: the quark spins are aligned in both cases and the only difference is that the net spin is parallel to the unit orbital angular momentum for the tensors and antiparallel for the scalars - which, in quark potential models, leads to minor differences in the masses[19]. However in the real world we expect the tensors to have narrower decay widths than the scalars because their decays into light pseudoscalars require non-zero angular momentum and corresponding near-threshold suppression factors. Thus they should be easier to identify experimentally - and that is their interest for us here.

Experimentally we find the following lightest tensor states[20]. There is an isoscalar $f_2(1270)$ with width, $\Gamma \sim 185$ MeV, a second isoscalar $f_2'(1525)$ ($\Gamma \sim 76$ MeV), an isovector $a_2(1320)$ ($\Gamma \sim 107$ MeV) and a strange isodoublet $K_2^*(1430)$ ($\Gamma \sim 100$ MeV) The first isoscalar decays mainly into pions while the second decays mainly into strange mesons. Thus it is natural to infer that the $f_2(1270)$ is mainly $u\bar{u}+d\bar{d}$ while the $f_2'(1525)$ is mainly $s\bar{s}$. We have a clear nonet of tensor mesons and we note that the splittings are exactly what one would expect from a mass-difference between strange and non-strange (constituent) quarks of ~ 100 MeV. Thus the lightest tensors provide a nice illustration of the quark model at its most successful: the mesons fall into $SU(3)$ multiplets with a modest symmetry breaking driven by the $m_s - m_n$ mass difference. (For convenience we shall adopt the shorthand notation $n\bar{n}$ for $\frac{1}{\sqrt{2}}(u\bar{u} + d\bar{d})$ in the following.)

We now turn to the 0^{++} mesons. The easiest such meson to see experimentally should be the strange isodoublet and the lightest such state turns out to be the $K_0^*(1430)$ with $\Gamma \sim 290$ MeV. Note that this is close in mass to the corresponding tensor although, as expected, its decay width is much larger. (This is what makes the scalars very much harder to identify experimentally.) There is also a candidate isovector $a_0(1450)$ ($\Gamma \sim 270$ MeV). In addition there are two isoscalars, the $f_0(1370)$ ($\Gamma \sim 300-500$ MeV) and the $f_0(1500)$ ($\Gamma \sim 120$ MeV). The former decays mainly into pions and so one would suppose it to be composed mainly of non-strange quarks.

So far all this looks much like the tensor nonet. However there is a puzzle. The $f_0(1500)$ does *not* have the obvious decays for a predominantly $s\bar{s}$ state. Moreover if we compare its width to the other scalars then we see that it is remarkably narrow – particularly for a state with, apparently, a large non-strange component. This motivates us to look for other nearby scalar states. There is evidence that the $f_J(1710)$ is, or contains, a 0^{++} state. Moreover the predominant 2-body decays of this state involve strange quarks ($K\bar{K}$ or $\eta\eta$). The state is relatively narrow $\Gamma \sim 175$ MeV. So if we include this state into our discussion, we no longer have an obvious $s\bar{s}$ problem. However now we have three isoscalar states in the 1370 – 1710 MeV mass region - too many for a quark model nonet! Since, as we have seen, lattice calculations predict a scalar glueball in the 1600 MeV mass region, with a width comparable to that which we observe for the $f_0(1500)$ and $f_J(1710)$, it is natural to conjecture that the three observed isoscalars are in fact the results of mixing between two quarkonium isoscalars and the lightest scalar glueball. This is the scenario explored in Refs.[23, 21, 22] with some differing assumptions.

All this might be a convincing picture if it were not for the existence of some lighter scalar states that we have so far failed to mention. These are the isoscalar $f_0(980)$ and the isovector $a_0(980)$. Both are narrow ($\Gamma \sim 40 - 100$ MeV). (There may also be an extremely wide isoscalar in the 400 – 1200 MeV mass range.) Since there is no nearby strange isodoublet, these states certainly do not fit into the usual quark picture where mesons should fall into approximate $SU(3)$ multiplets with modest symmetry breakings driven by the $m_s - m_n$ mass difference. In fact these mysterious states were interpreted some time ago[24] as being loosely-bound $K\bar{K}$ molecules - recall that they are narrow and occur very close to the $K\bar{K}$ threshold. If we accept this interpretation - and it has been widely accepted as being plausible - then we can ignore these states for our purposes and the interpretation of the previous paragraph remains. I should stress that alternative interpretations of this spectrum do exist, but I will ignore these here and instead refer you to Ref.[4] for a detailed discussion.

The picture we are thus led to is one where the $f_0(1370)$, the $f_0(1500)$ and the $f_0(1710)$ are the result of mixing between the glueball and the would-be $s\bar{s}$ and $n\bar{n}$ scalar quarkonia. We argued above that these quarkonia belong to the same nonet so the $s\bar{s}$ will be heavier than the $n\bar{n}$. Moreover the relative strengths of the glueball mixings will be determined on symmetry grounds.

What are the constraints on the mixing?

First the output masses of the mixing should correspond to the $f_0(1370)$, $f_0(1500)$ and $f_0(1710)$. The first state is very broad and so we shall allow its mass to lie in the region $M_1 \in [1.31, 1.40]$ GeV with a preference for 1.37 GeV. The second state is both quite narrow and well-defined and so we fix its mass to $M_2 = 1.5$ GeV. The third state appears to be again quite narrow but the precise location of the 0^{++} component is still quite uncertain. We shall consider the range $M_3 \in [1.64, 1.80]$ GeV with a preference for 1.71 GeV.

There are also some constraints on the input parameters. Given the lattice predictions, a generous range for the glueball mass would be $m_G \in [1.40, 1.80]$ GeV. We also have a qualitative constraint on the glueball-quarkonium matrix element of the Hamiltonian: it should be small, $\sim O(100)$ MeV or less. Similarly we expect a glueball to be narrow, and so any state with a large glueball component should be narrower than one would otherwise expect. As for the quarkonia, we expect $m_{s\bar{s}} > m_{n\bar{n}}$, with a mass difference in the $\sim 150 - 200$ MeV ball-park. And the mixing matrix elements should be close to their $SU(3)_{flavour}$ values. There are further, and important, constraints that arise from the observed decays of the three output states and again I refer to Ref.[4] for a more detailed discussion.

Here I shall give one example of what appears to be an acceptable mixing scheme. We assume that prior to mixing what we have are the 'bare' quarkonia with masses $m_{s\bar{s}} = 1.61$ GeV and $m_{n\bar{n}} = 1.36$ GeV and a bare glueball with mass $m_G = 1.48$ GeV These are not eigenstates of H and so the mass matrix will not be diagonal. We can write the latter as

$$\begin{pmatrix} m_G & z & \sqrt{2}z \\ z & m_{s\bar{s}} & 0 \\ \sqrt{2}z & 0 & m_{n\bar{n}} \end{pmatrix}$$

(The factor of $\sqrt{2}$ follows from the fact that $\frac{1}{\sqrt{2}}\{<u\bar{u}|+<d\bar{d}|\}H|G>= \sqrt{2}<u\bar{u}|H|G>$.) The physical masses, after mixing, are the eigenvalues of this mass matrix. With an acceptably weak mixing of $z = 62$ MeV we obtain as output masses 1.31, 1.50, 1.64 GeV for the $f_0(1370)$, $f_0(1500)$ and $f_0(1710)$, which is perfectly acceptable. The overlaps of the output states can be written in terms if the bare input states as follows:

i	f_{in}	f_{is}	f_{iG}
$f_0(1370)$	0.87	0.10	−0.49
$f_0(1500)$	0.48	−0.43	0.77
$f_0(1710)$	0.13	0.90	0.42

We see that most of the glueball resides in the $f_0(1500)$ and the $f_0(1710)$ is mainly an $s\bar{s}$ quarkonium. Such a mixing scheme is qualitatively compatible with the observed decays of the output states. This and other possible mixing scenarios are discussed in Ref.[4]. For some other recent discussions of mixing in this context see Ref.[21, 22].

Conclusions

We now know quite accurately the lightest scalar glueball mass in the pure gauge theory in units of the string tension. If we translate this into physical units, as described in this lecture, we find that the glueball should appear around $\sim 1.60 \pm 0.15$ GeV. This is just the mass range where one naively expects the two scalar flavour singlet quarkonia to lie. If they do then all these states will inevitably mix to some extent, even though we expect the dynamical mixing parameter to be weak. It is therefore intriguing that experimentally not only are there definitely states in this mass range, the $f_0(1370)$ and the $f_0(1500)$, but there is evidence for a third, the $f_0(1710)$. We presented a sample mixing scheme in which the glueball mainly resides in the $f_0(1500)$. This is not a new suggestion[21]. While it is still a little too early to come to a convincing conclusion, it certainly seems that it will not be very long before we are able to do so.

I have emphasised the scalar glueball because that is where most of the recent interest has been. What about the tensor and pseudoscalar? In looking for glueball candidates experimentalists naturally look for states that appear in 'gluon-rich' processes. For example in J/ψ decays, or in Pomeron-Pomeron collisions, or states that have large $\eta\eta$ decays. This picks out the $f_0(1500)$ and $f_0(1710)$ amongst the scalars. Amongst the tensors this picks out the $f_2(1900)$ and the $G(2150)$. These are, of course, in the right ball-park for the 2^{++} glueball as predicted by lattice calculations: $\sim 2.26 \pm 0.22$ GeV So here too, things look interesting.

With the pseudoscalar, on the other hand, we have little reason to feel smug. First, the lattice calculations are poor: a mass estimate of $\sim 2.16 \pm 0.32$ barely qualifies as an estimate at all. Moreover the obvious experimental candidate is the $\iota(1490)$ seen in J/ψ decays. We should however remember that this state is special: it has the quantum numbers of the vacuum topological charge. Which brings me smoothly to my next topic.

TOPOLOGICAL FLUCTUATIONS

As you know, SU(N) gauge fields in 3+1 dimensions possess[25] a topological charge. And it is the topological fluctuations of the gauge fields that are the reason why the η' has a mass ~ 1 GeV rather than being almost a Goldstone boson[26]. Moreover there is good reason to think that these fluctuations lead to the spontaneous breaking of chiral symmetry. The reason is that isolated instantons produce zero modes in the Dirac operator; these mix with each other, and shift away from zero, when the instantons are not isolated (as in the real vacuum). It is not hard to imagine that this might leave a non-zero density of modes close to zero, and this would suffice to break chiral symmetry (via the Banks-Casher formula[27]). This occurs explicitly in some instanton models (see

Shuryak's lectures at this School and Refs.[28, 30]) and has been observed in some lattice calculations[29]. It is also possible that instantons may affect the properties of some hadrons, as the near-zero modes may lead to large contributions to the valence-quark propagators. This is more speculative. All this to say that topology is interesting. In addition it is intrinsically non-perturbative. Here I will tell you some things that we have learned by studying topology in lattice gauge theory.

The first thing I need to address is the fact that when we discretise space-time we lose topology in a formal sense, since any field on a discrete set of points can be smoothly deformed to a trivial field. One should not get too excited about this; the same happens with dimensional regularisation. If the space-time dimension is not exactly 4 we have no topological winding. Since the theory is renormalisable we expect that, as the cut-off is removed, $a \to 0$, we recover all the properties of the continuum theory. We shall see how this occurs for topology in the lattice gauge theory.

So in this lecture I will address the following topics. First I discuss the basic ambiguity with defining topology on a lattice; and show why it does not really matter. Then I discuss the practical problem of calculating the topological charge of fields that have fluctuations on all length scales. This is a large subject and I will focus simply on the one technique that goes by the name of 'cooling'. I will then move onto the first bit of physics: the Witten-Veneziano fomula that relates the strength of the topological fluctuations in the pure gauge theory to the mass of the η' in QCD. Finally I will attempt to give some insight into the structure of the topological fluctuations in the vacuum.

A Basic Ambiguity and why it does not matter

Let me start by giving an explicit example. A lattice gauge field is defined by a set of SU(2) matrices on the links, l, of the lattice: $\{U_l\}$. (I will stick here to SU(2) for simplicity.) Consider now the following continuum gauge potential for an instanton of size ρ centered at $x = 0$:

$$A^I_\mu(x) = \frac{x^2}{x^2+\rho^2} g^{-1}(x)\partial_\mu g(x) \tag{12}$$

with

$$g(x) = \frac{x_0 + i x_j \sigma_j}{(x_\mu x_\mu)^{1/2}} \tag{13}$$

Let us now translate this field by $a/2$ in each direction so that it is centered at $\bar{x} = (a/2, a/2, a/2, a/2)$, i.e at the centre of a lattice hypercube. Define a lattice field by:

$$U^I_\mu(x) = \mathcal{P}\exp \int_x^{x+a\hat{\mu}} A^I_\mu(x)dx \tag{14}$$

For $\rho \gg a$ the lattice discretisation is very fine compared to the instanton core in which the instanton action and topological charge density, $Q(x) = \frac{1}{32\pi^2}\epsilon_{\mu\nu\rho\sigma} Tr\{F_{\mu\nu}(x)F_{\rho\sigma}(x)\}$, reside. In this case any reasonable definition of the topological charge (see below for an example) will assign a topological charge of $Q = 1$ to this lattice field. Suppose we now continuously reduce ρ. The lattice field will also vary continuously. Eventually we will have $\rho \ll a$. At this stage the instanton core, which is at the centre of a lattice hypercube, will be very far, in units of its size, from any of the lattice links. Thus even the nearest link matrices on the lattice will be arbitrarily close to pure gauge. That is to say, the field configuration will be the same as that due to a gauge singularity located at $x = \bar{x}$. In this case any reasonable definition of the topological charge will assign a topological charge of $Q = 0$ to this lattice field. Thus we have passed continuously from a field with $Q = 1$ to one with $Q = 0$.

Does this raise a fundamental problem with topology on the lattice? The answer is: not really. At least if we are interested, as here, with the limit $a \to 0$. Let me argue why this is so.

Consider the density of topological fluctuations as a function of their size ρ. This is only an unambiguous notion if $\rho \ll \xi_d$ where ξ_d is the typical dynamical length scale of the theory (e.g. 1 fermi in QCD or the inverse mass gap in the pure gauge theory). In that case we know the density of these 'instantons':

$$D(\rho)d\rho = \frac{d\rho}{\rho}\frac{1}{\rho^4}e^{-\frac{8\pi^2}{g^2(\rho)}} \ldots \tag{15}$$

where the '...' represent factors varying weakly with ρ. You recognise in this equation the scale-invariant integration measure; also a factor to account for the fact that a ball of volume ρ^4 can be placed in $1/\rho^4$ different ways in a unit volume; and finally a factor arising from the classical instanton action, $S_I = 8\pi^2/g^2$, with perturbative fluctuations promoting the bare g^2 to a running $g^2(\rho)$ in the usual way. This last sentence is the most important one for us here: if we substitute for $g^2(\rho)$ in eq. (15) we find that

$$D(\rho)d\rho \propto \rho^6 d\rho \qquad : SU(3) \tag{16}$$

with a power $\rho^{7/3}$ in the case of SU(2). So , because of the scale anomaly, the number of instantons rapidly vanishes as $\rho \to 0$ rather than diverging as $1/\rho^5$

On the lattice this density will change as follows if $a \ll \rho$ and $a \ll \xi_d$:

$$D(\rho) \to D(\rho) \times \{1 + 0(\frac{a^2}{\rho^2})\} \tag{17}$$

Now, suppose a lattice field configuration is to be smoothly deformed from $Q = 1$ to $Q = 0$. This requires a topological fluctuation to be squeezed out of the lattice, as described above. While we do not know much about the structure of the original fluctuation (it will typically be on a size scale $\sim \xi_d$ which is beyond the reach of our analytic techniques) we do know that if the lattice spacing is sufficiently small then to reach $\rho \sim a$ the 'instanton' will have to pass through sizes $\xi_d \gg \rho \gg a$ In this region the density is calculable as we saw above, with a probability that is very strongly suppressed; at least as $\sim (\rho/\xi_d)^6$ for SU(3). So the changing of Q is conditional upon the involvement of field configurations whose probability $\to 0$ as $a \to 0$. Thus, as we approach the continuum limit this lattice ambiguity vanishes very rapidly. We knew that this had to happen because the theory is renormalisable and the lattice is surely a good regulator. It is, however, nice to see it happen explicitly.

Cooling

Having convinced you that it makes sense to discuss topology on the lattice, I now want to discuss how you can calculate it. The obvious ways are three.

- Calculate the zero-modes of the Dirac operator $\not{D}[A]$. Q will equal the difference between the number of left and right handed zero-modes.
- Interpolate a smooth gauge potential. If $Q \neq 0$ and if space-time is compact (as it is here: a hypertorus) then we will need more than one patch. The net winding of the transition functions between the patches equals Q.

• We calculate the topological charge density $Q(x) = \frac{1}{32\pi^2}\epsilon_{\mu\nu\rho\sigma}Tr\{F_{\mu\nu}(x)F_{\rho\sigma}(x)\}$. Then $Q = \int Q(x)d^4x$

All these methods have been explored in lattice calculations. I will focus on the last because it is the simplest and because it immediately tells us something about the size and location of the core of the 'instanton' (since that is where $Q(x)$ is localised).

We need a lattice operator that becomes $Q(x)$ in the continuum limit. This is easy. Recall (see Chris Michael's lectures) that if we define the plaquette matrix, $U_{\mu\nu}(x)$, as the ordered product of link matrices around the corresponding plaquette of the lattice, then it is easy to see that $U_{\mu\nu}(x) = 1 + a^2F_{\mu\nu}(x) + ...$ and hence that[31]

$$Q_L(x) \equiv \frac{1}{32\pi^2}\epsilon_{\mu\nu\rho\sigma}Tr\{U_{\mu\nu}(x)U_{\rho\sigma}(x)\} = a^4Q(x) + 0(a^6). \tag{18}$$

If we apply this formula to an instanton of size ρ (discretised as described above) then we find, as expected, that $Q_L = \int Q_L(x)dx = 1 + 0(a^2/\rho^2)$.

Applied to the real vacuum however $Q_L(x)$ has problems. The operator is dimensionless and so $0(a^6)$ actually means terms like $\sim a^6F^3$, $\sim a^6FD^2F$ etc. For smooth fields these are indeed $0(a^6)$. However realistic fields (those that contribute to the path integral) have fluctuations all the way up to frequencies of $0(1/a)$. So if we take matrix elements of $Q_L(x)$ in the vacuum the contribution of the high frequency modes to the $0(a^6)$ terms will be $\delta Q_L(x) \sim a^6 \times 1/a^6 \sim 0(a^0)$. In practice this contribution is suppressed by some powers of β that can be calculated in perturbation theory (since these contributions are short-distance). Thus in the real world $Q_L(x)$ possesses interesting topological contributions that are of order $a^4 \propto e^{-c\beta}$ (since $g^2(a) \propto 1/\log(a\Lambda)$) and uninteresting ultraviolet contributions that are $\propto 1/\beta^n$. So as we approach the continuum limit, $\beta \to \infty$, the latter dominate and we are in trouble.

Actually things are a little worse than this. Like other composite lattice operators, $Q_L(x)$ possesses a multiplicative lattice renormalisation factor[32]: $Z_Q \simeq 1 + c'/\beta + ...$ This looks innocuous, and indeed in the continuum limit it obviously is. However it turns out that the value of c' is such that $Z_Q \ll 1$ precisely in the range of values of β where current lattice calculations are performed.

There are different ways to deal with these problems. I will describe a particularly simple technique[33]. The idea rests on the observation that the problems are all caused by the ultraviolet fluctuations on wavelengths $\sim a$. By contrast, if we are close to the continuum limit, the topology is on wavelengths $\rho \gg a$. One can therefore imagine taking the lattice fields and locally smoothing them over distances $\gg a$ but $\ll \rho$. Such a smoothing would erase the unwanted ultraviolet fluctuations while not significantly disturbing the physical topological charge fluctuations. One could then apply the operator $Q_L(x)$ to these 'cooled' fields to reveal the topological charge distribution of the vacuum.

How do we cool a lattice gauge field? The simplest procedure is to take the field and generate from it a new field by the standard Monte Carlo heat bath algorithm subject to one crucial modification: we always choose the new link matrix to locally minimise the plaquette action. Since $TrU_{\mu\nu}(x)$ measures the variations of the link matrices over a distance a, minimising the plaquette action is a very efficient way to erase the ultraviolet fluctuations. (Obviously there are many possible variations on this theme.)

Thus the notion is that we take our ensemble of N gauge fields, $\{U^{I=1,..,N}\}$, perform a suitable number of cooling sweeps on each one of these, so obtaining a corresponding ensemble $\{U_c^{I=1,...,N}\}$ of cooled fields, and then extract the desired topological properties from these cooled fields. As one might expect there are ambiguities for realistic values of a. As we cool, topological charges of opposite sign will gradually annihilate. This changes the topological charge density but not the total value of Q. Eventually this leads to a very dilute gas of instantons. As we cool even further these isolated instantons will gradually shrink and will eventually shrink within a hypercube and at this point even Q will change. (This is for a plaquette action on a large enough volume: other actions may have other effects.) Of course when an instanton becomes narrow it has a very peaked charge density and is impossible to miss. So we certainly know when it disappears out of the lattice and can, if we think it appropriate, correct for that. All this to say that cooling is a good way to calculate the total topological charge, but is not necessarily a reliable way to learn about the topological charge density.

Let me give you an example of how it works. I have produced a sequence of 90 Monte Carlo sweeps on a 12^4 lattice at $\beta = 2.5$. These field configurations are separated by just one heat bath sweep (i.e. each link matrix has been changed only once) so one expects the long distance physics on neighbouring field configurations to be almost identical. That is, the value of Q should change little. In Fig. 5 I show you the values

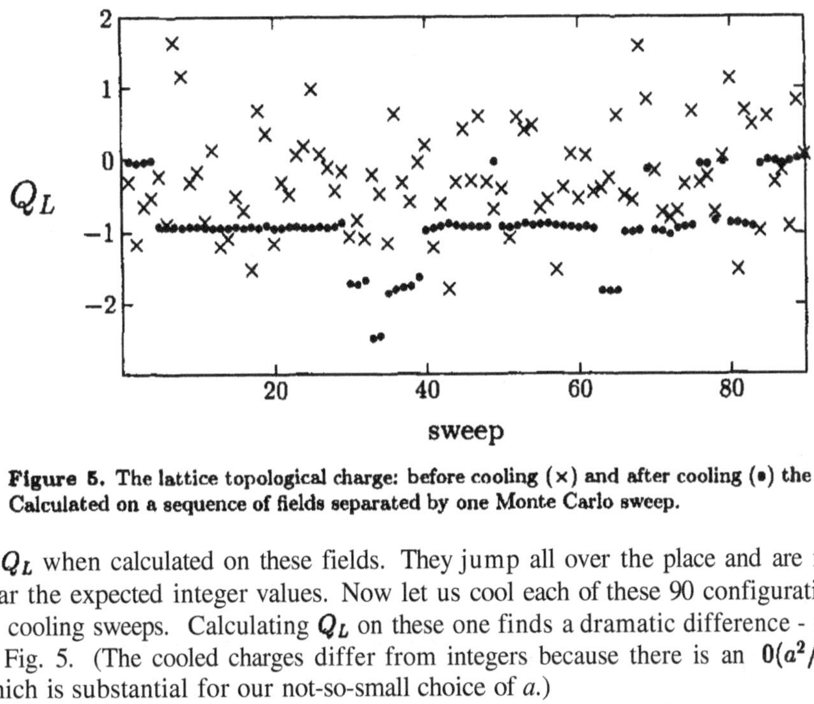

Figure 5. The lattice topological charge: before cooling (×) and after cooling (•) the fields. Calculated on a sequence of fields separated by one Monte Carlo sweep.

of Q_L when calculated on these fields. They jump all over the place and are nowhere near the expected integer values. Now let us cool each of these 90 configurations with 25 cooling sweeps. Calculating Q_L on these one finds a dramatic difference - as shown in Fig. 5. (The cooled charges differ from integers because there is an $0(a^2/\rho^2)$error which is substantial for our not-so-small choice of a.)

We now have a technique for calculating the topological charge of a lattice field. As we have seen, its application requires some care. Another reason to take care is the following. The probability of instantons with $\rho \sim a$ depends on the lattice discretisation. If the lattice action is much less than the continuum action then even when $a \to 0$ there may be a finite density (per unit physical volume) of these lattice artifacts. They might survive a couple of cooling sweeps. For realistic values of a the gap in scales between a and ξ_d is not very large and so such a narrow instanton, if sitting on a background field due to a physical but moderately narrow anti-instanton, might expand under cooling rather than shrinking. So it might occasionally survive our 20, or whatever, cooling sweeps and bias our calculations. Obviously this latter effect disappears as $a \to 0$. Another correction is due to the fact that with finite a one loses the tail of the continuum density that extends to $\rho \lesssim a$. This correction also disappears as $a \to 0$. To deal with these problems one needs to perform a suitable scaling analysis. For example suppose we define a topological charge $Q_L(\rho_c)$ which only includes topological charges for which $\rho \geq \rho_c = an_c$. As $a \to 0$ $Q_L(\rho_c)$ should become independent of ρ_c for a growing range $n_c^{min} \leq n_c \leq n_c^{max}$: with an_c^{min} large enough to exclude any $\rho \sim a$ artifacts and n_c^{max} growing exponentially with β. We shall demonstrate a simple calculation of this kind next.

The Topological Susceptibility

Since $< Q >= 0$ (we have no θ-term) the simplest quantity we can calculate is $< Q^2 >$. However there is a much better reason for calculating it: it is directly related[34, 35] to the η' mass:

$$\chi_t \equiv \frac{< Q^2 >}{volume} \simeq \frac{f_\pi^2}{2N_f}(m_{\eta'}^2 + m_\eta^2 - 2m_K^2) \tag{19}$$

If we put in the experimental numbers into the above formula, then we get

$$\chi_t \sim (180 \text{ MeV})^4 \tag{20}$$

In this relation the susceptibility, χ_t, is that of the pure gauge theory - which is something that we can calculate. We now describe such a calculation and see what happens.

The values of $< Q^2 >$ that I am going to use come from calculations that I have done in the past; for SU(3) they come from Refs.[36, 37] while for SU(2) they come from Refs.[38, 39, 40]. All these calculations have been performed using 20 to 25 cooling sweeps. The charge Q_L is then calculated on these cooled lattice fields. This charge is non-integer because the smallest instanton charges suffer significant $O(a^2/\rho^2)$ corrections. However such small instantons have very peaked densities and are easy to identify. One can then estimate the corrections due to these small instantons, and shift Q_L to the appropriate integer topological charge Q. This has been done in all these calculations.

In Fig. 6 I plot $\chi_t/\sqrt{\sigma}$ against the string tension, $a^2\sigma$ in lattice units. We expect

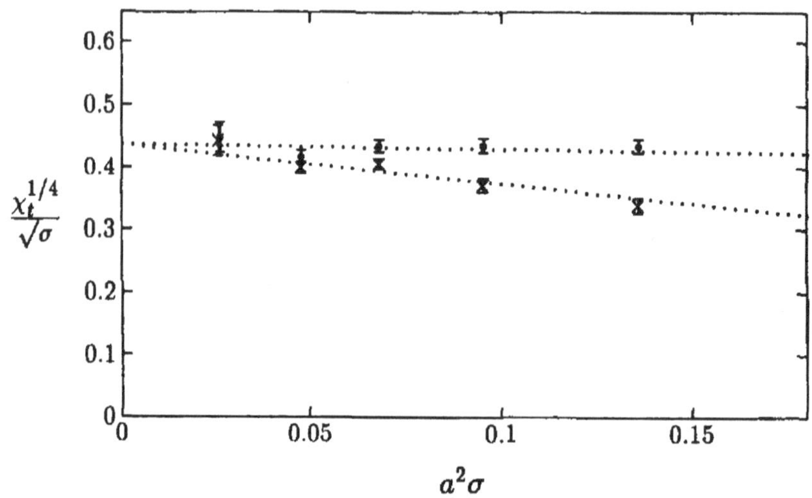

Figure 6. **The SU(3) topological susceptibility: from all peaks (•) and from peaks with $Q(x_{peak}) \leq 1/16\pi^2$ (×). Sample extrapolations to a common continuum limit are shown.**

the leading lattice corrections to this dimensionless mass ratio to be $0(a^2/\rho^2)$, as in eq. (1). So we would attempt a continuum extrapolation of the form

$$\frac{\chi_t(a)}{\sqrt{\sigma(a)}} = \frac{\chi_t(0)}{\sqrt{\sigma(0)}} + ca^2\sigma \tag{21}$$

which is a simple straight line on the plot. As we see the calculated values are consistent with this functional form. In addition to the susceptibility calculated from the total charge we also calculate Q with narrow instantons removed. The particular cut we have used is to remove all charges whose peak density is greater than $1/16\pi^2$. This corresponds to removing instantons with $\rho \lesssim 3a$. In doing this we are largely removing lattice artifacts with $\rho \sim a$ which, because of their environment, survive the cooling. Of course we are also removing a part of the small-ρ tail of the continuum density, $D(\rho)$. So we are not saying that this is necessarily a better measure of the continuum susceptibility. What we do wish to check is that both these susceptibilities are consistent when extrapolated to the continuum limit. And as we see in Fig. 6 this is indeed so. We extract:

$$\lim_{a\to 0} \frac{\chi_t}{\sqrt{\sigma}} = 0.437 \pm 0.020 \pm 0.015 \tag{22}$$

Here the first error is statistical and the second is a systematic error estimated using the two different extrapolations. If we now plug in our favourite value for the string tension, $\sqrt{\sigma} = 440$ MeV, we obtain

$$\chi_t = (192 \pm 12\ \mathrm{MeV})^4 \tag{23}$$

Of course we should really incorporate the uncertainty in the value of $\sqrt{\sigma}$, most of that coming from setting MeV units, and this would increase the error to ± 20 MeV. Irrespective of such details it is clear that the pure gauge theory susceptibility is indeed consistent with being large enough to drive the large η' mass.

In Fig. 7 I show the corresponding plot for the SU(2) susceptibility. Again we see that the two susceptibilities we calculate are consistent when extrapolated to the

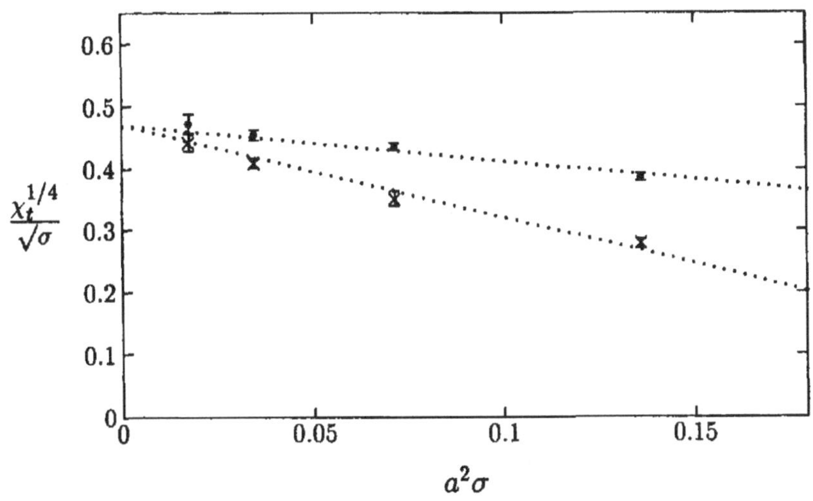

Figure 7. The SU(2) topological susceptibility: from all peaks (•) and from peaks with $Q(x_{peak}) \leq 1/16\pi^2$ (×). Sample extrapolations to a common continuum limit are shown.

continuum limit. Note that the difference between the two at corresponding values of a is much greater for $SU(2)$ than for SU(3). This is because the $SU(N)$ running coupling, $g^2(\rho)$, runs much faster to zero as N increases and so any finite-a ambiguities between physical and ultraviolet topological charges rapidly decrease. We extract:

$$\lim_{a\to 0} \frac{\chi_t}{\sqrt{\sigma}} = 0.470 \pm 0.011 \pm 0.015 \qquad : SU(2) \qquad (24)$$

which translates to

$$\chi_t = (207 \pm 9 \text{ MeV})^4 \qquad : SU(2) \qquad (25)$$

if we use $\sqrt{\sigma} = 440$ MeV. Unfortunately we do not know the η' etc. masses for the SU(2) theory so we cannot say if this is what is expected. And neither does it make much sense to introduce MeV units by using the SU(3) value of $\sqrt{\sigma}$ in this way. It does however allow me to compare to two recent estimates of[41] **200 ± 15 MeV** and of[42] **230 ± 10 MeV** which have been obtained using quite different methods, but a similar MeV scale. As we see, the values are all consistent - which is reassuring.

Vacuum Topological Structure

The calculation of Q on the lattice is relatively straightforward. If we continuously deform a continuum gauge field so as to minimise the action then we necessarily get driven to the semiclassical multi-instanton minimum. We do not change topological charge sectors since these are separated by infinite action barriers. Cooling is just a naive lattice version of this procedure and so it should work in this way for $a \to 0$. As we have seen, it does indeed seem to work well.

However there are other things we would like to know about the instanton density if we are interested in gauging the possible influence of instantons on chiral symmetry breaking, the quarkonium spectrum and related physics. For example a picture in which instantons have a mean size of $\sim 1/3$ fm and are relatively dilute produces a plethora of interesting physics[28]. Can we say something about such more detailed features of the vacuum topological structure?

This question involves both technical and conceptual ambiguities. There are two main technical problems. The first is that as we cool a gauge field its topological structure changes: instantons will change their sizes and nearby instantons and anti-instantons will annihilate. These quantities are not invariant under minimising the action; they are not even quasi-stable. So we must do as little cooling as possible and trust only those features that we find to be relatively insensitive to cooling. If the vacuum that one obtains is a dilute gas of instantons then there is no further difficulty. However what one might expect, and indeed finds, is a relatively dense gas of overlapping charges. This raises a difficult problem of pattern recognition, particularly for the larger instantons whose density, $Q_L(x)$ will be very small (it obviously varies as $1/\rho^4$) and which are most likely to overlap with several other large instantons. Such large instantons will, in any case, not be completely smooth, and may possess multiple peaks ('ripples') since we are trying to minimise the number of cooling sweeps used. These problems have been addressed in different ways by several recent calculations of this kind[43, 41, 42, 44, 45]. I think it is fair to say that all these should be regarded as exploratory. However, to whet your appetite I am going to show you some results from an analysis that I have been involved in[45]. What I am not going to do here is to tell you anything at all about the pattern recognition algorithms etc.

I mentioned a conceptual difficulty. Instantons are semiclassical objects. The real vacuum has fluctuations on all scales and there is no reason to think that the topological charge resides in instanton-like objects. Does it even make sense to talk of a distribution in ρ? This raises all kinds of interesting questions, including: do we care? After all, models such as that in Ref.[28] would not claim to be more than simplifications that are appropriate to the physics being considered. Perhaps we should regard a minimal amount of smoothening as performing such a simplification upon the fluctuating gauge fields? I am not going to address these questions any further here, but you should be aware of their existence.

Let me now move to some results of this kind of calculation[45]. The calculations are in the pure SU(3) gauge theory. We have used lattice fields that have been generated and stored (for other purposes) by the UKQCD Collaboration. We are performing calculations on $\mathbf{16^3 48}$ and $\mathbf{32^3 64}$ lattices at $\beta = \mathbf{6.0}$, $\mathbf{24^3 48}$ lattices at $\beta = \mathbf{6.2}$, and $\mathbf{32^3 64}$ lattices at $\beta = \mathbf{6.4}$. The point of the 2 lattice sizes at $\beta = \mathbf{6.0}$ is to check for finite volume effects - especially for the very large instantons. The various lattice spacings enable us to check whether the features we identify have the right scaling properties to survive into the continuum limit. In addition we do the calculations for various different numbers of cooling sweeps so that we can check whether these features are insensitive to cooling or not.

In Fig. 8 I show the size distribution, $D(\rho)$, with the size ρ expressed in units of $1/\sqrt{\sigma} \equiv 1/\sqrt{K}$. The peak is around $\rho \sim \mathbf{0.5}$ fm. This is not very sensitive to β or to the lattice size or to the number of cooling sweeps. The total number of charges is of course sensitive to the amount of cooling. We do, however, note that unless we go to a very large number of cools, the vacuum is dense with a great deal of overlap between the charges.

It is not straightforward to test for scaling, since a cooling sweep is intrinsically

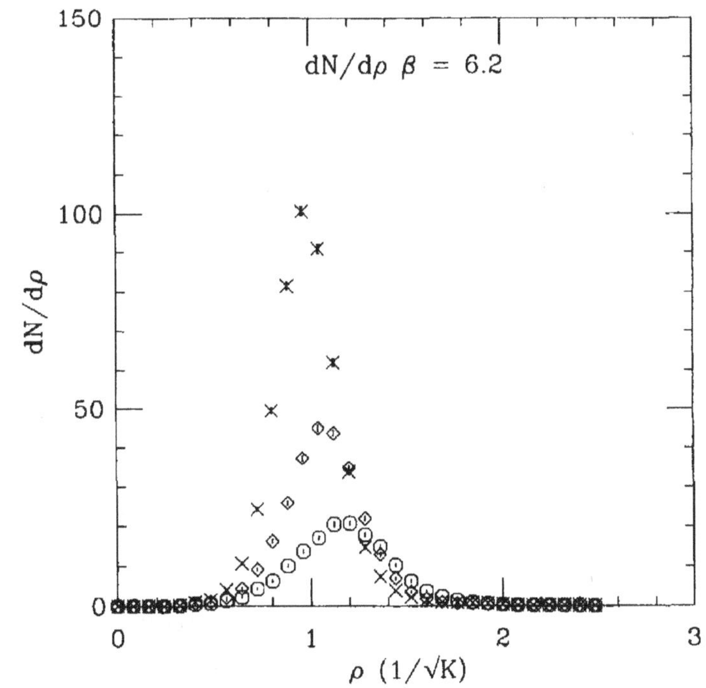

Figure 8. The number of topological charges versus size at $\beta = 6.2$: after 23 (×), 32 (◇) and 46 (○) (under-relaxed) cooling sweeps. The size is in units of $1/\sqrt{K}$ where K is the string tension.

non-scaling. What we do find is that if we tune the number of cooling sweeps so that the total number of charges is independent of β, then the detailed densities, $D(\rho)$, seem to scale as well.

Two other aspects of $D(\rho)$ are of particular interest: the fall-offs at small and large ρ. The former is interesting because it provides a check on our calculations: we know that $D(\rho) \propto \rho^6$ for $a \ll \rho \ll 1/\sqrt{\sigma}$. Of course we can only approach these severe inequalities by looking for the trend as we increase β. We do indeed find a tendency to approach something close to this functional behaviour. (Which, in any case, is modified by powers of log ρ.) By contrast large ρ is interesting because we do not know what happens to large instantons in a confining vacuum. If we used the semiclassical formula with a coupling that froze at large ρ then we would get $D(\rho) \propto 1/\rho^5$ as $a \to \infty$. What we find is a much stronger suppression:

$$D(\rho) \propto \frac{1}{\rho^{\sim 11}} \quad : \rho \gg \frac{1}{\sqrt{\sigma}} \tag{26}$$

Before closing this subject let me move to a quite different and more subtle aspect of the vacuum structure. In a dilute gas the correlation between the sign of a charge and the sign of the total charge will be independent of size ρ. In Fig. 9 I show the

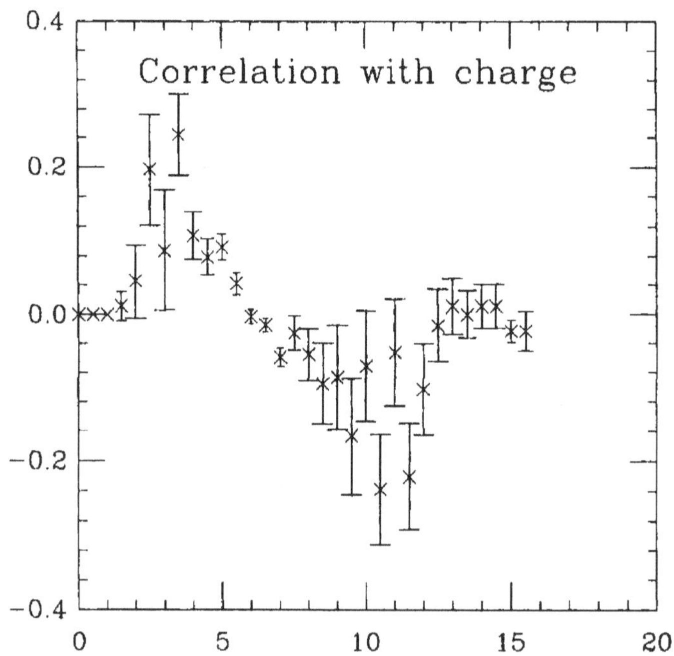

Figure 9. **The average correlation between the sign of a charge of size ρ and the sign of the total charge of the gauge field; at $\beta = 6.2$ after 23 cooling sweeps**

correlation we actually find. There is a striking size-dependence. Charges smaller than average tend to have the same sign as Q, larger charges tend to have the opposite charge. Since the sign of Q simply tells you which sign wins out on that configuration, this tells us that net charge of the small instantons wins out over the net charge of the larger instantons. This suggests a picture where the very large instantons are polarised, so that the small instantons sitting on this background tend to have the opposite charge throughout the volume. Since the charge of the small instantons wins out, this means that the large instantons are actually overscreened. Of course this would be a lot to glean from the one plot; however this intriguing picture is in fact supported by more detailed calculations. It leads to quite striking effects. Suppose for example we decide to calculate the total charge that includes instantons smaller than some value ρ_c: $Q_{\rho\leq\rho_c}$. The fluctuations are shown in Fig. 10. We note that if we were to limit ourselves to instantons with $\rho \leq 1/\sqrt{\sigma}$ then we would obtain a topological susceptibility that is ~ 10 times greater than the total susceptibility! It is clearly important to investigate the effect of these structures on quark propagators and hence on the observable physics.

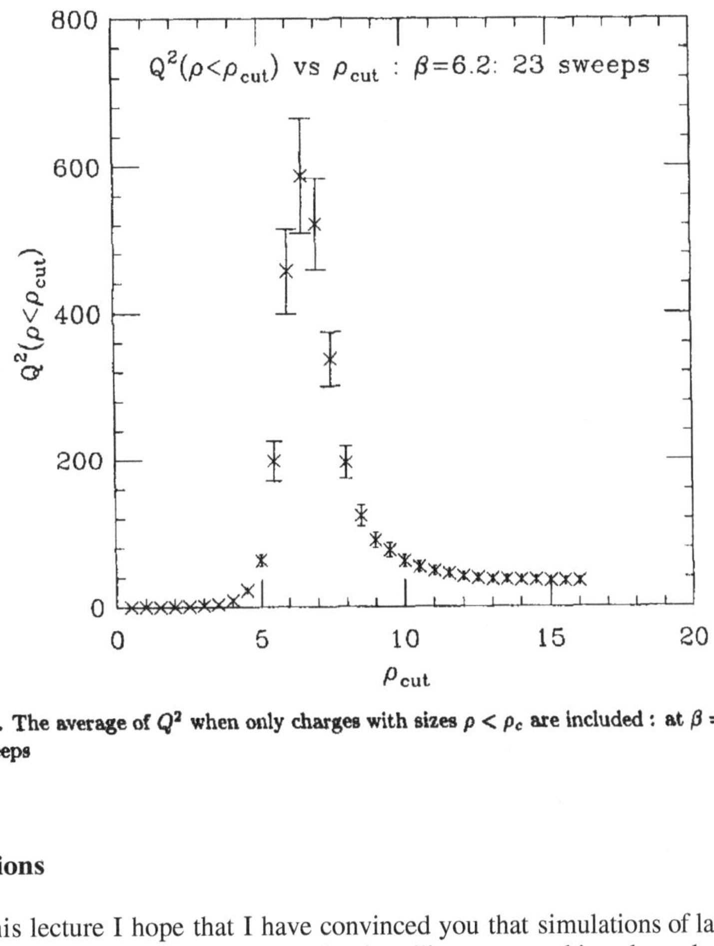

Figure 10. The average of Q^2 when only charges with sizes $\rho < \rho_c$ are included : at $\beta = 6.2$ after 23 cooling sweeps

Conclusions

In this lecture I hope that I have convinced you that simulations of lattice gauge theories can be both useful and interesting in telling us something about the topological fluctuations of the vacuum. Some of what I have shown you, particularly concerning the long distance polarisation of the vacuum, must be regarded as preliminary - to a potentially embarrassing degree. The calculations of the topological susceptibility, on the other hand, are now quite reliable. They show us that the topological vacuum fluctuations are indeed large enough to drive the large η' mass. The theoretical expectations with which we compared our result arise, most straightforwardly, from arguments about what happens in $SU(N)$ QCD at large N. This takes me smoothly to the final of my three topics in these lectures.

$SU(N_c)$ GAUGE THEORIES FOR ALL N_c

Quantum Chromodynamics is an $SU(3)$ gauge theory coupled to 3 lightish quark colour triplets. If we change the gauge group to $SU(2), SU(4), SU(5)$, ... then we obtain an infinite set of, a priori, quite different theories. However from an analysis of Fenyman diagrams to all orders[46] one finds that the $N_c \rightarrow \infty$ limit of such $SU(N_c)$ theories is smooth if we vary the coupling as $g^2 \propto 1/N_c$. This suggests that it should be possible to describe $SU(N_c)$ gauge theories as perturbations[46] in powers of $1/N_c$ around $SU(\infty)$, at least for large enough N_c. Moreover if one assumes confinement for all N_c, then one can easily show[46, 47] that the phenomenology of the $SU(\infty)$quark-gluon theory is strikingly similar to that of (the non-baryonic sector of) QCD. This makes it conceivable that the physically interesting $SU(3)$ theory could be largely understood by solving the much simpler $SU(\infty)$ theory. If all the $SU(N_c)$ theories down to $SU(3)$ can be treated in this way, then this represents an elegant and enormous theoretical simplification.

I do not have the time to review[48] this subject here, but let me at least indicate something of what is involved, albeit using arguments that lack any rigour. The constraint that $g^2 \propto 1/N_c$ is easy to motivate. Consider inserting a gluon loop into a gluon propagator. We have added two triple-gluon vertices and this gives a factor g^2. At the same time the sum over colour in the loop gives a factor of N_c. (Not N_c^2 because the colour of the incoming/outgoing gluon is fixed.) So we have a total factor $\sim g^2 N_c$. Now such loops can be inserted any number of times, so if we want a smooth large-N_c limit (at least in all-order perturbation theory) then we clearly need to impose $g^2 \propto 1/N_c$. Assume therefore that we have done so. Consider a typical meson decay, e.g. $\rho \rightarrow \pi\pi$. This requires the production of a $q\bar{q}$ pair. So the decay width contains a factor of $g^2 \propto 1/N_c$. But if we have confinement all the hadrons are colour singlets and so we do not acquire any compensating factors from summing over the colours of the decay products. Thus the decay width is suppressed by a factor $1/N_c$. That is, at large N_c hadrons do not decay. This is like a 'narrow-width' caricature of the real world. In fact this is a reasonable first approximation to the hadrons we know; mostly their widths are very much smaller than their masses. A similar argument tells us that mixings, e.g. between mesons and glueballs, are suppressed. This is reminiscent of the OZI rule discussed in my first lecture. This (and much more) suggests that the $SU(\infty)$ theory is indeed a first approximation to $SU(3)$.

All this motivates us to try and answer by explicit calculation some basic questions:

- does a non-perturbative calculation support the (all-orders) perturbative argument for a smooth $N_c \rightarrow \infty$ limit?
- is $N_c \rightarrow \infty$ confining?
- does such a limit really require $g^2 \propto 1/N_c$; and if so what does this mean when we have a running coupling?
- is $SU(N_c) \simeq SU(\infty)$ only for $N_c \gg 1$ or is it the case down to $N_c = 3$, or even $N_c = 2$?
- what is the $SU(\infty)$ mass spectrum?

There have been a number of interesting computational explorations of the lattice $SU(\infty)$ theory (for a review see Ref.[49]) based on the fact that it can be re-expressed as a single plaquette theory[50]. Unfortunately this scheme makes no statement about the size of the leading corrections to the $N_c = \infty$ limit, and so gives us no clue as to how close $N_c = 3$ is to $N_c = \infty$.

In this lecture I will describe an extremely straightforward approach to this problem. I will simply calculate the properties of $SU(N_c)$ gauge theories for several values of N_c and so determine explicitly how the physics varies as N_c increases. I will only look at the pure gauge theory, but there are good reasons for believing that the inclusion of quarks will not alter any of our conclusions (except in some obvious ways). Ideally I would like to present you with accurate calculations in 4 dimensions. Unfortunately, at present what I can provide you there is very rough and tentative. But I will make up for that by describing the results of the corresponding, but much more precise, calculation in 3 dimensions. Of course you will want to know what is the relevance of such a 'substitution'. I will come to that shortly.

My study was originally motivated by the observation that the $C = +$ sector of the light mass spectrum turned out to be quite similar[51, 52] in the $D = 2 + 1$ $SU(2)$ and[63] $SU(3)$ theories. (This also appears to be the case in $D = 3 + 1$, although there the comparison is weakened by the much larger errors.) One reason for this might be that both are close to the $N_c = \infty$ limit. In that case we would have an economical understanding of the spectra of $SU(N_c)$ gauge theories for all N_c: there is a common spectrum with small corrections.

A second, more practical reason for studying $N_c \to \infty$ was my interest in obtaining some model understanding of the structure of glueballs. Models are of interest even if they are very approximate in comparison with the results of simulations. A good model will embody the essential degrees of freedom in a problem and show how this leads to the main features of the physics, within some transparent and plausible approximation scheme. A model could provide the intuition necessary for an economical understanding of the role of glueballs in a wide range of contexts. For example, if one understands the structure of glueballs, one can make crude but reliable estimates of glueball-quarkonium mixing, of glueball decay, of the effects of dynamical quarks etc. In many cases the approximations made in the model include the neglect of decays and mixing. These are features of the $SU(\infty)$ rather than of the $SU(3)$ theory, and so it would be better to test the model against the spectrum of the former theory. One example that has been of particular interest to me is the flux tube model of glueballs[54, 55, 56]. The formulation of this model is identical for all $N_c > 2$. However, because the model does not incorporate the effects of glueball decay, it should presumably be tested against the $N_c \to \infty$ spectrum since it is only in that limit that there are no decays. It is also the case that many theoretical approaches are simpler in that limit. An example is provided by the recent progress in calculating the large N_c mass spectrum using light-front quantisation techniques[57].

The $D = 2 + 1$ analysis that I shall present here is based on my calculations over the last few years of the properties of $SU(N_c)$ gauge theories with $N_c = 2, 3, 4$ and 5. In $D = 3 + 1$ what I have done is to perform some $SU(4)$ calculations to supplement what is known about $SU(2)$ and $SU(3)$. My strategy is the very simple one of directly calculating the mass spectra of these theories and seeing whether they are approximately independent of N_c. The calculations are performed through the Monte Carlo simulation of the corresponding lattice theories, using the standard plaquette action. In the $D = 2 + 1$ case the calculations are very accurate and we are able to extrapolate our mass ratios to the continuum limit prior to the comparison. In the 3+1 dimensional case our $SU(4)$ calculations are not good enough for that, and our comparisons with $SU(2)$ and $SU(3)$ are correspondingly less precise. Some of the $SU(2)$ results have been published[51, 52] as have brief summaries of the results discussed here[58, 59]. A long paper is in preparation.

d=2+1 ~ d=3+1 ?

While one might naively expect that the $D = 2+1$ and $D = 3+1$ gauge theories would be so different as to make a unified treatment misleading, this is not in fact so. Theoretically the $D = 2+1$ theory shares with its $D = 3+1$ homologue four important properties.

• Both theories become free at short distances. In 3 dimensions the coupling, g^2, has dimensions of mass so that the effective dimensionless expansion parameter on a scale l will be

$$g_3^2(l) \equiv lg^2 \xrightarrow{l \to 0} 0 \tag{27}$$

In 4 dimensions the coupling is dimensionless and runs in a way we are all familiar with:

$$g_4^2(l) \simeq \frac{c}{\ln(l\Lambda)} \xrightarrow{l \to 0} 0 \tag{28}$$

In both cases the interactions vanish as $l \to 0$, although they do so much faster in the super-renormalisable $D = 2+1$ case than in the merely asymptotically free $D = 3+1$ case.

• Both theories become strongly coupled at large distances. This we see immediately by letting l increase in the above formulae. Thus in both cases the interesting physics is nonperturbative.

• In both theories the coupling sets the mass scale. In 3 dimensions it does so explicitly:

$$m_i = c_i g^2 \tag{29}$$

In 4 dimensions it does so through the phenomenon of dimensional transmutation: the classical scale invariance is anomalous, the coupling runs and this introduces a mass scale through the rate at which it runs:

$$m_i = c_i \Lambda \tag{30}$$

• Both theories confine with a linear potential. This is not something that we can prove by a simple argument. However lattice simulations provide convincing evidence that this is indeed the case. Note that although the $D = 2+1$ Coulomb potential is already confining, this is a weak logarithmic confinement, $V_C(r) \sim g^2 \ln(r)$, which has nothing to do with the nonperturbative linear potential, $V(r) \simeq \sigma r$, that one finds at large r.

In addition to these theoretical similarities, the calculated spectra also show some striking similarities.

• In both theories the lightest glueball is the scalar 0^{++} with a similar mass $m_{0^{++}} \sim 4\sqrt{\sigma}$. In the $C = +$ sector, the 2^{++} is the next lightest glueball (ignoring any excited scalars) with $m_{2^{++}}/m_{0^{++}} \sim 3/2$ in both cases.

All this motivates us to believe that a unified treatment makes sense.

Of course there are significant differences as well. For example:

∘ There are no instantons in $D = 2+1$ non-Abelian gauge theories. This probably implies quite different quark physics.

∘ The rotation group is Abelian.

∘ The details of the mass spectrum are very different in 3 and 4 dimensions. A particularly striking difference is that in 3 dimensions the spectrum exhibits parity doubling for states with non-zero angular momentum:

$$m_{J^+} = m_{J^-} \quad \text{if} \quad J \neq 0 \tag{31}$$

Table 1. The $D = 2+1$ $SU(N_c)$ confining string tension

N_c	$\sqrt{\sigma}/g^2$
2	0.3350 (15)
3	0.5530 (20)
4	0.7564 (45)

This follows from the fact that angular momentum flips sign under parity - which can be defined in $D = 3$ as $(x, y) \to (-x, y)$. The proof is elementary and I leave it as an exercise. Actually to show this you need the continuum rotation group. If you only have $\pi/2$ rotations then you find that you lose parity doubling for the $J = 2$ state. This can happen either because you are too close to the strong coupling limit or because the volume is too small and the rotation symmetry is broken by the boundary conditions. In either case this is a useful check: that we are effectively in the continuum limit and in an effectively infinite volume.

2+1 Dimensions

The calculations in 3 dimensions are performed in the same way as in 4 dimensions except that the computational problem is much more manageable. (Lattices grow as L^3 rather than as L^4.) The basic steps are just as outlined in my first lecture, and you will have to trust me that all the checks have been performed sufficiently carefully! After extrapolating to the continuum limit I obtain the string tension and a mass spectrum in units of g^2. Only the lightest portion of the mass spectrum is calculated, but that includes J^{PC} states for $J = 0, 1, 2$ and $P = \pm$ and $C = \pm$ as well as one or two further excited states in many cases. I do this for $SU(2)$, $SU(3)$, $SU(4)$ and $SU(5)$. In the case of $SU(2)$ there is no $C = -$ sector. In fact the main point of the $SU(5)$ calculation was to have 3 values of N_c for the $C = -$ states, so as to provide some control over their N_c dependence. The SU(5) calculation is recent and has not been incorporated into the analysis that follows.

I begin with the string tension, σ, since it turns out to be our most accurately calculated physical quantity. We use smeared Polyakov loops[38], to obtain $a^2\sigma$ for several values of the lattice spacing a. We then extrapolate the lattice results, using the asymptotic relation $\beta = 4/ag^2$, to obtain the continuum string tension in units of g^2:

$$\frac{\sqrt{\sigma}}{g^2} = \lim_{\beta\to\infty} \frac{\beta}{2N_c} a\sqrt{\sigma} \tag{32}$$

The results[58] for $SU(2)$, $SU(3)$ and $SU(4)$ in $D = 2+1$ are shown in Table 1 and are plotted in Fig. 11. We immediately see that there is an approximate linear rise with N_c and we find that we can obtain a good fit with

$$\frac{\sqrt{\sigma}}{g^2} = 0.1974(12)N_c - \frac{0.120(8)}{N_c}. \tag{33}$$

We obtain a similar behaviour with the light glueball masses (see below).

Some observations.

• For large N_c, eq. (33) tells us that $\sqrt{\sigma} \propto g^2 N_c$. That is to say, the overall mass scale of the theory, call it μ, is proportional to $g^2 N_c$. In other words, in units of the mass scale of the theory

$$g^2 \propto \frac{\mu}{N_c}. \tag{34}$$

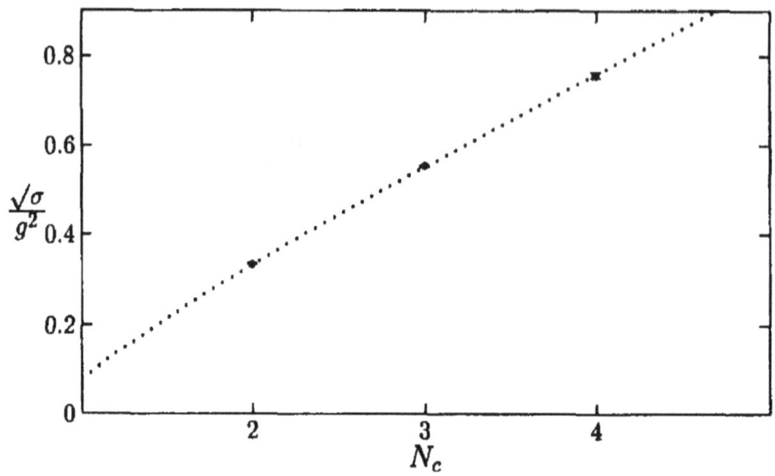

Figure 11. Continuum string tension versus number of colours in $D = 2 + 1$. Line is the fit in eq. (33).

While this coincides with the usual expectation based on an analysis of Feynman diagrams, we note that here the argument is fully non-perturbative.

• The string tension is non-zero for all N_c and, in particular, for $N_c \to \infty$ (when expressed in units of $g^2 N_c$ or the lightest glueball masses - see below). This confirms the basic assumption that needs to be made in 4 dimensions in order to extract the usual phenomenology of the large-N_c theory.

• In the pure gauge sector one expects[48] (again from an analysis of Feynman diagrams) that the first correction to the large-N_c limit will be $O(1/N_c^2)$ relative to the leading term. The fit in eq. (33) is indeed of this form. We note that if we try a fit with a $O(1/N_c)$ correction instead (which would be appropriate if we had quarks) then we obtain an unacceptably poor χ^2 (corresponding to a confidence level of only $\sim 2\%$ in contrast to the $\sim 45\%$ we obtain for the quadratic correction). We may regard this as providing some non-perturbative support for this diagram-based expectation.

• The coefficient of the correction term in eq. (33) is comparable to that of the leading term, suggesting an expansion in powers of $1/N_c$ that is rapidly convergent. Indeed one has to go to $N_c = 1$ before the correction term becomes comparable to the leading term. While the $SU(1)$ theory is completely trivial, we note that the $U(1)$ theory has a zero string tension (in the sense that $\sqrt{\sigma}/g^2 = 0$ in the continuum limit).

Let me now turn from the string tension to the mass spectrum. Recall that since we are in $D = 2 + 1$, states of opposite parity are degenerate as long as $J \neq 0$. This degeneracy is broken by lattice spacing and finite volume corrections. I will present the results separately for the $P = +$ and $P = -$ states so as to provide an explicit check on the presence of any such unwanted corrections.

I begin with the $C = +$ spectrum since the $SU(2)$ spectrum does not contain $C = -$ states. In this case we have masses for three values of N_c, and so can check how good is a fit of the kind in eq. (33). In Fig. 12 I plot the ratio $m_G/g^2 N_c$ against $1/N_c^2$ for a selection of the lightest states, G. On this plot a fit of the form in eq. (33) will be a straight line and I show the best such fits. As we can see, the data is consistent with such a $1/N_c^2$ correction being dominant for $N_c \geq 2$. However what is really striking is the lack of *any* apparent N_c dependence for the lightest 0^{++} and 2^{++} states.

Table 2. States with $C = +$ in $D = 2+1$: $R_\infty \equiv \lim_{N_c \to \infty} m_G/(g^2 N_c)$ and R_{slope} is the coefficient of the $1/N_c^2$ correction in eq. (35)

β	R_∞	R_{slope}
$\sqrt{\sigma}$	0.1974 (12)	-0.12 (1)
0^{++}	0.805 (13)	-0.06 (8)
0^{++*}	1.245 (27)	-0.41 (14)
0^{-+}	1.788 (88)	-0.48 (56)
2^{++}	1.333 (29)	-0.08 (18)
2^{-+}	1.340 (40)	-0.01 (24)
1^{++}	1.946 (75)	-0.59 (47)
1^{-+}	1.919 (115)	-0.18 (75)

In Table 2 I present the results of fitting the $C = +$ states to the form

$$\frac{m_G}{g^2 N_c} = R_\infty + \frac{R_{slope}}{N_c^2}. \tag{35}$$

where $R_\infty = \frac{m_G}{g^2 N_c}\Big|_{N_c=\infty}$. (Note that the errors on the slope and intercept are highly correlated.) The confidence levels of the fits are quite acceptable suggesting once again that for $N_c \geq 2$ a moderately sized correction of the form $\sim 1/N_c^2$ is all that is needed. Note that since the variation with N_c is small, the exact form of the correction used will not have a large impact on the extrapolation to $N_c = \infty$ (except in estimating the errors).

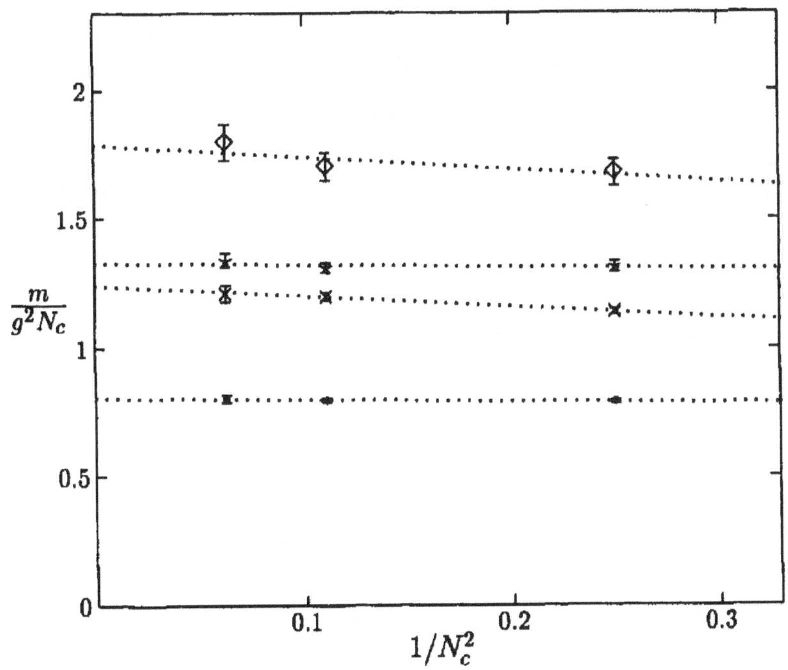

Figure 12. Some continuum glueball masses, in $D = 3$, for 2,3,4 colours: $0^{++}(\bullet)$, $0^{++*}(\times)$, $2^{++}(\star)$, $0^{-+}(\diamond)$ and linear fits.

Table 3. As in Table 2 but for states with $C = -$

G	R_∞	R_{slope}
0^{--}	1.18 (6)	0.1 (0.6)
0^{--*}	1.47 (10)	0.3 (1.1)
0^{+-}	1.98 (28)	-0.4 (2.7)
2^{--}	1.52 (14)	0.9 (1.4)
2^{+-}	1.58 (13)	-0.4 (1.3)
1^{--}	1.85 (15)	-0.3 (1.5)
1^{+-}	1.78 (23)	1.3 (2.3)

These calculations confirm my earlier claim that the physical mass scale at large N_c is $g^2 N_c$. So if we consider ratios of m_G to $\sqrt{\sigma}$ (as was explicitly done in Ref.[59]) we will find that they have finite non-zero limits as $N_c \to \infty$: that is to say, the large-N_c theory possesses linear confinement.

For the $C = -$ states we only have masses for 2 values of N_c and we cannot therefore check whether a fit of the form in eq. (35) is statistically favoured or not. However given that such a fit has proved accurate for the $C = +$ masses and for the string tension down to $N_c = 2$ it seems entirely reasonable to assume that it will be appropriate for $N_c \geq 3$ for the $C = -$ masses. Assuming this we obtain the results shown in Table 3 for the $N_c = \infty$ limit and for the coefficient of the first correction. The 'lever arm' on this extrapolation is, of course, shorter than for the $C = +$ states and that leads to correspondingly larger errors. This will be dramatically improved once I incorporate the $SU(5)$ spectrum into the analysis.

The results in the Tables provide us not only with values for the various mass ratios in the limit $N_c \to \infty$ but also, when inserted into eq. (33), predictions for *all* values of N_c.

Finally, I should remark that I have also calculated the deconfining temperature, T_c, for[60] $SU(2)$ and for[53] $SU(3)$. Extrapolating as in eq. (35), we find

$$\frac{T_c}{g^2 N_c} = 0.1745(52) + \frac{0.079(23)}{N_c^2}. \tag{36}$$

Of course, extrapolating from $N_c = 2, 3$ is much less reliable than extrapolating from $N_c = 3, 4$, and so this relation should be treated with some caution.

3+1 Dimensions

Our knowledge of 4 dimensional gauge theories is much less precise. As far as continuum properties are concerned, quantities that are known with reasonable accuracy include the string tension, the lightest scalar and tensor glueballs, the deconfining temperature and the topological susceptibility. As in 3 dimensions, the $SU(2)$ and $SU(3)$ values are within $\sim 20\%$ of each other, which encourages us to investigate the $SU(4)$ theory so as to see whether we are indeed 'close' to $N_c = \infty$. Of course these $SU(4)$ calculations are much slower than in $D = 2 + 1$ and the results I will present here are of a very preliminary nature[58].

I use the standard plaquette action, and so the first potential hurdle is the presence of the well-known bulk transition that occurs as we increase β from strong towards weak coupling. To locate this transition I have performed a scan on a 10^4 lattice and found that it occurred at $\beta = 10.4 \pm 0.1$. This corresponds to a rather large value of the lattice spacing, a, and so does not lie in the range of couplings within which we shall be working, i.e. β =10.7,10.9 and 11.1.

Table 4. $SU(4)$ in 4 dimensions; masses calculated at the values of β shown

β	$a\sqrt{\sigma}$	$am_{0^{++}}$	$am_{2^{++}}$
10.7	0.296 (14)	0.98 (17)	1.78 (34)
10.9	0.229 (7)	0.77 (8)	1.20 (10)
11.1	0.196 (7)	0.78 (6)	1.08 (10)

Our calculation consists of 4000, 6000 and 3000 sweeps on 10^4, 12^4 and 16^4 lattices at β =10.7, 10.9 and 11.1 respectively. Every fifth sweep we calculated correlations of (smeared) gluonic loops and from these we extracted the string tension and the masses of the lightest 0^{++} and 2^{++} particles, using standard techniques[38, 10]. These are presented in Table 4. We also calculated the topological susceptibility, $a^4\chi_t$. The charge Q was obtained using the cooling method discussed in the previous lecture. These calculations were performed every 50 sweeps. Overall this corresponds to rather small statistics and the errors are therefore unlikely to be very reliable.

We see from Table 4 that the most accurate physical quantity in our calculations is the string tension, σ. Can we learn from it how g^2 varies with N_c, just as we did in $D = 2+1$? We focus on a particular embodiment of this question: if we compare different $SU(N_c)$ theories at a value of a which is the same in physical units, i.e. for which $a\sqrt{\sigma}$ is the same, does the bare coupling vary as $1/N_c$, i.e. does $\beta \equiv 2N_c/g^2 \propto N_c^2$? Weperform this comparison for $\beta_4 = 10.9,\ 11.1$. (For convenience we shall label β by the value of N_c, i.e. we write it as β_{N_c}.) To find the corresponding values of β in $SU(2)$ and $SU(3)$ we simply interpolate between the values provided in (for example) Refs.[38, 10]. Doing so we find that the values of β corresponding to $\beta_4 = 10.9,\ 11.1$ are $\beta_3 \simeq 5.972(18),\ 6.071(24)$ and $\beta_2 \simeq 2.442(9),\ 2.485(11)$ respectively. If we simply scale β_4 by N_c^2 then what we would expect to obtain is $\beta_3 \simeq 6.131,\ 6.244$ and $\beta_2 \simeq 2.725,\ 2.775$ respectively. Superficially the numbers look to be in the right ballpark, but in fact the agreement is poor. For example $\beta_2 = 2.725$ and $\beta_2 = 2.442$ correspond to values of $a\sqrt{\sigma}$ that differ by about a factor of 3.

This disagreement should not, however, be taken too seriously, since it is well-known that the lattice bare coupling is a very poor perturbative expansion parameter. It is known that one can get a much better expansion parameter if one uses instead the mean-field improved coupling, g_I^2, obtained from g^2 by dividing it by the average plaquette[61], $< \frac{1}{N_c}TrU_p >$ Defining $\beta_{N_c}^I \equiv 2N_c/g_I^2(a)$we find that $\beta_4 = 10.9,\ 11.1$ correspond to $\beta_4^I = 6.215,\ 6.474$ respectively. Scaling β_4^I by N_c^2 we would expect the equivalent $SU(3)$ and $SU(2)$ couplings to be given by $\beta_3^I = 3.496,\ 3.642$ and $\beta_2^I = 1.554,\ 1.619$. What we actually find is that the equivalent couplings are $\beta_3^I \simeq 3.527(22),\ 3.649(28)$ and $\beta_2^I = 1.561(10),\ 1.613(12)$. The agreement is now excellent. That is to say, if the $SU(N_c)$ mean-field improved bare-coupling is defined on a length scale that is related to the physical length scale ($\sqrt{\sigma}$) by some constant factor, then it varies as $g^2 \propto 1/N_c$. This is, of course, the usual diagram-based expectation.

In Fig. 13 I plot the scalar and tensor glueball masses, in units of $\sqrt{\sigma}$, as a function of N_c. For $N_c = 2, 3$ we have used the continuum values. For $N_c = 4$ the calculations are not precise enough to permit an extrapolation to the continuum limit and so we simply present the values that we obtained at $\beta = 10.9$ and 11.1. (We do not use the $\beta = 10.7$ values since they have large errors and there is the danger that the scalar mass may be reduced by its proximity to the critical point at the end of the bulk transition line.) Although the $N_c = 4$ errors are quite large, it certainly seems that there is little variation with N_c for $N_c \geq 2$ and any dependence appears to be consistent with being

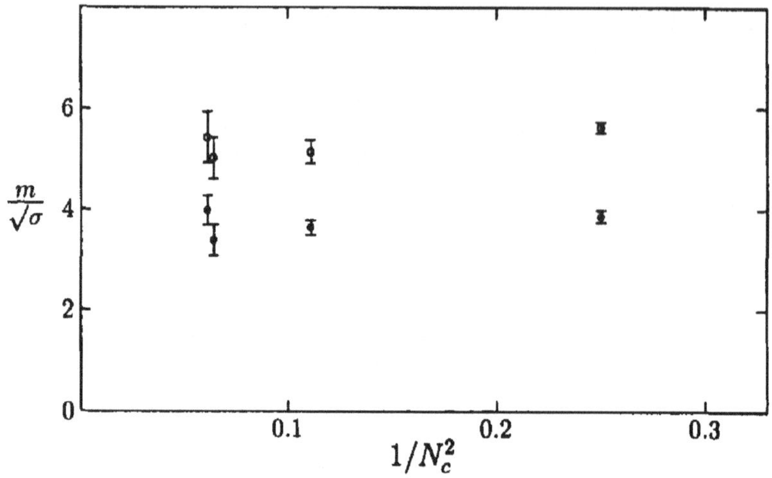

Figure 13. Lightest scalar (•) and tensor (○) glueball masses in $D = 4$. Continuum values for $N_c = 2, 3$ and lattice values ($\beta = 10.9$ and $\beta = 11.1$) for $N_c = 4$.

given by a simple $1/N_c^2$ correction. The fact that these mass ratios appear to have finite non-zero limits, implies that the **large-N_c** theory is confining.

As mentioned earlier we have also calculated the topological susceptibility. In Fig. 14 we plot the dimensionless ratio $\chi_t^{1/4}/\sqrt{\sigma}$ as a function of N_c. Once again the N_c=2 and 3 values are continuum extrapolations of lattice values, while in the case of $SU(4)$ we simply display the lattice values obtained at β=10.9 and 11.1. As we remarked earlier one expects, semiclassically, very few small instantons for $SU(4)$ and this is confirmed in our cooling calculations. This has the advantage that the lattice ambiguities that arise when instantons are not much larger than a are reduced as compared to $SU(3)$, and dramatically reduced as compared to $SU(2)$. This implies

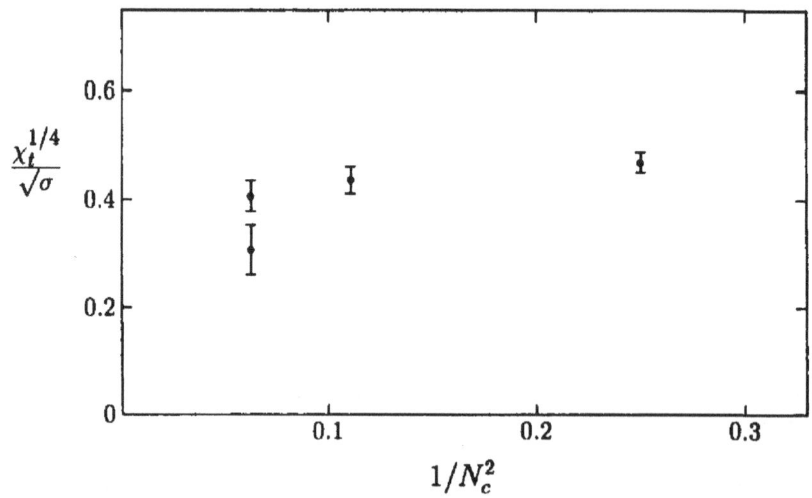

Figure 14. The topological susceptibility: continuum values for $N_c = 2, 3$ and lattice values ($\beta = 10.9$ and $\beta = 11.1$) for $N_c = 4$.

that the interesting large-N_c physics of topology (and the related meson physics) should be straightforward to study.

Conclusions

I have shown you the mass spectra and string tensions of gauge theories with $N_c = 2, 3, 4$ in 3 dimensions. We saw that there is only a small variation with N_c and that this can be accurately described by a modest $O(1/N_c^2)$ correction. That is to say, such theories are close to their $N_c = \infty$ limit for all values of $N_c \geq 2$. We also saw that the large-N_c theory is confining and that $g^2 \propto 1/N_c$ when expressed in physical units. This confirms, in a fully non-perturbative way, expectations arrived at from analyses of Feynman diagrams. It simultaneously provides a unified understanding of all our $SU(N_c)$ theories in terms of just the one theory, $SU(\infty)$, with modest corrections to it. In practical terms this means that, from the parameters in our Tables, we know the corresponding masses for *all* values of N_c

Our calculation in 4 dimensions, while quite preliminary, does suggest that the situation may be much the same there. This is obviously something that needs to be done much better. Of particular interest are topological fluctuations at large N_c: the instanton size density, $D(\rho)$, loses its small ρ tail and lattice topology should become unambiguous. At the same time it would be interesting to investigate the behaviour of hadrons composed of quarks: for example the η' mass should vanish as $N_c \to \infty$. There are many problems of real theoretical interest here. I hope some of you will choose to get involved.

Acknowledgements

My thanks to Pierre van Baal for inviting me to lecture at this School and for his efforts in making it such a success. I would also like to thank him, David Olive and Peter West for inviting me to the very stimulating Workshop at the Isaac Newton Institute that preceded the School and, finally, the Institute itself for providing such a splendid research environment.

REFERENCES

1. J. Sexton et al. Phys Rev. Letters, 75 (1995) 4563.
2. G. Bali et al. Phys. Lett. B309 (1993) 378.
3. H. Chen et al. Nucl. Phys. Proc. Suppl. B34 (1994) 357.
4. F. E. Close and M. Teper, in preparation.
5. M. Creutz, Quarks, Gluons and Lattices (Cambridge Univ. Press 1983).
6. I. Montvay and G. Münster, Quantum Fields on a Lattice (Cambridge Univ. Press 1994).
7. M. Lüscher, Comm. Math. Phys. 104 (1986) 177; 1983 Cargèse Lectures.
8. K. Symanzik, Nucl. Phys. B226 (1983) 187.
9. D. Perkins, Introduction to High Energy Physics (Addison-Wesley 1987).
10. C. Michael and M. Teper, Nucl. Phys. B314 (1989) 347.
11. Ph. de Forcrand et al. Phys. Lett. B152 (1985) 107.
12. Ph. de Forcrand et al. Phys. Lett. B160 (1985) 137; P. Bacilieri et al, Phys. Lett. B205 (1988) 535; S. Perantonis and C. Michael, Nucl. Phys. B347 (1990) 854; G. Bali and K. Schilling, Phys. Rev. D46 (1992) 2636; Phys. Rev. D47 (1993) 661; S. Booth et al, Nucl. Phys. B294 (1992) 385; C. Allton et al. Nucl. Phys. B407 (1993) 331; H. Wittig, Nucl. Phys. Proc. Suppl. B42 (1995) 288.
13. C. Morningstar and M. Peardon, Phys. Rev. D56 (1997) 4043.
14. F. Butler et al. Nucl. Phys. B430 (1994) 179.
15. R. Kenway et al. Nucl. Phys. Proc. Suppl. B53 (1997) 206; H. Shanahan et al, Phys. Rev. D55 (1997) 1548; S. Ryan, Ph.D. Thesis, Edinburgh 1996.

16. W. Lee and D. Weingarten, Nucl. Phys. Proc. Suppl. B53 (1997) 236.
17. P. Lacock et al. Phys. Rev. D54 (1996) 6997.
18. F. E. Close, Introduction to Quarks and Partons (Academic Press 1979).
19. S. Godfrey and N. Isgur, Phys. Rev. D32 (1985) 189.
20. Particle Data Group, Phys. Rev. D54 (1996) 1.
21. C. Amsler and F. Close, Phys. Rev. D53 (19%) 295.
22. D. Weingarten, Nucl. Phys. Proc. Suppl. B53 (1997) 232.
23. M. Genovese, Phys. Rev. D46 (1992) 5204.
24. R. Jaffe, Phys. Rev. D15 (1977) 267.
25. S. Coleman, Aspects of Symmetry, Ch.7 (Cambridge Univ. Press 1985).
26. G. 't Hooft, Physics Reports 142 (1986) 357.
27. T. Banks and A. Casher, Nucl. Phys. B169 (1980) 103.
28. T. Schaefer and E. Shuryak, hep-ph/9610451.
29. S. Hands and M. Teper, Nucl. Phys. B347 (1990) 819.
30. N. Dowrick and M. Teper, Nucl. Phys. Proc. Suppl. B42 (1995) 237.
31. P. di Vecchia et al, Nucl. Phys. B192 (1981) 392.
32. M. Campositrini et al, Phys. Lett. B212 (1988) 206.
33. M. Teper, Phys. Lett. B162 (1985) 357.
34. E. Witten, Nucl. Phys. B156 (1979) 269.
35. G. Veneziano, Nucl. Phys. B159 (1979) 213.
36. J. Hoek et al, Nucl. Phys. B288 (1987) 589.
37. M. Teper, Phys. Lett. B202 (1988) 553.
38. C. Michael and M. Teper, Nucl. Phys. B305 (1988) 453.
39. D. Pugh and M. Teper, Phys. Lett. B218 (1989) 326; B224 (1989) 159.
40. M. Teper, unpublished.
41. Ph. de Forcrand et al. Nucl. Phys. B499 (1997) 409.
42. T. DeGrand, A. Hasenfratz and T. Kovacs, hep-lat/9705009
43. C. Michael and P. Spencer, Phys. Rev. D52 (1995) 4691.
44. R. Brower et al. Nucl. Phys. Proc. Suppl. 53 (1997) 547.
45. D. Smith et al. hep-lat/9709128; in preparation.
46. G. 't Hooft, Nucl. Phys. B72 (1974) 461.
47. E. Witten, Nucl. Phys. B160 (1979) 57.
48. S. Coleman, 1979 Erice Lectures.
49. S.R. Das, Rev. Mod. Phys. 59 (1987) 235.
50. T. Eguchi and H. Kawai, Phys. Rev. Lett. 48 (1982) 1063.
51. M. Teper, Phys. Lett. B289 (1992) 115.
52. M. Teper, Phys. Lett. B311 (1993) 223.
53. M. Teper, in preparation.
54. N. Isgur and J. Paton, Phys. Rev. D31 (1985) 2910.
55. T. Moretto and M. Teper, hep-lat/9312035.
56. R. Johnson and M. Teper, hep-lat/9709083
57. F. Antonuccio and S. Dalley, Nucl. Phys. B461 (1996) 275.
58. M. Teper, Phys. Lett. B397 (1997) 223.
59. M. Teper, Nucl. Phys. (Proc. Suppl.) 53 (1997) 715.
60. M. Teper, Phys. Lett. B313 (1993) 417.
61. G. Parisi, in *High Energy Physics* - 1980(AIP 1981); G. Lepage and P. Mackenzie, Phys. Rev. D48 (1993) 2250.

QCD ON COARSE LATTICES

G. Peter Lepage*

Newman Lab. of Nuclear Studies, Cornell University
Ithaca, New York 14853, USA

INTRODUCTION

Lattice quantum chromodynamics is *the* fundamental theory of low-energy strong interactions. In principle, lattice QCD should tell us everything we want to know about hadronic spectra and structure, including such phenomenologically useful things as weak-interaction form factors, decay rates and deep-inelastic structure functions. In these lectures I discuss a revolutionary development that makes the techniques of lattice QCD much more widely accessible and greatly extends the range of problems that are tractable.

The basic approximation in lattice QCD is the replacement of continuous space and time by a discrete grid. The nodes or "sites" of the grid are separated by lattice spacing a, and the length of a side of the grid is L:

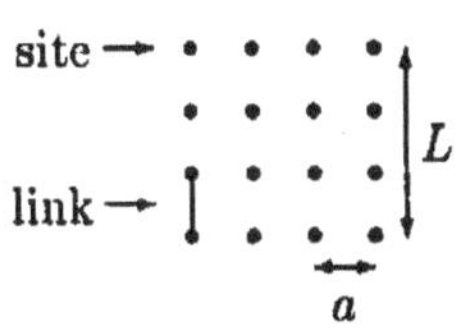

The quark and gluon fields from which the theory is built are specified only on the sites of the grid, or on the "links" joining adjacent sites; interpolation is used to find the fields between the sites. In this lattice approximation, the path integral, from which all quantum mechanical properties of the theory can be extracted, becomes an ordinary multidimensional integral where the integration variables are the values of the fields at each of the grid sites:

$$\int \mathcal{D}A_\mu \ldots \mathrm{e}^{-\int L dt} \longrightarrow \int \prod_{x_j \in \mathrm{grid}} \mathrm{d}A_\mu(x_j) \ldots \mathrm{e}^{-a\sum L_j}. \qquad (1)$$

Thus the problem of nonperturbative relativistic quantum field theory is reduced to one of numerical integration. The integral is over a large number of variables and so

*email: gpl@lns.mail.corell.edu

Monte Carlo methods are generally used in its evaluation. Note that the path integral uses euclidean time rather than ordinary minkowski time, where $t_{eucl} = i t_{mink}$; this removes a factor of i from the exponent, getting rid of high-frequency oscillations in the integrand that are hard to integrate.

Early enthusiasm for this approach to nonperturbative QCD gradually gave way to the sobering realization that very large computers would be needed for the numerical integration of the path integral — computers much larger than those that existed in the mid 1970's, when lattice QCD was invented. Much of a lattice theorist's effort in the first twenty years of lattice QCD was spent in accumulating computing time on the world's largest supercomputers, or in designing and building computers far larger than the largest commercially available computers. By the early 1990's, it was widely felt that teraflop computers costing tens of million of dollars would be essential to the final numerical solution of full QCD.

The revolutionary development I discuss here was the introduction of new techniques that allow one to do realistic numerical simulations of QCD on ordinary desktop workstations or even personal computers. To understand this development one must understand the factors that govern the cost of a full QCD simulation. This cost is governed by a formula like

$$\text{cost} \propto \left(\frac{L}{a}\right)^4 \left(\frac{1}{a}\right) \left(\frac{1}{m_\pi^2 a}\right) \tag{2}$$

where the first factor is just the number of lattice sites in the grid, and the remaining factors account for the "critical slowing down" of the algorithms used in the numerical integration. This formula shows that the single most important determinant of cost is the lattice spacing. The cost varies as the sixth power of $1/a$, suggesting that one ought to keep the lattice spacing as large as possible. Until very recently it was thought that lattice spacings as small as .05–.1 fm would be essential for reliable simulations of QCD. As I describe in these lectures, new simulation results based on new techniques indicate that spacings as large as .4 fm give accurate results. Assuming that the cost is proportional to $(1/a)^6$, these new simulations using coarse lattices should cost 10^3–10^6 times less than traditional simulations on fine lattices.

The computational advantages of coarse lattices are enormous and will certainly redefine numerical QCD: the simplest calculations can be done on a personal computer, while problems of unprecedented difficulty and precision can be tackled with large supercomputers. In these lectures I explain why we now think coarse lattices can be made to work well. I review the techniques from quantum field theory needed to redesign lattice QCD for coarse lattices. These techniques are based on ideas from renormalization theory and effective field theory. I begin by discussing the field-theoretic implications of discretizing space and time. Then I discuss in detail how to discretize the dynamics of gluons and light quarks. These sections present a mixture of very old and very new results. I end the lectures with a simple example of an effective field theory that illustrates the key field theoretic ideas underlying lattice QCD.

FIELD THEORY ON A LATTICE

In this section I discuss the factors that limit the maximum size of the lattice spacing. Replacing space-time by a discrete lattice is an approximation. If we make the lattice spacing too large, our answers will not be sufficiently accurate. For the purpose of these lectures, I define a "sufficiently accurate" simulation to be one that reproduces the low-energy properties of hadrons to within a few percent, which is very precise for low-energy strong-interaction physics. The issue then is: How large can we make a while keeping errors of order a few percent or less?

A nonzero lattice spacing results in two types of error: the error that arises when we replace derivatives in the field equations by finite-difference approximations, and the error due to the ultraviolet cutoff imposed by the lattice. I now discuss each of these in turn, and then focus on the key role played by perturbation theory in correcting for finite-a errors.

Approximate Derivatives

In the lattice approximation, field values are known only at the sites on the lattice. Consequently we approximate derivatives in the field equations by finite-differences that use only field values at the sites. This is completely conventional, and very familiar, numerical analysis. For example, the derivative of a field ψ evaluated at lattice site x_j is approximated by

$$\frac{\partial\psi(x_j)}{\partial x} \approx \Delta_x\psi(x_j) \tag{3}$$

where

$$\Delta_x\psi(x) \equiv \frac{\psi(x+a)-\psi(x-a)}{2a}. \tag{4}$$

It is easy to analyze the error associated with this approximation. Taylor's Theorem implies that

$$\begin{aligned} 2a\,\Delta_x\psi(x) &\equiv \psi(x+a)-\psi(x-a) \\ &= \left(e^{a\partial_x}-e^{-a\partial_x}\right)\psi(x) \end{aligned} \tag{5}$$

and therefore

$$\Delta_x\psi = \left(\partial_x + \frac{a^2}{6}\partial_x^3 + \mathcal{O}(a^4)\right)\psi. \tag{6}$$

Thus the relative error in $\Delta_x\psi$ is of order $(a/\lambda)^2$ where λ is the typical length scale in $\psi(x)$.

On coarse lattices we generally need more accurate discretizations than this one. These are easily constructed. For example from Eq. (6) it is obvious that

$$\frac{\partial\psi}{\partial x} = \Delta_x\psi - \frac{a^2}{6}\Delta_x^3\psi + \mathcal{O}(a^4), \tag{7}$$

which is a more accurate discretization. When one wishes to reduce the finite-a errors in a simulation, it is usually far more efficient to improve the discretization of the derivatives than to reduce the lattice spacing. For example, with just the first term in the approximation to $\partial_x\psi$, cutting the lattice spacing in half would reduce a 20% error to 5%; but the cost would increase by a factor of $2^6 = 64$ in a simulation where cost goes like $1/a^6$. On the other hand, including the a^2 correction to the derivative, while working at the larger lattice spacing, achieves the same reduction in error but with a cost increase of only a factor of 2.

Eq. (7) shows the first two terms of a systematic expansion of the continuum derivative in powers of a^2. In principle, higher-order terms can be included to obtain greater accuracy, but in practice the first couple of terms are sufficiently accurate for most purposes. Simple numerical experiments with a^3-accurate discretizations like this one show that only three or four lattice sites per bump in ψ are needed to achieve accuracies of a few percent or less. Since ordinary hadrons are approximately 1.8 fm in diameter, these experiments suggest that a lattice spacing of .4 fm would suffice for simulating these hadrons. However QCD is a quantum theory, and, as I discuss in the next section, quantum effects can change everything.

Ultraviolet Cutoff

The shortest wavelength oscillation that can be modeled on a lattice is one with wavelength $\lambda_{\rm min} = 2a$; for example, the function $\psi(x) = +1, -1, +1 \ldots$ for $x = 0, a, 2a \ldots$ oscillates with this wavelength. Thus gluons and quarks with momenta $k = 2\pi/\lambda$ larger than π/a are excluded from the lattice theory by the lattice; that is, the lattice functions as an ultraviolet cutoff. In simple classical field theories this is often irrelevant: short-wavelength ultraviolet modes are either unexcited or decouple from the long-wavelength infrared modes of interest. However, in a noisy nonlinear theory, like an interacting quantum field theory, ultraviolet modes strongly affect infrared modes by renormalizing masses and interactions. Thus we cannot simply discard all particles with momenta larger than π/a; we must somehow mimic their effects on infrared states. This we can do by modifying our discretized lagrangian.†

To see how we might mimic the effects of $k > \pi/a$ states on low momentum states, consider the scattering amplitude T for quark-quark scattering in one-loop perturbation theory (Fig. 1a). The difference between the correct amplitude T in the continuum theory and the cut-off amplitude $T^{(a)}$ in our lattice theory involves internal gluons with momentum k of order π/a or larger. This is because the classical theories agree at low momenta, and therefore all propagators and vertices agree there as well. Thus the loop contributions from low momenta will be the same in T and $T^{(a)}$, and cancel in the difference. Given that the external quarks have momenta p_i that are small relative to π/a, we can expand the difference $T - T^{(a)}$ in a Taylor series in $a\,p_i$ to obtain

$$\begin{aligned} T - T^{(a)} &= a^2\,c(a)\,\overline{u}(p_2)\gamma_\mu u(p_1)\,\overline{u}(p_4)\gamma^\mu u(p_3) \\ &+ a^2\,c_A(a)\,\overline{u}\gamma_\mu\gamma_5 u\,\overline{u}\gamma^\mu\gamma_5 u \\ &+ a^4\,d(a)\,(p_1 - p_2)^2\,\overline{u}\gamma_\mu u\,\overline{u}\gamma^\mu u \\ &+ \cdots . \end{aligned} \tag{8}$$

where the couplings c, $c_A \ldots$ are dimensionless functions of the cutoff. This difference is what is missing from the lattice theory; it is the contribution of the $k > \pi/a$ states excluded by the lattice.‡ The key observation is that we can reintroduce this high-k contribution into the lattice theory by adding new interactions to the lattice lagrangian:

$$\begin{aligned} \delta\mathcal{L}^{(a)}_{4q} &= \tfrac{1}{2}\,a^2\,c(a)\,\overline{\psi}\gamma_\mu\psi\,\overline{\psi}\gamma^\mu\psi \\ &+ \tfrac{1}{2}\,a^2\,c_A(a)\,\overline{\psi}\gamma_\mu\gamma_5\psi\,\overline{\psi}\gamma^\mu\gamma_5\psi \\ &+ a^4\,d(a)\,\overline{\psi}D^2\gamma_\mu\psi\,\overline{\psi}\gamma^\mu\psi \\ &+ \cdots . \end{aligned} \tag{9}$$

These new local interactions give the same contribution to $T(qq \to qq)$ in the lattice theory as the $k > \pi/a$ states do in the continuum theory. Although these correction terms are nonrenormalizable, they do not cause problems in our lattice theory because it

†The idea of modifying the lagrangian to compensate for a finite UV cutoff is central to chiral field theories for pions, nonrelativistic QED/QCD and all other effective field theories. The application to lattice field theories was pioneered in the form discussed here by Symanzik and is refered to as "Symanzik improvement" of lattice operators[1].

‡More precisely, this contribution is due to states above the cutoff, and to corrections needed to fix up the states below but sufficiently near the cutoff that they suffer severe lattice distortion.

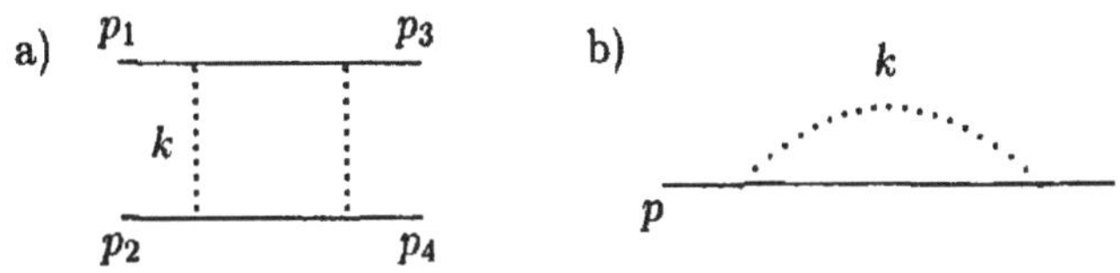

Figure 1. One-loop amplitudes contributing to: a) $qq \to qq$, and b) the quark self energy.

is cut off at π/a. On the contrary, they bring our lattice theory closer to the continuum without reducing the lattice spacing.

This simple analysis illustrates a general result of renormalization theory: one can mimic the effects of states excluded by a cutoff with extra local interactions in the cut-off lagrangian. The correction terms in $\delta\mathcal{L}^{(a)}$ are all local — that is, they are polynomials in the fields and in derivatives of the fields — since contributions omitted as a consequence of the cutoff can be Taylor-expanded in powers of ap_i. This is intuitively reasonable since these terms correct for contributions from intermediate states that are highly virtual and thus quite local in extent; the locality is a consequence of the uncertainty principle. The relative importance of the different interactions is determined by the number of powers of a in their coefficients, and that is determined by the dimensions of the operators involved: each term in $\delta\mathcal{L}^{(a)}$ must have total (energy) dimension four, and so if the operator in a particular term has dimension $n+4$ then the coefficient must contain a factor a^n. In principle there are infinitely many correction terms, but we need only include interactions with operators of dimension $n+4$ or less to achieve accuracy through order $(ap_i)^n$. In practice we only need precision to a given order n in ap_i, and there are only a finite number of local operators with dimension less than or equal to any given $n+4$.

Initially the $k > \pi/a$ contributions appear to be bad news. As we argued in the previous section, a^2 corrections are necessary for precision when a is large. However the quantum nature of our theory implies that there are a^2 terms, due to $k > \pi/a$ states, which depend in detail on the nature of the theory. Thus, for example, when we discretize the derivative in the QCD quark action we replace

$$\overline{\psi}\partial \cdot \gamma\psi \to \overline{\psi}\Delta \cdot \gamma\psi + a^2\, d(a)\, \overline{\psi}\Delta^{(3)} \cdot \gamma\psi + \cdots \tag{10}$$

where now $d(a)$ has a part, $-1/6$, from numerical analysis (equation Eq. (7)) plus a contribution that mimics the $k > \pi/a$ part of the quark self energy (Fig. 1b). The problem is that the latter contribution is, by its nature, completely theory specific; we cannot look it up in a book on numerical analysis. We must somehow solve the $k > \pi/a$ part of the field theory in order to calculate it; otherwise we are unable to correct our lagrangian and unable to use large lattice spacings.

The good news, for lattice QCD, is that $k > \pi/a$ QCD is perturbative provided a is small enough (because of asymptotic freedom). Then correction coefficients like $d(a)$ can be computed perturbatively, using Feynman diagrams, in an expansion in powers of $\alpha_s(\pi/a)$. Thus, for example, our corrected lattice action for quarks becomes

$$\begin{aligned}\mathcal{L}^{(a)} &= \overline{\psi}\,(\Delta \cdot \gamma + m(a))\,\psi \\ &+ d(a)\, a^2\, \overline{\psi}\Delta^{(3)} \cdot \gamma\psi \\ &+ c(a)\, a^2\, \overline{\psi}\gamma_\mu\psi\, \overline{\psi}\gamma^\mu\psi + \ldots\end{aligned} \tag{11}$$

where

$$d(a) = -\frac{1}{6} + d_1\, \alpha_s(\pi/a) + \cdots \tag{12}$$

$$c(a) = c_2\,\alpha_s^2(\pi/a) + \cdots \tag{13}$$

$$\cdots$$

are computed to whatever order in perturbation theory is necessary. In this way we use perturbation theory to, in effect, fill in the gaps between lattice points, allowing us to obtain continuum results without taking the lattice spacing to zero.

Perturbation Theory and Tadpole Improvement[2]

Improved discretizations and large lattice spacings are old ideas, pioneered by Wilson, Symanzik and others[3]. However, perturbation theory is essential; the lattice spacing a must be small enough so that $k \approx \pi/a$ QCD is perturbative. This was the requirement that drove lattice QCD towards very costly simulations with tiny lattice spacings. Traditional perturbation theory for lattice QCD begins to fail at distances of order 1/20 to 1/10 fm, and therefore lattice spacings must be at least this small before improved actions are useful. This seems very odd since phenomenological applications of continuum perturbative QCD suggest that perturbation theory works well down to energies of order 1 GeV, which corresponds to a lattice spacing of 0.6 fm. The breakthrough, in the early 1990's, was the discovery of a trivial modification of lattice QCD, called "tadpole improvement," that allows perturbation theory to work even at distances as large as[2, 4] 1/2 fm.

One can readily derive Feynman diagram rules for lattice QCD using the same techniques as in the continuum, but applied to the lattice lagrangian . The particle propagators and interaction vertices are usually complicated functions of the momenta that become identical to their continuum analogues in the low-momentum limit. All loop momenta are cut off at $k_\mu = \pm\pi/a$.

Testing perturbation theory is also straightforward. One designs short-distance quantities that can be computed easily in a simulation (i.e., in a Monte Carlo evaluation of the lattice path integral). The Monte Carlo gives the exact value which can then be compared with the perturbative expansion for the same quantity. An example of such a quantity is the expectation value of the Wilson loop operator,

$$W(C) \equiv \langle 0|\tfrac{1}{3}\,\mathrm{Re\,Tr\,P}\,e^{-ig\oint_C A\cdot dx}|0\rangle, \tag{14}$$

where A *is* the QCD vector potential, P denotes path ordering, and C is any small, closed path or loop on the lattice. $W(C)$ is perturbative for sufficiently small loops C. We can test the utility of perturbation theory over any range of distances by varying the loop size while comparing numerical Monte Carlo results for $W(C)$ with perturbation theory.

Fig. 2 illustrates the highly unsatisfactory state of traditional lattice-QCD perturbation theory. There I show the "Creutz ratio" of $2a \times 2a$, $2a \times a$ and $a \times a$ Wilson loops,

$$\chi_{2,2} \equiv -\ln\left(\frac{W(2a \times 2a)\,W(a \times a)}{W^2(2a \times a)}\right), \tag{15}$$

plotted versus the size $2a$ of the largest loop. Traditional perturbation theory (dotted lines) underestimates the exact result by factors of three or four for loops of order 1/2 fm; only when the loops are smaller than 1/20 fm does perturbation theory begin to give accurate results.

The problem with traditional lattice-QCD perturbation theory is that the coupling it uses is much too small. The standard practice was to express perturbative expansions of short-distance lattice quantities in terms of the bare coupling α_{lat} used in the lattice

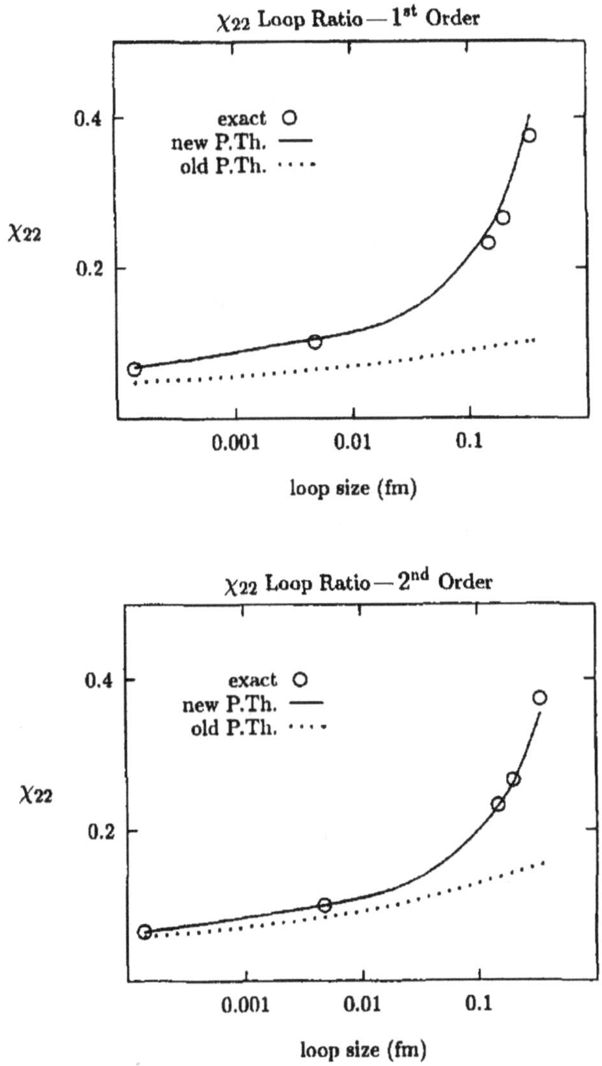

Figure 2. The χ_{22} Creutz ratio of Wilson loops versus loop size. Results from Monte Carlo simulations (exact), and from tadpole-improved (new) and traditional (old) lattice perturbation theory are shown.

lagrangian. This practice followed from the notion that the bare coupling in a cutoff theory is approximately equal to the running coupling evaluated at the cutoff scale, here $\alpha_s(\pi/a)$, and therefore that it is the appropriate coupling for quantities dominated by momenta near the cutoff. In fact the bare coupling in traditional lattice QCD is much smaller than true effective coupling at large lattice spacings: for example,

$$\alpha_{\rm lat} = \alpha_V(\pi/a) - 4.7\,\alpha_V^2 + \cdots \tag{16}$$

$$\leq \tfrac{1}{2}\alpha_V(\pi/a) \qquad \text{for } a > .1\,\text{fm} \tag{17}$$

where $\alpha_V(q)$ is a continuum coupling defined by the static-quark potential,

$$V_{Q\overline{Q}}(q) \equiv -4\,\pi\,C_F\frac{\alpha_V(q)}{q^2}. \tag{18}$$

Consequently $\alpha_{\rm lat}$ expansions, though formally correct, badly underestimate perturbative effects, and converge poorly.

The anomalously small bare coupling in the traditional lattice theory is a symptom of the "tadpole problem". As we discuss later, all gluonic operators in lattice QCD are built from the link operator

$$U_\mu(x) \equiv \mathrm{P}\,\mathrm{e}^{-\mathrm{i}\int_x^{x+a\hat\mu} gA\cdot dx} \approx \mathrm{e}^{-\mathrm{i}agA_\mu} \tag{19}$$

rather than from the vector potential A_μ. Thus, for example, the leading term in the lagrangian that couples quarks and gluons is $\overline{\psi}U_\mu\gamma_\mu\psi/a$. Such a term contains the usual $\overline{\psi}gA\cdot\gamma\psi$ vertex, but, in addition, it contains vertices with any number of additional powers of agA_μ. These extra vertices are irrelevant for classical fields since they are suppressed by powers of the lattice spacing. For quantum fields, however, the situation is quite different since pairs of A_μ's, if contracted with each other, generate ultraviolet divergent factors of $1/a^2$ that precisely cancel the extra a's. Consequently the contributions generated by the extra vertices are suppressed by powers of g^2 (not a), and turn out to be uncomfortably large. These are the tadpole contributions.

The tadpoles result in large renormalizations — often as large as a factor of two or three — that spoil naive perturbation theory, and with it our intuition about the connection between lattice operators and the continuum. However tadpole contributions are generically process independent and so it is possible to measure their contribution in one quantity and then correct for them in all other quantities.

The simplest way to do this is to cancel them out. The mean value,

$$u_0 \equiv \langle 0|\tfrac{1}{3}\,\mathrm{Re\,Tr}\,U_\mu|0\rangle_{\rm LG} \tag{20}$$

consists of only tadpoles and so we can largely cancel the tadpole contributions by dividing every link operator by u_0. That is, in every lattice operator we replace

$$U_\mu(x) \to \frac{U_\mu(x)}{u_0} \tag{21}$$

where u_0 is computed numerically in a simulation.

The mean link is gauge dependent and can be made arbitrarily small by changing the gauge. Thus we choose the gauge that maximizes $\langle \mathrm{Tr}\,U_\mu\rangle$, which happens to be Landau gauge (hence the subscript LG above). This gauge minimizes the tadpoles in u_0, pushing u_0 as close to 1 as possible; any tadpole contribution that remains cannot be a gauge artifact.

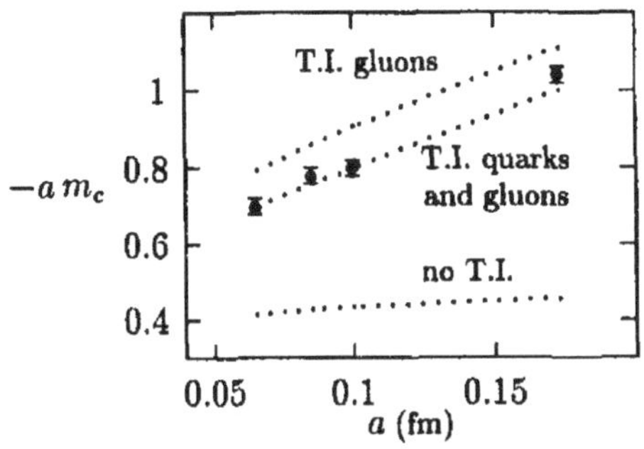

Figure 3. The critical bare quark mass for Wilson's lattice quark action versus lattice spacing. Monte Carlo data points are compared with perturbation theories in a theory with no tadpole improvement (T.I), tadpole-improved gluon dynamics, and tadpole-improved quark and gluon dynamics.

The u_0's cancel tadpole contributions, making lattice operators and perturbation theory far more continuum-like in their behavior. Thus, for example, the only change in the standard gluon action when it is tadpole-improved is that the new bare coupling α_{TI} is enhanced by a factor of $1/u_0^4$ relative to the coupling α_{lat} in the unimproved theory:

$$\alpha_{\mathrm{TI}} = \frac{\alpha_{\mathrm{lat}}}{u_0^4}. \tag{22}$$

Since $u_0^4 < .6$ when $a > .1\,\mathrm{fm}$, the tadpole-improved coupling is typically more than twice as large for coarse lattices. Expressing α_{TI} in terms of the continuum coupling α_V, we find that now our intuition is satisfied:

$$\alpha_{\mathrm{TI}} = \alpha_V(\pi/a) - .5\,\alpha_V^2 + \cdots \tag{23}$$

$$\approx \alpha_V(\pi/a). \tag{24}$$

Perturbation theory for the Creutz ratio Eq. (15) converges rapidly to the correct answer when it is reexpressed in terms of α_{TI}. An even better result is obtained if the expansion is reexpressed as a series in a coupling constant defined in terms of a physical quantity, like the static-quark potential, where that coupling constant is measured in a simulation. By measuring the coupling we automatically include any large renormalizations of the coupling due to tadpoles. It is important that the scale q^* at which the running coupling constant is evaluated be chosen appropriately for the quantity being studied[2, 4]. When these refinements are added, perturbation theory is dramatically improved, and, as illustrated in Fig. 2, is still quite accurate for loops as large as 1/2 fm.

This same conclusion follows from Fig. 3 which shows the value of the bare quark mass needed to obtain zero-mass pions using Wilson's lattice action for quarks. This quantity diverges linearly as the lattice spacing vanishes, and so should be quite perturbative. Here we see dramatic improvements as the tadpoles are removed first from the gluon action, through use of an improved coupling, and then also from the quark action.

The Creutz ratio and the critical quark mass are both very similar to the couplings we need to compute for improved lagrangians. Tadpole improvement has been very successful in a wide range of applications.

Summary

Asymptotic freedom implies that short-distance QCD is simple (perturbative) while long-distance QCD is difficult (nonperturbative). The lattice separates short from long distances, allowing us to exploit this dichotomy to create highly efficient algorithms for solving the entire theory: $k > \pi/a$ QCD is included via corrections $\delta\mathcal{L}$ to the lattice lagrangian that are computed using perturbation theory; $k < \pi/a$ QCD is handled nonperturbatively using Monte Carlo integration. Thus, while we wish to make the lattice spacing a as large as possible, we are constrained by two requirements. First a must be sufficiently small that our finite-difference approximations for derivatives in the lagrangian and field equations are sufficiently accurate. Second a must be sufficiently small that π/a is a perturbative momentum. Numerical experiments indicate that both constraints can be satisfied when $a \approx 1/2\,\mathrm{fm}$ or smaller, provided all lattice operators are tadpole improved.

It is important to remember that almost any perturbative analysis in QCD is contaminated by nonperturbative effects at some level. This will certainly be the case for the couplings in $\delta\mathcal{L}$. However we now have extensive experience showing that such nonperturbative effects are rarely significant for physical quantities at the distances relevant to our discussion. Should a situation arise where this is not the case (and one surely will, some day) we may have to supplement the perturbative approach outlined in this section with nonperturbative techniques. Such situations will not pose a problem if they are relatively rare, as now seems likely. Furthermore, as we discuss later, simple nonperturbative techniques already exist, should they be needed, for tuning the leading correction terms in both the gluon and quark actions.

GLUON DYNAMICS ON COARSE LATTICES

In this section I discuss first the construction of accurate discretizations of the classical theory of gluon dynamics. I then discuss the changes needed to make a quantum theory, and illustrate the discussion with Monte Carlo results. Finally I discuss actions for anisotropic lattices.

Classical Gluons

The continuum action for QCD is

$$S = \int d^4x\, \frac{1}{2} \sum_{\mu,\nu} \mathrm{Tr}\, F_{\mu\nu}^2(x) \tag{25}$$

where

$$F_{\mu\nu} \equiv \partial_\mu A_\nu - \partial_\nu A_\mu + ig[A_\mu, A_\nu] \tag{26}$$

is the field tensor, a traceless 3×3 hermitian matrix. The defining characteristic of the theory is its invariance with respect to gauge transformations where

$$F_{\mu\nu} \to \Omega(x)\, F_{\mu\nu}\, \Omega(x)^\dagger \tag{27}$$

and $\Omega(x)$ is an arbitrary x-dependent SU_3 matrix.

The standard discretization of this theory seems perverse at first sight. Rather than specifying the gauge field by the values of $A_\mu(x)$ at the sites of the lattice, the field is specified by variables on the links joining the sites. In the classical theory, the "link variable" on the link joining a site at x to one at $x + a\hat{\mu}$ is determined by the line integral of A_μ along the link:

$$U_\mu(x) \equiv \mathrm{P} \exp\left(-i \int_x^{x+a\hat{\mu}} gA \cdot \mathrm{d}y\right) \tag{28}$$

where the P-operator path-orders the A_μ's along the integration path. We use U_μ's in place of A_μ's on the lattice, because it is impossible to formulate a lattice version of QCD directly in terms of A_μ's that has exact gauge invariance. The U_μ's, on the other hand, transform very simply under a gauge transformation:

$$U_\mu(x) \to \Omega(x)\, U_\mu(x)\, \Omega(x + a\hat{\mu})^\dagger. \tag{29}$$

This makes it easy to build a discrete theory with exact gauge invariance.

A link variable $U_\mu(x)$ is represented pictorially by a directed line from x to $x + \hat{\mu}$, where this line is the integration path for the line integral in the exponent of $U_\mu(x)$:

$U_\mu(x)$

In the conjugate matrix $U_\mu^\dagger(x)$ the direction of the line integral is flipped and so we represent $U_\mu^\dagger(x)$ by a line going backwards from $x + \hat{\mu}$ to x:

$U_\mu^\dagger(x)$

A Wilson loop function,

$$W(C) \equiv \tfrac{1}{3} \mathrm{Tr}\, \mathrm{P} e^{-i \oint_C gA\cdot dx}, \tag{30}$$

for any closed path C built of links on the lattice can be computed from the path-ordered product of the U_μ's and $U_\mu^\dagger$'s associated with each link. For example, if C is the loop

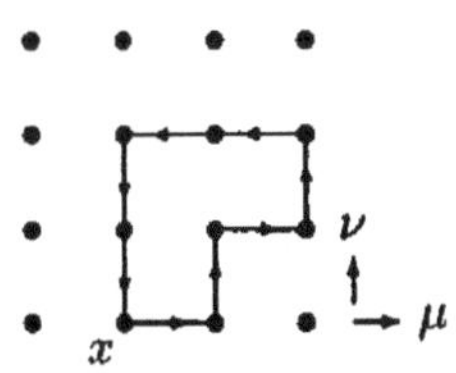

then

$$W(C) = \tfrac{1}{3} \mathrm{Tr}\left(U_\mu(x) U_\nu(x + a\hat{\mu}) \ldots U_\nu^\dagger(x)\right). \tag{31}$$

Such quantities are obviously invariant under arbitrary gauge transformations Eq. (29).

We must now build a lattice lagrangian from the link operators. We require that the lagrangian be gauge invariant, local, and symmetric with respect to axis interchanges (which is all that is left of Lorentz invariance). The most local nontrivial gauge invariant object one can build from the link operators is the "plaquette operator," which involves the product of link variables around the smallest square at site x in the $\mu\nu$ plane:

$$P_{\mu\nu}(x) \equiv \tfrac{1}{3} \mathrm{Re}\, \mathrm{Tr}\left(U_\mu(x) U_\nu(x + a\hat{\mu}) U_\mu^\dagger(x + a\hat{\nu}) U_\nu^\dagger(x)\right). \tag{32}$$

To see what this object is, consider evaluating the plaquette centered about a point x_0 for a very smooth weak classical A_μ field. In this limit,

$$P_{\mu\nu} \approx 1 \tag{33}$$

since

$$U_\mu \approx e^{-igaA_\mu} \approx 1. \tag{34}$$

Given that A_μ is slowly varying, its value anywhere on the plaquette should be accurately specified by its value and derivatives at x_0. Thus the corrections to Eq. (33) should be a polynomial in a with coefficients formed from gauge-invariant combinations of $A_\mu(x_0)$ and its derivatives: that is,

$$\begin{aligned} P_{\mu\nu} &= 1 - c_1 a^4 \operatorname{Tr}(gF_{\mu\nu}(x_0))^2 \\ &\quad - c_2 a^6 \operatorname{Tr}\left(gF_{\mu\nu}(x_0)(\mathrm{D}_\mu^2 + \mathrm{D}_\nu^2) gF_{\mu\nu}(x_0)\right) \\ &\quad + \mathcal{O}(a^8) \end{aligned} \tag{35}$$

where c_1 and c_2 are constants, and D_μ is the gauge-covariant derivative. The leading correction is order a^4 because $F_{\mu\nu}^2$ is the lowest-dimension gauge-invariant combination of derivatives of A_μ, and it has dimension 4. (There are no F^3 terms because $P_{\mu\nu}$ is invariant under $U_\mu \to U_\mu^\dagger$ or, equivalently, $F \to -F$.)

It is straightforward to find the coefficients c_1 and c_2. We need only examine terms in the expansion of $P_{\mu\nu}$ that are quadratic in A_μ; the cubic and quartic parts of $F_{\mu\mu}^2$ then follow automatically, by gauge invariance. Because of the trace, the path ordering is irrelevant to this order. Thus

$$\begin{aligned} P_{\mu\nu} &= \tfrac{1}{3} \operatorname{Re} \operatorname{Tr} \mathrm{P}\, e^{-i \oint_\square gA\cdot dx} \\ &= \tfrac{1}{3} \operatorname{Re} \operatorname{Tr} \left[1 - i \oint_\square gA \cdot dx - \tfrac{1}{2}\left(\oint_\square gA \cdot dx\right)^2 + \mathcal{O}(A^3)\right] \end{aligned} \tag{36}$$

where, by Stoke's Theorem,

$$\begin{aligned} \oint_\square A \cdot dx &= \int_{-a/2}^{a/2} dx_\mu dx_\nu \left[\partial_\mu A_\nu(x_0 + x) - \partial_\nu A_\mu(x_0 + x)\right] \\ &= \int_{-a/2}^{a/2} dx_\mu dx_\nu \left[F_{\mu\nu}(x_0) + (x_\mu \mathrm{D}_\mu + x_\nu \mathrm{D}_\mu) F_{\mu\nu}(x_0) + \cdots\right] \\ &= a^2 F_{\mu\nu}(x_0) + \frac{a^4}{24}(\mathrm{D}_\mu^2 + \mathrm{D}_\nu^2) F_{\mu\nu}(x_0) + \mathcal{O}(a^6, A^2). \end{aligned} \tag{37}$$

Thus $c_1 = 1/6$ and $c_2 = 1/72$ in Eq. (35).

The expansion in Eq. (35) is the classical analogue of an operator product expansion. Using this expansion, we find that the traditional "Wilson action" for gluons on a lattice,

$$S_{\mathrm{Wil}} = \beta \sum_{x,\mu>\nu} (1 - P_{\mu\nu}(x)) \tag{38}$$

where $\beta = 6/g^2$, has the correct limit for small lattice spacing up to corrections of order a^2:

$$S_{\mathrm{Wil}} = \int d^4x \sum_{\mu,\nu} \left\{\tfrac{1}{2} \operatorname{Tr} F_{\mu\nu}^2 + \frac{a^2}{24} \operatorname{Tr} F_{\mu\nu}(\mathrm{D}_\mu^2 + \mathrm{D}_\nu^2) F_{\mu\nu} + \cdots\right\}. \tag{39}$$

We can cancel the a^2 error in the Wilson action by adding other Wilson loops. For example, the $2a \times a$ "rectangle operator"

$$R_{\mu\nu} = \tfrac{1}{3}\,\mathrm{Re}\,\mathrm{Tr}$$ 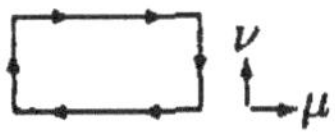 (40)

has expansion

$$R_{\mu\nu} = 1 - \frac{4}{6}\,a^4\,\mathrm{Tr}\,(gF_{\mu\nu})^2 - \frac{4}{72}\,a^6\,\mathrm{Tr}\,\left(gF_{\mu\nu}(4\,\mathrm{D}_\mu^2 + \mathrm{D}_\nu^2)gF_{\mu\nu}\right) - \cdots. \tag{41}$$

The mix of a^4 terms and a^6 terms in the rectangle is different from that in the plaquette. Therefore we can combine the two operators to obtain an improved classical lattice action that is accurate up to[6,7] $\mathcal{O}(a^4)$:

$$S_{\mathrm{classical}} \equiv -\beta \sum_{x,\mu>\nu} \left\{\frac{5P_{\mu\nu}}{3} - \frac{R_{\mu\nu} + R_{\nu\mu}}{12}\right\} + \mathrm{const} \tag{42}$$

$$= \int d^4x \sum_{\mu,\nu} \tfrac{1}{2}\,\mathrm{Tr}\,F_{\mu\nu}^2 + \mathcal{O}(a^4). \tag{43}$$

This process is the analogue of improving the derivatives in discretizations of non-gauge theories.[§]

Quantum Gluons[8]

In the previous section we derived an improved classical action for gluons that is accurate through order a^4. We now turn this into a quantum action. The most important step is to tadpole improve the action by dividing each link operator U_μ by the mean link u_0: for example, the action built of plaquette and rectangle operators becomes

$$S = -\beta \sum_{x,\mu>\nu} \left\{\frac{5}{3}\frac{P_{\mu\nu}}{u_0^4} - \frac{R_{\mu\nu} + R_{\nu\mu}}{12\,u_0^6}\right\}. \tag{44}$$

The u_0's cancel lattice tadpole contributions that otherwise would spoil weak-coupling perturbation theory in the lattice theory and undermine our procedure for improving the lattice discretization. Note that $u_0 \approx 3/4$ when $a = .4\,\mathrm{fm}$, and therefore the relative importance of the $R_{\mu\nu}$'s is larger by a factor of $1/u_0^2 \approx 2$ than without tadpole improvement. Without tadpole improvement, we cancel only half of the a^2 error.

The mean link u_0 is computed numerically by guessing a value for use in the action, measuring the mean link in a simulation, and then readjusting the value used in the action accordingly. This tuning cycle converges rapidly to selfconsistent values, and can be done very quickly using small lattice volumes. The u_0's depend only on lattice spacing, and become equal to one as the lattice spacing vanishes. The fourth root of the plaquette is often used in place of the mean link in the definition of u_0. The two agree well at small lattice spacings, but the mean link usually gives better results (that is, more convergent perturbation theory) for coarser lattices.

Tadpole improvement is the first step in a systematic procedure for improving the action. The next step is to add in renormalizations due to contributions from $k > \pi/a$ physics not already included in the tadpole improvement. The omission of this

[§] An important step that I have not discussed is to show that the gluon action is positive for any configuration of link variables. This guarantees that the classical ground state of the lattice action corresponds to $F_{\mu\nu} = 0$. See[7] for a detailed discussion.

physics induces $a^2\,\alpha_s(\pi/a)$ corrections,

$$
\begin{aligned}
\delta\mathcal{L} = {} & \alpha_s\, r_1\, a^2 \sum_{\mu,\nu} \mathrm{Tr}(F_{\mu\nu} \mathrm{D}_\mu^2 F_{\mu\nu}) \\
& + \alpha_s\, r_2\, a^2 \sum_{\mu,\nu} \mathrm{Tr}(\mathrm{D}_\mu F_{\nu\sigma} \mathrm{D}_\mu F_{\nu\sigma}) \\
& + \alpha_s\, r_3\, a^2 \sum_{\mu,\nu} \mathrm{Tr}(\mathrm{D}_\mu F_{\mu\sigma} \mathrm{D}_\nu F_{\nu\sigma}) \\
& + \cdots, \qquad (45)
\end{aligned}
$$

that must be removed. The last term is harmless; its coefficient can be set to zero by a change of field variable (in the path integral) of the form

$$A_\mu \to A_\mu + a^2\,\alpha_s\, f(\alpha_s) \sum_\nu \mathrm{D}_\nu F_{\nu\mu}. \qquad (46)$$

Since changing integration variables does not change the value of an integral, such field transformations must leave the physics unchanged.¶ Operators that can be removed by a field transformation are called "redundant." The other corrections are removed by renormalizing the coefficient of the rectangle operator $R_{\mu\nu}$ in the action, and by adding an additional operator. One choice for the extra operator is

$$C_{\mu\nu\sigma} \equiv \tfrac{1}{3}\,\mathrm{Re}\,\mathrm{Tr}$$ 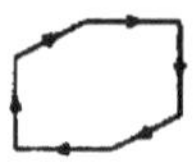 . (47)

Then the action, correct up to $\mathcal{O}(a^2\alpha_s^2, a^4)$, is[9]

$$S = -\beta \sum_{x,\mu>\nu} \left\{ \frac{5}{3}\frac{P_{\mu\nu}}{u_0^4} - r_g \frac{R_{\mu\nu} + R_{\nu\mu}}{12\,u_0^6} \right\} + c_g\,\beta \sum_{x,\mu>\nu>\sigma} \frac{C_{\mu\nu\sigma}}{u_0^6}, \qquad (48)$$

where

$$
\begin{aligned}
r_g &= 1 + .5\,\alpha_s(\pi/a) \qquad (49)\\
c_g &= .06\,\alpha_s(\pi/a). \qquad (50)
\end{aligned}
$$

These values are correct when u_0 is determined from the plaquette; the coefficients of α_s change slightly when the mean-link definition is used. (Of course, both definitions give the same result when the analysis is carried to all orders in α_s.)

The coefficients r_g and c_g are computed by "matching" physical quantities, like low-energy scattering amplitudes, computed using perturbation theory in the lattice theory with the analogous quantity in the continuum theory. The lattice result depends upon r_g and c_g; these parameters are tuned until the lattice amplitude agrees with the continuum amplitude to the order in a and α_s required:

$$T_{\mathrm{lat}}(r_g, c_g) = T_{\mathrm{contin}}. \qquad (51)$$

Note that tadpole improvement has a big effect on these coefficients. Without tadpole improvement, $r_g = 1+2\alpha_s$; that is, the coefficient of the radiative correction is four times larger. Tadpole improvement automatically supplies 75% of the one-loop contribution needed without improvement. Since $\alpha_s \approx 0.3$, the unimproved expansion for r_g is not particularly convergent. However, with tadpole improvement, the one-loop correction is only about 10–20% of r_g. Indeed, for most current applications, one-loop corrections to tadpole-improved actions are negligible.

¶One must, of course, include the jacobian for the transformation in the transformed path integral. This contributes only in one-loop order and higher; it has no effect on tree-level calculations.

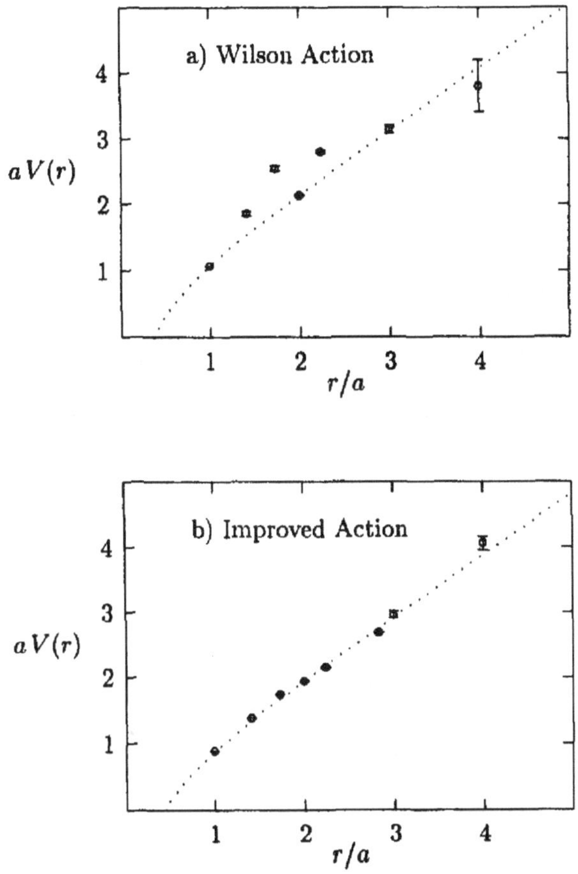

Figure 4. Static-quark potential computed on 6^4 lattices with $a \approx 0.4\,\text{fm}$ using the Wilson action and the improved action. The dotted line is the standard infrared parameterization for the continuum potential, $V(r) = Kr - \pi/12r + c$, adjusted to fit the on-axis values of the potential.

Tests of the Gluon Action

Having a procedure for systematically improving the gluon action, we need to establish whether the improvements really do improve simulations.

The effects of the $\mathcal{O}(a^2)$ improvements in the gluon action are immediately apparent if one compares the static-quark potentials computed with the Wilson action Eq. (38) and with the improved action[8] Eq. (48). Monte Carlo simulation results for these potentials are plotted in Fig. 4, together with the continuum potential. The Wilson action has errors as large as 40%. These are reduced to only 4–5% when the improved action is used instead.

The dominant errors in the uncorrected simulation of $V(r)$ reflect a failure of rotational invariance, which is expected since the a^2 error in the Wilson action Eq. (38) is neither Lorentz nor rotationally invariant. The points at $r = a, 2a, 3a \ldots$ are for separations that are parallel to an axis of the lattice, while the points at $r = \sqrt{2}a, \sqrt{3}a \ldots$ are for diagonal separations between the static quark and antiquark. The $\mathcal{O}(\alpha_s)$ corrections to the action have little effect on these results, suggesting that $\mathcal{O}(a^2\alpha_s)$ corrections are comparable to those of $\mathcal{O}(a^4)$ and therefore unimportant for most current applications. Note also that the rectangle term in the gluon action is the only $\mathcal{O}(a^2)$ term that affects rotation invariance. Thus one can tune its coefficient numerically, without resorting to perturbation theory, until rotation invariance is restored at large r (for example, until $V(2a,2a,a) = V(3a,0,0)$). Obviously, from our example, such nonperturbative tuning gives couplings that are consistent with perturbation theory.

Anisotropic Lattices

It is widely thought that Monte Carlo simulations on coarse lattices are noisier than those on fine lattices. Physics is extracted from simulations by computing Monte Carlo estimates of the correlations (Greens functions) between operators at widely separated times; for example, we compute masses by looking for the large-t plateau in the logarithmic derivative of a correlation function $G(t)$:

$$\frac{1}{G(t)}\frac{dG(t)}{dt} \to m \qquad \text{as } t \to \infty. \tag{52}$$

It is true that such measurements are noisier on conventional coarse lattices, especially for correlation functions where signal/noise vanishes exponentially quickly with increasing t. This is correct despite the fact that the ratio of signal to noise is approximately independent of lattice spacing when t is large; noise overcomes signal at some fixed physical distance, independent of the lattice spacing. The problem on coarse lattices is that this fixed distance might be only one or two lattice spacings long, since the lattice spacing is large. Consequently, without large Monte Carlo data sets, the statistically usable part of a mass plateau, for example, may be too short to allow identification or verification of the plateau. A lattice with small lattice spacing allows us to identify the plateau at smaller distances, with less noise, but at increased cost from the small lattice spacing. Since signal/noise decreases exponentially with increasing t, while simulation cost grows only as a (large) power of lattice spacing, one might decide that smaller lattice spacings are ultimately more efficient. The noise problem on coarse lattices can be greatly alleviated by the careful design of sources and sinks for correlation functions, but a far more efficient approach is to reduce the lattice spacing a_t in the time direction while retaining a large spacing a_s in spatial directions. Then more values of a correlation function can be measured at small t's, where relatively few Monte Carlo measurements are needed to suppress the noise. The small increase in simulation cost that results from a smaller a_t is more than offset by the exponential improvement in signal-to-noise at the smaller t's.

Anisotropic actions are easy to design. The standard improved gluon action becomes[10,11]

$$\begin{aligned} S_n = -\beta \sum_{x,s>s'} \frac{a_t}{a_s}\eta \left\{ \frac{5}{3}\frac{P_{s,s'}}{u_s^4} - \frac{1}{12}\frac{R_{ss,s'}}{u_s^6} - \frac{1}{12}\frac{R_{s's',s}}{u_s^6} \right\} \\ -\beta \sum_{x,s} \frac{a_s}{a_t}\eta^{-1} \left\{ \frac{4}{3}\frac{P_{s,t}}{u_s^2 u_t^2} - \frac{1}{12}\frac{R_{ss,t}}{u_s^4 u_t^2} \right\}, \end{aligned} \tag{53}$$

where u_s and u_t are the (Landau-gauge) spatial and temporal mean links, respectively, and η accounts for renormalization of the anisotropy by quantum effects. Since $a_t \ll a_s$, we have not bothered to correct for order-a_t^2 errors in this action; thus the action extends

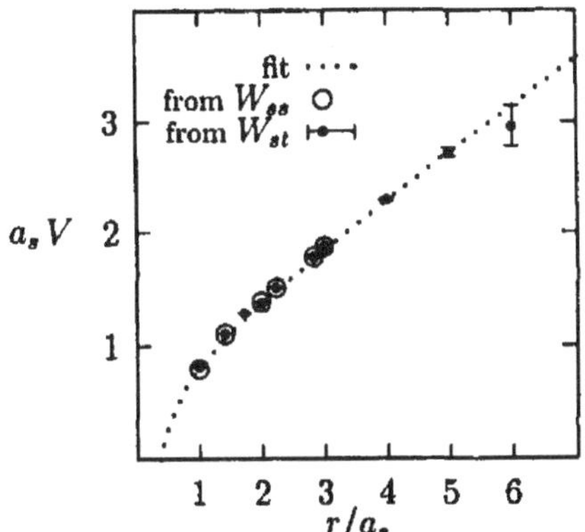

Figure 5. Static potential computed on an anisotropic lattice with $\eta\, a_t/a_s = 0.5$ and β=1.8. Potentials obtained from spatial-temporal loops (W_{st}) and from purely spatial loops (W_{ss}) are shown.

at most one step in the time direction and has no gluon ghosts. An a^2-improved version of Landau gauge is obtained by maximizing

$$\sum_{x\,\mu} \frac{1}{u_\mu a_\mu^2} \mathrm{ReTr} \left\{ U_\mu(x) - \frac{1}{16 u_\mu} U_\mu(x)\, U_\mu(x+\hat{\mu}) \right\}. \tag{54}$$

The anisotropic gluon action seems to have six parameters, but these appear in only three independent combinations,

$$\frac{\beta}{u_s^3 u_t}, \qquad \left(\frac{\eta a_t}{a_s}\right) \frac{u_t}{u_s}, \qquad u_s, \tag{55}$$

which control the spatial lattice spacing, the anisotropy, and the size of the order a_s^2 corrections, respectively. These can be tuned numerically by choosing β and $\eta a_t/a_s$, then determining u_t and u_s self-consistently, and finally measuring η. Alternatively one might compute the last three parameters using perturbation theory[12].

One can readily test an anisotropic simulation by showing that the (euclidean) time direction, with its smaller lattice spacing, is equivalent to any other direction. For example, in the same simulation, one can compute the same quantity twice, once taking the time direction to be along the time axis and once with, for example, the t and z axes interchanged. The two calculation should agree. Fig. 5 compares the static-quark potential computed on an anisotropic lattice in these two ways. Shifted to account for renormalization, the two potentials agree well for all r.

Anisotropic lattices with $a_t \ll a_s$ are useful for any simulation involving high-energy states, including glueballs, excited hadronic states, and high-momentum final states in form factors. Very impressive results have been obtained for the glueball spectrum using these techniques[13]. A variety of quark and gluon actions designed for anisotropic lattices are currently under development.

Summary

In this section I have outlined how one designs actions for simulating gluon dynamics on coarse lattices. The simulation results show that accurate simulations are possible even on lattices as coarse as a = .4fm using very simple actions. Tadpole improvement is essential, but one-loop radiative corrections to the action are not too important. Rotational invariance of the potential and, for anisotropic lattices, space-time interchange symmetry provide sensitive tests of improved gluon actions, and can be used to tune the leading correction terms.

The coarse lattices make simulations very fast. The simple simulation results shown here, as well as the glueball simulations in[13], were all generated using desktop workstations.

LIGHT QUARKS ON COARSE LATTICES[14]

In this section I review techniques for designing very accurate discretizations of the Dirac equation for quarks in a gluonic field.

The euclidean Dirac lagrangian in the continuum is

$$\mathcal{L} = \overline{\psi}_c \,(\mathrm{D}\cdot\gamma + m_c)\,\psi_c \tag{56}$$

where

$$\{\gamma_\mu, \gamma_\nu\} = 2\,\delta_{\mu\nu} \tag{57}$$

and I take

$$\gamma_t = \begin{pmatrix} 1 & 0 \\ 0 & -1 \end{pmatrix} \qquad \gamma_s = \begin{pmatrix} 0 & \sigma_s \\ \sigma_s & 0 \end{pmatrix} \qquad \gamma_5 \equiv \gamma_t\gamma_x\gamma_y\gamma_z = i\begin{pmatrix} 0 & 1 \\ -1 & 0 \end{pmatrix} \tag{58}$$

where, again, s and t as subscripts signify spatial and temporal indices respectively. The obvious discretization is

$$\mathcal{L}_{\mathrm{lat}} = \overline{\psi}_c \left(\Delta\cdot\gamma + m_c + \mathcal{O}(a^2)\right)\psi_c \tag{59}$$

where now Δ_μ is a (tadpole-improved) gauge-covariant derivative:

$$\Delta_\mu\psi(x) \equiv \frac{U_\mu(x)\,\psi(x+a\hat{\mu}) - U_\mu^\dagger(x-a\hat{\mu})\,\psi(x-a\hat{\mu})}{2\,a\,u_0}. \tag{60}$$

Unfortunately this simple lattice Dirac action has a very serious pathology; it suffers from "fermion doubling." This problem can be avoided by introducing a field transformation before discretizing the action. Starting with the continuum action, which we write as

$$\mathcal{L} = \overline{\psi}_c M_c \psi_c, \qquad M_c \equiv D\cdot\gamma + m_c, \tag{61}$$

we introduce a field transformation,

$$\psi = \Omega\psi_c \tag{62}$$

$$\Omega^2 = 1 - \tfrac{1}{2} r a (D\cdot\gamma - m_c) \tag{63}$$

to obtain a new lagrangian $\overline{\psi}M\psi$ where

$$\begin{aligned} M &= \Omega M_c \Omega \\ &= D\cdot\gamma + m_c\left(1 + \tfrac{1}{2} r a m_c\right) - \tfrac{1}{2} r a \left(D^2 + \tfrac{1}{2} g\sigma\cdot F\right). \end{aligned} \tag{64}$$

The third term in M, introduced by the field transformation, breaks the doubling symmetry when we discretize the action. Allowing errors only of order a^4 and higher, we obtain the "D234c" discretization[14] of M:

$$\begin{aligned} M_{\mathrm{D234}} \equiv\; & \Delta\cdot\gamma - c_3 \frac{a^2}{6}\Delta^{(3)}\cdot\gamma + m \\ & - \frac{ra}{2}\left(\Delta^{(2)} + c_F \frac{g\sigma\cdot F}{2} - c_4 \frac{a^2}{12}\Delta^{(4)}\right) \end{aligned} \tag{65}$$

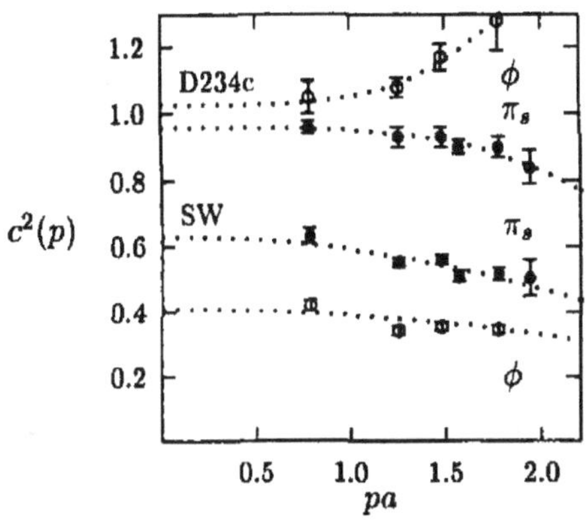

Figure 6. Speed of light squared from the quark dispersion relation using the D234c and SW actions, for pseudoscalar ("π_s") and vector ("ϕ") mesons made of strange quarks. Simulations are on isotropic lattices with $a \approx .4$ fm and an improved gluon action. Fits shown are to $c_0 + c_4(pa)^4$ for D234c and $c_0 + c_2(pa)^2$ for SW. In the continuum, $c^2 = 1$.

where $m \equiv m_c(1 + \frac{1}{2} r a m_c)$, $\Delta_\mu^{(n)}$ is the lattice discretization of the n'th gauge covariant derivative D_μ^n, and $F_{\mu\nu}$ is an a^2-improved discretization of the gluon field strength. All operators are tadpole improved; at tree level the couplings $c_F = c_3 = c_4 = 1$.This action with only the $\mathcal{O}(a)$ corrections is know as the Sheikholeslami-Wohlert (SW) action; with no correction terms, it is the Wilson (W) action. These last two actions have been extensively studied. The D234c action is quite new.

Tests of Quark Actions

The D234c action has been extensively tested on coarse lattices with[14] $a = 0.25$fm and 0.4 fm. This study focused on quark masses near the strange quark mass. Hadrons made of quarks with lighter masses are generally larger and lighter; consequently discretization errors are probably larger for the strange quark than for the other light quarks.

The leading correction in M_{D234c} beyond those in the more conventional SW formalism is

$$- c_3 \frac{a^2}{6} \sum_\mu \Delta_\mu^{(3)} \gamma_\mu . \tag{66}$$

This violates Lorentz invariance; it cancels similar violations in the leading-order terms. Thus we can test this correction by computing

$$c^2(\mathbf{p}) \equiv \frac{E_h^2(\mathbf{p}) - E_h^2(0)}{\mathbf{p}^2} \tag{67}$$

where $E_h(\mathbf{p})$ is the energy of hadron h at three-momentum $\mathbf{p}$. In a Lorentz invariant theory, $c(\mathbf{p})$ should equal 1, the speed of light, for all $\mathbf{p}$. Results at 0.4 fm are shown in Fig. 6 for both the D234c and SW actions. D234c is dramatically superior, deviating from 1 by only 3–4% at $\mathbf{p} = 0$. Also the dispersion relation is accurate to within 10% out to three-momenta of order $1.5/a$.

Table 1. Percentage change in hadron masses when the redundant coefficient r is changed from 1 to 2/3. The hadrons are all made out of strange quarks; π_s and N_s are analogues of the pion and nucleon. In the continuum, masses are r-independent.

Action:	Wilson	SW	D234c	D234c
a (fm):	0.4	0.4	0.4	0.25
π_s	16.0 (7) %	8 (2) %	2.0 (7) %	1.0 (1) %
ϕ	15.2 (11)	8 (2)	2.0 (5)	1.3 (2)
N_s	9.0 (14)	5 (2)	1.1 (6)	0.7 (2)
Ω	9.0 (20)	8 (2)	1.5 (6)	0.9 (2)

We could tune the coupling c_3 in the D234c action nonperturbatively until $c^2(0) = 1$ for the pseudoscalar, thereby restoring Lorentz symmetry through order a^2. Such tuning at .4 fm indicates that $c_3 \approx 1 + \alpha_s/2$, which is sufficiently close to $c_3 = 1$ that there is little need to bother. There is no value of c_3 for which both the pseudoscalar and vector have exactly $c^2 = 1$; uncorrected a^4 errors are larger than the radiative corrections to c_3 at this lattice spacing and quark mass.

There are two different definitions of a hadron's mass in lattice simulations. One is the "static mass," $E_h(0)$, and the other the "kinetic mass," which equals $E_h(0)/c^2(0)$. In D234c these two definitions agree to within 3-4%. In SW they differ by 40–60%. The kinetic and static masses in either formalism must be equal for zero mass mesons, because of the axis-interchange symmetry of the actions. The deviations seen here are because the strange quark is relatively massive at 0.4 fm; the ϕ mass, for example, is $2.1/a$ with D234c.

Another test of the D234c action involves varying the parameter r. The terms multiplied by r in the D234c action are approximately redundant at low momenta, since they originate from a field transformation. Thus, insofar as our discretization is accurate, hadron masses and other quantities are independent of r. Results from simulations both at the standard value $r = 1$ and at $r = 2/3$ are shown in Table 1. With D234c the hadron masses changed by no more than 1–2% at .4fm, and, as expected, less at .25 fm. This suggests that finite-a errors due to the r dependent terms in the action are of order 6% or less at .4fm. With SW, where the r terms are less highly corrected and therefore less redundant, shifts are of order 5–8% at .4fm, while they are 10–15% for the Wilson action, which is the least corrected.

The coefficients c_F, c_3, and c_4 are all renormalized by residual quantum effects, even after tadpole improvement. These can be varied in simulations to assess the importance of the potential corrections. Simulations show that only c_F has a strong effect on the static hadron masses. D234c with $c_F = 1$ gives errors in the ϕ mass of order 2% at 0.25fm and only 7% at 0.4fm; but a radiative correction of order $\pm\alpha_V/6$ in c_F would either remove these errors or double them. Thus to achieve precisions of order a couple percent at such large lattice spacings, we require a calculation of the order α_V correction to c_F. Work has begun on the perturbative evaluation of this correction; nonperturbative tuning is also possible[14]. Preliminary results suggest that radiative corrections are small.

These tests show that improved actions work very well for light quarks. Work continues on applications to phenomenology and on variations for anisotropic lattices. Simulations at .4 fm, with errors of order 5% or less, seem likely.

EFFECTIVE FIELD THEORY: AN EXERCISE[15]

Lattice QCD is a highly nontrivial example of an effective field theory. An effective field theory is a low-energy approximation to some more general theory. Here low-energy refers to wavelengths longer than the lattice spacing, while the more general theory is full continuum QCD, which is valid at wavelengths far smaller than any practical lattice spacing. A thorough understanding of effective field theories is essential for work in lattice QCD, and therefore it is useful to have simple examples of effective field theories with which to explore one's understanding. In this section I present a very simple example of an effective theory, one that uses nothing more sophisticated than the Schrödinger equation. This example, nevertheless, illustrates nonperturbative renormalization, effective field theories, operator product expansions, and many other crucial aspects of renormalization theory.

In this section, we design an effective Schrödinger theory that reproduces a given collection of low-energy data. We begin by discussing the data we will use. We then examine an obvious, well-known procedure for modelling such data, and its limitations. Next we construct a somewhat less obvious effective theory that overcomes these limitations. This modern approach allows us to model the data with arbitrary precision at low energies. It requires that we nonperturbatively renormalize the Schrödinger equation. The approach is numerical throughout, because this is the simplest approach to the particular problems under study.

Synthetic Data

To begin we need a collection of low-energy data that describe a particular physical system. While we could use real experimental data from a physically interesting system, it is better here to avoid the complexities associated with experimental error by generating "synthetic data."

I generated synthetic data for use in these lectures by inventing a simple physical system, and then solving the Schrödinger equation that describes it. I obtained binding energies, low-energy phase shifts, and matrix elements. For the system, I chose the familiar one-particle Coulombic atom, but with a short-range potential in $V_s(\mathbf{r})$ addition to the Coulomb potential:

$$H = \frac{\mathbf{p}^2}{2m} + V(\mathbf{r}) \tag{68}$$

where

$$V(r) = -\frac{\alpha}{r} + V_s(\mathbf{r}). \tag{69}$$

I arbitrarily set $m = 1$ and $\alpha = 1$.

For our purposes, the short-range potential may be anything; I made one up. The whole point of effective theories is that we can systematically design them directly from low-energy data, with no knowledge of the short-distance dynamics. Thus the form of $V_s(\mathbf{r})$ is irrelevant to our analysis. To underscore this point I will not reveal the functional form of the $V_s(\mathbf{r})$ I used. In what follows we need only know that it has finite range.

Such short-range interactions are common in real Coulombic atoms. The effect of the proton's finite size on the hydrogen atom's spectrum is an example. Another is the weak interaction between the electron and proton, which generates very short-range potentials.

Having chosen a particular $V_s(\mathbf{r})$, I wrote a simple computer code to numerically solve for the radial wavefunctions of energy eigenstates. I used this to compute a

Table 2. Binding energies for various S-wave eigenstates of the synthetic hamiltonian.

level	binding energy	level	binding energy
$1S$	1.28711542	$6S$	0.0155492598
$2S$	0.183325753	...	
$3S$	0.0703755485	$10S$	0.00534541931
$4S$	0.0371495726	$20S$	0.00129205010
$5S$	0.0229268241		

Table 3. S-wave phase shifts for the synthetic hamiltonian. Phase shifts are computed for $r=50$.

energy	phase shift	energy	phase shift
10^{-10}	−0.000421343353	.03	1.232867297
10^{-5}	−0.133227246	.07	−0.579619620
.001	−1.319383451	.1	−1.156444634
.003	0.900186195	.3	−0.106466466
.007	−0.146570028	.7	−1.457426179
.01	−0.654835316	1	1.160634967

variety of binding energies, phase shifts and matrix elements. Some of the S-state binding energies are listed in Table 2. Note that the energies would have been given by $-\alpha^2 m/2n^2 = -1/2n^2$, with $n = 1, 2 \ldots$, had there been no V_s. Thus the $1S$ energy is more than doubled by the short-range potential; V_s is not a small perturbation. Also the short-range nature of V_s is evident in this data since the energies approach the Coulombic energies for the very low-energy, large-n states.

Sample S-wave phase shifts are given in Table 3. These phase shifts depend upon the radius at which they are measured because of the long Coulomb tail in the potential $V(\mathbf{r})$; I arbitrarily chose $r=50$ for my phase-shift measurements, this being much larger than the Bohr radius, $1/\alpha m$, of my atom, and also much larger than the range of $V_s(\mathbf{r})$. (Alternatively, one can compute the phase shift with and without $V_s(\mathbf{r})$, and take the difference; the Coulomb divergence cancels in the difference.) I also computed $\langle \mathbf{p}^4 \rangle$ for several S-states, as well as the wavefunctions at the origin; these are tabulated in later sections.

The numerical analysis required to generate such data is minimal. The computation can easily be done using standard numerical analysis packages or symbolic manipulation programs on a personal computer.

Our challenge is to design a simple theory that reproduces our low-energy data with arbitrarily high precision. We must do this using only the data and knowledge of the long-range structure of the theory (that is, we are given m and α).

A Naive Approximation

A standard textbook approach to modelling our data is to approximate the unknown short-range potential by a delta function, whose effect is computed using first-

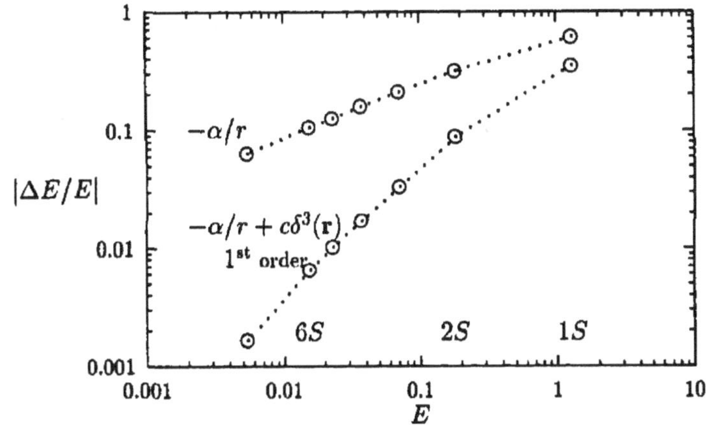

Figure 7. Relative errors in the S-wave binding energies are plotted versus binding energy for the Coulomb theory, and for the Coulomb theory augmented with a delta function in first-order perturbation theory.

order perturbation theory. The approximate hamiltonian is

$$H_{\text{app}} = \frac{\mathbf{p}^2}{2m} - \frac{\alpha}{r} + c\,\delta^3(\mathbf{r}), \tag{70}$$

where c is a parameter. Using first-order perturbation theory, the energy levels in our approximate theory are

$$\begin{aligned} E_n^{\text{app}} &= E_n^{\text{coul}} + c\left|\psi_n^{\text{coul}}(0)\right|^2 \\ &= -\frac{1}{2n^2} + c\frac{\delta_{l,0}}{\sqrt{\pi}\,n^3}. \end{aligned} \tag{71}$$

Our approximation has only a single parameter, c. This we must determine from the data. We do this by fitting formula (71) to our lowest-energy data, the 20S binding energy; this implies $c = -.5963$. We use the lowest-energy state because that is the state for which the replacement $V_s(\mathbf{r}) \to c\,\delta^3(\mathbf{r})$ is most accurate.

In Figure 7 I show the relative errors in several S-wave binding energies, plotted versus binding energy, obtained using just a Coulomb potential ($c = 0$) and using our approximate formula, Eq. (71), with $c = -.5963$. Both approximations become more accurate as the binding energy decreases, but adding the delta function in first order gives substantially better results. Our approximation to V_s is quite successful.

The limitations of this approximation become evident if we seek greater accuracy. We have made two approximations. First we used only first-order perturbation theory. There will be large contributions from second and higher orders in perturbation theory if the short-range potential is strong. So we might wish to compute the second-order contribution to E_n^{app}:

$$\sum_{m \neq n} \frac{\langle n|c\,\delta^3(\mathbf{r})|m\rangle\,\langle m|c\,\delta^3(\mathbf{r})|n\rangle}{E_n - E_m}. \tag{72}$$

Unfortunately this expression gives an infinite shift; the sum over scattering eigenstates diverges as scattering momentum $\mathbf{p} \to \infty$. The delta function is too singular to be meaningful beyond first-order perturbation theory.

The second approximation is replacing V_s by a delta function. The nature of this approximation can be appreciated by Fourier transforming V_s. Since V_s has a very short range, its transform $v_s(q^2)$ depends only weakly on the momentum transfer q. Thus we might try approximating v_s by the first few terms in its Taylor expansion when calculating low-energy matrix elements:

$$v_s(q^2) = v_s(0) + q^2 v_s'(0) + \qquad (73)$$

This is an economical parameterization of the short-distance dynamics since it replaces a function, $v_s(q^2)$, by a small set of numbers: $v_s(0)$, $v_s'(0)\ldots$. Keeping just the first term is equivalent to approximating $V_s(r)$ by a delta function. The additional terms correspond to derivatives of a delta function, and correct for the fact that the range of V_s is not infinitely small—the expansion is in powers of q times the range. This suggests that we might improve our approximation by adding a second parameter, again to be tuned to reproduce data:

$$V_s(\mathbf{r}) \rightarrow c\,\delta^3(\mathbf{r}) + d\,\nabla^2\delta^3(\mathbf{r}). \qquad (74)$$

Unfortunately the additional first-order shift in E_n^{app}, $\langle n|d\,\nabla^2\delta^3(\mathbf{r})|n\rangle$, is infinite. Again the delta function is too singular at short distances for such matrix elements to make sense.

Our first attempt to refine the approximate model has failed. We replaced the true, but unknown short-distance behavior of the potential by behavior that is pathologically singular as $r \rightarrow 0$. Conventional wisdom, often to be seen in books and even recent conference proceedings, is that our approximations must be abandoned once infinities appear, and that only knowledge of the true potential can get us beyond such difficulties. This wisdom is incorrect, as we now see.

Effective Theory

The infinities discussed in the previous section are exactly analogous to those found in relativistic quantum field theories. Among other things, they indicate that even low-energy processes are sensitive to physics at short distances. Modern renormalization theory, however, tells us that the low-energy (infrared) behavior of a theory is independent of the *details* of the short-distance (ultraviolet) dynamics. Insensitivity to the short-distance details means that there are infinitely many theories that have the same low-energy behavior; all are identical at large distances but each is quite different from the others at short distances. Thus we can generally replace the short-distance dynamics of a theory by something different, and perhaps simpler, without changing the low-energy behavior.

The freedom to redesign at short-distances allows us to create effective theories that model arbitrary low-energy data sets with arbitrary precision. There are three steps:

1. Incorporate the correct long-range behavior: The long-range behavior of the underlying theory must be known, and it must be built into the effective theory.

2. Introduce an ultraviolet cutoff to exclude high-momentum states, or, equivalently, to soften the short-distance behavior: The cutoff has two effects. First it excludes high-momentum states, which are sensitive to the unknown short-distance dynamics; only states that we understand are retained. Second, it makes all interactions regular at $r = 0$, thereby avoiding the infinities that plague the naive approach of the previous section.

3. Add local correction terms to the effective hamiltonian: These mimic the effects of the high-momentum states excluded by the cutoff in step 2. Each correction term consists of a theory-specific coupling constant, a number, multiplied by a theory-independent local operator. The correction terms systematically remove dependence on the cutoff. Their locality implies that only a finite number of corrections is needed to achieve any given level of precision.

We now apply this algorithm to design an effective hamiltonian that describes our data.

To begin, our effective theory is specified by a hamiltonian,

$$H_{\text{eff}} = \frac{\mathbf{p}^2}{2m} + V_{\text{eff}}(\mathbf{r}), \tag{75}$$

where the effective potential, V_{eff}, must become Coulombic at large r: $V_{\text{eff}}(r) \to -\alpha/r$, with $\alpha = 1$, for large r. We also need an ultraviolet cutoff. I chose to introduce a cutoff into the Coulomb potential through its Fourier transform:

$$\frac{1}{r} \quad \xrightarrow{\text{F.T.}} \quad \frac{4\pi}{q^2} \quad \xrightarrow{\text{cutoff}} \quad \frac{4\pi}{q^2}\,\mathrm{e}^{-q^2a^2/2} \quad \xrightarrow{\text{F.T.}} \quad \frac{\mathrm{erf}(r/\sqrt{2}a)}{r}, \tag{76}$$

where

$$\mathrm{erf}(x) = \frac{2}{\sqrt{\pi}} \int_0^x \mathrm{e}^{-t^2}\,dt \tag{77}$$

is the standard error function. The new, regulated potential is finite at $r = 0$, but goes to $1/r$ for $r \gg a$. It inhibits momentum transfers of order $\Lambda \equiv 1/a$ or larger. The exact form of the cutoff is irrelevant; there are infinitely many choices all of which give similar results.

Short-distance dynamics is explicitly excluded from the effective theory by the cutoff. We mimic the effects of the true short-distance structure by adding local correction terms. As discussed in the previous section, the low-momentum behavior of any short-range potential is efficiently described in terms of the Taylor expansion in momentum space. Transforming back to coordinate space gives a series that is a polynomial in the momentum operator $\mathbf{p} \equiv \boldsymbol{\nabla}/i$ multiplied by a delta function. We need an ultraviolet cutoff to avoid infinities, and therefore we smear the delta function over a volume whose radius is approximately the cutoff distance a. I chose a smeared delta function defined by

$$\delta_a^3(\mathbf{r}) \equiv \frac{\mathrm{e}^{-r^2/2a^2}}{(2\pi)^{3/2}\,a^3}, \tag{78}$$

but, again, the detailed structure of this function is irrelevant; other choices work just as well.

Remarkably, the structure of the correction terms is now completely determined, even though we have yet to examine the data. The effective potential must have the form

$$\begin{aligned} V_{\text{eff}}(\mathbf{r}) = {} & -\frac{\alpha}{r}\,\mathrm{erf}(r/\sqrt{2}a) + c\,a^2\,\delta_a^3(\mathbf{r}) \\ & + d_1\,a^4\,\nabla^2\delta_a^3(\mathbf{r}) + d_2\,a^4\,\boldsymbol{\nabla}\cdot\delta_a^3(\mathbf{r})\boldsymbol{\nabla} \\ & + \cdots + g\,a^{n+2}\,\nabla^n\delta_a^3(\mathbf{r}) + \cdots, \end{aligned} \tag{79}$$

where coupling constants c, d_1, $d_2 \ldots$ are dimensionless. It consists of a long-range part together with a series of local "contact" potentials. The contact terms are indistinguishable, to a low-momentum particle, from the true short-distance potential, provided the coupling constants are properly tuned. The potential is nonrenormalizable in the traditional sense, but that is not a problem here since the cutoff prevents infinities.

Generally the effective potential reflects the symmetries of the true theory. Here our effective potential is rotationally invariant because our data are. When the data are not rotationally invariant, additional terms like $a^3 \mathbf{g} \cdot \nabla \delta_a^3(\mathbf{r})$ must be included in $V_{\rm eff}$, where now the coupling constants include vectors, tensors... that characterize the rotational asymmetries of the true theory.

One might worry that our Taylor expansion of the short-distance dynamics would fail when we computed physical quantities like the scattering amplitude. This is because high-momentum states affect even low-energy processes through quantum fluctuations. For example, in the true theory the scattering amplitude is

$$\langle f|T(E)|i\rangle = \langle f|V|i\rangle + \sum_n \frac{\langle f|V|n\rangle\,\langle n|V|i\rangle}{E - E_n} + \cdots. \tag{80}$$

The sums over intermediate states in second-order and beyond include states with arbitrarily large momentum. A momentum-space Taylor expansion of the potential would not converge in matrix elements involving such states. Furthermore when we replace the exact potential V by our effective potential $V_{\rm eff}$, our ultraviolet cutoff in effect limits the sums to states with momenta less than the cutoff $1/a$; contributions from high-momentum intermediate states apparently are completely absent from our effective theory. Our effective theory is saved by the fact that high-momentum intermediate states are necessarily highly virtual if, as we assume, the initial and final states have a low energy. By the uncertainty principle, such states cannot propagate for long times or over large distances. Thus any contribution from these states is very local and can be incorporated into the effective theory using the same set of (smeared) delta-function potentials already included in $V_{\rm eff}$. In this way, the high-momentum states that are explicitly excluded, or badly distorted, by the cutoff are included implicitly through the coupling constants. Note that this means the coupling constants in our effective theory depend nonlinearly on the true potential $V(r)$.

While the true theory is obviously independent of the value of the cutoff a, results computed in the effective theory are only approximately a independent. The residual a dependence in such a result is typically a power series in qa, where q is the characteristic momentum associated with the process under study (for example, the initial momentum or the momentum transfer in a scattering amplitude). The contact terms remove these a-dependent errors order-by-order in a. Thus, for example, the $\mathcal{O}(a^2)$ term in $V_{\rm eff}$ removes errors of order $(qa)^2$ and is the most important correction. Errors of order a^4 are removed by the $\mathcal{O}(a^4)$ terms, of which there may be only two since there are only two independent ways of combining two momentum operators ∇/i with the smeared delta function. Obviously only a finite number of contact terms is needed to remove errors through any finite order in a.

The coupling constants vary with a since they must account for quantum fluctuations excluded by the cutoff. More or less is excluded as a is increased or decreased, and therefore the coupling constants must be adjusted to compensate. They are said to be "running coupling constants." When the short-distance potential is weak, the coupling to high-momentum intermediate states is weak and the coupling constants change only slowly with a.[‖] On the other hand, the coupling constants are often very

‖This behavior is modified if the cutoff distance a is very large. When a is larger than the range of the *long-range* potential, or when there is no long-range potential, the coefficients of the contact operators tend to go to a constant. Thus for example, c in $V_{\rm eff}$ will decrease like $1/a^2$ as a increases so that the coefficient of the delta function operator becomes constant.

cutoff dependent when the short-distance interactions are strong. The dependence of the effective theory on its coupling constants becomes highly nonlinear in this case. In fact this behavior can be so strong that the relative importance of different contact terms can change: an a^4 operator can act more like an a^2 operator, and vice versa. The physical systems I discuss here do not suffer from this complication, although it is easy to study the phenomenon by modifying the synthetic data appropriately.

Highly nonlinear behavior also results when the cutoff distance a is made too small. From the previous section we know that taking $a \to 0$ is a bad idea. In general it makes little sense to reduce a below the range r_s of the true potential. By assumption, our data involves energies that are too low—wavelengths that are too long—to probe the true structure of the theory at distances as small as r_s. When $a < r_s$, high-momentum states are included that are sensitive to structure at distances smaller than r_s. But the structure they see there is almost certainly wrong. Thus taking a smaller than r_s cannot improve results obtained from the effective theory. In fact, as the nonlinearities develop for small a's, results often degrade, or, in more extreme cases, the theory may become unstable or untunable.

Finally a comment on the physical significance of the cutoff: A common practice in applications of potential models, for example to nuclear or atomic physics, is to use a cutoff like our's to account for some physical effect, for example finite nuclear size. Then one must worry about whether the functional form of the cutoff is physically correct. We are *not* doing this here. Our cutoff is only a cutoff; physical effects like finite nuclear size are incorporated systematically through the contact terms. We need never worry, for example, about what the true charge distribution is inside a nucleus. Indeed this is the great strength of the effective theory. The effective theory gives us a simple, *universal* parameterization for the effects of short-distance structure. The form of the contact terms is theory independent. Only the numerical values of the coupling constants c, $d_1 \ldots$ are theory specific; for our (low-energy) purposes, everything that we need to know about the short-distance dynamics of the true theory is contained in these coupling constants. As discussed earlier, this situation is conceptually similar to a multipole analysis of a classical field. The couplings here are the analogues of the multipole moments, while the delta-function potentials are analogous to the multipoles that generate the field.

The exact form of the cutoff, and the exact definition of the smeared delta function are irrelevant. Different definitions give similar results for binding energies and phase shifts, but with different values for the coupling constants, and possibly quite different.

Tuning the Effective Theory; Results

We now tune the parameters of our effective theory so that it reproduces our low-energy data through order a^4. For simplicity, we examine only S-wave properties. Thus we need only the c and d_1 terms in V_{eff}, Eq. (79); we can drop the d_2 term in $V_{\text{eff}}(\mathbf{r})$ since it is important only for P-wave states.**

To generate the results discussed here I first chose a reasonable value for the cutoff distance: to begin with, a = 1, the Bohr radius for the Coulombic part of the interaction. Then I varied the coupling constants c and d_1 until the S-wave phase shifts from the effective hamiltonian agreed with the data at energies $E = 10^{-5}$ and 10^{-10}.

**It is obvious that the d_2 term couples only to P-waves in the limit $a \to 0$. When a is nonzero, however, this term has a small residual coupling to S-waves. This results in an interaction of order a^6, which may be ignored to the order we are working here. One can easily design contact terms that contribute only to a single channel in orbital angular momentum. For example, our smeared delta function could be replaced by a local potential that is separable.

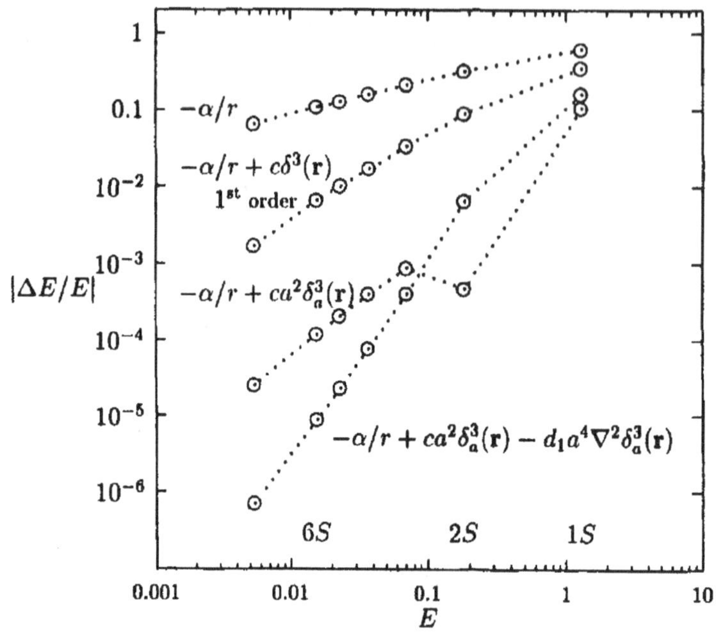

Figure 8. Relative errors in the S-wave binding energies are plotted versus binding energy for the Coulomb theory, the Coulomb theory augmented with a delta function in first-order perturbation theory, the nonperturbative effective theory through a^2, and the effective theory through a^4.

I found $c/4\pi = -3.18$ and $d/4\pi = -0.199$. Having found the coupling constants, I then generated binding energies, phase shifts, and matrix elements using the effective theory. I also generated results with only the a^2 correction $(d_1 = 0)$; the effective theory requires a different value of c in this case, since the two contact terms affect each other.

I tuned the couplings using phase shifts at very low energies. This was to minimize the effect of the a^6 errors, which arise because we truncated our effective potential at order a^4. These "truncation" errors are smallest for the lowest-energy data, since they are generally proportional to qa raised to some positive power. In general one should use the most infrared data available when tuning. Alternatively one might attempt a global fit to all of the data; but then it is crucial to give greater weight in the fit to low-energy data than to high-energy data: some estimate of the $(qa)^n$ error due to truncation must be added to the experimental errors used to weight the data in the fit[16].

So long as $a > 0$, $V_{\rm eff}(\mathbf{r})$ is a simple, nonsingular function of r. Consequently the effective theory is simple to solve numerically. To compute the results here, I reused the code that I wrote to generate the synthetic data, but with the true potential V(r) replaced by the effective potential $V_{\rm eff}(\mathbf{r})$. The effective theory is no harder to solve than the other; in particular, because of the cutoff, there are no infinities.

In Figure 8 the errors in the binding energies obtained from the tuned effective theory are compared with those obtained above from first-order perturbation theory. Even with only the a^2 correction there is a sizable reduction in errors, due to contributions from second and higher orders in the correction — our numerical solution of the effective theory is nonperturbative, and so includes all orders. The accuracy at low energies is further enhanced by the a^4 correction, which begins to account for the

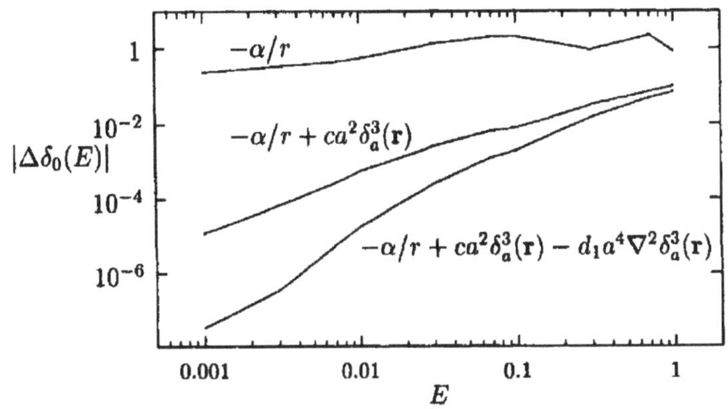

Figure 9. Errors in S-wave phase shifts (in radians) computed with the effective theory corrected through order a^2 and a^4. Values are for $a=1$ and are plotted versus energy. Results are also shown for the theory with no contact terms.

finite range of the short-range interaction. The crossing of the curves for the a^2 and a^4 theories around $E=0.1$ has no significance; the error in the a^2 theory changes sign there, passing through zero.

The phase shifts tell a similar story; see Figure 9. At low energies, the errors decrease steadily as the correction terms are added, order-by-order in a^2. The slope of the error curve changes as each new correction is added, getting steeper by one power of $E \propto p^2$ each time the order of the error is increased by a^2, as expected. Similar behavior is apparent in the previous figure, which shows the errors in the binding energies.

Both of these figures show that the correction terms have little effect at energies $E > 1$. At these energies, the particle's wavelength is sufficiently short that the particle can probe the detailed structure of $V_s(\mathbf{r})$. Effective theories are useless in this limit. To go much beyond $E=1$ in this example, one must somehow uncover the true short-distance structure of the theory.

I redid the analysis for a variety of different a's to explore cutoff dependence. Figure 10 shows how the error in the phase shift depends on a. As a is reduced from $10 \to 3 \to 1$, the errors decrease at all but the highest energies. For this range, a is larger than the range r_s of the true short-distance interaction, and therefore the errors are proportional to a power of qa, which decreases with decreasing a. The errors stop decreasing for $a < 1$, becoming almost insensitive to a. From our discussion in the previous section, this suggests that $r_s \approx 1$. The coupling constants also vary with a in the manner expected if $r_s = 1$. For example, with $d_1 = 0$, coupling constant $c/4\pi$ is roughly –3 and only weakly dependent on a when $a \gtrsim 1$. It grows rapidly as a is reduced below one; for example, $c=-74.0$ when $a=0.01$.

I also looked for multiple solutions for the coupling constants[17]. Setting $d_1 = 0$ and $a=1$, for example, I found that the phase shift at $E=10^{-10}$ is correct when $c/4\pi$ equals –12.4 or 1.6, as well as when it equals –3.18, the value used above. These extra solutions give reasonably good results for the phase shifts and binding energies, but not as good as our original value. The reason is that with these new c's the theory's short-distance structure is significantly modified. In particular, with $c=24.7$, there is an extra bound state with a very high binding energy (≈ 6). We do not expect our effective theory to work well at high energies, so there is nothing wrong with having the

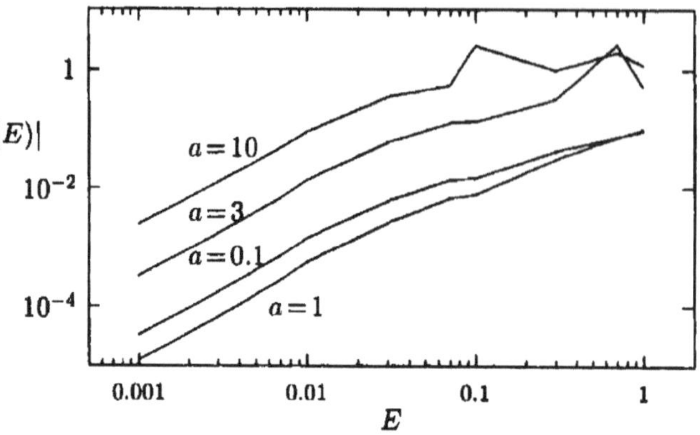

Figure 10. The dependence on cutoff of the S-wave phase shift errors plotted against energy is shown. Errors decrease as a is reduced until $a \approx 1$, beyond which point there is no improvement. The a^2 theory was used here ($d_1 = 0$).

extra state; the usual $1S$, $2S$... states are all still present, each with an extra node in its wavefunction at short distances. When $c/4\pi = 1.6$, the $1S$ state disappears, while the other states are missing a node at short distances.

It is remarkable that a simple two-parameter effective theory, tuned using only two very low energy phase shifts, can reproduce the full range of our data so successfully. The binding energy of the $2S$ state, for example, is accurately predicted by the effective theory to better than 1%, despite the fact that the cutoff scale $a = 1$ is almost a quarter the radius of the atom. Our most infrared data is reproduced to better than six digits.

Operators and the Operator Product Expansion

So far we have used our effective theory to compute binding energies and phase shifts. We now examine quantities that depend in detail on the wavefunctions. Consider, for example, the matrix element $\langle n|\mathbf{p}^4|n\rangle$, which might be important if we wished to include relativistic corrections in our potential model. In Table 4 values of this matrix element are listed for several S-states both for the true theory, and for our corrected theory (with $a = 1$). The values disagree by more than a factor of two, even for very low-energy states, despite the fact that the two theories agree on the corresponding binding energies to several digits.

The problem is that the operator in the effective theory that corresponds to $\mathbf{p}^4$ in the true theory is not $\mathbf{p}^4$. As is true of the hamiltonian, there are local corrections to $\mathbf{p}^4$ in the effective theory. Thus, for any S state, we expect

$$\langle \mathbf{p}^4 \rangle_{\text{true}} = Z\,\langle \mathbf{p}^4 \rangle_{\text{eff}} + \frac{\gamma}{a}\,\langle \delta_a^3(\mathbf{r}) \rangle_{\text{eff}} + \eta\, a\,\langle \nabla^2 \delta_a^3(\mathbf{r}) \rangle_{\text{eff}} + \mathcal{O}(a^3) \tag{81}$$

where Z, γ and η are dimensionless constants that are independent of the state. I tuned these constants so that Eq. (81) gave correct values for the $10S$, $15S$ and $20S$ bound states, and found $Z = 1$, $\gamma = -96.2$ and $\eta = -140.6$ when $a = 1$. This formula gives excellent results, as shown in Table 4. Just as for the binding energies, the relative error decreases steadily with decreasing energy.

This example illustrates the subtlety of the equivalence between the effective theory and the true theory. Although the infrared spectra and phase shifts are quite

Table 4. Expectation values of $\mathbf{p}^4$ for S-states in the true theory, and in the effective theory ($a=1$). These are compared with matrix elements of the renormalized operator in the effective theory that corresponds to $\mathbf{p}^4$ in the true theory.

level	$\langle \mathbf{p}^4 \rangle$	$\langle \mathbf{p}^4 \rangle_{\text{eff}}$	$\langle Z\mathbf{p}^4 + \gamma\delta_a^3/a + \cdots \rangle_{\text{eff}}$
$1S$	69	5.9	28
$2S$	5.50	1.7	5.34
$3S$	1.309	0.4	1.306
$4S$	0.5070	0.17	0.5068
$5S$	0.24740	0.08	0.24738
$6S$	0.138784	0.05	0.138780

Table 5. Wavefunctions evaluated at $r=0$ for S-states in the true theory, and in the effective theory ($a=1$). These are compared with the renormalized expression in the effective theory that corresponds to the wavefunction at the origin in the true theory.

level	$\psi(0)$	$\psi_{\text{eff}}(0)$	$\overline{\gamma} \int \psi_{\text{eff}} \delta_a^3 + \cdots$
$1S$	1.50	0.53	-3.4
$2S$	0.383	0.19	0.369
$3S$	0.1837	0.09	0.1830
$4S$	0.11353	0.06	0.11344
$5S$	0.079005	0.04	0.078986
$6S$	0.059031	0.03	0.059025

similar, the short-distance structure of corresponding wavefunctions in the two theories may be completely different. This means that matrix elements of operators sensitive to short distances, like $\mathbf{p}^4$, are not equal, even for very low-energy states. Renormalization theory, however, tells us how to define correction terms for the operators in the effective theory that correct for the differences in short-distance behavior between the two theories. Just as in the hamiltonian, the corrections are local operators with state-independent coupling constants, and, again, only a finite number of terms is needed to any given order in a.

Another example of the same idea is provided by the wavefunction at the origin, which is often needed in phenomenological work. In Table 5 values of the wave function at the origin are listed for several S-wave bound states. Again the effective theory gives results that are completely different from the true theory. Renormalization theory, however, implies that

$$\begin{aligned}\psi_{\text{true}}(r=0) = {} & \overline{\gamma} \int d^3r\, \psi_{\text{eff}}\, \delta_a^3(\mathbf{r}) \\ & + \overline{\eta}\, a^2 \int d^3r\, \psi_{\text{eff}}\, \nabla^2 \delta_a^3(\mathbf{r}) \\ & + \mathcal{O}(a^4) \end{aligned} \tag{82}$$

where $\overline{\gamma}$ and $\overline{\eta}$ are dimensionless, state-independent constants. For $a = 1$, I found $\overline{\gamma} = -28.53$ and $\overline{\eta} = -3.615$ by tuning this formula to give correct results for the $20S$ and $15S$ bound states. With these values, the results from the effective theory for

$\psi_{true}(0)$ are given in the last column of Table 5. Again the agreement is excellent for all but the most ultraviolet of the states, and it gets better as the states are more infrared—just as expected.

Both of these applications are closely related to the operator product expansion in quantum field theory and to factorization theorems of the sort used in perturbative QCD. Each quantity is a sum of terms, each of which factors into a long-distance state-dependent matrix element times a short-distance universal coefficient. This is exactly the structure exploited in lattice QCD to compute matrix elements of currents or four-quark operators, especially for applications in heavy-quark physics.

An Application: Nucleon-Nucleon Interactions

An example of an effective theory is the pion-nucleon theory of low-energy nuclear interactions. This problem is accurately described by a Schrödinger formalism, rather than a field theory, because pion retardation is negligible — that is, pion exchange is effectively instantaneous. The potential in our nucleon-nucleon hamiltonian can be determined by "matching" scattering amplitudes computed in the potential model with the same amplitudes computed using the chiral field theory, as above. There are both long-range parts, involving the exchange of one or more pions, and short-range or "contact" potentials like the delta function potentials above. The effective theory is an expansion in powers of the momentum divided by Λ_c, the momentum scale at which QCD color becomes apparent (500-1000 MeV).

The leading-order long-range potential is due to the exchange of a single pion; it gives

$$V_{1\pi} = \alpha_\pi\, \boldsymbol{\tau}_1\cdot\boldsymbol{\tau}_2\, \frac{\boldsymbol{\sigma}_1\cdot\boldsymbol{\nabla}\, \boldsymbol{\sigma}_2\cdot\boldsymbol{\nabla}}{m_\pi^2}\, v_\Lambda(r) \tag{83}$$

where $v_\Lambda(r)$ is a Yukawa potential at large r,

$$v_\Lambda(r) \xrightarrow{r\ \text{large}} \frac{e^{-m_\pi r}}{r}, \tag{84}$$

and the $\boldsymbol{\sigma}$'s and $\boldsymbol{\tau}$'s are Pauli matrix operators for the nucleon spin and isospin respectively. Chiral symmetry implies that the dimensionless coupling α_π is related to the pion decay constant, $f_\pi \approx 93$ MeV, the nucleon's axial vector coupling, $g_A \approx 1.25$, and the pion mass m_π:

$$\alpha_\pi \approx \frac{g_A^2\, m_\pi^2}{16\pi\, f_\pi^2} \approx 0.06\text{–}0.08. \tag{85}$$

The coupling is small because it is proportional to m_π^2/Λ_c^2, Λ_c being of order $4\sqrt{\pi} f_\pi$. Projecting the one-pion potential onto different spin-orbit eigenstates, we obtain

$$V_{1\pi} \to -\alpha_\pi\,[b\, v_\Lambda(r) + b_T\, v_T(r)] \tag{86}$$

where b and b_T are constants that depend upon the channel, and

$$v_T(r) \equiv \frac{1}{m_\pi^2}\left(v_\Lambda'' - \frac{v_\Lambda'}{r}\right) \tag{87}$$

$$\xrightarrow{r\ \text{large}} \frac{e^{-m_\pi r}}{r}\left[1 + \frac{3}{m_\pi r} + \frac{3}{m_\pi^2 r^2}\right]. \tag{88}$$

Here we drop a part of the potential that is indistinguishable from the contact terms we add below. We introduce an ultraviolet cutoff in the Fourier transform, as above, to obtain

$$v_\Lambda(r) \equiv \frac{1}{2r}\left[e^{-m_\pi r}\operatorname{erfc}\!\left(-(r\Lambda - m_\pi/\Lambda)/\sqrt{2}\right) - (r \to -r)\right], \tag{89}$$

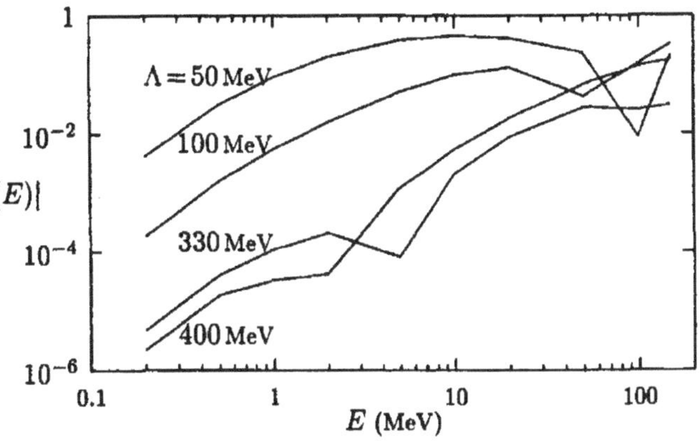

Figure 11. Errors in the $^1S_0$ phase shifts (in radians) versus energy for the full effective theory with different values of the cutoff Λ.

where $\mathrm{erfc}(x) = 1 - \mathrm{erf}(x)$, and $\Lambda = 1/a$ is the momentum cutoff.

Higher-order (in p/Λ) contributions to the long-range potential come from two-pion exchange. We will not include these here; they are described in the literature[18].

The short-distance terms in the potential are smeared delta functions, with $a = 1/\Lambda$, multiplied by powers of the momentum operator $\mathbf{p} = \nabla/i$, and the spin and isopin operators. It is convenient for our analysis to classify these by the spin-orbit channels that they affect.

The $^1S_0$ and $^3S_1$ channels have Λ^{-2} terms with no ∇'s; there is one for each channel, each with its own coupling constant. In order Λ^{-4} there are two contact potentials, one for each channel, proportional to $\mathbf{p}^2\delta^3_{1/\Lambda}(\mathbf{r}) + \delta^3_{1/\Lambda}(\mathbf{r})\mathbf{p}^2$. The form of this potential can be simplified by adding $-2\,\mathbf{p}\cdot\delta^3_{1/\Lambda}(\mathbf{r})\mathbf{p}$, which doesn't couple to S-states in this order; the resulting potential is then proportional to $(\nabla^2\delta^3_{1/\Lambda})$,just as before.

The P states do not couple to contact terms in $\mathcal{O}(\Lambda^{-2})$. Thus our effective theory predicts that P states are more accurately described by just the long-range potential than S-states. The leading P-wave contact terms are $\mathcal{O}(\Lambda^{-4})$, and, after projecting out the orbital and spin angular momenta, are all proportional to $\mathbf{p}\cdot\delta^3_{1/\Lambda}(\mathbf{r})\mathbf{p}$. This operator can be simplified by adding $-1/2(\mathbf{p}^2\delta^3_{1/\Lambda}(\mathbf{r}) + \delta^3_{1/\Lambda}(\mathbf{r})\mathbf{p}^2)$, which doesn't couple to P states in this order, to obtain again a potential proportional to $(\nabla^2\delta^3_{1/\Lambda})$. There is one such potential for each of the four spin-orbit P-wave channels, and each has its own coupling constant.

There are no contact terms in $\mathcal{O}(\Lambda^{-2}, \Lambda^{-4})$ that couple just to D states; thus D states should be even more accurately described by the long-range potential than P states. The only contact term in order Λ^{-4} that affects a D states is one that couples the $^3D_1$ and $^3S_1$ states.

There are nine contact terms in all through order Λ^{-4}: two Λ^{-2} terms and two Λ^{-4} terms for the S-states, four Λ^{-4} terms for the P states, and one Λ^{-4} term coupling S and D states. Each of these has its own coupling constant. This seems like a lot of coupling constants, but it is only one or two per channel. And so we analyze the np scattering data channel by channel.

We now compare this model with experimental data. For simplicity, we restrict our study to np scattering, thereby avoiding the problem of Coulomb distortion in pp scattering. We fit phase shifts for different channels, using phase-shift data from the

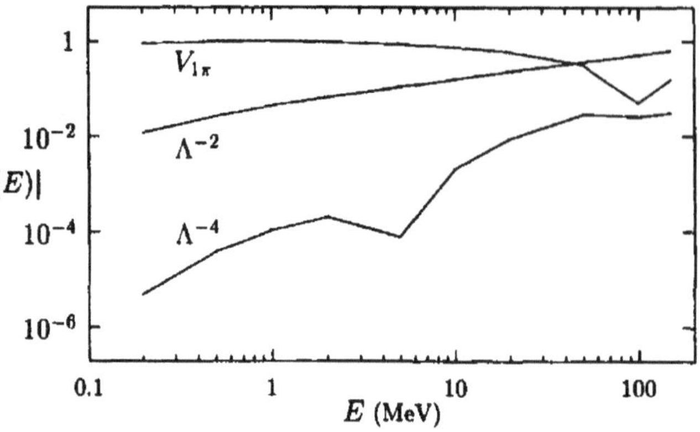

Figure 12. Errors in the $^1S_0$ phase shifts (in radians) versus energy for the effective theory through orders Λ^{-2} and Λ^{-4}. Results from the theory with just pion exchange ($V_{1\pi}$) are also shown. The cutoff was $\Lambda = 330$ MeV in each case.

multi-energy fits by the Nijmegen group[19].

The potential for $^1S_0$ phase shifts, including one-pion exchange and contact terms through order Λ^{-4}, is

$$V(r) = -\alpha_\pi v_\Lambda(r) + c\,\frac{\delta^3_{1/\Lambda}(\mathbf{r})}{\Lambda^2} - d\,\frac{\nabla^2\delta^3_{1/\Lambda}(\mathbf{r})}{\Lambda^4}, \tag{90}$$

where $\alpha_\pi \approx 0.075$ and $m_\pi \approx 140$ MeV. Coupling constants c and d must be tuned to the data. When I tried this, I used the Nijmegen phase shifts at $E = 0.05$ MeV and 0.1 MeV to tune c and d for Λ's ranging from 50 MeV to 1000 MeV. Here and elsewhere in this section E refers to the (nonrelativistic) center-of-mass energy.

In Figure 11 the errors obtained in the phase shift are plotted as a function of center-of-mass energy E for a variety of different values of the cutoff Λ. As in our synthetic example, the errors start out large when Λ is small, and decrease steadily until $\Lambda \approx 300$ MeV. No further improvement is seen for larger Λ's.

In Figure 12 the errors obtained from the Λ^{-2} theory $(d = 0)$ are compared with those from the Λ^{-4} theory. At low energies the errors in the Λ^{-2} theory grow approximately like $E^{0.5}$ with energy. The errors in the Λ^{-4} theory are much smaller, but grow like $E^{1.5}$ — that is faster by two powers of p, exactly as expected. So here, as in our synthetic data, we see clear evidence that the theory is improved by adding higher-order corrections, and that it is improved in just the manner predicted. This convinces me that the effective theory is working.

The curves in the logarithmic plots of $\Delta\delta_0(E)$ versus E dip at various different points. These dips result when the error changes sign. They should be ignored.

The last figure also shows errors for the effective theory with one-pion exchange but no contact terms ($V_{1\pi}$). Evidently one-pion exchange contributes little to this phase shift at most energies.

In Figure 13 the theoretical $^1S_0$ phase shifts are shown together with the data. The results from the full effective theory are almost independent of the cutoff. The variation at high energies gives an indication of the size of the Λ^{-6} corrections yet to be included.

Results for the $^1P_1$ and $^3D_2$ phase shifts are also shown in Figure 13. The effective

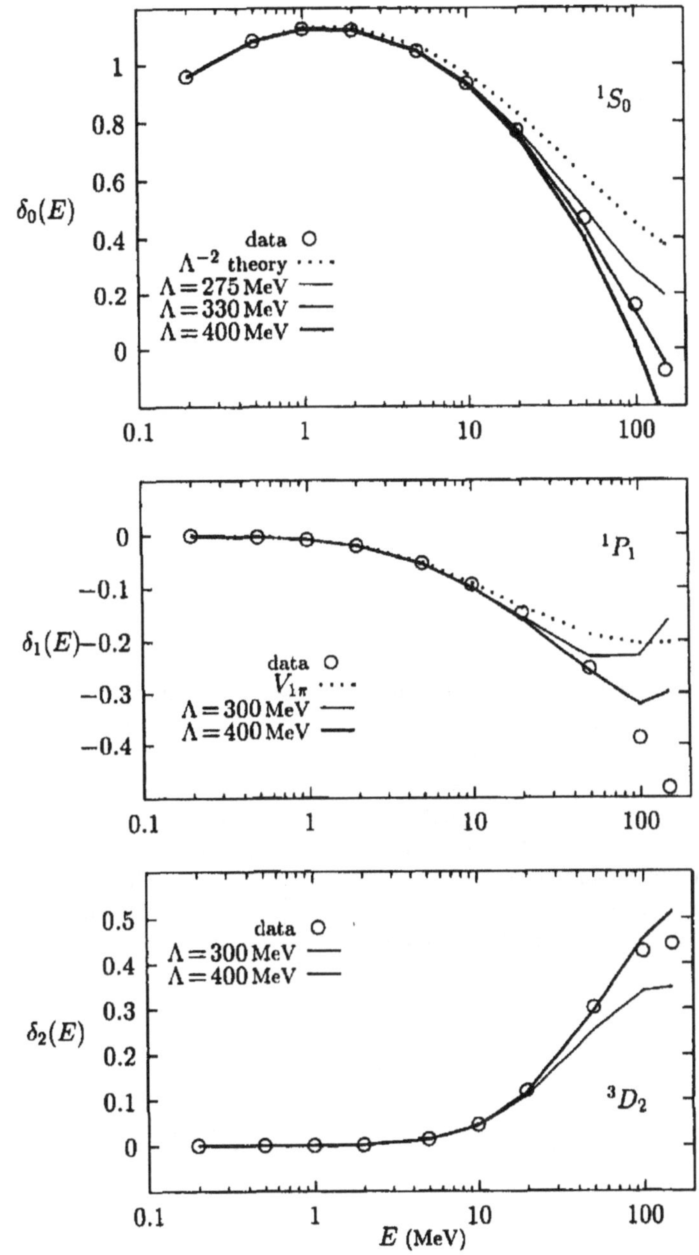

Figure 13. Phase shifts (in radians) for different channels from the effective theory plotted versus energy. Results are shown for various values of the cutoff Λ, and are compared with experiment. S-wave results are for the Λ^{-2} version of the theory ($d=0$, $\Lambda=200$ MeV) and for the full effective theory. P-wave results are for the uncorrected ($V_{1\pi}$) theory, and for the full effective theory. D-wave results are for the uncorrected one-pion theory with different values of Λ; there are no Λ^{-2} or Λ^{-4} corrections for this D-wave channel.

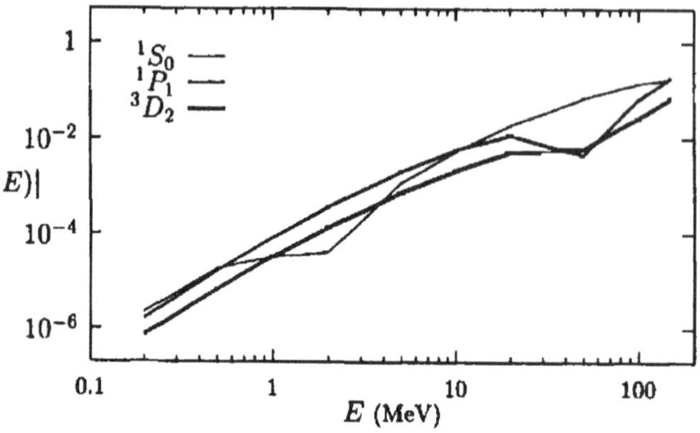

Figure 14. Errors in the phase shifts (in radians) for different channels from the effective theory corrected through order Λ^{-4}.

potentials for these two cases are, through order Λ^{-4},

$$V(^1P_1) = 3\,\alpha_\pi\, v_\Lambda(r) + c(^1P_1)\,\frac{\nabla^2\delta^3_{1/\Lambda}(\mathbf{r})}{\Lambda^4}, \tag{91}$$

and

$$V(^3D_2) = -\alpha_\pi\,[\,v_\Lambda(r) + 2\,v_T(r)]\,. \tag{92}$$

The data confirm that the long-range potential plays a much more important role for P-waves, which have no Λ^{-2} contact term, than for S-waves. And the D-wave analysis has no contact terms at all (until order Λ^{-6}). In both these cases the phase shifts are fairly linear in the couplings, unlike the low-energy S-wave phase shifts.

The relative errors are much smaller for our S-wave phase shifts than for the others. The absolute magnitude of the errors in each channel, however, is roughly the same when the effective theories for each channel are corrected to the same order in $1/\Lambda$. Also the energy dependence of these errors is the same. This is illustrated in Figure 14. It is exactly what is expected based upon the effective theory.

Effective theory does an excellent job of explaining nucleon-nucleon interactions — as it must.

Acknowledgements

These lectures are abridged and updated versions of lectures given at the Schladming Winter School[11] and at the Jorge André Swieca Summer School[15]. My work was supported by the National Science Foundation (USA) and by a fellowship from the Guggenheim Foundation.

REFERENCES

1. K. Symanzik, *Nucl. Phys.* **B226** (1983) 187.
2. G.P. Lepage and P.B. Mackenzie, *Phys. Rev.* **D48** (1993) 2250.
3. For an overview see K.G. Wilson, *Rev. Mod. Phys.* **55** (1983) 583.
4. G.P. Lepage, *Lattice QCD for Small Computers* in *The Building Blocks of Creation,* edited by S.Raby and T.Walker (World Scientific Press, Singapore, 1994).

5. See, for example, H. Kawai, R. Nakayama and K. Seo, *Nucl. Phys.* **B189** (1981) 40.
6. G.Curci, P. Menotti, and G.Paffuti, *Phys. Lett.* **130B** (1983) 205; Erratum: *ibid.* **135B** (1984) 516.
7. M. Lüscher and P. Weisz, *Comm. Math. Phys.* **97** (1985) 59.
8. M. Alford, W. Dimm, G.P. Lepage, G. Hockney and P. Mackenzie, *Phys. Lett.* **B361** (1995) 87.
9. M. Lüscher and P. Weisz, *Phys. Lett.* **158B**, (1985) 250, and references therein.
10. M. Alford, T. Klassen, G.P. Lepage, C. Morningstar and M. Peardon, in preparation.
11. For a review see G.P. Lepage, e-print hep-lat/9607076, in the proceedings of the 1996 Schladming School on *Perturbative and Nonperturbative Aspects of Quantum Field Theory.*
12. G.P. Lepage, e-print hep-lat/9707026, to be published in the proceedings of the International Workshop on Lattice QCD on Parallel Computers (Tsukuba, Japan, Mar 1997).
13. C. Morningstar and M. Peardon, *Phys. Rev.* **D56** (1997) 4043.
14. M. Alford, T. Klassen, and G.P. Lepage, e-print hep-lat/9709126, to be published in the proceedings of Lattice 97 (Edinburgh, Scotland, Jul 1997); *Nucl. Phys.* **B496** (1997) 377; and references therein.
15. G.P. Lepage, e-print nucl-th/9706029, to be published in the proceedings of the 8th Swieca Summer School: Nuclear Physics (Sao Paulo, Brazil, 1997).
16. This point is emphasized in E. Angelos, Cornell University Ph.D. Thesis, 1996.
17. I thank Martin Savage explaining the multiple solutions to me.
18. A particularly useful reference is C. Ordonez, L. Ray and U. van Kolck, *Phys. Rev.* **D53** (1995) 2086. They reference much of the older material as well.
19. Partial wave analysis of the Nijmegen University group, obtained from the World Wide Web page http://nn-online.sci.kun.nl/.

FINITE SIZE TECHNIQUES AND THE STRONG COUPLING CONSTANT

Peter Weisz

Max-Planck-Institut für Physik
Föhringer Ring 6, D-80805, Munich, Germany

1. INTRODUCTION

In these lectures I will mainly describe work done by the Alpha Collaboration to extract quantitative results in QCD from numerical simulations. Information on this collaboration, e.g. its members, physics program etc. can be obtained from the web page: www.ifh.de/computing/parallel/ape/alpha/alpha.html

One of the main goals of our collaboration is to determine the running coupling in the $\overline{\text{MS}}$-scheme[1] at energies around m_Z to a quotable accuracy of a few percent. Our emphasis is on the control of the systematic errors. So far this has only been achieved in pure Yang-Mills theory[2] and the method, which uses finite size techniques, will be described in section 3. The particular coupling defined there is based on the Schrödinger functional first investigated by Symanzik[3] which turns out to be a very powerful tool to probe the dynamics of quantum field theories (in particular non-abelian gauge theories).

Before going to the 4-dimensional theories I will first focus attention on the 2-d sigma-model. The reason for this is two-fold. Firstly it was in this model that we first tested the finite size scaling method mentioned above[4]. Secondly the extraction of physical results from numerical experiments in practice involves accepting some "conventional wisdom" as a plausible working hypothesis. This includes in particular 1) an assumption on the way the continuum limit is approached, and 2) the applicability of renormalized perturbation theory at high energies, the rational of which is the assumed property of asymptotic freedom. The validity of these hypotheses can be made more plausible by studies in simpler models. One can ask whether the validity of asymptotic freedom is at all phenomenologically relevant in its strict sense because it addresses the properties of the model at infinitely large energies, whereas QCD is (at best) an effective theory at intermediate high energies (at higher energies QCD can hardly be unraveled experimentally from the effects of other interactions). The question whether a particular non-perturbative field theory (e.g. QCD) is asymptotically free is however at the very least an interesting academic question.

In section 2 I will thus first introduce the recent work by Balog, Niedermaier and Hauer[5, 6, 7, 8, 9] on the form factor bootstrap approach[10, 11] in the 2-d O(3) non-linar sigma model because their results represent, in my opinion, the best evidence we have at present for the existence of a non-perturbatively defined field theory with asymptotic freedom. Then I will review the general strategy of the computation of a non-perturbatively defined running coupling using finite size recursive methods and a particular application thereof to the sigma-model[4].

The application of the finite size scaling method to the computation of a running coupling in full QCD will still take a few years to complete. Much preparation needs to be done, in particular development of efficient algorithms for simulating fermions and studies of improvement of the action to reduce cut-off effects. In section 4 I will review the Symanzik improvement program[12] applied to Wilson fermions[13], and some of the work done by the Alpha Collaboration to study improved operators and their non-perturbative renormalization which is required to extract physical matrix elements. In these applications the Schrödinger functional again plays a useful role.

The style of these lectures will be rather informal. The aim will be to emphasize the underlying ideas and issues; however, in so doing, many important technical details will be glossed over and for these the interested reader should refer to the original papers.

2. THE 2-d O(*N*) NON-LINEAR SIGMA MODEL

The "conventional wisdom"

The sigma model is a field theory in two dimensions formally described by the action

$$S = \frac{1}{2g_0^2} \int \mathrm{d}^2x (\partial_\mu \sigma^a)^2 \tag{1}$$

where the fields $\sigma^a(x)$, $a = 1, ..., N$ are restricted to lie on a sphere $\sigma^2 = 1$. The model is asymptotically free in renormalized perturbation theory i.e. renormalized running couplings go to zero as momenta $q \to \infty$ according to

$$g^2(q) \sim 1/b_0 \ln(q^2/\Lambda^2), \tag{2}$$

where Λ is the associated (scheme-dependent) lambda-parameter. The quantum model is supposed to have a mass gap with the lowest states above the vacuum an isovector multiplet of single particle states of mass m. In these respects the sigma-model resembles QCD and it is these features together with its comparative accessibility for numerical experiments which makes it a useful laboratory to study the "conventional wisdom". A wonderful property that makes it even more attractive is that the classical model possesses an infinite number of local[14] and non-local conservation laws[15]. Moreover these conservation laws carry over to the quantum theory[16, 17] and imply that in the scattering of two particles there is no particle production. Contrary to the situation in four dimensions this property does not imply that the theory has a trivial S-matrix. Indeed a proposal of the exact 2-2 particle S-matrix was given by the Zamolodchikov brothers[18]

$$S_{ik;jl}(p_1, p_2; p_1', p_2') = \delta(p_1 - p_1')\delta(p_2 - p_2')\sigma_2(\theta) \times$$
$$[\delta_{ij}\delta_{kl} - \frac{2\pi i}{N-2}\{\frac{1}{i\pi - \theta}\delta_{ik}\delta_{jl} + \frac{1}{\theta}\delta_{il}\delta_{jk}\}] + (i \leftrightarrow k, p_1 \leftrightarrow p_2), \tag{3}$$

a structure which also follows from the existence of the non-local conservation laws[16]. The multiparticle S-matrix elements are simply products of the 2-2 particle S-matrix elements. In Eq. (3) θ_i are rapidities $p_j = m \sinh \theta_j$ and $\theta = |\theta_1 - \theta_2|$. Properties of unitarity, analyticity and crossing together with the assumption of no bound states and the minimum number of poles and zeros in the physical strip fix the function σ_2[18]. The conjecture for the S-matrix already has an element of asymptotic freedom in the sense that the phase shifts approach zero logarithmically with the energy at high energy. Non-perturbative checks of the proposed S-matrix include agreement with the computation of the amplitudes in $1/N$ expansion to order N^{-2} and at low energy the phase shifts in the O(3) model have been measured by Lüscher and Wolff[19] using lattice simulations and found to be in good agreement with Eq. (3).

A further important result was obtained by Hasenfratz, Maggiore and Niedermayer[20] who established the exact ratio of the mass gap to the **Λ-parameter**

$$\frac{m}{\Lambda_{\overline{MS}}} = \frac{(8/e)^{1/(N-2)}}{\Gamma[1 + 1/(N-2)]}. \tag{4}$$

To obtain this result requires the ability to compute a physical quantity both perturbatively and non-perturbatively. The quantity they selected was the free energy of the model with a chemical potential h coupled to a Noether charge. For large h one can compute $F(h) - F(0)$ using re-normalized perturbation theory. For general h one can also use the thermodynamic Bethe ansatz (supplemented with some plausible assumptions) to derive an integral equation for $F(h)$. This equation, in which the scattering phase shift enters in the kernel, can be solved analytically at large h. A highly non-trivial consistency condition is that the analytic forms of the high h expansions obtained from the non-perturbative and perturbative computations match; in particular that the non-perturbative method reproduces the universal perturbative 1- and 2-loop beta-function coefficients.

Form factor bootstrap approach

Although the results above have not yet been rigorously derived they are already rather strong indication that conventional wisdom is internally consistent. Further stronger evidence in the O(3) case is supplied by the recent work of Balog and Niedermaier[8,9]. They have pursued the old so-called bootstrap program[10,11] which observed that in principle correlation functions of a local operators could be constructed if one knew all the matrix elements by saturation by a complete set of intermediate states. Of course the complication of such an explicit complete reconstruction is comparable to the computation of a partition function in statistical mechanics. However, as we shall see shortly, in certain energy regimes a good approximation to the correlation functions is obtained by truncating the sum over intermediate states.

Consider matrix elements of a local operator $\mathcal{O}$ of "spin" s and dimension Δ:

$$\langle\Omega|\mathcal{O}|n; i_n, \theta_n, ...\rangle = (m\sum_j e^{\theta_j})^{\Delta+s}(m\sum_k e^{-\theta_k})^{\Delta-s} f^{(n)}_{i_n,...,i_1}(\mathcal{O}; \theta_n, ...). \tag{5}$$

The matrix elements are obtained by solving a set of equations - the so-called Smirnov axioms[11] which have the following structure;
I.a (Watson's theorem):

$$f^{(n)}_{i_n,...,i_1}(\mathcal{O}; \theta_n, \theta_{n-1}, ...) = S^{j_{n-1}j_n}_{i_n,i_{n-1}}(\theta_{n,n-1}) f^{(n)}_{j_n,...,i_1}(\mathcal{O}; \theta_{n-1}, \theta_n, ...), \tag{6}$$

I.b (cyclicity) (for a recent derivation see ref.[21]):

$$f^{(n)}_{i_n,...,i_1}(\mathcal{O}; \theta_n + 2\pi i, \theta_{n-1}, ...) = f^{(n)}_{i_{n-1},...,i_1,i_n}(\mathcal{O}; \theta_{n-1}, ..., \theta_1, \theta_n), \tag{7}$$

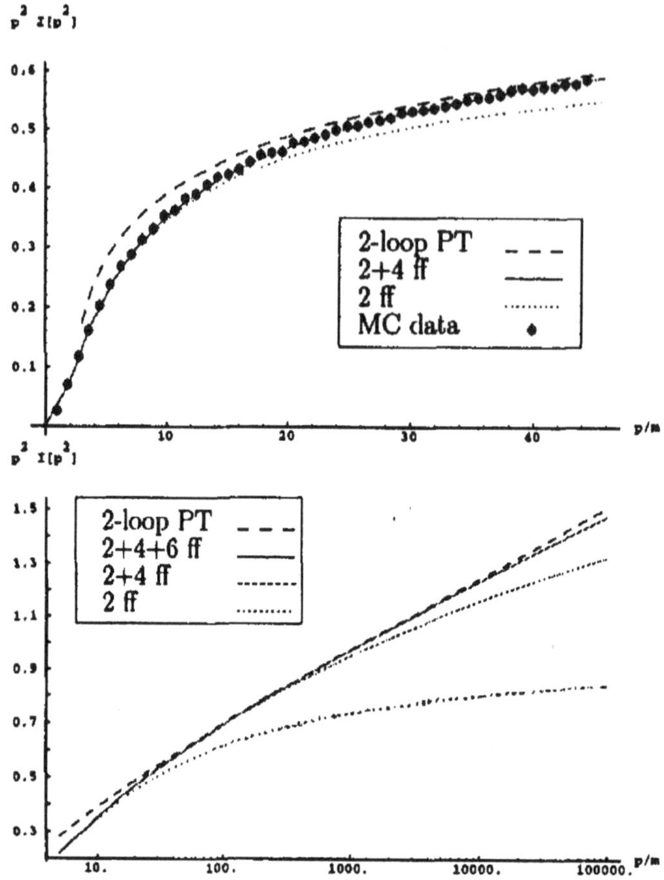

Figure 1. O(3) 2d σ-model current-current correlator (see Eq. (9)).

II. Kinematic residue:

$$\mathrm{Res}_{\theta_n=\theta_{n-1}+i\pi}[f^{(n)}_{i_n,\ldots,i_1}(\mathcal{O};\theta_n,..)] = \mathcal{T}[S]^{j_{n-2},\ldots,j_1}_{i_n,\ldots,i_1}\, f^{(n-2)}_{j_{n-2},\ldots,j_1}(\mathcal{O};\theta_{n-2},..), \tag{8}$$

where $\mathcal{T}$ is determined in terms of the S-matrix[8]. We do not give the explicit form here - the intention of writing down the equations is just to give an impression of the structure; there are two equations inter-relating a fixed particle number and one equation relating an n-particle form factor to a $n-2$ particle form factor. As one can imagine it is highly non-trivial to find a solution of this set of equations, but the problem becomes more tractable in the O(3) case where the S-matrix is a rational function of the rapidity difference θ. Even so the task is formidable but Balog, Niedermaier and Hauer[7] have succeeded to find solutions up to 6-particle matrix elements of the isovector $I = 1$ operators: the spin-field σ^a and the Noether current $j^a_\mu = g_0^{-2}\epsilon_{abc}\sigma^b\partial_\mu\sigma^c$ and the isoscalar $I = 0$ operators : the energy-momentum tensor $\theta_{\mu\nu} = g_0^{-2}[\partial_\mu\sigma^a\partial_\nu\sigma^a - \frac{1}{2}\eta_{\mu\nu}\partial_\rho\sigma^a\partial_\rho\sigma^a]$ and the topological density $k = \frac{1}{8\pi}\epsilon_{abc}\epsilon_{\mu\nu}\sigma^a\partial_\mu\sigma^b\partial_\nu\sigma^c$. Introducing the current-current 2-point function through

$$\int \mathrm{d}^2x e^{ipx}\langle j^a_\mu(x) j^b_\nu(0)\rangle = \delta^{ab} I(p^2)(p_\mu p_\nu - \delta_{\mu\nu}p^2), \tag{9}$$

a renormalized running coupling can be defined through $\bar{g}(p^2)^{-2} \propto p^2 I(p^2)$. Fig. 1 shows the approximation of including only the intermediate states containing up to 6 particles. There is impressive quantitative agreement with renormalized perturbation theory (using the Hasenfratz-Maggiore-Niedermayer result[20] (4) to convert the 2-loop perturbative result to a function of p/m) up to $p/m \sim 5000$, and up to $p/m \sim 40$ with results obtained from lattice simulations of the standard action[22]. A similar result holds for the spin-spin correlation function.

The Balog-Niedermaier scaling hypothesis

Consider the Källèn-Lehmann spectral decomposition[23]

$$I_O(p^2) = \int_0^\infty d\mu \frac{\rho_O(\mu)}{\mu^2 + p^2} \tag{10}$$

of the invariant amplitude defined by the two-point function of a local operator $\mathcal{O}$ (see e.g. Eq. (9)). From the form factor bootstrap one can show[8] that for $\mu \to \infty$ the n-particle contributions to the spectral function behave as

$$\rho_O^{(n)}(\mu) \sim \frac{A_O^{(n)}}{\mu(\ln \mu)^{4-2l}}, \tag{11}$$

whereas the behavior of the complete spectral function computed in renormalized perturbation theory is given by

$$\rho_1(\mu) \sim \frac{A_O}{\mu}\Big[1 + O\Big(\frac{1}{\ln \mu}\Big) + \ldots\Big], \tag{12}$$

$$\rho_0(\mu) \sim \frac{A_O}{\mu}\Big[\frac{1}{(\ln \mu)^2} + O\Big(\frac{\ln \ln \mu}{(\ln \mu)^3}\Big) + \ldots\Big], \tag{13}$$

for the isospin $l = 1, 0$ operators respectively. Note that each intermediate state with a finite number of particles gives a contribution to the current-current correlation function which is finite as $p \to \infty$. Hence for the asymptotic behavior predicted by perturbation theory to come about the sum of these contributions must diverge.

To get a grasp on the spectral functions involving an arbitrary number of particles Balog and Niedermaier observed that the spectral functions that they had computed so far possessed a remarkable scaling property. Defining new functions $Y_l^{(n)}$ through

$$\rho_l^{(n)}(\mu)/\mu = M_l^{(n)} Y_l^{(n)}\Big(\frac{\ln(\mu/m)}{\xi_l^{(n)}}\Big) \tag{14}$$

where $M_l^{(n)}, \xi_l^{(n)}$ are the values and positions of the maxima of $\rho_l^{(n)}(\mu)/\mu$, they observed that these rescaled functions are strikingly similar even for small n for fixed $l = 0, 1$ as is illustrated in Fig. 2. One can in these cases compare even and odd n (e.g. energy-momentum tensor and topological density spectral functions) since they are related by amazing clustering properties[8].

This similarity encouraged Balog and Niedermaier[9] to make the following scaling hypothesis for $n \to \infty$:

$$Y_l^{(n)} \to Y_l, \quad \xi_l^{(n)} \sim \xi_l n^{1+\alpha_l}, \quad M_l^{(n)} \sim M_l n^{\gamma_l}. \tag{15}$$

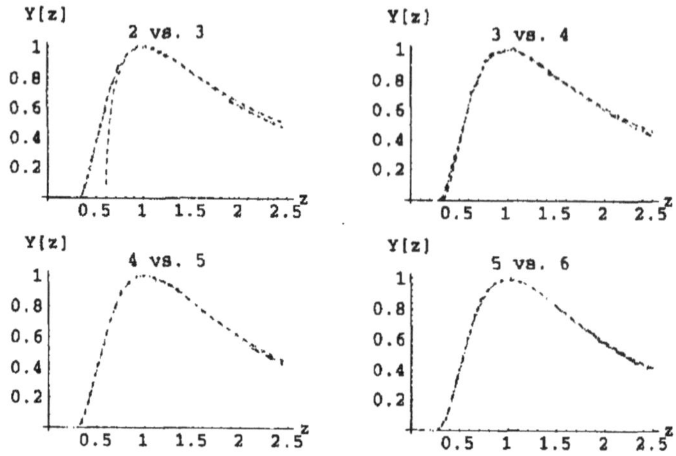

Figure 2. Comparison of consecutive O(3) Isospin 1 n-particle spectral functions.

Many of the exponents above are connected by consistency conditions and only one must be fitted numerically

$$\alpha_0 = \alpha_1 = \alpha = 0.273, \quad \gamma_0 = 3 + 2\alpha, \quad \gamma_1 = 1. \tag{16}$$

The consequences of this scaling hypothesis are far-reaching. Firstly if one couples the model to external fermions via the Noether current the n-particle production probability in the fermion scattering is directly related to the spectral function

$$P_{2n} \propto \mu\rho_1^{(2n)}(\mu), \quad \bar{n} \sim (\ln\mu)^{1/(1+\alpha)}. \tag{17}$$

Thus one finds the form of Koba-Nielsen-Olesen[24] scaling of multi-particle inclusive cross-sections $P_n = \sigma_n(s)/\sigma_{\rm tot}(s)$:

$$\bar{n}P_n \sim \psi(n/\bar{n}) \quad , \quad s \to \infty. \tag{18}$$

Secondly, ultraviolet properties are qualitatively consistent with renormalized perturbation theory and quantities accessible in perturbation theory are typically reproduced to better that 1%. For example e.g. the central charge is given by

$$c = 1.603 + 0.194 + 0.072 + \ldots = 1.997, \tag{19}$$

(where the first explicit contributions summing to 1.869 come from the 2,4 and 6 particle states and the rest of the sum obtained using the scaling hypothesis) which is within the systematic errors equal to the expected value[25] 2. Furthermore one finds

$$A^{\rm EM}/A^{\rm EM}_{\rm PT} = 0.99\,. \tag{20}$$

Moreover some normalizations not accessible to PT are fixed[9, 26] e.g.

$$A^{\rm curr} = \frac{\pi}{4}A^{\rm spin}, \quad A^{\rm EM} = 4A^{\rm top}\,. \tag{21}$$

Finally the topological density correlation function $\langle k(x)k(0)\rangle$ whose normalization in the form-factor bootstrap approach is fixed by matching with renormalized

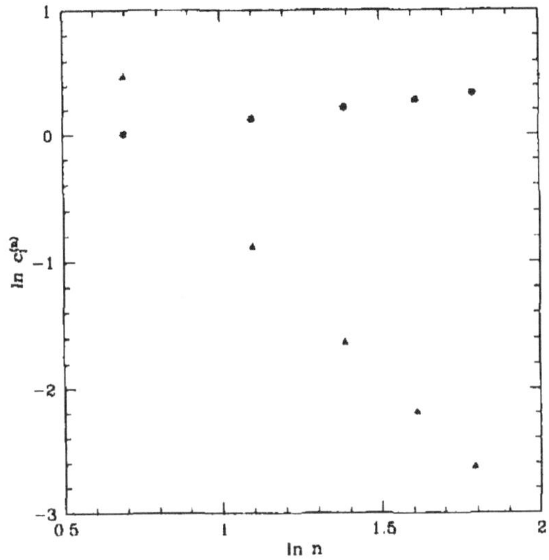

Figure 3. O(3) spectral function integrals $c_l^{(n)}$. The circles correspond to $l = 1$ and the triangles to $l = 0$.

perturbation theory at high energies, agree very well with that obtained by numerical simulations using a geometric definition of the topological charge density. The agreement is obtained only if one assumes that $k(x)$ is not multiplicatively renormalized - a property that would be anticipated for a topological object. Thus some remnant of topology survives although the topological susceptibility $\langle Q^2/V \rangle$ does not seem to have a continuum limit[27] ($\chi/m^2 \to \infty$ as $am \to 0$). The correlation function is everywhere negative for $x \neq 0$ and the susceptibility is rendered positive by the contact term at the origin. Whether a modified physically relevant definition of a topological susceptibility exists is an open question.

In a recent paper Patrascioiu and Seiler[28] have computed the current two-point function in the dodecahedron model (a model of 20 discrete spins arranged on the sphere) and found that up to $p/m = 80$ the amplitude cannot be distinguished from that of O(3) model (after including an overall normalization factor) and seems to have the same continuum limit (incidentally coinciding with the Balog-Niedermaier curve). Since they argue that the dodecahedron model is probably not asymptotically free they conclude the same should be true for the O(3) model. However another plausible interpretation is that their data just shows that for intermediate energies the dodecahedron model is a good "effective asymptotically free model". It is clear that one can not make any definitive statement on which scenario is correct using merely numerical experiments.

A very good indication of the validity of conventional wisdom is however obtained if one looks at Fig. 3 which is a plot of the integrals of the spectral densities

$$c_1^{(n)} = \int d\mu \rho_1^{(n)}(\mu), \quad c_0^{(n)} = 12\pi \int d\mu \rho_0^{(n)}(\mu). \tag{22}$$

In the scenario of Patrascioiu and Seiler, running couplings approach constants at infinity and hence they would require $c_1^{(n)} \to 0$. There is certainly no sign of this behavior and if it is true then it sets in at high n and hence at astronomic energy scales not accessible to numerical test. Conventional wisdom is thus "effectively true" in this model and any violation would be truly academic.

Measurement of a running coupling using lattice simulations

As in the sigma model a running coupling in QCD can also be defined through a two-point correlation function of currents e.g.

$$\frac{\sigma(e^+e^- \to \text{hadrons})}{\sigma(e^+e^- \to \mu^+\mu^-)} \equiv 3\sum_f Q_f^2\left(1 + \frac{1}{\pi}\alpha_R(q)\right). \tag{23}$$

However this and other couplings defined in infinite volume are extremely difficult to measure over a wide range of energies while controlling systematic errors at the same time. We will come back to this point in the next section. The proposal to overcome this is to use a recursive finite-size scaling method[4]. Here a running coupling is defined in finite volume $\bar{g}^2(L)$ and its evolution from small to large scales is done recursively. Contact with physical low energy scales is made at large L and at small L we can convert to other schemes e.g. $\overline{\text{MS}}$ using perturbation theory.

Here we outline how the general method proceeds. Consider a theory defined in a volume of finite extent L in some (not necessarily all) directions. Suppose we have defined an "admissible coupling" i.e. a (renormalized) observable $\bar{g}$ depending on L and equal to the bare coupling at tree level perturbation theory. Certainly in the continuum limit there will be a well defined curve $\bar{g}^2(2L)$ versus $\bar{g}^2(L)$ - the so called step scaling function. When we regularize the theory on the lattice the regulated form of the coupling will depend on the number M of points (in the finite directions) and on the bare coupling g_0. A given bare coupling corresponds to a given lattice spacing in units of some inverse physical mass. Thus if we simulate two lattices one with M points and another with $2M$ points at the same bare coupling we are investigating the system with some physical extent and the system with that extent doubled. If we measure $\bar{g}^2(M, g_0)$ and $\bar{g}^2(2M, g_0)$ for a rather random selection of input parameters M, g_0 we will, due to lattice artifacts, obtain a cloud of points (with their associated error bars) from which it would be difficult to precisely extract the continuum step scaling function.

To take the continuum limit one must proceed more systematically. The first step is to consider a sequence of lattices with points $M_1 < M_2 < M_3 ... < M_n$ and for each lattice size M_i determine (through numerical simulation and reweighting) the value of the bare coupling $g_0 = g_{i1}$ so that $\bar{g}$ stays fixed at some given value u_1. Now measure $\bar{g}$ at a sequence of lattices with double the points $2M_i$ but at the same values g_{i1} of the bare couplings. The results of these measurements v_{i1} will not be the same but they should approach a value v_1 corresponding to a point (u_1, v_1) on the continuum curve as $M \to \infty$. If we know the way the continuum limit is approached we can extrapolate v_{i1} without going to very large M_i. Symanzik proposed the limit is approached at a rate proportional to a^p i.e. as the physical length is being held fixed at a rate $\propto 1/M^p$, where p is an integer ≥ 1. It is to be stressed that this is a working hypothesis and it has to be tested. Symanzik's results are within the framework of all orders of perturbation theory but the structural statement is assumed to hold in the non-perturbative framework [the same holds true for renormalizability itself]. Once the continuum limit v_1 is determined we can repeat the procedure with a new fixed value u_2 say $u_2 = v_1$ and so on. In this way we determine a sequence of points on the step scaling function curve.

To carry out such a program and to get confidence that we are controlling the continuum limit it is clear that the coupling must have certain good properties. In

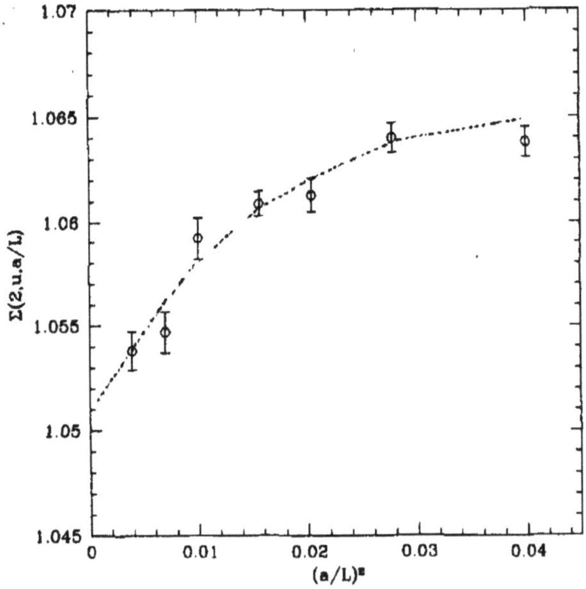

Figure 4. Approach to continuum limit of step scaling function $v = \Sigma(2, u, a/L)$ at $u = 0.9176$. The curve represents a second order fit polynomial in $(a/L)^2$.

particular it must be accurately calculable in numerical simulations. Consider the O(3) sigma-model on an infinite cylinder of circumference L, then a suitable coupling is defined through the (finite volume) mass gap

$$\bar{g}^2(L) = m(L)L. \tag{24}$$

Fig. 4 shows the data of one run of measurements[4], using the standard lattice action, of $\bar{g}^2(2L)$ with $\bar{g}^2(L)$ fixed at the value $u = 0.9176$. The data can certainly be fitted with a form as suggested by Symanzik (here the expected value of p is 2). One observes that in this case the continuum limit is approached from above. This is contrary to what happens at smaller renormalized couplings where perturbation theory predicts the limit to be approached from below. We are thus seeing some non-perturbative behavior at least on the level of cut-off effects. The same qualitative behavior at large coupling is also seen in the leading order of the $1/N$ expansion as illustrated in Fig. 5. Here all computations can be done analytically and one sees that the continuum limit is indeed approached at a rate $O(a^2)$ modulated by possible logarithmic factors.

By performing various steps of factor two we end up with the step scaling function over a wide range of energies. It remains to make contact with the low energy scale. For the particular quantity we are studying here this is rather easy. At a some large attained value of $\bar{g}^2(L)$ we simulate $\bar{g}^2(sL)$ for large s. The physical volume is now so big that we can expect finite volume corrections to be very small[29]

$$m(L) = m\left(1 + O(e^{-cmL})\right). \tag{25}$$

We finally obtain the running coupling over a wide range of energies as a function of mL. I will not show this plot here, but instead include Fig. 6 which shows approximations to m/Λ by taking 2,3 and 4 loop approximations to the Λ-parameter as a

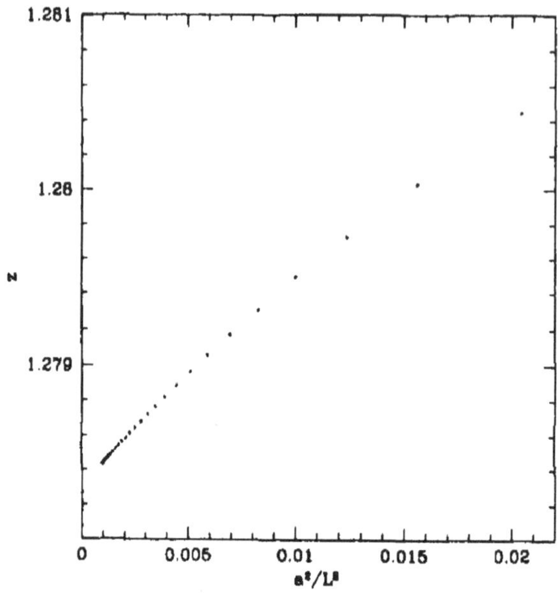

Figure 5. Approach to continuum limit of step scaling function $v = z' = \Sigma(2, u, a/L)$ at $u = 1$ in the limit $N = \infty$.

function of $\bar{g}(L)$. This is an update of the plot in the original paper[4] extended by the 4-loop computation of Shin[30] and new MC data at smaller $\bar{g}^2(L)$ of Hasenbusch[31]. Note that there is an indication of the 3,4-loop curves approaching the limit from below. The result is thus so far well consistent with the theoretical prediction[20, 32].

3. RUNNING COUPLINGS IN THE PURE YANG-MILLS THEORY

As mentioned previously one of the goals of the Alpha Collaboration is to obtain $\alpha_{\overline{\rm MS}}(m_Z)$ with a controlled error of say 1% i.e. without compromising approximations. To reach this goal we need at least four ingredients

- 1. An accurate determination of a reference low energy scale
- 2. A non-perturbatively defined running coupling $\alpha(q)$ accurately computable through numerical simulation over a wide range of q
- 3. A careful extrapolation to the continuum limit
- 4. The perturbative relation of our coupling to that of the $\overline{\rm MS}$-scheme to two loop order

$$\alpha_{\overline{\rm MS}}(sq) = \alpha(q) + c_1(s)\alpha(q)^2 + c_2(s)\alpha(q)^3 + \ldots \qquad (26)$$

where s is a so-called boost factor at our disposal.

Points 1-3 were encountered in the study of the sigma-model in the last section. Point 4 enters because once we have indeed checked that the chosen coupling runs at high energy according to the perturbative renormalization group, we can at the highest energies measured relate our coupling via perturbation theory to the $\overline{\rm MS}$-scheme. Since

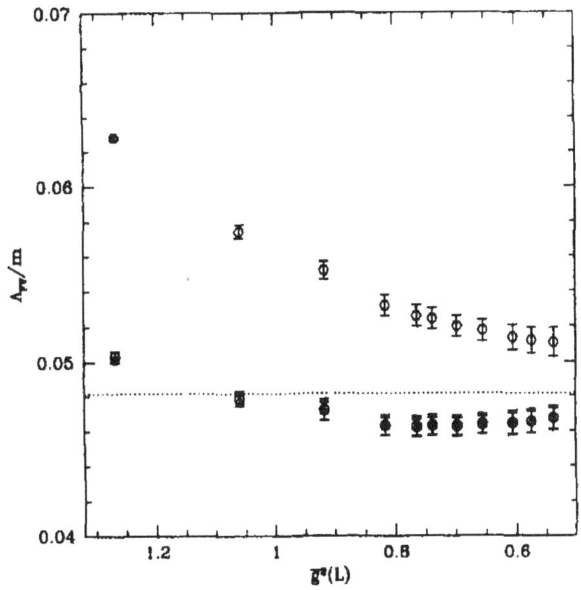

Figure 6. Approximations to m in units of the finite volume Λ-parameter, Λ_{FV}, in the O(3) model. The dotted line is the prediction of Hasenfratz, Maggiore and Niedermayer[20]. The upper points correspond to 2-loop and the lower points to 3,4-loop.

$\alpha_{\overline{MS}}(m_Z)$ is expected to be ~ 0.1, to achieve 1% accuracy we need the perturbative relation to two loops (at least).

As a first step we consider the pure SU(3) Yang-Mills theory. There are many definitions of running couplings (see e.g. ref.[33]). The basic difficulty is that for a given coupling its low- and high-energy properties must be related to each other. Consider as an example the coupling $\alpha_{q\bar{q}}$ defined in infinite volume through the force between static quarks at distance r

$$F_{q\bar{q}}(r) = \frac{4}{3}\frac{\alpha_{q\bar{q}}}{r^2}. \tag{27}$$

$\alpha_{q\bar{q}}$ is a "running" coupling at momentum $q = 1/r$ and its relation to the $\overline{\text{MS}}$-coupling is given by

$$\alpha_{\overline{MS}} = \alpha_{q\bar{q}} - 0.082 \times (\alpha_{q\bar{q}})^2 + \ldots \tag{28}$$

Now if we take for example a lattice with 32^4 points and require $L \geq 2\text{fm}$ to avoid large finite size effects the lattice spacing is $a \geq 0.06\text{fm}$. Further to avoid significant cutoff effects we should take $r \geq 3a$ (at the very least) and thus we are restricted to rather low energies $q \lesssim 2\text{GeV}$. The situation is illustrated in Fig. 7 of the force in pure SU(2) gauge theory. Although this plot is rather familiar to workers in lattice theory, I include it again here for the non-experts because it has some beautiful features. Firstly one sees that the force is consistent with tending to a constant K (the string tension) at large r - which is a manifestation of Wilson's criterion of static confinement[13]. Secondly there is evidence of the presence of a $1/r^2$ correction with a coefficient predicted by Lüscher[34] assuming the validity of an effective string picture. Perhaps surprising is that this long distance (i.e. non-perturbative) formula provides also a good quantitative description down to quite short distances.

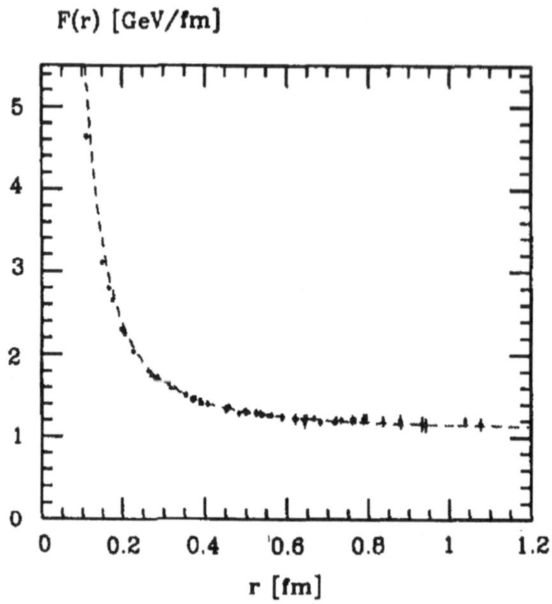

Figure 7. The force in pure SU(2) Yang-Mills theory, the dashed line represents Lüscher's force: $K + \frac{\pi}{12r^2}$.

Also we can see that Sommer's scale[35] r_0 defined as the distance where $r_0^2 F_{q\bar{q}}(r_0) = 1.65$ is rather precisely computable in present simulations and is hence a good low energy reference scale (see point 1 above) in pure gauge theories (and probably also in QCD). In nature (i.e. potential models applied to quarkonium) one finds $r_0 \sim 0.5\text{fm}$. (In pure SU(3) gauge theory $r_0\sqrt{K} = 1.18(2)$).

Now one could make (and of course it has been done) the daring assumption that for this coupling perturbative running sets in already at a low scale. But it is obviously better if we can avoid making such ad hoc assumptions.

As we saw illustrated in the sigma model, a way to overcome the difficulty is to define a coupling $\bar{g}(L)$ depending on the volume and use a finite step scaling technique. The finite volume is thus used as probe of the theory instead of being considered a source of systematic error. As stated previously it is important to make a good choice of the non-perturbatively defined $\bar{g}$ so that

- it is accurately computable by numerical simulation
- it has small cutoff effects
- it has an easy perturbative expansion

In our experience these properties are not easy to fulfill. In the next sub-section we will define a coupling through the Schrödinger functional but it is important to realize that there are many other proposals including in particular one which has been systematically studied[36, 37] which is defined by correlations of Polyakov loops in a volume with twisted boundary conditions.

The Schrödinger functional

Consider the SU(N) Yang-Mills theory on a "cylinder" $[0,T] \times S^1 \times S^1 \times S^1$. We impose periodic boundary conditions in the spatial directions and Dirichlet boundary conditions at the caps $t = 0, T$:

$$A_k(x)|_{x_0=0} = C_k(\mathbf{x}), \quad A_k(x)|_{x_0=T} = C'_k(\mathbf{x})\,. \tag{29}$$

The Schrödinger functional[3, 38, 39] is the partition function with these boundary conditions and formally given by the integral

$$\mathcal{Z}[C',C] = \int D[A]\, \mathrm{e}^{-S[A]}. \tag{30}$$

In the continuum one also has to integrate over gauge transformations of the boundary conditions at one end. Going to a Hamiltonian picture one obtains the quantum mechanical interpretation of the Schrödinger functional as a propagation kernel

$$\mathcal{Z}[C',C] = \langle C'|\mathrm{e}^{-\mathrm{H}T}\mathrm{P}|C\rangle \tag{31}$$

where H is the Hamilton operator and P the projector on gauge invariant states.

In the lattice formulation the gauge fields are replaced by a field of matrices $U(x,\mu) \in \mathrm{SU}(N)$, associated with the lattice links. On the boundary the matrices are related to the continuum fields C_k through

$$U(x,k)|_{x_0=0} = \mathcal{P}\exp\left\{a\int_0^1 \mathrm{d}t\, C_k(\mathbf{x} + (1-t)a\hat{\mathbf{k}})\right\}. \tag{32}$$

For lattice action we take

$$S[U] = \frac{1}{g_0^2}\sum_{\text{plaquettes } p} w(p)\,\mathrm{tr}\{1 - U(p)\} \tag{33}$$

i.e. the standard Wilson one plaquette action. The weights $w(p)$ which differ from 1 only for plaquettes touching the boundary are introduced for the possibility of diminishing some of the O(a) cutoff effects. The lattice regularized Schrödinger functional

$$\mathcal{Z}[C',C] = \int \prod_{x,\mu} \mathrm{d}U(x,\mu)\, \mathrm{e}^{-S[U]} \tag{34}$$

is a well defined gauge invariant functional of C, C'.

For the Schrödinger functional perturbation theory is the expansion about a "background field" i.e the field $B_\mu(x)$ with minimal action satisfying the prescribed boundary conditions. Consider for example constant Abelian boundary fields

$$C_k = \frac{i}{L} \times \mathrm{diag}\{\phi_1, \phi_2, \ldots, \phi_N\}, \qquad k = 1,2,3, \tag{35}$$

(here L is the circumference of the circles S^1) and similarly C', with angles ϕ_α satisfying

$$\phi_1 < \phi_2 < \ldots < \phi_N, \quad \phi_N - \phi_1 < 2\pi, \quad \sum_{\alpha=1}^{N} \phi_\alpha = 0\,. \tag{36}$$

In this case it is easy to sec that the induced background field is a constant color-electric field

$$B_0 = 0, \qquad B_k = \frac{x_0}{T}(C'_k - C_k) + C_k\,. \tag{37}$$

The lattice background field is simply $V_\mu(x) = e^{B_\mu(x)}$. For the specific boundary conditions Eqs. (35,36) one can show[39] that V is the unique global minimum of the action. This is an important result since it guarantees the stability of the perturbative expansion in this case.

The renormalization of the Schrödinger functional was first discussed by Symanzik[3]. He found that, in general, space-time boundaries with Dirichlet boundary conditions induce additional UV divergences, but that these can be cancelled by adding boundary counterterms to the action:

$$\int d^3\mathbf{x}\,\{\mathcal{L}_b(x)|_{x_0=0} + \mathcal{L}'_b(x)|_{x_0=T}\}\ . \tag{38}$$

Here $\mathcal{L}_b(x)$ is a linear combination of local fields of dimension $d \leq 3$, invariant under both spatial and internal symmetries. Indeed in scalar field theories one has boundary counterterms proportional to $\phi^2, \phi\partial_0\phi$. But in pure gauge theory there are no such (gauge invariant) fields which leads us to the conclusion that the effective action $\Gamma[B] - \Gamma[0] = \ln(\mathcal{Z}[B]/\mathcal{Z}[0])$ is finite after renormalization of the coupling (done exactly as in infinite volume). This has been checked explicitly to two loops PT[40, 41].

Now to define a running coupling one notes that

$$\Gamma[B] = \frac{1}{g_0^2}\Gamma_0[B] + \Gamma_1[B] + \ldots \propto \frac{1}{\bar{g}^2}\,. \tag{39}$$

To render the definition finite one makes the field B dependent on a parameter η and defines the coupling as the response of the system to a change of the boundary conditions

$$\bar{g}^{-2}(L) = \left(\frac{\partial\Gamma}{\partial\eta}\Big/\frac{\partial\Gamma_0}{\partial\eta}\right)\Bigg|_{\eta=0}\,. \tag{40}$$

For the simulation we set $T = L$ and chose constant Abelian boundary fields as in Eqs. (35,36) with a specific choice of the angles (for the case of SU(3))

$$(\phi_1, \phi_2, \phi_3) = \frac{1}{3}(-\pi, 0, \pi) + \frac{1}{2}(2, -1, -1)\,\eta\,, \tag{41}$$

and a similar expression for $(\phi'_1, \phi'_2, \phi'_3)$. The field (41) was found (after some trial) to represent a good choice in that the resulting coupling has a good signal and relatively small cutoff effects, thus satisfying our selection criteria above. Note that our observable is actually a set of plaquette expectation values at the boundary, which are accurately measurable. For large volumes our coupling is expected to diverge exponentially i.e. the observable becomes very small and hence accurate measurement is more difficult if the fluctuations are relatively large.

Now we have found a suitable definition of the coupling the finite step scaling analysis proceeds as in the sigma model as described in the previous section. An important step is, as we have stressed before, the extrapolation to the continuum limit. Here, if we follow Symanzik's ideas we expect the dominant cutoff effects to be O(a) and accordingly make a linear extrapolation of the data as illustrated in Fig. 8.

Having carried out this procedure during our investigations[2] in 1993 we ended up with Table 1. Note that by combining the evolution steps we obtained the evolution of $\bar{g}^2(L)$ from 1.243 to 3.474, corresponding to a total scale factor of about 25. In the meantime, in the process of computing the running quark mass in quenched QCD[42], we have refined this analysis and gone to smaller values of $\bar{g}$ and reached a scale factor of around 64.

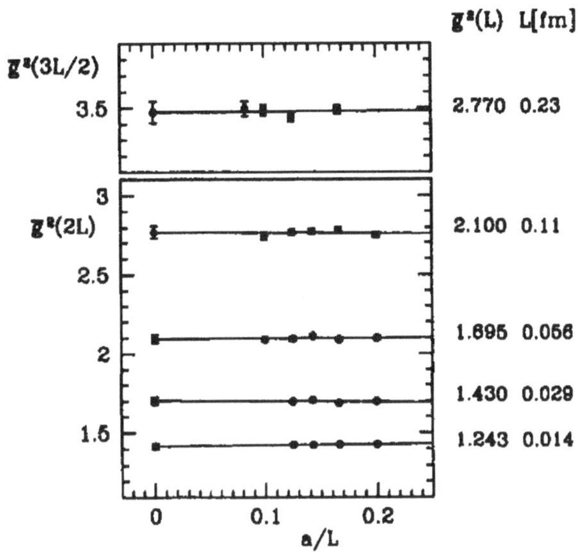

Figure 8. Taking the continuum limit of the step scaling function (SU(3)).

Table 1. Evolution of $\bar{g}^2(L)$ (SU(3)) in the continuum limit

s	$\bar{g}^2(L)$	$\bar{g}^2(sL)$
2	1.243	1.416(16)
2	1.430	1.703(24)
2	1.695	2.095(25)
2	2.100	2.771(41)
3/2	2.770	3.474(65)

Finally we must make contact with the low energy scales. Let us define

$$L_{\max} \equiv \text{value of } L \text{ where } \bar{g}^2(L) = 3.480\,. \tag{42}$$

The question is now: how large is $L_{\max}$ in physical units for example in terms of Sommer's scale r_0? To obtain this we note that both factors on the rhs of

$$L_{\max}/r_0 = (L_{\max}/a)(r_0/a)^{-1} \tag{43}$$

are functions of g_0 that can be determined separately. The procedure is then as follows. We choose a value of g_0 and compute r_0/a on a large lattice, avoiding finite-volume effects. Then at the same g_0 compute $\bar{g}^2(L)$ for a range of L/a and find $L_{\max}/a$ through interpolation. The results are not independent of g_0, but we expect

$$L_{\max}/r_0 = \text{constant} + \mathrm{O}(a/L_{\max}) \tag{44}$$

and this is born out by the data as illustrated in Fig. 9.

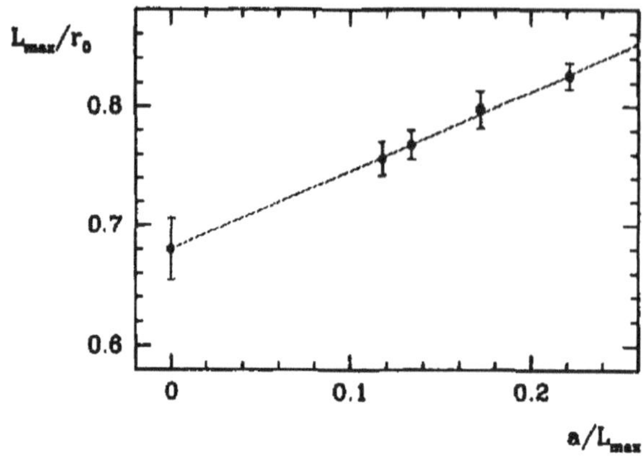

Figure 9. Extrapolation of L_{max}/r_0 to the continuum limit (SU(3)).

Fig. 10 shows the SF running coupling in pure SU(3) Yang-Mills theory[42] as a function of the energy calibrated by r_0. This is a truly impressive plot and clearly demonstrates that the data are well described by perturbative running at large energies. What is perhaps surprising is that the true non-perturbative running is well described by perturbative running right down to low energies. We emphasize that this is not a universal property of couplings; on the other hand it shows that some naturally defined couplings with this striking property exist.

The 3-loop beta-function is now known in this scheme[41] and using this we can extract a quite accurate value of the **Λ-parameter** in this scheme. Converting to the **$\overline{\text{MS}}$-scheme** we obtain in pure SU(3) Yang-Mills theory

$$\Lambda^{(0)}_{\overline{\text{MS}}} = 251(20)\text{MeV}. \tag{45}$$

Previous estimates using the potential[43, 44, 45] indeed produced similar values but the present analysis certainly has a much better control of the systematic errors.

Some estimates of the running coupling in full QCD using the assumption of perturbative running already at small scales have been made[46, 47, 48, 49, 50, 51, 52, 53, 54, 55, 56, 57]. The results are indeed consistent with experiment but it is surely very difficult to make a reliable estimate of the systematic errors. It will still take a few years to do a similarly good job on the running coupling in full QCD as we have now for the quenched theory. One requires accurate measurements which in turn require improved algorithms for fermions and also powerful computers. However the definition of the coupling introduced above can be readily extended to the theory with fermions since, as we shall see in the next section the Schrödinger functional is defined in the full theory with quarks[58]. In fact the Schrödinger functional has a wider range of applications and turns out to be a very useful tool for purposes of studying chiral symmetry and determining renormalization constants of composite operators.

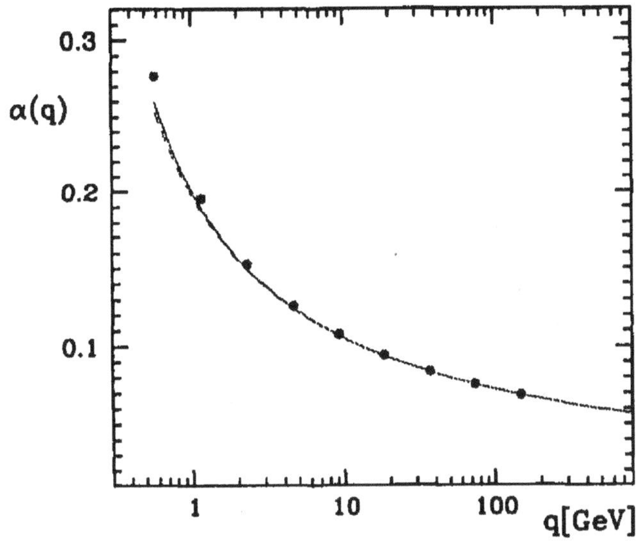

Figure 10. The SF running coupling in pure SU(3) Yang-Mills theory. The curves indicate 2-,3-loop running staring the integration at the highest energy.

4. IMPROVED FERMION ACTION AND NON-PERTURBATIVE RENORMALIZATION

If one constructs a lattice regulated free fermion action naively by just replacing the derivative in the Dirac operator by a simple difference operator one finds that the energy-momentum dispersion has extra minima and the continuum limit of the lattice theory contains 16 particles instead of 1. To lift this degeneracy (at the same time maintaining the usual classical continuum limit) Wilson[13] proposed to use an action which has an additional term of dimension 5

$$S = S_{\rm gluon} + S_{\rm quark}\,, \tag{46}$$

$$S_{\rm quark} = a^4 \sum_x \overline{\psi}(x)(D + m_0)\psi(x)\,. \tag{47}$$

Here m_0 is the bare quark mass and the Wilson-Dirac operator is given by

$$D = \frac{1}{2}\{\gamma_\mu(\nabla^*_\mu + \nabla_\mu) - a\nabla^*_\mu\nabla_\mu\}\,, \tag{48}$$

where ∇_μ, ∇^*_μ are gauge covariant forward and backward difference operators.

The Wilson action has many good properties e.g. a positive definite transfer matrix[59]. But it has one disadvantage - the Wilson term $\nabla^*_\mu\nabla_\mu$ violates chiral symmetry. This has several implications. Firstly chiral Ward identities are violated by terms of O(a) and thus low-energy theorems are generally affected. Secondly the quark mass requires additive renormalization (in addition to the customary multiplicative renormalization). Another consequence is that the axial current must be renormalized by a finite multiplicative renormalization constant and mixing of composite operators with different chiralities is not excluded. All these difficulties do not represent a fundamental problem but they do imply that, as indicated above, renormalization is more complicated and the continuum limit is reached only very slowly. Indeed, because of the presence of the Wilson term, cutoff effects are expected proportional to the lattice spacing a and this can be rather large for typical values of the Monte Carlo simulation parameters. Decreasing the lattice spacing at constant physical length scales means simulating lattices with a larger number of points and thus a significant increase in the cost (in time) of the simulation.

Diminishing cutoff effects

To make progress in controlling the cutoff effects other than the brute force method of simulating very large lattices, we need some theoretical understanding of cutoff effects in a general lattice theory. There are presently unfortunately very few rigorous non-perturbative statements in this respect. However there is a very plausible framework due to Symanzik[12], which we have already mentioned in previous sections, which he obtained through a systematic study of the continuum limit of lattice Feynman diagrams. Symanzik proposed that the structural properties that he found in perturbation theory should also hold at the non-perturbative level. To be precise he proposed that at energies well below the cutoff, the lattice theory is equivalent to an effective continuum theory with effective action

$$S_{\text{eff}} = S_0 + aS_1 + a^2 S_2 + \ldots \tag{49}$$

where S_i are integrals of local operators of dimension $4 + i$ having the symmetries of the lattice theory. In Wilson's formulation of QCD we expect the leading correction[60]

$$S_1 = \int \mathrm{d}^4x\, \mathcal{L}_1(x) \tag{50}$$

where $\mathcal{L}_1$ is a linear combination of local gauge invariant scalar fields of dimension 5. A linearly independent set of such operators is given by

$$\mathcal{O}_1 = \bar{\psi}\sigma_{\mu\nu}F_{\mu\nu}\psi, \tag{51}$$
$$\mathcal{O}_2 = \bar{\psi}D_\mu D_\mu\psi + \bar{\psi}\overleftarrow{D}_\mu\overleftarrow{D}_\mu\psi, \tag{52}$$
$$\mathcal{O}_3 = m\,\mathrm{tr}\left\{F_{\mu\nu}F_{\mu\nu}\right\}, \tag{53}$$
$$\mathcal{O}_4 = m\left\{\bar{\psi}\gamma_\mu D_\mu\psi - \bar{\psi}\overleftarrow{D}_\mu\gamma_\mu\psi\right\}, \tag{54}$$
$$\mathcal{O}_5 = m^2\bar{\psi}\psi. \tag{55}$$

In the effective theory renormalized local lattice fields $\phi_{\mathrm{R}} = Z_\phi\phi$ are represented by

$$\phi_{\text{eff}} = \phi_0 + a\phi_1 + a^2\phi_2 + \ldots \tag{56}$$

(where ϕ_i are local operators in the continuum theory) in the sense that renormalized lattice n-point correlation functions

$$G_n(x_1, \ldots, x_n) = (Z_\phi)^n \left\langle \phi(x_1)\ldots\phi(x_n)\right\rangle_{\text{con}} \tag{57}$$

at fixed separated points x_k are given by

$$G_n(x_1,\dots,x_n) = \langle\phi_0(x_1)\dots\phi_0(x_n)\rangle_{\rm con}$$
$$-a\int d^4y\,\langle\phi_0(x_1)\dots\phi_0(x_n)\mathcal{L}_1(y)\rangle_{\rm con}$$
$$+a\sum_{k=1}^{n}\langle\phi_0(x_1)\dots\phi_1(x_k)\dots\phi_0(x_n)\rangle_{\rm con} + \mathrm{O}(a^2)\,, \tag{58}$$

where the expectation values on the right hand side are taken in the continuum theory.

The effective action above contains a large number of operators even if we are only interested in describing the leading cutoff effects. It is important to remark that the situation is greatly simplified by restricting attention to on-shell quantities[60] since for this purpose some of the operators in the effective action become "redundant". This comes about by considering the field equations which imply for example

$$\mathcal{O}_4 + c\,\mathcal{O}_5 = 0, \qquad c = 2 + \mathrm{O}(g^2)\,, \tag{59}$$

which is true when inserted in a correlation function, up to contact terms which can be absorbed by a redefinition of ϕ_1 in Eq. (56). Thus $\mathcal{O}_4$ may be eliminated in favor of $\mathcal{O}_5$ in correlation functions with fields at non-zero physical distances from each other from which all on-shell data can be extracted. Similarly $\mathcal{O}_2$ can also be eliminated and we conclude that $\mathcal{L}_1$ can be expressed as a linear combination of $\mathcal{O}_1, \mathcal{O}_3, \mathcal{O}_5$ as far as on-shell quantities are concerned.

The implications of Symanzik's analysis is that it must be possible to construct an O(a) on-shell improved lattice action so that the S_1-term in its associated effective action vanishes by adding O(a) counterterms[12] δS of the form

$$\delta S = a^5\sum_x\left\{c_1\hat{\mathcal{O}}_1(x) + c_3\hat{\mathcal{O}}_3(x) + c_5\hat{\mathcal{O}}_5(x)\right\} \tag{60}$$

where $\hat{\mathcal{O}}_i$ are lattice counterparts of $\mathcal{O}_i$. The terms proportional to $\hat{\mathcal{O}}_3$ and $\hat{\mathcal{O}}_5$ in (60) may be dropped, because they simply correspond to a renormalization of the bare coupling and mass. An O(a) improved action in QCD, first proposed by Sheikholeslami and Wohlert[61], is thus given by

$$S_{\rm impr} = S_{\rm Wilson} + a^5\sum_x c_{\rm sw}\,\overline{\psi}(x)\frac{i}{4}\sigma_{\mu\nu}\hat{F}_{\mu\nu}(x)\psi(x) \tag{61}$$

where $\hat{F}_{\mu\nu}(x)$ is a lattice version of the field strength tensor (and the conventions for the Dirac matrices are as in ref.[60]). The coefficient $c_{\rm sw}(g_0)$ appearing here must be tuned to achieve cancellation of O(a) effects in on shell quantities. At small g_0 it can be computed perturbatively and to one-loop order PT

$$c_{\rm sw} = 1 + 0.2659\times g_0^2 + \mathrm{O}(g_0^4) \qquad (N=3,\ \text{any } N_{\rm f}). \tag{62}$$

This was first computed by Wohlert[62] by improving the residue of a pole in the S-matrix (in a volume finite in two directions with twisted boundary conditions), and later checked by improving certain correlation functions constructed using the Schrödinger functional[63] (see later).

The action can in principle be improved further to order a^2, but this involves introduction of many new operators of dimension 6. For the pure Yang-Mills theory the structure of the on-shell O(a^2) improved action is relatively simple[64], but for the theory with quarks the required set of operators contains also those involving products of 4 quark fields. The fact that the action is then no longer quadratic in the fermion fields causes extra complications for the numerical simulations and becomes impractical. Indeed, in general, improved actions are more difficult to simulate and one must ensure that the gain in reduction of cutoff effects is not swamped by the loss in simulation time. For the SW improved action above this problem is not severe because its structure is comparatively simple; with respect to the Wilson action it has just one extra term which involves the quark fields at the same space-time point.

Another effect of improving the action is that positivity of the transfer matrix is lost in general. Although positivity is a very nice property which one would have preferred to keep if possible it is not a necessity for the regularized theory. Indeed in the cases one can study explicitly one finds that the complex eigenvalues of the transfer matrix occur at energies of the order of the ultra-violet cutoff and hence not dangerous.

Improved operators

If one is only considering spectral quantities it is sufficient to improve the action. But if one is interested in matrix elements of renormalized operators one must also improve the corresponding lattice operators.

Consider for example in the theory with two degenerate flavors the following set of bare isovector operators bilinear in the quark fields:

$$V_\mu^a(x) = \overline{\psi}(x)\gamma_\mu \frac{1}{2}\tau^a \psi(x), \tag{63}$$

$$A_\mu^a(x) = \overline{\psi}(x)\gamma_\mu \gamma_5 \frac{1}{2}\tau^a \psi(x), \tag{64}$$

$$T_{\mu\nu}^a(x) = i\overline{\psi}(x)\sigma_{\mu\nu} \frac{1}{2}\tau^a \psi(x), \tag{65}$$

$$S^a(x) = \overline{\psi}(x) \frac{1}{2}\tau^a \psi(x), \tag{66}$$

$$P^a(x) = \overline{\psi}(x)\gamma_5 \frac{1}{2}\tau^a \psi(x), \tag{67}$$

where τ^a are the Pauli matrices. By inspection of symmetry properties the corresponding O(a)-improved (unrenormalized) fields are

$$(V_I)_\mu^a = V_\mu^a + c_V a \tilde{\partial}_\nu T_{\mu\nu}^a, \tag{68}$$

$$(A_I)_\mu^a = A_\mu^a + c_A a \tilde{\partial}_\mu P^a, \tag{69}$$

$$(T_I)_{\mu\nu}^a = T_{\mu\nu}^a + c_T a[\tilde{\partial}_\mu V_\nu^a - \tilde{\partial}_\nu V_\mu^a], \tag{70}$$

$$(P_I)^a = P^a, \tag{71}$$

$$(S_I)^a = S^a, \tag{72}$$

where $\tilde{\partial}_\mu$ denotes the symmetrized lattice derivative $\frac{1}{2}(\partial_\mu + \partial_\mu^*)$. The coefficients $c_A(g_0), c_V(g_0)$ and $c_T(g_0)$ must be chosen so as to cancel any remaining terms of O(a) in correlation functions involving the corresponding operators. To one-loop order PT[63, 65]

$$c_A = -0.005680(2) \times C_F g_0^2 + O(g_0^4), \tag{73}$$

$$c_V = -0.01225(1) \times C_F g_0^2 + O(g_0^4), \tag{74}$$

$$c_T = 0.00896(1) \times C_F g_0^2 + O(g_0^4), \tag{75}$$

(for any N_f) where $C_F = (N^2 - 1)/2N$.

Mass-independent renormalization schemes

Before going on to consider methods to determine improvement coefficients let us first discuss the renormalization scheme. For practical purposes it is convenient to use mass-independent renormalization schemes[66]. The only complication is that for compatibility with O(a) improvement mass-independent renormalization schemes require slight modifications from what is usually done and must have the form:

$$g_{\mathrm{R}}^2 = \tilde{g}_0^2 Z_{\mathrm{g}}(\tilde{g}_0^2, a\mu)\,, \tag{76}$$

$$m_{\mathrm{R}} = \widetilde{m}_{\mathrm{q}} Z_{\mathrm{m}}(\tilde{g}_0^2, a\mu)\,, \tag{77}$$

where the modified bare coupling and mass are defined by

$$\tilde{g}_0^2 = g_0^2\,(1 + b_{\mathrm{g}} a m_{\mathrm{q}})\,, \quad m_{\mathrm{q}} = m_0 - m_{\mathrm{c}}\,, \tag{78}$$

$$\widetilde{m}_{\mathrm{q}} = m_{\mathrm{q}}\,(1 + b_{\mathrm{m}} a m_{\mathrm{q}})\,. \tag{79}$$

To see how this more complicated structure comes about consider for example the "pole mass" of a free quark. It is given by

$$m_{\mathrm{p}} = \frac{1}{a}\ln(1 + a m_0) = m_{\mathrm{q}} - \frac{1}{2} a m_{\mathrm{q}}^2 + \ldots \tag{80}$$

and hence to tree level one immediately sees $b_{\mathrm{m}} = -1/2$.

To appreciate the necessity of the factor involving b_{g} in Eq. (78) consider the SF running coupling introduced in the last section. The perturbation expansion in the theory with N_{f} dynamical quarks

$$\bar{g}^2 = g_0^2 + [c^{(1,0)} + c^{(1,1)} N_{\mathrm{f}}] g_0^4 + \ldots \tag{81}$$

has been worked out to 1 loop in ref.[67]. The O(a) counterterms required for the improvement of the Schrödinger functional were taken into account in this calculation. If we insert the definition of the renormalized mass (with the value of b_{m} just obtained) $m_{\mathrm{R}} = m_{\mathrm{q}}(1 - \frac{1}{2} a m_{\mathrm{q}}) + O(g_0^2)$ the result assumes the form

$$c^{(1,0)} = \frac{11N}{24\pi^2}\ln(L/a) + k_1 + O(a^2), \tag{82}$$

$$c^{(1,1)} = -\frac{1}{12\pi^2}\ln(L/a) + k_2 + k_3 a m_{\mathrm{R}} + O(a^2), \tag{83}$$

where k_2 is a function of $m_{\mathrm{R}} L$ and

$$k_3 = 0.012000(2). \tag{84}$$

We can now express $\bar{g}^2$ as a series in the renormalized coupling g_{R}^2 by eliminating the bare coupling. The renormalization factor $Z_{\mathrm{g}}^{(1)}(a\mu)$ should be chosen so as to cancel the logarithmic divergence, but since it may not depend on the quark mass, the additional factor $1 + b_{\mathrm{g}}^{(1)} g_0^2 a m_{\mathrm{q}}$ is necessary to get rid of the term proportional to k_3.

Local operator renormalizations also involve extra factors, e.g. the renormalized improved isovector operators are given by

$$(X_{\mathrm{R}})^a = Z_{\mathrm{X}}(1 + b_{\mathrm{X}} a m_{\mathrm{q}})(X_{\mathrm{I}})^a, \tag{85}$$

for $X = V,A,T,S,P$. A summary of the tree level and one-loop improvement coefficients b_X known to date is given in Table 2[67, 65].

Table 2. Tree level and one-loop improvement coefficients b

X	$b_X^{(0)}$	$b_X^{(1)}$
g	0	$0.012000(2) \times N_f$
m	$-\frac{1}{2}$	$-0.07217(2) \times C_F$
ζ	$-\frac{1}{2}$	$-0.06738(4) \times C_F$
A	1	$0.11414(4) \times C_F$
V	1	$0.11492(4) \times C_F$
P	1	$0.11484(2) \times C_F$
T	1	$0.10434(5) \times C_F$

Non-perturbative determination of improvement coefficients

Although the improvement coefficients are computable in perturbation theory, in practice simulations are carried out at not so very small bare coupling and hence they should be determined non-perturbatively. There are a host of observables one could chose to compute and demand improvement thereof. To fix the O(a) improved Wilson action it is however natural to investigate the expressions associated with chiral symmetry. For example we can consider the PCAC relation on the lattice and tune c_{sw} and c_A so that the error term in

$$\langle \partial_\mu (A_R)^a_\mu(x)\, \mathcal{O}\rangle = 2m_R \langle (P_R)^a(x)\, \mathcal{O}\rangle + \mathrm{O}(a^p) \tag{86}$$

is reduced from $\mathrm{O}(a)$ to $\mathrm{O}(a^2)$.

To this end we define an "unrenormalized current quark mass" through

$$m = \frac{\langle \partial_\mu \{A^a_\mu(x) + c_A a \partial_\mu P^a(x)\}\mathcal{O}\rangle}{2\langle P^a(x)\, \mathcal{O}\rangle}\,. \tag{87}$$

The PCAC relation then implies

$$m = \frac{Z_P(1 + b_P a m_q)}{Z_A(1 + b_A a m_q)}\, m_R + \mathrm{O}(a^p)\,, \tag{88}$$

and the violation of chiral symmetry can be studied by computing m for various choices of $\mathcal{O}$. Note that the sources $\mathcal{O}$ need not be improved (but must be physically separated from x). Again there are many choices of $\mathcal{O}$ that one may consider but a very convenient set is obtained naturally in the framework of the Schrödinger fuctional.

When constructing the Schrödinger fuctional in QCD we impose the same boundary conditions for the gauge fields as we did in the theory without quarks, but we must also supplement boundary conditions for the quark fields. Since the Dirac equation is of first order, only half of the components of the quark and anti-quark fields can be prescribed. Appropriate boundary conditions are found to be[58]

$$\frac{1}{2}(1+\gamma_0)\psi(x) = \rho(\mathbf{x}),\ \overline{\psi}(x)(1+\gamma_0)\frac{1}{2} = \bar{\rho}(\mathbf{x}),\ \text{at } x_0 = 0, \tag{89}$$

and similar boundary conditions are imposed at $x_0 = T$ withfields ρ'. Notein particular that Dirichlet boundary conditions induce a frequency gap of order $1/L$ on the quark

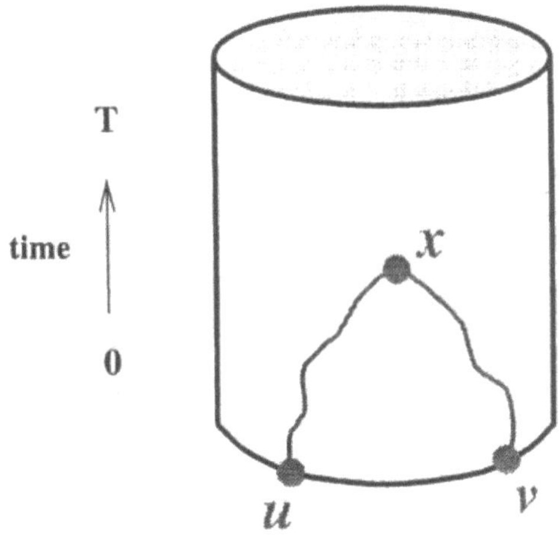

Figure 11. A sketch of a SF correlation function. The sources are at the boundary and the local operator at some point x well away from the boundary.

and gluon fields and hence there is no problem in performing numerical simulations for vanishing quark masses.

Now differentiation w.r.t. $\rho, \bar{\rho}$ produce sources at the boundary, e.g.

$$\mathcal{O}^a = a^6 \sum_{\mathbf{u},\mathbf{v}} \overline{\psi}(u)\gamma_5 \frac{1}{2}\tau^a \psi(v), \qquad u_0 = v_0 = 0\,, \tag{90}$$

and we can study correlation functions such as

$$\langle A_\mu^a(x)\mathcal{O}^a\rangle = \frac{1}{\mathcal{Z}} \int_{\text{fields}} A_\mu^a(x)\mathcal{O}^a\, \mathrm{e}^{-S} \tag{91}$$

illustrated schematically in Fig. 11.

The corresponding SF-mass $m = m_{\mathrm{SF}}$ must be independent of the boundary fields C, C', x_0 and the lattice sizes T and L, up to cutoff effects. We first inspect the PCAC relation at tree-level where all computations can be done analytically. Fig. 12 shows results on a 16×8^3 lattice with bare mass $am_0 = 0.01$. The large effects for the Wilson action are due to the fact that non-zero boundary values C and C' induce a background gauge field. Note that on adding the SW improvement term (with $c_{\mathrm{sw}} = 1$) the dependence on the boundary values is practically eliminated.

Turning to unimproved quenched QCD e.g. 16×8^3 lattice, $\beta = 6.4$, $a = 0.05$ fm (see Fig. 13) a strong violation of chiral symmetry is again manifest. The situation here can be improved by introducing the improvement terms in the action and axial vector current. This is illustrated in Fig. 14 where we have the same lattice as before, but $c_{\mathrm{sw}} = 1.60$, $c_{\mathrm{A}} = -0.027$. Similar results are obtained for larger lattices, other boundary values, etc. In each case we find that at lattice spacings $a \lesssim 0.1$ fm chiral symmetry is effectively restored provided c_{sw} and c_{A} are chosen appropriately.

A typical systematic non-perturbative determination of c_{sw} and c_{A} proceeds as follows[68]. Take three configurations of x_0, C, C', ... and require

$$m_1 = m_2 \quad \text{and} \quad m_2 = m_3\,. \tag{92}$$

For any other configuration Symanzik's analysis then guarantees that $m = m_1 + \mathrm{O}(a^2)$. Different "improvement conditions" yield values of c_{sw} and c_{A} that differ by terms

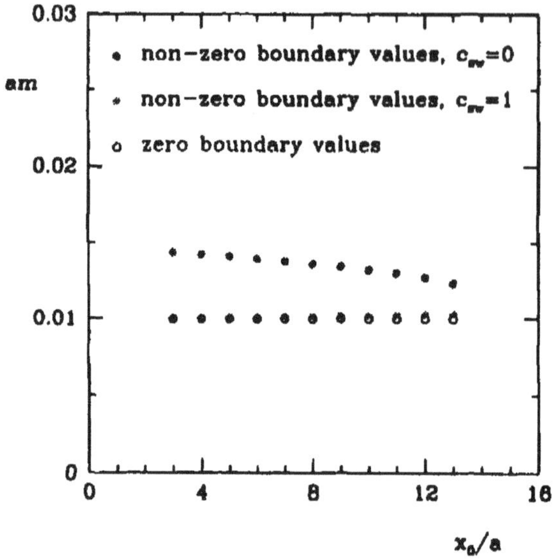

Figure 12. m_{SF} at tree-level. The $c_{sw} = 1$ results nearly fall on top of those for zero boundary values.

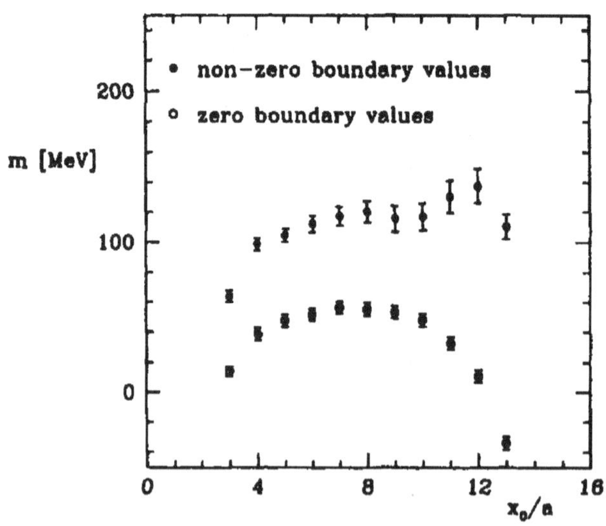

Figure 13. m_{SF} in unimproved quenched QCD.

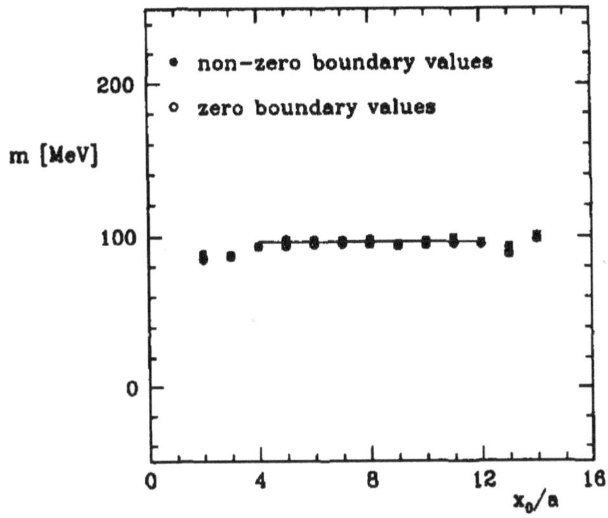

Figure 14. m_{SF} in improved quenched QCD.

of O(a). Thus improvement conditions should be chosen with care in order to avoid an artificial enhancement of O(a^2) cutoff effects, albeit this can be rather difficult to control in practice.

In Fig. 15 we show the results for c_{sw} in quenched QCD for a particular choice of improvement conditions[68]. The solid line is a parametrization of the data

$$c_{sw} = \frac{1 - 0.656\, g_0^2 - 0.152\, g_0^4 - 0.054\, g_0^6}{1 - 0.922\, g_0^2}, \qquad 0 \leq g_0 \leq 1. \tag{93}$$

This expression is now often taken for the definition of the improved action in various numerical simulations. A similar result is obtained for c_A but the analysis[68] is rather delicate and is now being reviewed.

Renormalization constants

For zero quark masses chiral symmetry is expected to become exact in the continuum limit. It is therefore natural to fix the renormalization constants Z_A, Z_V of the currents by imposing the continuum chiral Ward identities also at finite values of the cutoff[70, 71, 72].

In the case of the axial current the relevant Ward identity can be written in the form

$$\int_{\partial R} d\sigma_\mu(x) \epsilon^{abc} \langle A_\mu^a(x) A_\nu^b(y) Q^c \rangle = 2i \langle V_\nu^c(y) Q^c \rangle \tag{94}$$

where the integral is taken over the boundary of a region R containing the point y and Q^a is a source located outside R. In view of on-shell improvement it is important to note that all space-time arguments in Eq. (94) are well separated from one another.

As mentioned above, the Schrödinger functional is also a useful tool for the accurate determination of the renormalization constants of composite operators[73]. For the source

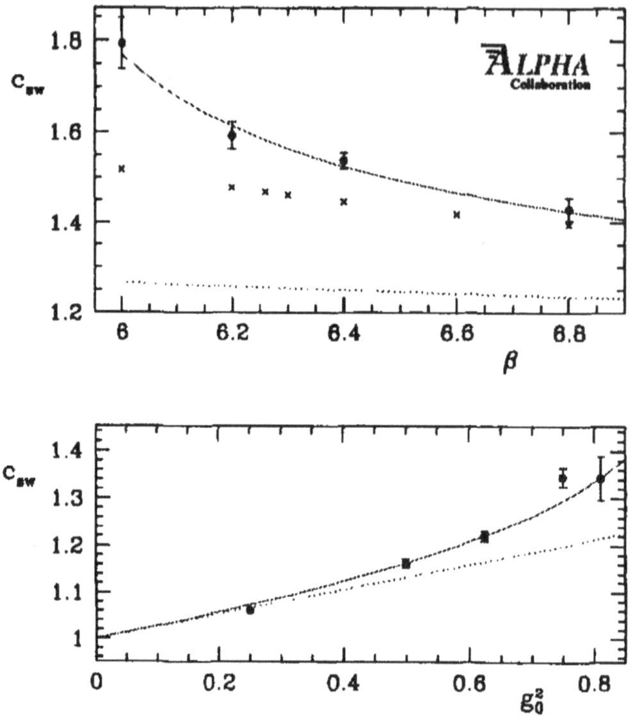

Figure 15. c_{sw} quenched QCD; the dotted line denotes one loop bare PT and the crosses correspond to tadpole improved PT[69].

in Eq. (94) one can take $Q^a = \epsilon^{abc}\mathcal{O}'^b\mathcal{O}^c$ with $\mathcal{O}^a$ as given in Eq. (90) and $\mathcal{O}'^a$ defined similarly with the primed fields. The region R is taken to be $R = \{x; t_1 \leq x_0 \leq t_2\}$.

Now introduce correlation functions:

$$f_{\mathrm{V}}^{\mathrm{R}}(x_0) = \frac{a^3}{6L^6}\sum_{\mathbf{x}} i\epsilon^{abc}\langle\mathcal{O}'^a(V_{\mathrm{R}})_0^b(x)\mathcal{O}^c\rangle, \tag{95}$$

$$f_{\mathrm{AA}}^{\mathrm{R}}(x_0, y_0) = \frac{a^6}{6L^6}\sum_{\mathbf{x},\mathbf{y}} \epsilon^{abe}\epsilon^{cde}\langle\mathcal{O}'^d(A_{\mathrm{R}})_0^a(x)(A_{\mathrm{R}})_0^b(y)\mathcal{O}^c\rangle, \tag{96}$$

$$f_1 = -\frac{1}{3L^6}\langle\mathcal{O}'^a\mathcal{O}^a\rangle. \tag{97}$$

At zero quark mass, and for the correct values of c_{sw} and c_{A} a lattice version of the chiral Ward identity Eq. (94) is

$$f_{\mathrm{AA}}^{\mathrm{R}}(x_0, y_0) = f_1 + O(a^2), \; x_0 > y_0. \tag{98}$$

Compared to Eq. (94) one has set $\nu = 0$ and included an additional summation over y, thus obtaining the isospin charge the action of which can be evaluated because of the exact isospin symmetry on the lattice. Further the conservation of the axial charge at $m_{\mathrm{q}} = 0$ was exploited.

Since the isospin symmetry remains unbroken for non-zero quark mass, one need not restrict the normalization of the vector current to the case $m_{\mathrm{q}} = 0$. Using the vector Ward identity analogous to Eq. (94) one obtains

$$f_{\mathrm{V}}^{\mathrm{R}}(x_0) = f_1 + O(a^2), \tag{99}$$

when c_{sw} is chosen correctly. Note that the improvement coefficient c_V is not needed here, because the tensor density does not contribute to the isospin charge.

Now to get precise definitions of the normalization constants starting from Eqs. (98) and (99) further specifications must be made. The following specific normalization conditions were chosen[73]:

$$f_V^R(T/2) = f_1, \; T = 2L, \tag{100}$$

$$f_{AA}^R(2T/3, T/3) = f_1, \; \text{at } m_q = 0, \; T = 9L/4, \tag{101}$$

where one also set the boundary fields to zero: $C = C' = \rho = \rho' = 0$. In order to guarantee that cutoff effects of matrix elements of renormalized currents vanish proportional to a^2 when approaching the continuum limit, one completes the normalization conditions by scaling L in units of a physical scale

$$L/a = 8 \text{ at } g_0 = 1, \quad L/r_0 = \text{constant}. \tag{102}$$

In the simulations the choices Eq. (102) are realized by smooth inter-/extrapolations.

Of course these explicit choices of the normalization conditions are rather arbitrary. One must just take care not to introduce large $O(a^2)$ effects through an "unfortunate" choice. A study of the cutoff effects in the lattice Ward identities in perturbation theory is often a good guide. Also in the course of the simulations it was checked that Z_V, Z_A do not change appreciably when parameters such as Eq. (102) are varied within reasonable limits.

Fig. 16 shows that the renormalization constants of the vector and axial vector currents are obtained rather accurately in this way[73]. The solid lines are fits by rational functions which incorporate the leading order perturbative results[74, 75, 76]

$$Z_V = 1 - 0.097072(1) \times C_F g_0^2 + O(g_0^4), \tag{103}$$

$$Z_A = 1 - 0.087344(1) \times C_F g_0^2 + O(g_0^4). \tag{104}$$

Since the condition Eq.(99) should hold also for non-vanishing quark mass one can also compute the coefficient b_V non-perturbatively. For quenched QCD the result[73] is shown in Fig. 17. The solid line (incorporating the known 1-loop coefficient) is

$$b_V = \frac{1 - 0.7613\, g_0^2 + 0.0012\, g_0^4 - 0.1136\, g_0^6}{1 - 0.9145\, g_0^2}, \qquad 0 \leq g_0 \leq 1 \tag{105}$$

Unfortunately the analogous coefficient b_A for the axial current cannot be obtained in the same way since the relevant Ward identity contains a physical mass dependence apart from the O(am) lattice artifact. A method of determining the combinations $b_A - b_P$ and $b_V - b_S$ in the quenched theory have been proposed[78, 79].

In this context it is instructive to note a rigorous result due to Lüscher[80] for the improvement coefficient b_s of the isoscalar scalar density $S(x) = \overline{\psi}(x)\psi(x)$. Let us work on an infinite lattice and consider the pseudoscalar density correlation function; at large x_0 we have

$$\sum_{\mathbf{x}} \langle P^a(x) P^a(0) \rangle \propto \exp(-x_0 m_\pi), \tag{106}$$

and differentiating with respect to the bare mass

$$\frac{\partial}{\partial m_0} \sum_{\mathbf{x}} \langle P^a(x) P^a(0) \rangle \propto \exp(-x_0 m_\pi) \Big\{ x_0 \frac{\partial m_\pi}{\partial m_0} + O(1) \Big\}, \; x_0 \to \infty. \tag{107}$$

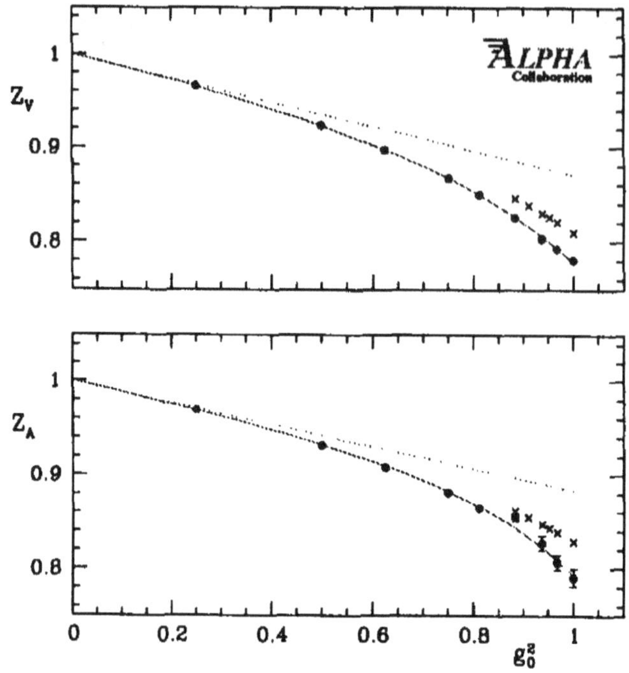

Figure 16. Non-perturbative determination of Z_A, Z_V for $g_0^2 \leq 1$ in the valence approximation.

On the other hand, from the functional integral one has

$$\frac{\partial}{\partial m_0} \sum_{\mathbf{x}} \langle P^a(x) P^a(0) \rangle = \sum_{\mathbf{x}} \sum_{y} \langle P^a(x) \big[S(y) - \langle S(y) \rangle \big] P^a(0) \rangle . \tag{108}$$

One now splits the above summation over y_0 into several ranges

$$y_0 \gg x_0, \; y_0 \sim x_0, \; x_0 > y_0 > 0, \; y_0 \sim 0, \; y_0 \ll 0 , \tag{109}$$

and observes that the only range which gives a contribution proportional to x_0 is the middle one. It then follows that

$$\frac{\partial}{\partial m_0} \sum_{\mathbf{x}} \langle P^a(x) P^a(0) \rangle \propto \langle \pi | S(0) - \langle S(0) | \pi \rangle \big[x_0 + O(1) \big] . \tag{110}$$

In other words, we have

$$\frac{\partial m_\pi}{\partial m_0} = \langle \pi | S - \langle S \rangle | \pi \rangle . \tag{111}$$

Finally one notes that

$$\frac{\partial \tilde{m}_0}{\partial m_0} = 1 + 2 b_{\mathrm{m}} a m_{\mathrm{q}} \tag{112}$$

and concludes that

$$\frac{\partial m_\pi}{\partial \tilde{m}_0} = \langle \pi | S - \langle S \rangle | \pi \rangle \big(1 + 2 b_{\mathrm{m}} a m_{\mathrm{q}} \big) . \tag{113}$$

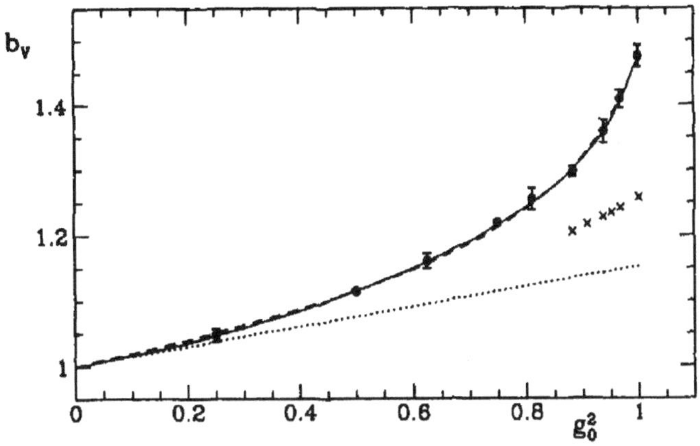

Figure 17. Non-perturbative determination of b_V for $g_0^2 \leq 1$ in the valence approximation (solid line). The dotted line indicates the perturbative estimate and the crosses the perturbative estimate using Parisi's coupling[77].

Since the left hand side is improved one deduces the relationship

$$b_s = -2b_m . \tag{114}$$

It should be emphasized that the scalar densities S and S^a may not have the same b's, although this should be the case in quenched QCD as has been explicitly verified to one-loop.

On-going computations

Various computations of improvement coefficients and renormalization constants are under way. Firstly to complete the definition of the renormalized SF-running mass m(L) one must specify a normalization condition for the pseudoscalar density. Defining

$$f_P^R(x_0) = -\frac{a^3}{3L^3}\sum_{\mathbf{x}}\langle (P_R)^a(x)\mathcal{O}^a\rangle, \tag{115}$$

one can for example impose the normalization condition

$$f_P^R(T/2)^2 = f_1, \text{ at } m_q = 0. \tag{116}$$

The corresponding non-perturbative running mass is now being measured in the quenched theory[42].

A computation of c_V appearing in the definition of the renormalized improved vector current is under way[81].

Finally the extension of non-perturbative improvement program to the theory with dynamical quarks is in principle straightforward. The first step - the non-perturbative determination of c_{sw} in the theory with two dynamical quarks is practically completed[82].

Acknowledgements

My thanks extend to all members of the Alpha Collaboration. In particular I would like to thank Martin Lüscher for sharing his deep insights and teaching me so much.

REFERENCES

1. W. A. Bardeen, A. Buras, D. W. Duke and T. Muta, Phys. Rev. D18 (1978) 3998
2. M. Lüscher, R. Sommer, P. Weisz and U. Wolff, Nucl. Phys. B413 (1994) 481
3. K. Symanzik, Nucl. Phys. B190 [FS3] 1981 1; M. Lüscher, Nucl. Phys. B254 (1985) 52
4. M. Lüscher, P. Weisz and U. Wolff, Nucl. Phys. B359 (1991) 221
5. J. Balog, Phys. Letts. B300 (1993) 145
6. J. Balog and T. Hauer, Phys. Letts. B337 (1994) 115
7. J. Balog, M. Niedermaier and T. Hauer, Phys. Letts. B386 (1996) 224
8. J. Balog and M. Niedermaier, *Off-shell solution of the O(3) NLS model. Beyond Monte Carlo and Perturbation Theory,* MPP-Ph/96-120; hep-th/9612039
9. J. Balog and M. Niedermaier, Phys. Rev. Lett. 78 (1997) 4151
10. M. Karowski and P. Weisz, Nucl. Phys. B139 (1978) 455
11. F.A. Smirnov, 'Form Factors in Completely Integrable Models of Quantum Field Theory', World Scientific, 1992
12. K. Symanzik, in: Mathematical problems in theoretical physics, eds. R. Schrader et al, Lecture Notes in Physics, Vol 153 (Springer New York, 1982); Nucl. Phys. B226 (1983) 187 and 205
13. K. Wilson, Phys. Rev. D10 (1974) 2445
14. H. Eichenherr and M. Forger, Comm. Math. Phys. 82 (1981) 227
15. K. Pohlmeyer, Comm. Math. Phys. 46 (1976) 207; M. Lüscher and K. Pohlmeyer, Nucl. Phys. B155 (1978) 46
16. M. Lüscher, Nucl. Phys. B135 (1978) 1; unpublished notes (1986)
17. D. Buchholz and J.T. **Łopuszański,** Lett. Math. Phys. 3 (1979) 175; D. Buchholz, J.T. **Łopuszański** and Sz. Rabsztyn, Nucl. Phys. B263 (1986) 155.
18. A. Zamolodchikov and Al.B. Zamolodchikov, Ann. Phys. 120 (1979) 253
19. M. Lüscher and U. Wolff, Nucl. Phys. B339 (1990) 222
20. P. Hasenfratz, M. Maggiore and F. Niedermayer, Phys. Letts. B245 (1990) 522; P. Hasenfratz and F. Niedermayer, Phys. Letts. B245 (1990) 529
21. M. Niedermaier, "A derivation of the cyclic form factor equation", MPI Preprint MPI-PhT/97-31
22. A. Patrascioiu and E. Seiler, "Two dimensional NLS Models: Orthodoxy and Heresy", in Fields and Particles, Schladming 1990, ed. H. Mitter and W. Schweiger; "The Difference between Abelian and Non-Abelian Models: Fact and Fancy", MPI-Ph/91-88; Phys. Rev. Lett 74 (1995) 1924
23. G. Källèn, Dan. mat. fys. Medd. **27** (1953) 1; H. Lehmann, Nuovo Cim. 11 (1954) 342
24. Z. Koba, H.B. Nielsen and P. Olesen, Nucl. Phys. B40 (1972) 317; A.M. Polyakov, Sov. Phys. JETP 32 (1971) 296; 33 (1971) 850
25. A. Zamolodchikov, JETP Lett. 3 (1986) 730; V. A. Fateev and Al. B. Zamolodchikov Phys. Lett. B271 (1991) 91
26. see also M. Campostrini, A. Pelissetto, P. Rossi and E. Vicari, "On the evaluation of universal non-perturbative constants in O(N) **σ-models",** hep-lat/9702010
27. M. Blatter, R. Burkhalter, P. Hasenfratz and F. Niedermayer, Phys. Rev. D53 (1996) 923
28. A. Patrascioiu and E. Seiler, "Is the 2D O(3) nonlinear **σ model** asymptotically free?", hep-lat/9706011
29. M. Lüscher, in Progress in Gauge Field Theory, G. 't Hooft eta, eds, (Plenum New York, 1984); Comm. Math. Phys. 104 (1986) 277; ibid 105 (1986) 153
30. D.-S. Shin, Nucl. Phys. B496 (1997) 408
31. M. Hasenbusch, Phys. Rev. D53 (1996) 3445
32. see also S. Caracciolo, R.G. Edwards, A. Pelissetto, A. Sokal, Phys. Rev. Letts. 75 (1995) 1891
33. P. Weisz, Nucl. Phys. B (Proc. Suppl.) 47 (1996) 71
34. M. Lüscher, Nucl. Phys. B180 [FS2] 1981 317
35. R. Sommer, Nucl. Phys. B411 (1994) 839

36. G. de Divitiis, R. Frezzotti, M. Guagnelli and R. Petronzio, Nucl. Phys. B422 (1994) 382; ibid. B433 (1995) 390
37. G. de Divitiis,R. Frezzotti, M. Guagnelli, M. Lüscher, R.Petronzio, R. Sommer, P. Weisz and U. Wolff (ALPHA collab.), Nucl. Phys. B437 (1995) 447
38. G. C. Rossi and M. Testa, Nucl. Phys. B163 (1980) 109; ibid B176 (1980) 477; ibid B237 (1984) 442
39. M. Lüscher, R. Narayanan, P. Weisz and U. Wolff, Nucl. Phys. B384 (1992) 168
40. R. Narayanan and U. Wolff, Nucl. Phys. B444 (1995) 425
41. A. Bode, Proceedings of the Lattice Conference, Edinburgh, July 1997
42. S. Capitani, M. Guagnelli, M. Lüscher, S. Sint, R. Sommer, P. Weisz and H. Wittig Non-perturbative quark mass renormalization, Proceedings of the Lattice Conference, Edinburgh, July 1997
43. C. Michael, Phys. Lett. B283 (1992) 103;
44. S. P. Booth, D. S. Henty, A. Hulsebos, A. C. Irving, C. Michael and P. W. Stephenson (UKQCD Collab.), Phys. Lett. B294 (1992) 385
45. G. S. Bali and K. Schilling, Phys. Rev. D47 (1993) 661; Nucl. Phys. B (Proc. Suppl.) 34 (1994) 147
46. A. X. El-Khadra, G. Hockney, A. S. Kronfeld and P. B. Mackenzie, Phys. Rev. Lett. 69 (1992) 729
47. A. X. El-Khadra, Nucl. Phys. B (Proc. Suppl.) 34 (1994) 141
48. G. P. Lepage and B. A. Thacker, Nucl. Phys. B (Proc. Suppl.) 4 (1988) 199; Phys. Rev. D43 (1991) 196
49. A. X. El-Khadra, PASCOS/Hopkins (1995) 127
50. G. P. Lepage, L. Magnea, C. Nakhleh, U. Magnea and K. Hornbostel, Phys. Rev. D46 (1992) 4052
51. P. McCallum and J. Shigemitsu, Nucl. Phys. B (Proc. Suppl.) 47 (1996) 409
52. C. T. H. Davies, K. Hornbostel, A. Langnau, G. P. Lepage, A. Lidsey, J. Shigemitsu and J. Sloan, Phys. Rev. D50 (1994) 6963
53. C. T. H. Davies, K. Hornbostel, G. P. Lepage, A. Lidsey, J. Shigemitsu and J. Sloan, Phys. Rev. D52 (1995) 6519; Phys. Rev. Lett. 73 (1994) 2654;Phys. Lett. B345 (1995) 42
54. M. Wingate, T. DeGrand, S. Collins and U. Heller, Nucl. Phys. B (Proc. Suppl.) 34 (1994) 373; Phys. Rev. D52 (1995) 307
55. S. Aoki, M. Fukugita, S. Hashimoto, N. Ishizuka, H. Hino, M. Okawa, T. Onogi, A. Ukawa Phys. Rev. Letts. 75 (1995) 22
56. S. M. Catteral, F. R. Devlin, I. T. Drummond and R. R. Horgan, Phys. Lett. B300 (1993) 393; Phys. Lett. B321 (1994) 246
57. M. Alford, W. Dimm, G. Hockney, G. P. Lepage, P. B. Mackenzie, Nucl. Phys. B (Proc. Suppl) 42 (1995) 787; Phys. Lett. B361 (1995) 87
58. S. Sint, Nucl. Phys. B421 (1994) 135, Nucl. Phys. B451 (1995) 416
59. M. Lüscher, Commun. Math. Phys. 54 (1977) 283
60. M. Lüscher, S. Sint, R. Sommer and P. Weisz, Nucl. Phys. B478 (1996) 365
61. B. Sheikholeslami and R. Wohlert, Nucl. Phys. B259 (1985) 572
62. R. Wohlert, Improved continuum limit lattice action for quarks, DESY preprint 87-069 (1987), unpublished
63. M. Lüscher and P. Weisz, Nucl. Phys. B479 (1996) 429
64. M. Lüscher and P. Weisz, Commun. Math. Phys. 97 (1985) 59; M. Lüscher, Improved lattice gauge theories, *in*: Critical phenomena, random systems, gauge theories (Les Houches, Session XLIII, 1984), ed. K. Osterwalder and R. Stora
65. S. Sint and P. Weisz, Nucl. Phys. B502 (1997) 251
66. K. Jansen, C. Liu, M. Lüscher, H. Simma, S. Sint, R. Sommer, P. Weisz and U. Wolff, Phys. Letts. B372 (1996) 275
67. S. Sint and R. Sommer, Nucl. Phys. B465 (1996) 71
68. M. Lüscher, S. Sint, R. Sommer, P. Weisz and U. Wolff, Nucl. Phys. B491 (1997) 323
69. G. P. Lepage and P. B. Mackenzie, Phys. Rev. D48 (1993) 2250
70. M. Boccicchio, L. Maiani, G. Martinelli, G. Rossi and M. Testa, Nucl. phys. B262 (1985) 331
71. L. Maiani and G. Martinelli, Phys. Lett. B178 (1986) 265
72. G. Martinelli, S. Petrarca, C. T. Sachrajda and A. Vladikas, Phys. Lett. B311 (1993) 241; E: B317 (1993) 660
73. M. Lüscher, S. Sint, R. Sommer and H. Wittig, Nucl. Phys. B491 (1997) 344

74. E. Gabrielli, G. Martinelli, C. Pittori, G. Heatlie and C. T. Sachrajda, Nucl. Phys. B362 (1991) 475
75. M. Göckeler, R. Horsley, E. M. Ilgenfritz, H. Oelrich, H. Perlt, P. Rakow, G. Schierholz, A. Schiller and P. Stephenson, Nucl. Phys. B (Proc. Suppl.) 53 (1997) 896
76. S. Sint, private notes (1993,1996)
77. G. Parisi, in: High-Energy Physics — 1980, XX. Int. Conf. Madison (1980), ed. L. Durand and L. G. Pondrom (American Institute of Physics, New York, 1981)
78. G. M. De Divitiis and R. Petronzio, Proceedings of the Lattice Conference, Edinburgh, July 1997
79. S. Sharpe, Proceedings of the Lattice Conference, Edinburgh, July 1997; G. Martinelli, G. C. Rossi, C. T. Sachrajda, S. Sharpe M. Talevi and M. Testa, hep-lat/9705018
80. M. Lüscher, private communication
81. M. Guagnelli and R. Sommer, Proceedings of the Lattice Conference, Edinburgh, July 1997, hep-lat/9709088
82. K. Jansen and R. Sommer, Proceedings of the Lattice Conference, Edinburgh, July 1997, hep-lat/9709022

CONTINUUM AND LATTICE COULOMB-GAUGE HAMILTONIAN

Daniel Zwanziger*

Physics Department, New York University
New York, NY 10003, USA

INTRODUCTION

In the Hamiltonian formalism one can calculate the spectrum directly[1, 2, 3, 4]. In the Coulomb gauge, the color form of Gauss's law, which would appear to be essential for confinement, is satisfied exactly. Moreover an obvious confinement mechanism suggests itself in the Coulomb gauge namely, a long-range instantaneous color-Coulomb potential.

This raises the question, can the Coulomb gauge be regularized? Certainly, on the lattice any configuration may be fixed to the Coulomb gauge. The lattice Coulomb hamiltonian which will be described in this lecture is an implementation of the Coulomb-gauge dynamics with lattice regularization. As to whether the Coulomb gauge is perturbatively renormalizable in the continuum theory, the question is under active investigation[5, 6]. The answer appears to be yes.

We review the canonical quantization of continuum Yang-Mills theory, and derive the continuum Coulomb-gauge Hamiltonian by a simplification of the Christ-Lee method. We then analogously derive, by a simple and elementary method, the lattice Coulomb-gauge Hamiltonian in the minimal Coulomb gauge (and in other Coulomb gauges) from the known Kogut-Susskind Hamiltonian.

CLASSICAL YANG-MILLS EQUATIONS

Classical Yang-Mills theory is designed to be invariant under the group of local gauge transformations $g(x)$. We shall consider the local $SU(N)$ theory, $g(x) \in SU(N)$. The quark field transforms covariantly, $\psi(x) \to {}^g\psi(x) = g(x)\psi(x)$. The gauge-covariant derivative $D = D(A)$, defined by $(D_\mu\psi)(x) = [\partial_\mu + t^a A^a_\mu(x)]\psi(x)$ also transforms covariantly, $(D_\mu\psi)(x) \to g(x)(D_\mu\psi)(x)$, provided that the connection A transforms according to

$$A_\mu(x) \to {}^gA_\mu(x) = g(x)\partial_\mu g^\dagger(x) + g(x)A_\mu(x)g^\dagger(x). \tag{1}$$

Here the anti-hermitian matrices t^a form a representation of the Lie algebra of SU(N), $[t^a, t^b] = f^{abc}t^c$, and $A_\mu \equiv t^a A^a_\mu$. The *local gauge principle* states that A and ${}^gA$ are

*email address: daniel.zwanziger@nyu.edu

physically identical $A \sim^{g} A$. The physical configuration space $\mathcal{Q} = \mathcal{A}/\mathcal{G}$ is the space of connections $\mathcal{A} = \{A\}$ modulo the local gauge transformations $\mathcal{G} = \{g\}$. This space is non-trivial because the local gauge transformation is non-linear.

The Yang-Mills field tensor,

$$F_{\mu\nu} = \partial_\mu A_\nu - \partial_\nu A_\mu + [A_\mu, A_\nu] = t^a F_{\mu\nu}, \tag{2}$$

transforms homogeneously,

$$F_{\mu\nu}(x) \to {}^g F^a_{\mu\nu}(x) = g(x) F_{\mu\nu}(x) g^\dagger(x). \tag{3}$$

The Yang-Mills action

$$S_{\rm YM} = g_0^{-2} \int d^4x (1/4)(-g^{\mu\kappa} g^{\nu\lambda}) F^a_{\mu\nu} F^a_{\kappa\lambda}, \tag{4}$$

is gauge-invariant and yields the field equations $D^\mu F_{\mu\nu} \equiv \partial^\mu F_{\mu\nu} + [A^\mu, F_{\mu\nu}] = 0$. Here the metric tensor is defined by $g_{\mu,\nu} = \mathbf{diag}(-1,1,1,1)$, and g_0 is the coupling constant. (If quarks were present we would have $D^\mu F_{\mu\nu} = -j_{\mathrm{qu},\nu}$.)

The color-electric and magnetic fields are defined by $E_i \equiv F_{i0} = -\dot{A}_i + D_i A_0$ and $B_i = F_{jk}$ for *i, j, k* cyclic.

GAUSS's LAW AND COLOR-COULOMB POTENTIAL

The time derivative of A_0 nowhere appears in the preceding equations, and the field equation which corresponds to variation with respect to A_0 is the time-independent constraint which constitutes the color Gauss law,

$$D_i E_i \equiv \nabla_i E_i + A_i \times E_i = 0 \quad (\text{or} = \rho_{\rm qu}), \tag{5}$$

where $(A \times E)^a \equiv f^{abc} A^b E^c$. This law may be used to fix the longitudinal part of the color-electric field.

We write

$$E_i = E_i^{\rm tr} - \nabla_i \Omega. \tag{6}$$

where $\nabla_i E_i^{\rm tr} = 0$, and Ω is the *color-Coulomb potential.* We have

$$D_i E_i = -D_i \nabla_i \Omega + A_i \times E_i^{\rm tr} = \rho_{\rm qu}. \tag{7}$$

The three-dimensional Faddeev-Popov operator is defined by

$$M = M(A) \equiv -D_i(A) \nabla_i \tag{8}$$

Gauss's law fixes the color-Coulomb potential

$$\Omega(x) = (M^{-1}\rho)(x) = \int d^3y\, M^{-1}(x, y; A) \rho(y). \tag{9}$$

This equation, which appears to be neither Lorentz nor gauge covariant, in fact holds in every Lorentz frame and every gauge. Here ρ is a color-charge density

$$\rho \equiv -A \times E^{\rm tr} + \rho_{\rm qu} \tag{10}$$

which does not contain the complete color-charge density of the gluons, $\rho_{\rm gl} = -A \times E$, but only the part that comes from $E^{\rm tr}$.

The color-Coulomb potential is propagated by the Green's function

$$M^{-1} = (-D_i \nabla_i)^{-1} = (-\vec{\nabla}^2 - A_i \times \nabla_i)^{-1} \tag{11}$$

of the three-dimensional Faddeev-Popov operator. The famous anti-screening property of non-Abelian gauge theory arises from this circumstance. For, whereas $(-\vec{\nabla}^2)$ is a positive operator, the second term $(-A_i \times \nabla_i)$ may be positive or negative. In the latter case there is a cancellation in the denominator which produces an enhanced color-Coulomb potential.

CLASSICAL CANONICAL FORMALISM

The Yang-Mills lagrangian is of the form

$$L_{\text{YM}} = \int d^3x \mathcal{L}_{\text{YM}} \tag{12}$$

where the Lagrangian density is

$$\mathcal{L}_{\text{YM}} = g_0^{-2}(1/2)(\vec{E}^2 - \vec{B}^2). \tag{13}$$

The time derivatives are contained in $E_i = -\dot{A}_i + D_i A_0$, and the canonical momentum densities are given by

$$\begin{aligned} \pi_i &= \partial\mathcal{L}/\partial\dot{A}_i = -g_0^{-2} E_i \\ \pi_0 &= \partial\mathcal{L}/\partial\dot{A}_0 = 0. \end{aligned} \tag{14}$$

The classical Hamiltonian

$$\begin{aligned} H &= \int d^3x(\pi_i \dot{A}_i - \mathcal{L}) \\ H &= \int d^3x \, [\, (1/2)(g_0^2\, \vec{\pi}^2 + g_0^{-2}\vec{B}^2) - A_0 D_i \pi_i \,] \end{aligned} \tag{15}$$

produces the equations of motion

$$\begin{aligned} \dot{A}_i &= \delta H/\delta\pi_i \Rightarrow \pi_i = -g_0^{-2} E_i \\ \dot{\pi}_i &= -\delta H/\delta A_i \Rightarrow D_0 E_i + \epsilon_{ijk} D_j B_k = 0, \end{aligned} \tag{16}$$

and the Gauss law constraint

$$0 = \delta H/\delta A_0 \Rightarrow D_i \pi_i = 0 \tag{17}$$

QUANTIZATION IN THE WEYL GAUGE

Before deriving the lattice Coulomb Hamiltonian, we shall derive the Christ-Lee Coulomb Hamiltonian for continuum gauge theory which it resembles. Christ and Lee start by canonical quantization in the Weyl gauge, $A_0 = 0$. This is not a complete gauge-fixing and allows time-independent local gauge transforamtions $g = g(\vec{x})$. In this gauge we have $E_i = -\dot{A}_i$, and the previous Hamiltonian reduces to

$$H_W = \int d^3x \; (1/2)(\, g_0^2\, \vec{\pi}^2 + g_0^{-2}\vec{B}^2 \,). \tag{18}$$

This Hamiltonian generates the previous equations of motion in the Weyl gauge, but not the Gauss law constraint $D_i \pi_i = 0$. In the classical theory this is imposed as an initial condition that holds at $t = 0$, and which is preserved by the equations of motion.

In the Christ-Lee method, one quantizes in the Weyl gauge by treating A and π as Cartesian variables, with canoncal equal-time commutation relations

$$[A_i^a(\vec{x}), A_j^b(\vec{y})] = [\pi_i^a(\vec{x}), \pi_j^b(\vec{y})] = 0 \tag{19}$$

$$[\pi_i^a(\vec{x}), A_j^b(\vec{y})] = -i\delta^{ab}\, \delta_{ij}\, \delta(\vec{x} - \vec{y}). \tag{20}$$

There is then no ordering problem in the Weyl Hamiltonian, nor in the definition of the left-hand side of Gauss's law

$$G^a(x) \equiv (D_i \pi_i)^a(x) = \nabla_i \pi_i^a + f^{abc} A_i^b(x) \pi_i^c(x). \tag{21}$$

Moreover with the above canonical commuation relations one observes that $G^a(x)$ is the generator of local time-independent gauge transformations,

$$[G^a(x), \pi_i^b(y)] = i\delta(x - y) f^{abc} \pi_i^c(y) \tag{22}$$

$$[G^a(x), A_i^b(y)] = -iD_i^{ab} \delta(x - y), \tag{23}$$

under which π transforms homogeneously. The $G^a(x)$ satisfy the Lie algebra of the time-independent local gauge group,

$$[G^a(x), G^b(y)] = i\delta(x - y) f^{abc} G^c(y) \tag{24}$$

and moreover these local time-independent gauge transformations are a symmetry of the Hamiltonian $[G^a(x), H_W] = 0$. Therefore it is consistent to impose Gauss's law as a subsidiary condition which must be satisfied for a wave function $\Psi = \Psi(A)$ to represent a physical state,

$$G^a(x)\Psi = 0. \tag{25}$$

This condition is also the statement that the wave function is invariant under time-independent local gauge transformations.

The inner product is defined by the functional integral

$$(\Phi, \Psi) = \int dA\, \Phi^*(A)\Psi(A), \tag{26}$$

and the Hamiltonian by $H_W = T + V$, where T and V correspond respectively to kinetic (electrical) and potential (magnetic) energies. Here

$$V = \int d^3x\, (1/2) g_0^{-2} [B_i^a(x)]^2 \tag{27}$$

acts by multiplication, and we define T by its expression as a quadratic form

$$(\Phi, T\Psi) = \int dA \int d^3x\, (g_0^2/2)\, \delta\Phi^*(A)/\delta A_i^a(x)\, \delta\Psi(A)/\delta A_i^a(x). \tag{28}$$

MINIMAL CONTINUUM COULOMB GAUGE

In the Christ-Lee method[7], one solves the Gauss law subsidiary condition explicitly to obtain the Hamiltonian in the physical subspace. In the original Christ-Lee paper, this is done by choosing coordinates in A-space adapted to the gauge orbit. A gauge orbit is a set of gauge-equivalent configurations $A \sim {}^gA$. (Here we refer only to time-independent local gauge transformations g.) This may be done by writing $A = {}^gA^{tr}$, where A^{tr} is transverse, $\nabla_i A_i^{tr} = 0$, and g runs over the set of all possible gauge transformations. Here the transverse configuration A^{tr} labels the gauge orbit and g labels the point on the orbit. One way to address the problem of Gribov copies[8], is to restrict A^{tr} to a subset Λ of transverse configurations Γ known as the fundamental modular region. For an alternative approach, see[9].

A convenient way to choose Λ is by means of a minimizing function

$$F_A(g) \equiv \| {}^gA \|^2 = \int d^3x |g\vec{\nabla}g^\dagger + g\vec{A}g^\dagger|^2. \tag{29}$$

The fundamental modular region A is defined to be the set of A that are an absolute minimum of this function,

$$\Lambda \equiv \{A : \| A \|^2 \leq \| {}^gA \|^2 \text{ for all } g\}. \tag{30}$$

With suitable assumptions on the space of configurations, one may show that such a minimum exists[10]. This definition (almost) uniquely fixes a single point on each gauge orbit, to within global gauge transformations, because the absolute minimum of the functional is unique apart from "accidental" degeneracies. These are degenerate (equal) absolute minima of the minimizing functional. They constitute the boundary of Λ. The fundamental modular region is obtained from Λ by topologically identifying the boundary points that are gauge-equivalent degenerate absolute minima. (For a more extensive discussion of Λ see[11].)

We may easily derive some properties of the fundamental modular region. At a relative or absolute minimum the functional is stationary to first order and its second-order variation is positive. With $g = \exp(\omega) = 1 + \omega + (1/2)\omega^2 + ...$ an elementary calculation gives

$$\| {}^gA \|^2 = \| A \|^2 - 2(\nabla_i A_i, \omega) + (1/2)(\omega, M(A)\omega) + \dots , \tag{31}$$

where $M(A) = -D_i(A)\nabla_i$ is the three-dimensional Faddeev-Popov operator. Since this must be positive for all ω, we obtain the Coulomb gauge condition $\nabla_i A_i = 0$. *In addition* we find that $M(A)$ is a positive operator, $(\omega, M(A)\omega) \geq 0$ for all ω and for all $A \in \Lambda$. These two conditions define the Gribov region Ω which is the set of all relative and absolute minima, of which the fundamental modular region Λ is a proper subset, $\Lambda \subset \Omega$. Note that because the Faddeev-Popov operator $M(A)$ is a positive operator for $A \in \Lambda$, $M(A)$ is invertible in the minimal Coulomb gauge and has non-negative eigenvalues.

CONTINUUM COULOMB-GAUGE HAMILTONIAN

In the present lecture we shall not explicitly change variables from A to the adapted coordinates A^{tr} and g, as was originally done by Christ and Lee[7], but instead use the more efficient Faddeev-Popov "trick" to obtain H_{coul}.

The Faddeev-Popov identity reads

$$1 = \int dg\, \delta_\Lambda(\nabla_i\, {}^g\!A)\, \det[M({}^g\!A)], \tag{32}$$

where $M(A) = -\nabla_i D_i(A)$ is the Faddeev-Popov operator. Here $\delta_\Lambda(\nabla_i A_i)$ is the restriction of $\delta(\nabla_i A_i)$ to the fundamental modular region Λ, namely $\delta_\Lambda(\nabla_i A_i) = \delta(\nabla_i A_i)\chi_\Lambda(A)$, where $\chi_\Lambda(A) = 1$ for (transverse) $A \in \Lambda$, and $\chi_\Lambda(A) = 0$ otherwise.

We require wave functions to be gauge invariant, $\Psi({}^g\!A) = \Psi(A)$, and we wish to express the inner-product of such wave-functions (Φ, Ψ) as an integral over the parameters $A^{tr} \in \Lambda$ that label the orbit space. For this purpose we insert the Faddeev-Popov identity into the inner product and obtain

$$\begin{aligned}
(\Phi, \Psi) &= \int dA\, \Phi^*(A)\Psi(A) \\
&= \int dA dg\, \delta_\Lambda(\nabla_i\, {}^g\!A_i)\, \det[M({}^g\!A)]\; \Phi^*(A)\Psi(A) \\
&= \int dA dg\, \delta_\Lambda(\nabla_i\, {}^g\!A_i)\, \det[M({}^g\!A)]\; \Phi^*({}^g\!A)\Psi({}^g\!A) \\
&= \int dA' dg\, \delta_\Lambda(\nabla_i A'_i)\, \det[M(A')]\; \Phi^*(A')\Psi(A') \\
&= N \int dA\, \delta_\Lambda(\nabla_i A_i)\, \det[M(A)]\; \Phi^*(A)\Psi(A) \\
&= N \int_\Lambda dA^{tr}\; \det[M(A^{tr})]\; \Phi^*(A^{tr})\Psi(A^{tr}).
\end{aligned} \tag{33}$$

Here $N = \int dg$ is the (infinite) volume of the gauge orbit which is an irrelevant normalization constant, and $\int_\Lambda dA^{tr}$ is the functional intregral over all transverse configurations in Λ. We have used the invariance of the measure dA under local gauge transformation, $dA = dA'$ for $A' = {}^g\!A_i$. These manipulations are somewhat formal in the continuum theory, but are well defined in lattice gauge theory.

We next express the Hamiltonian $H_W = T + V$ on the reduced space of wave functions that depend on A^{tr}. Because $V(A)$ is gauge invariant, it acts on the reduced space as the operator of multiplication by $V(A^{tr})$.

The expression for T as a quadratic form is particularly convenient for our purpose because it only involves first derivatives, and this allows us to apply Gauss's law. We have seen (22) that $\pi_i^a(x) = -i\delta/\delta A_i^a(x)$ transforms homogeneously under gauge transformations, so for gauge-invariant wave functions the quantity $\delta\Phi^*/\delta A_i^a(x)\; \delta\Psi/\delta A_i^a(x)$ is also gauge-invariant. Consequently when we insert the Faddeev-Popov identity into the quadratic form that defines T, we obtain, as for the inner product,

$$\begin{aligned}
(\Phi, T\Psi) = N &\int dA\, \delta_\Lambda(\nabla_i A_i)\, \det[M(A)] \\
\times &\int d^3x\, (g_0^2/2)\, \delta\Phi^*(A)/\delta A_i^a(x)\; \delta\Psi(A)/\delta A_i^a(x).
\end{aligned} \tag{34}$$

To proceed, we shall use Gauss's law $D_i\delta\Psi/\delta A_i = 0$ to express $\delta\Psi/\delta A_i$ as a derivative with respect to A^{tr}. For this purpose we write $E_i = E_i^{tr} - \nabla_i\Omega$, where $E_i = i\delta/\delta A_i$ and $\nabla_i E_i^{tr} = 0$. Here Ω is the color-Coulomb potential operator, and $E^{a,tr} = i\delta/\delta A_i^{a,tr}$. With $D_i E_i = \nabla_i E_i + A_i \times E_i$, Gauss's law, reads

$$-D_i\nabla_i\Omega\Psi = \rho\Psi. \tag{35}$$

where we have introduced the color-charge density operator corresponding to the dynamical degrees of freedom

$$\rho \equiv -A_i \times E_i^{tr}, \tag{36}$$

As in the classical theory, we may solve for $\Omega\Psi$ by inverting the Faddeev-Popov operator $M(A) = -D_i(A)\nabla_i$. We have $\Omega\Psi = M^{-1}\rho\Psi$, which gives

$$E_i\Psi = E_i^{tr}\Psi - \nabla_i M^{-1}(A)\,(-A_i \times E_i^{tr})\Psi. \tag{37}$$

Only derivatives with respect to A^{tr} now appear in the expression for the quadratic form T,

$$(\Phi, T\Psi) = N\int_\Lambda dA^{tr}\ \det[M(A^{tr})] \times \int d^3x\ (g_0^2/2)\ (E_i^a(x)\Phi)^*(A^{tr})\ (E_i^a(x)\Psi)(A^{tr}), \tag{38}$$

where $E\Psi$ and $E\Phi$ are defined by the preceding equation. The transverse and longitudinal parts of the electric field contribute separately to the kinetic energy,

$$(\Phi, T\Phi) = N\int_\Lambda dA^{tr}\ \det[M(A^{tr})]\int d^3x\ (g_0^2/2) \times [\,|E_i^{tr,a}(x)\Phi|^2 + |(\nabla_i M^{-1}\rho)^a(x)\Phi|^2\,]. \tag{39}$$

We have expressed T as a quadratic form on functions of transverse configurations, $A^{tr} \in \Lambda$ that parametrize the orbit space, and V as the operator of multiplication by $V(A^{tr})$. These expressions define the Hamiltonian in the minimal Coulomb gauge, $H_{coul} = T + V$, defined by the gauge choice $A_i = A_i^{tr} \in \Lambda$.

KOGUT-SUSSKIND LATTICE HAMILTONIAN

In a Euclidean lattice theory the Hamilton H may be derived from the partition function Z. The partition function is expressed in terms of the transfer matrix $\mathcal{T}$ by $Z = \lim_{N\to\infty} \mathrm{tr}\ \mathcal{T}^N$, where N is the number of Euclidean "time" slices. One chooses an asymmetric lattice with lattice unit a_0 in the Euclidean time direction and lattice unit a in spatial directions. Then $\mathcal{T} = \lim_{a_0\to 0} \exp(-a_0 H)$.

In[11] the lattice Coulomb-gauge Hamiltonian was derived directly from this formula, where the configurations were fixed to the lattice Coulomb gauge. In the present lecture we present a much briefer and elementary derivation. We shall start with the known Kogut-Susskind Hamiltonian[12], H_{KS}, which is the lattice analog of the Weyl Hamiltonian. We then fix the gauge, and solve the lattice form of the Gauss law constraint to obtain the lattice Coulomb-gauge Hamiltonian on the reduced space, just as we did in the continuum theory.

We designate points on the three-dimensional periodic cubic lattice by three-vectors with integer components, which we denote by $x, y, \ldots$. Gauge transformations are site variables, $g_x \in SU(N)$. The basic variables of the theory are the link variables $U_{(xy)} \in SU(N)$ defined for all links (xy) of the lattice, with $U_{(yx)} = U_{(xy)}^\dagger$. Local gauge transformations are defined by $U_{(xy)} \to {}^gU_{(xy)} = g_x U_{(xy)} g_y^\dagger$.

The $U_0 = 1$ gauge is the lattice analog of the Weyl $A_0 = 0$ gauge. In this gauge, the Hamiltonian obtained from Wilson's lattice action and the formula $\mathcal{T} = \lim_{a_0\to 0} \exp(-a_0 H)$ is

$$H_{KS} = T + V \tag{40}$$

where

$$T = \sum_{(xy)} (g_0^2/2a)\ J_{(xy)}^2$$
$$V = (2N/ag_0^2)\sum_p (1 - N^{-1}\mathrm{Re}\ \mathrm{tr} U_p). \tag{41}$$

Here a is the (spatial) lattice unit. Henceforth we set $a = 1$. In the last expression the sum extends over all plaquettes p of the (spatial) lattice, and U_p is the product of the link variables $U_{(xy)}$ on the links around the plaquette p. The potential energy operator is gauge invariant $V({}^gU) = V(U)$, as are physical wave functions, $\Psi({}^gU) = \Psi(U)$.

The variables $J_{(xy)}$ are electric field operators associated to each link of the lattice, defined as follows. Let the link variable $U_{(xy)}$ be parametrized by a set of variables $\theta^{\alpha}_{(xy)}$. Then, for each link (xy),

$$J_{(xy),a} \equiv i\, J^{\alpha}_{a}(\theta_{(xy)})\, \partial/\partial\theta^{\alpha}_{(xy)} \tag{42}$$

is the Lie generator of the $SU(N)$ group, that satisfies the commutation relations

$$[J_{(xy),a}, J_{(uv),b}] = i\delta_{(xy),(uv)} f^{abc} J_{(xy),c}. \tag{43}$$

For the SU(2) group, T would be the Hamiltonian for a set of non-interacting tops, located on every link of the lattice. The potential energy operator V provides an interaction between tops on the same plaquette.

Both T and V are gauge invariant, as is the inner product

$$(\Psi, \Phi) \equiv \int \prod_{(xy)} dU_{(xy)} \Psi^*(U)\Phi(U). \tag{44}$$

Here $dU_{(xy)}$ is the Haar measure on each link,

$$dU_{(xy)} \equiv \det \psi_{(xy)} \prod_{\alpha} d\theta^{\alpha}_{(xy)}, \tag{45}$$

where $\psi(\theta_{(xy)})$ is the inverse of the matrix $J(\theta_{(xy)})$,

$$J^{\alpha}_{a}(\theta_{(xy)})\psi^{b}_{\alpha}(\theta_{(xy)}) = \delta^{b}_{a}. \tag{46}$$

MINIMAL LATTICE COULOMB GAUGE

We wish to express the hamiltonian as on operator on the reduced space of gauge orbits. For this purpose we require a set of parameters that label the gauge orbits. A convenient way to identify the gauge orbits is to introduce a minimizing function, as in the continuum theory,

$$F_U(g) \equiv \sum_{x,i} \{ 1 - N^{-1}\mathrm{Re}\ \mathrm{tr}[({}^gU)_{x,i}] \}. \tag{47}$$

Here we have introduced the notation $U_{x,i} \equiv U_{(xy)}$ for $y = x + e_i$ where e_i is a unit vector in the positive i-direction. The gauge-transformed link variable is given by $({}^gU)_{x,i} = g_x U_{x,i} g^{\dagger}_{x+\hat{i}}$, where we have written $\hat{i} \equiv e_i$. The fundamental modular region Λ is defined to be the set of configurations U that are absolute minima of this function,

$$\Lambda \equiv \{U : F_U(1) \leq F_U(g) \text{ for all } g\}. \tag{48}$$

With this gauge choice, the link variables $U_{x,i}$ are made as close to unity as possible, equitably, over the whole lattice.

The advantage of this method is that we are assured that a minimizing configuration exists on each gauge orbit, because the minimizing function is defined on a compact space (a finite product of $SU(N)$ group manifolds). An alternative procedure that is sometimes followed is to posit a gauge condition. For example with the exponential mapping $U_{x,i} = \exp(A_{x,i})$, one may posit the condition $\sum_i(A_{x,i} - A_{x-\hat{\imath},i}) = 0$ which is a lattice analog of the continuum Coulomb-gauge transversality condition. An obstacle to this procedure is that one does not know whether every gauge orbit intersects this gauge-fixing surface. On the other hand for the purpose of formal expansions the alternative procedure may be used.

As in the continuum case, we observe that, at a relative or absolute minimum, the minimizing function is stationary to first order, and its second order variation is positive. In order to conveniently exploit these properties, we introduce a one-parameter subgroup of the local gauge group $g_x(s) = \exp(s\omega_x)$, where $\omega_x = t^a\omega_x^a$ is anti-hermitian, $\omega_x^\dagger = -\omega_x$. We write $F_U(s) \equiv F_U(g(s))$ and we have

$$dF_U(s)/ds = -N^{-1}\sum_{x,i} \mathrm{Re}\ \mathrm{tr}[\omega_x({}^gU)_{x,i} - ({}^gU)_{x,i}\omega_{x+\hat{\imath}}]$$

$$= N^{-1}\sum_{x,i} \mathrm{Re}\ \mathrm{tr}\{[\omega_{x+\hat{\imath}} - \omega_x]({}^gU)_{x,i}\} \tag{49}$$

$$= -(2N)^{-1}\sum_{x,i} \mathrm{tr}\{\omega_x[({}^gU)_{x,i} - ({}^gU)_{x,i}^\dagger - ({}^gU)_{x-\hat{\imath},i} + ({}^gU)_{x-\hat{\imath},i}^\dagger]\}. \tag{50}$$

At a relative or absolute minimum U we have $dF_U(s)/ds|_{s=0} = 0$. To express this condition compactly, we introduce the link variables

$$t^aA_{x,i}^a \equiv A_{x,i} \equiv (1/2)(U_{x,i} - U_{x,i}^\dagger)_{\mathrm{traceless}}. \tag{51}$$

We shall use these variables as coordinates to parametrize the $SU(N)$ group, and we write $U_{x,i} = U(A_{x,i})$. The manifold of the $SU(N)$ group requires more than one coordinate patch, but we shall not trouble to account for this explicitly. The formulas of the last section hold, with $\theta_{x,i}^a = A_{x,i}^a$. These coordinates agree with the exponential mapping, $U_{x,i} = \exp A_{x,i}$, to second order, for with the above definition one may show

$$U_{x,i} = 1 + A_{x,i} + (1/2)(A_{x,i}^2) + (6N)^{-1}\mathrm{tr}A^3 + O(A^4), \tag{52}$$

so these variables approach the continuum connection $A_i(x)$ in the continuum limit.

In terms of these variables, the condition $dF_U(s)/ds|_{s=0} = 0$ reads

$$\sum_x \mathrm{tr}[\omega_x \sum_i(-A_{x,i} + A_{x-\hat{\imath},i})] = 0. \tag{53}$$

Since this holds for arbitrary ω, we conclude that, at a minimum of the minimizing function, A satisfies

$$\sum_i (A_{x,i} - A_{x-\hat{\imath},i}) = 0. \tag{54}$$

This is the lattice analog of the transversality condition that characterizes the continuum Coulomb gauge, and we call our gauge choice the "minimal lattice Coulomb gauge". The lattice Coulomb-gauge condition is linear in the variables $A_{x,i}$, and it may be solved by lattice fourier transform.

To see the geometrical meaning of the gauge condition, and for future use, it is helpful to introduce some definitions. We define the lattice "gradient" by

$$(\nabla\omega)_{x,i} \equiv (\omega_{x+\hat{\imath}} - \omega_x), \tag{55}$$

It is a matrix that linearly maps site variables into link variables. The identity,

$$\sum_{x,i} \mathrm{tr}[(\omega_{x+i} - \omega_x)A_{x,i}] = \sum_x \mathrm{tr}[\omega_x \sum_i(-A_{x,i} + A_{x-\hat{\imath},i})], \tag{56}$$

holds for all ω and A. We write it

$$(\nabla\omega, A) = (\omega, \nabla^* A) \tag{57}$$

which defines the dual ∇^* of the lattice gradient. It is a matrix that linearly maps link variables into site variables, that is given explicitly by

$$(\nabla^* A)_x = -\sum_i (A_{x,i} - A_{x-\hat{i},i}), \tag{58}$$

and we have $\nabla^* = -$ (lattice "divergence"). These geometric definitions may be extended to arbitrary lattices that need not be cubic or periodic[11]. In the minimal Coulomb gauge, the lattice divergence of A vanishes, $(\nabla^* A)_x = 0$.

At a minimum of the minimizing function we also have the condition $d^2F_U(s)/ds^2|_{s=0} \geq 0$. From (49) we obtain

$$d^2F_U(s)/ds^2 = (2N)^{-1} \sum_{x,i} \mathrm{tr}\{[\omega_{x+\hat{i}} - \omega_x](d/ds)[({}^gU)_{x,i} - ({}^gU)^\dagger_{x,i}]\}. \tag{59}$$

With

$$A_{x,i}(s) \equiv A[({}^{g(s)}U)_{x,i}], \tag{60}$$

the positivity condition reads

$$d^2F_U(s)/ds^2|_{s=0} = N^{-1} \sum_{x,i} \mathrm{tr}\{[\omega_{x+\hat{i}} - \omega_x]\, dA_{x,i}(s)/ds\}|_{s=0} \geq 0. \tag{61}$$

To make explicit the geometrical meaning this condition, and for future use, we define the lattice gauge covariant "derivative" $D(A)$ by

$$[D(A)\omega]_{x,i} \equiv -dA_{x,i}(s)/ds|_{s=0}. \tag{62}$$

It is the matrix that maps an infinitesimal gauge transformation ω into the corresponding first-order change in the coordinates. (The minus sign is to be coherent with the continuum formula ${}^gA = A - D(A)\omega$.) Like the lattice gradient ∇, it maps site variables into link variables, and we have $D(0) = \nabla$. Its dual $D(A)^*$ maps link variables into site variables, and is defined by $(D^*(A)V, \omega) \equiv (V, D(A)\omega)$, for arbitary link variables V and site variables ω. For the coordinates A defined above, $D(A)$ is given explicitly by

$$[D(A)\omega]_{x,i} = t^b[D(A)\omega]^b_{x,i} = (1/2)[(U_{x,i}\omega_{x+\hat{i}} - \omega_x U_{x,i}) - (\mathrm{h.c.})]_{\mathrm{traceless}}, \tag{63}$$

where $U = U(A)$.

The positivity condition $d^2F_U(s)/ds^2|_{s=0} \geq 0$, which holds at absolute or relative minima of the minimizing function, thus reads

$$\sum_{x,i} (\nabla\omega)^b_{x,i} (D(U)\omega)^b_{x,i} \geq 0 \text{ for all } \omega. \tag{64}$$

We define the lattice Faddeev-Popov matrix

$$M(A) \equiv D^*(A)\nabla, \tag{65}$$

which maps site variables into site variables. For transverse A, one may verify that this matrix is symmetric,

$$D(A)^*\nabla = \nabla^* D(A) \quad \text{for } \nabla^* A = 0. \tag{66}$$

We conclude that at a minimum of the minimizing function, this matrix is also non-negative

$$M(A) \geq 0 \quad \text{for } A \in \Lambda. \tag{67}$$

It has a trivial null space consisting of x-independent eigenvectors, $\nabla\omega = 0$. This is a reflection of the fact that the minimal Coulomb gauge does not fix global (x-independent) gauge transformations. On the orthogonal subspace, $M(A)$ is strictly positive for configurations A that are interior points on the fundamental modular region Λ. (For additional properties of Λ, see[11].)

LATTICE COULOMB-GAUGE HAMILTONIAN

In the proceeding section we obtained a parametrization of the gauge orbit space by means of transverse configurations restricted to the fundamental modular region. We shall now calculate the lattice Coulomb-gauge Hamiltonian $H_{\rm coul}$ as the restriction of the Kogut-Susskind Hamiltonian $H_{\rm KS}$ to gauge-invariant wave functions, $\Phi({}^gU) = \Phi(U)$.

We represent states $\Phi(U)$ as functions of the coordinates, $\Phi = \Phi(A)$. The coordinates transform according to ${}^gA^b_{x,i} = A^b_{x,i}({}^gU)$ under local gauge transformation g_x. For an infinitesimal gauge transformation $g_x = \exp(\omega_x) = 1 + \omega_x$, we have

$${}^gA^b_{x,i} = A^b_{x,i} - [D(A)\omega]^b_{x,i}, \tag{68}$$

by the definition (62) of the lattice gauge-covariant derivative $D(A)$. With $\Phi({}^gA) = \Phi(A - D(A)\omega)$, we conclude that gauge-invariant wave functions, which satisfy $\Phi({}^gA) = \Phi(A)$ for all g, satisfy the first order differential equation,

$$\sum_{x,i}(D(A)\omega)^a_{x,i}\partial\Phi/\partial A^a_{x,i} = 0, \tag{69}$$

for all ω. This is equivalent to the condition

$$(D^*(A)\partial/\partial A)^b_x\Phi = 0 \tag{70}$$

for all lattice sites x, which is the lattice version of Gauss's law constraint. This equation holds for any coordinates $A_{x,i}$ on the group manifold, provided only that the lattice gauge-covariant derivative D is defined by (62).

We now proceed as in the continuum theory, and introduce the lattice Faddeev-Popov identity

$$1 = \int \prod_x dg_x {\prod_x}' \delta_\Lambda[(\nabla^* {}^gA)_x] \det M'({}^gA). \tag{71}$$

The δ-function δ_Λ is defined by strict analogy to the continuum theory. The product $\prod_x dg_x$ extends over all sites x of the lattice, where dg_x represents Haar measure. However the primed product, $\prod_x{}'\delta_\Lambda[(\nabla^*A)_x]$ extends over all but one site x_0 of the lattice because the lattice divergences satisfy the identity $\sum_x(\nabla^*A)_x = 0$, and are thus not all linearly independent. The primed matrix $M'(A)$ is the lattice Faddeev-Popov matrix $M(A) = D^*(A)\nabla$, with the rows and columns labelled by x_0 deleted. One may show that the integrand is independent of x_0 and that, apart from an overall normalization constant, $\det M'(A)$ for $\nabla^*A = 0$ is the determinant of $M(A)$ on the space orthogonal to its trivial null space. In the minimal Coulomb gauge $M'(A)$ is a strictly positive matrix, in the interior of the fundamental modular region Λ, and its determinant is positive.

Lattice gauge-covariant derivative

The above Faddeev-Popov identity also holds for other coordinates, such as that provided by the exponential mapping $U_{x,i} = \exp A_{x,i}$, with gauge condition $(\nabla^* A)_x = 0$. We now derive an explicit expression for the lattice gauge-covariant derivative $D(\theta)$ in any coordinate system $\theta^\alpha_{x,i}$, where $D(\theta)$ is defined by the linear change in the coordinates induced by an infinitesimal local gauge transformation. We start with the Maurer-Cartan differential

$$U_{x,i} dU^\dagger_{x,i} = t^a \psi^a_\alpha(\theta_{x,i}) d\theta^\alpha_{x,i}, \tag{72}$$

which relates infinitesimal changes in the coordinates $d\theta$ to infinitesimal changes dU in U. For an infinitesimal gauge transformation $g_x = 1 + \omega_x$ we have

$$\begin{aligned} {}^g U_{x,i} &= U_{x,i} + \omega_x U_{x,i} - U_{x,i}\omega_{x+\hat{i}} \\ dU_{x,i} &= \omega_x U_{x,i} - U_{x,i}\omega_{x+\hat{i}} \\ dU^\dagger_{x,i} &= -U^\dagger_{x,i}\omega_x + \omega_{x+\hat{i}} U^\dagger_{x,i} \\ U_{x,i} dU^\dagger_{x,i} &= U_{x,i}\omega_{x+\hat{i}} U^\dagger_{x,i} - \omega_x, \end{aligned} \tag{73}$$

where we have used $\omega^\dagger_x = -\omega_x$. The last expression represents the difference between the parallel-transport of ω, from $x + \hat{i}$ to x, and ω_x. By analogy with the continuum theory, it is natural to call this quantity the lattice gauge-covariant derivative in the *site* basis,

$$[\mathcal{D}(U)\omega]_{x,i} \equiv U_{x,i}\omega_{x+\hat{i}} U^\dagger_{x,i} - \omega_x. \tag{74}$$

If we choose a basis in the Lie algebra, $\omega_x = t^a \omega^a_x$, we have $U_{x,i} t^b U^\dagger_{x,i} = t^a R^{ab}(U)$, where $R^{ab}(U)$ is the adjoint representative of U. In the same Lie algebra basis we have the expansion $[\mathcal{D}(U)\omega]_{x,i} = t^a [\mathcal{D}(U)\omega]^a_{x,i}$. This gives the explicit expression for the lattice gauge-covariant derivative in the site basis

$$[\mathcal{D}(U)\omega]^a_{x,i} = R^{ab}(U)\omega^b_{x+\hat{i}} - \omega^a_x. \tag{75}$$

By analogy with the continuum formula for an infinitesimal gauge transformation ${}^g A = A - D(A)\omega$, we define the lattice gauge-covariant derivative in the *link* basis $D\omega$ as the first order change in the link coordinates induced by an infinitesimal gauge transformation

$$d\theta^\alpha_{x,i} \equiv [D(\theta)\omega]^\alpha_{x,i} \,. \tag{76}$$

The two quantities are related by use of the Maurer-Cartan differential

$$t^a \psi^a_\alpha(\theta_{x,i}) d\theta^\alpha_{x,i} = U_{x,i} dU^\dagger_{x,i} = t^a [\mathcal{D}(U)\omega]^a_{x,i} \tag{77}$$

which gives

$$\psi^a_\alpha(\theta_{x,i}) [D(\theta)\omega]^\alpha_{x,i} = [\mathcal{D}(U)\omega]^a_{x,i} \,. \tag{78}$$

The last formula may be inverted using $J^\beta_a(\theta_{x,i})\psi^a_\alpha(\theta_{x,i}) = \delta^\beta_\alpha$,

$$[D(\theta)\omega]^\alpha_{x,i} = J^\alpha_a(\theta_{xi}) [\mathcal{D}(\theta)\omega]^a_{x,i} \,, \tag{79}$$

and we obtain for the lattice gauge-covariant derivative in the link basis the explicit formula, valid in any coordinate system

$$[D(\theta)\omega]^\alpha_{x,i} = J^\alpha_a(\theta_{xi}) [R^{ab}(\theta)\omega^b_{x+\hat{i}} - \omega^a_x] \,, \tag{80}$$

where we have written $R^{ab}(\theta) = R^{ab}(U(\theta))$.

Hamiltonian

We shall now derive the lattice Coulomb-gauge Hamiltonian $H_{\rm coul}$ in any coordinate system with gauge condition $\sum_{x,i}(A_{x,i} - A_{x-\hat{i}})$. We insert the Faddeev-Popov identity into the formula for the inner-product (Φ, Ψ). Because the Haar measure $dU_{x,i} = \det \psi(A_{x,i}) dA_{x,i}$ is invariant under left and right group multiplication, it is invariant under local gauge transformation. Consequently we obtain, just as in the continuum theory,

$$(\Phi, \Psi) = N \int_\Lambda \prod_{x,i} dA_{x,i} \psi(A_{x,i}) \prod_x{}' \delta[(\nabla^* A)_x] \det M'(A) \Phi^*(A) \Psi(A), \tag{81}$$

where N is the finite volume of the local gauge group, and we have written ψ for $\det \psi$. The primed product $\prod_x{}'$ extends over all sites but one, as explained in the preceding section, and correspondingly for M'.

The gauge condition $\nabla^* A = 0$ may be solved by a fourier decomposition of $A_{x,i}$ on a finite, periodic, cubic lattice, using longitudinal and transverse polarization vectors for finite k, and including $k = 0$ or harmonic modes. This allows to write

$$A_{x,i,\alpha} = A^{\rm Tr}_{x,i,\alpha} - (\nabla \sigma_\alpha)_{x,i}, \tag{82}$$

where $(\nabla^* A^{\rm Tr}_\alpha)_x = 0$. Here $A^{\rm Tr}$ includes both transverse and harmonic modes, $A^{\rm Tr} = A^{\rm tr} + A^{\rm h}$. The integral over link variables which represents the inner product becomes an integral over fourier coefficients. Because of $\delta(\nabla^* A)$, only the transverse and harmonic modes survive. We express the result as

$$(\Phi, \Psi) = N \int_\Lambda dA^{\rm Tr} \prod_{x,i} \psi(A^{\rm Tr}_{x,i}) \det M'(A^{\rm Tr}) \Phi^*(A^{\rm Tr}) \Psi(A^{\rm Tr}), \tag{83}$$

where $dA^{\rm Tr}$ designates the integral over all transverse and harmonic modes.

We wish to express the Hamiltonian $H_{KS} = T + V$ as an operator on the reduced space of gauge orbits parametrized by $A^{\rm Tr} \in \Lambda$. The potential energy operator V acts in the larger space by multiplication by a gauge-invariant function $V({}^g A) = V(A)$. On the reduced space it acts simply by multiplication by $V = V(A^{\rm Tr})$.

The Lie generator J_a is symmetric with respect to Haar measure, so we may write the kinetic energy operator T in the Kogut-Susskind representation as the quadratic form

$$(\Phi, T\Psi) = \int \prod_{x,i} dU_{x,i} (g_0^2/2) \sum_{x,i,a} (\mathcal{E}_{x,i,a}\Phi)^* \, (\mathcal{E}_{x,i,a}\Psi). \tag{84}$$

Only first derivatives of the wave-function appear here, which will allow us to apply Gauss's law as in the continuum theory. Here we have written

$$\mathcal{E}_{x,i,a} \equiv J_{x,i,a} \tag{85}$$

in order to emphasize the analogy with the continuum electric field. The operator $\mathcal{E}_{x,i,a}$ represents the electric field in the site basis. We also introduce the electric field in the link basis

$$E_{x,i,\alpha} \equiv i\partial / \partial A^\alpha_{x,i}. \tag{86}$$

The two are related by

$$\mathcal{E}_{x,i,a} = J^\alpha_a(A_{x,i}) E_{x,i,\alpha}. \tag{87}$$

In this notation, Gauss's law constraint reads

$$(D(A)^* E)_x \Phi = 0. \tag{88}$$

The operator $\mathcal{E}_{x,i,a} = J_{x,i,a}$ transforms homogeneously under gauge transformation, so the integrand of the quadratic form T is gauge invariant. We insert the lattice Faddeev-Popov identity into this integral, and obtain

$$(\Phi, T\Psi) = N \int_\Lambda \prod_{x,i} dA_{x,i}\, \psi(A_{x,i}) \det M'(A)$$
$$\times \prod_x{}' \delta[(\nabla^* A)_x](g_0^2/2) \sum_{x,i,a} (\mathcal{E}_{x,i,a}\Phi)^* \, (\mathcal{E}_{x,i,a}\Psi). \tag{89}$$

We wish to use the Gauss law constraint to express $\mathcal{E}$ as a derivative operator that acts only on transverse variables. For this purpose we make a Fourier decomposition of the electric field $E_{x,i,a}$ on the periodic cubic lattice. For the $A^a_{x,i}$-field we write the fourier decomposition, as

$$A_X = \sum_K \phi_{X,K} \tilde{A}_K, \tag{90}$$

where X represents position and vector indices, and K represents wave vector and polarization indices. Here the $\phi_{X,K}$ form a real orthonormal basis, and the inversion reads

$$\tilde{A}_K = \sum_X \phi_{X,K} A_X. \tag{91}$$

We have

$$\frac{\partial}{\partial A_X} = \sum_K \frac{\partial \tilde{A}_K}{\partial A_X} \frac{\partial}{\partial \tilde{A}_K} = \sum_K \phi_{X,K} \frac{\partial}{\partial \tilde{A}_K}, \tag{92}$$

so the fourier decompositions of A_X and of $\frac{\partial}{\partial A_X}$ are the same. For the electric field E_X we make the decomposition onto the same basis

$$E_X = \sum_K \phi_{X,K} \tilde{E}_K. \tag{93}$$

Consequently, from

$$E_X = i \frac{\partial}{\partial A_X}, \tag{94}$$

we obtain

$$\tilde{E}_K = i \frac{\partial}{\partial \tilde{A}_K}. \tag{95}$$

We choose a polarization basis which is adapted to the decomposition into transverse and longitudinal and harmonic parts. Corresponding to the decomposition $A = A^{\text{Tr}} - \nabla\sigma$, we have

$$E = E^{\text{Tr}} - \nabla\Omega, \tag{96}$$

where $\nabla^* A^{\text{Tr}} = \nabla^* E^{\text{Tr}} = 0$, and we have introduced the lattice color-Coulomb potential operator Ω. It follows that if A^{Tr} has the fourier decomposition

$$A_X^{\text{Tr}} = \sum_{K \in T} \phi_{X,K} \tilde{A}_K, \tag{97}$$

then E^{Tr} has the fourier decomposition

$$E_X^{\text{Tr}} = i \sum_{K \in T} \phi_{X,K} \frac{\partial}{\partial \tilde{A}_K}, \tag{98}$$

where T is the set of transverse and harmonic modes K. This expression defines E^{Tr} as a differential operator that acts within the space of functions of A^{Tr}.

We shall now use the Gauss law constraint $(D^{*}E)_{x}\Phi = 0$ to determine the color-Coulomb potential $\Omega_{x}\Phi$ on the reduced subspace, parametrized by $A^{\mathrm{Tr}} \in \Lambda$. We write Gauss's law in the form

$$D^{*}(E^{\mathrm{Tr}} - \nabla\Omega)\Phi = 0, \tag{99}$$

which gives

$$D^{*}\nabla\,\Omega\,\Phi = \rho\,\Phi, \tag{100}$$

where

$$(\rho)_{x} \equiv (D^{*}E^{\mathrm{Tr}})_{x} = [(D-\nabla)^{*}E^{\mathrm{Tr}}]_{x}\,. \tag{101}$$

This operator represents the color-charge density carried by the dynamical degrees of freedom of the gluons. We shall only need to solve Gauss's law for $\Omega\Phi$ in the reduced space where the Faddeev-Popov matrix is symmetric in the minimal Coulomb gauge, $M(A^{\mathrm{Tr}}) = D^{*}(A^{\mathrm{Tr}})\nabla = \nabla^{*}D(A^{\mathrm{Tr}})$. Thus we have also

$$(\nabla^{*}D\,\Omega)_{x}\,\Phi = \rho_{x}\,\Phi. \tag{102}$$

Because $\sum_{x}(\nabla^{*}V)_{x} = 0$ for any link variable V, the last equation is consistent only if the total charge $Q \equiv \sum_{x}\rho_{x}$ of the state Φ vanishes. Thus for gauge-invariant states in the minimal Coulomb gauge we have the constraint on the reduced space

$$Q\Phi = 0. \tag{103}$$

In the preceding section we have seen that in the lattice Coulomb gauge $D^{*}(A^{\mathrm{Tr}})\nabla$ is also a strictly positive matrix for A^{Tr} in the interior of the fundamental modular region Λ. Consequently the equation for Ω always has a solution,

$$\Omega\,\Phi = (D^{*}\nabla)^{-1}\rho\,\Phi\,. \tag{104}$$

With $\rho = D^{*}(A^{\mathrm{Tr}})E^{\mathrm{Tr}}$, this expresses the color-Coulomb potential operator Ω on the reduced space as a derivative with respect to the components of A^{Tr}. We conclude that in the reduced space the color-electric field $E = E^{\mathrm{Tr}} - \nabla\Omega$ acts according to

$$E\Phi = E^{\mathrm{Tr}}\,\Phi - \nabla(D^{*}\nabla)^{-1}D^{*}E^{\mathrm{Tr}}\,\Phi, \tag{105}$$

where E^{Tr} is defined by its fourier expansion (98).

With this result we may express the kinetic energy operator T as a quadratic form on the reduced space,

$$(\Phi, T\Psi) = N\int_{\Lambda} dA^{\mathrm{Tr}} \prod_{x,i} \psi(A^{\mathrm{Tr}}_{x,i}) \det M'(A^{\mathrm{Tr}}) \times (g_0^2/2)\sum_{x,i,a}(\mathcal{E}_{x,i,a}\Phi)^{*}\,(\mathcal{E}_{x,i,a}\Psi), \tag{106}$$

where $\Phi = \Phi(A^{\mathrm{Tr}})$ and similarly for Ψ. Here $\mathcal{E}\Phi$ is defined by $\mathcal{E}_{x,i,a}\Phi \equiv J_{a}^{\alpha}(A^{\mathrm{Tr}}_{x,i})E_{x,i,\alpha}\Phi$, and $E\Phi$ is defined by the preceding equation. This completes the specification of the lattice Coulomb-gauge Hamiltonian, $H_{\mathrm{coul}} = T + V$.

CONCLUSION

We have derived by elementary methods the lattice Coulomb-gauge Hamilonian H_{coul} from the Kogut-Susskind Hamiltonian H_{KS}. The essential step was to write H_{KS}

as a quadratic form, involving only first derivatives, so Gauss's law $D^*E\Phi = 0$ could be used directly to solve for the color-Coulomb potential.

We have not addressed here invariance with respect to finite gauge transformations, which involves identification of points of the boundary of the fundamental region Λ, so that Gribov copies are avoided. These questions are addressed in[11], and have an important influence on the spectrum[1, 2, 3, 4].

However we believe that Gauss's law is essential to the phenomenon of confinement in QCD[5]. We have shown how, in Hamiltonian theory with lattice regularization, the imposition of gauge invariance in the form of Gauss's law leads to an instantaneous color-Coulomb potential in the lattice Coulomb-gauge Hamiltonian H_{coul}.

Acknowledgments

I wish to express my appreciation to the NATO Advanced Study Institute for its support, and to Pierre van Baal whose leadership and tireless efforts made the workshop a success. This research is supported in part by the National Science Foundation under grant No. PHY-9520978.

REFERENCES

1. M. Lüscher, Nucl. Phys. B219 (1983) 233.
2. R. E. Cutkosky, J. Math. Phys. 25 (1984) 939; R. E. Cutkosky and K. Wang, Phys. Rev. D37 (1988) 3024; R. E. Cutkosky, Czech J. Phys. 40 (1990) 252.
3. J. Koller and P. van Baal, Ann. Phys. 174 (1987) 288; J. Koller and P. van Baal, Nucl. Phys. B302 (1988) 1; P. van Baal, Acta Physica Pol. B20 (1989) 295; P. van Baal, Nucl. Phys. B351 (1991) 183; P. van Baal and N. D. Hari Dass, Nucl. Phys. B385 (1992) 185; P. van Baal and B. van den Heuvel, Nucl. Phys. B417 (1994) 215.
4. B. M. van den Heuvel, Non-perturbative phenomena in gauge theory on S^3, thesis, University of Leiden (1996).
5. D. Zwanziger, Isaac Newton Institute preprint NI97032.
6. L. Baulieu and D. Zwanziger, Renormalizable and Non-Covariant Gauges and Coulomb Gauge (in preparation).
7. N. Christ and T. D. Lee, Phys. Rev. D22 (1980) 939.
8. V. N. Gribov, Nucl. Phys. B 139 (1978) 1.
9. R. Friedberg, T. D. Lee, Y. Pang and H. C. Ren, Annals of Phys. 246 (1996) 381.
10. G. Dell'Antonio and D. Zwanziger, Commun. Math. Phys 138 (1991) 291.
11. D. Zwanziger, Nucl. Phys. B 485 (1997) 185 (hep-th/9603203).
12. J. Kogut and L. Susskind, Phys. Rev. D11 (1975) 395.

GRIBOV AMBIGUITIES AND THE FUNDAMENTAL DOMAIN

Pierre van Baal[1,2]

[1]Isaac Newton Institute for Mathematical Sciences
20 Clarkson Road, Cambridge CB3 0EH, UK
[2]Instituut-Lorentz for Theoretical Physics*, University of Leiden
P.O.Box 9506, NL-2300 RA Leiden, The Netherlands

DEDICATED TO THE MEMORY OF VLADIMIR GRIBOV

INTRODUCTION

Non-perturbative aspects are believed to play a crucial role in understanding the formation of a mass gap in the spectrum of excitations in a non-Abelian gauge theory, but despite much progress a simple explanation is still lacking. Here this problem will be addressed in a finite volume, where its size can be used as a control parameter, which is absent in infinite volumes. The starting point, as in solving a quantum mechanical problem, is finding a proper description of the classical potential and its minima.

Due to the non-Abelian nature of the theory not only the classical potential, but also the kinetic term in the Lagrangian or Hamiltonian, is more complicated. The latter is a manifestation of the non-trivial Riemannian geometry of the physical configuration space[1], formed by the set of gauge orbits $\mathcal{A}/\mathcal{G}$ ($\mathcal{A}$ is the collection of connections, $\mathcal{G}$ the group of local gauge transformations). Most frequently, coordinates of this orbit space are chosen by picking a representative gauge field on the orbit in a smooth and preferably unique way. It is by now well known that linear gauge conditions like the Landau or Coulomb gauge suffer from Gribov ambiguities[2]. The reason behind this is that topological obstructions prevent one from introducing affine coordinates[3] in a global way. In principle therefore, one can introduce different coordinate patches with transition functions to circumvent this problem[4]. One way to make this specific is to base the coordinate patches on the choice of a background gauge condition[5,6]. One could envisage to associate to each coordinate patch ghost fields and extend BRST symmetry to include fields with non-trivial "Grassmannian sections", although such a formulation is still in its infancy.

We will pursue, however, the issue of finding a fundamental domain for non-Abelian gauge theories[7] and its consequence for the glueball spectrum in intermediate volumes. The finite volume context allows us to make reliable statements on the non-perturbative

*Permanent address.

Confinement, Duality, and Nonperturbative Aspects of QCD
Edited by Pierre van Baal, Plenum Press, New York, 1998

contributions, because asymptotic freedom guarantees that at small volumes the effective coupling constant is small, such that high-momentum states can be treated perturbatively. Only the lowest (typically zero or near-zero momentum) states will be affected by non-perturbative corrections. We emphasize that it is essential that gauge invariance is implemented properly at all stages. We will describe the results mainly in the context of a Hamiltonian picture[8] with wave functionals on configuration space. Although rather cumbersome from a perturbative point of view, where the covariant path integral approach of Feynman is vastly superior, it provides more intuition on how to deal with non-perturbative contributions to observables that do not vanish in perturbation theory. An essential feature of the non-perturbative behaviour is that the wave functional spreads out in configuration space to become sensitive to its non-trivial geometry. If wave functionals are localised within regions much smaller than the inverse curvature of the field space, the curvature has no effect on the wave functionals. At the other extreme, if the configuration space has non-contractible circles, the wave functionals are drastically affected by the geometry, or topology, when their support extends over the entire circle. Instantons are of course the most important examples of this. Not only the vacuum energy is affected by these instantons, but also the low-lying glueball states and this is what we are after to describe accurately, albeit in sufficiently small volumes. The geometry of the finite volume, to be considered here, is the one of a three-torus[9, 5] and a three-sphere[10,11]. These lecture notes are an updated and extended version of ref [12].

COMPLETE GAUGE FIXING

An (almost) unique representative of the gauge orbit is found by minimising the L^2 norm of the vector potential along the gauge orbit[7, 13]

$$F_A(g) \equiv \|{}^gA\|^2 = -\int_M d^3x \ \mathrm{tr}\left(\left(g^{-1}A_ig + g^{-1}\partial_i g\right)^2\right), \qquad (1)$$

where the vector potential is taken anti-hermitian. Expanding around the minimum of eq. (1), writing $g(x) = \exp(X(x))$, one easily finds:

$$\|{}^gA\|^2 = \|A\|^2 + 2\int_M \mathrm{tr}\,(X\partial_iA_i) + \int_M \mathrm{tr}\,(X^\dagger FP(A)X)$$
$$+\frac{1}{3}\int_M \mathrm{tr}\,(X\,[[A_i,X],\partial_iX]) + \frac{1}{12}\int_M \mathrm{tr}\,([D_iX,X][\partial_iX,X]) + \mathcal{O}\left(X^5\right). \quad (2)$$

Where $FP(A)$ is the Faddeev-Popov operator $(\mathrm{ad}(A)X \equiv [A,X])$

$$FP(A) = -\partial_iD_i(A) = -\partial_i^2 - \partial_i\mathrm{ad}(A_i). \qquad (3)$$

At any local minimum the vector potential is therefore transverse, $\partial_iA_i = 0$, and $FP(A)$ is a positive operator. The set of all these vector potentials is by definition the Gribov region Ω. Using the fact that $FP(A)$ is linear in A, Ω is seen to be a convex subspace of the set of transverse connections Γ. Its boundary $\partial\Omega$ is called the Gribov horizon. At the Gribov horizon, the *lowest* eigenvalue of the Faddeev-Popov operator vanishes, and points on $\partial\Omega$ are hence associated with coordinate singularities. Any point on $\partial\Omega$ can be seen to have a finite distance to the origin of field space and in some cases even uniform bounds can be derived[14, 15].

The Gribov region is the set of *local* minima of the norm functional (3) and needs to be further restricted to the *absolute* minima to form a fundamental domain, which will be denoted by Λ. The fundamental domain is clearly contained within the Gribov region. To show that also Λ is convex we note that

$$\begin{aligned} \|{}^gA\|^2 - \|A\|^2 &= \int \operatorname{tr}\left(A_i^2\right) - \int \operatorname{tr}\left(\left(g^{-1}A_i g + g^{-1}\partial_i g\right)^2\right) \\ &= \int \operatorname{tr}\left(g^\dagger FP_f(A)\, g\right) \equiv \langle g, FP_f(A)\, g\rangle, \end{aligned} \tag{4}$$

where $FP_f(A)$ is the Faddeev-Popov operator generalised to the fundamental representation. At the critical points $A \in \Gamma$ of the norm functional (recall $\Gamma = \{A \in \mathcal{A} | \partial_i A_i = 0\}$) $FP_f(A)$, like $FP(A)$, is a hermitian operator. We can define Λ in terms of the absolute minima over $g \in \mathcal{G}$ of $\langle g, FP_f(A)\, g\rangle$

$$\Lambda = \{A \in \Gamma | \min_{g \in \mathcal{G}} \langle g, FP_f(A)\, g\rangle = 0\}. \tag{5}$$

Using that $FP_f(A)$ is linear in A and assuming that $A^{(1)}$ and $A^{(2)}$ are in Λ and therefore satisfy the equation $\min_{g \in \mathcal{G}} \langle g, FP_f(A)\, g\rangle = 0$, we find that $A = sA^{(1)} + (1-s)A^{(2)}$ satisfies the same identity for all $s \in [0,1]$ (such that both s and $(1-s)$ are positive). The line connecting two points in Λ, therefore lies within Λ.

If we would not specify anything further, as a convex space is contractible, the fundamental region could never reproduce the non-trivial topology of the configuration space. This means that Λ should have a boundary[16]. Indeed, as Λ is contained in Ω, this means Λ is also bounded in each direction. Clearly $A = 0$ is in the interior of Λ, which allows us to consider a ray extending out from the origin into a given direction, where it will have to cross the boundary of Λ and Ω. For any point along this ray in Λ, the norm functional is at its absolute minimum as a function of the gauge orbit. However, for points in Ω that are not also in Λ, the norm functional is necessarily at a relative minimum. The absolute minimum for this orbit is an element of Λ, but in general not along the ray. Continuity therefore tells us that at some point along the ray, this absolute minimum has to pass the local minimum. At the point they are exactly degenerate, there are two gauge equivalent vector potentials with the same norm, both at the absolute minimum. As in the interior the norm functional has a unique minimum, again by continuity, these two degenerate configurations have to both lie on the boundary of Λ. This is the generic situation.

It is important to note that $A = 0$ is a so-called reducible connection[17] which has a non-trivial stabiliser, i.e. a subgroup of the gauge group that leaves the connection invariant, in this case the set of constant gauge transformations. These reducible connections give rise to so-called orbifold singularities, that manifest themselves through curvature singularities. We know perfectly well how to deal with it by not fixing the constant gauge transformations. Indeed the norm functional is degenerate along the constant gauge transformations and the Coulomb gauge does not fix this gauge degree of freedom. We simply demand that the wave functional is in the singlet representation under the constant gauge transformations. The degeneracy of the norm functional along the constant gauge transformations gives rise to a trivial part in the kernel of the Faddeev-Popov operator (also directly seen by inspection, since it vanishes when acting on constant Lie-algebra elements that generate the constant gauge transformations). This trivial kernel is to be projected out, something that is possible in a finite volume. In this formalism there is a remnant gauge invariance G, which requires no gauge fixing since its volume is finite, but still needs to be divided out to get the proper identification

$$\Lambda/G = \mathcal{A}/\mathcal{G}. \tag{6}$$

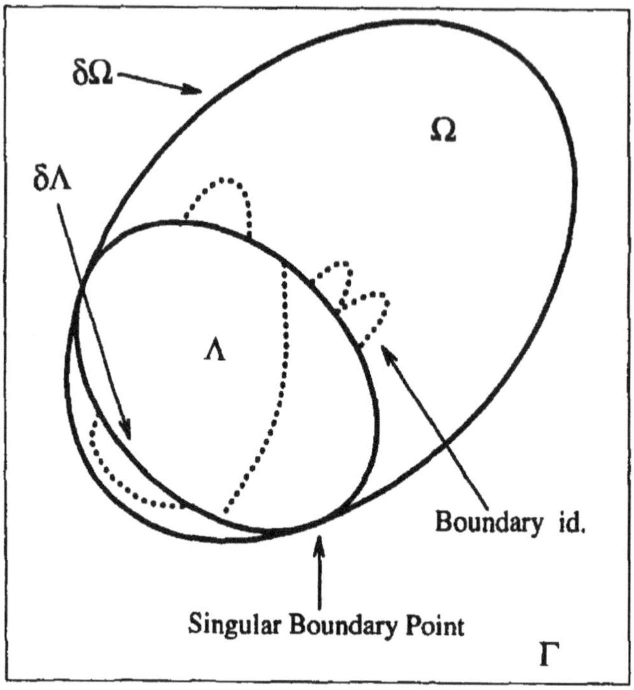

Figure 1. Sketch of the Gribov and fundamental regions. The dotted lines indicate the boundary identifications.

Here Λ is assumed to include the non-trivial boundary identifications. It is these boundary identifications that restore the non-trivial topology of $\mathcal{A}/\mathcal{G}$.

If the degeneracy at the boundary is continuous along non-trivial directions one necessarily has at least one non-trivial zero eigenvalue for *FP*(*A*) and the Gribov horizon will touch the boundary of the fundamental domain at these so-called singular boundary points. We sketch the general situation in figure 1. By singular we mean here a coordinate singularity. In principle, by choosing a different gauge fixing in the neighbourhood of these points one could resolve the singularity. If singular boundary points would not exist, all that would have been required is to complement the Hamiltonian in the Coulomb gauge with the appropriate boundary conditions in field space. Since the boundary identifications are by gauge transformations the boundary condition on the wave functionals is simply that they are identical under the boundary identifications, possibly up to a phase in case the gauge transformation is homotopically non-trivial, as will be discussed further on in more detail.

Unfortunately, the existence of non-contractible spheres[3] in the configuration space allows one to argue that singular boundary points are to be expected[16]. Consider the intersection of a *d*-dimensional non-contractible sphere with Λ. The part in the interior of Λ is contractible and it can only become non-contractible through the boundary identifications. The simplest way this can occur[16] is if for the (*d*-1)-dimensional intersection with the boundary, all points are to be identified. It would imply degeneracy of the the norm functional on a (*d*-1)-dimensional subspace, leading to at least *d*-1 zero-modes for the Faddeev-Popov operator. The intersection with the boundary of Λ can, however, exist of more than one connected component. In the case of two such components one can make a non-contractible sphere by identifying points of the (*d*-1)-dimensional

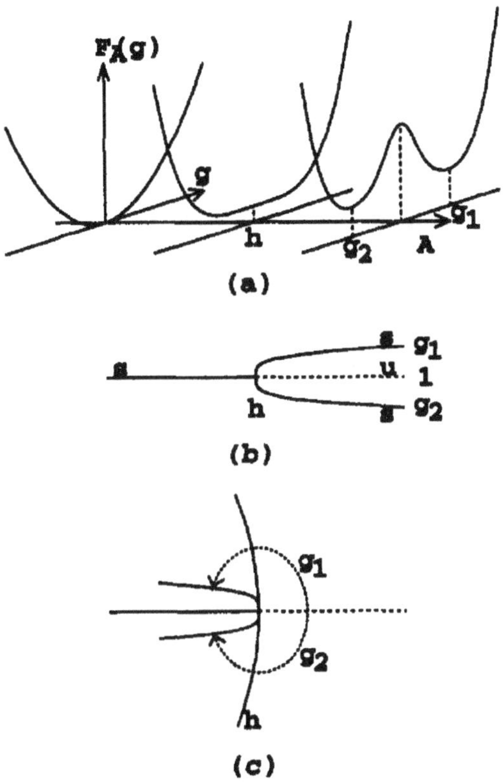

Figure 2. Sketch of a singular boundary point due to a bifurcation of the norm functional. It can be used to show that there are homotopically trivial gauge copies inside the Gribov horizon (h).

boundary intersection of the first connected component with that of the second and there is no necessity for a continuous degeneracy. Conversely, not all singular boundary points, even those associated with continuous degeneracies, need to be associated with non-contractible spheres.

When a singular boundary point is not associated to a continuous degeneracy, the norm functional undergoes a bifurcation moving from inside to outside the fundamental (and Gribov) region. The absolute minimum turns into a saddle point and two local minima appear, as indicated in figure 2. These are necessarily gauge copies of each other. The gauge transformation is homotopically trivial as it reduces to the identity at the bifurcation point, evolving continuously from there on. For reducible connections, that have a non-trivial stabiliser, this argument may be false[18], but examples of bifurcations at irreducible connections were explicitly found for S^3, see ref.[19] (app. A). We will come back to this.

Also Gribov's original arguments for the existence of gauge copies[2] (showing that points just outside the horizon are gauge copies of points just inside) can be easily understood from the perspective of bifurcations in the norm functional. It describes the generic case where the zero-mode of the Faddeev-Popov operator arises because of the coalescence of a local minimum with a saddle point with only one unstable direction. At the Gribov horizon the norm functional locally behaves in that case as X^3, with X the relevant zero eigenfunction of the Faddeev-Popov operator. The situation sketched in figure 2 corresponds to the case where the leading behaviour is like X^4. See ref.[16] for more details and a discussion of the Morse theory aspects that simplify the bifurcation analysis. As the Gribov region is associated with the local minima, and since the space of gauge transformations resembles that of a spin model, the analogy with spin glasses makes it unreasonable to expect that the Gribov region is free of further gauge copies. This will be illustrated by explicit examples. Unfortunately restrictions to a subset of the transverse gauge fields is a rather non-local procedure. This cannot be avoided since it reflects the non-trivial topology of field space.

GAUGE FIELDS ON THE THREE-TORUS

Homotopical non-trivial gauge transformations are in one to one correspondence with non-contractible loops in configuration space, which give rise to conserved quantum numbers. The quantum numbers are like the Bloch momenta in a periodic potential and label representations of the homotopy group of gauge transformations. On the fundamental domain the non-contractible loops arise through identifications of boundary points (as will be demonstrated quite explicitly for the torus in the zero-momentum sector). Although slightly more hidden, the fundamental domain will therefore contain all the information relevant for the topological quantum numbers. Sufficiently accurate knowledge of the boundary identifications will allow for an efficient and natural projection on the various superselection sectors (i.e. by choosing the appropriate "Bloch wave functionals"). All these features were at the heart of the finite volume analysis on the torus[5] and we see that they can in principle be naturally extended to the full theory, thereby including the desired θ dependence. In the next section this will be discussed in the context of the three-sphere. In ref.[6] we proposed formulating the Hamiltonian theory on coordinate patches, with homotopically non-trivial gauge transformations as transition functions. Working with boundary conditions on the boundary of the fundamental domain is easily seen to be equivalent and conceptually much simpler to formulate. If there would be no singular boundary points this would have provided a Hamiltonian formulation where all topologically non-trivial information can be encoded in the boundary conditions. Still, for the low-lying states in a finite volume, both on the three-torus and the three-sphere, singular boundary points will not play an important role in intermediate volumes.

Probably the most simple example to illustrate the relevance of the fundamental domain is provided by gauge fields on the torus in the abelian zero-momentum sector. For definiteness let us take $G = SU(2)$ and $A_i = i\frac{C_i}{2L}\tau_3$ (L is the size of the torus). These modes are dynamically motivated as they form the set of gauge fields on which the classical potential vanishes. It is called the vacuum valley (sometimes also referred to as toron valley) and one can attempt to perform a Born-Oppenheimer-like approximation for deriving an effective Hamiltonian in terms of these "slow" degrees of freedom. To find the Gribov horizon, one easily verifies that the part of the spectrum for $FP(A)$ that depends on $\vec{C}$, is given by $\lambda_{\vec{n}}^{gh}(\vec{C}) = 2\pi\vec{n}\cdot(2\pi\vec{n}\pm\vec{C})$, with $\vec{n}\neq\vec{0}$ an integer vector. The lowest eigenvalue therefore vanishes if $C_k = \pm 2\pi$. The Gribov region is therefore a cube with sides of length 4π, centred at the origin, specified by $|C_k| \leq 2\pi$ for all k, see figure 3.

The gauge transformation $g_{(k)} = \exp(\pi i x_k \tau_3/L)$ maps C_k to $C_k + 2\pi$, leaving the other components of $\vec{C}$ untouched. As $g_{(k)}$ is anti-periodic it is homotopically non-trivial (they are 't Hooft's twisted gauge transformations[20]). We thus see explicitly that gauge copies occur inside Ω, but furthermore the naive vacuum $A = 0$ has (many) gauge copies under these shifts of 2π that lie on the Gribov horizon. It can actually be shown for the Coulomb gauge that for any three-manifold, any Gribov copy by a

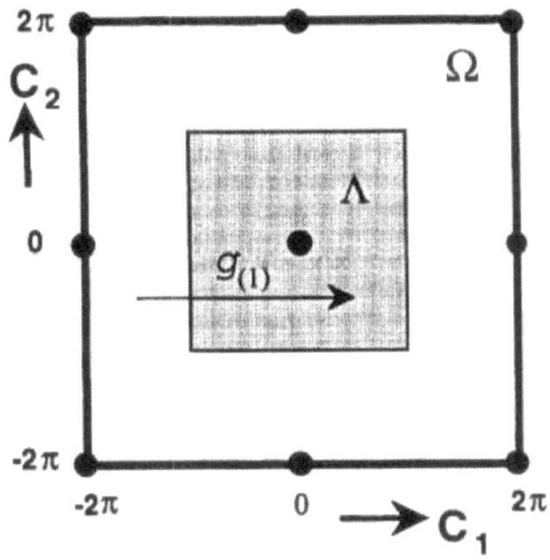

Figure 3. A two dimensional slice of the vacuum valley along the (C_1, C_2) plane. The fat square give the Gribov horizon, the grey square is the fundamental domain. The dots at the Gribov horizon are Gribov copies of the origin.

homotopically non-trivial gauge transformation of $A = 0$ will have vanishing Faddeev-Popov determinant[16]. Taking the symmetry under homotopically non-trivial gauge transformations properly into account is crucial for describing the non-perturbative dynamics and one sees that the singularity of the Hamiltonian at Gribov copies of $A = 0$, where the wave functionals are in a sense maximal, could form a severe obstacle in obtaining reliable results.

To find the boundary of the fundamental domain we note that the gauge copies $\vec{C} = (\pi, C_2, C_3)$ and $\vec{C} = (-\pi, C_2, C_3)$ have equal norm. The boundary of the fundamental domain, restricted to the vacuum valley formed by the abelian zero-momentum gauge fields, therefore occurs where $|C_k| = \pi$, well inside the Gribov region, see figure 3. The boundary identifications are by the homotopically non-trivial gauge transformations $g_{(k)}$. The fundamental domain, described by $|C_k| \leq \pi$, with all boundary points regular, has the topology of a torus. To be more precise, as the remnant of the constant gauge transformations (the Weyl group) changes $\vec{C}$ to $-\vec{C}$, the fundamental domain Λ/G restricted to the abelian constant modes is the orbifold T^3/Z_2. Generalisations to arbitrary gauge groups were considered in ref.[6]. (The fundamental domain turns out to coincide with the unit cell or "minimal" coordinate patch defined in ref.[6]).

Formulating the Hamiltonian on Λ, with the boundary identifications implied by the gauge transformations $g_{(k)}$, avoids the singularities at the Gribov copies of $A = 0$. "Bloch momenta" associated to the 2π shift, implemented by the non-trivial homotopy of $g_{(k)}$, label 't Hooft's electric flux quantum numbers[20] $\Psi(C_k = -\pi) = \exp(\pi i e_k)\Psi(C_k = \pi)$. Note that the phase factor is not arbitrary, but ± 1. This is because $g^2_{(k)}$ is homotopically trivial. In other words, the homotopy group of these anti-periodic gauge transformations is Z_2^3. Considering a slice of Λ can obscure some of the topological features. A loop that winds around the slice twice is contractible in Λ as soon as it is allowed to leave the slice. Indeed including the lowest modes transverse to this slice will make the Z_2 nature of the relevant homotopy group evident[5]. It should be mentioned that for the torus in the presence of fields in the fundamental representation (quarks), only periodic gauge transformations are allowed. In that case it is easily seen that the intersection of the fundamental domain with the constant abelian gauge fields is given by the domain $|C_k| \leq 2\pi$, whose boundary coincides with the Gribov horizon.

It is interesting to note that points on $\partial\Omega$ form an explicit example of a continuous degeneracy due to a non-contractible sphere[15].

In weak coupling Lüscher[9] showed unambiguously that the wave functionals are localised around $A = 0$, that they are normalisable and that the spectrum is discrete. In this limit the spectrum is insensitive to the boundary identifications (giving rise to a degeneracy in the topological quantum numbers). This is manifested by a vanishing electric flux energy, defined by the difference in energy of a state with $|\vec{e}| = 1$ and the vacuum state with $\vec{e} = \vec{0}$. Although there is no classical potential barrier to achieve this suppression, it comes about by a quantum induced barrier, in strength down by two powers of the coupling constant. This gives a suppression[21] with a factor $\exp(-S/g)$ instead of the usual factor of $\exp(-8\pi^2/g^2)$ for instantons[22]. Here $S = 12.4637$ is the action computed from the effective potential. At stronger coupling the wave functional spreads out over the vacuum valley and the boundary conditions drastically change the spectrum[5]. At this point the energy of electric flux suddenly switches on.

Integrating out the non-zero momentum degrees of freedom, for which Bloch degenerate perturbation theory provides a rigorous framework[23, 9], one finds an effective Hamiltonian. Near $A = 0$, due to the quartic nature of the potential energy $(F_{ij}^a)^2$ for the zero-momentum modes (the derivatives vanish and the field strength is quadratic in the field), there is no separation in time scales between the abelian and non-abelian modes. Away from $A = 0$ one could further reduce the dynamics to one along the vacuum valley, but near the origin this would be a singular decomposition (the adiabatic approximation breaks down). However, as long as the coupling constant is not too large, the wave functional can be reduced to a wave function on the vacuum valley near $\partial\Lambda$ where the boundary conditions can be implemented. These boundary conditions are formulated in a manner that preserves the invariance under constant gauge transformation and the effective Hamiltonian is solved by Rayleigh-Ritz (providing *also* lower bounds from the second moment of the Hamiltonian). The influence of the boundary conditions on the low-lying glueball states is felt as soon as the volume is bigger than an inverse scalar glueball mass. We summarise below the ingredients that enter the calculations.

The effective Hamiltonian is expressed in terms of the coordinates c_i^a, where $i = \{1,2,3\}$ is the spatial index ($c_0 = 0$) and $a = \{1,2,3\}$ is the SU(2)-colour index. These coordinates are related to the zero-momentum gauge fields through $A_i^a(x) = c_i^a/L$. We note that the field strength is given by $F_{ij}^a = -\varepsilon_{abd}c_i^b c_j^d/L^2$ and we introduce the gauge-invariant "radial" coordinate $r_i = \sqrt{\sum_a c_i^a c_i^a}$. The latter will play a crucial role in specifying the boundary conditions. For dimensional reasons the effective Hamiltonian is proportional to $1/L$. It will furthermore depend on L through the renormalised coupling constant ($g(L)$) at the scale $\mu = 1/L$. To one-loop order one has (for small L) $g(L)^2 = 12\pi^2/[-11\ln(\Lambda_{MS}L)]$. One expresses the masses and the size of the finite volume in dimensionless quantities, like mass-ratios and the parameter $z = mL$. In this way, the explicit dependence of g on L is irrelevant. This is also the preferred way of comparing results obtained within different regularisation schemes (i.e. dimensional and lattice regularisation). The effective Hamiltonian is now given by

$$
\begin{aligned}
L \cdot H_{eff}(c) = {} & \frac{g^2}{2(1+\alpha_1 g^2)} \sum_{i,a} \frac{\partial^2}{\partial c_i^{a\,2}} + \frac{1}{4}\left(\frac{1}{g^2} + \alpha_2\right) \sum_{ij,a} F_{ij}^{a\,2} \\
& + \gamma_1 \sum_i r_i^2 + \gamma_2 \sum_i r_i^4 + \gamma_3 \sum_{i>j} r_i^2 r_j^2 + \gamma_4 \sum_i r_i^6 + \gamma_5 \sum_{i \neq j} r_i^2 r_j^4 + \gamma_6 \prod_i r_i^2 \\
& + \alpha_3 \sum_{ijk,a} r_i^2 F_{jk}^{a\,2} + \alpha_4 \sum_{ij,a} r_i^2 F_{ij}^{a\,2} + \alpha_5 \det{}^2 c.
\end{aligned} \tag{7}
$$

We have organised the terms according to the importance of their contributions, ignoring terms quartic in the momenta. The first line gives (when ignoring $\alpha_{1,2}$) the lowest order effective Hamiltonian, whose energy eigenvalues are $\mathcal{O}(g^{2/3})$, as can be seen by rescaling c with $g^{2/3}$. Thus, in a perturbative expansion $c = \mathcal{O}(g^{2/3})$. The second line includes the vacuum-valley effective potential (i.e. the part that does not vanish on the set of abelian configurations). These two lines are sufficient to obtain the mass-ratios to an accuracy of better than 5%. The third line gives terms of $\mathcal{O}(g^4)$ in the effective potential, that vanish along the vacuum-valley. The coefficients (to two-loop order for γ_i) are

$$
\begin{aligned}
\gamma_1 &= -3.0104661\cdot 10^{-1} - (g/2\pi)^2 3.0104661\cdot 10^{-1}\ , \ \alpha_1 = +2.1810429\cdot 10^{-2},\\
\gamma_2 &= -1.4488847\cdot 10^{-3} - (g/2\pi)^2 9.9096768\cdot 10^{-3}\ , \ \alpha_2 = +7.5714590\cdot 10^{-3},\\
\gamma_3 &= +1.2790086\cdot 10^{-2} + (g/2\pi)^2 3.6765224\cdot 10^{-2}\ , \ \alpha_3 = +1.1130266\cdot 10^{-4},\\
\gamma_4 &= +4.9676959\cdot 10^{-5} + (g/2\pi)^2 5.2925358\cdot 10^{-5}\ , \ \alpha_4 = -2.1475176\cdot 10^{-4},\\
\gamma_5 &= -5.5172502\cdot 10^{-5} + (g/2\pi)^2 1.8496841\cdot 10^{-4}\ , \ \alpha_5 = -1.2775652\cdot 10^{-3},\\
\gamma_6 &= -1.2423581\cdot 10^{-3} - (g/2\pi)^2 5.7110724\cdot 10^{-3}\ .
\end{aligned}
\tag{8}
$$

The choice of boundary conditions, associated to each of the irreducible representations of the cubic group $O(3,\mathbb{Z})$ and to the states that carry electric flux[20], is best described by observing that the cubic group is the semidirect product of the group of coordinate permutations S_3 and the group of coordinate reflections Z_2^3. We denote the parity under the coordinate reflection $c_i^a \to -c_i^a$ by $p_i = \pm 1$ (*i* fixed). The electric flux quantum number for the same direction will be denoted by $q_i = \pm 1$. This is related to the more usual additive (mod 2) quantum number e_i by $q_j = \exp(i\pi e_j)$, Note that for SU(2) electric flux is invariant under coordinate reflections. If not all of the electric fluxes are identical, the cubic group is broken to $S_2 \times Z_2^3$, where $S_2(\sim Z_2)$ corresponds to interchanging the two directions with identical electric flux (unequal to the other electric flux). If all the electric fluxes are equal, the wave functions are irreducible representations of the cubic group. These are the four singlets $A^{\pm}_{1(2)}$, which are completely (anti-)symmetric with respect to S_3 and have each of the parities $p_i = \pm 1$. Then there are two doublets $E^{\pm}$, also with each of the parities $p_i = \pm 1$ and finally one has four triplets $T^{\pm}_{1(2)}$. Each of these triplet states can be decomposed into eigenstates of the coordinate reflections. Explicitly, for $T^{\pm}_{1(2)}$ we have one state that is (anti-)symmetric under interchanging the two- and three-directions, with $p_2 = p_3 = -p_1 = \mp 1$. The other two states are obtained through cyclic permutation of the coordinates. Thus, any eigenfunction of the effective Hamiltonian with specific electric flux quantum numbers q_i can be chosen to be an eigenstate of the parity operators p_i. The boundary conditions of these eigenfunctions $\Psi_{\vec{q},\vec{p}}(c)$ are simply given by

$$
\begin{aligned}
\Psi_{\vec{q},\vec{p}}(c)|_{r_i=\pi} &= 0 \ , \quad if \quad p_i q_i = -1\\
\frac{\partial}{\partial r_i}(r_i \Psi_{\vec{q},\vec{p}}(c))|_{r_i=\pi} &= 0 \ , \quad if \quad p_i q_i = +1
\end{aligned}
\tag{9}
$$

and one easily shows that with these boundary conditions the Hamiltonian is hermitian with respect to the innerproduct $<\Psi,\Psi'> = \int_{r_i\leq\pi} d^9c\Psi^*(c)\Psi'(c)$. For negative parity states ($\prod_i p_i = -1$) this description is, however, not accurate[24] as parity restricted to the vacuum valley is equivalent to a Weyl reflection (a remnant of the invariance under constant gauge transformations).

After correcting for lattice artefacts[25], the (semi-)analytic results agree extremely well with the best lattice data[26] (with statistical errors of 2% to 3%) up to a volume

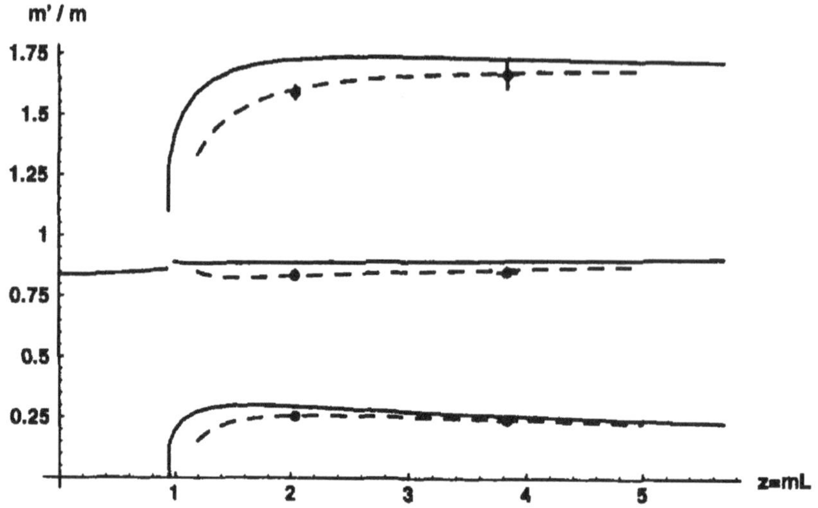

Figure 4. Mass ratios m'/m as a function of L in units of the inverse scalar glueball mass m^{-1}, $z = mL$. The analytic results are given by the full (continuum) and dashed (4^3 lattice) curves. Represented are the the square root of the energy of electric flux per unit length (bottom), the E tensor mass (middle) and the T_2 tensor mass (top). The latter two are (almost) degenerate below $z = 0.95$, which is also where the electric flux energy is exponentially suppressed. Where no error bars on the Monte Carlo data are visible they are smaller than the size of the data points.

of about .75 fermi, or about five times the inverse scalar glueball mass. In figure 4 we present the comparison for a lattice of spatial size 4^3. Monte Carlo data[26] are most accurate for this lattice size. For more detailed comparisons see ref.[25]. The analytic results below $z = 0.95$ are due to Lüscher and Münster[27], which is where the spectrum is insensitive to the identifications at the boundary of Λ.

Most conspicuously the tensor state in finite volumes is split in a doublet E, with a mass that is roughly 0.9 times the scalar A_1 mass and a triplet T_2 with a mass of roughly 1.7 times the scalar mass. Note that the multiplicity weighted average is approximately 1.4 times the scalar mass, agreeing well with what was found at large volumes from lattice data[26].

Apart from the corrections for the lattice artefacts, generalisation to SU(3) was established by Vohwinkel[28], with qualitatively similar results. In large volumes the rotational symmetry should be restored, as is observed from lattice simulations. The properties of the fundamental domain restricted to the zero-momentum modes for $SU(3)$ can be read off from the results in ref.[6]. In this reference also the generalisation to arbitrary gauge groups is discussed.

At large volumes extra degrees of freedom start to behave non-perturbatively. To demonstrate this, the minimal barrier height that separates two vacuum valleys that are related by gauge transformations with non-trivial winding number

$$\nu(g) = \frac{1}{24\pi^2} \int_M \mathrm{Tr}\left((g^{-1}dg)^3\right), \tag{10}$$

was found to be $72.605/Lg^2$, using the lattice approximation and carefully taking the continuum limit[29]. As long as the states under consideration have energies below this value, the transitions over this barrier can be neglected and the zero-momentum effective Hamiltonian provides an accurate description. One can now easily find for which

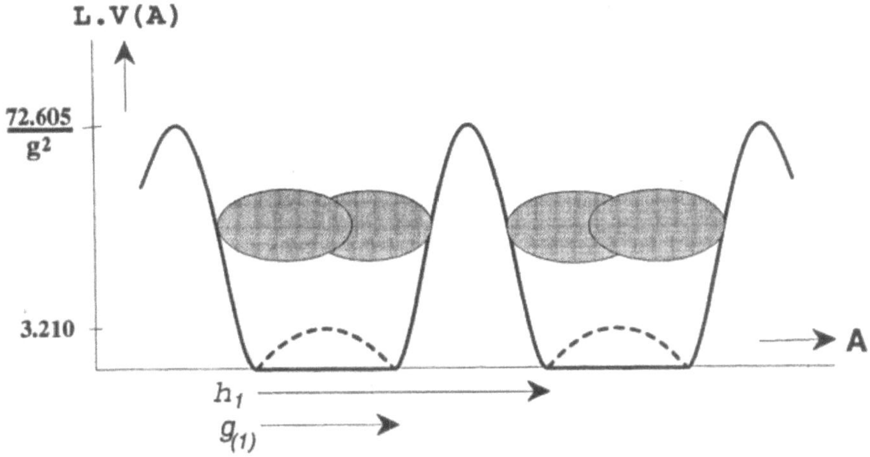

Figure 5. "Artistic impression" of the potential for the three-torus. Shown are two vacuum valleys, related to each other by a gauge transformation h_1 with winding number 1, whose shape in the direction perpendicular to the valley depends on the position along the valley. The induced one-loop effective potential, of height $3.210/L$, has degenerate minima related to each other by the anti-periodic gauge transformations $g_{(k)}$. The classical barrier, separating the two valleys, has a height $72.605/Lg^2$.

volume the energy of the level that determines the glueball mass (defined by the difference with the groundstate energy) starts to be of the order of this barrier height. This turns out to be the case for L roughly 5 to 6 times the correlation length set by the scalar glueball mass. The situation is sketched in figure 5.

We expect, as will be shown for the three-sphere, that the boundary of the fundamental domain along the path in field space across the barrier (which corresponds to the instanton path if we parametrise this path by Euclidean time t), occurs at the saddle point (which we call a finite volume sphaleron) in between the two minima. The degrees of freedom along this tunnelling path go outside of the space of zero-momentum gauge fields and if the energy of a state flows over the barrier, its wave functional will no longer be exponentially suppressed below the barrier and will in particular be non-negligible at the boundary of the fundamental domain. Boundary identifications in this direction of field space now become dynamically important too. The relevant "Bloch momentum" is in this case obviously the θ parameter, as wave functionals pick up a phase factor $e^{i\theta}$ under a gauge transformation with winding number one. For many of the intricacies in describing instantons on a torus we refer to ref.[30, 31, 32]. On the three-torus we have therefore achieved a self-contained picture of the low-lying glueball spectrum in intermediate volumes from first principles with *no free parameters*, apart from the overall scale.

GAUGE FIELDS ON THE THREE-SPHERE

The reason to consider the three-sphere lies in the fact that the conformal equivalence of $S^3 \times \mathbb{R}$ to $\mathbb{R}^4$ allows one to construct instantons explicitly[11, 33]. This greatly simplifies the study of how to formulate θ dependence in terms of boundary conditions on the fundamental domain, and indeed we will see that for S^3 simple enough results can be obtained to address this question[19, 34]. The disadvantage of the three-sphere is that in large volumes the corrections to the glueball masses are no longer exponential[9].

We will summarise the formalism that was developed in[11]. Alternative formulations, useful for diagonalising the Faddeev-Popov and fluctuation operators, were given in ref.[10]. We embed S^3 in $\mathbb{R}^4$ by considering the unit sphere parametrised by a unit vector n_μ. It is particularly useful to introduce the unit quaternions σ_μ and their conjugates $\bar{\sigma}_\mu = \sigma_\mu^\dagger$ by

$$\sigma_\mu = (id, i\vec{\tau}), \qquad \bar{\sigma}_\mu = (id, -i\vec{\tau}). \tag{11}$$

They satisfy the multiplication rules

$$\sigma_\mu \bar{\sigma}_\nu = \eta^\alpha_{\mu\nu}\sigma_\alpha, \qquad \bar{\sigma}_\mu \sigma_\nu = \bar{\eta}^\alpha_{\mu\nu}\sigma_\alpha, \tag{12}$$

where we used the 't Hooft η symbols[22], generalised slightly to include a component symmetric in μ and ν for $\alpha = 0$. We can use η and $\bar{\eta}$ to define orthonormal framings[35] of S^3, which were motivated by the particularly simple form of the instanton vector potentials in these framings. The framing for S^3 is obtained from the framing of $\mathbb{R}^4$ by restricting in the following equation the four-index α to a three-index a (for $\alpha = 0$ one obtains the normal on S^3):

$$e^\alpha_\mu = \eta^\alpha_{\mu\nu} n_\nu, \qquad \bar{e}^\alpha_\mu = \bar{\eta}^\alpha_{\mu\nu} n_\nu. \tag{13}$$

Note that e and $\bar{e}$ have opposite orientations. Each framing defines a differential operator and associated (mutually commuting) angular momentum operators $\vec{L}_1$ and $\vec{L}_2$:

$$\partial^i = e^i_\mu \frac{\partial}{\partial x^\mu}, \quad L_1^i = \frac{i}{2}\partial^i \quad , \quad \bar{\partial}^i = \bar{e}^i_\mu \frac{\partial}{\partial x^\mu}, \quad L_2^i = \frac{i}{2}\bar{\partial}^i. \tag{14}$$

It is easily seen that $\vec{L}_1^2 = \vec{L}_2^2$, which has eigenvalues $l(l+1)$, with $l = 0, \frac{1}{2}, 1, \cdots$.

The (anti-)instantons[36] in these framings, obtained from those on $\mathbb{R}^4$ by interpreting the radius in $\mathbb{R}^4$ as the exponential of the time t in the geometry $S^3 \times \mathbb{R}$, become

$$A_0 = \frac{\vec{\varepsilon}\cdot\vec{\sigma}}{2(1+\varepsilon\cdot n)}, \qquad \vec{A} = \frac{\vec{\sigma}\wedge\vec{\varepsilon} - (u+\varepsilon\cdot n)\vec{\sigma}}{2(1+\varepsilon\cdot n)}, \tag{15}$$

where

$$u = \frac{2s^2}{1+b^2+s^2}, \qquad \varepsilon_\mu = \frac{2sb_\mu}{1+b^2+s^2}, \qquad s = \lambda e^t. \tag{16}$$

Here $\vec{\varepsilon}$ and $\vec{A}$ are defined with respect to the framing e^a_μ for instantons and with respect to the framing $\bar{e}^a_\mu$ for anti-instantons. The instanton describes tunnelling from $A = 0$ at $t = -\infty$ to $A_a = -\sigma_a$ at $t = \infty$, over a potential barrier at $t = 0$ that is lowest when $b_\mu \equiv 0$. This configuration corresponds to a sphaleron[37], i.e. the vector potential $A_a = -\frac{1}{2}\sigma_a$ is a saddle point of the energy functional with one unstable mode, corresponding to the direction (u) of tunnelling. At $t = \infty$, $A_a = -\sigma_a$ has zero energy and is a gauge copy of $A_a = 0$ by a gauge transformation $g = n\cdot\bar{\sigma}$ with winding number one.

We will be concentrating our attention to the modes that are degenerate in energy to lowest order with the modes that describe tunnelling through the sphaleron and "anti-sphaleron". The latter is a gauge copy by a gauge transformation $g = n\cdot\sigma$ with winding number –1 of the sphaleron. The two dimensional space containing the tunnelling paths through these sphalerons is consequently parametrised by u and v through

$$A_\mu(u,v) = \left(-ue^a_\mu - v\bar{e}^a_\mu\right)\frac{\sigma_a}{2}. \tag{17}$$

The gauge transformation with winding number −1 is easily seen to map $(u,v)=(w,0)$ into $(u,v)=(0,2-w)$. The 18 dimensional space is defined by

$$A_\mu(c,d)=\left(c_i^a e_\mu^i + d_j^a \bar{e}_\mu^j\right)\frac{\sigma_a}{2}=A_i(c,d)e_\mu^i. \tag{18}$$

The c and d modes are mutually orthogonal and satisfy the Coulomb gauge condition:

$$\partial_i A_i(c,d)=0. \tag{19}$$

This space contains the (u,v) plane through $c_i^a=-u\delta_i^a$ and $d_i^a=-v\delta_i^a$. The significance of this 18 dimensional space is that the energy functional[11]

$$\mathcal{V}(c,d)\equiv-\int_{S^3}\frac{1}{2}\operatorname{tr}(F_{ij}^2)=\mathcal{V}(c)+\mathcal{V}(d)+\frac{2\pi^2}{3}\left\{(c_i^a)^2(d_j^b)^2-(c_i^a d_j^a)^2\right\}, \tag{20}$$

$$\mathcal{V}(c)=2\pi^2\left\{2(c_i^a)^2+6\det c+\frac{1}{4}[(c_i^a c_i^a)^2-(c_i^a c_j^a)^2]\right\}, \tag{21}$$

is degenerate to second order in c and d. Indeed, the quadratic fluctuation operator $\mathcal{M}$ in the Coulomb gauge, defined by

$$-\int_{S^3}\frac{1}{2}\operatorname{tr}(F_{ij}^2)=\int_{S^3}\operatorname{tr}(A_i\mathcal{M}_{ij}A_j)+\mathcal{O}\left(A^3\right),$$
$$\mathcal{M}_{ij}=2\vec{L}_1^2\delta_{ij}+2\left(\vec{L}_1+\vec{S}\right)_{ij}^2,\quad S_{ij}^a=-i\varepsilon_{aij}, \tag{22}$$

has $A(c,d)$ as its eigenspace for the (lowest) eigenvalue 4. These modes are consequently the equivalent of the zero-momentum modes on the torus, with the difference that their zero-point frequency does not vanish.

$FP_f(A)$ in eq. (4) is defined as a hermitian operator acting on the vector space $\mathcal{L}$ of functions g over S^3 with values in the space of the quaternions $\mathbb{H}=\{q_\mu\sigma_\mu|q_\mu\in\mathbb{R}\}$. The gauge group $\mathcal{G}$ is contained in $\mathcal{L}$ by restricting to the unit quaternions: $\mathcal{G}=\{g\in\mathcal{L}|g=g_\mu\sigma_\mu, g_\mu\in\mathbb{R}, g_\mu g_\mu=1\}$. For arbitrary gauge groups $\mathcal{L}$ is defined as the algebra generated by the identity and the (anti-hermitian) generators of the algebra. When minimising the same functional over the larger space $\mathcal{L}$ one obviously should find a smaller space $\tilde{\Lambda}\subset\Lambda$. Since $\mathcal{L}$ is a linear space $\tilde{\Lambda}$ can also be specified by the condition that $FP_f(A)$ be positive,

$$\tilde{\Lambda}=\{A\in\Gamma|\langle g,FP_f(A)\,g\rangle\geq 0,\ \forall g\in\mathcal{L}\}. \tag{23}$$

As for the Gribov horizon, the boundary of $\tilde{\Lambda}$ is therefore determined by the location where the lowest eigenvalue vanishes. For the (c,d) space it can be shown[19] that the boundary $\partial\tilde{\Lambda}$ will touch the Gribov horizon $\partial\Omega$. This establishes the existence of singular points on the boundary of the fundamental domain due to the inclusion $\tilde{\Lambda}\subset\Lambda\subset\Omega$. By showing that the fourth order term in eq. (2) is positive (see app. A of ref.[19]) this is seen to correspond to the situation as sketched in figure 2.

To simplify the notation, write $FP_f=FP_{\frac{1}{2}}$ and $FP=FP_1$, with the indices related to the isospin. The associated generators are

$$\vec{T}_{\frac{1}{2}}=\tfrac{1}{2}\vec{\tau}\qquad\text{and}\qquad\vec{T}_1=\tfrac{1}{2}\mathrm{ad}\vec{\tau}. \tag{24}$$

One can now make convenient use of the $SU(2)^3$ symmetry generated by $\vec{L}_1$, $\vec{L}_2$ and $\vec{T}_t$ to calculate explicitly the spectrum of $FP_t(A)$. One has

$$FP_t(A(c,d))=4\vec{L}_1^2-\frac{2}{t}c_i^aT_t^aL_1^i-\frac{2}{t}d_i^aT_t^aL_2^i, \tag{25}$$

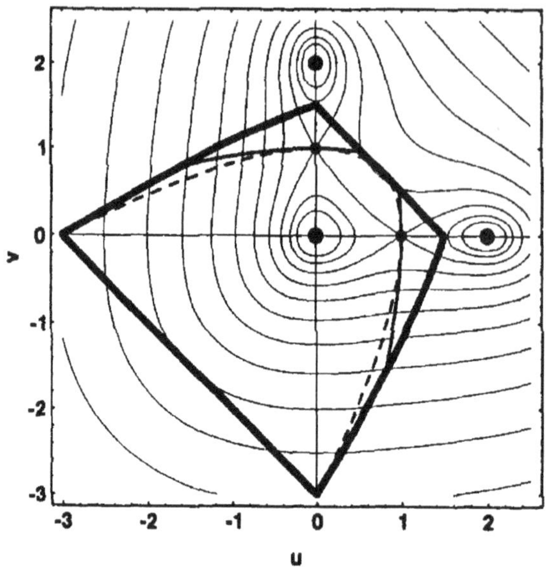

Figure 6. Location of the classical vacua (large dots), sphalerons (smaller dots), the Gribov horizon (fat sections), the boundary of $\tilde{\Lambda}$ (dashed curves) and part of the boundary of the fundamental domain (full curves). Also indicated are the lines of equal potential in units of 2^n times the sphaleron energy.

which commutes with $\vec{L}_1^2 = \vec{L}_2^2$, but for arbitrary (c, d) there are in general no other commuting operators (except for a charge conjugation symmetry when $t = \frac{1}{2}$). Restricting to the (u, v) plane one easily finds that

$$FP_t(A(u,v)) = 4\vec{L}_1^2 + \frac{2}{t}u\vec{L}_1 \cdot \vec{T}_t + \frac{2}{t}v\vec{L}_2 \cdot \vec{T}_t, \tag{26}$$

which also commutes with the total angular momentum $\vec{J}_t = \vec{L}_1 + \vec{L}_2 + \vec{T}_t$ and is easily diagonalized. Figure 6 summarises the results for this (u, v) plane and also shows the equal-potential lines as well as exhibiting the multiple vacua and the sphalerons. As it is easily seen that the two sphalerons are gauge copies (by a unit winding number gauge transformation) with equal norm, they lie on $\partial\Lambda$, which can be extend by perturbing around these sphalerons[38].

To obtain the result for general (c, d) one can use the invariance under rotations generated by $\vec{L}_1$ and $\vec{L}_2$ and under constant gauge transformations generated by $\vec{T}_t$, to bring c and d to a standard form, or express $\det\left(FP_t(A(c,d))|_{t=\frac{1}{2}}\right)$, which determines the locations of $\partial\Omega$ and $\partial\tilde{\Lambda}$, in terms of invariants. We define the matrices X and Y by $X_b^a = (cc^t)_b^a$ and $Y_b^a = (dd^t)_b^a$, which allows us to find

$$\begin{aligned}\det\left(FP_{\frac{1}{2}}(A(c,d))|_{t=\frac{1}{2}}\right) = [&81 - 18\,\mathrm{Tr}\,(X+Y) + 24(\det c + \det d) \\ &-(\mathrm{Tr}\,(X-Y))^2 + 2\,\mathrm{Tr}\,((X-Y)^2)]^2.\end{aligned} \tag{27}$$

The two-fold multiplicity is due to charge conjugation symmetry. The expression for $t = 1$, that determines the location of the Gribov horizon in the (c, d) space, is given in app. B of ref.[19]. If we restrict to $d = 0$ the result simplifies considerably. In that case one can bring c to a diagonal form $c_i^a = x_i\delta_i^a$. The rotational and gauge symmetry reduce to permutations of the x_i and simultaneous changes of the sign of two of the x_i.

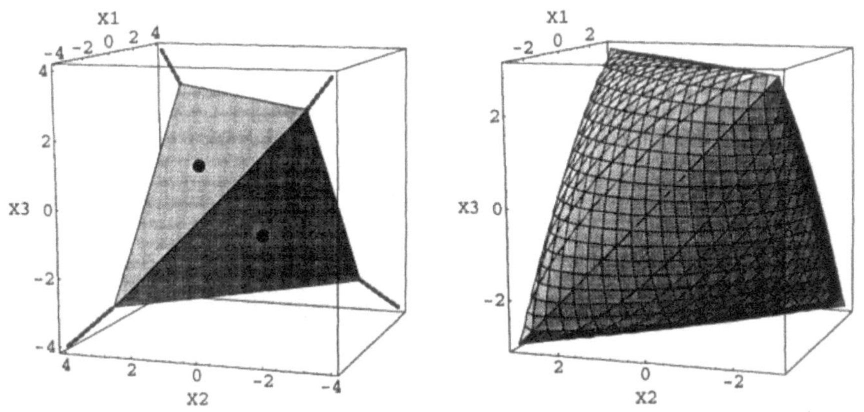

Figure 7. The fundamental domain (left) for constant gauge fields on S^3, with respect to the "instanton" framing e^a_μ, in the diagonal representation $A_a = x_a\sigma_a$ (no sum over a). By the dots on the faces we indicate the sphalerons, whereas the dashed lines represent the symmetry axes of the tetrahedron. To the right we display the Gribov horizon, which encloses the fundamental domain, coinciding with it at the singular boundary points along the edges of the tetrahedron.

Table 1. Numerical coefficients for $\mathcal{V}^{(1)}_{\text{eff}}$

$\kappa_1 = -0.2453459985179565$	$\kappa_2 = 3.66869179814223$
$\kappa_3 = 0.500703203096610$	$\kappa_4 = -0.839359633413003$
$\kappa_5 = -0.849965412245339$	$\kappa_6 = -0.06550330854836428$
$\kappa_7 = -0.3617122159967145$	$\kappa_8 = -2.295356861354712$

One easily finds the invariant expression ($\mathrm{Tr}\,(X) = \sum_i x_i^2$ and $\det c = \prod_i x_i$)

$$\det\left(FP_1(A(c,0))|_{l=\frac{1}{4}}\right) = (2\det c - 3\,\mathrm{Tr}\,(X) + 27)^4. \tag{28}$$

In figure 7 we present the results for Λ and Ω. In this particular case, where $d = 0$, Λ coincides with $\tilde{\Lambda}$, a consequence of the convexity and the fact that both the sphalerons (indicated by the dots) and the edges of the tetrahedron lie on $\partial\Lambda$, the latter also lying on $\partial\Omega$. It is essential that the sphalerons do not lie on the Gribov horizon and that the potential energy near $\partial\Omega$ is relatively high. This is why we can take the boundary identifications near the sphalerons into account without having to worry about singular boundary points, as long as the energies of the low-lying states will be not much higher than the energy of the sphaleron. It allows one to study the glueball spectrum as a function of the CP violating angle θ, but more importantly it incorporates for $\theta = 0$ the noticeable influence of the barrier crossings, i.e. of the instantons.

An effective Hamiltonian for the c and d modes is derived from the one-loop effective action[34]. To lowest order it is given by

$$H = -\frac{g^2(R)}{4\pi^2 R}\left(\left(\frac{\partial}{\partial c_i^a}\right)^2 + \left(\frac{\partial}{\partial d_i^a}\right)^2\right) + \frac{1}{g^2(R)R}\mathcal{V}(c,d) + \frac{1}{R}\mathcal{V}^{(1)}_{\text{eff}}(c,d), \tag{29}$$

where $g(R)$ *is* the running coupling constant (related to the MS running coupling by a finite renormalisation, such that kinetic term above has no corrections). The one-loop correction to the effective potential[34] is given by (for κ_i see table 1):

$$\mathcal{V}^{(1)}_{\text{eff}}(c,d) = \mathcal{V}^{(1)}_{\text{eff}}(c) + \mathcal{V}^{(1)}_{\text{eff}}(d) + \kappa_7(c_i^a)^2(d_j^b)^2 + \kappa_8(c_i^a d_j^a)^2, \tag{30}$$
$$\mathcal{V}^{(1)}_{\text{eff}}(c) = \kappa_1(c_i^a)^2 + \kappa_2\det c + \kappa_3(c_i^a c_i^a)^2 + \kappa_4(c_i^a c_j^a)^2 + \kappa_5(c_i^a)^2\det c + \kappa_6(c_i^a c_i^a)^3.$$

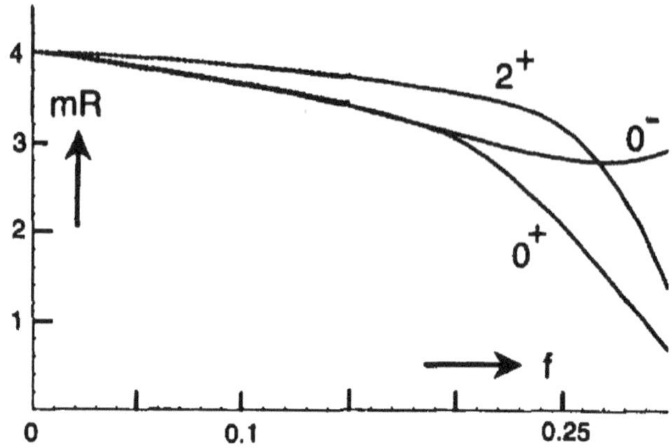

Figure 8. The full one-loop results for the masses of scalar, tensor and odd glueballs on S^3 as a function of $f = g^2(R)/2\pi^2$ for $\theta = 0$. The dashed lines correspond to the perturbative result.

Errors due to an adiabatic approximation are not necessarily suppressed by powers of the coupling constant. Nevertheless, one expects to achieve an approximate understanding of the non-perturbative dynamics in this way[34].

The boundary conditions are chosen so as to coincide with the appropriate boundary conditions near the sphalerons, but such that the gauge and (left and right) rotational invariances are not destroyed. Projections on the irreducible representations of these symmetries turned out to be essential to reduce the size of the matrices to be diagonalised in a Rayleigh-Ritz analysis. Remarkably all this could be implemented in a tractable way[34]. Results are summarised in figure 8. One of the most important features is that the 0^- glueball is (slightly) lighter than the 0^+ in perturbation theory, but when including the effects of the boundary of the fundamental domain, setting in at $f \equiv g^2/2\pi^2 \sim 0.2$, the $0^-/0^+$ mass ratio rapidly increases. Beyond $f \sim 0.28$ it can be shown that the wave functionals start to feel parts of the boundary of the fundamental domain which the present calculation is not representing properly[34]. This value of f corresponds to a circumference of roughly 1.3 fm, when setting the scale as for the torus, assuming the scalar glueball mass in both geometries at this intermediate volume to coincide.

DISCUSSION

We have analysed in detail the boundary of the fundamental domain for SU(2) gauge theories on the three-torus and three-sphere. It is important to note that it is necessary to divide $\mathcal{A}$ by the set of *all* gauge transformations, including those that are homotopically non-trivial, to get the physical configuration space. All the non-trivial topology is then retrieved by the identifications of points on the boundary of the fundamental domain.

As we already mentioned in the introduction, the knowledge of the boundary identifications is important in the case that the wave functionals spread out in configuration space to such an extent that they become sensitive to these identifications. This happens at large volumes, whereas at very small volumes the wave functional is localised around $A = 0$ and one need not worry about these non-perturbative effects. That these effects can be dramatic, even at relatively small volumes (above a tenth of a fermi across), was demonstrated for the case of the torus. However, for that case the structure of the fundamental domain (restricted to the abelian zero-energy modes) is a hypercube and deviates considerably from the fundamental domain of the three-sphere. Nevertheless, the spectrum for the sphere is compatible with that for the torus in volumes around one fermi across[39], with $m(2^+)/m(0^+) \sim 1.5$ and $m(0^-)/m(0^+) \sim 1.7$.

It should be noted that the shape of Λ is independent of L if the gauge field is expressed in units of $1/L$. Suppose that the coupling constant will grow without bound. This would make the potential irrelevant and makes the wave functional spread out over the whole of field space (which could be seen as a strong coupling expansion). If the kinetic term would have been trivial the wave functionals would be "plane waves" on a space with complicated boundary conditions. In that case it seems unavoidable that the infinite volume limit would depend on the geometry (like T^3 or S^3) that is scaled-up to infinity. With the non-triviality of the kinetic term this conclusion cannot be readily made and our present understanding only allows comparison in volumes around one cubic fermi. However, one way to avoid this undesirable dependence on the geometry is that the vacuum is unstable against domain formation. As periodic subdivisions are space filling on a torus, this seems to be the preferred geometry to study domain formation. In a naive way it will give the correct string tension (flux conservation tells us to "string" the domains that carry electric flux) and tensor to scalar mass ratio (averaging over the orientations of the domains is expected to lead to a multiplicity weighted average of the T_2 and E masses). Furthermore, the natural dislocations of such a domain picture are gauge dislocations. The point-like gauge dislocations in four dimensions are instantons and in three dimensions they are monopoles. Their density is expected to be given roughly as one per domain (with a volume of around 0.5 cubic fermi). Also the coupling constant will stop running at the scale of the domain size. We have discussed this elsewhere and refer the reader to refs.[6, 32, 40] for further details.

REFERENCES

1. O. Babelon and C. Viallet, *Comm.Math.Phys.* 81:515 (1981).
2. V. Gribov, *Nucl.Phys.* B139:1 (1978).
3. I. Singer, *Comm.Math.Phys.* 60:7 (1978).
4. W. Nahm, *in:* "IV Warsaw Sym.Elem.Part.Phys," 1981, Z.Ajduk, ed. (1981) p.275.
5. J. Koller and P. van Baal, *Nucl. Phys.* B302:1 (1988); P. van Baal, *Acta Phys. Pol.* B20:295 (1989).
6. P. van Baal, *in:* "Probabilistic Methods in Quantum Field Theory and Quantum Gravity," ed. P.H. Damgaard e.a., Plenum Press, New York (1990) p31; *Nucl.Phys.* B(Proc.Suppl.)20:3 (1991).
7. M.A. Semenov-Tyan-Shanskii and V.A. Franke, *Zapiski Nauchnykh Seminarov Leningradskogo Otdeleniya Matematicheskogo Instituta im. V.A. Steklov AN SSSR* 120:159 (1982). Translation: Plenum Press, New York (1986) p.999.
8. N.M. Christ and T.D. Lee, *Phys. Rev.* D22:939 (1980).
9. M. Lüscher, *Nucl. Phys.* B219:233 (1983).
10. R.E. Cutkosky, *J. Math. Phys.* 25:939 (1984) 939; R.E. Cutkosky and K. Wang, *Phys. Rev.* D37:3024 (1988); R.E. Cutkosky, *Czech. J. Phys.* 40:252 (1990).
11. P. van Baal and N. D. Hari Dass, *Nucl.Phys.* B385:185 (1992).
12. P. van Baal, *in:* "Non-perturbative approaches to Quantum Chromodynamics", D. Diakonov, ed., Gatchina, 1995, pp.4-23.
13. G. Dell'Antonio and D. Zwanziger, *in:* "Probabilistic Methods in Quantum Field Theory and Quantum Gravity," ed. P.H. Damgaard e.a., (Plenum Press, New York, 1990) p07; G. Dell'Antonio and D. Zwanziger, *Comm. Math. Phys.* 138:291 (1991).
14. G. Dell'Antonio and D. Zwanziger, *Nucl.Phys.* B326:333 (1989).
15. D. Zwanziger, *Nucl. Phys.* B378:525 (1992).
16. P. van Baal, *Nucl.Phys.* B369:259 (1992).

17. S. Donaldson and P. Kronheimer, The geometry of four manifolds (Oxford University Press, 1990); D. Freed and K. Uhlenbeck, Instantons and four-manifolds, M.S.R.I. publ. Vol. I (Springer, New York, 1984).
18. P. van Baal, *in*: Proceedings of the International Symposium on Advanced Topics of Quantum Physics, eds. J.Q. Liang, e.a., Science Press (Beijing, 1993), p.133, hep-lat/9207029.
19. P. van Baal and B. van den Heuvel, *Nucl.Phys.* B417:215 (1994).
20. G. 't Hooft, *Nucl. Phys.* B153:141 (1979).
21. P. van Baal and J. Koller, *Ann. Phys. (N.Y.)* 174:299 (1987).
22. G. 't Hooft, *Phys.Rev.* D14:3432 (1976).
23. C. Bloch, *Nucl. Phys.* 6:329 (1958).
24. C. Vohwinkel, *Phys. Lett.* B213:54 (1988).
25. P. van Baal, *Phys. Lett.* 224B:397 (1989); *Nucl.Phys.* B(Proc.Suppl)17:581 (1990); *Nucl. Phys.* B351:183 (1991).
26. C. Michael, G.A. Tickle and M.J. Teper, *Phys. Lett.* 207B:313 (1988); C. Michael, *Nucl. Phys.* B329:225 (1990).
27. M. Lüscher and G. Münster, *Nucl. Phys.* B232:445 (1984).
28. C. Vohwinkel, *Phys. Rev. Lett.* 63:2544 (1989).
29. M. García Pérez and P. van Baal, *Nucl. Phys.* B429:451 (1994).
30. M. García Pérez, A. González-Arroyo, J. Snippe and P. van Baal, *Nucl. Phys.* B413:535 (1994); *Nucl. Phys.* B(Proc.Suppl)34:222 (1994).
31. P. van Baal, *Nucl. Phys.* B(Proc.Suppl)49:238 (1996), hep-th/9512223.
32. P. van Baal, *The QCD vacuum,* review at Lattice'97 (Edinburgh, 22-26 July 1996), hep-lat/9709066.
33. Y. Hosotani, *Phys. Lett.* 147B:44 (1984).
34. B.M van den Heuvel, *Nucl. Phys.* B(Proc.Suppl.)42:823 (1995); *Phys. Lett.* B368:124 (1996); B386:233 (1996); *Nucl. Phys.* B488:282 (1997).
35. M. Lüscher, *Phys. Lett.* B70:321 (1977).
36. A. Belavin, A. Polyakov, A. Schwarz and Y. Tyupkin, *Phys. Lett.* 59B:85 (1975); M. Atiyah, V. Drinfeld, N. Hitchin and Yu. Manin, *Phys. Lett.* 65A:185 (1978).
37. F. R. Klinkhamer and M. Manton, *Phys. Rev.* D30:2212 (1984).
38. P. van Baal and R.E. Cutkosky, *Int. J. Mod. Phys.* A(Proc. Suppl.)3A:323 (1993).
39. C. Michael and M. Teper, *Phys.Lett.* B199:95 (1987).
40. J. Koller and P. van Baal, *Nucl. Phys.* B(Proc.Suppl)4:47 (1988).

PERFECT ACTIONS

Peter Hasenfratz*

Institute for Theoretical Physics
University of Bern, Sidlerstrasse 5
CH-3012 Bern, Switzerland

INTRODUCTION

This lecture note is a pedagogical introduction to the theoretical background and properties of perfect actions. It is intended to be largely self-contained. The general aspects of lattice regularization and Wilson's renormalization group[1] and in particular those which are closely related to our subject will be explicitly introduced and discussed. The reader might even consider the topic of perfect actions as a pretext to learn about these powerful theoretical and practical tools in quantum field theories. I hope, however, that the unexpected and amazing properties of the perfect actions will catch the readers' fantasy also.

The very definition of a quantum field theory requires a regularization. As it is well known, all the regularizations break some of the symmetries of the underlying classical theory. They introduce a new scale (the cut-off) which is actually reflected in the final predictions even after the regularization is removed: naive dimensional analysis breaks down, anomalous dimensions are created, dynamical mass generation might occur, etc. On the other hand, beyond these exciting physical effects, the new scale creates dirt, cut-off artifacts everywhere which have to be eliminated in a theoretically and practically difficult limiting process. On the tree level (in the classical theory) the new introduced scale creates dirt only, nothing else. If the classical theory is scale invariant (like the Yang-Mills theory or massless QCD) the cut-off artifacts destroy the most interesting and relevant classical properties of the model like the existence of scale invariant classical solutions or fermionic zero modes and index theorems.

Regularizations often break other symmetries also. The lattice, the only known non-perturbative regularization, breaks Euclidean rotation symmetry (Lorentz symmetry) which is less of a headache, and also chiral symmetry which creates a difficult tuning and renormalization problem. Chiral symmetry awakes deep theoretical issues (anomalies, no-go theorems) culminating in problems concerning chiral gauge theories.

On the other hand, perfect actions[2] are perfect. We shall call a lattice regularized local action *classically perfect* if its classical predictions (independently whether the lattice is fine or coarse, whether the resolution is good or bad) agree with those of the

*Work supported in part by Schweizerischer Nationalfonds

continuum. The *quantum perfect action* does the same for all the physical questions in the quantum theory. That such actions exist might seem to be surprising. Their existence is closely related to renormalization group theory. As we shall see, they have beautiful properties which one would not expect a lattice action can have.

The classically perfect action has scale invariant instanton solutions, has no topological artifacts, satisfies the index theorem concerning the fermionic zero modes on the lattice, preserves some (presumably, all) important physical consequences of chiral symmetry in the quantum theory and is expected to reduce the cut-off effects significantly even in quantum simulations.

In this note we shall mainly discuss the theoretical properties of the classically perfect actions. These actions are the fixed points of renormalization group transformations in asymptotically free theories and are determined by classical field theory equations. They are not abstract theoretical constructions only. Actually, most of the effort during the last four years was directed towards constructing and parametrizing them explicitly and testing their performance in numerical simulations[3]. We shall not discuss these developments here at all beyond the remark that the classically perfect action gives very good results in quantum simulations including questions related to topology. This way, this lecture note presents the first part of a longer story. Even this first part is a torso: no discussion will be given on the recent developments concerning chiral symmetry.

We close this introduction with a list of content. Unlike in the main text, the sections and subsections are numbered in the list below. These numbers will be used later for reference.

LATTICE REGULARIZATION AND THE CONTINUUM LIMIT

In this Section we summarize some of the basic notions concerning lattice regularization, continuum limit and the relation between Quantum Field Theories (QFTs) and critical phenomena in classical statistical mechanics. Readers, who are familiar with these basics are advised to move directly to Section 2.4 on locality.

QFTs defined by path integrals in Euclidean space

In these lectures we shall consider QFTs in their Euclidean formulation and use path integrals. Using path integrals to describe quantum theories not only provides an appealing physical picture for the quantum evolution as a sum over classical paths (in quantum mechanics) or sum over classical configurations (in QFT), but gives a powerful framework for analytic manipulations and opens the way for numerical calculations also. In addition, it helps to understand the deep relation between QFTs and critical phenomena in classical statistical mechanics[4, 5].

The formal relation between QFTs and classical statistical physics is easy to see. Consider an n-point function in a scalar QFT in Minkowski space

$$\langle 0|T\big(\hat{\phi}(x_1)\hat{\phi}(x_2)\dots\hat{\phi}(x_n)\big)|0\rangle\,, \tag{1}$$

where $\hat{\Phi}(x)$ is the field operator. In the path integral language this expression is given by

$$\frac{1}{Z}\int D\phi\,\phi(x_1)\phi(x_2)\dots\phi(x_n)\exp\Big(i\int dtd\mathbf{x}\ \mathcal{L}(\phi,\partial_\mu\phi)\Big), \tag{2}$$

where $Z = \int D\phi \exp(i\int dtd\mathbf{x}\,\mathcal{L})$ provides for the correct normalization, $x = (t,\mathbf{x}) = (t, x_1, \dots, x_{d_s})$, and the Lagrangian is defined as

$$\mathcal{L}_{Mink} = \frac{1}{2}\partial_0\phi\partial_0\phi - \sum_{i=1}^{d_s}\frac{1}{2}\partial_i\phi\partial_i\phi - V(\phi)\,. \tag{3}$$

The relation we are looking for becomes clear if we go to Euclidean space by rotating time to imaginary time

$$t = x_0 \rightarrow -ix_d \tag{4}$$

It can be shown[6] that in every order of perturbation theory the analytic continuation in eq. (4) (Wick rotation) is possible due to the specific pole structure of the Feynman propagators. It is an assumption that this remains true beyond perturbation theory. Using

$$i\int dtd\mathbf{x} \rightarrow \int d^dx, \tag{5}$$

and

$$\mathcal{L}_{Mink} \to -\Big(\sum_{\mu=1}^{d} \frac{1}{2}\partial_\mu\phi\partial_\mu\phi + V(\phi)\Big) = -\mathcal{L}_{Eucl}, \tag{6}$$

we get

$$Z = \int D\phi \exp\Big(-\int d^dx\, \mathcal{L}_{Eucl}(\phi, \partial\phi)\Big), \tag{7}$$

$$\langle\phi(x_1)\ldots\phi(x_n)\rangle = \frac{1}{Z}\int D\phi\,\phi(x_1)\ldots\phi(x_n)\exp(-\int d^dx\,\mathcal{L}_{Eucl}). \tag{8}$$

These equations can be interpreted as the partition function and the correlation function of a system in classical statistical mechanics. The quantum theory of fields in d_s space dimensions is transformed into classical statistical mechanics of fields in $d = d_s + 1$ Euclidean dimensions. The Euclidean action plays the role of E/k_BT, where E is the classical energy of the d dimensional configuration.

Actually, the relation between QFTs and classical statistical systems is deeper and more specific than the observation above. We shall return to this question in Sect.2.3

Lattice regularization

Some of the mathematical operations which enter the definition of a QFT require a careful limiting procedure. In field theory, the variables are associated with space-time points. As can be seen from the form of Lagrangian's of different classical FT's (see, for example eq. (3)), these variables have some kind of self-interaction, whereas the elementary interaction between different degrees of freedom is over infinitesimal distances as expressed by derivatives. Already in the classical theory, the definition of a derivative requires the temporary introduction of a finite increment (of the argument of the function considered) which disappears at the end by some limiting procedure

$$f'(x) = \lim_{a\to 0}\frac{f(x+a)-f(x)}{a}. \tag{9}$$

Defining the derivative, which has the important role of providing communication between different field variables, is a much less trivial problem in QFTs. Like in eq. (9), the very definition of a QFT requires the temporary introduction of a defining framework called regularization, which disappears from the theory by a limiting process. The way to introduce and remove the regularization is a highly non-trivial problem.

We shall use lattice regularization in these lectures. A hypercubic lattice is introduced in the $d = d_s + 1$ dimensional Euclidean space. Scalar and fermion field variables live on the lattice points, vector fields on the connecting links. Derivatives are replaced by some kind of finite difference operation.

It will be convenient to work with dimensionless quantities. In the following, all the fields, momenta, masses, couplings and other possible parameters will be defined dimensionless by absorbing appropriate powers of the lattice unit a in their definition. For any quantity Q_{df} with dimension s (measured in mass units) we define

$$Q_{df} \to Q = a^{-s}Q_{df}. \tag{10}$$

For a scalar field $\phi(x)$ we write

$$\phi(x) \to \phi_n \tag{11}$$

where n is a d-dimensional vector of integers. The simple choice for the finite difference operation

$$\partial_\mu \phi(x) \to \nabla_\mu \phi_n = \phi_{n+\hat{\mu}} - \phi_n \tag{12}$$

leads to the following lattice regularized form of the classical Euclidean Lagrangian

$$A = \int d^d x\, \mathcal{L} \to \sum_n \sum_\mu \frac{1}{2} \nabla_\mu \phi_n \nabla_\mu \phi_n + \sum_n V(\phi_n)\,. \tag{13}$$

Having a finite lattice unit a the lattice defines a short distance regularization. Going over to momentum space, in the Fourier integral $\exp(ipn)$ enters which is periodic under $p_\mu \to p_\mu + 2\pi$. This constrains the momentum to the Brillouin zone $-\pi \geq p_\mu < \pi$. Therefore, the lattice provides a momentum cut-off also.

The lattice regularization has several advantages over other conventional regularizations used in perturbation theory. In general, in problems which require numerical analysis (e.g. integrals, differential equations,...), the standard procedure is to introduce meshes. Also the path integral can be defined in a natural way by introducing meshes. The path integral regularized this way becomes a set of integrals and is ready for numerical procedures in case of non-perturbative problems. Actually, lattice regularization is the only known non-perturbative regularization. It is a very natural regularization also for theories with gauge invariance or with constrained variables (like the non-linear σ-model).

Undoubtedly, lattice regularization has some disadvantages also. It breaks certain symmetries, as every regularization does. But the lattice has special problems (although not unsurmountable) with chiral symmetry and tough problems with chiral gauge theories. The discretization of space-time breaks Lorentz symmetry (O(d) symmetry in Euclidean space), but, as we shall see later (Sect. 8.1), the remaining cubic symmetry is sufficient to have it restored automatically as the regularization is removed (at least in asymptotically free theories). The lack of Poincaré symmetry creates problems, however, in theories with supersymmetry.

Although the lattice is used, in general, to investigate non-perturbative problems, for certain questions (renormalization of operators, finding improved actions or comparing lattice results with those obtained in other regularizations) it is unavoidable to use lattice perturbation theory. Unfortunately, perturbation theory on the lattice is technically somewhat cumbersome.

Some aspects of lattice regularization will be discussed in Sect. 5. For further reading we refer to textbooks[7].

The continuum limit of QFTs versus critical phenomena in statistical mechanics

Assume, we can solve our lattice regularized QFT at some set of (bare) couplings and masses in the Lagrangian. The predicted spectrum will contain dimensionless numbers, since all our quantities were made dimensionless by absorbing appropriate powers of the lattice unit a (Sect. 2.2). The mass of an excitation is given by a number M, the corresponding dimensionful mass $M_{df}, [M_{df}] = g$ and correlation length $\xi_{df}, [\xi_{df}] = cm$ are given by

$$M_{df} = M\frac{1}{a} \qquad \xi_{df} = \frac{1}{M_{df}} = \frac{1}{M}a = \xi a\,. \tag{14}$$

The correlation length defined above is the length scale over which the massive particle can propagate with a significant amplitude. For a generic value of the couplings and masses in the Lagrangian the predicted M will be an $O(1)$ number, the mass of the particle is of the order of the cut-off $1/a$, the correlation length is $O(a)$. This corresponds to a very poor resolution, the situation is far from the continuum limit. It requires a careful tuning of a certain number of parameters in the Lagrangian to make the predicted M's in the low lying spectrum to become very small numbers which correspond to physical length scales much larger than the lattice unit a. In this limit the lattice becomes very fine and the presence of the regularization will not distort the physical results (like mass ratios) anymore. This is the process of removing the lattice regularization which leads to the continuum limit. Theories in which this can be achieved by tuning a finite number of parameters (and by fixing the normalization of fields appropriately) are called renormalizable. This criterion is identical to that introduced in perturbation theory.

In the continuum limit the correlation length of physical excitations is much larger than the lattice unit a. But how large is it in Fermi? This can not be predicted, the absolute scale should come from observations. If the theory is expected to describe the physics of hadrons, for example, then the lowest lying particle with spin=1/2, electric charge=1, baryon number=1, strangeness=0,... should be identified with the proton having a mass of 940 MeV. This fixes the physical length scale and relative to that unit the lattice constant goes to zero in the continuum limit.

In Sect. 2.2 we observed a formal relation between QFTs and classical statistical systems. We see now that the continuum limit of a QFT corresponds to a statistical system with a correlation length which is much larger than the lattice unit a. This is the case for critical statistical systems. In critical solid state problems the lattice unit is fixed (typically $O(\text{Å})$) and the correlation length becomes macroscopical. This is a matter of choosing the absolute scale, however, and it does not reduce the extent of the deep analogy between the two subjects.

Locality

A recurrent issue in our discussion will be the locality of the action density (Lagrangian density). The laws of classical physics (from Newton through Maxwell to Einstein) are expressed in terms of (partial) differential equations (rather than, say, integro-differential equations). They correspond to actions, where the interaction between variables in different space-time points extends over infinitesimal distances as expressed by derivatives. This continues to be true in quantum systems also. No symmetry principles would prevent us to add a term to the Lagrangian in eq. (3) like

$$\mathcal{L}(x) = \ldots + \int d^d y \; f(x-y)\phi(x+\frac{y}{2})\phi(x-\frac{y}{2}), \tag{15}$$

where $f(z)$ is an even function of z and has an extension λ, for example $f(z) \sim \exp(-z^2/2\lambda^2)$, where λ is finite when measured in Fermi. One can imagine functions also which decay only as a power in z^2. Nevertheless, such models found their place neither theoretically, nor experimentally.

In a QFT with non-local interactions renormalizability (in the sense discussed in Sect. 2.3) will be lost. In addition, a vital and beautiful property of local QFTs (shared by critical statistical systems), called universality will be lost also. Universality means that the physical predictions become independent of the microscopic details of the Lagrangian, i.e. they are not sensitive to the detailed form of the interaction at the cut-off scale. Since our knowledge on interactions much above the scales of present experiments is very limited, universality is vital to preserve the predictive power of a QFT.

On the lattice we shall call an action density local, if it has an extension of $O(a)$. Typically, it is an exponential function of the distance $|n - n'|$ between the variables at n and n': $\sim \exp(-\gamma|n - n'|)$, $\gamma = O(1)$. In the continuum limit $(a \to 0)$, the extension of the action density measured in physical units goes to zero.

A Lagrangian which has nearest-neighbour interactions only or interactions which are identically zero beyond a few lattice units is certainly local, but no physical principle requires to have this extreme case. If this were necessary to assure the above mentioned nice properties of a QFT (or of a critical statistical system), then no experiment on ferromagnets close to the Curie temperature would observe universality. Really, the elementary interaction between the magnetic moments in the crystal decays rapidly, but certainly does not become identically zero beyond a few lattice spacings.

In the following we shall call an action where the couplings become identically zero beyond a certain $O(a)$ range ultralocal. We shall also use the expression 'short ranged' for local actions where the exponential decay goes with a large γ.

RENORMALIZATION GROUP

A QFT is defined over a large span of scales from low (relative to the cut-off) physical scales up to the cut-off which goes to infinity in the continuum limit. Although field variables associated with very high scales do influence the physical predictions through a complicated cascade process, no physical question involves them directly. Their presence and indirect influence makes it difficult to establish an intuitive connection between the form of the interaction and the final expected predictions. The presence of a large number of degrees of freedom makes the problem technically difficult also. It is, therefore a natural idea to integrate them out in the path integral. This process, which reduces the number of degrees of freedom, taking into account their effect on the remaining variables exactly, is called a renormalization group transformation[1,8] (RGT).

In this Section we introduce some of the basic notions related to RG theory and set the notations. Since these lectures are mainly concerned with asymptotically free (AF) QFTs, our discussion will be biased by this goal.

Renormalization group transformation

For simplicity consider a scalar field theory regularized on the lattice in d Euclidean dimensions as discussed in Section 2.2. The RGT averages out the short distance fluctuations, i.e. the fluctuations over distances $O(a)$. The lattice is indexed by n, the field associated with the point n *is* denoted by ϕ_n. We introduce a blocked lattice with lattice unit $a' = 2a$ whose points are labeled by n_B and the associated block field is denoted by χ_{n_B}. The block variable χ_{n_B} is an average of the original ϕ fields in the neighbourhood of the point n_B:

$$\chi_{n_B} = b \sum_n \omega(2n_B - n)\phi_n \,, \tag{16}$$

where ω specifies the weights of the averaging whose normalization is chosen as

$$\sum_n \omega(2n_B - n) = 1 \,. \tag{17}$$

The significance of the scale factor b in eq. (16) will become clear later. A trivial choice for ω might be to take $\omega(2n_B - n) = 2^{-d}$ if n is in the hypercube whose center is indexed by n_B and zero otherwise.

We integrate out now the original ϕ_n variables keeping the block averages fixed:

$$\exp\left(-\mathcal{A}'(\chi)\right) = \prod_n \int d\phi_n \prod_{n_B} \delta\left(\chi_{n_B} - b\sum_n \omega(2n_B - n)\phi_n\right) \exp\left(-\mathcal{A}(\phi)\right). \tag{18}$$

The new action $\mathcal{A}'(\chi)$ describes the interaction between the block variables. The partition function remains unchanged

$$Z' = \prod_{n_B} \int d\chi_{n_B} \exp\left(-\mathcal{A}'(\chi)\right) = \prod_n \int d\phi_n \exp\left(-\mathcal{A}(\phi)\right) = Z. \tag{19}$$

The long-distance behaviour† of Green's functions, and so the spectrum and other low energy properties of the system is expected to remain unchanged as well. On the other hand, the lattice unit is increased by a factor of 2: $a' = 2a$. Since the dimensions are carried by the lattice unit (Sect. 2.2), the dimensionless correlation length, which is measured in the actual lattice distance, is reduced by a factor of 2:

$$\xi \to \frac{\xi}{2}. \tag{20}$$

By eliminating the fluctuations on the shortest scales we reduced the number of variables by a factor of 2^d, while taking their effect into account by changing the action $\mathcal{A} \to \mathcal{A}'$ appropriately. Although the lattice unit is increased by a factor of 2, and so the resolution becomes worse, the long-distance behaviour remains unchanged. In particular, no new cut-off effects are generated in the predictions for physical quantities. Iterating this RGT, the goal formulated at the beginning of Sect. 3 is achieved.

We close this subsection by generalizing the RGT of eq. (18) slightly[9] (a step which will be very useful later):

$$\exp\left(-\mathcal{A}'(\chi)\right) = \mathcal{N}(\kappa)\prod_n \int d\phi_n \exp\left[-\mathcal{A}(\phi) - \kappa\sum_{n_B}\left(\chi_{n_B} - b\sum_n \omega(2n_B - n)\phi_n\right)^2\right], \tag{21}$$

where κ is a free parameter. For $\kappa \to \infty$ eq. (21) goes over to eq. (18), for finite κ the block average is allowed to fluctuate around χ (slightly, if κ is not too small). Therefore, the parameter κ specifies the stiffness of the RG averaging. $\mathcal{N}$ is a normalization factor (which is field independent, trivial in this case) introduced to keep eq. (19) valid.

Constraints on the block transformation from symmetries

It is largely arbitrary how the block averages in a RGT are constructed, but the procedure should conform with the intuitive goal of a RGT: it should lead to the elimination of the short distance fluctuations. Identifying, for example, the block variable with one of the original fine variables in the block ('decimation') is not really an averaging, and although it is a legal transformation of the path integral, it will not lead to useful results in general.

Beyond this general requirement, symmetries put further constraints on the form of the block transformation. Denote the fields on the fine and coarse lattice by ϕ and χ, respectively and write the RGT in the form

$$\exp\left(-\mathcal{A}'(\chi)\right) = \int D\phi \prod_{\text{blocks}} f(\chi_{\text{block}};\phi) \exp\left(-\mathcal{A}(\phi)\right). \tag{22}$$

† 'Long-distance' is meant here and elsewhere in this work as 'many lattice units', and it might mean short, or long physical distances.

Here ϕ, χ can be scalar, vector or fermion fields and f defines how the coarse field in the block χ_{block} is constructed from the (neighbouring) ϕ fields.

Assume, there is a symmetry transformation $\phi \to R\phi$ under which the action and the measure are invariant: $\mathcal{A}(R\phi) = \mathcal{A}(\phi)$, $D(R\phi) = D\phi$. The blocked action $\mathcal{A}'(\chi)$ will inherit this symmetry, $\mathcal{A}'(R\chi) = \mathcal{A}'(\chi)$, if the averaging function satisfies

$$f(R\chi_{block};\phi) = f(\chi_{block};R^{-1}\phi)\,, \tag{23}$$

as it is seen easily by changing integration variables in eq. (22).

In case of gauge symmetries we shall require somewhat more: we shall not only require that the effect of a gauge transformation on the coarse lattice on f is equivalent to a gauge transformation on the fine lattice, eq. (23), but also the other way around. This constraint implies that gauge equivalent fine configurations contribute equally to the path integral in eq. (22).

A basic assumption of the RG theory

Consider a step of the RGT (eq. (18) or eq. (21)) for the case where the parameters of $\mathcal{A}(\phi)$ were chosen so that the original system was close to the continuum (or, in the language of statistical physics, close to criticality). As eq. (21) shows, performing a RGT on a system with action $\mathcal{A}(\phi)$ leads to a path integral where $\mathcal{A}$ is replaced by:

$$\mathcal{A}(\phi) + \kappa \sum_{n_B} \Big(\chi_{n_B} - b\sum_{n} \omega(2n_B - n)\phi_n\Big)^2 . \tag{24}$$

For this path integral the field χ enters as an external field. Even if $\mathcal{A}(\phi)$ was carefully tuned to criticality, the system in eq. (24) is not expected to be critical. Actually, this is a basic assumption of the RG theory: the path integral entering a step of the RGT defines a non-critical problem with short-range interactions only. This is what one expects intuitively: the χ fields constrain the ϕ**-averages** in their neighbourhood and disrupt the long-range fluctuations. This assumption leads to the important conclusion that the new action $\mathcal{A}'(\chi)$ will be local. Really, if the r.h.s. of eq. (21) describes a system with short-range fluctuations only, the generated interaction between distant χ fields is expected to be negligibly small.

Another consequence of this basic assumption is that the path integral in eq. (18) or eq. (21) is a technically much simpler problem than the path integral of the original system. Saying differently: to perform a RG step is not an easy problem in general, but it is much simpler than to solve the original (near) critical theory. This is what makes the RG theory a practically useful idea.

Fixed Point (FP) and the behaviour in the vicinity of a FP

In general, the transformed action $\mathcal{A}'(\chi)$ in eq. (21) will contain all kinds of interactions, even if the original action $\mathcal{A}(\phi)$ had a simple form. It is useful to introduce a sufficiently general interaction space to describe the actions generated by the RGT, and write

$$\mathcal{A}(\phi) = \sum_{\alpha} K_\alpha \theta_\alpha(\phi), \tag{25}$$

where $\theta_\alpha(\phi)$, $\alpha = 1, 2, \ldots$ denotes the different interaction terms like (for scalar fields)

$$\theta_1 = \sum_{n,\mu} \nabla_\mu\phi_n \nabla_\mu\phi_n\,, \qquad \theta_2 = \sum_n \phi_n^2\,, \ldots\,, \; \theta_i = \sum_{n,\mu} \phi_n^4 \nabla_\mu\phi_n \nabla_\mu\phi_n, \ldots \tag{26}$$

and K_α are the corresponding dimensionless couplings. The transformed action $\mathcal{A}'(\chi)$ is also expanded in terras of these interaction terms

$$\mathcal{A}'(\chi) = \sum_\alpha K'_\alpha \theta_\alpha(\chi) \,. \tag{27}$$

The RGT induces a motion in the coupling constant space: $\{K_\alpha\} \to \{K'_\alpha\}$. Under repeated RGTs a coupling constant flow is generated

$$\{K_\alpha\} \to \{K'_\alpha\} \to \{K''_\alpha\} \to \dots \,, \tag{28}$$

while the (dimensionless) correlation length is reduced with every step (eq. (20))

$$\xi \to \frac{1}{2}\xi \to \frac{1}{4}\xi \to \dots \,. \tag{29}$$

It might happen that certain points in the coupling constant space are reproduced by the RGT

$$\{K^*_\alpha\} \to \{K^*_\alpha\} \,. \tag{30}$$

A point with this property is called a fixed point (FP) of the RGT and the corresponding action

$$\mathcal{A}^{FP}(\phi) = \sum_\alpha K^*_\alpha \theta_\alpha(\phi) \tag{31}$$

is the FP action. FP actions play an important role in QFTs in general and will play an important role in our later discussion of the perfect actions.

Eq. (28) and eq. (29) imply that at the FP, $\xi = \infty$ or $\xi = 0$. We will be interested in FPs with $\xi = \infty$. The set of points in the coupling constant space where $\xi = \infty$, forms a hypersurface, which is called the critical surface. As eq. (29) shows, an RGT drives the point $\{K_\alpha\}$ away from the critical surface, except when $\{K_\alpha\}$ is on the critical surface.

Let us consider now the behaviour of the flow under the RGT in the vicinity[10] of a FP. Take a point $\{K_\alpha\}$ with $\Delta K_\alpha = K_\alpha - K^*_\alpha$ small. Under the RGT we have

$$\Delta K_\alpha \to \Delta K'_\alpha = K'_\alpha - K^*_\alpha \,. \tag{32}$$

Expanding $\Delta K'_\alpha = \Delta K'_\alpha(\{K\})$ around $\{K^*\}$

$$\Delta K'_\alpha = \Delta K'_\alpha(\{K^*\}) + \sum_\beta \frac{\partial}{\partial K_\beta} \Delta K'_\alpha(\{K\})_{|K=K^*} \Delta K_\beta + O\big((\Delta K)^2\big) \,, \tag{33}$$

and observing that the first term on the r.h.s. is zero, we get the following linearized RG equation

$$\Delta K'_\alpha = \sum_\beta T_{\alpha\beta} \Delta K_\beta , \tag{34}$$

with

$$T_{\alpha\beta} = \frac{\partial}{\partial K_\beta} \Delta K'_\alpha(\{K\})_{|K=K^*} \,. \tag{35}$$

Let us denote the eigenvectors and eigenvalues of the matrix T by h_a and λ_a, respectively:

$$\sum_\beta T_{\alpha\beta} h^a_\beta = \lambda^a h^a_\alpha \,, \; a = 1, 2, \dots \,. \tag{36}$$

The eigenvectors h^a define the eigenoperators

$$h^a(\phi) = \sum_\alpha h^a_\alpha \theta_\alpha(\phi) . \tag{37}$$

These operators define a new basis in terms of which the action can be expanded

$$\mathcal{A}(\phi) = \mathcal{A}^{FP}(\phi) + \sum_a c^a h^a(\phi), \tag{38}$$

where c^a are the corresponding expansion coefficients (couplings) which are small if $\mathcal{A}$ is close to $\mathcal{A}^{FP}$. Repeated application of the RGT gives

$$\mathcal{A}(\phi) \to \mathcal{A}^{FP}(\phi) + \sum_a \lambda^a c^a h^a(\phi) \to \mathcal{A}^{FP}(\phi) + \sum_a (\lambda^a)^2 c^a h^a(\phi) \to \dots , \tag{39}$$

i.e., the coupling c^a goes over to $(\lambda^a)^n c^a$ after n RGT steps. For $|\lambda^a| > 1$ ($|\lambda^a| < 1$) the coupling $|c^a|$ is increasing (decreasing) under repeated RG steps. The corresponding interaction h^a is called relevant (irrelevant). For $|\lambda^a| = 1$ the fate of the operator (called marginal) is decided by the higher order corrections suppressed in eq. (33).

RG flows of a Yang-Mills theory

The following discussion applies also for other AF theories, like the $O(N)$ non-linear σ-model or the CP^n model in $d = 2$.

The classical action of an $SU(N)$ Yang-Mills theory has the form

$$\beta \mathcal{A}_{cont} = \frac{\beta}{2N} \int_x F^a_{\mu\nu}(x) F^a_{\mu\nu}(x) , \quad a = 1, \dots , N^2 - 1 , \tag{40}$$

where $\beta = 2N/g^2$ and $F^a_{\mu\nu}$ is the colour field strength tensor. Any lattice representation of this action should go over to the form in eq. (40) for slowly changing smooth fields. Even if we use a gauge symmetric discretization (as we always do), there is a large freedom in writing down a local action with this property on the lattice. As we discussed in Sect. 3.4, we have to introduce an infinite dimensional coupling constant space if we want to follow the change of the action under RGTs. We write

$$\beta \mathcal{A}_{latt} = \beta \mathcal{A}_{latt}(K_1, K_2, \dots), \tag{41}$$

where we indicated the coupling constant dependence of the action only. Under a RGT we have: $\{\beta, K_1, K_2, ...\} \to \{\beta', K_1', K_2', ...\}$.

The expected flow diagram is sketched in fig. 1. In the $\beta = \infty$ hyperplane, which is the critical surface, there is a FP with coordinates $K_1^*, K_2^*, \dots$. We shall suppress the index 'latt' and introduce the notation $\mathcal{A}(K_1^*, K_2^*, ...) = \mathcal{A}^{FP}$. All the eigenoperators which lie in the $\beta = \infty$ hypersurface are irrelevant. There is one marginal direction which is pointing out of this surface. Actually, this direction is marginal in the linear approximation only, it becomes weakly relevant by the higher order corrections to eq. (33).

Consider the point $\{\beta, K_1, K_2, ...\}$, β very large. Under RGTs this point runs rapidly towards the FP (the largest irrelevant eigenvalue is 1/4), then slowly moves away from it along the weakly relevant direction tracing a trajectory, called the renormalized trajectory.

Consider the interaction corresponding to the FP, $\mathcal{A}^{FP}$, multiply it by β: $\beta\mathcal{A}^{FP}$ and allow β to move away from ∞. This is a straight line (the dashed line in fig. 1.)

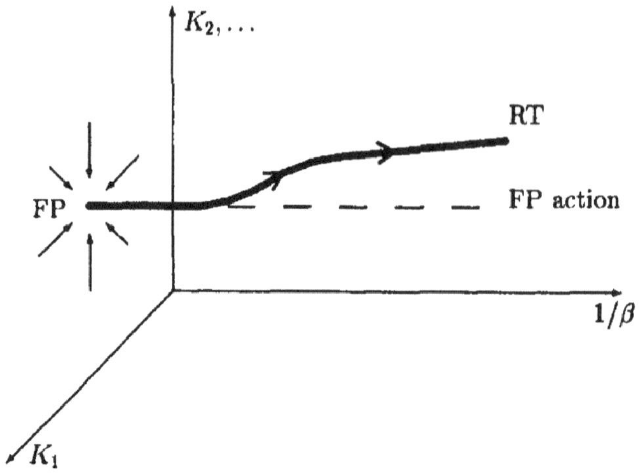

Figure 1. Flow of the couplings under RG transformations in the O(N) non-linear σ-model.

which is not a RG flow, but defines an action for every value of β. We shall call this action the FP action.

We postpone until Sects. 7 and 8 the discussion on the arguments supporting the structure of the flow diagram in fig. 1. We mention here only that theories with this flow diagram can be renormalized by tuning a single coupling constant and the resulting renormalized theory is universal (independent of the value of the other couplings in the (bare) action) . The continuum renormalized theory is obtained as $g \to 0$ ($\beta \to \infty$) in which limit the correlation length goes to infinity, the lattice unit $a \to 0$, the resolution becomes infinitely good and the cut-off artifacts disappear from the predictions.

PERFECT ACTIONS IN AF THEORIES

We shall call a lattice regularized local action *classically perfect* if its classical predictions (independently whether the lattice is fine or coarse, whether the resolution is good or bad) agree with those of the continuum. The *quantum perfect action* does the same for all the physical questions in the quantum theory. That such actions exist might seem to be surprising. As we shall see, they do exist and have beautiful properties which one would not expect a lattice action can have[2, 3, 11, 12, 13, 14].

We shall argue now that the actions defined by the points of the renormalized trajectory (RT) in fig. 1 define quantum perfect actions. It means that by taking an arbitrary point $(\beta, K_1, K_2, \ldots) \in RT$, where $g^2 = 2N/\beta$ need not be small, the corresponding action $\mathcal{A}(\beta, K_1, K_2, \ldots)$ will produce quantum results for physical questions (mass ratios, for example) which are exactly the same as in the continuum limit. This is surprising, since in the point $(\beta, K_1, K_2, \ldots)$ the correlation length is not large, the lattice unit a is not small, the resolution is not fine. Nevertheless, the statement is true.

The argument goes as follows[2]. At any given β, the point of the RT is connected to the infinitesimal neighbourhood of the FP by (infinitely many steps of) exact RG transformations. Since each step increases the lattice unit by a factor of 2, *any* distance at the given β (even 1 lattice unit) corresponds to a long distance close to the FP. The infinitesimal neighbourhood of the FP is in the continuum limit, there are no cut-off effects there at long distances. On the other hand, for all the questions which can be formulated in terms of the degrees of freedom after the transformation we get the same answer as before the transformation. Thus, there are no lattice artifacts at the given β on the RT at any distances.

We shall also argue that the FP action, as defined in Sect. 3.5 (fig. 1) is a classically perfect action. We are not able to demonstrate this statement so compactly as that for the quantum perfect action before. In the following sections we shall go over the different classical properties of continuum Yang-Mills, QCD and other AF theories and show one by one that they are reproduced by the FP action independently of the lattice resolution.

The detailed form of the FP action and the RT depend on the form and parameters of the block transformation. The theoretical properties of the FP action and RT are, however, independent of these details. In this sense, any RGT which has a local FP is suitable. On the other hand, different local FP actions (belonging to different block transformations) have different extensions and this has an important practical significance when it comes to parametrization and simulation. In order to find short ranged actions (Sect. 2.4), different block transformations were tested and optimized[3].

LATTICE REGULARIZATION AND EXPLICIT RGTs IN DIFFERENT AF THEORIES

It is time now to come down from generalities to concrete examples. We shall discuss how the lattice regularization is set up and give explicit RGTs in different AF theories. First, the *O(N)* non-linear **σ-model** and the *SU(N)* Yang-Mills theory will be considered. Then, as a preparation for QCD, the special difficulties of lattice regularized Dirac fermions will be illustrated by treating the case of free spin-1/2 fermions. We close this section with QCD.

In all these theories we write down the 'standard' lattice action which is the simplest discretization of the continuum action satisfying the basic requirements. These are the actions which - due to their technical simplicity - were used in most of the numerical simulations. These actions are simple, but they represent poorly most of the properties even of the classical field theory if the resolution is not sufficiently fine. As we discussed in Sect. 4, in the framework of the RG theory actions can be defined, which have beautiful properties. For later use we define explicit RGTs in all the theories considered in this section.

O(N) non-linear σ-model in d=2

The classical action in the continuum has the form

$$\beta \mathcal{A}_{cont}(\mathbf{S}) = \frac{\beta}{2} \int_x \partial_\mu \mathbf{S}(x) \partial_\mu \mathbf{S}(x) , \tag{42}$$

where S is an .N-component vector satisfying the constraint $\mathbf{S}^2(x) = 1$, $\beta = 1/g$. Writing

$$\mathbf{S}(x) = \begin{pmatrix} \sqrt{1-\boldsymbol{\pi}(x)^2} \\ \boldsymbol{\pi}(x) \end{pmatrix} \tag{43}$$

and expanding in the fluctuations $\boldsymbol{\pi}(x)$ (it has $N-1$ components) a systematic perturbation theory can be set up. The theory is AF in the coupling g for $N \geq 3$. It is believed (and this is strongly supported by different theoretical and numerical results) that a non-zero mass is generated dynamically, the spectrum contains a massive $O(N)$ multiplet. The σ-model is discussed in different modern textbooks on QFT and statistical systems[5,15,16]

Looking for classical solutions with finite action, the spin vectors $\mathbf{S}$ should be parallel at large distances. For $N = 3$ this leads to an $S_2 \to S_2$ mapping between the coordinate and group spaces. The $O(3)$ model has a non-trivial topology, there exist scale invariant instanton solutions with an action $\mathcal{A} = 4\pi|Q|$, where Q is the topological charge of the solution.

The $O(3)$ model shows many analogies[17,16] with the Yang-Mills theory in $d = 4$. It offers the possibility to test non-perturbative ideas and numerical methods.

On the lattice, the scalar field $\mathbf{S}$ lives on the lattice points. The simplest realization of the action has the form ('standard action')

$$\beta \mathcal{A}_{st}(\mathbf{S}) = \frac{\beta}{2} \sum_{n,\mu} \nabla_\mu \mathbf{S}_n \nabla_\mu \mathbf{S}_n \,, \tag{44}$$

where ∇_μ is defined in eq. (12) and $\mathbf{S}_n^2 = 1$, On smooth configurations $\mathcal{A}_{st}$ goes over to $\mathcal{A}_{cont}$ – a property we shall require for any lattice action.

Consider RGTs with a scale factor of 2, forming averages out of the four fine spins in a block. The block spin will be denoted by $\mathbf{R}_{n_B}$, $\mathbf{R}_{n_B}^2 = 1$. The RGT has the form

$$\exp\bigl(-\beta' \mathcal{A}'(\mathbf{R})\bigr) = \int D\mathbf{S} \exp\bigl[-\beta\bigl(\mathcal{A}(\mathbf{S}) + T_\sigma(\mathbf{R}, \mathbf{S})\bigr)\bigr] \,, \tag{45}$$

where T_σ defines the averaging process and we used the notation

$$\int D\mathbf{S} = \prod_n \int d\mathbf{S}_n \delta(\mathbf{S}^2 - 1) \,. \tag{46}$$

We shall give two explicit examples for the averaging which were studied in the context of perfect actions[2,19]. Both RGTs are $O(N)$ invariant, satisfying eq. (23) for global $O(N)$ transformations. In the first case:

$$T_{\sigma 1}(\mathbf{R}, \mathbf{S}) = \sum_{n_B} \bigl(-\kappa_\sigma \mathbf{R}_{n_B} \mathbf{\Sigma}_{n_B} + \mathcal{N}(|\mathbf{\Sigma}_{n_B}|)\bigr) \,, \tag{47}$$

where $\mathbf{\Sigma}_{n_B}$ is the sum of the four S spins in the block n_B

$$\mathbf{\Sigma}_{n_B} = \sum_{n \in n_B} \mathbf{S}_n \,, \tag{48}$$

and $\mathcal{N}$ assures the correct normalization, eq. (19). We get

$$\mathcal{N}(|\mathbf{\Sigma}_{n_B}|) = \frac{1}{\beta} \ln Y_N(\beta\kappa |\mathbf{\Sigma}_{n_B}|) \,, \tag{49}$$

where $Y_N(z)$ is related to the modified Bessel function (some of its properties are summarized in appendix F in[20], for example), specifically $Y_3(z) \sim \sin(z)/z$. For $\kappa \to \infty$, the block transformation goes over to a δ-function constraining the average to lie parallel to the coarse spin

$$\frac{\mathbf{\Sigma}_{n_B}}{|\mathbf{\Sigma}_{n_B}|} = \mathbf{R}_{n_B}, \qquad (\kappa = \infty) \,. \tag{50}$$

It will be useful, however, to keep k finite and optimize its value to get a short ranged FP action.

Modifying slightly the averaging function T

$$T_{\sigma 2} = 2\kappa_\sigma \sum_{n_B} \Big(\mathbf{R}_{n_B} - \frac{\mathbf{\Sigma}_{n_B}}{|\mathbf{\Sigma}_{n_B}|} \Big)^2 , \tag{51}$$

we can make the technically cumbersome normalizing function $\mathcal{N}$ constant. Really, the integral

$$\int d\mathbf{R}\, \delta(\mathbf{R}^2 - 1) \exp(4\beta\kappa_\sigma \mathbf{R}\mathbf{e}) , \tag{52}$$

with $\mathbf{e} = \mathbf{\Sigma}/|\mathbf{\Sigma}|$, depends on the length of $\mathbf{e}$ only, therefore it is an $\mathbf{S}$-independent constant.

SU(N) Yang-Mills theory in d=4

The classical action in the continuum has the form

$$\beta \mathcal{A}^{YM}_{cont}(A_\mu) = \frac{\beta}{4N} \int_x Tr(F_{\mu\nu} F_{\mu\nu}) , \tag{53}$$

where $\beta = 2N/g^2$, $F_{\mu\nu} = F^a_{\mu\nu} T^a, a = 1, ..., N^2 - 1$ and T^a are the generators of the colour group *SU(N)*, $Tr(T^a T^b) = 1/2\delta_{ab}$.
The classical theory has scale invariant instanton solutions[18] with an action $\mathcal{A} = 8\pi^2|Q|$.

The quantum theory is AF in the coupling g. It is believed (and this is supported by many, but non-rigorous results) that non-perturbative effects create a confining potential between static sources in the fundamental colour representation (static quarks) and the spectrum contains massive colour singlet excitations (glueballs) only.

On the lattice, the vector gauge field lives on the links of the hypercubic lattice[21, 7]. The convenient variables, in terms of which gauge invariance is easily kept on the lattice are not the vector potentials themselves, but their exponentiated forms. Consider a static quark and antiquark in the points x and y, respectively (in the continuum). The form

$$\bar{\psi}(x)\, \mathcal{P} \exp\Big(i \int_x^y A_\mu(z) dz_\mu\Big)\, \psi(y) , \tag{54}$$

where the integral runs along some path between x and y and $\mathcal{P}$ denotes path ordering, is gauge invariant. The gauge part in eq. (54), which leads the colour flux between the sources, is an element of the *SU(N)* group rather than that of the algebra. Going to the lattice and placing x, y in the endpoints of the link $(n, n + \hat{\mu})$, we get

$$\bar{\psi}_n\, U_\mu(n)\, \psi_{n+\hat{\mu}} , \tag{55}$$

where $U_\mu(n)$ is related to the lattice vector potential as

$$U_\mu(n) = \exp(iaA_\mu(n)) . \tag{56}$$

The variables of the lattice regularized Yang-Mills theory are the link matrices $U_\mu(n)$ The link matrix associated with the same link but directed oppositely: $(n + \hat{\mu}, n)$ is $U^\dagger_\mu(n)$. Under a gauge transformation the link matrix transforms as

$$U_\mu(n) \to V(n) U_\mu(n) V^\dagger(n + \hat{\mu}) , \tag{57}$$

where $V(n) \in SU(N)$ is the gauge transformation in the point n. Consequently, the trace of the product of link matrices along a closed path on the lattice is gauge invariant.

The simplest lattice realization of the Yang-Mills action has the form ('Wilson action')

$$\beta \mathcal{A}_W(U) = \frac{\beta}{N} \sum_{plaq} (N - ReTr\, U_p) , \tag{58}$$

where the sum is over the plaquettes and U_p is the product of four directed link matrices around the plaquette p. For smooth gauge fields eq. (58) goes over to eq. (53), as it should.

We shall consider RGTs with a scale factor of 2. The blocked link variable $V_\mu(n_B) \in SU(N)$, which lives on the coarse lattice with lattice unit $a' = 2a$, is coupled to a local average of the original link variables. The blocking kernel $T(V, U)$ enters the RGT as in the σ-model

$$\exp\left(-\beta' \mathcal{A}'(V)\right) = \int DU \exp\left[-\beta\left(\mathcal{A}(U) + T_g(V,U)\right)\right] , \tag{59}$$

where

$$\int DU = \prod_{n,\mu} \int dU_\mu(n) . \tag{60}$$

The kernel T_g is taken in the form

$$T_g(V,U) = \sum_{n_B,\mu} \left[-\frac{\kappa_g}{N} ReTr \left(V_\mu(n_B) Q^\dagger_\mu(n_B) \right) + \mathcal{N}\left(Q_\mu(n_B) \right) \right] . \tag{61}$$

The $N \times N$ complex matrix $Q_\mu(n_B)$ is an average of the fine link variables in the neighbourhood of the coarse link $l_B = (n_B, \mu)$. $\mathcal{N}(Q_\mu(n_B))$ is fixed by the normalization condition, eq. (19):

$$\exp\left(\mathcal{N}(Q)\right) = \int dW \exp\left(\frac{\beta \kappa_g}{N} ReTr\, (WQ^\dagger)\right), \qquad W \in SU(N) . \tag{62}$$

We shall give two examples which were studied in detail before[13, 22]:

RGT type A (fig. 2):

$$\begin{aligned} Q_\mu(n_B) = {} & (1-6c) U_\mu(n) U_\mu(n+\hat\mu) + \\ & c \sum_{\nu \neq \mu} \left[U_\nu(n) U_\mu(n+\hat\nu) U_\mu(n+\hat\mu+\hat\nu) U^\dagger_\nu(n+2\hat\mu) + \right. \\ & \left. U^\dagger_\nu(n-\hat\nu) U_\mu(n-\hat\nu) U_\mu(n+\hat\mu-\hat\nu) U_\nu(n+2\hat\mu-\hat\nu) \right] , \end{aligned} \tag{63}$$

where c, the relative weight of the staples versus the central link, is a tunable parameter (see, last paragraph of Sect. 4).

Figure 2. The renormalization group transformation of type A.

The averaging process in eq. (63) is not a very good one: there are many link variables on the fine lattice which do not contribute to any of the Q's. Experience shows that poor averaging might lead to a FP which is local, but not sufficiently short ranged or might not even have a FP.

RGT type B:

The second example provides for a better averaging. Instead of using just simple staples in constructing $Q_\mu(n_B)$ it builds also 'diagonal staples'. Let us define first the matrices $W^m(n,n')$ connecting the sites n and n', where $n'_\mu - n_\mu = 0,\ |n'_\nu - n_\nu| \leq 1$ (for any ν):

$$W^{(0)}(n,n) = 1\,, \qquad W^{(1)}(n,n+\hat{\nu}) = U_\nu(n)\,,$$
$$W^{(2)}(n,n+\hat{\nu}+\hat{\rho}) = \frac{1}{2}\left(U_\nu(n)U_\rho(n+\hat{\nu}) + U_\rho(n)U_\nu(n+\hat{\rho})\right)$$
$$W^{(3)}(n,n+\hat{\nu}+\hat{\rho}+\hat{\lambda}) = \frac{1}{6}\left(U_\nu(n)U_\rho(n+\hat{\nu})U_\lambda(n+\hat{\nu}+\hat{\rho}) + \text{permutations}\right)\,. \quad (64)$$

Here ν, ρ and λ go over all (positive and negative) directions different from μ and from each other (with $U_{-\nu}(n) = U^\dagger_\nu(n-\hat{\nu})$). Values of $W^{(m)}(n,n')$ not indicated are taken to be zero. $W^{(2)}(n,n')$ represents the 'planar diagonal link', $W^{(3)}(n,n')$ the spatial one. In eq. (64) the sums are taken over all shortest paths leading to the endpoint n' of the corresponding diagonal. A fuzzy link operator is constructed then as

$$\mathcal{W}_\mu(n) = \sum_{m=0}^{3}\sum_{n'} c_m W^{(m)}(n,n')U_\mu(n')W^{(m)}(n'+\hat{\mu},n+\hat{\mu}). \quad (65)$$

The coefficients c_m are free parameters subject to the constraint

$$c_0 + 6\,c_1 + 12\,c_2 + 8\,c_3 = 1. \quad (66)$$

assuring that for a trivial configuration $Q_\mu(n_B)$ is equal to the unit matrix. Finally, the matrix $Q_\mu(n_B)$ is the product of two fuzzy link operators connecting the points $2n_B$ and $2n_B + 2\hat{\mu}$ on the fine lattice

$$Q_\mu(n_B) = \mathcal{W}_\mu(2n_B)\mathcal{W}_\mu(2n_B+\hat{\mu})\,. \quad (67)$$

Free fermion fields in **d=4**

The action of spin 1/2 Dirac fermions in the continuum has the form

$$\mathcal{A}_{cont}(\bar{\psi},\psi) = \int_x \bar{\psi}(x)(\gamma_\mu\partial_\mu + m)\psi(x)\,, \quad (68)$$

where $\{\gamma_\mu,\gamma_\nu\} = 2\delta_{\mu\nu}$.

On the lattice, the fermion field lives on the lattice points. The most general action which is invariant under cubic rotation, reflection, permutation of the coordinate axes and discrete translations, reads

$$\mathcal{A}(\bar{\psi},\psi) = \sum_{n,n'} \bar{\psi}_n\left(f^\mu(n-n')\gamma_\mu + f(n-n')\right)\psi_{n'}\,, \quad (69)$$

where the local functions f^μ and f have the following reflection properties

$$f^\mu(-r) = -f^\mu(r), \qquad f(-r) = f(r). \quad (70)$$

A naive discretization of eq. (68) leads to

$$\mathcal{A}_{naive}(\bar{\psi},\psi) = \frac{1}{2}\sum_{n,\mu}(\bar{\psi}_n\gamma_\mu\psi_{n+\hat{\mu}} - \bar{\psi}_{n+\hat{\mu}}\gamma_\mu\psi_n) + \sum_n m\bar{\psi}_n\psi_n\,, \quad (71)$$

corresponding to

$$f^\mu(r) = 1/2(\delta_{r+\hat\mu,0} - \delta_{r-\hat\mu,0}), \qquad f(r) = m\delta_{r,0}, \tag{72}$$

or, in Fourier space

$$f^\mu(k) = i\sin(k_\mu), \qquad f(k) = m. \tag{73}$$

The momentum space propagator has the form

$$D(k) = \frac{-i\sum_\mu \sin(k_\mu)\gamma_\mu + m}{\sum_\mu \sin^2(k_\mu) + m^2}. \tag{74}$$

The poles of the denominator determine the spectrum. Since $\sin(k_\mu) = 0$ at $k_\mu = 0$ and $k_\mu = \pi$, the propagator in eq. (74) has 16 poles defining 16 fermion species with the correct relativistic dispersion relation $E = -ik_4 = (\vec{k}^2 + m^2)^{1/2}$ in the continuum limit. This is the species doubling problem specific to fermions on the lattice.

For $m = 0$, the naive action in eq. (71) is chiral invariant, i.e. it remains unchanged under the transformation

$$\psi_n \to \exp(i\alpha\gamma_5)\,\psi_n, \qquad \bar\psi_n \to \bar\psi_n \exp(i\alpha\gamma_5). \tag{75}$$

This symmetry requires $f = 0$ in eq. (69) and leads to a propagator $(\sum_\mu f^\mu \gamma_\mu)^{-1}$. For small k, $f^\mu(k) \sim ik_\mu$ since the lattice action should reproduce the continuum action in this limit. In addition, $f^\mu(k)$ is a periodic function (Sect. 2.2). It follows then that $-if^\mu(k) > 0$ for k_μ small and positive and < 0 for k_μ close to, but less than 2π If $f^\mu(k)$ is a continuous function, it should have a zero in between. This intuitive argument implies a close relation between the chiral symmetry of the action and the doubling problem: if the action is chiral symmetric, we have species doubling[23].

The arguments above use certain assumptions and it is useful to think it over, which of them one would be willing to give up. The periodicity in k might be avoided using a random lattice. The related theoretical and technical difficulties do not make this possibility attractive. We assumed that $-if^\mu(k)$ is a real function, which corresponds to hermiticity of the action in Minkowski space. If we do not want to see ghosts in the spectrum it is better to keep this. We used translation symmetry (f_μ depends only on $n - n'$, eq. (69)) related to momentum conservation - we want to keep this also‡. Finally, we might drop the assumption that $f^\mu(k)$ is a continuous function. This would imply loosing the locality of $f^\mu(n - n')$ and of the action. This is the most horrible of all the possibilities.

There remains to violate the chiral symmetry of the action. It is well known in perturbation theory in models without chiral symmetry (scalar QFT) or when non-chiral invariant regularization is used (QED, for example, with Pauli-Villars regularization) that radiative corrections induce an additive mass renormalization and so, a tuning problem. This will be the generic situation in our case also. However, as we mentioned in the introduction, the FP action offers an elegant solution to the tuning and other related problems[3].

A simple way to kill the doublers is to add to the action an appropriate non-chiral invariant term which disappears in the continuum limit. Wilson suggested[24] ('Wilson fermion action') to add to eq. (71)

$$\frac{r}{2}\sum_{n,\mu}\left(2\bar\psi_n\psi_n - \bar\psi_n\psi_{n+\hat\mu} - \bar\psi_{n+\hat\mu}\psi_n\right) \tag{76}$$

‡I do not know whether it is possible to break hermiticity or translation symmetry so that the symmetry comes back in the continuum limit but the doublers are killed

with $r \neq 0$ a free parameter. This term gives a contribution to $f(k)$

$$f(k) = m + 2r \sum_{\mu} \sin^2(\frac{k_\mu}{2}) . \tag{77}$$

The extra term gives an $O(1/a)$ mass to the doublers, but it is suppressed around the $k \sim 0$ pole. This procedure can be generalized to interactive theories also and this is the method which is used in most of the QCD calculations.

Consider now RGTs on a d-dimensional hypercubic lattice with a scale factor of 2. Write the RGT in the form

$$\exp\Big(-\mathcal{A}'(\bar{\chi}, \chi)\Big) = \int D\bar{\psi} D\psi \exp\Big[-\Big(\mathcal{A}(\bar{\psi}, \psi) + T_f(\bar{\chi}, \chi; \bar{\psi}, \psi)\Big)\Big] \tag{78}$$

with

$$\int D\bar{\psi} D\psi = \prod_n \int d\bar{\psi}_n d\psi_n , \tag{79}$$

where $\bar{\psi}, \psi$ are Grassmann variables. Consider blocking kernels of the form

$$T_f(\bar{\chi}, \chi; \bar{\psi}, \psi) = \kappa_f \sum_{n_B} \Big(\bar{\chi}_{n_B} - b_f \sum_n \bar{\psi}_n \omega(2n_B - n)\Big) \times \Big(\chi_{n_B} - b_f \sum_n \omega(2n_B - n)\psi_n\Big) . \tag{80}$$

We shall assume that the averaging function ω is diagonal in Dirac space and real. Observe that T_f is not chiral invariant.

We shall consider two different block transformations[12, 25, 26]. The first is just the blocking introduced for a scalar field in Sect. 3.1 and also for the σ-model in Sect. 5.1. We define $\omega(2n_B - n) = 2^{-d}$ if n is in the hypercube whose center in indexed by n_B and zero otherwise.

Keeping in mind QCD, the blocking in the second example has a form which is easy to make gauge invariant in the presence of gauge interactions. Place the block fields in the even points of the fine lattice (n is even if all its coordinates are even). A simple choice for the averaging function ω in this case is

$$\omega(0,0,\dots,0) = c_0, \qquad \omega(\pm 1, 0, \dots, 0) = \dots = \omega(0,0,\dots,\pm 1) = c_1, \dots, \omega(\pm 1, \pm 1, \dots, \pm 1) = c_d . \tag{81}$$

The parameters $c_0, c_1, \dots, c_d$ satisfy the normalization condition, eq. (17),

$$c_0 + 2d\, c_1 + 2d(d-1)\, c_2 + \dots + 2^d c_d = 1 . \tag{82}$$

Observe that the fine field ψ_n contributes to several block averages, in general. We shall call a blocking with this property 'overlapping'. The block transformations in eqs. (47,51) and the first example above are 'non-overlapping', while those in eqs. (63,67, 81) are overlapping.

There is a natural symmetric choice for the c_i parameters in eq. (81):

$$c_i = 2^{-d-i}, \qquad i = 0, 1, \dots, d . \tag{83}$$

In this case, the sum of contributions of ψ_n to all the block fields χ is independent of n, the averaging is 'flat'.

QCD

The action is the sum of the Yang-Mills action and the action of quarks in interaction with the colour gauge field

$$\mathcal{A}^{QCD}_{cont}(\bar{\psi},\psi,A_\mu) = \beta\mathcal{A}^{FP}_{cont}(A_\mu) + \int_x \bar{\psi}(x)\Big[\big(\gamma_\mu(\partial_\mu - iA_\mu(x)\big) + m\Big]\psi(x)\,. \qquad (84)$$

The quark fields carry Dirac, colour and flavour indices, the gauge field does not know about flavour, the mass (matrix) does not know about colour. For $N = 3$, this theory is believed to be the theory of the hadrons. It should explain the masses, widths and other static properties of hadrons, scattering events, all kinds of hadronic matrix elements, the behaviour of hadronic matter at finite temperature and density, spontaneous chiral symmetry breaking, the η' problem, etc. Most of these problems are non-perturbative.

The simplest lattice regularized QCD action, which we shall call the 'Wilson action' is the sum of the gauge action eq. (58) and the fermionic action, eq. (71) plus eq. (76), made gauge invariant. Take $r = 1$, which is the preferred value (for technical reasons) in simulations

$$\mathcal{A}^{QCD}_W(\bar{\psi},\psi,U) = \frac{\beta}{N}\sum_{plaq}(N - Re Tr\, U_p) +$$
$$\frac{1}{2}\sum_{n,\mu}\Big(\bar{\psi}_n(\gamma_\mu - 1)U_\mu(n)\psi_{n+\hat{\mu}} - \bar{\psi}_{n+\hat{\mu}}(\gamma_\mu + 1)U^\dagger_\mu(n)\psi_n\Big) + \sum_n \bar{\psi}_n(m+4)\psi_n\,. \qquad (85)$$

We write the RGT in the form

$$\exp\Big[-\big(\beta'\mathcal{A}'_g(V) + \mathcal{A}'_f(\bar{\chi},\chi,V)\big)\Big] =$$
$$\int D\bar{\psi}D\psi DU \exp\Big[-\beta\big(\mathcal{A}_g(U) + T_g(V,U)\big) + \big(\mathcal{A}_f(\bar{\psi},\psi,U) + T_f(\bar{\chi},\chi;\bar{\psi},\psi,U)\big)\Big]\,. \qquad (86)$$

The gauge kernel T_g is defined in eq. (61), while the fermion kernel is the gauge invariant version of eq. (80)

$$T_f(\bar{\chi},\chi;\bar{\psi},\psi,U) = \kappa_f\sum_{n_B}\Big(\bar{\chi}_{n_B} - b_f\sum_n \bar{\psi}_n\omega^\dagger(U)_{n,n_B}\Big)$$
$$\times\Big(\chi_{n_B} - b_f\sum_n \omega(U)_{n_B,n}\psi_n\Big)\,. \qquad (87)$$

In eq. (87), the averaging function ω is made gauge invariant by connecting n_B with the point n by a product of U matrices. This can be done without problems if n_B overlaps with one of the fine lattice points as in the second example in Sect. 5.3, eq. (81). Beyond hypercubic symmetry the choice of these paths is part of the freedom we have in defining an RGT. A simple choice is to take the shortest connecting paths. For $U \to 1$, $\omega(U)_{n_B,n} \to \omega(2n_B - n)$ defined in eq. (80), which satisfies the normalization condition, eq. (17). This condition gives a meaning to b_f in eq. (87). As we shall see (Sect. 8), in a free field theory b_f is determined by the engineering dimension of the field. This should be the case also in the classical theory even in the presence of interaction, since the corrections to b_f are related to the anomalous dimension of the field generated by quantum fluctuations[1,8,5]. We shall return to this point later.

THE SADDLE POINT EQUATION FOR THE FIXED POINT

For free field theories with a quadratic blocking kernel, like those in eq. (80) or in eq. (21), the RGT is reduced to Gaussian integrals. The integrals can be performed exactly and the problem can be investigated in every detail (Sect. 8). In general, however, the path integral of a RGT is a highly non-trivial problem.

A basic observation to proceed is that in AF theories the FP lies at $\beta = \infty$ and in this limit the path integral can be calculated in the saddle point approximation[2]. This leads to an equation in classical field theory which determines the FP action. We shall derive this equation first in the non-linear **σ-model** and then in QCD.

The FP of the O(N) non-linear σ-model

As eqs. (47,49) or eq. (51) show, for $\beta \to \infty$ the blocking kernel is proportional to β. For a given fixed coarse field configuration $\mathbf{R}$ we can perform the path integral

$$\int DS \exp\left[-\beta\left(\mathcal{A}(\mathbf{S}) + T_\sigma(\mathbf{R},\mathbf{S})\right)\right] \tag{88}$$

in the saddle point approximation leading to the classical equation

$$\mathcal{A}'(\mathbf{R}) = \min_{\{\mathbf{S}\}}\left[\mathcal{A}(\mathbf{S}) + T_\sigma^\infty(\mathbf{R},\mathbf{S})\right], \tag{89}$$

where T_σ^∞ is the $\beta = \infty$ limit of the blocking kernel. The FP of the transformation is determined by the equation

$$\mathcal{A}^{FP}(\mathbf{R}) = \min_{\{\mathbf{S}\}}\left[\mathcal{A}^{FP}(\mathbf{S}) + T_\sigma^\infty(\mathbf{R},\mathbf{S})\right]. \tag{90}$$

Write down this equation explicitly for the block transformation defined in eqs. (47,49)

$$\mathcal{A}^{FP}(\mathbf{R}) = \min_{\{\mathbf{S}\}}\left[\mathcal{A}^{FP}(\mathbf{S}) - \kappa_\sigma \sum_{n_B}\left(\mathbf{R}_{n_B}\mathbf{\Sigma}_{n_B} - |\mathbf{\Sigma}_{n_B}|\right)\right], \tag{91}$$

where we have used $\ln Y_N(z) = z(1 + O(\ln z/z))$ for large z and $\mathbf{\Sigma}_{n_B}$ is defined in eq. (48).

Observe that eq. (88) reduces to the saddle point equation, eq. (89), for any configuration $\mathbf{R}$. If the configuration $\mathbf{R}$ is strongly fluctuating then the minimizing configuration $\mathbf{S}_{min}$ will not be smooth either. In general, eqs. (89,90) and their solutions have nothing to do with perturbation theory. A starting condition for the FP equation, eq. (90), is that $\mathcal{A}^{FP}$ goes over to the classical action $\mathcal{A}_{cont}(\mathbf{R})$, eq. (42), for very smooth $\mathbf{R}$ configurations.

The FP of QCD

In the $\beta \to \infty$ limit ($V, \bar{\chi}, \chi$ fixed), the Boltzmann factor on the r.h.s. of eq. (86) is dominated by the gauge part. The integral over the U field is saturated by the minimizing configuration $U_{min} = U_{min}(V)$ leading to the equation

$$\mathcal{A}_g'(V) = \min_{\{U\}}\left[\mathcal{A}_g(U) + T_g^\infty(V,U)\right]. \tag{92}$$

The gauge part of the FP is determined by

$$\mathcal{A}_g^{FP}(V) = \min_{\{U\}}\left[\mathcal{A}_g^{FP}(U) + T_g^\infty(V,U)\right]. \tag{93}$$

Using the explicit form of the blocking kernel T_g in eq. (61) we obtain

$$\mathcal{A}_g^{FP}(V) = \min_{\{U\}}\left(\mathcal{A}_g^{FP}(U) - \frac{\kappa_g}{N}\sum_{n_B,\mu}\left[ReTr\left(V_\mu(n_B)Q_\mu^\dagger(n_B)\right) - f\left(Q_\mu(n_B)\right)\right]\right), \tag{94}$$

where

$$f(Q) = \max_W \left[ReTr(WQ^\dagger) \right], \qquad W \in SU(N) . \tag{95}$$

Eq. (95) follows from eq. (62) in the $\beta \to \infty$ limit.

The remaining fermionic integral has the form

$$\exp\left(-\mathcal{A}'_f(\bar{\chi},\chi,V)\right) = \int D\bar{\psi} D\psi \exp\left[-\left(\mathcal{A}_f(\bar{\psi},\psi,U_{min}) + T_f(\bar{\chi},\chi;\bar{\psi},\psi,U_{min})\right)\right], \tag{96}$$

where $U_{min}(V)$ is the minimizing configuration from eq. (92). If the action $\mathcal{A}_f$ is quadratic in the fermion fields, then the blocked action $\mathcal{A}'_f$ is quadratic also, since the kernel T_f is quadratic (eq. (87)) and the integral in eq. (96) is Gaussian. Write the fermion action in the following general form

$$\mathcal{A}_f(\bar{\psi},\psi,U) = \sum_{n,n'} \bar{\psi}_n h_{n,n'}(U) \psi'_n , \tag{97}$$

where $h(U)_{n,n'}$ carries Dirac, colour and flavour indices which are not indicated explicitly. Using eq. (87) and performing the Gaussian integral in eq. (96) we get

$$h'(V)^{-1}_{n_B,n'_B} = \frac{1}{\kappa_f}\delta_{n_B,n'_B} + b_f^2 \sum_{n,n'} \omega(U)_{n_B,n} h(U)^{-1}_{n,n'} \omega(U)^\dagger_{n',n'_B} . \tag{98}$$

The FP satisfies the equation

$$h^{FP}(V)^{-1}_{n_B,n'_B} = \frac{1}{\kappa_f}\delta_{n_B,n'_B} + b_f^2 \sum_{n,n'} \omega(U)_{n_B,n} h^{FP}(U)^{-1}_{n,n'} \omega(U)^\dagger_{n',n'_B} . \tag{99}$$

Eqs. (94,99) determine the FP action of QCD. Gaussian integrals of c-number fields are equivalent to minimization. This can be generalized to Gaussian Grassmann integrals to express eq. (99) as

$$\mathcal{A}_f^{FP}(\bar{\chi},\chi,V) = \left[\mathcal{A}_f^{FP}(\bar{\psi},\psi,U) + T_f(\bar{\chi},\chi;\bar{\psi},\psi,U)\right]_{|\bar{\psi}_{st},\psi_{st}} \tag{100}$$

where $\psi_{st}, \bar{\psi}_{st}$ make the r.h.s. stationary, $\delta/\delta\psi$ [...]=0, $\delta/\delta\bar{\psi}$ [...]=0. In eq. (100), $U = U(V)$ and the fields $\bar{\chi},\chi,\bar{\psi},\psi$ are c-number fields.

THE FP ACTION FOR WEAK FIELDS

We have claimed in Sect. 4 that the FP action is classically perfect: it reproduces all the essential physical properties of the continuum classical action on the lattice even if the resolution of the lattice is poor. The proof of this statement follows directly from the FP equations (eq. (91) for the σ-**model** or eqs. (94,99) for QCD) without solving them explicitly. Before presenting these arguments let us study the FP equations in a limit where analytic results can be obtained.

Consider first the FP equation in the *O(N)* non-linear σ-**model,** eq. (91). Take a configuration where the spins fluctuate around the first axis in *O(N)* space

$$\mathbf{R}_{n_B} = \begin{pmatrix} \sqrt{1-\chi^2_{n_B}} \\ \chi_{n_B} \end{pmatrix} , \tag{101}$$

where χ_{n_B} has $(N-1)$ components. Assume, these fluctuations are 'weak', $|\chi_{n_B}| \ll 1$. In this case, the minimizing field $\mathbf{S}_{min}(\mathbf{R})$ will also fluctuate around the first axis, eq. (43), and these fluctuations will also be weak, $|\pi_{n_B}| \ll 1$. For weak fields eq. (91) can be expanded in powers of the fluctuations and one can study the solution order by order.

There is a difference between the 'weak field' introduced here and the 'smooth field' used repeatedly earlier. A smooth field changes slowly in configuration space and is dominated by small k values in Fourier space. In a weak field, the size of deviations from the classical vacuum configuration is small, but the field might change rapidly in configuration space, so higher momentum components might be important in Fourier space.

In leading order of the weak field expansion, the FP equation is quadratic in the fluctuations. Using translation symmetry we can write

$$\mathcal{A}^{FP}(\mathbf{R}) = \frac{1}{2} \sum_{n_B, r_B} \rho(r_B) \chi_{n_B} \chi_{n_B+r_B} + O(\chi^4), \tag{102}$$

where the unknown couplings, to be determined by the FP equation, are denoted by $\rho(r_B)$. Expanding also the blocking kernel in eq. (91), we get at the quadratic level

$$\frac{1}{2} \sum_{n_B, r_B} \rho(r_B) \chi_{n_B} \chi_{n_B+r_B} = \min_{\{\pi\}} \Big[\frac{1}{2} \sum_{n,r} \rho(r) \pi_n \pi_{n+r} + 2\kappa_\sigma \sum_{n_B} (\chi_{n_B} - \frac{1}{4} \sum_{n \in n_B} \pi_n)^2 \Big]. \tag{103}$$

Using the trivial relation between Gaussian integrals and minimization we observe that eq. (103) is just the FP equation of a free scalar theory:

$$c \exp\Big(-\frac{1}{2} \sum_{n_B, r_B} \rho(r_B) \chi_{n_B} \chi_{n_B+r_B}\Big) = \prod_n \int d\pi_n \exp\Big[-\frac{1}{2} \sum_{n,r} \rho(r) \pi_n \pi_{n+r} - 2\kappa_\sigma \sum_{n_B} (\chi_{n_B} - \frac{1}{4} \sum_{n \in n_B} \pi_n)^2 \Big]. \tag{104}$$

The same steps can be used for the Yang-Mills or QCD FP actions. In quadratic order the Yang-Mills FP equation, eq. (94), goes over to the FP problem of a free (Abelian) gauge theory, while the fermion part of the QCD FP action, eq. (99), becomes equivalent to the FP equation of free Dirac fermions. In the next section we give a general solution of the free field FP problem.

THE FREE FIELD FP PROBLEM

As we discussed in Sect. 7, in leading order of the weak field expansion the FP equations of AF theories are reduced to that of free fields. We shall study the free field problem in this section.

In all our previous considerations we applied lattice regularization. For free field theories without gauge symmetry we might also use a simple momentum space cut-off, which leads quickly to some results we already referred to in the previous sections[1,8,27]. We present these results first, and then turn to the problem of FP actions on the lattice.

Scalar field with momentum space cut-off

We work in Fourier space and constrain all the momenta to be smaller than Λ^{cut}. As always, we use dimensionless quantities, the dimensions are carried by the cut-off. We write the quadratic action in the general form

$$\mathcal{A}(\phi) = \frac{1}{2}\int_{|q|\leq 1} \phi(-q)\,\rho(q)\,\phi(q)\,, \tag{105}$$

and perform a RGT by integrating out the fields in the momentum range $1/2 < |q| \leq 1$. We write

$$\phi(q) = \phi_0(q) + \phi_1(q)\,, \tag{106}$$

where $\phi_0(q)$ and $\phi_1(q)$ are non-zero only for $|q| \leq 1/2$ and $1/2 < |q| \leq 1$, respectively. In the RGT we have to integrate out $\phi_1(q)$, i.e. perform the path integral

$$\exp\bigl(-\tilde{\mathcal{A}}(\phi_0)\bigr) = \prod_{\frac{1}{2}<|q|\leq 1} \int d\phi_1(q)\ \exp\bigl(\mathcal{A}(\phi_0+\phi_1)\bigr)\,. \tag{107}$$

Using the quadratic action in eq. (105) we get

$$\mathcal{A}(\phi_0+\phi_1) = \frac{1}{2}\int_{|q|\leq\frac{1}{2}} \phi(-q)\rho(q)\phi(q) + \frac{1}{2}\int_{\frac{1}{2}<|q|\leq 1} \phi(-q)\rho(q)\phi(q)\,. \tag{108}$$

Writing eq. (108) into eq. (105), the ϕ_0-dependent part can be brought out of the integral, while the ϕ_1-integral gives a field independent, irrelevant constant. By rescaling and relabeling the ϕ_0 field as

$$\phi'(q') = \zeta^{-1}\phi_0(q) \qquad q' = 2q\,, \qquad 0 \leq |q'| \leq 1\,, \tag{109}$$

we get

$$\tilde{\mathcal{A}}(\phi_0) = \zeta^2\frac{1}{2}\int_{|q|\leq\frac{1}{2}} \phi'(-2q)\,\rho(q)\,\phi'(2q) =$$

$$\zeta^2 2^{-d}\frac{1}{2}\int_{|q|\leq 1} \phi'(-q)\,\rho(q/2)\,\phi'(q) = \mathcal{A}'(\phi')\,. \tag{110}$$

The role of the rescaling factor ζ in eq. (109) is the same as that of b in eqs. (16,18,...), whose significance we shall see in a moment. As eq. (105) and eq. (110) show, under the RGT ρ transforms as

$$\rho(q) \to \zeta^2 2^{-d}\rho(q/2)\,. \tag{111}$$

Expand ρ in powers of momenta

$$\rho(q) = \alpha_0 + \alpha_2\sum_\mu q_\mu q_\mu + \alpha_4\bigl(\sum_\mu q_\mu q_\mu\bigr)^2 + \alpha_6\bigl(\sum_\mu q_\mu q_\mu\bigr)^3 \ldots\,. \tag{112}$$

For α_0 small, the pole of the propagator $1/\rho(q)$ is close to $q = 0$ and $\alpha_0 = m^2$, where m is the mass of the free particle. We shall require that the action density before and after the transformation goes over into the form of classical field theory for small q-values, i.e. we fix the coefficient of $\sum_\mu q_\mu q_\mu$ in ρ to be 1. This condition fixes ζ

$$\zeta^2 2^{-d}\sum_\mu \frac{q_\mu}{2}\frac{q_\mu}{2} = \sum_\mu q_\mu q_\mu \to \zeta = 2^{\frac{d+2}{2}}\,. \tag{113}$$

Eq. (111) gives then the transformation rules for the other couplings

$$\alpha_0 \to 2^2\,\alpha_0\,, \qquad \alpha_4 \to 2^{-2}\,\alpha_4\,, \qquad \alpha_6 \to 2^{-3}\,\alpha_6\,,\ldots\,. \tag{114}$$

Eq. (114) shows that the operator ϕ^2 (the coupling α_0 or m^2) is relevant with an eigenvalue $\lambda_0 = 4$, all the other operators (couplings) which contain four or more derivatives are irrelevant. The largest irrelevant eigenvalue is 1/4.

Replacing the region of allowed momenta of a cut-off sphere by a cut-off hypercube, the regularization will violate rotation symmetry. Only hypercubic symmetry remains, like on the lattice. In this case we have to allow in eq. (112) terms like $\beta_4 \sum_\mu (q_\mu q_\mu q_\mu q_\mu)$, which is not $O(d)$ invariant, but hypercubic symmetric. The coupling β_4 is also irrelevant, $\beta_4 \to 1/4\beta_4$ under the RGT. That means that rotational symmetry is restored as the RGT is iterated, as we go towards the infrared.

If we start with $\alpha_0 = 0$, we are and remain on the critical surface and run rapidly towards the FP which, in this case, has the simple form

$$\mathcal{A}^{FP}(\phi) = \frac{1}{2}\int_{|q|\le 1} \phi(-q)q^2\phi(q)\,, \tag{115}$$

describing a free massless scalar particle. As we mentioned before, the form of the FP depends on the block transformation (typically, it will look more complicated on the lattice), but the conclusions concerning the relevant, marginal and irrelevant eigenvalues is independent of that.

Consider now small perturbations around the FP, eq. (115), like $\lambda_4\phi^4$, $\lambda_6\phi^6,\cdots$, with $\lambda_4, \lambda_6, \cdots$ small, as discussed in Sect. 3.4

$$\begin{aligned}\mathcal{A}(\phi) = \mathcal{A}^{FP}(\phi) + \frac{\alpha_0}{2}\int_q \phi(-q)\phi(q) + \lambda_4 \int_{q_1,\ldots,q_4} \phi(q_1)\ldots\phi(q_4)\delta(q_1+\ldots+q_4) + \\ \lambda_6 \int_{q_1,\ldots,q_6} \phi(q_1)\ldots\phi(q_6)\delta(q_1+\ldots+q_6) + \ldots\,,\end{aligned} \tag{116}$$

where only three of the perturbations are written out explicitly. Using this action $\mathcal{A}$ in eq. (107), expanding the exponent in $\alpha_0, \lambda_4, \lambda_6 \ldots$ up to linear order, performing the Gaussian ϕ_0-integrals and rescaling and relabeling as in eq. (107), we get

$$\begin{aligned}\alpha_0 &\to (\alpha_0 + 6\,c\lambda_4 + 15\,c^2\lambda_6 + \ldots)\zeta^2 2^{-d}\,,\\ \lambda_4 &\to (\lambda_4 + 15\,c^2\lambda_6 + \ldots)\zeta^4 2^{-3d}\,,\\ \lambda_6 &\to (\lambda_6 + \ldots)\zeta^6 2^{-5d}\,,\end{aligned} \tag{117}$$

with

$$c = \int_{|q|=\frac{1}{2}}^{|q|=1} \frac{1}{q^2}\,. \tag{118}$$

The corresponding T-matrix (eq. (35)) is a triangular matrix, the eigenvalues λ are just the diagonal elements

$$\begin{aligned}\lambda_1 &= 2^2\,, & h^{(1)} &= \int_x \phi^2(x)\,,\\ \lambda_2 &= 2^{4-d}\,, & h^{(2)} &= \int_x \left[\phi^4(x) + \alpha\phi^2(x)\right],\\ \lambda_3 &= 2^{6-2d}\,, & h^{(3)} &= \int_x \left[\phi^6(x) + \beta_1\phi^4(x) + \beta_2\phi^2(x)\right].\end{aligned} \tag{119}$$

As eq. (119) shows, there is a simple relation between the eigenvalue λ_θ and the dimension d_θ of the operator $\theta(x)$ which has the highest dimension in the corresponding eigenoperator *h:*

$$\lambda_\theta = 2^{d-d_\theta} \tag{120}$$

In the eigenvector h the highest dimensional operator $\theta(x)$ is mixed with lower dimensional operators.

In $d = 4$, operators with dimension less than, equal, larger than 4 are relevant, marginal and irrelevant, respectively. This result is also valid in gauge and fermion theories. In *SU(N)* Yang-Mills theory the lowest dimensional gauge invariant operator is $\mathrm{Tr}(F_{\mu\nu}F_{\mu\nu})$, which is marginal.

Lattice FP actions in different free field theories

Consider first a real scalar field. Denote the fields on the coarse and on the fine lattice by χ and ϕ, respectively. The lattice units are a_0 and $a_1 = a_0/2$. Assume, we are in the FP, and perform a RGT:

$$c\exp\Big(-\frac{1}{2}\sum_{n_B,r_B}\rho(r_B)\chi_{n_B}\chi_{n_B+r_B}\Big) =$$
$$\prod_n\int d\phi_n \exp\Big[-\frac{1}{2}\sum_{n,r}\rho(r)\phi_n\phi_{n+r} - \frac{1}{2}\kappa\sum_{n_B}(\chi_{n_B} - b\sum_n\omega(2n_B - n)\phi_n)^2\Big]. \tag{121}$$

Multiply this equation by $\chi_{n'_B}\chi_{n''_B}$ and integrate over the χ fields. Using eq. (19) we obtain

$$\langle\chi_{n'_B}\chi_{n''_B}\rangle = b^2\sum_{n',n''}\omega(2n'_B - n')\omega(2n''_B - n'')\langle\phi_{n'}\phi_{n''}\rangle + \frac{1}{\kappa}\delta_{n'_B,n''_B}\,. \tag{122}$$

In Fourier space eq. (122) has the form

$$\frac{1}{\tilde\rho(q_B)} = b^2\frac{1}{2^d}\sum_l\tilde\omega\Big(\frac{q_B}{2}+\pi l\Big)\frac{1}{\tilde\rho(\frac{q_B}{2}+\pi l)}\tilde\omega^*\Big(\frac{q_B}{2}+\pi l\Big) + \frac{1}{\kappa},$$
$$l = (l_1,\ldots,l_d)\,,\, l_\mu = 0,1\,. \tag{123}$$

The Fourier transformation is defined as

$$f(n) = \int_{-\pi}^{\pi}\cdots\int_{-\pi}^{\pi}\frac{d^dq}{(2\pi)^d}\exp(iqn)\tilde f(q)\,. \tag{124}$$

We demand, as discussed in Sect. 8.1, that any action should go over into the classical continuum action for smooth fields. That requires that $\rho(q_B) \to q_B^2 + O(q_B^4)$ for small q_B. (The FP action is on the critical surface, the mass parameter α_0 in eq. (112) is zero.) Considering eq. (123) in the $q_B \to 0$ limit, the $l \neq 0$ contribution on the r.h.s. can be neglected and we obtain

$$\frac{1}{q_B^2} = b^2\frac{1}{2^d}|\tilde\omega(0)|^2\frac{1}{(q_B/2)^2}\,. \tag{125}$$

The normalization condition, eq. (17), gives $\tilde\omega(0) = 1$ and so

$$b = 2^{-\frac{d-2}{2}}\,. \tag{126}$$

We shall use the relation of eq. (122) recursively, i.e. we connect the propagator $\langle\phi_{n'}\phi_{n''}\rangle$ on the r.h.s. of eq. (122) with the propagator on the lattice with unit $a_2 = a_0/4$. Denote the field on this lattice by η and the lattice points by n_f:

$$\langle\chi_{n'_B}\chi_{n''_B}\rangle = b^4 \sum_{n',n'_f} \omega(2n'_B - n')\omega(2n' - n'_f) \times \sum_{n'',n''_f} \omega(2n''_B - n'')\omega(2n'' - n''_f)\langle\eta_{n'_f}\eta_{n''_f}\rangle + \frac{1}{\kappa}\left[\delta_{n'_B,n''_B} + b^2\sum_n \omega(2n'_B - n)\omega(2n''_B - n)\right]. \quad (127)$$

Iterating this equation further, after j steps one arrives at a lattice with lattice unit $a_j = a_0/2^j$. For $j \to \infty$, the lattice will be infinitely fine and the propagator for this lattice can be replaced by the propagator in the continuum. This way, the FP propagator $\langle\chi_{n'_B}\chi_{n''_B}\rangle$ will be expressed explicitly in terms of the propagator in the continuum and of the product and sum of the averaging function ω. The inverse propagator gives the FP action.

If the block transformation is non-overlapping (see the paragraph after eq. (82)), the sum in the last term of eq. (127) is $\sim 1/\kappa\delta_{n'_B,n''_B}$. This remains true in every order of the iteration. Although for overlapping transformations this is not true anymore, the $\sim 1/\kappa$ contribution to the propagator remains local. The non-locality of the propagator is carried completely by the first term in eq. (127).

Without going through the detailed derivation, we quote the final results only. Let us start with an example: take the simple non-overlapping transformation, where $\omega(2n_B - n) = 2^{-d}$ if n is in the hypercube whose center is indexed by n_B and zero otherwise (Sect. 3.1). The FP propagator in Fourier space has the form[9, 28, 29]

$$\frac{1}{\tilde{\rho}(q)} = \sum_{l\in Z^d} \frac{1}{(q + 2\pi l)^2} \prod_\mu \frac{\sin^2(\frac{q_\mu}{2} + \pi l_\mu)}{(\frac{q_\mu}{2} + \pi l_\mu)^2} + \frac{4}{3\kappa}, \quad (128)$$

where the summation is over integer vectors $l = (l_1, \cdots, l_d)$ and $(q+2\pi l)^2 = \sum_\mu (q_\mu + 2\pi l_\mu)^2$. The $\sim 1/\kappa$ term in eq. (128) is a constant in Fourier space, hence $\sim \delta_{n,n'}$ inconfiguration space, which is the general result for non-overlapping transformations. The second factor under the sum in eq. (128) is regular, it is the Fourier transform of the product of the local averaging function ω. The pole singularities come entirely from the $(q+2\pi l)^{-2}$ piece. These poles define the spectrum of the FP theory on the lattice. Fix the spatial part of the momentum to be $\mathbf{q} = \mathbf{p}, p_i \in (-\pi, \pi)$, and find the pole singularities in the complex q_d-plane. There is an infinite sequence of poles on the imaginary axis

$$q_d = \pm i\,|\mathbf{p} + 2\pi\mathbf{l}|, \qquad \mathbf{l} \in Z^{d-1}. \quad (129)$$

Denoting $\mathbf{p} + 2\pi\mathbf{l}$ by k, $k_i \in (-\infty, \infty)$, we get the exact relativistic dispersion relation of a massless particle

$$-i\, q_d = E(\mathbf{k}) = |\mathbf{k}|. \quad (130)$$

Through the l-summation in eq. (128), the full exact continuum spectrum is reproduced (the momentum k is not constrained to the Brillouin zone), even though we are on the lattice, which resolves the large k waves very poorly. This result is to be compared with the dispersion relation of the standard nearest-neighbour action (eq. (13) with $V = 0$). If k is close to the end of the Brillouin zone, the dispersion relation is wrong by a factor of 2 or more.

Table 1. The couplings of the free field FP action at a distance $r = (r_0, r_1)$ for the optimal choice of the block transformation with κ=4. Note that in our convention, for the standard action the only non-vanishing entry in this list would be $\rho_{ST}(1,0) = -1$.

r	$\rho(r)$	r	$\rho(r)$
(1,0)	-0.61802	(4,0)	$-2.632 \cdot 10^{-6}$
(1,1)	-0.19033	(4,1)	$7.064 \cdot 10^{-7}$
(2,0)	$-1.998 \cdot 10^{-3}$	(4,2)	$1.327 \cdot 10^{-6}$
(2,1)	$-6.793 \cdot 10^{-4}$	(4,3)	$-7.953 \cdot 10^{-7}$
(2,2)	$1.625 \cdot 10^{-3}$	(4,4)	$6.895 \cdot 10^{-8}$
(3,0)	$-1.173 \cdot 10^{-4}$	(5,0)	$-8.831 \cdot 10^{-8}$
(3,1)	$1.942 \cdot 10^{-5}$	(5,1)	$3.457 \cdot 10^{-8}$
(3,2)	$5.232 \cdot 10^{-5}$	(5,2)	$3.491 \cdot 10^{-8}$
(3,3)	$-1.226 \cdot 10^{-5}$	(5,3)	$-3.349 \cdot 10^{-8}$
		(5,4)	$8.408 \cdot 10^{-9}$
		(5,5)	$-1.657 \cdot 10^{-10}$

Performing the sum in eq. (128) and calculating $\rho(r)$ by Fourier transformation is an easy numerical exercise. The function $\rho(r)$ and hence the FP action is local, as expected: it decays exponentially $\sim \exp(-\gamma|r|)$, as discussed in Sect. 2.4. The size of γ which defines how short ranged the action is, depends on the free parameter κ. The value $\kappa \approx 4$ is optimal. For $d = 2$, we have $\gamma \approx 3.4$ for $\kappa = 4$. In this case $\rho(r)$ is strongly dominated by the nearest neighbour and diagonal couplings as table 1 shows[2].

Actually, $\rho(r)$ is a perfect discretization of the continuum Laplacian. The question raised by this observation is, whether the FP idea can be used in solving partial differential equations, like the Navier-Stokes equation, numerically[30] .

Let us present now the general result. The fields can be scalars, fermions or vectors. For the blocking kernel we write

$$\kappa \sum_{n_B} \Big(\bar{\chi}^{a}_{n_B} - b \sum_{n,\beta} \bar{\phi}^{\beta}_{n} \omega^{\dagger}_{\beta,a}(2n_B - n)\Big)\Big(\chi^{a}_{n_B} - b \sum_{n,\beta} \omega_{a,\beta}(2n_B - n)\phi^{\beta}_{n}\Big)\,, \tag{131}$$

where the bar above the fields denotes complex or Dirac conjugation[§], the index a can be a Lorentz or Dirac index[¶]. The FP propagator is given by

$$\tilde{D}^{FP}(q) = \lim_{n\to\infty} \tilde{D}^{(n)}(q)\,, \tag{132}$$

where

$$\tilde{D}^{(n)}(q) = (2^{-d}b^2)^n \sum_{l_\mu=0}^{2^n-1} \tilde{\Omega}^{(n)}\Big(\frac{q+2\pi l}{2^n}\Big)\tilde{D}^{(0)}\Big(\frac{q+2\pi l}{2^n}\Big)\tilde{\Omega}^{\dagger(n)}\Big(\frac{q+2\pi l}{2^n}\Big) + \frac{1}{\kappa}\sum_{j=0}^{n-1}\tilde{Q}^{(j)}(q)\,, \tag{133}$$

where Ω and Q are defined as

$$\tilde{\Omega}^{(j)} = \tilde{\omega}(2^{-1})\tilde{\omega}(2^{-2})\ldots\tilde{\omega}(2^{-j})\,, \tag{134}$$

$$\tilde{Q}^{(j)}(q) = (2^{-d}b^2)^j \sum_{l_\mu=0}^{2^j-1} \tilde{\Omega}^{(j)}\Big(\frac{q+2\pi l}{2^j}\Big)\tilde{\Omega}^{\dagger(j)}\Big(\frac{q+2\pi l}{2^j}\Big)\,, \tag{135}$$

[§]For a real scalar or vector field $\bar{\phi} = \phi$ and we include a factor of 1/2 both in the action and in the kernel, as in eq. (121).

[¶]We shall consider the block transformation only where ω is diagonal in the Dirac indices.

and the rescaling factor b is given by

$$b = 2^{-d_\phi}, \tag{136}$$

where d_ϕ is the dimension of the field ϕ. For non-overlapping transformations $\tilde{Q}^{(j)}(q)$ is independent of q, in the overlapping case it is an analytic function of q. For illustration, consider free Dirac fermions in $d = 4$ using the overlapping block transformation defined by the eqs. (81,83). In this case[26]

$$\begin{aligned}
\tilde{\omega}(q) &= \prod_\mu \cos^2\left(\frac{q_\mu}{2}\right), \\
\tilde{\Omega}^{(j)}(q) &= \prod_\mu 2^{-2j} \frac{\sin^2(\frac{q_\mu}{2})}{\sin^2(\frac{q_\mu}{2^{j+1}})}, \\
\tilde{Q}^{(j)}(q) &= \frac{2^{-j}}{81} \prod_\mu \left[(1+4^{-j})\cos(q_\mu) + (2+4^{-j})\right].
\end{aligned} \tag{137}$$

For the initial fermion propagator $D^{(0)}$ one might take the Wilson fermion propagator (Sect. 5.3). Eq. (133) and eq. (137) gives the FP fermion propagator

$$D^{FP}(q) = -i\alpha^\mu(q)\gamma_\mu + \beta(q), \tag{138}$$

where

$$\alpha^\mu(q) = \sum_{l \in Z^d} \frac{q_\mu + 2\pi l_\mu}{(q+2\pi l)^2} \prod_\nu \frac{\sin^4(\frac{q_\nu}{2})}{(\frac{q_\nu}{2} + \pi l_\nu)^4} \tag{139}$$

and

$$\begin{aligned}
\beta(q) = \frac{1}{81\kappa}\Big[c_1 \prod_\mu \cos(q_\mu) &+ c_2\big(\cos(q_1)\cos(q_2)\cos(q_3) + \ldots \\
+ (\cos(q_2)\cos(q_3)\cos(q_4)\big) &+ c_3\big(\cos(q_1)\cos(q_2) + \ldots + \cos(q_3)\cos(q_4)\big) + \\
& c_4 \sum_\mu \cos(q_\mu) + c_5\Big].
\end{aligned} \tag{140}$$

The constants $c_1, \ldots, c_5$ result from different geometric sums and have the value $c_1 = 0.592581$, $c_2 = 1.388406$, $c_3 = 3.349502$, $c_4 = 8.336553$, $c_5 = 102.4105571$. This FP propagator has all the features we expected. The spectrum is given by the $\sim \alpha^\mu\gamma_\mu$ part and it is exact. The $\sim 1/\kappa$ part is analytic in q-space and ultralocal (Sect. 2.4) in configuration space. The propagator is not chiral symmetric. The chiral symmetry breaking term is $\sim 1/\kappa$ comes entirely from the non-chiral invariant block transformation.

INSTANTON SOLUTIONS AND THE TOPOLOGICAL CHARGE ON THE LATTICE

In this section we shall consider the $O(3)$ non-linear σ-model in $d = 2$. The arguments and results can be immediately generalized to Yang-Mills theories in $d = 4$.

As we discussed in Sect. 5.1, the configurations of the $O(3)$ model fall into different topological sectors characterized by an integer number Q, the topological charge of the configuration. While configurations from the same topological sector can be continuously deformed into each other, this is not possible for configurations with a different charge Q. The charge Q is the number of times the internal sphere is wrapped as the coordinate sphere is traversed. It may be defined as

$$Q = \frac{1}{8\pi} \int d^2x \epsilon_{\mu\nu} \mathbf{S}(\partial_\mu \mathbf{S} \times \partial_\nu \mathbf{S}) , \tag{141}$$

and it is related to the action by the inequality

$$\mathcal{A} \geq 4\pi |Q| . \tag{142}$$

If, for a given configuration the equality is satisfied, the configuration minimizes the action for the given topological charge Q and is therefore a solution of the equations of motion.

Instanton solutions of the FP action

Replacing the continuous Euclidean space by a discrete set of points, the notion of a 'continuous deformation of the configuration' is lost and the definition of the topological charge becomes problematic. In addition, since the lattice introduces a scale (lattice unit a), classical scale invariance is broken and no scale invariant (instanton) solutions are expected to exist.

Of coarse, if the radius ρ of the instanton is much larger than the lattice unit a, the values of the continuum solution in the lattice points define a configuration which is 'almost' a solution of the lattice Euler-Lagrange equations. The action of this quasi-solution will be

$$\mathcal{A} = 4\pi |Q| \left(1 + c\frac{a^2}{\rho^2} + O(\frac{a^4}{\rho^4})\right) . \tag{143}$$

The size and sign of c depends on the details of the form of the lattice action. Depending on the sign of c, large instantons shrink or grow to get a smaller action.

An even more serious problem is the possible existence of dislocations, topological artifacts, which are small, $O(a)$, objects but contribute to the topological charge and have an action below $4\pi|Q|$.

There are many high energy physicists who believe that fluctuating instanton configurations play an important role in QCD, especially for questions where chiral symmetry is relevant. If a discretized action describes the classical topological properties of the theory poorly, then it might happen that continuum quantum physics will be reproduced on very fine lattices only. This would increase the numerical difficulties which are not small anyhow.

The FP actions offer a solution to this problem[2,13,31,32,33,34]. We show now that the Euler-Lagrange equations of the FP action have scale invariant instanton solutions. We demonstrate first the following statement[2]:

If the configuration $\mathbf{R}$ on the coarse lattice satisfies the FP classical equations and it is a local minimum of $\mathcal{A}^{FP}$ (R), then the configuration $\mathbf{S}_{min}$ (R) on the finer lattice which minimizes the r.h.s. of the FP equation eq. (91) satisfies the FP equations as well. In addition, the value of the action remains unchanged, $\mathcal{A}^{FP}(\mathbf{S}_{min}) = \mathcal{A}^{FP}(\mathbf{R})$.

This statement is easy to show. Since $\mathbf{R}$ is a solution of the classical equations of motion $\delta\mathcal{A}^{FP}/\delta\mathbf{R} = 0$ and the configuration $\mathbf{S} = \mathbf{S}_{min}$ (R) should satisfy the equation

$$\mathbf{\Sigma}_{n_B} = \lambda_{n_B} \mathbf{R}_{n_B} , \tag{144}$$

for any n_B. Here $\mathbf{\Sigma}_{n_B}$ is the sum of the four $\mathbf{S}_n$ spins in the block, eq. (48). Eq. (144) follows by taking the variation $\delta/\delta\mathbf{R}$ of the FP equation, eq. (91). Since we assumed that the configuration R is a local minimum of $\mathcal{A}^{FP}$, we have $\lambda_{n_B} \geq 0$. On the other hand, the $\sim \kappa_\sigma$ term in eq. (91)

$$-\kappa_\sigma \sum_{n_B} \left(\mathbf{R}_{n_B} \mathbf{\Sigma}_{n_B} - |\mathbf{\Sigma}_{n_B}| \right), \tag{145}$$

takes its absolute minimum (zero) on the configuration $\mathbf{S}_{min}$(R) satisfying eq. (144). Since $\mathbf{S}_{min}$(R) is the minimum of the sum of eq. (145) and , $\mathcal{A}^{FP}$(S), it follows that $\mathbf{S}_{min}$(R) is a stationary point of $\mathcal{A}^{FP}$:

$$\left(\frac{\delta \mathcal{A}^{FP}(\mathbf{S})}{\delta \mathbf{S}} \right)_{|\mathbf{S}=\mathbf{S}_{min}} = 0\,. \tag{146}$$

Since eq. (145) is zero if eq. (144) is satisfied, we get $\mathcal{A}^{FP}$(R) $=$ $\mathcal{A}^{FP}(\mathbf{S}_{min})$, as we wanted to show.

According to this statement, if $\mathcal{A}^{FP}$ has an instanton solution of size ρ, then there exists an instanton solution of size 2p, $2^2\rho, \ldots$ with the value of the action being exactly 4π for all these instantons (scale invariance). The value 4π follows from the fact that very large instantons are smooth on the lattice and then any valid lattice action gives the continuum value.

It is important to observe that the reverse of the statement is not true: if the configuration **S** is a solution, then the configuration **R**, where $\mathbf{R}_{n_B} = \mathbf{\Sigma}_{n_B}/|\mathbf{\Sigma}_{n_B}|$, is not necessarily a solution. The proof fails because for this configuration R the minimizing configuration is not necessarily equal to **S** itself. One can show that **S** is a minimum, but not always the absolute minimum. Actually, this is the mechanism which prevents the existence of arbitrarily small instanton solutions on the lattice.

To prove the statement above we only had to show that the blocking kernel $T_\sigma^{(\infty)}$(R, S) in the FP equation, eq. (90), is zero at its absolute minimum in **R**. This is true in all the cases we considered (σ-model, Yang-Mills theory) simply by construction: the normalization condition of the kernel reads

$$\int D\mathbf{R} \exp\left(-\beta T_\sigma(\mathbf{R}, \mathbf{S})\right) = 1\,, \tag{147}$$

which implies in the $\beta \to \infty$ limit that $\min_{\mathbf{R}} T_\sigma^{(\infty)}$(R, S) $=$ 0. This can be seen explicitly in eq. (91) and eq. (51) for two different RGTs in the σ-model, and in eqs. (94,95) in the Yang-Mills theory.

The way this theorem works for the instanton solutions in the $O(3)$ σ-model and in the *SU(2)* Yang-Mills theory has been tested in detail numerically. We shall return to the results after discussing the construction of the topological charge Q.

Fixed point operators

Until now we have discussed the construction of the FP action only. We can add to the action different operators O_i multiplied by infinitesimal sources c_i, and consider the RGT

$$\begin{aligned} &\exp\left[-\beta' \mathcal{A}'(\mathbf{R}) - c_1' O_1'(\mathbf{R}) - c_2' O_2'(\mathbf{R}) - \ldots\right] = \\ &\qquad \int D\mathbf{S} \exp\left[-\beta\left(\mathcal{A}(\mathbf{S}) + T_\sigma(\mathbf{R}, \mathbf{S})\right) - c_1 O_1(\mathbf{S}) - c_2 O_2(\mathbf{S}) - \ldots\right]. \end{aligned} \tag{148}$$

Start with the FP action $\mathcal{A}^{FP}$(S) and take the $\beta \to \infty$ limit. Using eq. (90) we obtain

$$c_1' O_1'(\mathbf{R}) = c_1 O_1(\mathbf{S}_{min}(\mathbf{R})), \qquad c_2' O_2'(\mathbf{R}) = c_2 O_2(\mathbf{S}_{min}(\mathbf{R})), \ldots\,. \tag{149}$$

The FP form of the operator is reproduced by the RGT:

$$\lambda_1 O_1^{FP}(\mathbf{R}) = O_1^{FP}(\mathbf{S}_{min}(\mathbf{R}))\,, \qquad \lambda_2 O_2^{FP}(\mathbf{R}) = O_2^{FP}(\mathbf{S}_{min}(\mathbf{R}))\,,\ldots\,, \tag{150}$$

with $\lambda_i = c_i'/c_i$. The analysis above is similar to the general discussion in Sect. 3.4 or Sect. 8.1, except that here we also allow operators which one can not consider as part of the action (for example local operators which are not summed over the lattice points). As we discussed in Sect. 8.1, the eigenvalues λ are determined by simple dimensional analysis if the FP is the Gaussian FP (which is the case for free fields and for AF theories).

Let us consider examples. Denote the field variables of a free scalar theory in d dimension by ϕ_n and χ_{n_B} on the fine and coarse lattice, respectively. Consider a RGT with scale factor 2. The FP field at the point n_B is constructed out of the χ fields in the local neighbourhood of n_B and has the transformation law

$$2^{-\frac{d-2}{2}}\,\varphi^{FP}(\chi; n_B) = \varphi^{FP}(\phi_{min}(\chi); n = 2n_B)\,, \tag{151}$$

where $\phi_{min}(\chi)$ is the minimizing configuration of the quadratic FP equation of the free field theory, Sect. 7.

As a second example, consider a current $J_\mu(n)$ in QCD. The FP current is a local combination of the fermion fields connected by the product of gauge matrices to assure gauge invariance. The details are determined by the equation

$$2^{-3}\,J_\mu(\bar{\chi}, \chi, V; n_B) = J_\mu\Big(\bar{\psi}_{st}(\bar{\chi}, U_{min}), \psi_{st}(\chi, U_{min}), U_{min}(V); n = 2n_B\Big)\,, \tag{152}$$

where U_{min} is the minimizing configuration of the FP equation, eq. (94), while $\bar{\psi}_{st}, \psi_{st}$ are the configurations which make the r.h.s. of eq. (100) stationary.

Only a few cases were treated explicitly until now: the FP free field eq. (151), the FP Polyakov **loop**[13] and the FP topological **charge** [31, 32], which will be discussed in the next subsection.

The FP topological charge

There are many ways to discretize the continuum definition of the topological charge, eq. (141). One might replace the derivatives by finite differences taking care of the basic symmetries of the expression. This is the so called 'field theoretical' definition. The corresponding topological charge on the lattice will not be an integer number. This creates immediately a complicated renormalization problem since there will be contributions to this operator even in perturbation theory. An alternative possibility is to construct an operator on the lattice which gives an integer number on any configuration ('geometric definition') and so it is protected in perturbation theory. This definition assigns, however, a non-zero topological charge to different topological dislocations of size *O(a)* which have nothing to do with instantons and which heavily distort the results for quantities related to topology.

The FP topological charge used together with the FP action avoids these difficulties.
Since Q is a dimensionless quantity it satisfies the FP equation:

$$Q^{FP}(\mathbf{R}) = Q^{FP}(\mathbf{S}_{min}(\mathbf{R}))\,. \tag{153}$$

One can solve this equation the following way. Let us denote the minimizing configuration associated with R by $\mathbf{S}^{(1)}_{min}(\mathbf{R})$ ($\equiv \mathbf{S}_{min}$(R) in eq. (153)). Consider now the FP equation, eq. (90), with $\mathbf{S}^{(1)}_{min}$ on the l.h.s . The corresponding minimizing configuration $\mathbf{S}^{(2)}_{min} = \mathbf{S}^{(2)}_{min}(\mathbf{S}_{min}(\mathbf{R}))$ lives on a lattice whose lattice unit is a factor of 2^2 smaller than that of the lattice of **R**. Iterating eq. (153) this way, we get

$$Q^{FP}(\mathbf{R}) = Q^{FP}(\mathbf{S}^{(j)}_{min}(\mathbf{R})) , \tag{154}$$

where $\mathbf{S}^{(j)}_{min}(\mathbf{R})$ lives on a very fine lattice (containing big instantons with small fluctuations) if j is large. On such configurations any lattice definition Q_{latt} gives the correct result

$$Q^{FP}(\mathbf{R}) = \lim_{j\to\infty} Q_{latt}(\mathbf{S}^{(j)}_{min}(\mathbf{R})) . \tag{155}$$

In practice the $j = 2$ value agrees with the $j = \infty$ limit if the geometrical definition is used[31, 33] for Q_{latt}. The construction of Q^{FP} goes the same way in Yang-Mills theories[32, 35, 36].

There are no topological artifacts if FP operators are used

We show now that using the FP action and the FP topological charge, the action of any configuration (classical solution or not) satisfies

$$\mathcal{A}^{FP}(\mathbf{R}) \geq 4\pi|Q| , \tag{156}$$

like in the continuum classical theory. There are no configurations with topological charge whose action is below that of the instanton solutions.

In the equation $\mathcal{A}^{FP}(\mathbf{R}) = \mathcal{A}^{FP}(\mathbf{S}_{min}) + T^{(\infty)}_{\sigma}(\mathrm{R}, \mathbf{S}_{min})$ we can express $\mathcal{A}^{FP}(\mathbf{S}_{min})$ again in terms of the sum of the action and the kernel on the next finer lattice, etc. Iterating further we get arbitrarily close to the continuum, where the inequality eq. (156) is certainly satisfied. Since the kernels are non-negative (Sect. 9.1), eq. (156) follows.

Numerical tests on the classical solutions

As we discussed in Sect. 9.1, if the FP action has an instanton solution with size ρ then there is certainly a solution with size 2ρ. The opposite is not always true: if there is a solution **S** with size ρ, then the coarse configuration $\mathbf{R}_{n_B} = \Sigma_{n_B}/|\Sigma_{n_B}|$ does not necessarily satisfy the equations of motion. This happens if in the minimization problem, eq. (90), more than one local minima occurs and the absolute minimum does not agree with the starting **S**. It is expected that this happens if $\rho = O(a)$ and the would be instanton with $\rho' = \rho/2$ 'falls through the lattice'. The theorem leaves open when this happens.

Detailed numerical studies of classical solutions show[31, 33, 35, 36] that the radius ρ_{min} of the smallest instanton which exists on the lattice is less than the lattice unit a both for spin and for gauge models (for the O(3) model with the kernel eq. (47), for example, $\rho_{min} \approx 0.7$a). For instantons with $\rho_{min} < \rho < \infty$, the value the action is exactly $4\pi|Q|$, where Q is the topological charge given by the operator Q^{FP}.

FERMIONIC ZERO MODES AND THE INDEX THEOREM ON THE LATTICE

Consider the continuum eigenvalue problem of the Dirac operator of massless fermions in the presence of a background gauge field

$$\gamma_\mu\big(\partial_\mu - i\,A_\mu(x)\big)\psi^{(\lambda)}(x) = \lambda\psi^{(\lambda)}(x) . \tag{157}$$

If the background gauge field has a non-zero topological charge Q, then there necessarily exist zero eigenvalues. The corresponding eigenvectors are γ_5 eigenstates: $\gamma_5\psi^{(0)}_{L,R} = \pm\psi^{(0)}_{L,R}$. If we denote by n_L and n_R the number of left and right handed eigenvalues, we have

$$n_R - n_L = Q\,, \tag{158}$$

which is a special case of the Atiyah-Singer index theorem[37]. It is believed by many that eq. (158) has important consequences on the low energy properties of QCD. A possible intuitive picture of a typical gauge configuration in QCD is that of a gas or liquid of instantons and anti-instantons with quantum fluctuations. Over such a configuration, the Dirac operator is expected to have a large number of quasi-zero modes which could be responsible for spontaneous chiral symmetry breaking.

Standard lattice formulations violate the conditions of the index theorem in all possible ways: the topology of the gauge sector is in a bad shape, the Dirac operator breaks chiral symmetry (Sect. 5.3, 5.4) and the discretization errors distort the spectrum.

We shall show now that using the FP action the index theorem for the zero modes of the Dirac operator remains valid even for coarse lattices[38]. This statement is based on the FP equations for the gauge action (eq. (93)) and for the Dirac operator (eq. (100)).

Consider a gauge configuration V on a lattice with lattice unit a_0. Assume that V has a topological charge Q. As discussed in Sect. 9.3, the FP topological charge is defined through a sequence of minimizing configurations $U^{(1)}_{min}(V), U^{(2)}_{min}(U^{(1)}_{min}(\wedge))$i Here $U^{(j)}_{min}$ lives on a lattice with lattice unit $a_j = a_0/2^j$. By construction all these configurations have the same topological charge Q, which is the charge of $U^{(\infty)}_{min}$ in the continuum, at the end of the sequence.

Assume that the FP Dirac operator h^{FP} has a zero eigenvalue for the background field V

$$\sum_{n'_B} h^{FP}(V)_{n_B,n'_B}\chi_{n'_B} = 0\,. \tag{159}$$

Eq. (159) is the Dirac equation, i.e. the classical Euler-Lagrange equation of motion. We proceed now as in the case of the classical solutions of the Yang-Mills theory, Sect. 9. Taking the variation of eq. (100) with respect to $\bar{\chi}_{n_B}$ and using eqs. (97,159,87) we find that ψ^{st}_n, which makes the r.h.s. of eq. (100) stationary, satisfies the Dirac equation for the background field $U^{(1)}_{min}(V)$. Since eq. (100) is quadratic in the fermion field, the stationary configuration is unique. Hence the opposite statement is also true: if ψ_n is a solution of the Dirac equation for the background field $U^{(1)}_{min}(V)$ then $\chi_{n_B} = b_f \sum_n \omega(U)_{n_B,n}\psi_n$ is a solution on the coarse lattice for the configuration V.

This result can be iterated towards finer lattices. Having a zero mode eigenfunction ψ^{st}_n for $U^{(1)}_{min}$ there is a zero mode on the next finer lattice for the configuration $U^{(2)}_{min}$, etc. At every step we can also move in the opposite direction. Going up, we finally get to a very smooth configuration for which the Atiyah-Singer theorem is applicable. This way we get that for each gauge configuration of the sequence, the FP Dirac eigenvalue problem has the same number of zero modes and with the same helicities as in the continuum. The statement for the helicities follows from the relation between the eigenvectors of the i-th and $(i-1)$-th step

$$\psi^{st}_{n_{(i-1)}} = b_f \sum_{n_{(i)}} \omega(U^{(i)}_{min})_{n_{(i-1)},n_{(i)}}\psi^{st}_{n_{(i)}}\,, \tag{160}$$

and from the fact that ω is diagonal in Dirac space (Sect. 5.3).

REFERENCES

1. K. G. Wilson and J. B. Kogut, Phys. Rep. C12 (1974) 75;
K. G. Wilson, Rev. Mod. Phys. 47 (1975) 773, *ibid.,* 55 (1983) 583;
K. G. Wilson, Adv. Math. 16 (1975) 176, 444.
2. P. Hasenfratz and F. Niedermayer, Nucl. Phys. B414 (1994) 785;
3. For a recent summary on the perfect actions, see P. Hasenfratz, hep-lat/9709110.
4. V. N. Gribov and A. A. Migdal, Sov. Phys. JETP 28 (1968) 784;
A. A. Migdal, Sov. Phys. JETP 32 (1971) 552;
A. M. Polyakov, Sov. Phys. JETP 29 (1969) 533;
M. A. Moore, Nuovo Cim. 3 (1972) 275.
5. G. Parisi, Statistical Field Theory, Benjamin/Cummings (1988);
J. Zinn-Justin, Quantum Field Theory and Critical Phenomena, Oxford Univ. Press, (1993).
6. J. Schwinger, Phys. Rev. 115 (1959) 721;
G. C. Wick, Phys. Rev. 96 (1954) 1124.
7. M. Creutz, Quarks, Gluons and Lattices, Cambridge Univ. Press (1983);
I. Montvay and G. Münster, Quantum Fields on the Lattice, Cambridge Univ. Press (1994).
8. S. K. Ma, Rev. Mod. Phys. 45 (1973) 589;
S. K. Ma, Modern Theory of Critical Phenomena, Benjamin Reading, MA (1976);
L. L. Kadanoff, Rev. Mod. Phys. 49 (1977) 267;
Th. Niemeyer and J. M. J. van Leeuwen, Phys. Rev. Lett. 31 (1973) 1411.
9. T. L. Bell and K. G. Wilson, Phys. Rev. B10 (1974) 3935.
10. F. Wegner, Phys. Rev. B5 (1972) 4529.
11. P. Hasenfratz, Nucl. Phys. B(Proc.Suppl.) 34 (1994) 3;
T. DeGrand, A. Hasenfratz, P. Hasenfratz, F. Niedermayer and U.-J. Wiese,
Nucl. Phys. B(Proc.Suppl.) 42 (1995) 67;
F. Niedermayer, Nucl. Phys. B(Proc.Supph.) 53 (1997) 56.
12. U.-J. Wiese, Phys. Lett. B315 (1993) 417;
W. Bietenholz and U.-J. Wiese, Nucl. Phys. B(Proc.Suppl.) 34 (1994) 516.
13. T. DeGrand, A. Hasenfratz, P. Hasenfratz and F. Niedermayer,
Nucl. Phys. B454 (1995) 587.
14. W. Bietenholz, R. Brower, S. Chandrasekharan and U.-J. Wiese,
Nucl. Phys. B(Proc.Suppl.) 53 (1997) 921.
15. M. E. Peskin and D. V. Schroeder, An Introduction to Quantum Field Theory,
Addison-Wesley (1995).
16. A. M. Polyakov, Gauge Fields and Strings, Harwood (1987).
17. A. A. Migdal, Sov. Phys. JETP 42 (1975) 413; 42 (1976) 743;
L. Kadanoff, Ann. Phys.(NY) 100 (1976) 359; Rev. Mod. Phys. 49 (1977) 267.
18. R. Rajaraman, Solitons and Instantons, North-Holland, Amsterdam (1982).
19. P. Hasenfratz and F. Niedermayer, hep-latt/9706002.
20. P. Hasenfratz and H. Leutwyler, Nucl. Phys. B343 (1990) 241.
21. K. G. Wilson, Phys. Rev. D10 (1974) 2445;
R. Balian, J.-M. Drouffe and C. I. Itzykson, Phys. Rev. D10 (1974) 3376.
22. M. Blatter and F. Niedermayer, Nucl. Phys. B482 (1996) 286.
23. N. B. Nielsen and M. Ninomiya, Phys. Lett. B105 (1981) 219; Nucl. Phys. B185 (1981) 20.
24. K. G. Wilson, in New Phenomena in Subnuclear Physics, ed. A. Zichichi,
Plenum Press, New York (1997).
25. T. DeGrand, A. Hasenfratz, P. Hasenfratz, P. Kunszt and F. Niedermayer,
Nucl. Phys. B(Proc.Suppl.) 53 (1997) 942.
26. P. Kunszt, hep-lat/9706019.
27. S. H. Shenker, in Recent Development in Field Theory and Statistical Mechanics,
eds. J. B. Zuber and R. Stora, North Holland (1984).
28. Y. Iwasaki, Nucl. Phys. B314 (1989) 63.
29. K. Kawedzki and A. Kupiainen, Comm. Math. Phys. 77 (1980) 31.
30. E. Katz and U.-J. Wiese, comp-gas/9709001.
31. F. Niedermayer, Nucl. Phys. B(Proc.Suppl.) 34 (1994) 513;
M. Blatter, R. Burkhalter, P. Hasenfratz and F. Niedermayer, Phys. Rev. D53 (1996) 923.
32. T. DeGrand, A. Hasenfratz and D. Zhu, Nucl. Phys. B475 (1996) 321;
Nucl. Phys. B478 (1996) 349; Nucl. Phys. B(Proc.Suppl.) 53 (1997) 945.
33. R. Burkhalter, Phys. Rev. D54 (1996) 4121.

34. W. Bietenholz, R. Brower, S. Chandrasekharan and U.-J. Wiese, hep-lat/9704015.
35. T. DeGrand, A. Hasenfratz and T. Kovacs, hep-lat/9705009.
36. P. Farchioni and A. Papa, hep-lat/9709012 .
37. M. Atiyah and I. M. Singer, Ann. Math. 93 (1971) 139.
38. P. Hasenfratz and F. Niedermayer, unpublished.

NONPERTURBATIVE FLOW EQUATIONS, LOW–ENERGY QCD, AND THE CHIRAL PHASE TRANSITION

D.–U. Jungnickel and C. Wetterich*

Institut für Theoretische Physik
Universität Heidelberg, Philosophenweg 16
69120 Heidelberg, Germany

EFFECTIVE AVERAGE ACTION

Quantum chromodynamics (QCD) describes qualitatively different physics at different length scales. At short distances the relevant degrees of freedom are quarks and gluons which can be treated perturbatively. At long distances we observe hadrons, and an essential part of the dynamics can be encoded in the masses and interactions of mesons. Any attempt to deal with this situation analytically and to predict the meson properties from the short distance physics (as functions of the strong gauge coupling α_s and the current quark masses m_q) has to bridge the gap between two qualitatively different effective descriptions. Two basic problems have to be mastered for an extrapolation from short distance QCD to mesonic length scales:

- The effective couplings change with scale. This does not only concern the running gauge coupling, but also the coefficients of non–renormalizable operators as, for example, four quark operators. Typically, these non–renormalizable terms become important in the momentum range where α_s is strong and deviate substantially from their perturbative values. Consider the four–point function which results after integrating out the gluons. For heavy quarks it contains the information about the shape of the heavy quark potential whereas for light quarks the complicated spectrum of light mesons and chiral symmetry breaking are encoded in it. At distance scales around 1 fin one expects that the effective action resembles very little the form of the classical QCD action which is relevant at short distances.

- Not only the couplings, but even the relevant variables or degrees of freedom are different for long distance and short distance QCD. It seems forbiddingly difficult to describe the low-energy scattering of two mesons in a language of quarks and gluons only. An appropriate analytical field theoretical method should be capable of introducing field variables for composite objects such as mesons.

A conceptually very appealing idea for our task is the block–spin action[1,2]. It realizes

*Lectures presented by Christof Wetterich

that physics with a given characteristic length scale l is conveniently described by a functional integral with an ultraviolet (UV) cutoff Λ for the momenta. Here Λ should be larger than l^{-1} but not necessarily by a large factor. The Wilsonian effective action S_Λ^W replaces then the classical action in the functional integral. It is obtained by integrating out the fluctuations with momenta $q^2 \gtrsim \Lambda^2$. An exact renormalization group equation[2,3,4,5,6] describes how S_Λ^W changes with the UV cutoff Λ.

We will use here the somewhat different but related concept of the effective average action[7] Γ_k which, in the language of statistical physics, is a coarse grained free energy with coarse graining scale k. The effective average action is based on the quantum field theoretical concept of the effective action[8] Γ which is obtained by integrating out all quantum fluctuations. The effective action contains all information about masses, couplings, form factors and so on, since it is the generating functional of the *1PI* Green functions. The field equations derived from Γ are exact including all quantum effects. For a field theoretical description of thermal equilibrium this concept is easily generalized to a temperature dependent effective action which includes now also the thermal fluctuations. In statistical physics Γ describes the free energy as a functional of some convenient (space dependent) order parameter, for instance the magnetization. In particular, the behavior of Γ for a constant order parameter (the effective potential) specifies the equation of state. The effective average action Γ_k is a simple generalization of the effective action, with the distinction that only quantum fluctuations with momenta $q^2 \gtrsim k^2$ are included. This can be achieved by introducing an explicit infrared cutoff $\sim k$ in the functional integral defining the partition function (or the generating functional for the n-point functions). Typically, this IR-cutoff is quadratic in the fields and modifies the inverse propagator, for example by adding a mass–like term $\sim k^2$. The effective average action can then be defined in complete analogy to the effective action (via a Legendre transformation of the logarithm of the partition function). The mass-like term in the propagator suppresses the contributions from the small momentum modes with $q^2 \lesssim k^2$ and Γ_k accounts effectively only for the fluctuations with $q^2 \gtrsim k^2$.

Following the behavior of Γ_k for different k is like looking at the world through a microscope with variable resolution: For large k one has a very precise resolution $\sim k^{-1}$ but one also studies effectively only a small volume $\sim k^{d}$. Taking in QCD the coarse graining scale k much larger than the confinement scale guarantees that the complicated nonperturbative physics does not play a role yet. In this case, Γ_k will look similar to the classical action, typically with a running gauge coupling evaluated at the scale k. (This does not hold for Green functions with much larger momenta $p^2 \gg k^2$ where the relevant IR cutoff is p, and the effective coupling is $\alpha_s(p)$.) For lower k the resolution is smeared out and the detailed information of the short distance physics can be lost. (Again, this does not concern Green functions at high momenta.) On the other hand, the "observable volume" is increased and long distance aspects such as collective phenomena become visible. In a theory with a physical UV cutoff Λ we may associate Γ_Λ with the classical action S since no fluctuations are effectively included. By definition, the effective average action equals the effective action for $k = 0$, $\Gamma_0 = \Gamma$, since the infrared cutoff is absent. Thus Γ_k interpolates between the classical action S and the effective action Γ as k is lowered from Λ to zero. The ability to follow the evolution $k \to 0$ is equivalent to the ability to solve the quantum field theory.

For a formal description we will consider in the first two sections a model with real scalar fields χ^a (the index a labeling internal degrees of freedom) in d Euclidean dimensions with classical action $S[\chi]$. We define the generating functional for the connected Green functions by

$$W_k[J] = \ln \int \mathcal{D}\chi \exp\left\{-S_k[\chi] + \int d^dx J_a(x)\chi^a(x)\right\} \tag{1}$$

where we have added to the classical action an IR cutoff $\Delta_k S$

$$S_k[\chi] = S[\chi] + \Delta S_k[\chi] \tag{2}$$

which is quadratic in the fields and conveniently formulated in momentum space

$$\Delta S_k[\chi] = \frac{1}{2}\int \frac{d^dq}{(2\pi)^d} R_k(q^2)\chi_a(-q)\chi^a(q) \,. \tag{3}$$

Here J_a are the usual scalar sources introduced to define generating functionals and $R_k(q^2)$ denotes an appropriately chosen (see below) IR cutoff function. We require that $R_k(q^2)$ vanishes rapidly for $q^2 \gg k^2$ whereas for $q^2 \ll k^2$ it behaves as $R_k(q^2) \simeq k^2$. This means that all Fourier components of χ^a with momenta smaller than the IR cutoff k acquire an effective mass $m_{\text{eff}} \simeq k$ and therefore decouple while the high momentum components of χ^a should not be affected by R_k. The classical fields

$$\Phi^a \equiv \langle \chi^a \rangle = \frac{\delta W_k[J]}{\delta J_a} \tag{4}$$

now depend on k. In terms of W_k the effective average action is defined via a Legendre transformation

$$\Gamma_k[\Phi] = -W_k[J] + \int d^dx J_a(x)\Phi^a(x) - \Delta S_k[\Phi] \,. \tag{5}$$

In order to define a reasonable coarse grained free energy we have subtracted in Eq.(5) the infrared cutoff piece. This guarantees that the only difference between Γ_k and Γ is the effective IR cutoff in the fluctuations. Furthermore, this has the consequence that Γ_k does not need to be convex whereas a pure Legendre transform is always convex by definition. (The coarse grained free energy becomes convex[9] only for $k \to 0$.) This is very important for the description of phase transitions, in particular first order ones. One notes

$$\begin{aligned} \lim_{k\to 0} R_k(q^2) &= 0 \quad &\Rightarrow \quad \lim_{k\to 0} \Gamma_k[\Phi] &= \Gamma[\Phi] \\ \lim_{k\to \Lambda} R_k(q^2) &= \infty \quad &\Rightarrow \quad \lim_{k\to \Lambda} \Gamma_k[\Phi] &= S[\Phi] \end{aligned} \tag{6}$$

for a convenient choice of R_k like

$$R_k(q^2) = Z_{\Phi,k} q^2 \frac{1 + e^{-q^2/k^2} - e^{-q^2/\Lambda^2}}{e^{-q^2/\Lambda^2} - e^{-q^2/k^2}} \,. \tag{7}$$

Here $Z_{\Phi,k}$ denotes the wave function renormalization to be specified below and we will often use for R_k the limit $\Lambda \to \infty$

$$R_k(q^2) = \frac{Z_{\Phi,k} q^2}{e^{q^2/k^2} - 1} \,. \tag{8}$$

We note that the property $\Gamma_\Lambda = S$ is not essential since the short distance laws may be parameterized by Γ_Λ as well as by S. In addition, for momentum scales much smaller than Λ universality implies that the precise form of Γ_Λ is irrelevant, up to the values of a few relevant renormalized couplings.

A few properties of the effective average action are worth mentioning:

1. All symmetries of the model which are respected by the IR cutoff $\Delta_k S$ are automatically symmetries of Γ_k. In particular, this concerns translation and rotation invariance and one is not plagued by many of the problems encountered by a formulation of the block-spin action on a lattice.

2. In consequence, Γ_k can be expanded in terms of invariants with respect to these symmetries with couplings depending on k. For the example of a scalar theory one may use a derivative expansion ($\rho = \Phi^a\Phi_a/2$)

$$\Gamma_k = \int d^dx \left\{ U_k(\rho) + \frac{1}{2} Z_{\Phi,k}(\rho)\partial^\mu\Phi_a\partial_\mu\Phi^a + \dots \right\} \tag{9}$$

and expand further in powers of ρ

$$\begin{aligned} U_k(\rho) &= \frac{1}{2}\bar{\lambda}_k\,(\rho - \rho_0(k))^2 + \frac{1}{6}\bar{\gamma}_k\,(\rho - \rho_0(k))^3 + \dots \\ Z_{\Phi,k}(\rho) &= Z_{\Phi,k}(\rho_0) + Z'_{\Phi,k}(\rho_0)\,(\rho - \rho_0) + \dots \end{aligned} \tag{10}$$

where ρ_0 denotes the (k-dependent) minimum of the effective average potential $U_k(\rho)$. We see that Γ_k describes infinitely many running couplings. ($Z_{\Phi,k}$ in Eq.(7) can be identified with $Z_{\Phi,k}(\rho_0)$.)

3. There is no problem incorporating chiral fermions since a chirally invariant cutoff R_k can be formulated[10,11].

4. Gauge theories can be formulated along similar lines[12,13, 14,15,16,17] even though $\Delta_k S$ may not be gauge invariant. In this case the usual Ward identities receive corrections for which one can derive closed expressions[15]. These corrections vanish for $k \to 0$.

5. The high momentum modes are very effectively integrated out because of the exponential decay of R_k for $q^2 \gg k^2$. Nevertheless, it is sometimes technically easier to use a cutoff without this fast decay property (e.g. $R_k \sim k^2$ or $R_k \sim k^4/q^2$). In the latter case one has to be careful with possible remnants of an incomplete integration of the short distance modes. Also our cutoff does not introduce any non–analytical behavior as would be the case for a sharp **cutoff**[7].

6. Despite a similar spirit and many analogies there remains also a conceptual difference to the Wilsonian effective action S_Λ^W. The Wilsonian effective action describes a set of different actions (parameterized by Λ) for one and the same theory — the *n–point* functions are independent of A and have to be computed from S_Λ^W by further functional integration. In contrast, Γ_k describes the effective action for different theories — for any value of k the effective average action is related to the generating functional of a theory with a different action $S_k = S + \Delta_k S$. The n – point functions depend on k. The Wilsonian effective action does not generate the $1PI$ Green functions[19].

7. Because of the incorporation of an infrared cutoff, Γ_k is closely related to an effective action for averages of fields[7], where the average is taken over a volume $\sim k^d$.

EXACT RENORMALIZATION GROUP EQUATION

The dependence of Γ_k on the coarse graining scale k is governed by an exact renormalization group equation (ERGE)[20]

$$\partial_t \Gamma_k[\Phi] = \frac{1}{2} \mathrm{Tr} \left\{ \left[\Gamma_k^{(2)}[\Phi] + R_k \right]^{-1} \partial_t R_k \right\} . \tag{11}$$

Here $t = \ln(k/\Lambda)$ with some arbitrary momentum scale Λ, and the trace includes a momentum integration as well as a summation over internal indices, $\mathrm{Tr} = \int \frac{d^dq}{(2\pi)^d} \sum_a$. The second functional derivative $\Gamma_k^{(2)}$ denotes the *exact* inverse propagator

$$\left[\Gamma_k^{(2)}\right]_{ab}(q, q') = \frac{\delta^2 \Gamma_k}{\delta\Phi^a(-q)\delta\Phi^b(q')} . \tag{12}$$

The flow equation Eq.(11) can be derived from Eq.(5) in a straightforward way using

$$\begin{aligned}\partial_t \Gamma_k|_\Phi &= -\partial_t W_k|_J - \partial_t \Delta_k S[\Phi] \\ &= \frac{1}{2}\mathrm{Tr}\left\{\partial_t R_k \left(\langle\Phi\Phi\rangle - \langle\Phi\rangle\langle\Phi\rangle\right)\right\} \\ &= \frac{1}{2}\mathrm{Tr}\left\{\partial_t R_k W_k^{(2)}\right\}\end{aligned} \tag{13}$$

and

$$\begin{gathered} W^{(2)}_{k,ab}(q, q') = \frac{\delta^2 W_k}{\delta J^a(-q)\delta J^b(q')} \\ \frac{\delta^2 W_k}{\delta J_a(-q)\delta J_b(q')} \frac{\delta^2\left(\Gamma_k + \Delta_k S\right)}{\delta\Phi_b(-q')\delta\Phi_c(q'')} = \delta_{ac}\delta_{qq''} . \end{gathered} \tag{14}$$

It has the form of a renormalization group improved one-loop expression[7]. Indeed, the one-loop formula for Γ_k reads

$$\Gamma_k[\Phi] = S[\Phi] + \frac{1}{2}\mathrm{Tr}\ln\left(S^{(2)}[\Phi] + R_k\right) \tag{15}$$

with $S^{(2)}$ the second functional derivative of the *classical* action, similar to Eq.(12). (Remember that $S^{(2)}$ is the field dependent classical inverse propagator. Its first and second derivative with respect to the fields describe the classical three– and four–point vertices, respectively.) Taking a t–derivative of Eq.(15) gives a one–loop flow equation very similar to Eq.(11) with $\Gamma_k^{(2)}$ replaced by $S^{(2)}$. It may seem surprising, but it is nevertheless true, that the renormalization group improvement $S^{(2)} \to \Gamma_k^{(2)}$ promotes the one–loop flow equation to an exact nonperturbative flow equation which includes the effects from all loops as well as all contributions which are non–analytical in the couplings like instantons, etc.! For practical computations it is actually often quite convenient to write the flow equation Eq. (11) as a formal derivative of a renormalization group improved one–loop expression

$$\partial_t \Gamma_k = \frac{1}{2}\mathrm{Tr}\,\tilde{\partial}_t \ln\left(\Gamma_k^{(2)} + R_k\right) \tag{16}$$

with $\tilde{\partial}_t$ acting only on R_k and not on $\Gamma_k^{(2)}$, i.e. $\tilde{\partial}_t = (\partial R_k/\partial t)(\partial/\partial R_k)$. Flow equations for n–point functions follow from appropriate functional derivatives of Eq.(11) or Eq.(16) with respect to the fields. For their derivation it is sufficient to evaluate the corresponding one–loop expressions (with the vertices and propagators derived from Γ_k) and then to take a formal $\tilde{\partial}_t$-derivative. (If the one–loop expression is finite or properly regularized the $\tilde{\partial}_t$-derivative can be taken after the evaluation of the trace.) This permits the use of (one–loop) Feynman diagrams and standard perturbative techniques in many circumstances. Most importantly, it establishes a very direct connection between the solution of flow–equations and perturbation theory. If one uses on the right hand side of Eq.(11) a truncation for which the propagator and vertices appearing in

$\Gamma_k^{(2)}$ are replaced by the ones derived from the classical action, but with running k–dependent couplings, and then expands the result to lowest non–trivial order in the coupling constants one recovers standard renormalization group improved one–loop perturbation theory. The formal solution of the flow equation can also be employed for the development of a systematically resummed perturbation theory[21].

For a choice of the cutoff function similar to Eq.(8) the momentum integral contained in the trace on the right hand side of the flow equation is both infrared and ultraviolet finite. Infrared finiteness arises through the presence of the infrared regulator $\sim R_k$. We note that all eigenvalues of the matrix $\Gamma_k^{(2)} + R_k$ must be positive semi-definite. The proof follows from the observation that the functional $\Gamma_k + \Delta_k S$ is convex since it is obtained from W_k by a Legendre transform. On the other hand, ultraviolet finiteness is related to the fast decay of $\partial_t R_k$ for $q^2 \gg k^2$. This expresses the fact that only a narrow range of fluctuations with $q^2 \simeq k^2$ contributes effectively if the infrared cutoff k is lowered by a small amount. If for some other choice of R_k the right hand side of the flow equation would not remain UV finite this would indicate that the high momentum modes have not yet been integrated out completely in the computation of Γ_k. Since the flow equation is manifestly finite this can be used to define a regularization scheme. The "ERGE–scheme" is specified by the flow equation, the choice of R_k and the "initialcondition" Γ_Λ. This is particularly important for gauge theories where other regularizations in four dimensions and in the presence of chiral fermions are difficult to construct. For gauge theories Γ_Λ has to obey appropriately modified Ward identities. In the context of perturbation theory a first proposal how to regularize gauge theories by use of flow equations can be found in[13,14]. We note that in contrast to previous versions of exact renormalization group equations there is no need in the present formulation to construct an ultraviolet momentum cutoff — a task known to be very difficult in non–Abelian gauge theories.

Despite the conceptual differences between the Wilsonian effective action S_Λ^W and the effective average action Γ_k the exact flow equations describing the Λ-dependence of S_Λ^W and the k–dependence of Γ_k are simply related. Polchinski's continuum version of the Wilsonian flow equation[5] can be transformed into Eq.(11) by means of a Legendre transform and a suitable variable redefinition[22].

Even though intuitively simple, the replacement of the (RG–improved) classical propagator by the full propagator turns the solution of the flow equation Eq.(11) into a difficult mathematical problem: The evolution equation is a functional differential equation. Once Γ_k is expanded in terms of invariants (e.g. Eqs.(9), (10)) this is equivalent to a coupled system of non-linear partial differential equations for infinitely many couplings. General methods for the solution of functional differential equations are not developed very far. They are mainly restricted to iterative procedures that can be applied once some small expansion parameter is identified. This covers usual perturbation theory in the case of a small coupling, the 1/N–expansion or expansions in the dimensionality $4 - d$ or $2 - d$. It may also be extended to less familiar expansions like a derivative expansion which is related in critical three dimensional scalar theories to a small anomalous dimension. In the absence of a clearly identified small parameter one nevertheless needs to truncate the most general form of Γ_k in order to reduce the infinite system of coupled differential equations to a (numerically) manageable size. This truncation is crucial. It is at this level that approximations have to be made and, as for all nonperturbative analytical methods, they are often not easy to control. The challenge for nonperturbative systems like low momentum QCD is to find flow equations which (a) incorporate all the relevant dynamics such that neglected effects make only small changes, and (b) remain of manageable size. The difficulty with the first task is a reliable estimate of the error. For the second task the main limitation is a practical restriction for numerical solutions of differential equations to functions depending only on a small number of variables. The existence of an exact functional differential flow equation is a very useful starting point and guide for this task. At this point the precise form of the exact flow equation is quite important. Furthermore, it can be used for systematic expansions through enlargement of the truncation and for an error estimate in this way. Nevertheless, this is not all. Usually, physical insight into a model is necessary to device a useful nonperturbative truncation!

So far, two complementary approaches to nonperturbative truncations have been explored: an expansion of the effective Lagrangian in powers of derivatives ($\rho \equiv \frac{1}{2}\Phi_a\Phi^a$)

$$\Gamma_k[\Phi] = \int d^dx \left\{ U_k(\rho) + \frac{1}{2} Z_{\Phi,k}(\Phi)\partial_\mu\Phi^a\partial^\mu\Phi_a + \frac{1}{4} Y_{\Phi,k}(\rho)\partial_\mu\rho\partial^\mu\rho + \mathcal{O}(\partial^4) \right\} \tag{17}$$

or one in powers of the fields

$$\Gamma_k[\Phi] = \sum_{n=0}^{\infty} \frac{1}{n!} \int \left(\prod_{j=0}^{n} d^dx_j \left[\Phi(x_j) - \Phi_0\right] \right) \Gamma_k^{(n)}(x_1, \ldots, x_n)\,. \tag{18}$$

If one chooses Φ_0 as the k–dependent VEV of Φ, the series Eq.(18) starts effectively at n = 2. The flow equations for the $1PI$ n–point functions $\Gamma_k^{(n)}$ are obtained by functional differentiation of Eq.(11). Such flow equations have been discussed earlier from a somewhat different **viewpoint**[4]. They can also be interpreted as a differential form of Schwinger–Dyson **equations**[23].

The formation of mesonic bound states, which typically appear as poles in the (Minkowskian) four-quark Green function, is most efficiently described by expansions like Eq.(18). This is also the form needed to compute the nonperturbative momentum dependence of the gluon propagator and the heavy quark **potential**[16, 17, 18]. On the other hand, a parameterization of Γ_k as in Eq.(17) seems particularly suited for the study of phase transitions. The evolution equation for the average potential U_k follows by evaluating Eq.(17) for constant Φ, In the limit where the Φ-dependence of $Z_{\Phi,k}$ is neglected and $Y_{\Phi,k} = 0$ one **finds**[7] for the O(N)–symmetric scalar model

$$\partial_t U_k(\rho) = \frac{1}{2} \int \frac{d^dq}{(2\pi)^d} \frac{\partial R_k}{\partial t} \left(\frac{N-1}{Z_{\Phi,k}q^2 + R_k + U_k'} + \frac{1}{Z_{\Phi,k}q^2 + R_k(q) + U_k' + 2\rho U_k''} \right) \tag{19}$$

with $U_k' \equiv \frac{\partial U_k}{\partial \rho}$, etc. One observes the appearance of ρ-dependent mass terms in the effective propagators of the right hand side of Eq.(19). Once $\eta_\Phi = -\partial_t \ln Z_{\Phi,k}$ is **determined**[7] in terms of the couplings parameterizing U_k this is a partial differential equation, for a function U_k depending on two variables k and ρ which can be solved **numerically**[24, 25, 26, 27]. (The Wilson–Fisher fixed point relevant for a second order phase transition ($d = 3$) corresponds to a scaling **solution**[28, 29] where $\partial_t U_k = 0$.) A suitable truncation of a flow equation of the type Eq.(17) will play a central role in the description of chiral symmetry breaking below.

It should be mentioned at this point that the weakest point in the ERGE approach seems to be a reliable estimate of the truncation error in a nonperturbative context. This problem is common to all known analytical approaches to nonperturbative phenomena and appears often even within systematic (perturbative) expansions. One may hope that the existence of an exact flow equation could also be of some help for error estimates. An obvious possibility to test a given truncation is its enlargement to include more variables — for example, going one step higher in a derivative expansion. This is similar to computing higher orders in perturbation theory and limited by practical considerations. As an alternative, one may employ different truncations of comparable size — for instance, by using different definitions of retained couplings. A comparison of the results can give a reasonable picture of the uncertainty if the used set of truncations is wide enough. In this context we should also note the dependence of the results on the choice of the cutofffunction $R_k(q)$. Of course, for $k \to 0$ the physics should not depend on a particular choice of R_k and, in fact, it does not for full solutions of Eq.(11). Different choices of R_k just correspond to different trajectories in the space of effective average actions along which the unique IR limit $\Gamma[\Phi]$ is reached. Once approximations are used to solve the ERGE Eq.(11), however, not only the trajectory but also its end point will depend on the precise definition of the function R_k. This is very similar to the renormalization scheme dependence usually encountered in perturbative computations of Green functions. One may use this scheme dependence as a tool to study the robustness of a given approximation scheme.

Before applying a new nonperturbative method to a complicated theory like QCD it should be tested for simpler models. A good criterion for the capability of the ERGE to deal with nonperturbative phenomena concerns the critical behavior in three dimensional scalar theories. In a first step the well known results of other methods for the critical exponents have been reproduced within a few percent accuracy[28]. The ability of the method to produce new results has been demonstrated by the computation of the critical equation of state for Ising and Heisenberg models[25] which has been verified by lattice simulations[30]. This has been extended to first order transitions in matrix models[27] or for the Abelian Higgs model relevant for superconductors[31, 26]. Analytical investigations of the high temperature phase transitions in $d = 4$ scalar theories *(O(N)*–models) have correctly described the second order nature of the in contrast to earlier attempts within high temperature perturbation theory.

For an extension of the flow equations to Abelian and non–Abelian gauge theories we refer the reader to[12, 13, 14, 15, 16, 17, 18]. The other necessary piece for a description of low-energy QCD, namely the transition from fundamental (quark and gluon) degrees of freedom to composite (meson) fields within the framework of the ERGE can be found in[33].. We will describe the most important aspects of this formalism for mesons below.

CHIRAL SYMMETRYBREAKING IN QCD

The strong interaction dynamics of quarks and gluons at short distances or high energies is successfully described by quantum chromodynamics (QCD). One of its most striking features is asymptotic freedom[34] which makes perturbative calculations reliable in the high energy regime. On the other hand, at scales around a few hundred MeV confinement sets in. As a consequence, the low–energy degrees of freedom in strong interaction physics are mesons, baryons and glueballs rather than quarks and gluons. When constructing effective models for these IR degrees of freedom one usually relies on the symmetries of QCD as a guiding principle, since a direct derivation of such models from QCD is still missing. The most important symmetry of QCD, its local color $SU(3)$ invariance, is of not much help here, since the IR spectrum appears to be color neutral. When dealing with bound states involving heavy quarks the so called "heavy quark symmetry" may be invoked to obtain approximate symmetry relations between IR observables[35]. We will rather focus here on the light scalar and pseudoscalar meson spectrum and therefore consider QCD with only the light quark flavors u, d and s. To a good approximation the masses of these three flavors can be considered as small in comparison with other typical strong interaction scales. One may therefore consider the chiral limit of QCD (vanishing current quark masses) in which the classical QCD Lagrangian does not couple left– and right–handed quarks. It therefore exhibits a global chiral invariance under $U_L(N) \times U_R(N) = SU_L(N) \times SU_R(N) \times U_V(1) \times U_A(1)$ where N denotes the number of massless quarks ($N = 2$ or 3) which transform as

$$\psi_R \equiv \frac{1-\gamma_5}{2}\psi \longrightarrow \mathcal{U}_R\psi_R \; ; \quad \mathcal{U}_R \in U_R(N)$$
$$\psi_L \equiv \frac{1+\gamma_5}{2}\psi \longrightarrow \mathcal{U}_L\psi_L \; ; \quad \mathcal{U}_L \in U_L(N) \, . \tag{20}$$

Even for vanishing quark masses only the diagonal $SU_V(N)$ vector-like subgroup can be observed in the hadron spectrum ("eightfold way"). The symmetry $SU_L(N) \times SU_R(N)$ must therefore be spontaneously broken to $SU_V(N)$

$$SU_L(N) \times SU_R(N) \longrightarrow SU_{L+R}(N) \equiv SU_V(N) \, . \tag{21}$$

Chiral symmetry breaking is one of the most prominent features of strong interaction dynamics and phenomenologically well established[36], though a rigorous derivation of this phenomenon starting from first principles is still missing. In particular, the chiral symmetry breaking Eq.(21) predicts for $N = 3$ the existence of eight light parity–odd (pseudo–)Goldstone bosons: π^0, $\pi^\pm$, K^0, $\overline{K}^0$, $K^\pm$ and η. Their comparably small masses are a consequence of the explicit chiral symmetry breaking due to small but non-vanishing current quark masses. The axial Abelian subgroup $U_A(1) = U_{L-R}(1)$ is broken in the quantum theory by an anomaly of the axial-vector current. This breaking proceeds without the occurrence of a Goldstone boson[37]. Finally, the $U_V(1) = U_{L+R}(1)$ subgroup corresponds to baryon number conservation.

The light pseudoscalar and scalar mesons are thought of as color neutral quark-antiquark bound states $\Phi^{ab} \sim \overline{\psi}_L^b \psi_R^a$, $a, b = 1, \ldots, N$, which therefore transform under chiral rotations Eq.(20) as

$$\Phi \longrightarrow \mathcal{U}_R \Phi \mathcal{U}_L^\dagger \, . \tag{22}$$

Hence, the chiral symmetry breaking pattern Eq.(21) is realized if the meson potential develops a VEV

$$\left\langle \Phi^{ab} \right\rangle = \overline{\sigma}_0 \delta^{ab} \; ; \quad \overline{\sigma}_0 \neq 0 \, . \tag{23}$$

One of the most crucial and yet unsolved problems of strong interaction dynamics is to derive an effective field theory for the mesonic degrees of freedom directly from QCD which exhibits this behavior.

A SEMI-QUANTITATIVE PICTURE

Before turning to a quantitative description of chiral symmetry breaking using flow equations it is perhaps helpful to give a brief overview of the relevant scales which appear in relation to this phenomenon and the physical degrees of freedom associated to them. Some of this will be explained in more detail in the remainder of these lectures whereas other parts are rather well established features of strong interaction physics.

At scales above approximately 1.5 GeV, the relevant degrees of freedom of strong interactions are quarks and gluons and their dynamics appears to be well described by perturbative QCD. At somewhat lower energies this changes dramatically. Quark and gluon bound states form and confinement sets in. Concentrating on the physics of scalar and pseudoscalar mesons[†] there are three important momentum scales which appear to be rather well separated:

[†]One may assume that all other bound states are integrated out. We will comment on this issue below.

- The compositeness scale k_Φ at which mesonic $\overline{\psi}\psi$ bound states form because of the increasing strength of the strong interaction. It will turn out to be somewhere in the range (600 – 700) MeV.

- The chiral symmetry breaking scale $k_{\chi SB}$ at which the chiral condensate $\left\langle \overline{\psi}^b \psi^a \right\rangle$ or $\left\langle \Phi^{ab} \right\rangle$ assumes a non-vanishing value, therefore breaking chiral symmetry according to Eq.(21). This scale is found to be around (400 – 500) MeV. For k below $k_{\chi SB}$ the quarks acquire constituent masses $M_q \simeq 350\,\mathrm{MeV}$ due to their Yukawa coupling to the chiral condensate Eq.(23).

- The confinement scale $\Lambda_{QCD} \simeq 200\,\mathrm{MeV}$ which corresponds to the Landau pole in the perturbative evolution of the strong coupling constant α_s. In our context, this is the scale where possible deviations of the effective quark propagator from its classical form and multi-quark interactions not included in the meson physics may become very important.

For scales k in the range $k_{\chi SB} \lesssim k \lesssim k_\Phi$ the most relevant degrees of freedom are mesons and quarks. Typically, the dynamics in this range is dominated by the strong Yukawa coupling h between quarks and mesons: $h^2/(4\pi) \gg \alpha_s$. One may therefore assume that the dominant QCD effects are included in the meson physics and consider a simple model of quarks and mesons only. As one evolves to scales below $k_{\chi SB}$ the Yukawa coupling decreases whereas α_s increases. Of course, getting closer to Λ_{QCD} it is no longer justified to neglect the QCD effects which go beyond the dynamics of effective meson degrees of freedom. On the other hand, the final IR value of the Yukawa coupling h is fixed by the typical values of constituent quark masses $M_q \simeq 350\,\mathrm{MeV}$ to be $h^2/(4\pi) \simeq 4.5$. One may therefore speculate that the domination of the Yukawa interaction persists down to scales $k \simeq M_q$ at which the quarks decouple from the evolution of the mesonic degrees of freedom altogether due to their mass. Of course, details of the gluonic interactions are expected to be crucial for an understanding of quark and gluon confinement. Strong interaction effects may dramatically change the momentum dependence of the quark n–point functions for k around Λ_{QCD}. Yet, as long as one is only interested in the dynamics of the mesons one is led to expect that these effects are quantitatively not too important. Because of the effective decoupling of the quarks and therefore the whole colored sector the details of confinement have only little influence on the mesonic flow equations for $k \lesssim \Lambda_{QCD}$. We conclude that there are good prospects that the meson physics can be described by an effective action for mesons and quarks for $k < k_\Phi$. The main part of the work presented here is concerned with this effective quark meson model.

In order to obtain this effective action at the compositeness scale k_Φ from short distance QCD two steps have to be carried out. In a first step one computes at the scale $k_p \simeq 1.5\,\mathrm{GeV}$ an effective action involving only quarks. This step integrates out the gluon degrees of freedom in a "quenched approximation". More precisely, one solves a truncated flow equation for QCD with quark and gluon degrees of freedom in presence of an effective infrared cutoff $k_p \simeq 1.5\,\mathrm{GeV}$ in the quark propagators. The exact flow equation to be used for this purpose is obtained by lowering the infrared cutoff R_k for the gluons to zero while keeping the one for the quarks fixed. Subsequently, the gluons are eliminated by solving the field equations for the gluon fields as functionals of the quarks. This will result in a non–trivial momentum dependence of the quark propagator and effective non–local four and higher quark interactions. Because of the infrared cutoff k_p the resulting effective action for the quarks resembles closely the one for heavy quarks (at least for Euclidean momenta). The dominant effect is the appearance of an effective quark potential (similar to the one for the charm quark) which describes the effective four–quark interactions. For the effective quark action at k_p we only retain this four-quark interaction in addition to the two–point function, while neglecting n–point functions involving six and more quarks.

For typical momenta larger than Λ_{QCD} a reliable computation of the effective quark action should be possible by using in the quark–gluon flow equation a relatively simple truncation. The result[18] for the Fourier transform of the potential $V(q^2)$ is shown in figure 1. (See ref[18] for the relation between the four-quark interaction and the heavy quark potential which involves a rescaling in dependence on k_p.) A measure for the truncation uncertainties is the parameter α_* which corresponds to the value to which an appropriately defined running strong coupling α_s evolves for $k \to 0$. We see that these uncertainties affect principally the low-q^2 region. For high values of q^2 the potential is very close to the perturbative two–loop potential whereas for intermediate q^2 relevant for quarkonium spectra it is quite close to phenomenologically acceptable potentials (e.g. the Richardson potential[38]). The inverse quark propagator is found in this computation to remain very well approximated by the simple classical momentum dependence $\not{q}$.

In the second step one has to lower the infrared cutoff in the effective non–local quark model in order to extrapolate from k_p to k_Φ. This task can be carried out by means of the exact flow equation for quarks only, starting at k_p with an initial value $\Gamma_{k_p}[\psi]$ as obtained after integrating out the gluons. For fermions the trace in Eq.(11) has to be replaced by a supertrace in order to account for the minus sign related to Grassmann variables[10]. A first investigation in this direction[33] has used a truncation with a chirally invariant four quark interaction whose most general momentum dependence was retained

$$\begin{aligned}\Gamma_k = & \int \frac{d^4p}{(2\pi)^4} \overline{\psi}^i_a(p) Z_{\psi,k}(p) \left[\not{p}\,\delta^{ab} + m^{ab}(p)\gamma_5 + i\tilde{m}^{ab}(p)\right] \psi_{ib}(p) \\ & + \frac{1}{2}\int \left(\prod_{l=1}^{4} \frac{d^4p_l}{(2\pi)^4}\right) (2\pi)^4\,\delta(p_1+p_2-p_3-p_4) \\ & \times \lambda_k^{(\psi)}(p_1,p_2,p_3,p_4) \left\{\left[\overline{\psi}^i_a(-p_1)\psi_i^b(p_2)\right]\left[\overline{\psi}^j_b(p_4)\psi_j^a(-p_3)\right]\right. \\ & \left. - \left[\overline{\psi}^i_a(-p_1)\gamma_5\psi_i^b(p_2)\right]\left[\overline{\psi}^j_b(p_4)\gamma_5\psi_j^a(-p_3)\right]\right\}. \end{aligned} \tag{24}$$

Here i,j run from one to N_c which is the number of quark colors. The indices a,b label the different light quark flavors and run from 1 to N. The matrices m and $\tilde{m}$ are hermitian and $m + i\tilde{m}\gamma_5$ forms therefore the most general quark mass matrix. (Our chiral conventions[10] where the hermitean part of the mass matrix is multiplied by γ_5 may be somewhat unusual but they are quite convenient for Euclidean calculations.) The ansatz Eq.(24) does not correspond to the most general chirally invariant four-quark interaction. It neglects similar interactions in the ρ–mesonand pomeron channels which are also obtained from a Fierz transformation of the heavy quark potential[16]. With $V(q^2)$ the heavy quark potential in a Fourier representation, the initial value at $k_p = 1.5\,\text{GeV}$ was taken as $(\hat{Z}_{\psi,k} = Z_{\psi,k}(p^2 = -k_p^2))$

$$\lambda_{k_p}^{(\psi)}(p_1,p_2,p_3,p_4)\hat{Z}_{\psi,k_p}^{-2} = \frac{1}{2}V((p_1-p_3)^2) = \frac{2\pi\alpha_s}{(p_1-p_3)^2} + \frac{8\pi\lambda}{((p_1-p_3)^2)^2}\,. \tag{25}$$

This corresponds to an approximation by a one gluon exchange term $\sim \alpha_s(k_p)$ and a string tension $\lambda \simeq 0.18\,\text{GeV}^2$ and is in reasonable agreement with the form computed recently[18] from the solution of flow equations (see fig. 1). In the simplified

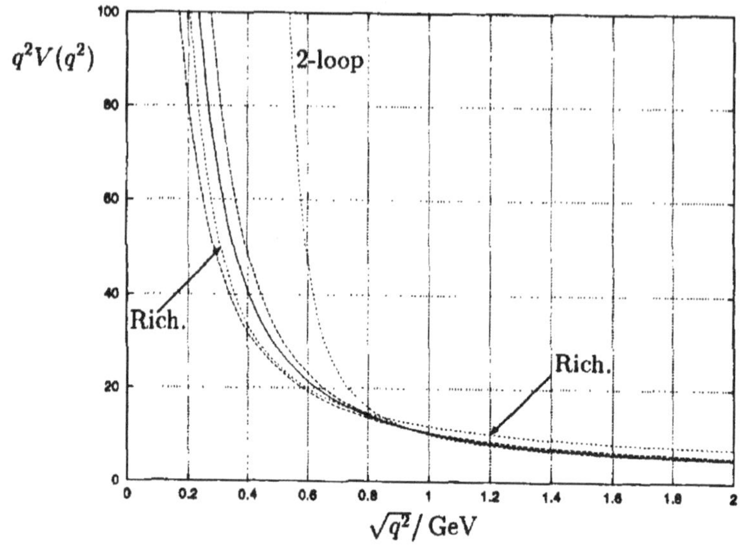

Figure 1. Heavy quark potential for different values of α_*. The solid line corresponds to $\alpha_* = 1.5$ whereas the dashed lines are for $\alpha_* = 1$ (lower dashed line) and $\alpha_* = 2$ (upper dashed line). Also shown are the two loop potential and the Richardson fit as dotted lines.

ansatz Eq.(25) the string tension introduces a second scale in addition to k_p and it becomea clear that the incorporation of gluon fluctuations is a crucial ingredient for the emergence of mesonic bound states. For a more precise treatment[18] of the four–quark interaction at the scale k_Φ this second scale is set by the running of α_s or Λ_{QCD}.

The evolution equation for the function $\lambda_k^{(\psi)}$ for $k < k_p$ can be derived from the fermionic version of Eq.(11) and the truncation Eq.(24). Since $\lambda_k^{(\psi)}$ depends on six independent momentum invariants it is a partial differential equation for a function depending on seven variables and has to be solved numerically[33]. The "initial value" Eq.(25) corresponds to the t–channel exchange of a "dressed" colored gluonic state and it is by far not clear that the evolution of $\lambda_k^{(\psi)}$ will lead at lower scales to a momentum dependence representing the exchange of colorless mesonic bound states. Yet, at the compositeness scale

$$k_\Phi \simeq 630\,\mathrm{MeV} \tag{26}$$

one finds an approximate factorization

$$\lambda_{k_\Phi}^{(\psi)}(p_1, p_2, p_3, p_4) = g(p_1, p_2)\tilde{G}(s)g(p_3, p_4) + \dots \tag{27}$$

which indicates the formation of mesonic bound states. Here $g(p_1, p_2)$ denotes the amputated Bethe–Salpeter wave function and $\tilde{G}(s)$ is the mesonic bound state propagator displaying a pole–like structure in the s–channel if it is continued to negative $s = (p_1 + p_2)^2$. The dots indicate the part of $\lambda_k^{(\psi)}$ which does not factorize and which will be neglected in the following. In the limit where the momentum dependence of g and $\tilde{G}$ is neglected we recover the four-quark interaction of the Nambu–Jona-Lasinio model[39,40]. It is therefore not surprising that our description of the dynamics for $k < k_\Phi$ will parallel certain aspects of other investigations of this model, even though we are not bound to the approximations used typically in such studies (large–N_c expansion, perturbative renormalization group, etc.).

It is clear that for scales $k \lesssim k_\Phi$ a description of strong interaction physics in terms of quark fields alone would be rather inefficient. Finding physically reasonable truncations of the effective average action should be much easier once composite fields for the mesons are introduced. The exact renormalization group equation can indeed be supplemented by an exact formalism for the introduction of composite field variables or, more generally, a change of variables[33]. For our purpose, this amounts in practice to inserting at the scale k_Φ the identities

$$
\begin{aligned}
1 &= \text{const} \int \mathcal{D}\sigma_A \\
&\times \exp\left\{-\operatorname{tr}\left(\sigma_A^\dagger - K_A^\dagger \tilde{G} - m_A^\dagger - \mathcal{O}^\dagger \tilde{G}\right) \frac{1}{2\tilde{G}} \left(\sigma_A - \tilde{G}K_A - m_A - \tilde{G}\mathcal{O}\right)\right\} \\
1 &= \text{const} \int \mathcal{D}\sigma_H \qquad\qquad (28) \\
&\times \exp\left\{-\operatorname{tr}\left(\sigma_H^\dagger - K_H^\dagger \tilde{G} - m_H^\dagger - \mathcal{O}^{(5)\dagger} \tilde{G}\right) \frac{1}{2\tilde{G}} \left(\sigma_H - \tilde{G}K_H - m_H - \tilde{G}\mathcal{O}^{(5)}\right)\right\}
\end{aligned}
$$

into the functional integral which formally defines the quark effective average action. Here we have used the shorthand notation $\operatorname{tr} A^\dagger G B \equiv \int \frac{d^d q}{(2\pi)^d} A_a^*(q) G^{ab}(q) B_b(q)$, and $K_{A,H}$ are sources for the collective fields $\sigma_{A,H}$ which correspond in turn to the anti-hermitian and hermitian parts of the meson field Φ. They are associated to the fermion bilinear operators $\mathcal{O}[\psi]$, $\mathcal{O}^{(5)}[\psi]$ whose Fourier components read

$$
\begin{aligned}
\mathcal{O}^a{}_b(q) &= -i \int \frac{d^4 p}{(2\pi)^4} g(-p, p+q) \overline{\psi}^a(p) \psi_b(p+q) \\
\mathcal{O}^{(5)a}{}_b(q) &= -\int \frac{d^4 p}{(2\pi)^4} g(-p, p+q) \overline{\psi}^a(p) \gamma_5 \psi_b(p+q) \,. \qquad (29)
\end{aligned}
$$

The choice of $g(-p, p+q)$ as the bound state wave function renormalization and of $\tilde{G}(q)$ as its propagator guarantees that the four-quark interaction contained in Eq.(28) cancels the dominant factorizing part of the QCD–induced non-local four–quark interaction Eqs.(24), (27). In addition, one may choose

$$
\begin{aligned}
m^T_{Hab} &= m_{ab}(0) g^{-1}(0,0) Z_{\psi,k_\Phi}(0) \\
m^T_{Aab} &= \tilde{m}_{ab}(0) g^{-1}(0,0) Z_{\psi,k_\Phi}(0) \qquad (30)
\end{aligned}
$$

such that the explicit quark mass term cancels out for $q = 0$. The remaining quark bilinear is $\sim m(q) - m(0) Z_{\psi,k_\Phi}(0) g(-q,q) / [Z_{\psi,k_\Phi}(q) g(0,0)]$. It vanishes for zero momentum and will be neglected in the following. Without loss of generality we can take m real and diagonal and $\tilde{m} = 0$.

In consequence, we have replaced at the scale k_Φ the effective quark action Eq.(24) with Eq.(27) by an effective quark meson action given by

$$
\begin{aligned}
\hat{\Gamma}_k &= \Gamma_k - \frac{1}{2} \int d^4 x \operatorname{tr}\left(\Phi^\dagger \jmath + \jmath^\dagger \Phi\right) \\
\Gamma_k &= \int d^4 x U_k(\Phi, \Phi^\dagger) + \int \frac{d^4 q}{(2\pi)^d} \left\{ Z_{\Phi,k}(q) q^2 \operatorname{tr}\left[\Phi^\dagger(q)\Phi(q)\right] + Z_{\psi,k}(q) \overline{\psi}_a(q) \gamma^\mu q_\mu \psi^a(q) \right. \\
&\left. + \int \frac{d^4 p}{(2\pi)^d} \bar{h}_k(-q, q-p) \overline{\psi}^a(q) \left(\frac{1+\gamma_5}{2} \Phi_{ab}(p) - \frac{1-\gamma_5}{2} \Phi^\dagger_{ab}(-p) \right) \psi^b(q-p) \right\} . \qquad (31)
\end{aligned}
$$

At the scale k_Φ the inverse scalar propagator is related to $\tilde{G}(q)$ in Eq.(27) by

$$
\tilde{G}^{-1}(q^2) = 2\overline{m}^2_{k_\Phi} + 2 Z_{\Phi,k_\Phi}(q) q^2 \,. \qquad (32)
$$

This fixes the term in U_{k_Φ} which is quadratic in Φ to be positive, $U_{k_\Phi} = \overline{m}^2_{k_\Phi} \operatorname{tr} \Phi^\dagger\Phi + \ldots$. The higher order terms in U_{k_Φ} cannot be determined in the approximation Eq.(24) since they correspond to terms involving six or more quark fields. The initial value of the Yukawa coupling corresponds to the "quark wave function in the meson" in Eq.(27), i.e.

$$\overline{h}_{k_\Phi}(-q, q-p) = g(-q, q-p) \tag{33}$$

which can be normalized with $\overline{h}_{k_\Phi}(0,0) = g(0,0) = 1$. We observe that the explicit chiral symmetry breaking from non-vanishing current quark masses appears now in the form of a meson source term with

$$\jmath = 2\overline{m}^2_{k_\Phi} Z_{\psi,k_\Phi}(0) g^{-1}(0,0)\,(m_{ab} + i\tilde{m}_{ab}) = 2Z_{\psi,k_\Phi}\overline{m}^2_{k_\Phi} \operatorname{diag}(m_u, m_d, m_s)\,. \tag{34}$$

This induces a non–vanishing $\langle\Phi\rangle$ and an effective quark mass M_q through the Yukawa coupling. We note that the current quark mass m_q and the constituent quark mass $M_q \sim \overline{h}_k\langle\Phi\rangle$ are identical at the scale k_Φ. (By solving the field equation for Φ as a functional of $\overline{\psi}$, ψ (with $U_k = \overline{m}^2_k \operatorname{tr}\Phi^\dagger\Phi$) one recovers from Eq.(31) the effective quark action Eq.(24). For a generalization beyond the approximation of a four–quark interaction or a quadratic potential see ref[52].) Spontaneous chiral symmetry breaking can be described in this language by a non-vanishing $\langle\Phi\rangle$ in the limit $\jmath \to 0$. Because of spontaneous chiral symmetry breaking the constituent quark mass M_q can differ from zero even for $m_q = 0$. It is a nice feature of our formalism that it provides for a unified description of the concepts of the current and the constituent quark masses. As long as the effective average potential U_k has a unique minimum at $\Phi = 0$ there is simply no difference between the two. The running of the current quark mass in the pure quark model should be equivalent in the quark meson language to the running of $\overline{h}_k\langle\Phi\rangle(k)$. (A verification of this property would actually provide a good check for the truncation errors.) Nevertheless, the formalism is now adapted to account for the quark mass contribution from chiral symmetry breaking since the absolute minimum of U_k may be far from the perturbative one for small k. We also note that Γ_k in Eq.(31) is chirally invariant. The explicit chiral symmetry breaking in $\hat{\Gamma}_k$ appears only in the form of a linear source term which is independent of k and does not affect the flow of Γ_k. The flow equation for Γ_k therefore respects chiral symmetry even in the presence of quark masses. This leads to a considerable simplification.

At the scale k_Φ the propagator $\tilde{G}$ and the wave function $g(-q, q-p)$ should be optimized for a most complete elimination of terms quartic in the quark fields. In the present context we will, however, neglect the momentum dependence of $Z_{\psi,k}$, $Z_{\Phi,k}$ and $\overline{h}_k$. The mass $\overline{m}_{k_\Phi}$ was found in[33] for the simple truncation Eq.(24) with $Z_{\psi,k_\Phi} = 1$, $m = \tilde{m} = 0$ to be $\overline{m}_{k_\Phi} \simeq 120\,\text{MeV}$. In view of the possible large truncation errors we will take this only as an order of magnitude estimate. Below we will consider the range $\overline{m}_{k_\Phi} = (45 - 120)\,\text{MeV}$ for which chiral symmetry breaking can be obtained in a two flavor model. Furthermore, we will assume, as usually done in large -N_c computations within the NJL–model, that $Z_{\Phi,k_\Phi} \equiv Z_{\Phi,k_\Phi}(q=0) \ll 1$. The quark wave function renormalization $Z_{\psi,k} \equiv Z_{\psi,k}(q = 0)$ is set to one at the scale k_Φ for convenience. For $k < k_\Phi$ we will therefore study an effective action for quarks and mesons in the truncation

$$\Gamma_k = \int d^4x \Big\{ Z_{\psi,k}\overline{\psi}_a i\partial\!\!\!/ \psi^a + Z_{\Phi,k} \operatorname{tr}\left[\partial_\mu\Phi^\dagger\partial^\mu\Phi\right] + U_k(\Phi, \Phi^\dagger) + \overline{h}_k\overline{\psi}^a\left(\frac{1+\gamma_5}{2}\Phi_{ab} - \frac{1-\gamma_5}{2}(\Phi^\dagger)_{ab}\right)\psi^b \Big\} \tag{35}$$

with compositeness conditions

$$\begin{aligned}
U_{k_\Phi}(\Phi,\Phi^\dagger) &= \overline{m}^2_{k_\Phi}\,\mathrm{tr}\,\Phi^\dagger\Phi - \frac{1}{2}\overline{\nu}_{k_\Phi}\left(\det\Phi + \det\Phi^\dagger\right)\\
&\quad + \frac{1}{2}\overline{\lambda}_{1,k_\Phi}\left(\mathrm{tr}\,\Phi^\dagger\Phi\right)^2 + \frac{N-1}{4}\overline{\lambda}_{2,k_\Phi}\,\mathrm{tr}\left(\Phi^\dagger\Phi - \frac{1}{N}\mathrm{tr}\,\Phi^\dagger\Phi\right)^2 + \ldots\\
\overline{m}^2_{k_\Phi} &\equiv \frac{1}{2\tilde{G}(0)} \simeq (45\,\mathrm{MeV})^2 - (120\,\mathrm{MeV})^2\\
\overline{h}_{k_\Phi} &= Z_{\psi,k_\Phi} = 1\\
Z_{\Phi,k_\Phi} &\ll 1\,. \qquad\qquad (36)
\end{aligned}$$

As a consequence, the initial value of the renormalized Yukawa coupling which is given by $h_{k_\Phi} = \overline{h}_{k_\Phi} Z^{-1}_{\psi,k_\Phi} Z^{-1/2}_{\Phi,k_\Phi}$ is large! Note that we have included in the potential an explicit $U_A(1)$ breaking term[‡] $\sim \overline{\nu}_k$ which mimics the effect of the chiral anomaly of QCD to leading order in an expansion of the effective potential in powers of Φ. Because of the infrared stability of the evolution of Γ_k which will be discussed below the precise form of the potential, i.e. the values of the quartic couplings $\overline{\lambda}_{i,k_\Phi}$ and so on, will turn out to be unimportant.

We have refrained here for simplicity from considering four quark operators with vector and pseudo-vector spin structure. Their inclusion is straightforward and would lead to vector and pseudo-vector mesons in the effective action Eq.(35). We will concentrate first on two flavors and consider only the two limiting cases $\overline{\nu}_k = 0$ and $\overline{\nu}_k \to \infty$. We also omit first the explicit quark masses and study the chiral limit $\jmath = 0$. Because of the positive mass term $\overline{m}^2_{k_\Phi}$ one has at the scale k_Φ a vanishing expectation value $\langle\Phi\rangle = 0$ (for $\jmath = 0$). There is no spontaneous chiral symmetry breaking at the compositeness scale. This means that the mesonic bound states at k_Φ and somewhat below are not directly connected to chiral symmetry breaking.

The question remains how chiral symmetry is broken. We will try to answer it by following the evolution of the effective potential U_k from k_Φ to lower scales using the exact renormalization group method outlined above with the compositeness conditions Eq.(36) defining the initial values. In this context it is important that the formalism for composite fields[33] also induces an infrared cutoff in the meson propagator. The flow equations are therefore exactly of the form Eq.(11) (except for the supertrace), with quarks and mesons treated on an equal footing. In fact, one would expect that the large renormalized Yukawa coupling will rapidly drive the scalar mass term to negative values as the IR cutoff k is lowered[10]. This will then finally lead to a potential minimum away from the origin at some scale $k_{\chi SB} < k_\Phi$ such that $\langle\Phi\rangle \neq 0$. The ultimate goal of such a procedure, besides from establishing the onset of chiral symmetry breaking, would be to extract phenomenological quantities, like f_π or meson masses, which can be computed in a straightforward manner from Γ_k in the IR limit $k \to 0$.

At first sight, a reliable computation of $\Gamma_{k\to 0}$ seems a very difficult task. Without a truncation Γ_k is described by an infinite number of parameters (couplings, wave function renormalizations, etc.) as can be seen if Γ_k is expanded in powers of fields and derivatives. For instance, the pseudoscalar and scalar meson masses are obtained as the poles of the exact propagator, $\lim_{k\to 0}\Gamma^{(2)}_k(q)|_{\Phi=\langle\Phi\rangle}$, which receives formally contributions from terms in Γ_k with arbitrarily high powers of derivatives and the expectation value σ_0. Realistic nonperturbative truncations of Γ_k which reduce the problem to a manageable size are crucial. We will argue in the following that there may be a twofold solution to this problem:

‡The anomaly term in the fermionic effective average action has been computed in[41].

- Due to an IR fixed point structure of the flow equations in the symmetric regime, i.e. for $k_{\chi SB} < k < k_\Phi$, the values of many parameters of Γ_k for $k \to 0$ will be approximately independent of their initial values at the compositeness scale k_Φ. For small enough Z_{Φ,k_Φ} only a few relevant parameters $(\overline{m}^2_{k_\Phi}, \overline{\nu}_{k_\Phi})$ need to be computed accurately from QCD. They can alternatively be determined from phenomenology.
- One can show that physical observables like meson masses, decay constants, etc., can be expanded in powers of the quark masses within the linear meson model[42, 43, 44]. This is similar to the way it is usually done in chiral perturbation theory[36]. To a given finite order of this expansion only a finite number of terms of a simultaneous expansion of Γ_k in powers of derivatives and Φ are required if the expansion point is chosen properly.

In combination, these two results open the possibility for a perhaps unexpected degree of predictive power within the linear meson model.

We wish to stress, though, that a perturbative treatment of the model at hand, e.g., using perturbative RG techniques, cannot be expected to yield reliable results. The renormalized Yukawa coupling is expected to be large at the scale k_Φ. Even the IR value of h is still relatively big

$$\lim_{k\to 0} h = \frac{2M_q}{f_\pi} \simeq 7.5 \tag{37}$$

and h increases with increasing k. The dynamics of the linear meson model is therefore clearly nonperturbative for all scales $k \leq k_\Phi$.

FLOW EQUATIONS FOR THE LINEAR QUARK MESON MODEL

We will next turn to the ERGE analysis of the linear meson model which was introduced in the last section[§]. In a first approach we will attack the problem at hand by truncating Γ_k in such a way that it contains all perturbatively relevant and marginal operators, i.e. those with canonical dimensions. This has the advantage that the flow of a small number of couplings permits quantitative insight into the relevant mechanisms, e.g., of chiral symmetry breaking. More quantitative precision will be obtained once we generalize our truncation (see below) for the $O(4)$–model by allowing for the most general effective average potential U_k. The effective potential U_k is a function of only four $SU_L(N) \times SU_R(N)$ invariants for $N = 3$:

$$\begin{aligned}
\rho &= \mathrm{tr}\,\Phi^\dagger\Phi \\
\tau_2 &= \frac{N}{N-1}\,\mathrm{tr}\left(\Phi^\dagger\Phi - \frac{1}{N}\rho\right)^2 \\
\tau_3 &= \mathrm{tr}\left(\Phi^\dagger\Phi - \frac{1}{N}\rho\right)^3 \\
\xi &= \det\Phi + \det\Phi^\dagger\,.
\end{aligned} \tag{38}$$

The invariant τ_3 is only independent for $N \geq 3$. For $N = 2$ it can be eliminated by a suitable combination of τ_2 and ρ. The additional $U_A(1)$ breaking invariant $\omega = i(\det\Phi - \det\Phi^\dagger)$ is CP violating and may therefore appear only quadratically in U_k. It is

[§]For a study of chiral symmetry breaking in QED using related exact renormalization group techniques see ref[45].

straightforward to see that ω^2 is expressible in terms of the invariants Eq.(38). We may expand U_k as a function of these invariants around its minimum, i.e. $\rho = \rho_0 \equiv N\overline{\sigma}_0^2$, $\xi = \xi_0 = 2\overline{\sigma}_0^N$ and $\tau_2 = \tau_3 = 0$ where

$$\overline{\sigma}_0 \equiv \frac{1}{N}\operatorname{tr}\langle\Phi\rangle \ . \tag{39}$$

The expansion coefficients are the k–dependent couplings of the model. In our first version we only keep couplings of canonical dimension $d_c \leq 4$. This yields in the chirally symmetric regime, i.e., for $k_{\chi SB} \leq k \leq k_\Phi$ where $\overline{\sigma}_0 = 0$

$$U_k = \overline{m}_k^2\rho + \frac{1}{2}\overline{\lambda}_{1,k}\rho^2 + \frac{N-1}{4}\overline{\lambda}_{2,k}\tau_2 - \frac{1}{2}\overline{\nu}_k\xi \tag{40}$$

whereas in the SSB regime for $k \leq k_{\chi SB}$ we have

$$U_k = \frac{1}{2}\overline{\lambda}_{1,k}\left[\rho - N\overline{\sigma}_{0,k}^2\right]^2 + \frac{N-1}{4}\overline{\lambda}_{2,k}\tau_2 + \frac{1}{2}\overline{\nu}_k\left[\overline{\sigma}_{0,k}^{N-2}\rho - \xi\right] \ . \tag{41}$$

Before continuing to compute the nonperturbative beta functions for these couplings it is worthwhile to pause here and emphasize that naively (perturbatively) irrelevant operators can by no means always be neglected. The most prominent example for this is QCD itself. It is the very assumption of our treatment of chiral symmetry breaking (substantiated by the results of[33]) that the momentum dependence of the coupling constants of some six-dimensional quark operators $(\overline{\psi}\psi)^2$ develop poles in the s–channel indicating the formation of mesonic bound states. On the other hand, it is quite natural to assume that Φ^6 or Φ^8 operators are not really necessary to understand the properties of the potential in a neighborhood around its minimum. Yet, truncating higher dimensional operators does not imply the assumption that the corresponding coupling constants are small. In fact, this could only be expected as long as the relevant and marginal couplings are small as well. What is required, though, is that their *influence* on the evolution of those couplings kept in the truncation, for instance, the set of equations Eq.(43) below, is small. A comparison with the results of the later sections for the full potential U_k (i.e., including arbitrarily many couplings in the formal expansion) will provide a good check for the validity of this assumption. In this context it is perhaps also interesting to note that the truncation Eq.(35) includes the known one-loop beta functions of a small coupling expansion as well as the leading order result of the large–N_c expansion of the $U_L(N) \times U_R(N)$ model[46]. This should provide at least some minimal control over this truncation, even though we believe that our results are significantly more accurate.

Inserting the truncation Eqs.(35), (40), (41) into Eq.(11) reduces this functional differential equation for infinitely many variables to a finite set of ordinary differential equations. This yields, in particular, the beta functions for the couplings $\overline{\lambda}_{1,k}$, $\overline{\lambda}_{2,k}$, $\overline{\nu}_k$ and $\overline{m}_k^2$ or $\overline{\sigma}_{0,k}$. Details of the calculation can be found in[47]. We will refrain here from presenting the full set of flow equations but rather illustrate the main results with a few examples. Defining dimensionless renormalized VEV and coupling constants

$$\begin{aligned}
\kappa &= Z_{\Phi,k}N\overline{\sigma}_{0,k}^2k^{-2}\\
h^2 &= \overline{h}_k^2 Z_{\Phi,k}^{-1}Z_{\psi,k}^{-2}\\
\lambda_i &= \overline{\lambda}_{i,k}Z_{\Phi,k}^{-2}\ ; \quad i = 1,2\\
\nu &= \overline{\nu}_k Z_{\Phi,k}^{-\frac{N}{2}}k^{N-4}
\end{aligned} \tag{42}$$

one finds, e.g., for the spontaneous symmetry breaking (SSB) regime and $v_k = 0$

$$
\begin{aligned}
\frac{d}{dt}\kappa &= -(2+\eta_\Phi)\kappa + \frac{1}{16\pi^2}\Big\{N^2 l_1^4(0;\eta_\Phi) + 3 l_1^4(2\lambda_1\kappa;\eta_\Phi) \\
&\quad + (N^2-1)\left[1+\frac{\lambda_2}{\lambda_1}\right] l_1^4(\lambda_2\kappa;\eta_\Phi) - 4N_c\frac{h^2}{\lambda_1} l_1^{(F)4}(\frac{1}{N}h^2\kappa;\eta_\psi)\Big\} \\
\frac{d}{dt}\lambda_1 &= 2\eta_\Phi\lambda_1 + \frac{1}{16\pi^2}\Big\{N^2\lambda_1^2 l_2^4(0;\eta_\Phi) + 9\lambda_1^2 l_2^4(2\lambda_1\kappa;\eta_\Phi) \\
&\quad + (N^2-1)\left[\lambda_1+\lambda_2\right]^2 l_2^4(\lambda_2\kappa;\eta_\Phi) - 4\frac{N_c}{N}h^4 l_2^{(F)4}(\frac{1}{N}h^2\kappa;\eta_\psi)\Big\} \\
\frac{d}{dt}\lambda_2 &= 2\eta_\Phi\lambda_2 + \frac{1}{16\pi^2}\Big\{\frac{N^2}{4}\lambda_2^2 l_2^4(0;\eta_\Phi) + \frac{9}{4}(N^2-4)\lambda_2^2 l_2^4(\lambda_2\kappa;\eta_\Phi) \\
&\quad - \frac{1}{2}N^2\lambda_2^2 l_{1,1}^4(0,\lambda_2\kappa;\eta_\Phi) + 3[\lambda_2+4\lambda_1]\lambda_2 l_{1,1}^4(2\lambda_1\kappa,\lambda_2\kappa;\eta_\Phi) \\
&\quad - 8\frac{N_c}{N}h^4 l_2^{(F)4}(\frac{1}{N}h^2\kappa;\eta_\psi)\Big\} \\
\frac{d}{dt}h^2 &= [d-4+2\eta_\psi+\eta_\Phi]h^2 - \frac{4}{N}v_d h^4\Big\{N^2 l_{1,1}^{(FB)d}(\frac{1}{N}\kappa h^2,0;\eta_\psi,\eta_\Phi) \\
&\quad - (N^2-1) l_{1,1}^{(FB)d}(\frac{1}{N}\kappa h^2,\kappa\lambda_2;\eta_\psi,\eta_\Phi) \\
&\quad - l_{1,1}^{(FB)d}(\frac{1}{N}\kappa h^2, 2\kappa\lambda_1;\eta_\psi,\eta_\Phi)\Big\}
\end{aligned}
\tag{43}
$$

Here $\eta_\Phi = -\frac{d}{dt}\ln Z_{\Phi,k}$, $\eta_\psi = -\frac{d}{dt}\ln Z_{\psi,k}$ are the meson and quark anomalous dimensions, respectively[47]. The symbols l_n^d, l_{n_1,n_2}^d and $l_n^{(F)d}$ denote bosonic and fermionic mass threshold functions, respectively, which are defined in[47]. They describe the decoupling of massive modes and provide an important nonperturbative ingredient. For instance, the bosonic threshold functions

$$
l_n^d(w;\eta_\Phi) = \frac{n+\delta_{n,0}}{4}v_d^{-1}k^{2n-d}\int\frac{d^dq}{(2\pi)^d}\frac{1}{Z_{\Phi,k}}\frac{\partial R_k}{\partial t}\frac{1}{[P(q^2)+k^2w]^{n+1}} \tag{44}
$$

involve the inverse average propagator $P(q^2) = q^2 + Z_{\Phi,k}^{-1}R_k(q^2)$ where the infrared cutoff is manifest. These functions decrease $\sim w^{-(n+1)}$ for $w \gg 1$. Since typically $w = M^2/k^2$ with M a mass of the model, the main effect of the threshold functions is to cut off fluctuations of particles with masses $M^2 \gg k^2$. Once the scale k is changed below a certain mass threshold, the corresponding particle no longer contributes to the evolution and decouples smoothly.

Within our truncation the beta functions Eq.(43) for the dimensionless couplings look almost the same as in one–loop perturbation theory. There are, however, two major new ingredients which are crucial for our approach: First, there is a new equation for the running of the mass term in the symmetric regime or for the running of the potential minimum in the regime with spontaneous symmetry breaking. This equation is related to the quadratic divergence of the mass term in perturbation theory and does not appear in the Callan–Symanzik[48] or Coleman– Weinberg[49] treatment of the renormalization group. Obviously, these equations are the key for a study of the onset of spontaneous chiral symmetry breaking as k is lowered from k_Φ to zero. Second, the most important nonperturbative ingredient in the flow equations for the dimensionless Yukawa and scalar couplings is the appearance of effective mass threshold functions like

Eq.(44) which account for the decoupling of modes with masses larger than k. Their form is different for the symmetric regime (massless fermions, massive scalars) or the regime with spontaneous symmetry breaking (massive fermions, massless Goldstone bosons). Without the inclusion of the threshold effects the running of the couplings would never stop and no sensible limit $k \to 0$ could be obtained because of unphysical infrared divergences. The threshold functions are not arbitrary but have to be computed carefully. The mass terms appearing in these functions involve the dimensionless couplings. Expanding the threshold functions in powers of the mass terms (or the dimensionless couplings) makes their non–perturbative content immediately visible. It is these threshold functions which will make it possible below to use one–loop type formulae for the necessarily nonperturbative computation of the critical behavior of the (effectively) three–dimensional $O(4)$–symmetric scalar model.

THE CHIRAL ANOMALY AND THE $O(4)$–MODEL

We have seen how the mass threshold functions in the flow equations describe the decoupling of heavy modes from the evolution of Γ_k as the IR cutoff k is lowered. In the chiral limit with two massless quark flavors ($N = 2$) the pions are the massless Goldstone bosons. Below, once we will consider the flow of the full effective average potential without resorting to a quartic truncation, we will also consider the more general case of non-vanishing current quark masses. For the time being, the effect of the physical pion mass of $m_\pi \simeq 140\,\text{MeV}$, or equivalently of the two small but non-vanishing current quark masses, can easily be mimicked by stopping the flow of Γ_k at $k = m_\pi$ by hand. This situation changes significantly once the strange quark is included. Now the η and the four K mesons appear as additional massless Goldstone modes in the spectrum. They would artificially drive the running of Γ_k at scales $m_\pi \lesssim k \lesssim 500\,\text{MeV}$ where they should already be decoupled because of their physical masses. It is therefore advisable to focus on the two flavor case $N = 2$ as long as the chiral limit of vanishing current quark masses is considered.

It is straightforward to obtain an estimate of the (renormalized) coupling ν which parameterizes the explicit $U_A(1)$ breaking due to the chiral anomaly. From Eq.(41) we find

$$m_{\eta'}^2 = \frac{N}{2}\nu\sigma_0^{N-2} \simeq 1\,\text{GeV} \tag{45}$$

which translates for $N = 2$ for $k \to 0$ into

$$\nu(k \to 0) \simeq 1\,\text{GeV}\,. \tag{46}$$

This suggests that $\nu \to \infty$ can be considered as a realistic limit. An important simplification occurs for $N = 2$ and $\nu \to \infty$, related to the fact that for $N = 2$ the chiral group $SU_L(2) \times SU_R(2)$ is (locally) isomorphic to $O(4)$. Thus, the complex (2,2) representation Φ of $SU_L(2) \times SU_R(2)$ may be decomposed into two vector representations, (σ, π^k) and (η', a^k) of $O(4)$:

$$\Phi = \frac{1}{2}(\sigma - i\eta') + \frac{1}{2}\left(a^k + i\pi^k\right)\tau_k\,. \tag{47}$$

For $\nu \to \infty$ the masses of the η' and the a^k diverge and these particles decouple. We are then left with the original $O(4)$ symmetric linear σ-model of Gell–Mann and Levy[50] coupled to quarks. The flow equations of this model have been derived previously[10, 7] for the truncation of the effective action used here. For mere comparison we also consider the opposite limit $\nu \to 0$. Here the η' meson becomes an additional Goldstone boson in the chiral limit which suffers from the same problem as the K and the η in the case $N = 3$. Hence, we may compare the results for two different approximate limits of the effects of the chiral anomaly:

- the $O(4)$ model corresponding to $N = 2$ and $\nu \to \infty$
- the $U_L(2) \times U_R(2)$ model corresponding to $N = 2$ and $\nu = 0$.

For the reasons given above we expect the first situation to be closer to reality. In this case we may imagine that the fluctuations of the kaons, η, η' and the scalar mesons (as well as vector and pseudovector mesons) have been integrated out in order to obtain the initial values of Γ_{k_Φ} — in close analogy to the integration of the gluons for the effective quark action $\Gamma_{k_\varphi}[\psi]$ discussed above. We will keep the initial values of the couplings $\overline{m}^2_{k_\Phi}$, $\overline{\lambda}_{1,k_\Phi}$ and Z_{Φ,k_Φ} as free parameters. Our results should be quantitatively accurate to the extent to which the local polynomial truncation is a good approximation.

INFRARED STABILITY

Eq.(43) and the corresponding set of flow equations for the symmetric regime constitute a coupled system of ordinary differential equations which can be integrated numerically. Similar equations can be computed for the $O(4)$ model where $N = 2$ and the coupling λ_2 is absent. We have neglected in a first step the dependence of all threshold functions appearing in the flow equations on the anomalous dimensions. This dependence will be taken into account below once we abandon the quartic potential approximation. The most important result is that chiral symmetry breaking indeed occurs for a wide range of initial values of the parameters including the presumably realistic case of large renormalized Yukawa coupling and a bare mass $\overline{m}_{k_\Phi}$ of order 100 MeV. A typical evolution of the renormalized meson mass m with k is plotted in fig. 2. Driven by the strong Yukawa coupling, m decreases rapidly and goes through

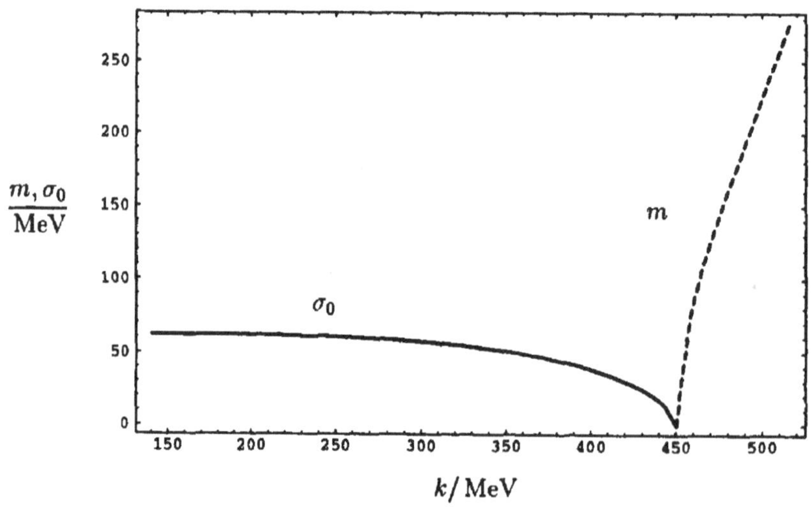

Figure 2. Evolution of the renormalized mass m in the symmetric regime (dashed line) and the vacuum expectation value $\sigma_0 = Z^{1/2}_{\Phi,k}\overline{\sigma}_{0,k}$ of the scalar field in the SSB regime (solid line) as functions of k for the $U_L(2) \times U_R(2)$ model. Initial values at $k = k_\Phi$ are $\lambda_{i,I} = 0$, $h^2_I = 300$ and $\overline{m}_{k_\Phi} = 63\,\text{MeV}$.

zero at a scale $k_{\chi SB}$ not far below k_Φ. Here the system enters the SSB regime and a non-vanishing (renormalized) VEV σ_0 for the meson field Φ develops. The evolution of σ_0 with k turns out to be reasonably stable already before scales $k \simeq m_\pi$ where the evolution is stopped. We take this result as an indication that our truncation of the effective action Γ_k leads at least qualitatively to a satisfactory description of chiral symmetry breaking. The reason for the relative stability of the IR behavior of the VEV (and all other couplings) is that the quarks acquire a constituent mass $M_q = h\sigma_0 \simeq 350\,\mathrm{MeV}$ in the SSB regime. As a consequence they decouple once k becomes smaller than M_q and the evolution is then dominantly driven by the massless Goldstone bosons. This is also important in view of potential confinement effects expected to become important for the quark n–point functions for k around $\Lambda_{QCD} \simeq 200\,\mathrm{MeV}$. Since confinement is not explicitly included in our truncation of Γ_k, one might be worried that such effects could spoil our results completely. Yet, as discussed in some more detail above, only the colored quarks should feel confinement and they are no longer important for the evolution of the meson couplings for k around 200 MeV. One might therefore hope that a precise treatment of confinement is not crucial for this approach to chiral symmetry breaking.

Most importantly, one finds that the system of flow equations exhibits a partial IR fixed point in the symmetric phase. As already pointed out one expects Z_{Φ,k_Φ} to be rather small. In turn, one may assume that, at least for the initial range of running in the symmetric regime the mass parameter $m^2 \sim Z_{\Phi,k}^{-1}$ is large. This means, in particular, that all threshold functions with arguments $\sim m^2$ may be neglected in this regime. As a consequence, the flow equations simplify considerably. We find, for instance, for the $U_L(2) \times U_R(2)$ model

$$
\begin{aligned}
\frac{d}{dt}\tilde{\lambda}_1 &\equiv \frac{d}{dt}\frac{\lambda_1}{h^2} \simeq \frac{N_c}{4\pi^2}h^2\left[\frac{1}{2}\tilde{\lambda}_1 - \frac{1}{N}\right] \\
\frac{d}{dt}\tilde{\lambda}_2 &\equiv \frac{d}{dt}\frac{\lambda_2}{h^2} \simeq \frac{N_c}{4\pi^2}h^2\left[\frac{1}{2}\tilde{\lambda}_2 - \frac{2}{N}\right] \\
\frac{d}{dt}h^2 &\simeq \frac{N_c}{8\pi^2}h^4 \\
\eta_\Phi &\simeq \frac{N_c}{8\pi^2}h^2\,, \quad \eta_\psi \simeq 0\,.
\end{aligned}
\tag{48}
$$

This system possesses an attractive IR fixed point for the quartic scalar self interactions

$$
\tilde{\lambda}_{1*} = \frac{1}{2}\tilde{\lambda}_{2*} = \frac{2}{N}\,. \tag{49}
$$

Furthermore it is exactly soluble[47]. Because of the strong Yukawa coupling the quartic couplings $\tilde{\lambda}_1$ and $\tilde{\lambda}_2$ generally approach their fixed point values rapidly, long before the systems enters the broken phase ($m \to 0$) and the approximation of large m breaks down. In addition, for large initial values h_I^2 the Yukawa coupling at the scale $k_{\chi SB}$ where m vanishes (or becomes small) only depends on the initial value $\overline{m}_{k_\Phi}$. Hence, the system is approximately independent in the IR upon the initial values of λ_1, λ_2 and h^2, the only "relevant" parameter being $\overline{m}_{k_\Phi}$. (Once quark masses and a proper treatment of the chiral anomaly are included for $N = 3$ one expects that m_q and ν are additional relevant parameters. Their values may be fixed by using the masses of π, K and η' as phenomenological input.) In other words, the effective action looses almost all its "memory" in the far IR of where in the UV it came from. This feature of the flow equations leads to a perhaps surprising degree of predictive power. In addition, also the dependence of $f_\pi = 2\sigma_0$ on $\overline{m}_{k_\Phi}$ is not very strong for a large range in $\overline{m}_{k_\Phi}$, as shown in fig. 3. The relevant parameter $\overline{m}_{k_\Phi}$ can be fixed by using the constituent quark mass $M_q \equiv h\sigma_0 \simeq 350\,\mathrm{MeV}$ as a phenomenological input. One obtains for the $O(4)$–model

$$\frac{\overline{m}^2_{k_\Phi}}{k^2_\Phi} \simeq 0.02\,. \tag{50}$$

The resulting value for the decay constant is

$$f_\pi = (91 - 100)\,\mathrm{MeV} \tag{51}$$

for $h_I^2 = 10^4 - 300$. It is striking that this comes close to the real value $f_\pi = 92.4\,\mathrm{MeV}$ but we expect that the uncertainty in the determination of the compositeness scale k_Φ and the truncation errors exceed the influence of the variation of h_I^2. We have furthermore used this result for an estimate of the chiral condensate:

$$\left\langle\overline{\psi}\psi\right\rangle \equiv -\overline{m}^2_{k_\Phi} f_\pi Z^{-1/2}_{\Phi,k=0} A \simeq -(195\,\mathrm{MeV})^3 \tag{52}$$

where the factor $A \simeq 1.7$ accounts for the change of the normalization scale of $\left\langle\overline{\psi}\psi\right\rangle$ from k_Φ to the commonly used value 1 GeV. Our value is in reasonable agreement with results from sum rules[51]. This is non-trivial since not only $\overline{m}_{k_\Phi}$ and f_π enter but also the IR value $Z_{\Phi,k=0}$. Integrating Eq.(48) for η_Φ one finds

$$Z_{\Phi,k} = Z_{\Phi,k_\Phi} + \frac{N_c}{8\pi^2}\ln\frac{k_\Phi}{k}\,. \tag{53}$$

Thus $Z_{\Phi,k}$ will indeed be practically independent of its initial value Z_{Φ,k_Φ} already after some running as long as Z_{Φ,k_Φ} is small compared to 0.01.

The alert reader may have noticed that the beta functions Eq.(48) correspond exactly to those obtained in the one–quark–loop approximation or, in other words, to the leading order in the large–N_c expansion for the Nambu–Jona-Lasinio model[46]. The fixed point Eq.(49) is then nothing but the large–N_c boundary condition on the evolution of λ_1 and λ_2 in this model. Yet, we wish to stress that nowhere we have

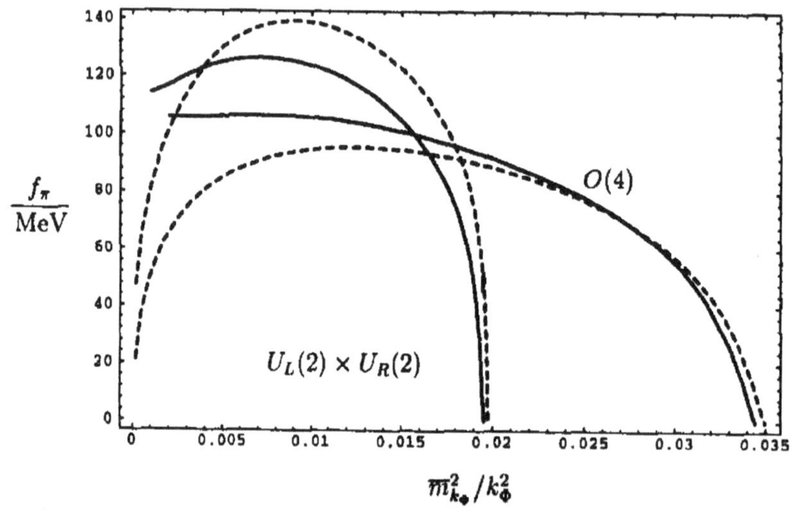

Figure 3. The pion decay constant f_π as a function of $\overline{m}^2_{k_\Phi}/k^2_\Phi$ for $k_\Phi = 630\,\mathrm{MeV}$ and initial values $\lambda_{i,I} = 0$ and $h_I^2 = 300$ (solid line) as well as $h_I^2 = 10^4$ (dashed line).

made the assumption that N_c is a large number. On the contrary, the physical value $N_c = 3$ suggests that the large–N_c expansion should a priori only be trusted on a quantitatively rather crude level. The reason why we expect Eq.(48) to nevertheless give rather reliable results is based on the fact that for small Z_{Φ,k_Φ} all (renormalized) meson masses are much larger than the scale k for the initial part of the running. This implies that the mesons are effectively decoupled and their contribution to the beta functions is negligible leading to the one–quark–loop approximation. Yet, already after some relatively short period of running the renormalized meson masses approach zero and our approximation of neglecting mesonic threshold functions breaks down. Hence, the one–quark–loop approximation is reasonable only for scales close to k_Φ but is bound to become inaccurate around $k_{\chi SB}$ and in the SSB regime. What is important in our context is not the numerical value of the partial fixed points Eq.(49) but rather their mere existence and the presence of a large coupling h^2 driving the $\tilde{\lambda}_i$ fast towards them. This is enough for the IR values of all couplings to become almost independent of the initial values $\lambda_{i,I}$. Similar features of IR stability are expected if the truncation is enlarged, for instance, to a more general form of the effective potential U_{k_Φ} as will be discussed in the next section.

FLOW EQUATIONS FOR THE SCALAR POTENTIAL

In this and the remaining sections we consider the $O(4)$–symmetric quark meson model without truncating its effective average potential to a polynomial form[52]. A comparison with the results of the last sections will give us a feeling for the size of the errors induced by the quartic truncation used so far. Furthermore, such an approach is well suited for a study of the chiral phase transition close to which the form of the potential deviates substantially from a polynomial.

It is convenient to work with dimensionless and renormalized variables therefore eliminating all explicit k–dependence. With $(t = \ln(k/k_\Phi))$

$$u(t,\tilde{\rho}) \equiv k^{-d} U_k(\rho) \,, \quad \tilde{\rho} \equiv Z_{\Phi,k} k^{2-d} \rho \tag{54}$$

and using Eq.(35) as a first truncation of the effective average action Γ_k one obtains the flow equation in arbitrary dimensions d

$$\begin{aligned} \frac{\partial}{\partial t} u &= -du + (d - 2 + \eta_\Phi)\, \tilde{\rho} u' \\ &+ 2v_d \left\{ 3 l_0^d(u'; \eta_\Phi) + l_0^d(u' + 2\tilde{\rho} u''; \eta_\Phi) - 2^{\frac{d}{2}+1} N_c l_0^{(F)d}\left(\frac{1}{2}\tilde{\rho} h^2; \eta_\psi\right) \right\} . \end{aligned} \tag{55}$$

Here $v_d^{-1} \equiv 2^{d+1} \pi^{d/2} \Gamma(d/2)$ and primes denote derivatives with respect to $\tilde{\rho}$. We will always use in the following for the number of quark colors $N_c = 3$. Eq.(55) is a partial differential equation for the effective potential $u(t,\tilde{\rho})$ which has to be supplemented by the flow equation for the Yukawa coupling and expressions for the anomalous dimensions η_Φ, η_ψ. The definition of the threshold functions l_n^d, $l_n^{(F)d}$ can be found in[47].

The dimensionless renormalized expectation value $k = 2k^{2-d} Z_{\Phi,k} \overline{\sigma}_{0,k}^2$, with $\overline{\sigma}_{0,k}$ the k–dependent VEV of Φ, may be computed for each k directly from the condition

$$u'(t,\kappa) = \frac{\jmath}{\sqrt{2\kappa}} k^{-\frac{d+2}{2}} Z_{\Phi,k}^{-1/2} \equiv \epsilon_g \tag{56}$$

where[52]

$$\jmath = 2 \overline{m}_{k_\Phi,k}^2 \hat{m} \tag{57}$$

and $\hat{m} = (m_u + m_d)/2$ denotes the average light current quark mass normalized at k_Φ. Note that $\kappa \equiv 0$ in the symmetric regime for vanishing source term. Eq.(56) allows us to follow the flow of κ according to

$$\frac{d}{dt}\kappa = \frac{\kappa}{\epsilon_g + 2\kappa\lambda}\left\{ [\eta_\Phi - d - 2]\,\epsilon_g - 2\frac{\partial}{\partial t}u'(t,\kappa)\right\} \tag{58}$$

with $\lambda \equiv u''(t,\kappa)$. We define the Yukawa coupling for $\tilde{\rho} = \kappa$ and its flow equation reads[47]

$$\begin{aligned} \frac{d}{dt}h^2 &= (d - 4 + 2\eta_\psi + \eta_\Phi)\,h^2 - 2v_d h^4 \left\{ 3l_{1,1}^{(FB)d}(\frac{1}{2}h^2\kappa, \epsilon_g; \eta_\psi, \eta_\Phi) \right. \\ &- \left. l_{1,1}^{(FB)d}(\frac{1}{2}h^2\kappa, \epsilon_g + 2\lambda\kappa; \eta_\psi, \eta_\Phi)\right\}. \end{aligned} \tag{59}$$

Similarly, the scalar and quark anomalous dimensions are infered from

$$\begin{aligned} \eta_\Phi &\equiv -\frac{d}{dt}\ln Z_{\Phi,k} = 4\frac{v_d}{d}\left\{ 4\kappa\lambda^2 m_{2,2}^d(\epsilon_g, \epsilon_g + 2\lambda\kappa; \eta_\Phi) \right. \\ &+ \left. 2^{\frac{d}{2}} N_c h^2 m_4^{(F)d}(\frac{1}{2}h^2\kappa; \eta_\psi)\right\}, \\ \eta_\psi &\equiv -\frac{d}{dt}\ln Z_{\psi,k} = 2\frac{v_d}{d}h^2\left\{ 3m_{1,2}^{(FB)d}(\frac{1}{2}h^2\kappa, \epsilon_g; \eta_\psi, \eta_\Phi) \right. \\ &+ \left. m_{1,2}^{(FB)d}(\frac{1}{2}h^2\kappa, \epsilon_g + 2\lambda\kappa; \eta_\psi, \eta_\Phi)\right\}, \end{aligned} \tag{60}$$

which is a linear set of equations for the anomalous dimensions. The definitions of the threshold functions $l_{n_1,n_2}^{(FB)d}$, m_{n_1,n_2}^d, $m_4^{(F)d}$ and $m_{n_1,n_2}^{(FB)d}$ are again specified in[47].

The flow equations (55), (58)—(60), constitute a coupled system of ordinary and partial differential equations which can be integrated numerically[53, 54]. Here we take the effective current quark mass dependence of h, $Z_{\Phi,k}$ and $Z_{\psi,k}$ into account by stopping the evolution according to Eqs.(59), (60), evaluated for the chiral limit, below the pion mass m_π.

Similarly to the case of the quartic truncation of the effective average potential described in the preceding section one finds for $d = 4$ that chiral symmetry breaking indeed occurs for a wide range of initial values of the parameters. These include the presumably realistic case of large renormalized Yukawa coupling and a bare mass $\bar{m}_{k_\Phi}$ of order 100 MeV.

Most importantly, the approximate partial IR fixed point behavior encountered for the quartic potential approximation before, carries over to the truncation of Γ_k which maintains the full effective average potential[52]. To see this explicitly we study the flow equations (55), (58)—(60) subject to the condition $Z_{\Phi,k_\Phi} \ll 1$. For the relevant range of $\tilde{\rho}$ both $u'(t,\tilde{\rho})$ and $u'(t,\tilde{\rho}) + 2\tilde{\rho}u''(t,\tilde{\rho})$ are then much larger than $\tilde{\rho}h^2(t)$ and we may therefore neglect in the flow equations all scalar contributions with threshold functions involving these large masses. This yields the simplified equations ($d = 4, v_4^{-1} = 32\pi^2$)

$$\begin{aligned} \frac{\partial}{\partial t}u &= -4u + (2 + \eta_\Phi)\,\tilde{\rho}u' - \frac{N_c}{2\pi^2}l_0^{(F)4}(\frac{1}{2}\tilde{\rho}h^2), \\ \frac{d}{dt}h^2 &= \frac{N_c}{8\pi^2}h^4, \\ \eta_\Phi &= \frac{N_c}{8\pi^2}h^2, \quad \eta_\psi = 0. \end{aligned} \tag{61}$$

Again, this approximation is only valid for the initial range of running below k_Φ before the (dimensionless) renormalized scalar mass squared $u'(t,\tilde{\rho}=0)$ approaches zero near the chiral symmetry breaking scale. The system Eq.(61) is exactly soluble. We find

$$\begin{aligned} h^2(t) &= Z_\Phi^{-1}(t) = \frac{h_I^2}{1-\frac{N_c}{8\pi^2}h_I^2 t}\,, \quad Z_\psi(t)=1\,, \\ u(t,\tilde{\rho}) &= e^{-4t}u_I(e^{2t}\tilde{\rho}\frac{h^2(t)}{h_I^2}) - \frac{N_c}{2\pi^2}\int_0^t dr e^{-4r} l_0^{(F)4}(\frac{1}{2}h^2(t)\tilde{\rho}e^{2r})\,. \end{aligned} \tag{62}$$

(The integration over r on the right hand side of the solution for u can be carried out by first exchanging it with the one over momentum implicit in the definition of the threshold function $l_0^{(F)4}$.) Here $u_I(\tilde{\rho}) \equiv u(0,\tilde{\rho})$ denotes the effective average potential at the compositeness scale and h_I^2 is the initial value of h^2 at k_Φ, i.e. for $t=0$. For simplicity we will use an expansion of the initial value effective potential $u_I(\tilde{\rho})$ in powers of $\tilde{\rho}$ around $\tilde{\rho}=0$

$$u_I(\tilde{\rho}) = \sum_{n=0}^{\infty}\frac{u_I^{(n)}(0)}{n!}\tilde{\rho}^n \tag{63}$$

even though this is not essential for the forthcoming reasoning. Expanding also $l_0^{(F)4}$ in Eq.(62) in powers of its argument one finds for $n>2$

$$\frac{u^{(n)}(t,0)}{h^{2n}(t)} = e^{2(n-2)t}\frac{u_I^{(n)}(0)}{h_I^{2n}} + \frac{N_c}{\pi^2}\frac{(-1)^n(n-1)!}{2^{n+2}(n-2)}l_n^{(F)4}(0)\left[1-e^{2(n-2)t}\right]\,. \tag{64}$$

For decreasing $t\to-\infty$ the initial values $u_I^{(n)}$ become rapidly unimportant and $u^{(n)}/h^{2n}$ approaches a fixed point. For $n=2$, i.e., for the quartic coupling, one finds

$$\frac{u^{(2)}(t,0)}{h^2(t)} = 1 - \frac{1-\frac{u_I^{(2)}(0)}{h_I^2}}{1-\frac{N_c}{8\pi^2}h_I^2 t} \tag{65}$$

leading to the fixed point value $(u^{(2)}/h^2)_* = 1$ already encountered in Eq.(49). As a consequence of this fixed point behavior the system looses all its "memory" on the initial values $u_I^{(n\geq 2)}$ at the compositeness scale k_Φ! This typically happens before the approximation $u'(t,\tilde{\rho}), u'(t,\tilde{\rho})+2\tilde{\rho}u''(t,\tilde{\rho}) \gg \tilde{\rho}h^2(t)$ breaks down and the solution Eq.(62) becomes invalid. Furthermore, the attraction to partial infrared fixed points continues also for the range of k where the scalar fluctuations cannot be neglected anymore. As for the quartic truncation the initial value for the bare dimensionless mass parameter

$$\frac{u_I'(0)}{h_I^2} = \frac{\overline{m}_{k_\Phi}^2}{k_\Phi^2} \tag{66}$$

is never negligible. (In fact, using the values for $\overline{m}_{k_\Phi}^2$ and k_Φ computed previously[33] for a pure quark effective action as described above one obtains $\overline{m}_{k_\Phi}^2/k_\Phi^2 \simeq 0.036$.) For large h_I (and dropping the constant piece $u_I(0)$) the solution Eq.(62) therefore behaves with growing $|t|$ as

$$\begin{aligned} Z_\Phi(t) &\simeq -\frac{N_c}{8\pi^2}t\,, \quad Z_\psi(t)=1\,, \\ h^2(t) &\simeq -\frac{8\pi^2}{N_c t}\,, \\ u(t,\tilde{\rho}) &\simeq \frac{u_I'(0)}{h_I^2}e^{-2t}h^2(t)\tilde{\rho} - \frac{N_c}{2\pi^2}\int_0^t dr e^{-4r} l_0^{(F)4}(\frac{1}{2}h^2(t)\tilde{\rho}e^{2r})\,. \end{aligned} \tag{67}$$

Table 1. The table shows the dependence on the constituent quark mass M_q of the input parameters k_Φ, $\overline{m}^2_{k_\Phi}/k^2_\Phi$ and $\jmath$ as well as some of our "predictions". The phenomenological input used here besides M_q is $f_\pi = 92.4\,\mathrm{MeV}$, $m_\pi = 135\,\mathrm{MeV}$.The first line corresponds to the values for M_q and λ_I used in the remainder of this work. The other three lines demonstrate the insensitivity of our results with respect to the precise values of these parameters.

$\frac{M_q}{\mathrm{MeV}}$	$\frac{\lambda_I}{h_I^2}$	$\frac{k_\Phi}{\mathrm{MeV}}$	$\frac{\overline{m}^2_{k_\Phi}}{k^2_\Phi}$	$\frac{\jmath^{1/3}}{\mathrm{MeV}}$	$\frac{\hat{m}(k_\Phi)}{\mathrm{MeV}}$	$\frac{\hat{m}(1\,\mathrm{GeV})}{\mathrm{MeV}}$	$\frac{\langle\overline{\psi}\psi\rangle(1\,\mathrm{GeV})}{\mathrm{MeV}^3}$	$\frac{f_\pi^{(0)}}{\mathrm{MeV}}$
303	1	618	0.0265	66.8	14.7	11.4	$-(186)^3$	80.8
300	0	602	0.026	66.8	15.8	12.0	$-(183)^3$	80.2
310	0	585	0.025	66.1	16.9	12.5	$-(180)^3$	80.5
339	0	552	0.0225	64.4	19.5	13.7	$-(174)^3$	81.4

In other words, for $h_I \to \infty$ the IR behavior of the linear quark meson model will depend (in addition to the value of the compositeness scale k_Φ and the quark mass $\hat{m}$) only on one parameter, $\overline{m}^2_{k_\Phi}$. We have numerically verified this feature by starting with different values for $u_I^{(2)}(0)$. Indeed, the differences in the physical observables were found to be small. This IR stability of the flow equations leads to a perhaps surprising degree of predictive power: Not only the scalar wave function renormalization but even the full effective potential $U(\rho) = \lim_{k\to 0} U_k(\rho)$ is (approximately) fixed for $Z_{\Phi,k_\Phi} \ll 1$ once $\overline{m}_{k_\Phi}$ is known! For definiteness we will perform our numerical analysis of the full system of flow equations (55), (58)—(60) with the idealized initial value $u_I(\tilde{\rho}) = u_I'(0)\tilde{\rho}$ in the limit $h_I^2 \to \infty$. It should be stressed, though, that deviations from this idealization will lead only to small numerical deviations in the IR behavior of the linear quark meson model as long as the condition $Z_{\Phi,k_\Phi} \ll 1$ holds, say for[42] $h_I \gtrsim 15$.

With this knowledge at hand we may now fix the remaining three parameters of our model, k_Φ, $\overline{m}^2_{k_\Phi}$ and $\hat{m}$ by using $f_\pi = 92.4\,\mathrm{MeV}$, the pion mass $M_\pi = 135\,\mathrm{MeV}$ and the constituent quark mass M_q as phenomenological input. This approach differs from that of the preceding sections where we took an earlier determination of k_Φ as input for the computation of f_π. It will be better suited for precision estimates of the high temperature behavior in the following sections. Because of the uncertainty regarding the precise value of M_q we give in tab. 1 the results for several values of M_q. The first line of tab. 1 corresponds to the choice of M_q and $\lambda_I \equiv u_I''(\kappa)$ which we will use for the forthcoming analysis of the model at finite temperature. As argued analytically above the dependence on the value of λ_I is weak for large enough h_I as demonstrated numerically by the second line. Moreover, we notice that our results, and in particular the value of $\jmath$, are rather insensitive with respect to the precise value of M_q. It is remarkable that the values for k_Φ and $\overline{m}_{k_\Phi}$ are not very different from those computed[33] from four–quark interactions as described above. As compared to the analysis of the preceding sections the present truncation of Γ_k is of a higher level of accuracy. We now consider an arbitrary form of the effective average potential instead of a polynomial approximation and we have included the pieces in the threshold functions which are proportional to the anomalous dimensions. It is encouraging that the results are rather robust with respect to these improvements.

Once the parameters k_Φ, $\overline{m}^2_{k_\Phi}$ and $\hat{m}$ are fixed there are a number of "predictions" of the linear meson model which can be compared with the results obtained by other methods or direct experimental observation. First of all one may compute the value of $\hat{m}$ at a scale of 1 GeV which is suitable for comparison with results obtained from chiral perturbation theory[55] and sum rules[51]. For this purpose one has to account for the running of this quantity with the normalization scale from k_Φ as given in tab. 1

to the commonly used value of 1 GeV: $\hat{m}(1\,\mathrm{GeV}) = A^{-1}\hat{m}(k_\Phi)$. A reasonable estimate of the factor A is obtained from the three loop running of $\hat{m}$ in the $\overline{MS}$ scheme[51]. For $M_q \simeq 300\,\mathrm{MeV}$ corresponding to the first two lines in tab. 1 its value is $A \simeq 1.3$. The results for $\hat{m}(1\,\mathrm{GeV})$ are in acceptable agreement with recent results from other methods[55, 51] even though they tend to be somewhat larger. Closely related to this is the value of the chiral condensate

$$\left\langle\overline{\psi}\psi\right\rangle(1\,\mathrm{GeV}) \equiv -A\overline{m}_{k_\Phi}^2\left[f_\pi Z_{\Phi,k=0}^{-1/2} - 2\hat{m}\right] . \tag{68}$$

These results are quite non–trivial since not only f_π and $\overline{m}_{k_\Phi}^2$ enter but also the computed IR value $Z_{\Phi,k=0}$. We emphasize in this context that there may be substantial corrections both in the extrapolation from k_Φ to 1 GeV and because of the neglected influence of the strange quark which may be important near k_Φ. These uncertainties have only little effect on the physics at lower scales as relevant for our analysis of the temperature effects. Only the value of $\jmath$ which is fixed by m_π enters here.

A further more qualitative test concerns the mass of the sigma resonance or radial mode in the limit $k \to 0$ whose renormalized mass squared is given by

$$m_\sigma^2 = Z_{\Phi,k_\Phi}^{-1/2}\frac{\overline{m}_{k_\Phi}^2\hat{m}}{\sigma_0} + 4\lambda\sigma_0^2 . \tag{69}$$

From our numerical analysis we obtain $\lambda_{k=0} \simeq 20$ which translates into $m_\sigma \simeq 430\,\mathrm{MeV}$. One should note, though, that this result is presumably not very accurate as we have employed in this work the approximation of using the Goldstone boson wave function renormalization constant also for the radial mode. Furthermore, the explicit chiral symmetry breaking contribution to m_σ^2 is certainly underestimated as long as the strange quark is neglected. In any case, we observe that the sigma meson is significantly heavier than the pions. This is a crucial consistency check for the linear quark meson model. A low sigma mass would be in conflict with the numerous successes of chiral perturbation theory[36] which requires the decoupling of all modes other than the Goldstone bosons in the IR–limit of QCD. The decoupling of the sigma meson is, of course, equivalent to the limit $\lambda \to \infty$ which formally describes the transition from the linear to the non–linear sigma model and which appears to be reasonably well realized by the large IR–values of λ obtained in our analysis. We also note that the issue of the sigma mass is closely connected to the value of $f_\pi^{(0)}$, the value of f_π in the chiral limit $\hat{m} = 0$ also given in tab. 1. To lowest order in $(f_\pi - f_\pi^{(0)})/f_\pi$ or, equivalently, in $\hat{m}$ one has

$$f_\pi - f_\pi^{(0)} = \frac{\jmath}{Z_\Phi^{1/2} m_\sigma^2} = \frac{f_\pi m_\pi^2}{m_\sigma^2} . \tag{70}$$

A larger value of m_σ would therefore reduce the difference between $f_\pi^{(0)}$ and f_π.

In fig. 4 we show the dependence of the pion mass and decay constant on the average current quark mass $\hat{m}$. These curves depend very little on the values of the initial parameters as demonstrated in tab. 1 by $f_\pi^{(0)}$. We observe a relatively large difference of 12 MeV between the pion decay constants at $\hat{m} = \hat{m}_{\mathrm{phys}}$ and $\hat{m} = 0$. According to Eq.(70) this difference is related to the mass of the sigma particle and will be modified in the three flavor case. We will later find that the critical temperature T_c for the second order phase transition in the chiral limit is almost independent of the initial conditions. The values of $f_\pi^{(0)}$ and T_c essentially determine the non–universal amplitudes in the critical scaling region (see below). In summary, we find that the behavior of our model for small k is quite robust as far as uncertainties in the initial conditions at the scale k_Φ are concerned. We will see that the difference of observables between non–vanishing and vanishing temperature is entirely determined by the flow of couplings in the range $0 < k \lesssim 3T$.

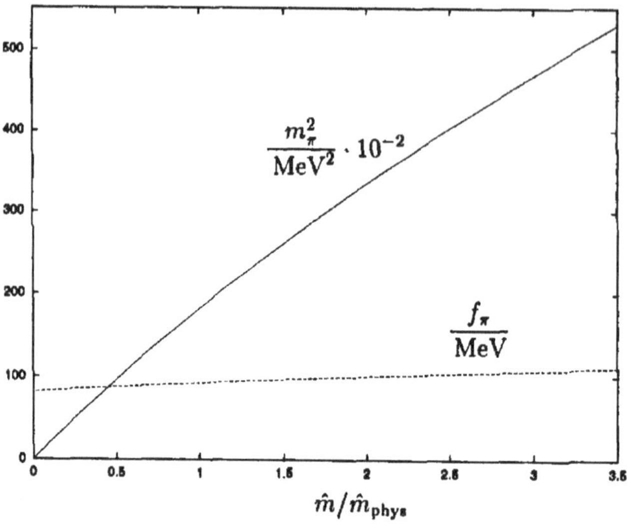

Figure 4. The plot shows m_π^2 (solid line) and f_π (dashed line) as functions of the current quark mass $\hat{m}$ in units of the physical value $\hat{m}_{\mathrm{phys}}$.

CHIRAL PHASE TRANSITION OF TWO FLAVOR QCD

Strong interactions in thermal equilibrium at high temperature T — as realized in early stages of the evolution of the Universe — differ in important aspects from the well tested vacuum or zero temperature properties. A phase transition at some critical temperature T_c or a relatively sharp crossover may separate the high and low temperature physics[56]. Many experimental activities at heavy ion colliders[57] search for signs of such a transition. It was realized early that the transition should be closely related to a qualitative change in the chiral condensate according to the general observation that spontaneous symmetry breaking tends to be absent in a high temperature situation. A series of stimulating contributions[58, 59, 60] pointed out that for sufficiently small up and down quark masses, m_u and m_d, and for a sufficiently large mass of the strange quark, m_s, the chiral transition is expected to belong to the universality class of the $O(4)$ Heisenberg model. This means that near the critical temperature only the pions and the sigma particle play a role for the behavior of condensates and long distance correlation functions. It was suggested[59,60] that a large correlation length may be responsible for important fluctuations or lead to a disoriented chiral condensate[61]. This was even related[59,60] to the spectacular "Centauro events"[62] observed in cosmic rays. The question how small m_u and m_d would have to be in order to see a large correlation length near T_c and if this scenario could be realized for realistic values of the current quark masses remained, however, unanswered. The reason was the missing link between the universal behavior near T_c and zero current quark mass on one hand and the known physical properties at $T = 0$ for realistic quark masses on the other hand.

It is the purpose of the remaining sections of these lectures to provide this link[52]. We present here the equation of state for two flavor QCD within an effective quark meson model. The equation of state expresses the chiral condensate $\langle\overline{\psi}\psi\rangle$ as a function of temperature and the average current quark mass $\hat{m} = (m_u + m_d)/2$. This connects explicitly the universal critical behavior for $T \to T_c$ and $\hat{m} \to 0$ with the temperature dependence for a realistic value $\hat{m}_{\text{phys}}$. Since our discussion covers the whole temperature range $0 \leq T \lesssim 1.7\,T_c$ we can fix $\hat{m}_{\text{phys}}$ such that the (zero temperature) pion mass is $m_\pi = 135\,\text{MeV}$. The condensate $\langle\overline{\psi}\psi\rangle$ plays here the role of an order parameter. Fig. 5 shows our results for $\langle\overline{\psi}\psi\rangle(T, \hat{m})$: Curve (*a*) gives the temperature dependence of $\langle\overline{\psi}\psi\rangle$ in the chiral limit $\hat{m} = 0$. Here the lower curve is the full result for arbitrary *T* whereas the upper curve corresponds to the universal scaling form of the equation of state for the *O*(4) Heisenberg model. We see perfect agreement of both curves for *T* sufficiently close to $T_c = 100.7\,\text{MeV}$. This demonstrates the capability of our method to cover the critical behavior and, in particular, to reproduce the critical exponents of the *O*(4)–model. We have determined (see below) the universal critical equation of state as well as the non–universal amplitudes. This provides the full functional dependence of $\langle\overline{\psi}\psi\rangle(T, \hat{m})$ for small $T - T_c$ and $\hat{m}$. The curves (*b*), (*c*) and (*d*) are for non–vanishing values of the average current quark mass $\hat{m}$. Curve (c) corresponds to $\hat{m}_{\text{phys}}$ or, equivalently, $m_\pi(T = 0) = 135\,\text{MeV}$. One observes a crossover in the range !$T = (1.2 - 1.5)T_c$. The *O*(4) universal equation of state (upper curve) gives a reasonable approximation in this temperature range. The transition turns out to be much less dramatic than for $\hat{m} = 0$. We have also plotted in curve (*b*) the results for comparably small quark masses $\simeq 1\,\text{MeV}$, i.e. $\hat{m} = \hat{m}_{\text{phys}}/10$, for which the $T = 0$ value of m_π equals 45 MeV. The crossover is considerably sharper but a substantial deviation from the chiral limit remains even for such small values of $\hat{m}$. In order to facilitate comparison with lattice simulations which are typically performed for larger values of m_π we also present results for $m_\pi(T = 0) = 230\,\text{MeV}$ in curve *(d)*. One may define a "pseudocritical temperature" T_{pc} associated to the smooth crossover as the inflection point of $\langle\overline{\psi}\psi\rangle(T)$ as often done in lattice simulations. Our results for this definition of T_{pc} are denoted by $T_{pc}^{(1)}$ and are presented in tab. 2 for the four different values of $\hat{m}$ or, equivalently,

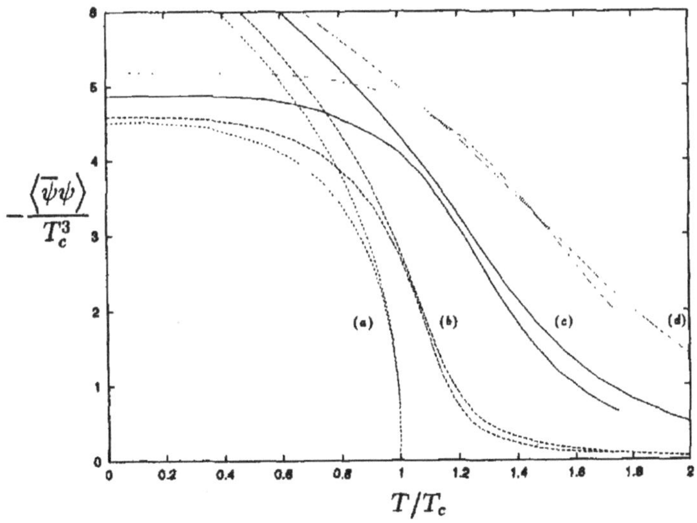

Figure 5. The plot shows the chiral condensate $\langle\overline{\psi}\psi\rangle$ as a function of temperature T. Lines (a), (b), (c), (d) correspond at zero temperature to $m_\pi = 0, 45\,\text{MeV}, 135\,\text{MeV}, 230\,\text{MeV}$, respectively. For each pair of curves the lower one represents the full T-dependence of $\langle\overline{\psi}\psi\rangle$ whereas the upper one shows for comparison the universal scaling form of the equation of state for the $O(4)$ Heisenberg model. The critical temperature for zero quark mass is $T_c = 100.7\,\text{MeV}$. The chiral condensate is normalized at a scale $k_\Phi \simeq 620\,\text{MeV}$.

Table 2. The table shows the critical and "pseudocritical" temperatures for various values of the zero temperature pion mass. Here $T_{pc}^{(1)}$ is defined as the inflection point of $\langle\overline{\psi}\psi\rangle(T)$ whereas $T_{pc}^{(2)}$ is the location of the maximum of the sigma correlation length.

$\frac{m_\pi}{\text{MeV}}$	0	45	135	230
$\frac{T_{pc}^{(1)}}{\text{MeV}}$	100.7	$\simeq 110$	$\simeq 130$	$\simeq 150$
$\frac{T_{pc}^{(2)}}{\text{MeV}}$	100.7	113	128	—

$m_\pi(T=0)$. The value for the pseudocritical temperature for $m_\pi = 230\,\text{MeV}$ compares well with the lattice results for two flavor QCD (cf. the discussion below). One should mention, though, that a determination of T_{pc} according to this definition is subject to sizeable numerical uncertainties for large pion masses as the curve in fig. 5 is almost linear around the inflection point for quite a large temperature range. A difficult point in lattice simulations with large quark masses is the extrapolation to realistic values of m_π or even to the chiral limit. Our results may serve here as an analytic guide. The overall picture shows the approximate validity of the $O(4)$ scaling behavior over a large temperature interval in the vicinity of and above T_c once the (non–universal) amplitudes are properly computed.

A second important result of our investigations is the temperature dependence of the space–like pion correlation length $m_\pi^{-1}(T)$. (We will often call $m_\pi(T)$ the temperature dependent pion mass since it coincides with the physical pion mass for $T = 0$.) The plot for $m_\pi(T)$ in fig. 6 again shows the second order phase transition in the chiral limit $\hat{m} = 0$. For $T < T_c$ the pions are massless Goldstone bosons whereas for $T > T_c$ they form together with the sigma particle a degenerate vector of $O(4)$ whose mass increases as a function of temperature. For $\hat{m} = 0$ the behavior for small positive $T - T_c$ is characterized by the critical exponent ν, i.e. $m_\pi(T) = (\xi^+)^{-1} T_c\,((T-T_c)/T_c)^\nu$ and we

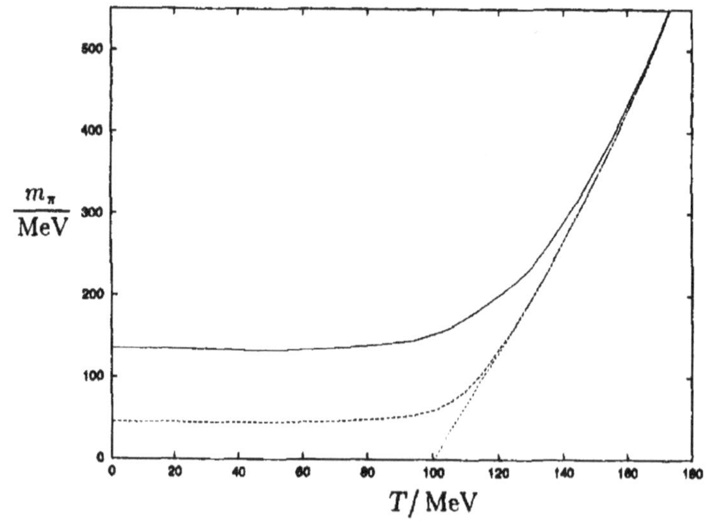

Figure 6. The plot shows m_π as a function of temperature T for three different values of the average light current quark mass $\hat{m}$. The solid line corresponds to the realistic value $\hat{m} = \hat{m}_{\text{phys}}$ whereas the dotted line represents the situation without explicit chiral symmetry breaking, i.e., $\hat{m} = 0$. The intermediate, dashed line assumes $\hat{m} = \hat{m}_{\text{phys}}/10$.

obtain $\nu = 0.787$, $\xi^+ = 0.270$., For $\hat{m} > 0$ we find that $m_\pi(T)$ remains almost constant for $T \lesssim T_c$ with only a very slight dip for T near $T_c/2$. For $T > T_c$ the correlation length decreases rapidly and for $T \gg T_c$ the precise value of $\hat{m}$ becomes irrelevant. We see that the universal critical behavior near T_c is quite smoothly connected to $T = 0$. The full functional dependence of $m_\pi(T, \hat{m})$ allows us to compute the overall size of the pion correlation length near the critical temperature and we find $m_\pi(T_{pc}) \simeq 1.7 m_\pi(0)$for the realistic value $\hat{m}_{phys}$. This correlation length is even smaller than the vacuum ($T = 0$) one and shows no indication for strong fluctuations of pions with long wavelength. It would be interesting to see if a decrease of the pion correlation length at and above T_c is experimentally observable. It should be emphasized, however, that a tricritical behavior with a massless excitation remains possible for three flavors. This would not be characterized by the universal behavior of the $O(4)$–model. We also point out that the present investigation for the two flavor case does not take into account a speculative "effective restoration" of the axial $U_A(1)$ symmetry at high temperature[58, 63]. We will comment on these issues in the last section. In the next sections we will describe the formalism which leads to these results.

THERMAL EQUILIBRIUM AND DIMENSIONAL REDUCTION

The extension of flow equations to thermal equilibrium situations at non–vanishing temperature T is straightforward[32]. In the Euclidean formalism non-zero temperature results in (anti–)periodic boundary conditions for (fermionic) bosonic fields in the Euclidean time direction with periodicity[64] $1/T$. This leads to the replacement

$$\int \frac{d^d q}{(2\pi)^d} f(q^2) \rightarrow T \sum_{l \in \mathbb{Z}} \int \frac{d^{d-1}\vec{q}}{(2\pi)^{d-1}} f(q_0^2(l) + \vec{q}^{\,2}) \tag{71}$$

in the trace in Eq.(11) when represented as a momentum integration, with a discrete spectrum for the zero component

$$q_0(l) = \begin{cases} 2l\pi T & \text{for bosons} \\ (2l+1)\pi T & \text{for fermions}. \end{cases} \tag{72}$$

Hence, for $T > 0$ a four–dimensional QFT can be interpreted as a three–dimensional model with each bosonic or fermionic degree of freedom now coming in an infinite number of copies labeled by $l \in \mathbb{Z}$ (Matsubara modes). Each mode acquires an additional temperature dependent effective mass term $q_0^2(l)$. In a high temperature situation where all massive Matsubara modes decouple from the dynamics of the system one therefore expects to observe an effective three–dimensional theory with the bosonic zero modes as the only relevant degrees of freedom. In other words, if the characteristic length scale associated with the physical system is much larger than the inverse temperature the compactified Euclidean "time" dimension cannot be resolved anymore. This phenomenon is known as "dimensional reduction"[65].

The formalism of the effective average action automatically provides the tools for a smooth decoupling of the massive Matsubara modes as the scale k is lowered from $k \gg T$ to $k \ll T$. It therefore allows us to directly link the low–T, four–dimensional QFT to the effective three–dimensional high–T theory. The replacement Eq.(71) in Eq.(11) manifests itself in the flow equations (55), (58)—(60) only through a change to T–dependent threshold functions. For instance, the dimensionless functions $l_n^d(w; \eta_\Phi)$

defined in Eq.(44) are replaced by

$$l_n^d(w,\frac{T}{k};\eta_\Phi) \equiv \frac{n+\delta_{n,0}}{4} v_d^{-1} k^{2n-d} T \sum_{l\in\mathbb{Z}} \int \frac{d^{d-1}\vec{q}}{(2\pi)^{d-1}} \left(\frac{1}{Z_{\Phi,k}} \frac{\partial R_k(q^2)}{\partial t}\right) \frac{1}{[P(q^2)+k^2 w]^{n+1}} \tag{73}$$

where $q^2 = q_0^2 + \vec{q}^2$ and $q_0 = 2\pi l T$. A list of the various temperature dependent threshold functions appearing in the flow equations can be found[52]. There we also discuss some subtleties regarding the definition of the Yukawa coupling and the anomalous dimensions for $T \neq 0$. In the limit $k \gg T$ the sum over Matsubara modes approaches the integration over a continuous range of q_0 and we recover the zero temperature threshold function $l_n^d(w;\eta_\Phi)$. In the opposite limit $k \ll T$ the massive Matsubara modes($l \neq 0$) are suppressed and we expect to find a $d-1$ dimensional behavior of l_n^d. In fact, one obtains from Eq.(73)

$$\begin{aligned} l_n^d(w,T/k;\eta_\Phi) &\simeq l_n^d(w;\eta_\Phi) && \text{for } T \ll k\,, \\ l_n^d(w,T/k;\eta_\Phi) &\simeq \frac{T}{k}\frac{v_{d-1}}{v_d} l_n^{d-1}(w;\eta_\Phi) && \text{for } T \gg k\,. \end{aligned} \tag{74}$$

For our choice of the infrared cutoff function R_k, Eq.(8), the temperature dependent Matsubara modes in $l_n^d(w,T/k;\eta_\Phi)$ are exponentially suppressed for $T \ll k$ whereas the behavior is more complicated for other threshold functions appearing in the flow equations (55), (58)—(60). Nevertheless, all bosonic threshold functions are proportional to T/k for $T \gg k$ whereas those with fermionic contributions vanish in this limit¶. This behavior is demonstrated in fig. 7 where we have plotted the quotients $l_1^4(w,T/k)/l_1^4(w)$ and $l_1^{(F)4}(w,T/k)/l_1^{(F)4}(w)$ of bosonic and fermionic threshold functions, respectively. One observes that for $k \gg T$ both threshold functions essentially behave as for zero temperature. For growing T or decreasing k this changes as more and more Matsubara modes decouple until finally all massive modes are suppressed. The bosonic threshold function l_1^4 shows for $k \ll T$ the linear dependence on T/k derived in Eq.(74). In particular, for the bosonic excitations the threshold function for $w \ll 1$ can be approximated with reasonable accuracy by $l_n^4(w;\eta_\Phi)$ for $T/k < 0.25$ and by $(4T/k)l_n^3(w;\eta_\Phi)$ for $T/k > 0.25$. The fermionic threshold function $l_1^{(F)4}$ tends to zero for $k \ll T$ since there is no massless fermionic zero mode, i.e. in this limit all fermionic contributions to the flow equations are suppressed. On the other hand, the fermions remain quantitatively relevant up to $T/k \simeq 0.6$ because of the relatively long tail in fig. 7b. The transition from four to three–dimensional threshold functions leads to a *smooth dimensional reduction* as k is lowered from $k \gg T$ to $k \ll T$! Whereas for $k \gg T$ the model is most efficiently described in terms of standard four–dimensional fields Φ a choice of rescaled three–dimensional variables $\Phi_3 = \Phi/\sqrt{T}$ becomes better adapted for $k \ll T$. Accordingly, for high temperatures one will use a potential

$$u_3(t,\tilde{\rho}_3) = \frac{k}{T} u(t,\tilde{\rho})\,; \quad \tilde{\rho}_3 = \frac{k}{T}\tilde{\rho}\,. \tag{75}$$

In this regime $\Gamma_{k\to 0}$ corresponds to the free energy of classical statistics and $\Gamma_{k>0}$ is a classical coarse grained free energy.

For our numerical calculations at non-vanishing temperature we exploit the discussed behavior of the threshold functions by using the zero temperature flow equations in the range $k \geq 10T$. For smaller values of k we approximate the infinite Matsubara sums (cf. Eq.(73)) by a finite series such that the numerical uncertainty at $k = 10T$ is better than 10^{-4}. This approximation becomes exact in the limit $k \ll 10T$.

¶For the present choice of R_k the temperature dependence of the threshold functions is considerably smoother than in that of previous investigations[32].

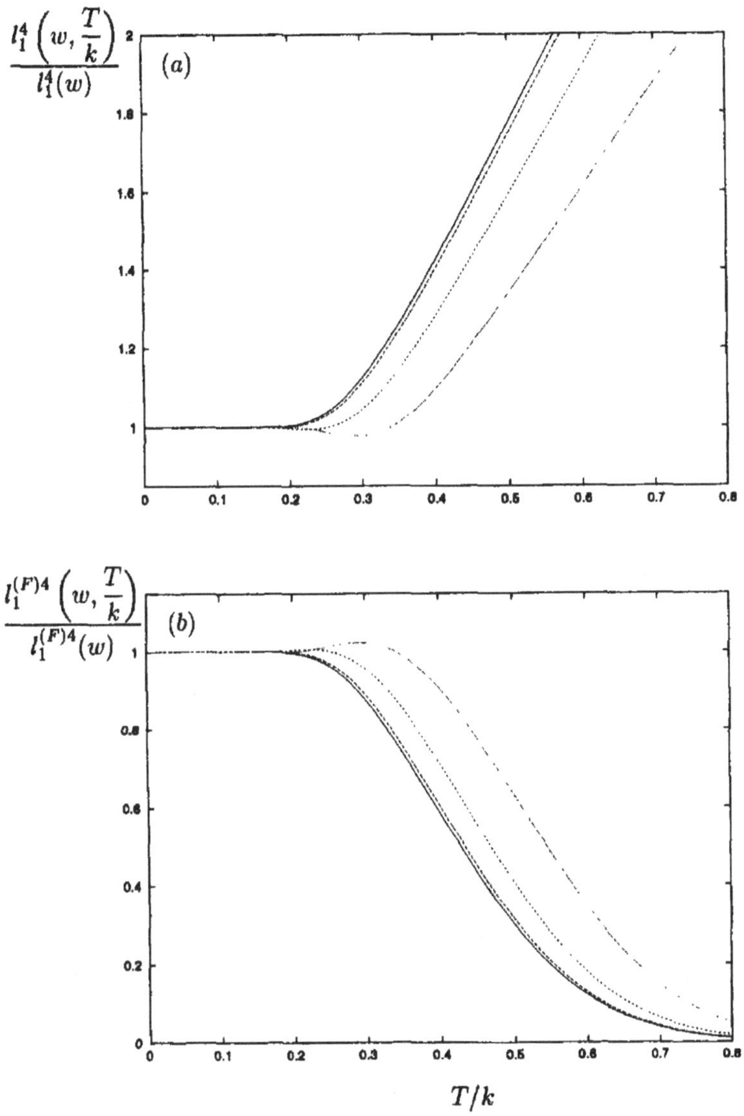

Figure 7. The plot shows the temperature dependence of the bosonic (a) and the fermionic (b) threshold functions $l_1^4(w, T/k)$ and $l_1^{(F)4}(w, T/k)$, respectively, for different values of the dimensionless mass term w. The solid line corresponds to $w = 0$ whereas the dotted ones correspond to $w = 0.1$, $w = 1$ and $w = 10$ with decreasing size of the dots. For $T \gg k$ the bosonic threshold function becomes proportional to T/k whereas the fermionic one tends to zero. In this range the theory with properly rescaled variables behaves as a classical three–dimensional theory.

THE QUARK MESON MODEL AT $T \neq 0$

So far we have considered the relevant fluctuations that contribute to the flow of Γ_k in dependence on the scale k. In a thermal equilibrium situation Γ_k also depends on the temperature T and one may ask for the relevance of thermal fluctuations at a given scale k. In particular, for not too high values of T the "initial condition" Γ_{k_Φ} for the solution of the flow equations should essentially be independent of temperature.

This will allow us to fix Γ_{k_Φ} from phenomenological input at $T = 0$ and to compute the temperature dependent quantities in the infrared $(k \to 0)$. We note that the thermal fluctuations which contribute to the r.h.s. of the flow equation for the meson potential Eq.(55) are effectively suppressed for $T \lesssim k/4$ as discussed in detail in the last section. Clearly for $T \gtrsim k_\Phi/3$ temperature effects become important at the compositeness scale. We expect the linear quark meson model with a compositeness scale $k_\Phi \simeq 600\,\mathrm{MeV}$ to be a valid description for two flavor QCD below a temperature of about 170 MeV. We note that there will be an effective temperature dependence of Γ_{k_Φ} induced by the fluctuations of other degrees of freedom besides the quarks, the pions and the sigma which are taken into account here. We will comment on this issue in the last section. For realistic three flavor QCD the thermal kaon fluctuations will become important for $T \gtrsim 170\,\mathrm{MeV}$.

We compute the quantities of interest for temperatures $T \lesssim 170\,\mathrm{MeV}$ by numerically solving the T–dependent version of the flow equations[52] (55), (58)—(60) by lowering k from k_Φ to zero. For this range of temperatures we use the initial values as given in the first line of tab. 1. This corresponds to choosing the zero temperature pion mass and the pion decay constant ($f_\pi = 92.4\,\mathrm{MeV}$ for $m_\pi = 135\,\mathrm{MeV}$) as phenomenological input. The only further input is the constituent quark mass M_q which we vary in the range $M_q \simeq 300 - 350\,\mathrm{MeV}$. We observe only a minor dependence of our results on M_q for the considered range of values. In particular, the value for the critical temperature T_c of the model remains almost unaffected by this variation.

We have plotted in fig. 8 the renormalized expectation value $2\sigma_0$ of the scalar field as a function of temperature for three different values of the average light current quark mass $\hat{m}$. (We remind that $2\sigma_0(T = 0) = f_\pi$.) For $\hat{m} = 0$ the order parameter σ_0 of chiral symmetry breaking continuously goes to zero for $T \to T_c = 100.7\,\mathrm{MeV}$ characterizing the phase transition to be of second order. The universal behavior of the model for small $T - T_c$ and small $\hat{m}$ is discussed in more detail in the following section. We point out that the value of T_c corresponds to $f_\pi^{(0)} = 80.8\,\mathrm{MeV}$, i.e. the value of the pion decay constant for $\hat{m} = 0$, which is significantly lower than $f_\pi = 92.4\,\mathrm{MeV}$ obtained for the realistic value $\hat{m}_{\mathrm{phys}}$. If we would fix the value of the pion decay constant to be 92.4 MeV

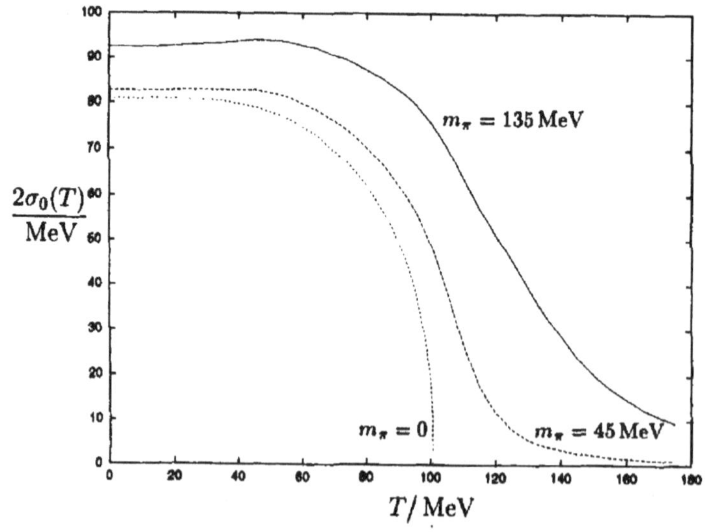

Figure 8. The expectation value $2\sigma_0$ is shown as a function of temperature T for three different values of the zero temperature pion mass.

also in the chiral limit ($\hat{m} = 0$), the value for the critical temperature would raise to 115 MeV. The nature of the transition changes qualitatively for $\hat{m} \neq 0$ where the second order transition is replaced by a smooth crossover. The crossover for a realistic $\hat{m}_{\rm phys}$ or $m_\pi(T = 0) = 135\,{\rm MeV}$ takes place in a temperature range $T \simeq (120 - 150)\,{\rm MeV}$. The middle curve in fig. 8 corresponds to a value of $\hat{m}$ which is only a tenth of the physical value, leading to a zero temperature pion mass $m_\pi = 45\,{\rm MeV}$. Here the crossover becomes considerably sharper but there remain substantial deviations from the chiral limit even for such small quark masses $\hat{m} \simeq 1\,{\rm MeV}$. The temperature dependence of m_π has already been mentioned (see fig. 6) for the same three values of $\hat{m}$. As expected, the pions behave like true Goldstone bosons for $\hat{m} = 0$, i.e. their mass vanishes for $T \leq T_c$. Interestingly, m_π remains almost constant as a function of T for $T < T_c$ before it starts to increase monotonically. We therefore find for two flavors no indication for a substantial decrease of m_π around the critical temperature.

The dependence of the mass of the sigma resonance m_σ on the temperature is displayed in fig. 9 for the above three values of $\hat{m}$. In the absence of explicit chiral symmetry breaking, $\hat{m} = 0$, the sigma mass vanishes for $T \leq T_c$. For $T < T_c$ this is a consequence of the presence of massless Goldstone bosons in the Higgs phase which drive the renormalized quartic coupling λ to zero. In fact, λ runs linearly with k for $T \gtrsim k/4$ and only evolves logarithmically for $T \lesssim k/4$. Once $\hat{m} \neq 0$ the pions acquire a mass even in the spontaneously broken phase and the evolution of λ with k is effectively stopped at $k \sim m_\pi$. Because of the temperature dependence of $\sigma_{0,k=0}$ (cf. fig. 8) this leads to a monotonically decreasing behavior of m_σ with T for $T \lesssim T_c$. This changes into the expected monotonic growth once the system reaches the symmetric phase for $T > T_c$. For low enough $\hat{m}$ one may use the minimum of $m_\sigma(T)$ for an alternative definition of the (pseudo-)critical temperature denoted as $T_{pc}^{(2)}$. Tab. 2 in the introduction shows our results for the pseudocritical temperature for different values of $\hat{m}$ or, equivalently, $m_\pi(T = 0)$. For a zero temperature pion mass $m_\pi = 135\,{\rm MeV}$ we find $T_{pc}^{(2)} = 128\,{\rm MeV}$. At larger pion masses of about 230 MeV we observe no longer a characteristic minimum for m_σ apart from a very broad, slight dip at $T \simeq 90\,{\rm MeV}$. A comparison of our results with lattice data is given below in the next section.

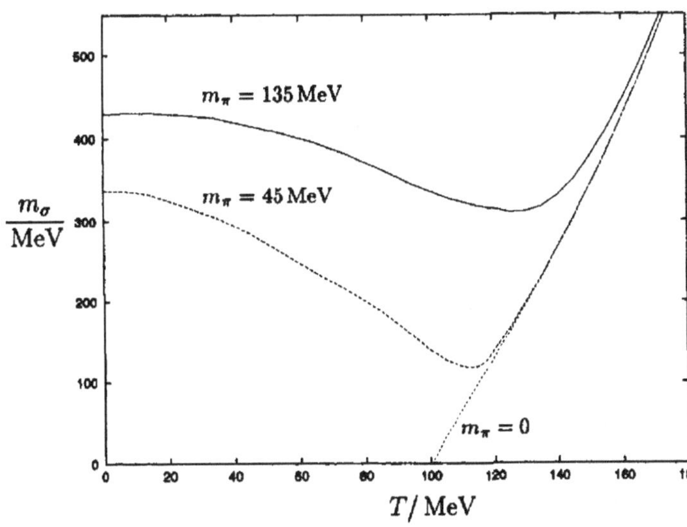

Figure 9. The plot shows the m_σ as a function of temperature T for three different values of the zero temperature pion mass.

Our results for the chiral condensate $\langle\overline{\psi}\psi\rangle$ as a function of temperature for different values of the average current quark mass are presented in fig. 5. We will compare $\langle\overline{\psi}\psi\rangle(T,\hat{m})$ with its universal scaling form for the $O(4)$ Heisenberg model in the following section.

Our ability to compute the complete temperature dependent effective meson potential U is demonstrated in fig. 10 where we display the derivative of the potential with respect to the renormalized field $\phi_R = (Z_\Phi\rho/2)^{1/2}$, for different values of T. The curves cover a temperature range $T = (5 - 175)$ MeV. The first one to the left corresponds to $T = 175$ MeV and neighboring curves differ in temperature by $\Delta T = 10\,\mathrm{MeV}$. One observes only a weak dependence of $\partial U(T)/\partial\phi_R$ on the temperature for $T \lesssim 60\,\mathrm{MeV}$. Evaluated for $\phi_R = \sigma_0$ this function connects the renormalized field expectation value with $m_\pi(T)$, the source $\jmath$ and the mesonic wave function renormalization $Z_\Phi(T)$ according to

$$\frac{\partial U(T)}{\partial\phi_R}(\phi_R = \sigma_0) = \frac{2\jmath}{Z_\Phi^{1/2}(T)} = 4\sigma_0(T)m_\pi^2(T)\ . \tag{76}$$

We close this section with a short assessment of the validity of our effective quark meson model as an effective description of two flavor QCD at non–vanishing temperature. The identification of qualitatively different scale intervals which appear in the context of chiral symmetry breaking, as presented in the preceding sections for the zero temperature case, can be generalized to $T \neq 0$: For scales below k_Φ there exists a hybrid description in terms of quarks and mesons. For $k_{\chi SB} \leq k \lesssim 600\,\mathrm{MeV}$ chiral symmetry remains unbroken where the symmetry breaking scale $k_{\chi SB}(T)$ decreases with increasing temperature. Also the constituent quark mass decreases with T. The running Yukawa coupling depends only mildly on temperature for $T \lesssim 120\,\mathrm{MeV}$. (Only near the critical temperature and for $\hat{m} = 0$ the running is extended because of massless pipn fluctuations.) On the other hand, for $k \lesssim 4T$ the effective three–dimensional gauge coupling increases faster than at $T = 0$ leading[12] to an increase of $\Lambda_{QCD}(T)$ with T. As k gets closer to the scale $\Lambda_{QCD}(T)$ it is no longer justified to neglect in the quark

$\partial U(T)/\partial\phi_R$ / MeV^3

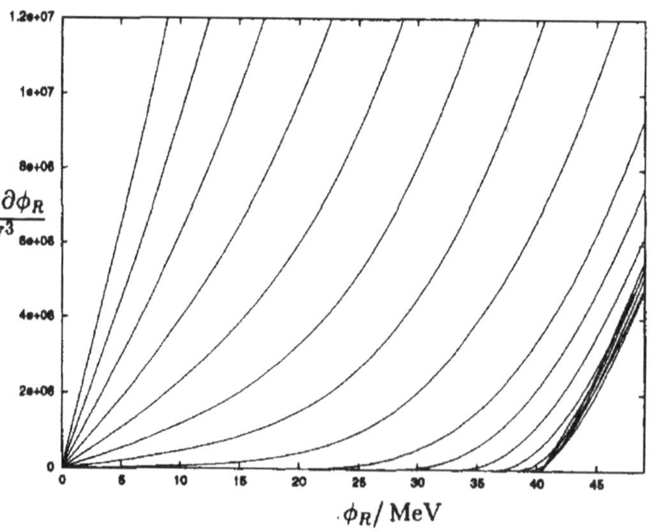

ϕ_R/ MeV

Figure 10. The plot shows the derivative of the meson potential $U(T)$ with respect to the renormalized field $\phi_R = (Z_\Phi\rho/2)^{1/2}$ for different values of T. The first curve on the left corresponds to $T = 175\,\mathrm{MeV}$. The successive curves to the right differ in temperature by $\Delta T = 10\,\mathrm{MeV}$ down to $T = 5\,\mathrm{MeV}$.

sector confinement effects which go beyond the dynamics of our present quark meson model. Here it is important to note that the quarks remain quantitatively relevant for the evolution of the meson degrees of freedom only for scales $k \gtrsim T/0.6$ (cf. fig. 7). In the limit $k \ll T/0.6$ all fermionic Matsubara modes decouple from the evolution of the meson potential according to the temperature dependent version of Eq.(55). Possible sizeable confinement corrections to the meson physics may occur if $\Lambda_{QCD}(T)$ becomes larger than the maximum of $M_q(T)$ and $T/0.6$. This is particularly dangerous for small $\hat{m}$ in a temperature interval around T_c. Nevertheless, the situation is not dramatically different from the zero temperature case since only a relatively small range of k is concerned. We do not expect that the neglected QCD non–localities lead to qualitative changes. Quantitative modifications, especially for small $\hat{m}$ and $|T - T_c|$ remain possible. This would only effect the non–universal amplitudes which will be discussed in the next section. The size of these corrections depends on the strength of (non–local) deviations of the quark propagator and the Yukawa coupling from the values computed in the quark meson model.

CRITICAL BEHAVIOR NEAR THE CHIRAL PHASE TRANSITION

In this section we study the linear quark meson model in the vicinity of the critical temperature T_c close to the chiral limit $\hat{m} = 0$. In this region we find that the sigma mass m_σ^{-1} is much larger than the inverse temperature T^{-1}, and one observes an effectively three–dimensional behavior of the high temperature quantum field theory. We also note that the fermions are no longer present in the dimensionally reduced system as has been discussed above. We therefore have to deal with a purely bosonic $O(4)$–symmetric linear sigma model. At the phase transition the correlation length becomes infinite and the effective three–dimensional theory is dominated by classical statistical fluctuations. In particular, the critical exponents which describe the singular behavior of various quantities near the second order phase transition are those of the corresponding classical system.

Many properties of this system are universal, i.e. they only depend on its symmetry ($O(4)$), the dimensionality of space (three) and its degrees of freedom (four real scalar components). Universality means that the long–range properties of the system do not depend on the details of the specific model like its short distance interactions. Nevertheless, important properties as the value of the critical temperature are non–universal. We emphasize that although we have to deal with an effectively three-dimensional bosonic theory, the non–universal properties of the system crucially depend on the details of the four–dimensional theory and, in particular, on the fermions.

Our aim is a computation of the critical equation of state which relates for arbitrary T near T_c the derivative of the free energy or effective potential U to the average current quark mass $\hat{m}$. The equation of state then permits to study the temperature and quark mass dependence of properties of the chiral phase transition.

At the critical temperature and in the chiral limit there is no scale present in the theory. In the vicinity of T_c and for small enough $\hat{m}$ one therefore expects a scaling behavior of the effective average potential u_k and accordingly a universal scaling form of the equation of state[28]. There are only two independent scales close to the transition point which can be related to the deviation from the critical temperature, $T - T_c$, and to the explicit symmetry breaking through the quark mass $\hat{m}$. As a consequence, the properly rescaled potential can only depend on one scaling variable. A possible choice for the parameterization of the rescaled "unrenormalized" potential is the use of the

Table 3. The table shows the critical exponents corresponding to the three-dimensional $O(4)$–Heisenberg model. Our results are denoted by "average action" whereas "3d PT" labels the exponents obtained from a 3d perturbative expansion using Pade techniques[70, 71]. The bottom line contains lattice Monte Carlo results.[69]

	ν	γ	δ	β	η
average action	0.787	1.548	4.80	0.407	0.0344
3d PT	0.73(2)	1.44(4)	4.82(5)	0.38(1)	0.03(1)
MC	0.7479(90)	1.477(18)	4.851(22)	0.3836(46)	0.0254(38)

Widom scaling variable[66]

$$x = \frac{(T - T_c)/T_c}{(2\overline{\sigma}_0/T_c)^{1/\beta}} . \tag{77}$$

Here β is the critical exponent of the order parameter $\overline{\sigma}_0$ in the chiral limit $\hat{m} = 0$ (see Eq.(81)). With $U'(\rho = 2\overline{\sigma}_0^2) = \jmath/(2\overline{\sigma}_0)$ the Widom scaling form of the equation of state reads[66]

$$\frac{\jmath}{T_c^3} = \left(\frac{2\overline{\sigma}_0}{T_c}\right)^{\delta} f(x) \tag{78}$$

where the exponent δ is related to the behavior of the order parameter according to Eq.(83). The equation of state Eq.(78) is written for convenience directly in terms of four–dimensional quantities. They are related to the corresponding effective variables of the three–dimensional theory by appropriate powers of T_c. The source $\jmath$ is determined by the average current quark mass $\hat{m}$ as $\jmath = 2\overline{m}_{k_\Phi}^2 \hat{m}$. The mass term at the compositeness scale, $\overline{m}_{k_\Phi}^2$, also relates the chiral condensate to the order parameter according to $\langle\overline{\psi}\psi\rangle = -2\overline{m}_{k_\Phi}^2(\overline{\sigma}_0 - \hat{m})$. The critical temperature of the linear quark meson model was found above to be $T_c = 100.7\,\text{MeV}$.

The scaling function f is universal up to the model specific normalization of x and itself. Accordingly, all models in the same universality class can be related by a rescaling of $\overline{\sigma}_0$ and $T - T_c$. The non–universal normalizations for the quark meson model discussed here are defined according to

$$f(0) = D \quad , \qquad f(-B^{-1/\beta}) = 0 \,. \tag{79}$$

We find $D = 1.82 \cdot 10^{-4}, B = 7.41$ and our result for β is given in tab. 3. Apart from the immediate vicinity of the zero of $f(x)$ we find the following two parameter fit for the scaling function[25],

$$\begin{aligned} f_{\rm fit}(x) \;&=\; 1.816 \cdot 10^{-4}(1 + 136.1\,x)^2\,(1 + 160.9\,\theta\,x)^{\Delta} \\ &\quad\; (1 + 160.9\,(0.9446\,\theta^{\Delta})^{-1/(\gamma-2-\Delta)}\,x)^{\gamma-2-\Delta} \end{aligned} \tag{80}$$

to reproduce the numerical results for f and df/dx at the 1 – 2% level with $\theta = 0.625$ (0.656), $\Delta = -0.490$ (-0.550) for $x > 0$ ($x < 0$) and γ as given in tab. 3. The universal properties of the scaling function can be compared with results obtained by other

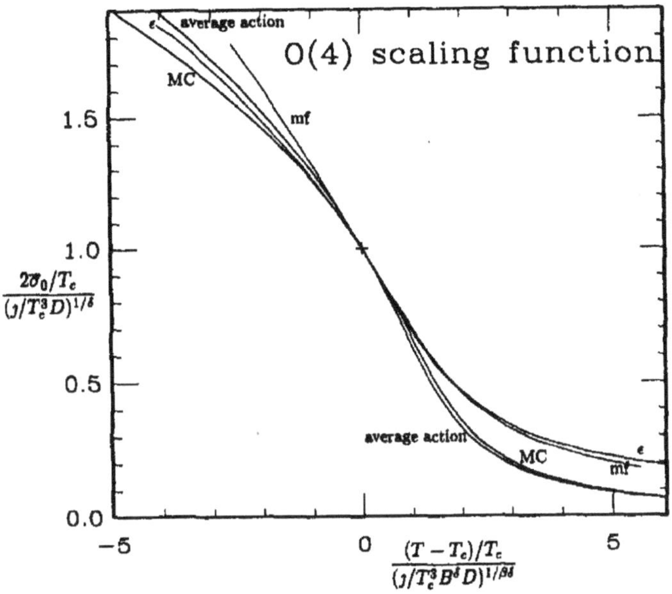

Figure 11. The figure shows a comparison of our results, denoted by "average action", with results of other methods for the scaling function of the three–dimensional $O(4)$ Heisenberg model. We have labeled the axes for convenience in terms of the expectation value $\bar{\sigma}_0$ and the source $\jmath$ of the corresponding four–dimensional theory. The constants B and D specify the non–universal amplitudes of the model (cf. Eq.(79). The curve labeled by "MC" represents a fit to lattice Monte Carlo data. The second order epsilon expansion[68] and mean field results are denoted by "ϵ" and "mf", respectively. Apart from our results the curves are taken from[67].

methods for the three–dimensional $O(4)$ Heisenberg model. In fig. 11 we display our results along with those obtained from lattice Monte Carlo simulation[67], second order epsilon expansion[68] and mean field theory. We observe a good agreement of average action, lattice and epsilon expansion results within a few per cent for $T < T_c$. Above T_c the average action and the lattice curve go quite close to each other with a substantial deviation from the epsilon expansion and mean field scaling function. (We note that the question of a better agreement of the curves for $T < T_c$ or $T > T_c$ depends on the chosen non–universal normalization conditions for x and f (cf. Eq.(79)).)

Before we use the scaling function $f(x)$ to discuss the general temperature and quark mass dependent case, we consider the limits $T = T_c$ and $\hat{m} = 0$, respectively. In these limits the behavior of the various quantities is determined solely by critical amplitudes and exponents. In the spontaneously broken phase $(T < T_c)$ and in the chiral limit we observe that the renormalized and unrenormalized order parameters scale according to

$$\begin{aligned} \frac{2\sigma_0(T)}{T_c} &= (2E)^{1/2}\left(\frac{T_c - T}{T_c}\right)^{\nu/2}, \\ \frac{2\bar{\sigma}_0(T)}{T_c} &= B\left(\frac{T_c - T}{T_c}\right)^{\beta}, \end{aligned} \tag{81}$$

respectively, with $E = 0.814$ and the value of B given above. In the symmetric phase the renormalized mass $m = m_\pi = m_\sigma$ and the unrenormalized mass $\bar{m} = Z_\Phi^{1/2} m$ behave as

$$\frac{m(T)}{T_c} = (\xi^+)^{-1}\left(\frac{T-T_c}{T_c}\right)^{\nu},$$
$$\frac{\overline{m}(T)}{T_c} = (C^+)^{-1/2}\left(\frac{T-T_c}{T_c}\right)^{\gamma/2}, \tag{82}$$

where $\xi^+ = 0.270$, $C^+ = 2.79$. For $T = T_c$ and non–vanishing current quark mass we have

$$\frac{2\overline{\sigma}_0}{T_c} = D^{-1/\delta}\left(\frac{\jmath}{T_c^3}\right)^{1/\delta} \tag{83}$$

with the value of D given above.

Though the five amplitudes E, B, ξ^+, C^+ and D are not universal there are ratios of amplitudes which are invariant under a rescaling of $\overline{\sigma}_0$ and $T - T_c$. Our results for the universal amplitude ratios are

$$R_\chi = C^+DB^{\delta-1} = 1.02\,,$$
$$\tilde{R}_\xi = (\xi^+)^{\beta/\nu}D^{1/(\delta+1)}B = 0.852\,, \tag{84}$$
$$\xi^+E = 0.220\,.$$

Those for the critical exponents are given in tab. 3. Here the value of η is obtained from the temperature dependent version of Eq.(60) at the critical temperature[52]. For comparison tab. 3 also gives the results[70, 71] of a $3d$ perturbative computation as well as lattice Monte Carlo results[69] which have been used for the lattice form of the scaling functionin fig. 11.[‖] There are only two independent amplitudes and critical exponents, respectively. They are related by the usual scaling relations of the three–dimensional scalar $O(N)$–model[71] which we have explicitly verified by the independent calculation of our exponents.

We turn to the discussion of the scaling behavior of the chiral condensate $\langle\overline{\psi}\psi\rangle$ for the general case of a temperature and quark mass dependence. In fig. 5 we have displayed our results for the scaling equation of state in terms of the chiral condensate

$$\langle\overline{\psi}\psi\rangle = -\overline{m}_{k_\Phi}^2 T_c\left(\frac{\jmath/T_c^3}{f(x)}\right)^{1/\delta} + \jmath \tag{85}$$

as a function of $T/T_c = 1 + x(\jmath/T_c^3 f(x))^{1/\beta\delta}$ for different quark masses or, equivalently, different values of $\jmath$. The curves shown in fig. 5 correspond to quark masses $\hat{m} = 0$, $\hat{m} = \hat{m}_{\rm phys}/10$, $\hat{m} = \hat{m}_{\rm phys}$ and $\hat{m} = 3.5\hat{m}_{\rm phys}$ or, equivalently, to zero temperature pion masses $m_\pi = 0$, $m_\pi = 45\,{\rm MeV}$, $m_\pi = 135\,{\rm MeV}$ and $m_\pi = 230\,{\rm MeV}$, respectively (cf. fig. 4). One observes that the second order phase transition with a vanishing order parameter at T_c for $\hat{m} = 0$ is turned into a smooth crossover in the presence of non-zero quark masses.

The scaling form Eq.(85) for the chiral condensate is exact only in the limit $T \to T_c$, $\jmath \to 0$. It is interesting to find the range of temperatures and quark masses for which $\langle\overline{\psi}\psi\rangle$ approximately shows the scaling behavior Eq.(85). This can be infered from a comparison (see fig. 5) with our full non–universal solution for the T and $\jmath$ dependence of $\langle\overline{\psi}\psi\rangle$. For $m_\pi = 0$ one observes approximate scaling behavior for temperatures $T \gtrsim 90\,{\rm MeV}$. This situation persists up to a pion mass of $m_\pi = 45\,{\rm MeV}$. Even for the realistic case, $m_\pi = 135\,{\rm MeV}$, and to a somewhat lesser extent for $m_\pi = 230\,{\rm MeV}$ the

‖See also ref[72] and references therein for a recent calculation of critical exponents using similar methods as in this work. For high precision estimates of the critical exponents see also refs[73, 74].

scaling curve reasonably reflects the physical behavior for $T \gtrsim T_c$. For temperatures below T_c, however, the zero temperature mass scales become important and the scaling arguments leading to universality break down.

The above comparison may help to shed some light on the use of universality arguments away from the critical temperature and the chiral limit. One observes that for temperatures above T_c the scaling assumption leads to quantitatively reasonable results even for a pion mass almost twice as large as the physical value. This in turn has been used for two flavor lattice QCD as theoretical input to guide extrapolation of results to light current quark masses. From simulations based on a range of pion masses $0.3 \lesssim m_\pi/m_\rho \lesssim 0.7$ and temperatures $0 < T \lesssim 250\,\mathrm{MeV}$ a "pseudocritical temperature" of approximately 140 MeV with a weak quark mass dependence is reported[75]. Here the "pseudocritical temperature" T_{pc} is defined as the inflection point of $\langle\overline{\psi}\psi\rangle$ as a function of temperature. The values of the lattice action parameters used in[75] with $N_t = 6$ were $a\hat{m} = 0.0125$, $6/g^2 = 5.415$ and $a\hat{m} = 0.025$, $6/g^2 = 5.445$. For comparison with lattice data we have displayed in fig. 5 the temperature dependence of the chiral condensate for a pion mass $m_\pi = 230\,\mathrm{MeV}$. From the free energy of the linear quark meson model we obtain in this case a pseudocritical temperature of about 150 MeV in reasonable agreement with the results of ref[75]. In contrast, for the critical temperature in the chiral limit we obtain $T_c = 100.7\,\mathrm{MeV}$. This value is considerably smaller than the lattice results of about (140 –150) MeV obtained by extrapolating to zero quark mass in ref[75]. We point out that for pion masses as large as 230 MeV the condensate $\langle\overline{\psi}\psi\rangle(T)$ is almost linear around the inflection point for quite a large range of temperature. This makes a precise determination of T_c somewhat difficult. Furthermore, fig. 5 shows that the scaling form of $\langle\overline{\psi}\psi\rangle(T)$ underestimates the slope of the physical curve. Used as a fit with T_c as a parameter this can lead to an overestimate of the pseudocritical temperature in the chiral limit. We also mention here the results of ref[76]. There two values of the pseudocritical temperature, $T_{pc} = 150(9)\,\mathrm{MeV}$ and $T_{pc} = 140(8)$, corresponding to $a\hat{m} = 0.0125$, $6/g^2 = 5.54(2)$ and $a\hat{m} = 0.00625$, $6/g^2 = 5.49(2)$, respectively, (both for $N_t = 8$) were computed. These values show a somewhat stronger quark mass dependence of T_{pc} and were used for a linear extrapolation to the chiral limit yielding $T_c = 128(9)\,\mathrm{MeV}$.

The linear quark meson model exhibits a second order phase transition for two quark flavors in the chiral limit. As a consequence the model predicts a scaling behavior near the critical temperature and the chiral limit which can, in principle, be tested in lattice simulations. For the quark masses used in the present lattice studies the order and universality class of the transition in two flavor QCD remain a partially open question. Though there are results from the lattice giving support for critical scaling[77,78] there are also recent simulations with two flavors that reveal significant finite size effects and problems with $O(4)$ scaling[79,80].

ADDITIONAL DEGREES OF FREEDOM

So far we have investigated the chiral phase transition of QCD as described by the linear $O(4)$–model containing the three pions and the sigma resonance as well as the up and down quarks as degrees of freedom. Of course, it is clear that the spectrum of QCD is much richer than the states incorporated in our model. It is therefore important to ask to what extent the neglected degrees of freedom like the strange quark, strange (pseudo)scalar mesons, (axial)vector mesons, baryons, etc., might be important for the chiral dynamics of QCD. Before doing so it is perhaps instructive to first look into the opposite direction and investigate the difference between the linear quark meson model described here and chiral perturbation theory based on the non–linear sigma model[36]. In some sense, chiral perturbation theory is the minimal model of chiral symmetry breaking containing only the Goldstone degrees of freedom. By construction it is therefore only valid in the spontaneously broken phase and can not be expected to yield realistic results for temperatures close to T_c or for the symmetric phase. However, for small temperatures (and momentum scales) the non–linear model is expected to describe the low–energy and low–temperature limit of QCD reliably as an expansion in powers of the light quark masses. For vanishing temperature it has been demonstrated recently[42, 43, 44] that the results of chiral perturbation theory can be reproduced within the linear meson model once certain higher dimensional operators in its effective action are taken into account for the three flavor case. Moreover, some of the parameters of chiral perturbation theory $(L_4, \ldots, L_8)$ can be expressed and therefore also numerically computed in terms of those of the linear model. For non–vanishing temperature one expects agreement only for low T whereas deviations from chiral perturbation theory should become large close to T_c. Yet, even for $T \ll T_c$ small quantitative deviations should exist because of the contributions of (constituent) quark and sigma meson fluctuations in the linear model which are not taken into account in chiral perturbation theory.

From[81] we infer the three–loop result for the temperature dependence of the chiral condensate in the chiral limit for N light flavors

$$\begin{aligned} \left\langle \overline{\psi}\psi \right\rangle (T)_{\chi PT} &= \left\langle \overline{\psi}\psi \right\rangle_{\chi PT} (0) \left\{ 1 - \frac{N^2-1}{N} \frac{T^2}{12F_0^2} - \frac{N^2-1}{2N^2} \left(\frac{T^2}{12F_0^2} \right)^2 \right. \\ &\left. + \; N(N^2-1) \left(\frac{T^2}{12F_0^2} \right)^3 \ln \frac{T}{\Gamma_1} \right\} + \mathcal{O}(T^8) \, . \end{aligned} \tag{86}$$

The scale Γ_1 can be determined from the D–wave isospin zero $\pi\pi$ scattering length and is given by $\Gamma_1 = (470 \pm 100)\,\text{MeV}$. The constant F_0 is (in the chiral limit) identical to the pion decay constant $F_0 = f_\pi^{(0)} = 80.8\,\text{MeV}$. In fig. 12 we have plotted the chiral

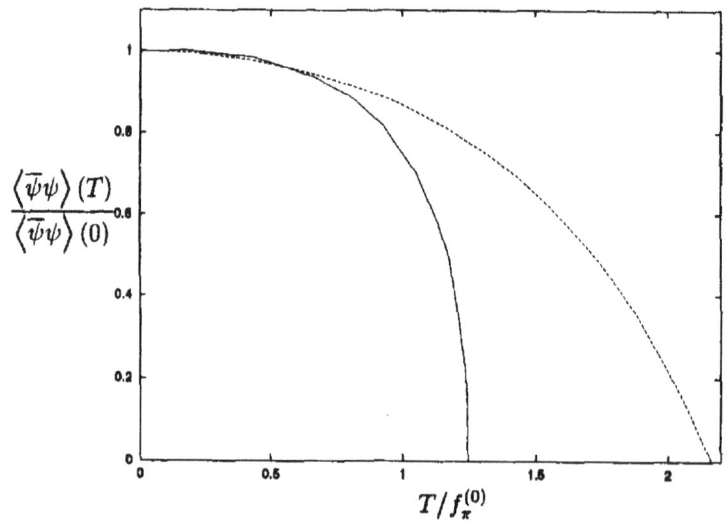

Figure 12. The plot displays the chiral condensate $\left\langle \overline{\psi}\psi \right\rangle$ as a function of $T/f_\pi^{(0)}$. The solid line corresponds to our results for vanishing average current quark mass $\hat{m} = 0$ whereas the dashed line shows the corresponding three–loop chiral perturbation theory result for $\Gamma_1 = 470\,\text{MeV}$.

condensate as a function of T/F_0 for both, chiral perturbation theory according to Eq.(86) and for the linear quark meson model. As expected the agreement for small T is very good. Nevertheless, the anticipated small numerical deviations present even for $T \ll T_c$ due to quark and sigma meson loop contributions are manifest. For larger values of T, say for $T \gtrsim 0.8 f_\pi^{(0)}$ the deviations become significant because of the intrinsic inability of chiral perturbation theory to correctly reproduce the critical behavior of the system near its second order phase transition.

Within the language of chiral perturbation theory the neglected effects of thermal quark fluctuations may be described by an effective temperature dependence of the parameter $F_0(T)$. We notice that the temperature at which these corrections become important equals approximately one third of the constituent quark mass $M_q(T)$ or the sigma mass $m_\sigma(T)$, respectively, in perfect agreement with fig. 7. As suggested by this figure the onset of the effects from thermal fluctuations of heavy particles with a T–dependent mass $m_H(T)$ is rather sudden for $T \gtrsim m_H(T)/3$. These considerations also apply to our two flavor quark meson model. Within full QCD we expect temperature dependent initial values at k_Φ.

The dominant contribution to the temperature dependence of the initial values presumably arises from the influence of the mesons containing strange quarks as well as the strange quark itself. Here the quantity $\overline{m}^2_{k_\Phi}$ seems to be the most important one. (The temperature dependence of higher couplings like $\lambda(T)$ is not very relevant if the IR attractive behavior remains valid, i.e. if Z_{Φ,k_Φ} remains small for the range of temperatures considered. We neglect a possible T–dependence of the current quark mass $\hat{m}$.) In particular, for three flavors the potential U_{k_Φ} contains a term

$$-\frac{1}{2}\overline{\nu}_{k_\Phi}\left(\det\Phi + \det\Phi^\dagger\right) = -\overline{\nu}_{k_\Phi}\varphi_s\Phi_{uu}\Phi_{dd} + \ldots \tag{87}$$

which reflects the axial $U_A(1)$ anomaly. It yields a contribution to the effective mass term proportional to the expectation value $\langle\Phi_{ss}\rangle \equiv \varphi_s$, i.e.

$$\Delta\overline{m}^2_{k_\Phi} = -\frac{1}{2}\overline{\nu}_{k_\Phi}\varphi_s\,. \tag{88}$$

Both, $\overline{\nu}_{k_\Phi}$ and φ_s, depend on T. We expect these corrections to become relevant only for temperatures exceeding $m_K(T)/3$ or $M_s(T)/3$. We note that the temperature dependent kaon and strange quark masses, $m_K(T)$ and $M_s(T)$, respectively, may be somewhat different from their zero temperature values but we do not expect them to be much smaller. A typical value for these scales is around 500 MeV. Correspondingly, the thermal fluctuations neglected in our model should become important for $T \gtrsim 170\,\mathrm{MeV}$. It is even conceivable that a discontinuity appears in $\varphi_s(T)$ for sufficiently high T (say $T \simeq 170\,\mathrm{MeV}$). This would be reflected by a discontinuity in the initial values of the $O(4)$–model leading to a first order transition within this model. Obviously, these questions should be addressed in the framework of the three flavor $SU_L(3) \times SU_R(3)$ quark meson model. Work in this direction is in progress.

We note that the temperature dependence of $\overline{\nu}(T)\varphi_s(T)$ is closely related to the question of an effective high temperature restoration of the axial $U_A(1)$ symmetry[58, 63]. The η' mass term is directly proportional to this combination[42], $m^2_{\eta'}(T) - m^2_\pi(T) \simeq \frac{3}{2}\overline{\nu}(T)\varphi_s(T)$. Approximate $U_A(1)$ restoration would occur if $\varphi_s(T)$ or $\overline{\nu}(T)$ would decrease sizeably for large T. For realistic QCD this question should be addressed by a three flavor study. Within two flavor QCD the combination $\overline{\nu}_k\varphi_s$ is replaced by an effective anomalous mass term $\overline{\nu}^{(2)}_k$. The temperature dependence of $\overline{\nu}^{(2)}(T)$ could be studied by introducing quarks and the axial anomaly in the two flavor matrix model of ref[54]. We add that this question has also been studied within full two flavor QCD in lattice simulations[79, 82, 83]. So far there does not seem to be much evidence for a restoration of the $U_A(1)$ symmetry near T_c but no final conclusion can be drawn yet.

To summarize, we have found that the effective two flavor quark meson model presumably gives a good description of the temperature effects in two flavor QCD for a temperature range $T \lesssim 170\,\mathrm{MeV}$. Its reliability should be best for low temperature where our results agree with chiral perturbation theory. However, the range of validity is considerably extended as compared to chiral perturbation theory and includes, in particular, the critical temperature of the second order phase transition in the chiral limit. We have explicitly connected the universal critical behavior for small $|T - T_c|$ and small current quark masses with the renormalized couplings at $T = 0$ and realistic quark masses. The main quantitative uncertainties from neglected fluctuations presumably concern the values of $f_\pi^{(0)}$ and T_c which, in turn, influence the non–universal amplitudes B and D in the critical region. We believe that our overall picture is rather solid. Where applicable our results compare well with numerical simulations of full two flavor QCD.

CONCLUSIONS

Our conclusions may be summarized in the following points:

1. The connection between short distance perturbative QCD and long distance meson physics by analytical methods seems to emerge step by step. The essential ingredients are nonperturbative flow equations as approximations of exact renormalization group equations and a formalism which allows a change of variables by introducing meson–like composite fields.

2. The relevance of meson–like $\overline{\psi}\psi$ composite objects is established in this framework. A typical scale where the mesonic bound states appear is the compositeness scale $k_\Phi \simeq (600-700)\,\mathrm{MeV}$. The occurrence of chiral symmetry breaking depends on the effective meson mass $\overline{m}_{k_\Phi}$ at the compositeness scale. For a certain range of values of the ratio $\overline{m}_{k_\Phi}^2/k_\Phi^2$ spontaneous chiral symmetry breaking is induced by quark fluctuations. A definite analytical establishment of spontaneous chiral symmetry breaking from "first principles" (i.e., short distance QCD) still awaits a reliable calculation of this ratio.

3. Phenomenologically, the ratio $\overline{m}_{k_\Phi}^2/k_\Phi^2$ may be determined from the value of the constituent quark mass in units of the pion decay constant f_π. For this value the ratios f_π/k_Φ and $\left\langle\overline{\psi}\psi\right\rangle/k_\Phi^3$ can be computed. Together with an earlier estimate of k_Φ this yields rather encouraging results: $f_\pi \simeq 100\,\mathrm{MeV}$ and $\left\langle\overline{\psi}\psi\right\rangle \simeq -(190\,\mathrm{MeV})^3$ for two flavor QCD. It will be very interesting to generalize these results to the realistic three flavor case.

4. The formalism presented here naturally leads to an effective linear quark meson model for the description of mesons below the compositeness scale k_Φ. For this model the standard non–linear sigma model of chiral perturbation theory emerges as a low energy approximation due to the large sigma mass.

5. The old puzzle about the precise connection between the current and the constituent quark mass reveals new interesting aspects in this formalism. In the context of the effective average action one may define these masses by the quark propagator at zero or very small momentum, either in the quark gluon picture (current quark mass) or the effective quark meson model (constituent quark mass). There

is no conceptual difference between the two situations. As a function of k the running quark mass smoothly interpolates between the standard current quark mass m_q for high k and the standard constituent quark mass M_q for low k. (This holds at least as long as the minimum of the effective scalar potential is unique.) A rapid quantitative change occurs for $k \simeq (400 - 500)$ MeV because of the onset of chiral symmetry breaking. For $k = 0$ one expects this behavior to carry over to the momentum dependence of the quark propagator: For small q^2 the inverse quark propagator is dominated by M_q. In contrast, for high q^2 the constant term in the inverse propagator is reduced to m_q since q^2 replaces k^2 as an effective infrared cutoff. One expects a smooth interpolation between the two limits and it would be interesting to know the form of the propagator for momenta in the transition region. Furthermore, the symmetry breaking source term which determines the pion mass in the linear or non–linear meson model can be related to the current quark mass at the compositeness scale. This constitutes a bridge between the low energy meson properties and the running quark mass at a scale which is not too far from the validity of perturbation theory.

6. A particular version of the Nambu–Jona-Lasinio model appears in our formalism as a limiting case (infinitely strong renormalized Yukawa coupling at the compositeness scale k_Φ). Here the ultraviolet cutoff which is implicit in the NJL model is dictated by the momentum dependence of the infrared cutoff function R_k in the effective average action. The characteristic cutoff scale is k_Φ, In this context our results can be interpreted as an approximative solution of the NJL model. Our method includes many contributions beyond the leading order contribution of the $1/N_c$ expansion.

7. Based on a satisfactory understanding of the meson properties in the vacuum we have described their behavior in a thermal equilibrium situation. Our method should remain valid for temperature below ~ 170 MeV. This extends well beyond the validity of chiral perturbation theory. In particular, in the chiral limit of vanishing quark masses we can describe the universal critical behavior near a second order phase transition of two flavor QCD. The universal behavior is quantitatively connected to observed quantities at zero temperature and realistic quark masses.

Acknowledgements

We thank J. Berges and B. Bergerhoff for collaboration on many subjects covered by these lectures. We also wish to express our gratitude to the organizers of the *NATO Advanced Study Institute: Confinement, Duality and Non–perturbative Aspects of QCD* for their excellent organization and for providing a most stimulating environment.

REFERENCES

1. L.P. Kadanoff, *Physica* **2** (1966) 263.
2. K. G. Wilson, Phys. Rev. **B4** (1971) 3174; K. G. Wilson and I. G. Kogut, Phys. Rep. **12** (1974) 75.
3. F. Wegner and A. Houghton, Phys. Rev. **A8** (1973) 401; F.J. Wegner in *Phase Transitions and Critical Phenomena,* vol. 6, eds. C. Domb and M.S. Greene, Academic Press (1976); J.F. Nicoll and T.S. Chang, Phys. Lett. **62A** (1977) 287.
4. S. Weinberg in *Critical Phenomena for Field Theorists,* Erice Subnucl. Phys. (1976) 1.
5. J. Polchinski, Nucl. Phys. **B231** (1984) 269.
6. A. Hasenfratz and P. Hasenfratz, Nucl. Phys. **B270** (1986) 685; P. Hasenfratz and J. Nager, Z. Phys. **C37** (1988) 477.

7. C. Wetterich, Nucl. Phys. **B352** (1991) 529; Z. Phys. **C57** (1993) 451; C60 (1993) 461.
8. J. Schwinger, Proc. Nat. Acad. Sc. **37** (1951) 452; J. Goldstone, A. Salam and S. Weinberg, Phys. Rev. **127** (1962) 965; G. Jona–Lasinio, Nuovo Cim. **34** (1964) 1790; K. Symanzik, Comm. Math. Phys. **16** (1970) 48; J. Iliopoulos, C. Itzykson and A. Martin, Rev. Mod. Phys. **47** (1975) 165.
9. A. Ringwald and C. Wetterich, Nucl. Phys. **B334** (1990) 506; N. Tetradis and C. Wetterich, Nucl. Phys. **B383** (1992) 197.
10. C. Wetterich, Z. Phys. **C48** (1990) 693; S. Bornholdt and C. Wetterich, Z. Phys. **C58** (1993) 585.
11. J. Comelas, Y. Kubyshin and E.Moreno, Nucl. Phys. **B490** (1997) 653 (hep-th/9512086).
12. M. Reuter and C. Wetterich, Nucl. Phys. **B391** (1993) 147; B408 (1993) 91; B417 (1994) 181; B427 (1994) 291; preprint HD-THEP-94-39 (hep-th/9708051).
13. C. Becchi, Lecture given at the Parma Theoretical Physics Seminar, Sep 1991, preprint GEF-TH-96-11 (hep-th/9607188).
14. M. Bonini, M. D'Attanasio, and G. Marchesini, Nucl. Phys. **B418** (1994) 81; B421 (1994) 429.
15. U. Ellwanger, Phys. Lett. **B335** (1994) 364; U. Ellwanger, M. Hirsch and A. Weber, Z. Phys. **C69** (1996) 687.
16. C. Wetterich, preprint HD–THEP–95–2 (hep-th/9501119); Z. Phys. **C72** (1996) 139 (hep-ph/9604227).
17. U. Ellwanger, M. Hirsch and A. Weber, preprint LPTHE Orsay 96–50 (hep-ph/9606468).
18. B. Bergerhoff and C. Wetterich, preprint HD–THEP–97–46 (hep-ph/9708425).
19. G. Keller, C. Kopper and M. Salmhofer, Helv. Phys. Acta **65** (1992) 32; G. Keller and G. Kopper, Phys. Lett. **273B** (1991) 323.
20. C. Wetterich, Phys. Lett. **301B** (1993) 90.
21. C. Wetterich, Mod. Phys. Lett. **A11** (1996) 2573 (hep-th/9602039).
22. M. Bonini, M. D'Attanasio, and G. Marchesini, Nucl. Phys. **B409** (1993) 441; U. Ellwanger, Z. Phys. **C58** (1993) 619; C. Wetterich, Int. J. Mod. Phys. **A9** (1994) 3571.
23. F.J. Dyson, Phys. Rev. **75** (1949) 1736; J. Schwinger, Proc. Nat. Acad. Sc. **37** (1951) 452.
24. J. Adams, J. Berges, S. Bornholdt, F. Freire, N. Tetradis and C. Wetterich, Mod. Phys. Lett. **A10** (1995) 2367.
25. J. Berges, N. Tetradis and C. Wetterich, Phys. Rev. Lett. **77** (1996) 873.
26. N. Tetradis, Nucl. Phys. **B488** (1997) 92 (hep-ph/9608272)
27. J. Berges and C. Wetterich, Nucl. Phys. **B487** (1997) 675 (hep-th/9609019).
28. N. Tetradis and C. Wetterich, Nucl. Phys. **B422** [FS] (1994) 541.
29. T. Morris, Phys. Lett. **B329** (1994) 241.
30. M.M. Tsypin, Phys. Rev. Lett. **73** (1994) 2015; preprint FIAN/TD-23/95 (hep-lat/9601021).
31. B. Bergerhoff, F. Freire, D. Litim, S. Lola, C. Wetterich, Phys. Rev. **B53** (1996) 5734; B. Bergerhoff, D. Litim, S. Lola and C.Wetterich, Int. J. Mod. Phys. **A11** (1996) 4273.
32. N. Tetradis and C. Wetterich, Nucl. Phys. **B398** (1993) 659.
33. U. Ellwanger and C. Wetterich, Nucl. Phys. **B423** (1994) 137.
34. D. J. Gross and F. A. Wilczek, Phys. Rev. Lett. **30** (1973) 1343; H. D. Politzer, Phys. Rev. Lett. **30** (1973) 1346.
35. See for instance M. Neubert, Phys. Rep. **245** (1994) 259.
36. J. Gasser and H. Leutwyler, Phys. Rep. **C87** (1982) 77; Nucl. Phys. **B250** (1985) 465; for a review see for instance H. Leutwyler, lectures given at the *Hadroni 96* workshop, Gramado, Brazil, 10 – 14 April 1994 (hep-ph/9406283).
37. G. 't Hooft, Phys. Rep. 142 (1986) 357.
38. J. L. Richardson, Phys. Lett. **82B** (1979), 272.
39. Y. Nambu and G. Jona–Lasinio, Phys. Rev. **122** (1961) 345.
40. J. Bijnens, Phys. Rep. **265** (1996) 369; T. Hatsuda and T. Kunihiro, Phys. Rep. **247** (1994) 221.
41. J. Pawlowski, preprint FSUJ-TPI-05-96 (hep-th/9605037).
42. D.–U. Jungnickel and C. Wetterich, preprint HD–THEP–96–19 to appear in Z. Phys. **C76** (1997) (hep-ph/9606483).
43. D.–U. Jungnickel and C. Wetterich, Phys. Lett. **389B** (1996) 600 (hep-ph/9607411).
44. D.–U. Jungnickel and C. Wetterich, preprint HD–THEP–97–9 to appear in Z. Phys. **C77** (1997) (hep-ph/9704345).
45. K.–I. Aoki, K. Morikawa, J.–I. Sumi, H. Terao and M. Tomoyose, Progr. Theor. Phys. **97** (1997) 479 (hep-ph/9612459).
46. W.A. Bardeen, C.T. Hill and D.–U. Jungnickel, Phys. Rev. **D49** (1994) 1437.

47. D.–U. Jungnickel and C. Wetterich, Phys. Rev. **D53** (1996) 5142.
48. C.G. Callan, Phys. Rev. **D2** (1970) 1541; K. Symanzik, Comm. Math. Phys. **18** (1970) 227.
49. S. Coleman and E. Weinberg, Phys. Rev. **D7** (1973) 1888.
50. M. Gell–Mann and M. Levy, Nuovo Cim. **16** (1960) 705.
51. M. Jamin and M. Münz, Z. Phys. **66** (1995) 633; K.G. Chetyrkin, C.A. Dominguez, D. Pirjol and K. Schilcher Phys. Rev. **D51** (1995) 5090; J. Bijnens, J. Prades and E. de Rafael, Phys. Lett. **B348** (1995) 226.
52. J. Berges, D.–U. Jungnickel and C. Wetterich, preprint HD–THEP–97–20 (hep-ph/9705474).
53. J. Adams, J. Berges, S. Bornholdt, F. Freire, N. Tetradis and C. Wetterich, Mod. Phys. Lett. **A10** (1995) 2367 (hep-th/9507093).
54. J. Berges and C. Wetterich, Nucl. Phys. **B487** (1997) 675 (hep-th/9609019).
55. H. Leutwyler, preprint HEPPH–9609467 (hep-ph/9609467).
56. For a recent review see H. Meyer–Ortmanns, Rev. Mod. Phys. **68** (1996) 473 (hep-lat/9608098).
57. Proceedings of Quark Matter '96, Nucl. Phys. **A610** (1996).
58. R.D. Pisarski and F. Wilczek, Phys. Rev. **D29** (1984) 338.
59. K. Rajagopal and F. Wilczek, Nucl. Phys. B399 (1993) 395. Nucl. Phys. **B399** (1993) 577.
60. K. Rajagopal, in *Quark-Gluon Plasma 2*, edited by R. Hwa, World Scientific (1995) (hep-ph/9504310).
61. A. A. Anselm, Phys. Lett. **217B** (1988) 169; A. A. Anselm and M. G. Ryskin, Phys. Lett. **266B** (1991) 482; J.–P. Blaizot and A. Krzywicki, Phys. Rev. **D46** (1992) 246; J. D. Bjorken, Int. J. Mod. Phys. **A7** (1992) 4189; Acta Phys. Pol. B23 (1992) 561; K. L. Kowalski and C. C. Taylor, preprint CWRUTH-92-6 (1992) (hep-ph/9211282); J. D. Bjorken, K. L. Kowalski and C. C. Taylor, in Proceedings of Les Rencontres de Physique de la Vallée D'Aoste, La Thuile, SLAC-PUB-6109 (1993); and in Proceedings of the workshop on Physics at Current Accelerators and the Supercollider, Argonne (1993) hep-ph/9309235.
62. C.M.G. Lattes, V. Fujimoto and S. Hasegawa, Phys. Rep. **65** (1980) 151.
63. E. Shuryak, Comm. Nucl. Part. Phys. **21** (1994) 235.
64. See for instance J. Kapusta, *Finite Temperature Field Theory,* Cambridge University Press (1989).
65. P. Ginsparg, Nucl. Phys. **B170** (1980) 388; T. Appelquist and R. Pisarski, Phys. Rev. **D23** (1981) 2305; S. Nadkarni, Phys. Rev. **D27** (1983) 917; N.P. Landsman, Nucl. Phys. **B322** (1989) 498.
66. B. Widom, J. Chem. Phys. **43** (1965) 3898.
67. D. Toussaint, Phys. Rev. **D55** (1997) 362 (hep-lat/9607084).
68. E. Brezin, D.J. Wallace and K.G. Wilson, Phys. Rev. **B7** (1973) 232.
69. K. Kanaya and S. Kaya, Phys. Rev. **D51** (1995) 2404.
70. G. Baker, D. Meiron and B. Nickel, Phys. Rev. **B17** (1978) 1365.
71. J. Zinn–Justin, *Quantum Field Theory and Critical Phenomena* (Oxford University Press, 1993).
72. T.R. Morris and M.D. Turner, preprint SHEP–97–06 (hep-th/9704202).
73. P. Butera and M. Comi, Phys. Rev. **B52** (1995) 6185.
74. T. Reisz, Phys. Lett. **360B** (1995) 77 (hep-lat/9507011).
75. C. Bernard et al., Phys. Rev. **D55** (1997) 6861 (hep-lat/9612025); C. Bernard et al., Nucl. Phys. Proc. Suppl. **53** (1997) 442 (hep-lat/9608026).
76. S. Gottlieb et al., Phys. Rev. **D55** (1997) 6852 (hep-lat/9612020).
77. F. Karsch, Phys. Rev. **D49** (1994) 3791; F. Karsch and E. Laermann, Phys. Rev. D50 (1994) 6954.
78. Y. Iwasaki, K. Kanaya, S. Kaya and T. Yoshie, Phys. Rev. Lett. **78** (1997) 179 (hep-lat/9609022).
79. G. Boyd, F. Karsch, E. Laermann and M. Oevers, Talk given at 10th International Conference on Problems of Quantum Field Theory, Alushta, Ukraine, 13-17 May 1996 (hep-lat/9607046).
80. A. Ukawa, Nucl. Phys. Proc. Suppl. **53** (1997) 106 (hep-ph9612011); S. Aoki, T. Kaneda, A. Ukawa and T. Umemura, Nucl. Phys. Proc. Suppl. 63 (1997) 438 (hep-lat/9612010).
81. J. Gasser and H. Leutwyler, Phys. Lett. **184B** (1987) 83; H. Leutwyler, Nucl. Phys. Proc. Suppl. **B4** (1988) 248.
82. C. Bernard et al., Phys. Rev. Lett. **78** (1997) 598 (hep-lat/9611031).
83. J.B. Kogut, J.F. Lagae and D.K. Sinclair, Nucl. Phys. Proc. Suppl. **53** (1997) 269 (hep-lat/9608128).

LIGHT-FRONT QCD: A CONSTITUENT PICTURE OF HADRONS

Robert J. Perry

Department of Physics, The Ohio State University
Columbus, Ohio, 43210, USA

MOTIVATION AND STRATEGY

We seek to derive the structure of hadrons from the fundamental theory of the strong interaction, QCD. Our work is founded on the hypothesis that a constituent approximation can be *derived* from QCD, so that a relatively small number of quark *and gluon* degrees of freedom need be explicitly included in the state vectors for low-lying hadrons.[1] To obtain a constituent picture, we use a Hamiltonian approach in light-front coordinates.[2]

I do not believe that light-front Hamiltonian field theory is extremely useful for the study of low energy QCD unless a constituent approximation can be made, and I do not believe such an approximation is possible unless cutoffs that it violate manifest gauge invariance and covariance are employed. Such cutoffs *inevitably* lead to relevant and marginal effective interactions *(i.e.,* counterterms) that contain functions of longitudinal momenta.[3,4] It is it not possible to renormalize light-front Hamiltonians in any useful manner without developing a renormalization procedure that can produce these non-canonical counterterms.

The line of investigation I discuss has been developed by a small group of theorists who are working or have worked at Ohio State University and Warsaw University.[1,3–19] Ken Wilson provided the initial impetus for this work, and at a very early stage outlined much of the basic strategy we employ.[15]

I make no attempt to provide enough details to allow the reader to start doing light-front calculations. The introductory article by Harindranath[16] is helpful in this regard. An earlier version of these lectures[8] also provides many more details.

A Constituent Approximation Depends on Tailored Renormalization

If it is possible to derive a constituent approximation from QCD, we can formulate the hadronic bound state problem as a set of coupled few-body problems. We obtain the states and eigenvalues by solving

$$H_\Lambda \mid \Psi_\Lambda\rangle = E \mid \Psi_\Lambda\rangle, \tag{1}$$

where,

$$\mid \Psi_\Lambda\rangle = \phi^\Lambda_{q\bar{q}} \mid q\bar{q}\rangle + \phi^\Lambda_{q\bar{q}g} \mid q\bar{q}g\rangle + \cdots. \tag{2}$$

Confinement, Duality, and Nonperturbative Aspects of QCD
Edited by Pierre van Baal, Plenum Press, New York, 1998

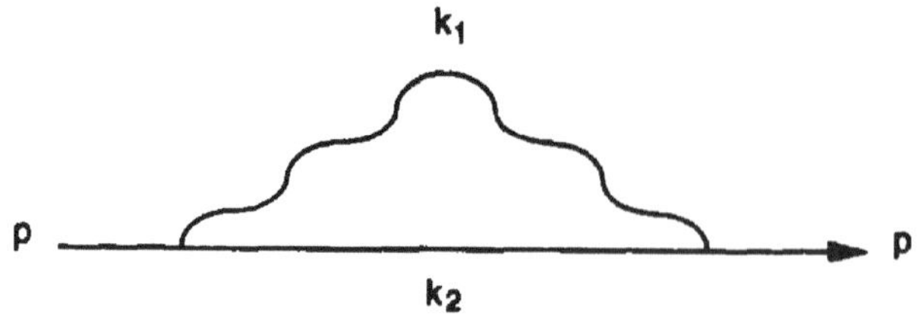

Figure 1. **The second-order shift in the self-energy of a bare electron due to mixing with electron-photon states.**

where I use shorthand notation for the Fock space components of the state. The full state vector includes an infinite number of components, and in a constituent approximation we truncate this series. We derive the Hamiltonian from QCD, so we must allow for the possibility of constituent gluons. I have indicated that the Hamiltonian and the state both depend on acutoff, Λ, which is critical for the approximation.

This approach has no chance of working without a renormalization scheme *tailored to it.* Much of our work has focused on the development of such a renormalization scheme.[3,4,17–19] In order to understand the constraints that have driven this development, seriously consider under what conditions it might be possible to truncate the above series without making an arbitrarily large error in the eigenvalue. I focus on the eigenvalue, because it will certainly not be possible to approximate all observable properties of hadrons (*e.g.,* wee parton structure functions) this way.

For this approximation to be valid, *all* many-body states must approximately decouple from the dominant few-body components. We know that even in perturbation theory, high energy many-body states do not simply decouple from few-body states. In fact, the errors from simply discarding high energy states are infinite. In second-order perturbation theory, for example, high energy photons contribute an arbitrarily large shift to the mass of an electron. This second-order effect is illustrated in Figure 1, and the precise interpretation for this light-front time-ordered diagram will be given below. The solution to this problem is well-known, renormalization. We *must* use renormalization to move the effects of high energy components in the state to effective interactions* in the Hamiltonian.

It is difficult to see how a constituent approximation can emerge using any regularization scheme that does not employ a cutoff that either removes degrees of freedom or removes direct couplings between degrees of freedom. A Pauli-Villars "cutoff," for example, drastically increases the size of Fock space and destroys the hermiticity of the Hamiltonian.

In the best case scenario we expect the cutoff to act like a resolution. If the cutoff is increased to an arbitrarily large value, the resolution increases and instead of seeing a few constituents we resolve the substructure of the constituents and the few-body approximation breaks down. As the cutoff is lowered, this substructure is removed from the state vectors, and the renormalization procedure replaces it with effective interactions in the Hamiltonian. Any "cutoff" that does not remove this substructure from the states is of no use to us.

*These include one-body operators that modify the free dispersion relations.

This point is well-illustrated by the QED calculations discussed below[8, 10, 11]. There is a window into which the cutoff must be lowered for the constituent approximation to work. If the cutoff is too large, atomic states must explicitly include photons. After the cutoff is lowered to a value that can be self-consistently determined *a-posteriori,* photons are removed from the states and replaced by the Coulomb interaction and relativistic corrections. The cutoff cannot be lowered too far using a perturbative renormalization group, hence the window.

Thus, if we remove high energy degrees of freedom, or coupling to high energy degrees of freedom, we should encounter self-energy shifts leading to effective one-body operators, vertex corrections leading to effective vertices, and exchange effects leading to explicit many-body interactions not found in the canonical Hamiltonian. We naively expect these operators to be local when acting on low energy states, because simple uncertainty principle arguments indicate that high energy virtual particles cannot propagate very far. Unfortunately this expectation is indeed naive, and at best we can hope to maintain transverse locality.[4] I will elaborate on this point below. The study of perturbation theory with the cutoffs we must employ makes it clear that it is *not enough* to adjust the canonical couplings and masses to renormalize the theory. It is not possible to make significant progress towards solving light-front QCD without fully appreciating this point.

Low energy many-body states do not typically decouple from low energy few-body states. The worst of these low energy many-body states is the vacuum. This is what drives us to use light-front coordinates.[2] Figure 2 shows a pair of particles being produced out of the vacuum in equal-time coordinates t and z. The transverse components x and y are not shown, because they are the same in equal-time and light-front coordinates. The figure also shows light-front time,

$$x^+ = t + z \,, \tag{3}$$

and the light-front longitudinal spatial coordinate,

$$x^- = t - z \,. \tag{4}$$

In equal-time coordinates it is kinematically possible for virtual pairs to be produced from the vacuum (although relevant interactions actually produce three or more particles from the vacuum), as long as their momenta sum to zero so that three-momentum is conserved. Because of this, the state vector for a proton includes an arbitrarily large number of particles that are disconnected from the proton. The only constraint imposed by relativity is that particle velocities be less than or equal to that of light.

In light-front coordinates, however, we see that all allowed trajectories lie in the first quadrant. In other words, light-front longitudinal momentum, p^+ (conjugate to x^- since $a \cdot b = \frac{1}{2}(a^+b^- + a^-b^+) - \mathbf{a}_\perp \cdot \mathbf{b}_\perp$), is always positive,

$$p^+ \geq 0 \,. \tag{5}$$

We exclude $p^+ = 0$, forcing the vacuum to be trivial because it is the only state with $p^+ = 0$. Moreover, the light-front energy of a free particle of mass m is

$$p^- = \frac{\mathbf{p}_\perp^2 + m^2}{p^+} \,. \tag{6}$$

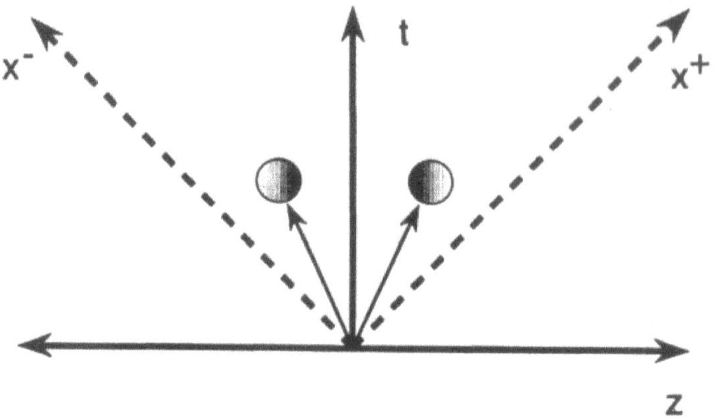

Figure 2. Light-front coordinates. Light-front 'time' is $x^+ = t + z$, and the light-front longitudinal spatial coordinate is $x^- = t - z$.

This implies that all free particles with zero longitudinal momentum have infinite energy, unless their mass and transverse momentum are identically zero. Replacing such particles with effective interactions should be reasonable.

- Is the vacuum really trivial
- What about confinement?
- What about chiral symmetry breaking?
- What about instantons?
- What about the job security of theorists who study the vacuum?

The question of how one should treat "zero modes," degrees of freedom (which may be constrained) with identically zero longitudinal momentum, divides the light-front community. Our attitude is that explicitly including zero modes defeats the purpose of using light-front coordinates, and we do not believe that significant progress will be made in this direction, at least not in 3 + 1 dimensions.

The vacuum in our formalism is trivial. We are forced to work in the "hidden symmetry phase" of the theory, and to introduce effective interactions that reproduce all effects associated with the vacuum in other formalisms.[1, 20, 21] The simplest example of this approach is provided by a scalar field theory with spontaneous symmetry breaking. It is possible to shift the scalar field and deal explicitly with a theory containing symmetry breaking interactions. In the simplest case ϕ^3 is the only relevant or marginal symmetry breaking interaction, and one can simply tune this coupling to the value corresponding to spontaneous rather than explicit symmetry breaking. Ken Wilson and I have also shown that in such simple cases one can use coupling coherence to fix the strength of this interaction so that tuning is not required.[3]

I will make an additional drastic assumption in these lectures, an assumption that Ken Wilson does not believe will hold true. I will assume that all effective interactions we require are local in the transverse direction. If this is true, there are a finite number of relevant and marginal operators, although each contains a function of longitudinal momenta that must be determined by the renormalization procedure.† There are *many more* relevant and marginal operators in the renormalized light-front Hamiltonian than in $\mathcal{L}_{QCD}$. If transverse locality is violated, the situation is much worse than this.

The presence of extra relevant and marginal operators that contain functions tremendously complicates the renormalization problem, and a common reaction to this problem is denial, which may persist for years. However, this situation may make possible tremendous simplifications in the final nonperturbative problem. For example, few-body operators must produce confinement manifestly!

Confinement cannot require particle creation and annihilation, flux tubes, etc. This is easily seen using a variational argument. Consider a color neutral quark-antiquark pair that are separated by a distance *R*, which is slowly increased to infinity. Moreover, to see the simplest form of confinement assume that there are no light quarks, so that the energy should increase indefinitely as they are separated if the theory possesses confinement. At each separation the gluon components of the state adjust themselves to minimize the energy. But this means that the expectation value of the Hamiltonian for a state with no gluons must exceed the energy of the state with gluons, and therefore must diverge even more rapidly than the energy of the true ground state. This means that there must be a two-body confining interaction in the Hamiltonian. If the renormalization procedure is unable to produce such confining two-body interactions, the constituent picture will not arise.

Manifest gauge invariance and manifest rotational invariance require all physical states to contain an arbitrarily large number of constituents. Gauge invariance is not manifest since we work in light-cone gauge with the zero modes removed; and it is easy to see that manifest rotational invariance requires an infinite number of constituents. Rotations about transverse axes are generated by dynamic operators in interacting light-front field theories, operators that create and annihilate particles.[2, 22] No state with a finite number of constituents rotates into itself or transforms as a simple tensor under the action of such generators. These symmetries seem to imply that we can not obtain a constituent approximation.

To cut this Gordian knot we employ cutoffs that violate gauge invariance and covariance, symmetries which then must be restored by effective interactions, and which need not be restored exactly. A familiar example of this approach is supplied by lattice gauge theory, where rotational invariance is violated by the lattice.

Simple Strategy

We have recently employed a conceptually simple strategy to complete bound state calculations. The first step is to use a perturbative similarity renormalization group[17, 18, 19] and coupling coherence[3, 4] to find the renormalized Hamiltonian as an expansion in powers of the canonical coupling:

$$H^{\Lambda} = h_0 + g_{\Lambda} h_1^{\Lambda} + g_{\Lambda}^2 h_2^{\Lambda} + \cdots \tag{7}$$

We compute this series to a finite order, and to date have not required any *ad hoc* assumptions to uniquely fix the Hamiltonian. No operators are added to the hamiltonian, so the hamiltonian is completely determined by the underlying theory to this order.

The second step is to employ bound state perturbation theory to solve the eigenvalue problem. The complete Hamiltonian contains every interaction (although each

†These functions imply that there are effectively an infinite number of relevant and marginal operators; however, their dependence on fields and transverse momenta is extremely limited.

is cut off) contained in the canonical Hamiltonian, and many more. We separate the Hamiltonian,

$$H^\Lambda = H_0^\Lambda + V^\Lambda , \tag{8}$$

treating H_0^Λ nonperturbatively and computing the effects of V^Λ in bound state perturbation theory. We must choose H_0^Λ and Λ so that H_0^Λ is *manageable* and to minimize corrections from higher orders of V^Λ within a constituent approximation.

If a constituent approximation is valid *after* Λ is lowered to a critical value which must be determined, we may be able to move all creation and annihilation operators to V^Λ. H_0^Λ will include many-body interactions that do not change particle number, and these interactions should be primarily responsible for the constituent bound state structure.

There are several obvious flaws in this strategy. Chiral symmetry-breaking operators, which must be included in the Hamiltonian since we work entirely in the hidden symmetry phase of the theory, do not appear at any finite order in the coupling. These operators must simply be added and tuned to fit spectra or fixed by a non-perturbative renormalization procedure.[1, 7, 9] In addition, there are perturbative errors in the strengths of all operators that do appear. We know from simple scaling arguments[23] that when Λ is in the scaling regime:

- small errors in relevant operators exponentiate in the output,
- small errors in marginal operators produce comparable errors in output,
- small errors in irrelevant operators tend to decrease exponentially in the output.

This means that even if a relevant operator appears (*e.g.,* a constituent quark or gluon mass operator), we may need to tune its strength to obtain reasonable results. We have not had to do this, but we have recently studied some of the effects of tuning a gluon mass operator.[14]

To date this strategy has produced well-known results in QED[8, 10, 11] through the Lamb shift, and reasonable results for heavy quark bound states in QCD.[8, 12, 13, 14] The primary objective of the remainder of these lectures is to review these results. I first use the Schwinger model as an illustration of a light-front bound state calculation. This model does not require renormalization, so before turning to QED and QCD I discuss the renormalization procedure that we have developed.

LIGHT-FRONT SCHWINGER MODEL

In this section I use the Schwinger model, massless QED in 1 + 1 dimensions, to illustrate the basic strategy we employ *after* we have computed the renormalized Hamiltonian. No model in 1 + 1 dimensions illustrates the renormalization problems we must solve before we can start to study QCD_{3+1}.

The Schwinger model can be solved analytically.[24, 25] Charged particles are confined because the Coulomb interaction is linear and there is only one physical particle, a massive neutral scalar particle with no self-interactions. The Fock space content of the physical states depends crucially on the coordinate system and gauge, and it is only in light-front coordinates that a simple constituent picture emerges.[26]

The Schwinger model was first studied in Hamiltonian light-front field theory by Bergknoff.[26] My description of the model follows his closely, and I recommend his paper to the reader. Bergknoff showed that the physical boson in the light-front massless

Schwinger model in light-cone gauge is a pure electron-positron state. This is an amazing result in a strong-coupling theory of massless bare particles, and it illustrates how a constituent picture may arise in QCD. The electron-positron pair is confined by the linear Coulomb potential. The light-front kinetic energy vanishes in the massless limit, and the potential energy is minimized by a wave function that is flat in momentum space, as one might expect since a linear potential produces a state that is as localized as possible (given kinematic constraints due to the finite velocity of light) in position space.

In order to solve this theory I must first set up a large number of details. I recommend that for a first reading these details be skimmed, because the general idea is more important than the detailed manipulations. The Lagrangian for the theory is

$$\mathcal{L} = \overline{\psi}(i \not\partial - m)\psi - e\overline{\psi}\gamma_\mu\psi A^\mu - \frac{1}{4}F_{\mu\nu}F^{\mu\nu} , \tag{9}$$

where $F_{\mu\nu}$ is the electromagnetic field strength tensor. I have included an electron mass, m, which is taken to zero later. I choose light-cone gauge,

$$A^+ = 0 . \tag{10}$$

In this gauge we avoid ghosts, so that the Fock space has a positive norm. This is absolutely essential if we want to apply intuitive techniques from many-body quantum mechanics.

Many calculations are simplified by the use of a chiral representation of the Dirac gamma matrices, so in this section I will use:

$$\gamma^0 = \begin{pmatrix} 0 & -i \\ i & 0 \end{pmatrix} , \quad \gamma^1 = \begin{pmatrix} 0 & i \\ i & 0 \end{pmatrix} , \tag{11}$$

which leads to the light-front coordinate gamma matrices,

$$\gamma^+ = \begin{pmatrix} 0 & 0 \\ 2i & 0 \end{pmatrix} , \quad \gamma^- = \begin{pmatrix} 0 & -2i \\ 0 & 0 \end{pmatrix} . \tag{12}$$

In light-front coordinates the fermion field ψ contains only one dynamical degree of freedom, rather than two. To see this, first define the projection operators,

$$\Lambda_+ = \frac{1}{2}\gamma^0\gamma^+ = \begin{pmatrix} 1 & 0 \\ 0 & 0 \end{pmatrix} \quad \Lambda_- = \frac{1}{2}\gamma^0\gamma^- = \begin{pmatrix} 0 & 0 \\ 0 & 1 \end{pmatrix} . \tag{13}$$

Using these operators split the fermion field into two components,

$$\psi = \psi_+ + \psi_- = \Lambda_+\psi + \Lambda_-\psi . \tag{14}$$

The two-component Dirac equation in this gauge is

$$\left(\frac{i}{2}\gamma^+\partial^- + \frac{i}{2}\gamma^-\partial^+ - m - \frac{e}{2}\gamma^+A^-\right)\psi = 0 ; \tag{15}$$

which can be split into two one-component equations,

$$i\partial^-\tilde{\psi}_+ = -im\tilde{\psi}_- + eA^-\tilde{\psi}_+ , \tag{16}$$

$$i\partial^+\tilde{\psi}_- = im\tilde{\psi}_+ . \tag{17}$$

Here $\tilde{\psi}_\pm$ refers to the non-zero component of $\psi_\pm$.

The equation for ψ_+ involves the light-front time derivative, ∂^-; so ψ_+ is a dynamical degree of freedom that must be quantized. On the other hand, the equation for ψ_- involves only spatial derivatives, so ψ_- is a constrained degree of freedom that should be eliminated in favor of ψ_+. Formally,

$$\tilde{\psi}_- = \frac{m}{\partial^+}\tilde{\psi}_+ \,. \tag{18}$$

This equation is not well-defined until boundary conditions are specified so that ∂^+ can be inverted. I will eventually define this operator in momentum space using a cutoff, but I want to delay the introduction of a cutoff until a calculation requires it.

I have chosen the gauge so that $A^+ = 0$, and the equation for A^- is

$$-\frac{1}{4}(\partial^+)^2 A^- = e\psi_+^\dagger\psi_+ \,. \tag{19}$$

A^- is also a constrained degree of freedom, and we can formally eliminate it,

$$A^- = -\frac{4e}{(\partial^+)^2}\psi_+^\dagger\psi_+ \,. \tag{20}$$

We are now left with a single dynamical degree of freedom, ψ_+, which we can quantize at $x^+ = 0$,

$$\{\psi_+(x^-), \psi_+^\dagger(y^-)\} = \Lambda_+\delta(x^- - y^-) \,. \tag{21}$$

We can introduce free particle creation and annihilation operators and expand the field operator at $x^+ = 0$,

$$\tilde{\psi}_+(x^-) = \int_{k^+>0} \frac{dk^+}{4\pi}\left[b_k e^{-ik\cdot x} + d_k^\dagger e^{ik\cdot x}\right] , \tag{22}$$

with,

$$\{b_k, b_p^\dagger\} = 4\pi\delta(k^+ - p^+) \,. \tag{23}$$

In order to simplify notation, I will often write k to mean k^+. If I need $k^- = m^2/k^+$, I will provide the superscript.

The next step is to formally specify the Hamiltonian. I start with the canonical Hamiltonian,

$$H = \int dx^-\left(H_0 + V\right) , \,. \tag{24}$$

$$H_0 = \psi_+^\dagger\left(\frac{m^2}{i\partial^+}\right)\psi_+ \,, \tag{25}$$

$$V = -2e^2\psi_+^\dagger\psi_+\left(\frac{1}{\partial^+}\right)^2\psi_+^\dagger\psi_+ \,. \tag{26}$$

To actually calculate:

- replace ψ_+ with its expansion in terms of b_k and d_k,
- normal-order,
- throw away constants,
- drop all operators that require b_0 and d_0.

The free part of the Hamiltonian becomes

$$H_0 = \int_{k>0} \frac{dk}{4\pi} \left(\frac{m^2}{k}\right) (b_k^\dagger b_k + d_k^\dagger d_k) \,. \tag{27}$$

When V is normal-ordered, we encounter new one-body operators,

$$H_0' = \frac{e^2}{2\pi} \int_{k>0} \frac{dk}{4\pi} \left[\int_{p>0} dp \left(\frac{1}{(k-p)^2} - \frac{1}{(k+p)^2} \right) \right] (b_k^\dagger b_k + d_k^\dagger d_k) \,. \tag{28}$$

This operator contains a divergent momentum integral. From a mathematical point of view we have been sloppy and need to carefully add boundary conditions and define how ∂^+ is inverted. However, I want to apply physical intuition and even though no physical photon has been exchanged to produce the initial interaction, I will act as if a photon has been exchanged and everywhere an 'instantaneous photon exchange' occurs I will cut off the momentum. In the above integral I insist,

$$|k - p| > \epsilon \,. \tag{29}$$

Using this cutoff we find that

$$H_0' = \frac{e^2}{\pi} \int \frac{dk}{4\pi} \left(\frac{1}{\epsilon} - \frac{1}{k} + \mathcal{O}(\epsilon) \right) (b_k^\dagger b_k + d_k^\dagger d_k) \,. \tag{30}$$

Comparing this result with the original free Hamiltonian, we see that a divergent mass-like term appears; but it does not have the same dispersion relation as the bare mass. Instead of depending on the inverse momentum of the fermion, it depends on the inverse momentum cutoff, which cannot appear in any physical result. There is also a finite shift in the bare mass, with the standard dispersion relation.

The normal-ordered interactions are

$$\begin{aligned} V' = 2e^2 \int \frac{dk_1}{4\pi} \cdots \frac{dk_4}{4\pi} 4\pi\delta(k_1 + k_2 - k_3 - k_4) \\ \Big\{ -\frac{2}{(k_1 - k_3)^2} b_1^\dagger d_2^\dagger d_4 b_3 + \frac{2}{(k_1 + k_2)^2} b_1^\dagger d_2^\dagger d_3 b_4 \\ - \frac{1}{(k_1 - k_3)^2} (b_1^\dagger b_2^\dagger b_3 b_4 + d_1^\dagger d_2^\dagger d_3 d_4) + \cdots \Big\} \,. \end{aligned} \tag{31}$$

I do not display the interactions that involve the creation or annihilation of electron/positron pairs, which are important for the study of multiple boson eigenstates.

The first term in V' is the electron-positron interaction. The longitudinal momentum cutoff I introduced above requires $|k_1 - k_3| > \epsilon$, so in position space we encounter a potential which I will naively define with a Fourier transform that ignores the fact that the momentum transfer cannot exceed the momentum of the state,

$$\begin{aligned} v(x^-) &= 4q_1 q_2 \int_{-\infty}^{\infty} \frac{dk}{4\pi} \frac{1}{k^2} \, \theta(|k| - \epsilon) \exp(-\frac{i}{2} k x^-) \\ &= \frac{q_1 q_2}{\pi} \left[\frac{2}{\epsilon} - \frac{\pi}{2} |x^-| + \mathcal{O}(\epsilon) \right] . \end{aligned} \tag{32}$$

This potential contains a linear Coulomb potential that we expect in two dimensions, but it also contains a divergent constant that is negative for unlike charges and positive for like charges.

In charge neutral states the infinite constant in V' is *exactly* canceled by the divergent 'mass' term in H_0'. This Hamiltonian assigns an infinite energy to states with net charge, and a finite energy as $\epsilon \to 0$ to charge zero states. This does not imply that charged particles are confined, but the linear potential prevents charged particles from moving to arbitrarily large separation except as charge neutral states. The confinement mechanism I propose for QCD in 3+1 dimensions shares many features with this interaction.

I would also like to mention that even though the interaction between charges is long-ranged, there are no van der Waals forces in 1+1 dimensions. It is a simple geometrical calculation to show that all long range forces between two neutral states cancel exactly. This does not happen in higher dimensions, and if we use long-range two-body operators to implement confinement we must also find many-body operators that cancel the strong long-range van der Waals interactions.

Given the complete Hamiltonian in normal-ordered form we can study bound states. A powerful tool for the initial study of bound states is the variational wave function. In this case, we can begin with a state that contains a single electron-positron pair,

$$|\Psi(P)\rangle = \int_0^P \frac{dp}{4\pi} \phi(p) b_p^\dagger d_{P-p}^\dagger |0\rangle \,. \tag{33}$$

The norm of this state is

$$\langle \Psi(P')|\Psi(P)\rangle = 4\pi P \delta(P' - P) \left\{ \frac{1}{P} \int_0^P \frac{dp}{4\pi} |\phi(p)|^2 \right\} , \tag{34}$$

where the factors outside the brackets provide a covariant plane wave normalization for the center-of-mass motion of the bound state, and the bracketed term should be set to one.

The expectation value of the one-body operators in the Hamiltonian is

$$\langle \Psi | H_0 + H_0' | \Psi \rangle = \frac{1}{P} \int \frac{dk}{4\pi} \left[\frac{m^2 - e^2/\pi}{k} + \frac{m^2 - e^2/\pi}{P-k} + \frac{2e^2}{\pi\epsilon} \right] |\phi(k)|^2 \,, \tag{35}$$

and the expectation value of the normal-ordered interactions is

$$\langle \Psi | V' | \Psi \rangle = -\frac{4e^2}{P} \int' \frac{dk_1}{4\pi} \frac{dk_2}{4\pi} \left[\frac{1}{(k_1 - k_2)^2} - \frac{1}{P^2} \right] \phi^*(k_1)\phi(k_2) \,, \tag{36}$$

where I have dropped the overall plane wave norm. The prime on the last integral indicates that the range of integration in which $|k_1 - k_2| < \epsilon$ must be removed. By expanding the integrand about $k_1 = k_2$, one can easily confirm that the $1/\epsilon$ divergences cancel.

With $m = 0$ the energy is minimized when

$$\phi(k) = \sqrt{4\pi} \,, \tag{37}$$

and the invariant-mass is

$$M^2 = \frac{e^2}{\pi} \,. \tag{38}$$

This type of simple analysis can be used to show that this electron-positron state is actually the *exact ground* state of the theory with momentum P, and that bound states do not interact with one another.

The primary purpose of introducing the Schwinger model is to illustrate that bound state center-of-mass motion is easily separated from relative motion in light-front coordinates, and that standard quantum mechanical techniques can be used to analyze the relative motion of charged particles once the Hamiltonian is found. It is intriguing that even when the fermion is massless, the states are constituent states in light-cone gauge and in light-front coordinates. This is not true in other gauges and coordinate systems. The success of light-front field theory in 1+1 dimensions can certainly be downplayed, but it should be emphasized that no other method on the market is as powerful for bound state problems in 1+1 dimensions.

The most significant barriers to using light-front field theory to solve low energy QCD are not encountered in 1+1 dimensions. The Schwinger model is super-renormalizable, so we completely avoid serious ultraviolet divergences. There are no transverse directions, and we are not forced to introduce a cutoff that violates rotational invariance, because there are no rotations. Confinement results from the Coulomb interaction, and chiral symmetry is not spontaneously broken. This simplicity disappears in realistic 3 + 1-dimensional calculations, which is one reason there are so few 3 + 1-dimensional light-front field theory calculations.

LIGHT-FRONT RENORMALIZATION GROUP: SIMILARITY TRANSFORMATION AND COUPLING COHERENCE

As argued above, in 3 + 1 dimensions we must introduce a cutoff on energies, Λ, and we never perform explicit bound state calculations with Λ anywhere near its continuum limit. In fact, we want to let Λ become as small as possible. In my opinion, any strategy for solving light-front QCD that requires the cutoff to explicitly approach infinity in the nonperturbative part of the calculation is useless. Therefore, we must set up and solve

$$P_\Lambda^- \mid \Psi_\Lambda(P)\rangle = \frac{\mathbf{P}_\perp^2 + M^2}{P^+} \mid \Psi_\Lambda(P)\rangle \, . \tag{39}$$

Physical results, such as the mass, M, can not depend on the arbitrary cutoff, Λ, *even* as Λ approaches the scale of interest. This means that P_Λ^- and $\mid \Psi_\Lambda\rangle$ must depend on the cutoff in such a way that $\langle \Psi_\Lambda \mid P_\Lambda^- \mid \Psi_\Lambda\rangle$ does not. Wilson based the derivation of his renormalization group on this observation,[27, 28, 29, 23] and we use Wilson's renormalization group to compute P_Λ^-.

It is difficult to even talk about how the Hamiltonian depends on the cutoff without having a means of changing the cutoff. If we can change the cutoff, we can explicitly watch the Hamiltonian's cutoff dependence change and fix its cutoff dependence by insisting that this change satisfy certain requirements (*e.g.*, that the limit in which the cutoff is taken to infinity exists). We introduce an operator that changes the cutoff,

$$H(\Lambda_1) = T[H(\Lambda_0)] \, , \tag{40}$$

where I assume that $\Lambda_1 < \Lambda_0$. To simplify the notation, I will let $H(\Lambda_l) = H_l$. To renormalize the hamiltonian we study the properties of the transformation.

Figure 3 displays two generic cutoffs that might be used. Traditionally theorists have used cutoffs that remove high energy states, as shown in Figure 3a. This is the type of cutoff Wilson employed in his initial work[27] and I have studied its use in light-front field theory.[4] When a cutoff on energies is reduced, all effects of couplings eliminated must be moved to effective operators. As we will see explicitly below, when these effective operators are computed perturbatively they involve products of matrix elements divided by energy denominators. Expressions closely resemble those encountered in standard perturbation theory, with the second-order operator involving terms of the form

$$\delta V_{ij} \sim \frac{\langle \phi_i \mid V \mid \phi_k\rangle\langle \phi_k \mid V \mid \phi_j\rangle}{\epsilon_i - \epsilon_k} \, . \tag{41}$$

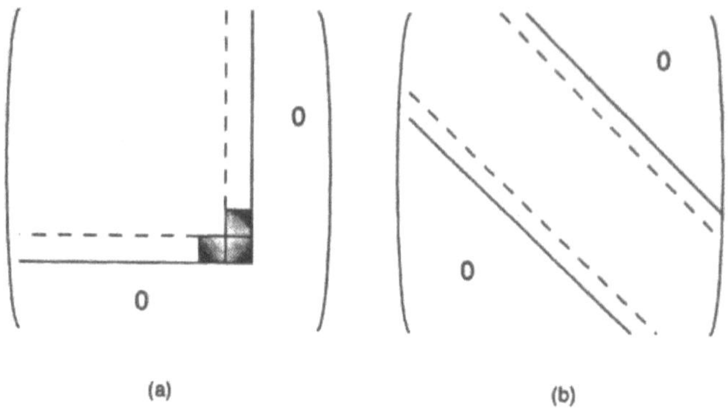

Figure 3. Two ways to run a cutoff on free energy. In (a) a cutoff on the magnitude of the energy is lowered from the solid to the dashed lines, with problems resulting from the removed shaded region. In (b) a cutoff on how far off diagonal matrix elements appear is lowered from the dashed to the solid lines.

This new effective interaction replaces missing couplings, so the states ϕ_i and ϕ_j are retained and the state ϕ_k is one of the states removed. The problem comes from the shaded, lower right-hand corner of the matrix, where the energy denominator vanishes for states at the corner of the remaining matrix. In this corner we should use nearly degenerate perturbation theory rather than perturbation theory, but to do this requires solving high energy many-body problems nonperturbatively before solving the low energy few-body problems.

An alternative cutoff, which does not actually remove any states and which can be run by a similarity transformation‡ is shown in Figure 3b. This cutoff removes couplings between states whose free energy differs by more than the cutoff. The advantage of this cutoff is that the effective operators resulting from it contain energy denominators which are never smaller than the cutoff, so that a perturbative approximation for the effective Hamiltonian may work well. I discuss a conceptually simple similarity transformation that runs this cutoff below.

Given a cutoff and a transformation that runs the cutoff, we can discuss how the Hamiltonian depends on the cutoff by studying how it changes with the cutoff. Our objective is to find a "renormalized" Hamiltonian, which should give the same results as an idealized Hamiltonian in which the cutoff is infinite and which displays all of the symmetries of the theory. To state this in simple terms, consider a sequence of Hamiltonians generated by repeated application of the transformation,

$$H_0 \rightarrow H_1 \rightarrow H_2 \rightarrow \cdots . \tag{42}$$

What we really want to do is fix the final value of Λ at a reasonable hadronic scale, and let Λ_0 approach infinity. In other words, we seek a Hamiltonian that survives an infinite number of transformations. In order to do this we need to understand what happens when the transformation is applied to a broad class of Hamiltonians.

Perturbative renormalization group analyses typically begin with the identification of at least one fixed point, H^*. A *fixed point* is defined to be any Hamiltonian that

‡In deference to the original work I will call this a similarity transformation even though in all cases of interest to us it is a unitary transformation.

satisfies the condition

$$H^* = T[H^*] . \tag{43}$$

For perturbative renormalization groups the search for such fixed points is relatively easy. If H^* contains no interactions (*i.e.,* no terms with a product of more than two field operators), it is called *Gaussian.* If H^* has a massless eigenstate, it is called *critical.* If a Gaussian fixed point has no mass term, it is a *critical Gaussian* fixed point. If it has a mass term, this mass must typically be infinite, in which case it is a *trivial Gaussian* fixed point. In lattice QCD the trajectory of renormalized Hamiltonians stays 'near' a critical Gaussian fixed point until the lattice spacing becomes sufficiently large that a transition to strong-coupling behavior occurs. If H^* contains only weak interactions, it is called *near-Gaussian,* and we may be able to use perturbation theory both to identify H^* and to accurately approximate 'trajectories' of Hamiltonians near H^*. Of course, once the trajectory leaves the region of H^*, it is generally necessary to switch to a non-perturbative calculation of subsequent evolution.

Consider the immediate 'neighborhood' of the fixed point, and assume that the trajectory remains in this neighborhood. This assumption must be justified a *posteriori,* but if it is true we should write

$$H_l = H^* + \delta H_l , \tag{44}$$

and consider the trajectory of small deviations δH_l.

As long as δH_l is 'sufficiently small,' we can use a perturbative expansion in powers of δH_l, which leads us to consider

$$\delta H_{l+1} = L \cdot \delta H_l + N[\delta H_l] . \tag{45}$$

Here L is the linear approximation of the full transformation in the neighborhood of the fixed point, and $N[\delta H_l]$ contains all contributions to δH_{l+1} of $\mathcal{O}(\delta H_l^2)$ and higher.

The object of the renormalization group calculation is to compute trajectories and this requires a representation for δH_l. The problem of computing trajectories is one of the most common in physics, and a convenient basis for the representation of δH_l is provided by the eigenoperators of L, since L dominates the transformation near the fixed point. These eigenoperators and their eigenvalues are found by solving

$$L \cdot O_m = \lambda_m O_m . \tag{46}$$

If H^* is Gaussian or near-Gaussian it is usually straightforward to find L, and its eigenoperators and eigenvalues. This is not typically true if H^* contains strong interactions, but in QCD we hope to use a perturbative renormalization group in the regime of asymptotic freedom, and the QCD ultraviolet fixed point is apparently a critical Gaussian fixed point. For light-front field theory this linear transformation is a scaling of the transverse coordinate, the eigenoperators are products of field operators and transverse derivatives, and the eigenvalues are determined by the transverse dimension of the operator. All operators can include both powers and inverse powers of longitudinal derivatives because there is no longitudinal locality.[4]

Using the eigenoperators of L as a basis we can represent δH_l,

$$\delta H_l = \sum_{m \in R} \mu_{m_l} O_m + \sum_{m \in M} g_{m_l} O_m + \sum_{m \in I} w_{m_l} O_m . \tag{47}$$

Here the operators O_m with $m \in R$ are *relevant* (*i.e.,* $\lambda_m > 1$), the operators O_m with $m \in M$ are *marginal* (*i.e.,* $\lambda_m = 1$), and the operators with $m \in I$ are either *irrelevant*

($i.e.$, $\lambda_m < 1$) or become irrelevant after many applications of the transformation. The motivation behind this nomenclature is made clear by considering repeated application of L, which causes the relevant operators to grow exponentially, the marginal operators to remain unchanged in strength, and the irrelevant operators to decrease in magnitude exponentially. There are technical difficulties associated with the symmetry of L and the completeness of the eigenoperators that I ignore.[30]

For the purpose of illustration, let me assume that $\lambda_m = 4$ for all relevant operators, and $\lambda_m = 1/4$ for all irrelevant operators. The transformation can be represented by an infinite number of coupled, nonlinear difference equations:

$$\mu_{m_{l+1}} = 4\mu_{m_l} + N_{\mu_m}[\mu_{m_l}, g_{m_l}, w_{m_l}] , \tag{48}$$

$$g_{m_{l+1}} = g_{m_l} + N_{g_m}[\mu_{m_l}, g_{m_l}, w_{m_l}] , \tag{49}$$

$$w_{m_{l+1}} = \frac{1}{4}w_{m_l} + N_{w_m}[\mu_{m_l}, g_{m_l}, w_{m_l}] . \tag{50}$$

Sufficiently near a critical Gaussian fixed point, the functions N_{μ_m}, N_{g_m}, and N_{w_m} should be adequately approximated by an expansion in powers of μ_{m_l}, g_{m_l}, and w_{m_l}. The assumption that the Hamiltonian remains in the neighborhood of the fixed point, so that all μ_{m_l}, g_{m_l}, and w_{m_l} remain small, must be justified *a posteriori*. Any precise definition of the neighborhood of the fixed point within which all approximations are valid must also be provided *a posteriori*.

Wilson has given a general discussion of how these equations can be solved,[23] but I will use coupling coherence[3, 4] to fix the Hamiltonian. This is detailed below, so at this point I will merely state that coupling coherence allows us to fix all couplings as functions of the canonical couplings and masses in a theory. The renormalization group equations specify how all of the couplings run, and coupling coherence uses this behavior to fix the strength of all of the couplings. But the first step is to develop a transformation.

Similarity Transformation

Stan Glazek and Ken Wilson studied the problem of small energy denominators which are apparent in Wilson's first complete non-perturbative renormalization group calculations,[28] and realized that a similarity transformation which runs a different form of cutoff (as discussed above) avoids this problem.[17, 18, 19] Independently, Wegner[32] developed a similarity transformation which is easier to use than that of Glazek and Wilson.

In this section I want to give a simplified discussion of the similarity transformation, using sharp cutoffs that must eventually be replaced with smooth cutoffs which require a more complicated formalism.

Suppose we have a Hamiltonian,

$$H^{\Lambda_0} = H_0^{\Lambda_0} + V^{\Lambda_0} , \tag{51}$$

where $H_0^{\Lambda_0}$ is diagonal. The cutoff Λ_0 indicates that $\langle\phi_i|V^{\Lambda_0}|\phi_j\rangle = 0$ if $|E_{0i} - E_{0j}| > \Lambda_0$. I should note that Λ_0 is defined differently in this section from later sections. We want to use a similarity transformation, which automatically leaves all eigenvalues and other physical matrix elements invariant, that lowers this cutoff to Λ_1. This similarity transformation will constitute the first step in a renormalization group transformation, with the second step being a rescaling of energies that returns the cutoff to its original numerical value.[§]

The transformed Hamiltonian is

$$H^{\Lambda_1} = e^{iR}(H_0^{\Lambda_0} + V^{\Lambda_0})e^{-iR} , \tag{52}$$

where R is a hermitian operator. If H^{Λ_0} is already diagonal, $R = 0$. Thus, if R has an expansion in powers of V, it starts at first order and we can expand the exponents in powers of R to find the perturbative approximation of the transformation.

We must adjust R so that the matrix elements of H^{Λ_1} vanish for all states that satisfy $\Lambda_1 < |E_{0i} - E_{0j}| < \Lambda_0$. We insist that this happens to each order in perturbation theory. Consider such a matrix element,

$$\begin{aligned}\langle\phi_i|H^{\Lambda_1}|\phi_j\rangle &= \langle\phi_i|e^{iR}(H_0^{\Lambda_0} + V^{\Lambda_0})e^{-iR}|\phi_j\rangle \\ &= \langle\phi_i|(1 + iR + \cdots)(H_0^{\Lambda_0} + V^{\Lambda_0})(1 - iR - \cdots)|\phi_j\rangle \\ &= \langle\phi_i|H_0^{\Lambda_0}|\phi_j\rangle + \langle\phi_i|V^{\Lambda_0} + i\left[R, H_0^{\Lambda_0}\right]|\phi_j\rangle + \cdots .\end{aligned} \tag{53}$$

The last line contains all terms that appear in first-order perturbation theory. Since $\langle\phi_i|H_0^{\Lambda_0}|\phi_j\rangle = 0$ for these off-diagonal matrix elements, we can satisfy our new constraint using

$$\langle\phi_i|R|\phi_j\rangle = \frac{i\langle\phi_i|V^{\Lambda_0}|\phi_j\rangle}{E_{0j} - E_{0i}} + \mathcal{O}(V^2) . \tag{54}$$

This fixes the matrix elements of R when $\Lambda_1 < |E_{0i} - E_{0j}| < \Lambda_0$ to first order in V^{Λ_0}. I will assume that the matrix elements of R for $|E_{0i} - E_{0j}| < \Lambda_1$ are zero to first order in V, and fix these matrix elements to second order below.

Given R we can compute the nonzero matrix elements of H^{Λ_1}. To second order in V^{Λ_0} these are

$$\begin{aligned}H_{ab}^{\Lambda_1} &= \langle\phi_a|H_0 + V + i\left[R, H_0\right] + i\left[R, V\right] - \frac{1}{2}\left\{R^2, H_0\right\} + RH_0R + \mathcal{O}(V^3)|\phi_b\rangle \\ &= \langle\phi_a|H_0 + V + i\left[R_2, H_0\right]|\phi_b\rangle \\ &\quad -\frac{1}{2}\sum_k \Theta_{ak}\Theta_{kb}V_{ak}V_{kb}\left[\frac{1}{E_{0k} - E_{0a}} + \frac{1}{E_{0k} - E_{0b}}\right] \\ &\quad -\sum_k\left[\Theta_{ak}(1 - \Theta_{kb})\frac{V_{ak}V_{kb}}{E_{0k} - E_{0a}} + \Theta_{kb}(1 - \Theta_{ak})\frac{V_{ak}V_{kb}}{E_{0k} - E_{0b}}\right] + \mathcal{O}(V^3) \\ &= \langle\phi_a|H_0 + V + i\left[R_2, H_0\right]|\phi_b\rangle \\ &\quad +\frac{1}{2}\sum_k \Theta_{ak}\Theta_{kb}V_{ak}V_{kb}\left[\frac{1}{E_{0k} - E_{0a}} + \frac{1}{E_{0k} - E_{0b}}\right] \times \\ &\qquad \left[\theta(|E_{0a} - E_{0k}| - |E_{0b} - E_{0k}|) - \theta(|E_{0b} - E_{0k}| - |E_{0a} - E_{0k}|)\right] \\ &\quad -\sum_k V_{ak}V_{kb}\left[\Theta_{ak}\frac{\theta(|E_{0a} - E_{0k}| - |E_{0b} - E_{0k}|)}{E_{0k} - E_{0a}} + \right. \\ &\qquad \left.\Theta_{kb}\frac{\theta(|E_{0b} - E_{0k}| - |E_{0a} - E_{0k}|)}{E_{0k} - E_{0b}}\right] .\end{aligned} \tag{55}$$

I have dropped the Λ_0 superscript on the right-hand side of this equation and used subscripts to indicate matrix elements. The operator Θ_{ij} is one if $\Lambda_1 < |E_{0i} - E_{0j}| < \Lambda_0$

[§]The rescaling step is not essential, but it avoids exponentials of the cutoff in the renormalization group equations.

and zero otherwise. It should also be noted that V_{ij} is zero if $|E_{0i} - E_{0j}| > \Lambda_0$. All energy denominators involve energy differences that are at least as large as Λ_1, and this feature persists to higher orders in perturbation theory; which is the main motivation for choosing this transformation. R_2 is second-order in V, and we are still free to choose its matrix elements; however, we must be careful not to introduce small energy denominators when choosing R_2.

The matrix element $\langle\phi_a|i\big[R_2, H_0\big]|\phi_b\rangle = i(E_{0b} - E_{0a})\langle\phi_a|R_2|\phi_b\rangle$ must be specified. I will choose this matrix element to cancel the first sum in the final right-hand side of Eq. (55). This choice leads to the same result one obtains by integrating a differential transformation that runs a step function cutoff. To cancel the first sum in the final right-hand side of Eq. (55) requires

$$\langle\phi_a|R_2|\phi_b\rangle = \frac{i}{2}\sum_k \Theta_{ak}\Theta_{kb}V_{ak}V_{kb}\left[\frac{1}{E_{0k} - E_{0a}} + \frac{1}{E_{0k} - E_{0b}}\right] \times \left[\theta(|E_{0a} - E_{0k}| - |E_{0b} - E_{0k}|) - \theta(|E_{0b} - E_{0k}| - |E_{0a} - E_{0k}|)\right] . \quad (56)$$

No small energy denominator appears in R_2 because it is being used to cancel a term that involves a large energy difference. If we tried to use R_2 to cancel the remaining sum also, we would find that it includes matrix elements that diverge as $E_{0a} - E_{0b}$ goes to zero, and this is not allowed.

The non-vanishing matrix elements of H^{Λ_1} are now completely determined to $\mathcal{O}(V^2)$,

$$H_{ab}^{\Lambda_1} = \langle\phi_a|H_0 + V^{\Lambda_0}|\phi_b\rangle - \sum_k V_{ak}V_{kb}\left[\Theta_{ak}\frac{\theta(|E_{0a} - E_{0k}| - |E_{0b} - E_{0k}|)}{E_{0k} - E_{0a}} + \Theta_{kb}\frac{\theta(|E_{0b} - E_{0k}| - |E_{0a} - E_{0k}|)}{E_{0k} - E_{0b}}\right] . \quad (57)$$

Let me again mention that $V_{ij}^{\Lambda_0}$ is zero if $|E_{0i} - E_{0j}| > \Lambda_0$, so there are implicit cutoffs that result from previous transformations.

As a final word of caution, I should mention that the use of step functions produces long-range pathologies in the interactions that lead to infrared divergences in gauge theories. We must replace the step functions with smooth functions to avoid this problem. This problem will not show up in any calculations detailed in these lectures, but it does affect higher order calculations in QED[10, 11] and QCD.[31]

Light-Front Renormalization Group

In this section I use the similarity transformation to form a perturbative light-front renormalization group for scalar field theory. If we want to stay as close as possible to the canonical construction of field theories, we can:

- Write a set of 'allowed' operators using powers of derivatives and field operators.
- Introduce 'free' particle creation and annihilation operators, and expand all field operators in this basis.
- Introduce cutoffs on the Fock space transition operators.

Instead of following this program I will skip to the final step, and simply write a Hamiltonian to initiate the analysis.

$$
\begin{aligned}
H = & \int d\tilde{q}_1\, d\tilde{q}_2\, (16\pi^3)\delta^3(q_1 - q_2)\, u_2(-q_1, q_2)\, a^\dagger(q_1)a(q_2) \\
& +\frac{1}{6}\int d\tilde{q}_1\, d\tilde{q}_2\, d\tilde{q}_3\, d\tilde{q}_4\, (16\pi^3)\delta^3(q_1 + q_2 + q_3 - q_4)\Theta(q_4^- - q_3^- - q_2^- - q_1^-) \\
& \qquad u_4(-q_1, -q_2, -q_3, q_4)\, a^\dagger(q_1)a^\dagger(q_2)a^\dagger(q_3)a(q_4) \\
& +\frac{1}{4}\int d\tilde{q}_1\, d\tilde{q}_2\, d\tilde{q}_3\, d\tilde{q}_4\, (16\pi^3)\delta^3(q_1 + q_2 - q_3 - q_4)\Theta(q_4^- + q_3^- - q_2^- - q_1^-) \\
& \qquad u_4(-q_1, -q_2, q_3, q_4)\, a^\dagger(q_1)a^\dagger(q_2)a(q_3)a(q_4) \\
& +\frac{1}{6}\int d\tilde{q}_1\, d\tilde{q}_2\, d\tilde{q}_3\, d\tilde{q}_4\, (16\pi^3)\delta^3(q_1 - q_2 - q_3 - q_4)\Theta(q_4^- + q_3^- + q_2^- - q_1^-) \\
& \qquad u_4(-q_1, q_2, q_3, q_4)\, a^\dagger(q_1)a(q_2)a(q_3)a(q_4) \\
& + \quad \mathcal{O}(\phi^6)\,, \qquad (58)
\end{aligned}
$$

where,

$$
d\tilde{q} = \frac{dq^+ d^2q_\perp}{16\pi^3 q^+}\,, \qquad (59)
$$

$$
q_i^- = \frac{q_{i\perp}^2}{q_i^+}\,, \qquad (60)
$$

and,

$$
\Theta(Q^-) = \theta\left(\frac{\Lambda^2}{P^+} - |Q^-|\right)\,. \qquad (61)
$$

I assume that no operators break the discrete $\phi \to -\phi$ symmetry. The functions u_2 and u_4 are not yet determined. If we assume locality in the transverse direction, these functions can be expanded in powers of their transverse momentum arguments. Note that to specify the cutoff both transverse and longitudinal momentum scales are required, and in this case the longitudinal momentum scale is independent of the particular state being studied. Note also that P^+ breaks longitudinal boost invariance and that a change in P^+ can be compensated by a change in Λ^2. This may have important consequences, because the Hamiltonian should be a fixed point with respect to changes in the cutoff's longitudinal momentum scale, since this scale invariance is protected by Lorentz covariance.[4]

I have specified the similarity transformation in terms of matrix elements, and will work directly with matrix elements, which are easily computed in the free particle Fock space basis. In order to study the renormalization group transformation I will assume that the Hamiltonian includes only the interactions shown above. A single transformation will produce a Hamiltonian containing products of arbitrarily many creation and annihilation operators, but it is not necessary to understand the transformation in full detail.

I will *define* the full renormalization group transformation as: (i) a similarity transformation that lowers the cutoff in Θ, (ii) a rescaling of all transverse momenta that returns the cutoff to its original numerical value, (iii) a rescaling of the creation and annihilation operators by a constant factor ζ, and (iv) an overall constant rescaling of the Hamiltonian to absorb a multiplicative factor that results from the fact that it has the dimension of transverse momentum squared. These rescaling operations are introduced so that it may be possible to find a fixed point Hamiltonian that contains interactions.[23]

To find the critical Gaussian fixed point we need to study the linearized approximation of the full transformation, as discussed above. In general the linearized approximation can be extremely complicated, but near a critical Gaussian fixed point it is particularly simple in light-front field theory with zero modes removed, because tadpoles are excluded. We have already seen that the similarity transformation does not produce any first order change in the Hamiltonian (see Eq. (57)), so the first order change is determined entirely by the rescaling operation. If we let

$$\Lambda_1 = \eta \Lambda_0 \,, \tag{62}$$

$$\mathbf{p}_{i\perp} = \eta \mathbf{p}'_{i\perp} \,, \tag{63}$$

and,

$$a_p = \zeta a_{p'} \,, \quad a_p^\dagger = \zeta a_{p'}^\dagger \,, \tag{64}$$

to first order the transformed Hamiltonian is

$$\begin{aligned}
H = \quad & \zeta^2 \int d\tilde{q}_1\, d\tilde{q}_2\, (16\pi^3)\delta^3(q_1 - q_2)\, u_2(q_i^+, \eta q_i^\perp)\, a^\dagger(q_1)a(q_2) \\
+ & \frac{1}{6}\eta^4\zeta^4 \int d\tilde{q}_1\, d\tilde{q}_2\, d\tilde{q}_3\, d\tilde{q}_4\, (16\pi^3)\delta^3(q_1 + q_2 + q_3 - q_4)\Theta(q_4^- - q_3^- - q_2^- - q_1^-) \\
& \qquad u_4(q_i^+, \eta q_i^\perp)\, a^\dagger(q_1)a^\dagger(q_2)a^\dagger(q_3)a(q_4) \\
+ & \frac{1}{4}\eta^4\zeta^4 \int d\tilde{q}_1\, d\tilde{q}_2\, d\tilde{q}_3\, d\tilde{q}_4\, (16\pi^3)\delta^3(q_1 + q_2 - q_3 - q_4)\Theta(q_4^- + q_3^- - q_2^- - q_1^-) \\
& \qquad u_4(q_i^+, \eta q_i^\perp)\, a^\dagger(q_1)a^\dagger(q_2)a(q_3)a(q_4) \\
+ & \frac{1}{6}\eta^4\zeta^4 \int d\tilde{q}_1\, d\tilde{q}_2\, d\tilde{q}_3\, d\tilde{q}_4\, (16\pi^3)\delta^3(q_1 - q_2 - q_3 - q_4)\Theta(q_4^- + q_3^- + q_2^- - q_1^-) \\
& \qquad u_4(q_i^+, \eta q_i^\perp)\, a^\dagger(q_1)a(q_2)a(q_3)a(q_4) \\
+ \quad & \mathcal{O}(\phi^6) \,,
\end{aligned} \tag{65}$$

where I have simplified my notation for the arguments appearing in the functions u_i. An overall factor of η^2 that results from the engineering dimension of the Hamiltonian has been removed. The Gaussian fixed point is found by insisting that the first term remains constant, which requires

$$\zeta^2 u_2^*(q^+, \eta \mathbf{q}_\perp) = u_2^*(q^+, \mathbf{q}_\perp) \,. \tag{66}$$

I have used the fact that u_2 actually depends on only one momentum other than the cutoff.

The solution to this equation is a monomial in $\mathbf{q}_\perp$ which depends on ζ,

$$u_2^*(q^+, \mathbf{q}_\perp) = f(q^+)(\mathbf{q}_\perp)^n \,, \quad \zeta = \left(\frac{1}{\eta}\right)^{(n/2)} \,. \tag{67}$$

The solution depends on our choice of n and to obtain the appropriate free particle dispersion relation we need to choose $n = 2$, so that

$$u_2^*(q^+, \mathbf{q}_\perp) = f(q^+)\mathbf{q}_\perp^2 \,, \quad \zeta = \frac{1}{\eta} \,. \tag{68}$$

$f(q^+)$ is allowed because the cutoff scale P^+ allows us to form the dimensionless variable q^+/P^+, which can enter the one-body operator. We will see that this happens in QED and QCD. Note that the constant in front of each four-point interaction becomes one, so that their scaling behavior is determined entirely by u_4. If we insist on transverse locality (which may be violated because we remove zero modes), we can expand u_4 in powers of its transverse momentum arguments, and discover powers of $\eta q_i^{\perp}$ in the transformed Hamiltonian. Since we are lowering the cutoff, $\eta < 1$, and each power of transverse momentum will be suppressed by this factor. This means increasing powers of transverse momentum are increasingly *irrelevant.*

I will not go through a complete derivation of the eigenoperators of the linearized approximation to the renormalization group transformation about the critical Gaussian fixed point, but the derivation is simple.[4] Increasing powers of transverse derivatives and increasing powers of creation and annihilation operators lead to increasingly irrelevant operators. The irrelevant operators are called 'non-renormalizable' in old-fashioned Feynman perturbation theory. Their magnitude decreases at an exponential rate as the cutoff is lowered, which means that they *increase* at an exponential rate as the cutoff is raised and produce increasingly large divergences if we try to follow their evolution perturbatively in this exponentially unstable direction.

The only relevant operator is the mass operator,

$$\mu^2 \int d\tilde{q}_1\, d\tilde{q}_2\, (16\pi^3)\delta^3(q_1 - q_2)\;\; a^\dagger(q_1)a(q_2)\,, \tag{69}$$

while the fixed point Hamiltonian is marginal (of course), and the operator in which $u_4 = \lambda$ (a constant) is marginal. A ϕ^3 operator would also be relevant.

The next logical step in a renormalization group analysis is to study the transformation to second order in the interaction, keeping the second-order corrections from the similarity transformation. I will compute the correction to u_2 to this order and refer the interested reader to Ref. 4 for more complicated examples.

The matrix element of the one-body operator between single particle states is

$$\langle p'|h|p\rangle = \langle 0|\; a(p')\; h\; a^\dagger(p)\; |0\rangle = (16\pi^3)\delta^3(p' - p)\; u_2(-p,p)\,. \tag{70}$$

Thus, we easily determine u_2 from the matrix element. It is easy to compute matrix elements between other states. We computed the matrix elements of the effective Hamiltonian generated by the similarity transformation when the cutoff is lowered in Eq. (57), and now we want to compute the second-order term generated by the four-point interactions above. There are additional corrections to u_2 at second order in the interaction if u_6, *etc.* are nonzero.

Before rescaling we find that the transformed Hamiltonian contains

$$\begin{aligned}(16\pi^3)&\delta^3(p' - p)\; \delta v_2(-p,p) = \\ &\frac{1}{3!}\int d\tilde{k}_1 d\tilde{k}_2 d\tilde{k}_3\; \theta(\Lambda_0 - |p^- - k_1^- - k_2^- - k_3^-|) \\ &\theta(|p^- - k_1^- - k_2^- - k_3^-| - \eta\Lambda_0)\; \frac{\langle p'|V|k_1,k_2,k_3\rangle\langle k_1,k_2,k_3|V|p\rangle}{p^- - k_1^- - k_2^- - k_3^-}\,,\end{aligned} \tag{71}$$

where $p^- = \mathbf{p}_\perp^2/p^+$, *etc.* One can readily verify that

$$\langle p'|V|k_1,k_2,k_3\rangle = (16\pi^3)\delta^3(p' - k_1 - k_2 - k_3)\; \Theta(p^- - k_1^- - k_2^- - k_3^-)\; u_4(-p',k_1,k_2,k_3)\,, \tag{72}$$

$$\langle k_1,k_2,k_3|V|p\rangle = (16\pi^3)\delta^3(p - k_1 - k_2 - k_3)\; \Theta(p^- - k_1^- - k_2^- - k_3^-)\; u_4(-k_1,-k_2,-k_3,p)\,. \tag{73}$$

This leads to the result,

$$\begin{aligned}\delta v_2(-p,p) = \frac{1}{3!}\int &d\tilde{k}_1 d\tilde{k}_2 d\tilde{k}_3\; (16\pi^3)\delta^3(p - k_1 - k_2 - k_3) \\ &\theta(\Lambda_0 - |p^- - k_1^- - k_2^- - k_3^-|)\; \theta(|p^- - k_1^- - k_2^- - k_3^-| - \eta\Lambda_0) \\ &\frac{u_4(-p,k_1,k_2,k_3)\; u_4(-k_1,-k_2,-k_3,p)}{p^- - k_1^- - k_2^- - k_3^-}\,.\end{aligned} \tag{74}$$

To obtain δu_2 from δv_2 we must rescale the momenta, the fields, and the Hamiltonian. The final result is

$$\delta u_2(-p,p) = \frac{1}{3!}\int d\tilde{k}_1 d\tilde{k}_2 d\tilde{k}_3\, (16\pi^3)\delta^3(p-k_1-k_2-k_3)$$
$$\theta\left(\frac{\Lambda_0}{\eta} - |p^- - k_1^- - k_2^- - k_3^-|\right)\theta(|p^- - k_1^- - k_2^- - k_3^-| - \Lambda_0)$$
$$\frac{u_4(p',k_1',k_2',k_3')\, u_4(-k_1',-k_2',-k_3',p')}{p^- - k_1^- - k_2^- - k_3^-}, \tag{75}$$

where $p^{+'} = p^+$, $p^{\perp'} = \eta p^\perp$, $k_i^{+'} = k_i^+$, and $k_i^{\perp'} = \eta k_i^\perp$.

Coupling Coherence

The basic mathematical idea behind coupling coherence was first formulated by Oehme, Sibold, and Zimmerman.[33] They were interested in field theories where many couplings appear, such as the standard model, and wanted to find some means of reducing the number of couplings. Wilson and I developed the ideas independently in an attempt to deal with the functions that appear in marginal and relevant light-front operators.[3]

The puzzle is how to reconcile our knowledge from covariant formulations of QCD that only one running coupling constant characterizes the renormalized theory with the appearance of new counterterms and functions required by the light-front formulation. What happens in perturbation theory when there are effectively an infinite number of relevant and marginal operators? In particular, does the solution of the perturbative renormalization group equations require an infinite number of independent counterterms (i.e., independent functions of the cutoff)? Coupling coherence provides the conditions under which a finite number of running variables determines the renormalization group trajectory of the renormalized Hamiltonian. To leading nontrivial orders these conditions are satisfied by the counterterms introduced to restore Lorentz covariance in scalar field theory and gauge invariance in light-front gauge theories. In fact, the conditions can be used to determine all counterterms in the Hamiltonian, including relevant and marginal operators that contain functions of longitudinal momentum fractions; and with no direct reference to Lorentz covariance, this symmetry is restored to observables by the resultant counterterms in scalar field theory.[4]

A coupling-coherent Hamiltonian is analogous to a fixed point Hamiltonian, but instead of reproducing itself exactly it reproduces itself in form with a limited number of independent running couplings. If g_Λ is the only independent coupling in a theory, in a coupling-coherent Hamiltonian *all other couplings are invariant functions of* g_Λ, $f_i(g_\Lambda)$. The couplings $f_i(g_\Lambda)$ depend on the cutoff only through their dependence on the running coupling g_Λ, and in general we demand $f_i(0) = 0$. This boundary condition on the dependent couplings is motivated in our calculations by the fact that it is the combination of the cutoff and the interactions that force us to add the counterterms we seek, so the counterterms should vanish when the interactions are turned off,

Let me start with a simple example in which there is a finite number of relevant and marginal operators *ab initio,* and use coupling coherence to discover when only one or two of these may independently run with the cutoff. In general such conditions are met only when an underlying symmetry exists.

Consider a theory in which two scalar fields interact,

$$V(\phi) = \frac{\lambda_1}{4!}\phi_1^4 + \frac{\lambda_2}{4!}\phi_2^4 + \frac{\lambda_3}{4!}\phi_1^2\phi_2^2\,. \tag{76}$$

Under what conditions will there be fewer than three independent running coupling constants? We can use a simple cutoff on Euclidean momenta, $q^2 < \Lambda^2$. Letting $t = \ln(\Lambda/\Lambda_0)$, the Gell-Mann–Low equations are

$$\frac{\partial\lambda_1}{\partial t} = 3\zeta\lambda_1^2 + \frac{1}{12}\zeta\lambda_3^2 + \mathcal{O}(2\text{ loop}) , \tag{77}$$

$$\frac{\partial\lambda_2}{\partial t} = 3\zeta\lambda_2^2 + \frac{1}{12}\zeta\lambda_3^2 + \mathcal{O}(2\text{ loop}) , \tag{78}$$

$$\frac{\partial\lambda_3}{\partial t} = \frac{2}{3}\zeta\lambda_3^2 + \zeta\lambda_1\lambda_3 + \zeta\lambda_2\lambda_3 + \mathcal{O}(2\text{ loop}) ; \tag{79}$$

where $\zeta = \hbar/(16\pi^2)$. It is not important at this point to understand how these equations are derived.

First suppose that λ_1 and λ_2 run separately, and ask whether it is possible to find $\lambda_3(\lambda_1, \lambda_2)$ that solves Eq. (79). To one-loop order this leads to

$$\left(3\lambda_1^2 + \frac{1}{12}\lambda_3^2\right)\frac{\partial\lambda_3}{\partial\lambda_1} + \left(3\lambda_2^2 + \frac{1}{12}\lambda_3^2\right)\frac{\partial\lambda_3}{\partial\lambda_2} = \frac{2}{3}\lambda_3^2 + \lambda_1\lambda_3 + \lambda_2\lambda_3 . \tag{80}$$

If λ_1 and λ_2 are independent, we can equate powers of these variables on each side of Eq. (80). If we allow the expansion of λ_3 to begin with a constant, we find a solution to Eq. (80) in which all powers of λ_1 and λ_2 appear. In this case a constant appears on the right-hand-sides of Eqs. (77) and (78), and there will be no Gaussian fixed points for λ_1 and λ_2. We are generally not interested in the possibility that a counterterm does not vanish when the canonical coupling vanishes, so we will simply discard this solution both here and below. We are interested in the conditions under which one variable ceases to be independent, and the appearance of such an arbitrary constant indicates that the variable remains independent even though its dependence on the cutoff is being reparameterized in terms of other variables.

If we do not allow a constant in the solution, we find that $\lambda_3 = \alpha\lambda_1 + \beta\lambda_2 + \mathcal{O}(\lambda^2)$. When we insert this in Eq. (80) and equate powers on each side, we obtain three coupled equations for α and β. These equations have no solution other than $\alpha = 0$ and $\beta = 0$, so we conclude that if λ_1 and λ_2 are independent functions of t, λ_3 will also be an independent function of t unless the two fields decouple.

Assume that there is only one independent variable, $\tilde{\lambda} = \lambda_1$, so that λ_2 and λ_3 are functions of $\tilde{\lambda}$. In this case we obtain two coupled equations,

$$\left(3\tilde{\lambda}^2 + \frac{1}{12}\lambda_3^2\right)\frac{\partial\lambda_2}{\partial\tilde{\lambda}} = 3\lambda_2^2 + \frac{1}{12}\lambda_3^2 , \tag{81}$$

$$\left(3\tilde{\lambda}^2 + \frac{1}{12}\lambda_3^2\right)\frac{\partial\lambda_3}{\partial\tilde{\lambda}} = \frac{2}{3}\lambda_3^2 + \tilde{\lambda}\lambda_3 + \lambda_2\lambda_3 . \tag{82}$$

If we again exclude a constant term in the expansions of λ_2 and λ_3 we find that the only non-trivial solutions to leading order are $\lambda_2 = \tilde{\lambda}$, and either $\lambda_3 = 2\tilde{\lambda}$ or $\lambda_3 = 6\tilde{\lambda}$. If $\lambda_3 = 2\tilde{\lambda}$,

$$V(\phi) = \frac{\tilde{\lambda}}{4!}\left(\phi_1^2 + \phi_2^2\right)^2 , \tag{83}$$

and we find the $O(2)$ symmetric theory. If $\lambda_3 = 6\tilde{\lambda}$,

$$V(\phi) = \frac{\tilde{\lambda}}{2\cdot 4!}\left[(\phi_1 + \phi_2)^4 + (\phi_1 - \phi_2)^4\right] , \tag{84}$$

and we find two decoupled scalar fields. Therefore, λ_2 and λ_3 do not run independently with the cutoff if there is a symmetry that relates their strength to λ_1.

The condition that a limited number of variables run with the cutoff does not only reveal symmetries broken by the regulator, it may also be used to uncover symmetries that are broken by the vacuum. I will not go through the details, but it is straightforward to show that in a scalar theory with a ϕ^3 coupling, this coupling can be fixed as a function of the ϕ^2 and ϕ^4 couplings only if the symmetry is spontaneously broken rather than explicitly broken.[3]

This example is of some interest in light-front field theory, because it is difficult to reconcile vacuum symmetry breaking with the requirements that we work with a trivial vacuum and drop zero-modes in practical non-perturbative Hamiltonian calculations. Of course, the only way that we can build vacuum symmetry breaking into the theory without including a nontrivial vacuum as part of the state vectors is to include symmetry breaking interactions in the Hamiltonian and work in the hidden symmetry phase. The problem then becomes one of finding all necessary operators without sacrificing predictive power. The renormalization group specifies what operators are relevant, marginal, and irrelevant; and coupling coherence provides one way to fix the strength of the symmetry-breaking interactions in terms of the symmetry-preserving interactions. This does not solve the problem of how to treat the vacuum in light-front QCD by any means, because we have only studied perturbation theory; but this result is encouraging and should motivate further investigation.

For the QED and QCD calculations discussed below, I need to compute the hamiltonian to second order, while the canonical coupling runs at third order. To determine the generic solution to this problem, I present an oversimplified analysis in which there are three coupled renormalization group equations for the independent marginal coupling (g), in addition to dependent relevant (μ) and irrelevant (w) couplings.

$$\mu_{m_{l+1}} = 4\mu_{m_l} + c_\mu g_l^2 + \mathcal{O}(g_l^3) , \tag{85}$$

$$g_{m_{l+1}} = g_{m_l} + c_g g_l^3 + \mathcal{O}(g_l^4) , \tag{86}$$

$$w_{m_{l+1}} = \frac{1}{4} w_{m_l} + c_w g_l^2 + \mathcal{O}(g_l^3) . \tag{87}$$

I assume that $\mu_l = \zeta g_l^2 + \mathcal{O}(g_l^3)$ and $w_l = \eta g_l^2 + \mathcal{O}(g_l^3)$, which satisfy the conditions of coupling coherence. Substituting into the renormalization group equations yields (dropping all terms of $\mathcal{O}(g_l^3)$),

$$\zeta g_{l+1}^2 = \zeta g_l^2 = 4\zeta g_l^2 + c_\mu g_l^2 , \tag{88}$$

$$\eta g_{l+1}^2 = \eta g_l^2 = \frac{1}{4}\eta g_l^2 + c_w g_l^2 . \tag{89}$$

The solutions are $\zeta = -\frac{1}{3}c_\mu$ and $\eta = \frac{4}{3}c_w$.

These are *exactly* the coefficients in a Taylor series expansion for $\mu(g)$ and $w(g)$ that reproduce themselves. This observation suggests an alternative way to find the coupling coherent Hamiltonian without explicitly setting up the renormalization group equations. Although this method is less general, we only need to find what operators must be added to the Hamiltonian so that at $\mathcal{O}(g^2)$ it reproduces itself, with the only change being the change in the specific value of the cutoff. Coupling coherence allows us to substitute the running coupling in this solution, but it is not until third order that we would explicitly see the coupling run. This is how I will fix the QED and QCD hamiltonians to second order.

QED and QCD Hamiltonians

In order to derive the renormalized QED and QCD Hamiltonians I must first list the canonical Hamiltonians. I follow the conventions of Brodsky and **Lepage.**[34] There is no need to be overly rigorous, because coupling coherence will fix any perturbative errors. I recommend the papers of Zhang and Harindranath[6] for a more detailed discussion from a different point of view. The reader who is not yet concerned with details can skip this section.

I will use $A^+ = 0$ gauge, and I drop zero modes. I use the Bjorken and Drell conventions for gamma matrices. The gamma matrices are

$$\gamma^0 = \begin{pmatrix} 1 & 0 \\ 0 & -1 \end{pmatrix} , \quad \gamma^k = \begin{pmatrix} 0 & \sigma_k \\ -\sigma_k & 0 \end{pmatrix} , \tag{90}$$

where σ_k are the Pauli matrices. This leads to

$$\gamma^+ = \begin{pmatrix} 1 & \sigma_3 \\ -\sigma_3 & -1 \end{pmatrix} , \quad \gamma^- = \begin{pmatrix} 1 & -\sigma_3 \\ \sigma_3 & -1 \end{pmatrix} . \tag{91}$$

Useful identities for many calculations are $\gamma^+\gamma^-\gamma^+ = 4\gamma^+$, and $\gamma^-\gamma^+\gamma^- = 4\gamma^-$.

The operator that projects onto the dynamical fermion degree of freedom is

$$\Lambda_+ = \frac{1}{2}\gamma^0\gamma^+ = \frac{1}{4}\gamma^-\gamma^+ = \frac{1}{2}\begin{pmatrix} 1 & \sigma_3 \\ \sigma_3 & 1 \end{pmatrix} , \tag{92}$$

and the complement projection operator is

$$\Lambda_- = \frac{1}{2}\gamma^0\gamma^- = \frac{1}{4}\gamma^+\gamma^- = \frac{1}{2}\begin{pmatrix} 1 & -\sigma_3 \\ -\sigma_3 & 1 \end{pmatrix} . \tag{93}$$

The Dirac spinors $u(p,\sigma)$ and $v(p,\sigma)$ satisfy

$$(\not p - m)u(p,\sigma) = 0 , \quad (\not p + m)v(p,\sigma) = 0 , \tag{94}$$

and,

$$\overline{u}(p,\sigma)u(p,\sigma') = -\overline{v}(p,\sigma)v(p,\sigma') = 2m\delta_{\sigma\sigma'} , \tag{95}$$

$$\overline{u}(p,\sigma)\gamma^\mu u(p,\sigma') = \overline{v}(p,\sigma)\gamma^\mu v(p,\sigma') = 2p^\mu\delta_{\sigma\sigma'} , \tag{96}$$

$$\sum_{\sigma=\pm\frac{1}{2}} u(p,\sigma)\overline{u}(p,\sigma) = \not p + m , \quad \sum_{\sigma=\pm\frac{1}{2}} v(p,\sigma)\overline{v}(p,\sigma) = \not p - m . \tag{97}$$

There are only two physical gluon (photon) polarization vectors, $\epsilon_{1\perp}$ and $\epsilon_{2\perp}$; but it is sometimes convenient (and dangerous once covariance and gauge invariance are violated) to use ϵ^μ, where

$$\epsilon^+ = 0 , \quad \epsilon^- = \frac{2\mathbf{q}_\perp \cdot \epsilon_\perp}{q^+} . \tag{98}$$

It is often possible to avoid using an explicit representation for $\epsilon_\perp$, but completeness relations are required,

$$\sum_\lambda \epsilon_\perp^\mu(\lambda)\epsilon_\perp^{*\nu}(\lambda) = -g_\perp^{\mu\nu} , \tag{99}$$

so that,

$$\sum_\lambda \epsilon^\mu(\lambda)\epsilon^{*\nu}(\lambda) = -g_\perp^{\mu\nu} + \frac{1}{q^+}(\eta^\mu q_\perp^\nu + \eta^\nu q_\perp^\mu) + \frac{\mathbf{q}_\perp^2}{(q^+)^2}\eta^\mu\eta^\nu , \tag{100}$$

where $\eta^+ = \eta^1 = \eta^2 = 0$ and $\eta^- = 2$. One often encounters diagrammatic rules in which the gauge propagator is written so that it looks covariant; but this is dangerous in loop calculations because such expressions require one to add and subtract terms that contain severe infrared divergences.

The QCD Lagrangian density is

$$\mathcal{L} = -\frac{1}{2}TrF^{\mu\nu}F_{\mu\nu} + \overline{\psi}\big(i\not{D} - m\big)\psi\,, \tag{101}$$

where $F^{\mu\nu} = \partial^\mu A^\nu - \partial^\nu A^\mu + ig[A^\mu, A^\nu]$ and $iD^\mu = i\partial^\mu - gA^\mu$. The SU(3) gauge fields are $A^\mu = \sum_a A_a^\mu T^a$, where T^a are one-half the Gell-Mann matrices, λ^a, and satisfy $Tr\ T^aT^b = 1/2\ \delta^{ab}$ and $[T^a, T^b] = if^{abc}T^c$.

The dynamical fermion degree of freedom is $\psi_+ = \Lambda_+\psi$, and this can be expanded in terms of plane wave creation and annihilation operators at $x^+ = 0$,

$$\psi_+^r(x) = \sum_{\sigma=\pm1/2}\int_{k^+>0}\frac{dk^+d^2k_\perp}{16\pi^3k^+}\Big[b^r(k,\sigma)u_+(k,\sigma)e^{-ik\cdot x} + d^{r\dagger}(k,\sigma)v_+(k,\sigma)e^{ik\cdot x}\Big]\,, \tag{102}$$

where these field operators satisfy

$$\Big\{\psi_+^r(x), \psi_+^{s\dagger}(y)\Big\}_{x^+=y^+=0} = \Lambda_+\delta_{rs}\delta^3(x-y)\,, \tag{103}$$

and the creation and annihilation operators satisfy

$$\Big\{b^r(k,\sigma), b^{s\dagger}(p,\sigma')\Big\} = \Big\{d^r(k,\sigma), d^{s\dagger}(p,\sigma')\Big\} = 16\pi^3k^+\delta_{rs}\delta_{\sigma\sigma'}\delta^3(k-p)\,. \tag{104}$$

The indices r and s refer to SU(3) color. In general, when momenta are listed without specification of components, as in $\delta^3(p)$, I am referring to p^+ and $\mathbf{p}_\perp$.

The transverse dynamical gluon field components can also be expanded in terms of plane wave creation and annihilation operators,

$$A_\perp^{ic}(x) = \sum_\lambda\int_{k^+>0}\frac{dk^+d^2k_\perp}{16\pi^3k^+}\Big[a^c(k,\lambda)\epsilon_\perp^i(\lambda)e^{-ik\cdot x} + a^{c\dagger}(k,\lambda)\epsilon_\perp^{i*}(\lambda)e^{ik\cdot x}\Big]\,. \tag{105}$$

The superscript i refers to the transverse dimensions x and y, and the superscript c is for SU(3) color. If required the physical polarization vector can be represented

$$\epsilon_\perp(\uparrow) = -\frac{1}{\sqrt{2}}(1,i)\,,\quad \epsilon_\perp(\downarrow) = \frac{1}{\sqrt{2}}(1,-i)\,. \tag{106}$$

The quantization conditions are

$$\Big[A_\perp^{ic}(x), \partial^+A_\perp^{jd}(y)\Big]_{x^+=y^+=0} = i\delta^{ij}\delta_{cd}\delta^3(x-y)\,, \tag{107}$$

$$\Big[a^c(k,\lambda), a^{d\dagger}(p,\lambda')\Big] = 16\pi^3k^+\delta_{\lambda\lambda'}\delta_{cd}\delta^3(k-p)\,. \tag{108}$$

The classical equations for $\psi_- = \Lambda_-\psi$ and A^- do not involve time-derivatives, so these variables can be eliminated in favor of dynamical degrees of freedom. This formally yields

$$\begin{aligned}\psi_- &= \frac{1}{i\partial^+}\Big[i\alpha_\perp\cdot\mathbf{D}_\perp + \beta m\Big]\psi_+\\ &= \tilde{\psi}_- - \frac{g}{i\partial^+}\alpha_\perp\cdot\mathbf{A}_\perp\psi_+\,,\end{aligned} \tag{109}$$

where the variable $\tilde{\psi}_-$ is defined on the second line to separate the interaction-dependent part of ψ_-; and

$$A^{a-} = \frac{2}{i\partial^+} i\partial_\perp \cdot \mathbf{A}_\perp^a + \frac{2igf^{abc}}{(i\partial^+)^2}\left\{(i\partial^+ A_\perp^{bi})A_\perp^{ci} + 2\psi_+^\dagger T^a \psi_+\right\}$$
$$= \tilde{A}^{a-} + \frac{2igf^{abc}}{(i\partial^+)^2}\left\{(i\partial^+ A_\perp^{bi})A_\perp^{ci} + 2\psi_+^\dagger T^a \psi_+\right\}, \tag{110}$$

where the variable $\tilde{A}^-$ is defined on the second line to separate the interaction-dependent part of A^-.

Given these replacements, we can follow a canonical procedure to determine the Hamiltonian. This path is full of difficulties that I ignore, because ultimately I will use coupling coherence to refine the definition of the Hamiltonian and determine the non-canonical interactions that are inevitably produced by the violation of explicit covariance and gauge invariance. For my purposes it is sufficient to write down a Hamiltonian that can serve as a starting point:

$$H = H_0 + V\,, \tag{111}$$

$$H_0 = \int d^3x \left\{Tr(\partial_\perp^i A_\perp^j \partial_\perp^i A_\perp^j) + \psi_+^\dagger(i\alpha_\perp^i \partial_\perp^i + \beta m)\psi_+\right\}$$
$$= \sum_{colors} \int \frac{dk^+ d^2k_\perp}{16\pi^3 k^+}\left\{\sum_\lambda \frac{\mathbf{k}_\perp^2}{k^+} a^\dagger(k,\lambda)a(k,\lambda)\right.$$
$$\left. + \sum_\sigma \frac{\mathbf{k}_\perp^2 + m^2}{k^+}\Big(b^\dagger(k,\sigma)b(k,\sigma) + d^\dagger(k,\sigma)d(k,\sigma)\Big)\right\}. \tag{112}$$

In the last line the 'self-induced inertias' (*i.e.*, one-body operators produced by normal-ordering V) are not included. It is difficult to regulate the field contraction encountered when normal-ordering in a manner exactly consistent with the cutoff regulation of contractions encountered later. Coupling coherence avoids this issue and produces the correct one-body counterterms with no discussion of normal-ordering required.

The interactions are complicated and are most easily written using the variables, $\tilde{\psi} = \tilde{\psi}_- + \psi_+$, and $\tilde{A}$, where $\tilde{A}^+ = 0$, $\tilde{A}^-$ is defined above, and $\tilde{A}_\perp^i = A_\perp^i$. Using these variables we have

$$V = \int d^3x \Bigg\{ g\bar{\tilde{\psi}}\gamma_\mu \tilde{A}^\mu \tilde{\psi} + 2g\, Tr\Big(i\partial^\mu \tilde{A}^\nu \big[\tilde{A}_\mu, \tilde{A}_\nu\big]\Big)$$
$$-\frac{g^2}{2} Tr\Big(\big[\tilde{A}^\mu, \tilde{A}^\nu\big]\big[\tilde{A}_\mu, \tilde{A}_\nu\big]\Big) + g^2 \bar{\tilde{\psi}}\gamma_\mu \tilde{A}^\mu \frac{\gamma^+}{2i\partial^+}\gamma_\nu \tilde{A}^\nu \tilde{\psi}$$
$$+\frac{g^2}{2}\bar{\psi}\gamma^+ T^a \psi \frac{1}{(i\partial^+)^2}\bar{\psi}\gamma^+ T^a \psi$$
$$-g^2 \bar{\psi}\gamma^+ \left(\frac{1}{(i\partial^+)^2}\big[i\partial^+ \tilde{A}^\mu, \tilde{A}_\mu\big]\right)\psi$$
$$+g^2\, Tr\left(\big[i\partial^+ \tilde{A}^\mu, \tilde{A}_\mu\big]\frac{1}{(i\partial^+)^2}\big[i\partial^+ \tilde{A}^\nu, \tilde{A}_\nu\big]\right) \Bigg\}. \tag{113}$$

The commutators in this expression are SU(3) commutators only. The potential algebraic complexity of calculations becomes apparent when one systematically expands every term in V and replaces:

$$\tilde{\psi}^- \to \frac{1}{i\partial^+}\Big(i\alpha_\perp \cdot \partial_\perp + \beta m\Big)\psi_+\,, \tag{114}$$

$$\tilde{A}^- \rightarrow \frac{2}{i\partial^+} i\partial_\perp \cdot \mathbf{A}_\perp \, ; \tag{115}$$

and then expands ψ_+ and $\mathbf{A}_\perp$ in terms of creation and annihilation operators. It rapidly becomes evident that one should avoid such explicit expansions if possible.

LIGHT-FRONT QED

In this section I will follow the strategy outlined in the first section to compute the positronium spectrum. I will detail the calculation through the leading order Bohr results[8] and indicate how higher order calculations proceed.[10, 11]

The first step is to compute a renormalized cutoff Hamiltonian as a power series in the coupling *e*. Starting with the canonical Hamiltonian as a 'seed,' this is done with the similarity renormalization group[17, 18] and coupling coherence.[3, 4] The result is an apparently unique perturbative series,

$$H_N^\Lambda = h_0 + e_\Lambda h_1 + e_\Lambda^2 h_2 + \cdots + e_\Lambda^N h_N \, . \tag{116}$$

Here e_Λ is the running coupling constant, and all remaining dependence on Λ in the operators h_i must be explicit. In principle I must also treat m_Λ, the electron running mass, as an independent function of Λ; but this will not affect the results to the order I compute here. We must calculate the Hamiltonian to a fixed order, and systematically improve the calculation later by including higher order terms.

Having obtained the Hamiltonian to some order in *e*, the next step is to split it into two parts,

$$H^\Lambda = \mathcal{H}_0^\Lambda + \mathcal{V}^\Lambda \, . \tag{117}$$

As discussed before, $\mathcal{H}_0^\Lambda$ must be accurately solved non-perturbatively, producing a zeroth order approximation for the eigenvalues and eigenstates. The greatest ambiguities in the calculation appear in the choice of $\mathcal{H}_0^\Lambda$, which requires one of science's most powerful computational tools, trial and error.

In QED and QCD I *assume* that all interactions in $\mathcal{H}_0^\Lambda$ preserve particle number, with all interactions that involve particle creation and annihilation in $\mathcal{V}^\Lambda$. This assumption is consistent with the original hypothesis that a constituent picture will emerge, but it should emerge as a valid approximation.

The final step before the loop is repeated, starting with a more accurate approximation for H^Λ, is to compute corrections from $\mathcal{V}^\Lambda$ in bound state perturbation theory. There is no reason to compute these corrections to arbitrarily high order, because the initial Hamiltonian contains errors that limit the accuracy we can obtain in bound state perturbation theory.

In this section I: (i) compute H^Λ to $\mathcal{O}(e^2)$, (ii) assume the cutoff is in the range $\alpha m^2 < \Lambda^2 < \alpha^2 m^2$ for non-perturbative analyses, (iii) include the most infrared singular two-body interactions in $\mathcal{H}_0^\Lambda$, and (iv) estimate the binding energy for positronium to $\mathcal{O}(\alpha^2 m)$.

Since $\mathcal{H}_0^\Lambda$ is assumed to include interactions that preserve particle number, the zeroth order positronium ground state will be a pure electron-positron state. We only need one- and two-body interactions; i.e., the electron self-energy and the electron-positron interaction. The canonical interactions can be found in Eq. (113), and the second-order change in the Hamiltonian is given in Eq. (57). The shift due to the bare electron mixing with electron-photon states to lowest order (see Figure 1) is

$$\delta\Sigma_p = e^2 \int \frac{dk_1^+ d^2k_{1\perp}}{16\pi^3} \frac{\theta(k_1^+)\theta(k_2^+)}{k_1^+ k_2^+} \frac{\overline{u}(p) D_{\mu\nu}(k_1)\gamma^\mu(\not{k}_2 + m)\gamma^\nu u(p)}{p^- - k_1^- - k_2^-} \, , \tag{118}$$

where,

$$k_2^+ = p^+ - k_1^+ \ , \quad \mathbf{k}_{2\perp} = \mathbf{p}_\perp - \mathbf{k}_{1\perp} \ , \tag{119}$$

$$k_i^- = \frac{k_{i\perp}^2 + m_i^2}{k_i^+} \ , \tag{120}$$

$$D_{\mu\nu}(k) = -g_{\perp\mu\nu} + \frac{k_\perp^2}{(k^+)^2}\eta_\mu\eta_\nu + \frac{1}{k^+}\Big(k_{\perp\mu}\eta_\nu + k_{\perp\nu}\eta_\mu\Big) \ , \tag{121}$$

$$\eta_\mu a^\mu = a^+ \ . \tag{122}$$

I have not yet displayed the cutoffs. To evaluate the integrals it is easiest to use Jacobi variables x and $\mathbf{s}$ for the relative electron-photon motion,

$$k_1^+ = xp^+ \ , \quad \mathbf{k}_{1\perp} = x\mathbf{p}_\perp + \mathbf{s} \ , \tag{123}$$

which implies

$$k_2^+ = (1-x)p^+ \ , \quad \mathbf{k}_{2\perp} = (1-x)\mathbf{p}_\perp - \mathbf{s} \ . \tag{124}$$

The second-order change in the electron self-energy becomes

$$\delta\Sigma_p = -\frac{e_\Lambda^2}{p^+}\int \frac{dx d^2s}{16\pi^3}\theta\left(y\Lambda_0^2 - \frac{s^2+x^2m^2}{x(1-x)}\right)\theta\left(\frac{s^2+x^2m^2}{x(1-x)} - y\Lambda_1^2\right)$$
$$\left(\frac{1}{s^2+x^2m^2}\right)\overline{u}(p,\sigma)\Big\{2(1-x)\not p - 2m +$$
$$\frac{\gamma^+}{p^+}\left[\frac{2s^2}{x^2} + \frac{2s^2}{1-x} + \frac{x(2-x)m^2}{1-x}\right]\Big\}u(p,\sigma) \ , \tag{125}$$

where $y = p^+/P^+$.

It is straightforward in this case to determine the self-energy required by coupling coherence. Since the electron-photon coupling does not run until third order, to second order the self-energy must exactly reproduce itself with $\Lambda_0 \to \Lambda_1$. For the self-energy to be finite we must assume that $\delta\Sigma$ reduces a positive self-energy, so that

$$\Sigma_{coh}^\Lambda(p) = \frac{e_\Lambda^2}{p^+}\int \frac{dx d^2s}{16\pi^3}\theta(yx-\epsilon)\theta\left(y\Lambda^2 - \frac{s^2+x^2m^2}{x(1-x)}\right)\left(\frac{1}{s^2+x^2m^2}\right)$$
$$\overline{u}(p,\sigma)\Big\{2(1-x)\not p - 2m + \frac{\gamma^+}{p^+}\left[\frac{2s^2}{x^2} + \frac{2s^2}{1-x} + \frac{x(2-x)m^2}{1-x}\right]\Big\}u(p,\sigma)$$
$$= \frac{e_\Lambda^2}{8\pi^2 p^+}\Big\{2y\Lambda^2\ln\left(\frac{y^2\Lambda^2}{(y\Lambda^2+m^2)\epsilon}\right) - \frac{3}{2}y\Lambda^2 + \frac{1}{2}\frac{ym^2\Lambda^2}{y\Lambda^2+m^2}$$
$$+ 3m^2\ln\left(\frac{m^2}{y\Lambda^2+m^2}\right)\Big\} + \mathcal{O}(\epsilon/y) \ . \tag{126}$$

I have been forced to introduce *a second cutoff,*

$$xp^+ > \epsilon P^+ \ , \tag{127}$$

because after the $\mathbf{s}$ integration is completed we are left with a logarithmically divergent x integration. Other choices for this second infrared cutoff are possible and lead to similar results. This second cutoff must be taken to zero and no new counterterms can be added to the Hamiltonian, so all divergences must cancel before it is taken to zero.

The electron and photon (quark and gluon) 'mass' operators, are a function of a longitudinal momentum scale introduced by the cutoff, and there is an exact scale invariance required by longitudinal boost invariance. Here I mean by 'mass operator' the one-body operator when the transverse momentum is zero, even though this does not agree with the free mass operator because it includes longitudinal momentum dependence. The cutoff violates boost invariance and the mass operator is required to restore this symmetry.

We must interpret this new infrared divergence, because we have no choice about whether it is in the Hamiltonian if we use coupling coherence. We can only choose between putting the divergent operator in $\mathcal{H}_0^\Lambda$ or in $\mathcal{V}^\Lambda$. I make different choices in QED and QCD, and the arguments are based on physics.

The *divergent* electron 'mass' is a complete lie. We encounter a term proportional to $e_\Lambda^2 \Lambda^2 \ln(1/\epsilon)/P^+$ when the scale is Λ; however, we can reduce this scale as far as we please in perturbation theory. Photons are massless, so the electron will continue to dress itself with small-x photons to arbitrarily small Λ. Since I believe that this divergent self-energy is exactly canceled by mixing with small-x photons, and that this mixing can be treated perturbatively in QED, I simply put the divergent electron self-energy in $\mathcal{V}^\Lambda$, which is treated perturbatively.

There are two time-ordered diagrams involving photon exchange between an electron with initial momentum p_1 and final momentum p_2, and a positron with initial momentum k_1 and final momentum k_2. These are shown in Figure 4, along with the instantaneous exchange diagram. Using Eq. (57), we find the required matrix element of δH,

$$\delta H = -\frac{e^2}{q^+} D_{\mu\nu}(q)\overline{u}(p_2,\sigma_2)\gamma^\mu u(p_1,\sigma_1)\overline{v}(k_1,\lambda_1)\gamma^\nu v(k_2,\lambda_2)$$
$$\theta(|q^+| - \epsilon P^+)\theta\left(\frac{\Lambda_0^2}{P^+} - |\, p_1^- - p_2^- - q^- \,|\right)\theta\left(\frac{\Lambda_0^2}{P^+} - |\, k_2^- - k_1^- - q^- \,|\right)$$
$$\left[\frac{\theta(|p_1^- - p_2^- - q^-| - \Lambda_1^2/\mathcal{P}^+)\ \theta(|p_1^- - p_2^- - q^-| - |k_2^- - k_1^- - q^-|)}{p_1^- - p_2^- - q^-}\right.$$
$$\left. + \frac{\theta(|k_2^- - k_1^- - q^-| - \Lambda_1^2/\mathcal{P}^+)\ \theta(|k_2^- - k_1^- - q^-| - |p_1^- - p_2^- - q^-|)}{k_2^- - k_1^- - q^-}\right]$$
$$\theta\left(\frac{\Lambda_1^2}{P^+} - |\, p_1^- + k_1^- - p_2^- - k_2^- \,|\right), \tag{128}$$

where $q^+ = p_1^+ - p_2^+$, $\mathbf{q}_\perp = \mathbf{p}_{1\perp} - \mathbf{p}_{2\perp}$, and $q^- = q_\perp^2/q^+$.

I have used the second cutoff on longitudinal momentum that I was forced to introduce when computing the change in the self-energy. We will see in the section on confinement that it is essential to include this cutoff everywhere consistently. In QED this point is not immediately important, because all infrared singular interactions, including the infrared divergent self-energy, are put in $\mathcal{V}^\Lambda$ and treated perturbatively. Divergences from higher orders in $\mathcal{V}^\Lambda$ cancel.[10, 11]

To determine the interaction that must be added to the Hamiltonian to maintain coupling coherence, we must again find an interaction that when added to δV reproduces itself with $\Lambda_0 \to \Lambda_1$ everywhere. The coupling coherent interaction generated by the first terms in δH are not uniquely determined at this order. There is some ambiguity because we can obtain coupling coherence either by having δH increase the strength of an operator by adding additional phase space strength, or we can have δH reduce the strength of an operator by subtracting phase space strength. The ambiguity is resolved in higher orders, so I will simply state the result. If an instantaneous photon-exchange interaction is present in H, δH cancels part of this marginal operator and increases the

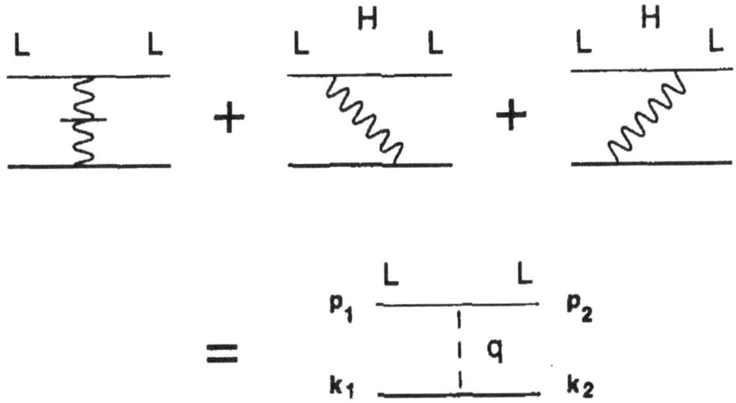

Figure 4. Effective two-body interaction between low energy constituents resulting from: (i) the canonical instantaneous exchange interaction, and (ii) the elimination of direct coupling between low energy two-body states and high energy states containing an additional gauge particle.

strength of a new photon-exchange interaction. This new interaction reproduces the effects of high energy photon exchange removed by the cutoff. The result is

$$V^{\Lambda}_{coh} = -\frac{e_{\Lambda}^2}{q^+} D_{\mu\nu}(q)\overline{u}(p_2,\sigma_2)\gamma^{\mu}u(p_1,\sigma_1)\overline{v}(k_1,\lambda_1)\gamma^{\nu}v(k_2,\lambda_2)$$
$$\theta(|q^+| - \epsilon P^+)\,\theta\left(\frac{\Lambda^2}{P^+} - |\,p_1^- + k_1^- - p_2^- - k_2^-\,|\right)$$
$$\left[\frac{\theta(|p_1^- - p_2^- - q^-| - \Lambda^2/\mathcal{P}^+)\;\theta(|p_1^- - p_2^- - q^-| - |k_2^- - k_1^- - q^-|)}{p_1^- - p_2^- - q^-} + \frac{\theta(|k_2^- - k_1^- - q^-| - \Lambda^2/\mathcal{P}^+)\;\theta(|k_2^- - k_1^- - q^-| - |p_1^- - p_2^- - q^-|)}{k_2^- - k_1^- - q^-}\right]. \tag{129}$$

This matrix element exactly reproduces photon exchange above the cutoff. The cutoff removes the direct coupling of electron-positron states to electron-positron-photon states whose energy differs by more than the cutoff, and coupling coherence dictates that the result of this mixing should be replaced by a direct interaction between the electron and positron. We could obtain this result by much simpler means at this order by simply demanding that the Hamiltonian produce the 'correct' scattering amplitude at $\mathcal{O}(e^2)$ with the cutoffs in place. Of course, this procedure requires us to provide the 'correct' amplitude, but this is easily done in perturbation theory.

V^{Λ}_{coh} is non-canonical, and we will see that it is responsible for producing the Coulomb interaction. We need some guidance to decide which irrelevant operators are most important. We find *a posteriori* that differences of external transverse momenta, and differences of external longitudinal momenta are both proportional to α. This allows us to identify the dominant operators by expanding in powers of these *implicit* powers of α. This indicates that it is the *most infrared singular* part of V_{coh} that is important. As explained above, this operator receives substantial strength only from the exchange of photons with small longitudinal momentum; so we expect inverse q^+ dependence to indicate 'strong' interactions between low energy pairs. So the part of

V_{coh} that is included in $\mathcal{H}_0^\Lambda$ is

$$\tilde{V}^\Lambda_{coh} = -e_\Lambda^2 \frac{q_\perp^2}{(q^+)^3} \overline{u}(p_2,\sigma_2)\gamma^+ u(p_1,\sigma_1)\overline{v}(k_1,\lambda_1)\gamma^+ v(k_2,\lambda_2)$$
$$\theta(|q^+| - \epsilon P^+)\, \theta\left(\frac{\Lambda^2}{P^+} - |\, p_1^- + k_1^- - p_2^- - k_2^- \,|\right)$$
$$\left[\frac{\theta(|p_1^- - p_2^- - q^-| - \Lambda^2/\mathcal{P}^+)\ \theta(|p_1^- - p_2^- - q^-| - |k_2^- - k_1^- - q^-|)}{p_1^- - p_2^- - q^-}\right.$$
$$\left. + \frac{\theta(|k_2^- - k_1^- - q^-| - \Lambda^2/\mathcal{P}^+)\ \theta(|k_2^- - k_1^- - q^-| - |p_1^- - p_2^- - q^-|)}{k_2^- - k_1^- - q^-}\right]$$
$$= -4e_\Lambda^2 \sqrt{p_1^+ p_2^+ k_1^+ k_2^+} \frac{q_\perp^2}{(q^+)^3} \delta_{\sigma_1\sigma_2}\delta_{\lambda_1\lambda_2}$$
$$\theta(|q^+| - \epsilon P^+)\, \theta\left(\frac{\Lambda^2}{P^+} - |\, p_1^- + k_1^- - p_2^- - k_2^- \,|\right)$$
$$\left[\frac{\theta(|p_1^- - p_2^- - q^-| - \Lambda^2/\mathcal{P}^+)\ \theta(|p_1^- - p_2^- - q^-| - |k_2^- - k_1^- - q^-|)}{p_1^- - p_2^- - q^-}\right.$$
$$\left. + \frac{\theta(|k_2^- - k_1^- - q^-| - \Lambda^2/\mathcal{P}^+)\ \theta(|k_2^- - k_1^- - q^-| - |p_1^- - p_2^- - q^-|)}{k_2^- - k_1^- - q^-}\right] . \tag{130}$$

The Hamiltonian is almost complete to second order in the electron-positron sector, and only the instantaneous photon exchange interaction must be added. The matrix element of this interaction is

$$V_{instant} = -e_\Lambda^2 \left(\frac{1}{q^+}\right)^2 \overline{u}(p_2,\sigma_2)\gamma^+ u(p_1,\sigma_1)\overline{v}(k_1,\lambda_1)\gamma^+ v(k_2,\lambda_2)$$
$$\times\, \theta\left(\frac{\Lambda^2}{P^+} - |\, p_1^- + k_1^- - p_2^- - k_2^- \,|\right)$$
$$= -4e_\Lambda^2 \sqrt{p_1^+ p_2^+ k_1^+ k_2^+} \left(\frac{1}{q^+}\right)^2 \delta_{\sigma_1\sigma_2}\delta_{\lambda_1\lambda_2}$$
$$\times\, \theta\left(\frac{\Lambda^2}{P^+} - |\, p_1^- + k_1^- - p_2^- - k_2^- \,|\right) . \tag{131}$$

The only cutoff that appears is the cutoff directly run by the similarity transformation that prevents the initial and final states from differing in energy by more than Λ^2/P^+.

This brings us to a final subtle point. Since there are no cutoffs in $V_{instant}$ that directly limit the momentum exchange, the matrix element diverges as $q^+ \rightarrow 0$. Consider $\tilde{V}_{coh}$ in this same limit,

$$\tilde{V}^\Lambda_{coh} \rightarrow 4e_\Lambda^2 \sqrt{p_1^+ p_2^+ k_1^+ k_2^+} \left(\frac{1}{q^+}\right)^2 \delta_{\sigma_1\sigma_2}\delta_{\lambda_1\lambda_2}$$
$$\times\, \theta\left(|\, q^- \,| - \frac{\Lambda^2}{P^+}\right) \theta\left(\frac{\Lambda^2}{P^+} - |\, p_1^- + k_1^- - p_2^- - k_2^- \,|\right) . \tag{132}$$

This means that as $q^+ \rightarrow 0$, V_{coh} partially screens $V_{instant}$, leaving the original operator multiplied by $\theta(\Lambda^2/P^+ - |q^-|)$. However, even after this partial screening, the matrix elements of the remaining part of $V_{instant}$ between bound states diverge and we must introduce the same infrared cutoff used for the self-energy to regulate these divergences.

This is explicitly shown in the section on confinement. However, all divergences from $V_{instant}$ are exactly canceled by the exchange of massless photons, which persists to arbitrarily small cutoff. This cancellation is exactly analogous to the cancellation of the infrared divergence of the self-energy, and will be treated in the same way. The portion of $V_{instant}$ that is not canceled by $\tilde{V}_{coh}$ will be included in $\mathcal{V}^\Lambda$, the perturbative part of the Hamiltonian. We will not encounter this interaction until we also include photon exchange below the cutoff perturbatively, so all infrared divergences should cancel in this bound state perturbation theory. I repeat that this is not guaranteed for arbitrary choices of $\mathcal{H}_0^\Lambda$, and we are not free to simply cancel these divergent interactions with counterterms because coupling coherence completely determines the Hamiltonian.

We now have the complete interaction that I include in $\mathcal{H}_0^\Lambda$. Letting $\mathcal{H}_0^\Lambda = h_0 + v_0^\Lambda$, where h_0 is the free hamiltonian, I add parts of $V_{instant}$ and V_{coh} to obtain

$$v_0^\Lambda = -4e_\Lambda^2 \sqrt{p_1^+ p_2^+ k_1^+ k_2^+}\,\delta_{\sigma_1\sigma_2}\delta_{\lambda_1\lambda_2}\theta\left(\frac{\Lambda^2}{\mathcal{P}^+} - |\, p_1^- + k_1^- - p_2^- - k_2^- \,|\right) \times$$
$$\Bigg\{\left[\frac{q_\perp^2}{(q^+)^3}\frac{1}{p_1^- - p_2^- - q^-} + \frac{1}{(q^+)^2}\right]\theta(|p_1^- - p_2^- - q^-| - \Lambda^2/\mathcal{P}^+)$$
$$\times\, \theta(|p_1^- - p_2^- - q^-| - |k_2^- - k_1^- - q^-|)$$
$$+\left[\frac{q_\perp^2}{(q^+)^3}\frac{1}{k_2^- - k_1^- - q^-} + \frac{1}{(q^+)^2}\right]\theta(|k_2^- - k_1^- - q^-| - \Lambda^2/\mathcal{P}^+)$$
$$\times\, \theta(|k_2^- - k_1^- - q^-| - |p_1^- - p_2^- - q^-|)\Bigg\}\,. \qquad (133)$$

In order to present an analytic analysis I will make assumptions that can be justified *a posteriori*. First I will assume that the electron and positron momenta can be arbitrarily large, but that in low-lying states their relative momenta satisfy

$$|\mathbf{p}_\perp - \mathbf{k}_\perp| \ll m\,, \qquad (134)$$

$$|p^+ - k^+| \ll p^+ + k^+\,. \qquad (135)$$

It is essential that the condition for longitudinal momenta not involve the electron mass, because masses have the scaling dimensions of transverse momenta and not longitudinal momenta. As above, I use p for the electron momenta and k for the positron momenta. To be even more specific, I will assume that

$$|\mathbf{p}_\perp - \mathbf{k}_\perp| \sim \alpha m\,, \qquad (136)$$

$$|p^+ - k^+| \sim \alpha(p^+ + k^+)\,. \qquad (137)$$

This allows us to use power counting to evaluate the perturbative strength of operators for small coupling, which may prove essential in the analysis of QCD.[1] Note that these conditions allow us to infer

$$|\mathbf{p}_{1\perp} - \mathbf{p}_{2\perp}| \sim \alpha m\,, \qquad (138)$$

$$|p_1^+ - p_2^+| \sim \alpha(p^+ + k^+)\,. \qquad (139)$$

Given these order of magnitude estimates for momenta, we can drastically simplify the free energies in the kinetic energy operator and the energy denominators in v_0^Λ. We can use transverse boost invariance to choose a frame in which

$$p_i^+ = y_i \mathcal{P}^+\,,\quad p_{i\perp} = \kappa_i\,,\quad k_i^+ = (1 - y_i)\mathcal{P}^+\,,\quad k_{i\perp} = -\kappa_i\,, \qquad (140)$$

so that

$$P^+(p_1^- - p_2^- - q^-) = \frac{\kappa_1^2+m^2}{y_1} - \frac{\kappa_2^2+m^2}{y_2} - \frac{(\kappa_1-\kappa_2)^2}{y_1-y_2}$$
$$= -4m^2(y_1-y_2) - \frac{(\kappa_1-\kappa_2)^2}{y_1-y_2} + \mathcal{O}(\alpha^2 m^2) . \qquad (141)$$

To leading order all energy denominators are the same. Each energy denominator is $\mathcal{O}(\alpha m^2)$, which is large in comparison to the binding energy we will find. This is important, because the bulk of the photon exchange that is important for the low energy bound state involves intermediate states that have larger energy than the differences in constituent energies in the region of phase space where the wave function receives most of its strength. This allows us to use a perturbative renormalization group to compute the dominant effective interactions.

There are similar simplifications for all energy denominators. After making these approximations we find that the matrix element of v_0^Λ is

$$v_0^\Lambda = 4e_\Lambda^2 \sqrt{y_1 y_2 (1-y_1)(1-y_2)}\; \theta\Big(4m^2|y_1-y_2| + \frac{(\kappa_1-\kappa_2)^2}{|y_1-y_2|} - \Lambda^2\Big)$$
$$\theta\Big(\Lambda^2 - 4\,|\,\kappa_1^2 + 4m^2(1-2y_1)^2 - \kappa_2^2 - 4m^2(1-2y_2)^2\,|\Big)$$
$$\frac{(\kappa_1-\kappa_2)^2}{(y_1-y_2)^2}\left[\frac{1}{4m^2(y_1-y_2)^2+(\kappa_1-\kappa_2)^2} - \frac{1}{(\kappa_1-\kappa_2)^2}\right]$$
$$= -16e_\Lambda^2 m^2 \sqrt{y_1 y_2 (1-y_1)(1-y_2)}\; \theta\Big(4m^2|y_1-y_2| + \frac{(\kappa_1-\kappa_2)^2}{|y_1-y_2|} - \Lambda^2\Big)$$
$$\theta\Big(\Lambda^2 - 4\,|\,\kappa_1^2 + 4m^2(1-2y_1)^2 - \kappa_2^2 - 4m^2(1-2y_2)^2\,|\Big)$$
$$\left[\frac{1}{4m^2(y_1-y_2)^2+(\kappa_1-\kappa_2)^2}\right] . \qquad (142)$$

In principle the electron-positron annihilation graphs should also be included at this order, but the resultant effective interactions do not diverge as $q^+ \to 0$, so I include such effects perturbatively in $\mathcal{V}^\Lambda$.

At this point we can complete the zeroth order analysis of positronium using the state,

$$|\Psi(P)\rangle = \sum_{\sigma\lambda} \int \frac{dp^+ d^2p_\perp}{16\pi^3 p^+} \frac{dk^+ d^2k_\perp}{16\pi^3 k^+} \sqrt{p^+k^+}\, 16\pi^3 \delta^3(P-p-k)$$
$$\phi(p,\sigma;k,\lambda) b^\dagger(p,\sigma) d^\dagger(k,\lambda)|0\rangle , \qquad (143)$$

where $\phi(p,\sigma;k,\lambda)$ is the wave function for the relative motion of the electron and positron, with the center-of-mass momentum being P. We need to choose the longitudinal momentum appearing in the cutoff, and I will use the natural scale P^+. The matrix element of $\mathcal{H}_0^\Lambda$ is

$$\langle\Psi(P)|\mathcal{H}_0^\Lambda|\Psi(P')\rangle = 16\pi^3\delta^3(P-P') \times$$
$$\Big\{ \int \frac{dy d^2\kappa}{16\pi^3}\Big[4m^2 + 4\kappa^2 + 4m^2(1-2y)^2\Big]|\phi(\kappa,y)|^2$$
$$- 16e^2m^2 \int \frac{dy_1 d^2\kappa_1}{16\pi^3}\frac{dy_2 d^2\kappa_2}{16\pi^3}\theta\Big(4m^2|y_1-y_2| + \frac{(\kappa_1-\kappa_2)^2}{|y_1-y_2|} - \Lambda^2\Big)$$
$$\theta\Big(\Lambda^2 - 4\,|\,\kappa_1^2 + 4m^2(1-2y_1)^2 - \kappa_2^2 - 4m^2(1-2y_2)^2\,|\Big)$$
$$\left[\frac{1}{4m^2|y_1-y_2|^2+(\kappa_1-\kappa_2)^2}\right]\phi^*(\kappa_2,y_2)\phi(\kappa_1,y_1)\Big\} . \qquad (144)$$

I have chosen a frame in which $P_\perp = 0$ and used the Jacobi coordinates defined above, and indicated only the electron momentum in the wave function since momentum conservation fixes the positron momentum. I have also dropped the spin indices because the interaction in $\mathcal{H}_0^\Lambda$ is independent of spin.

If we vary this expectation value subject to the constraint that the wave function is normalized we obtain the equation of motion,

$$\begin{aligned} M^2\phi(\kappa_1, y_1) &= (4m^2 - 4mE + E^2)\phi(\kappa_1, y_1) \\ &= \left[4m^2 + 4\kappa_1^2 + 4m^2(1-2y_1)^2\right]\phi(\kappa_1, y_1) \\ &\quad -16e^2m^2 \int \frac{dy_2 d^2\kappa_2}{16\pi^3}\theta\Big(4m^2|y_1 - y_2| + \frac{(\kappa_1-\kappa_2)^2}{|y_1-y_2|} - \Lambda^2\Big) \\ &\quad \theta\Big(\Lambda^2 - 4\mid \kappa_1^2 + 4m^2(1-2y_1)^2 - \kappa_2^2 - 4m^2(1-2y_2)^2 \mid\Big) \\ &\quad \left[\frac{1}{4m^2|y_1-y_2|^2 + (\kappa_1-\kappa_2)^2}\right]\phi(\kappa_2, y_2) \,. \end{aligned} \tag{145}$$

E is the binding energy, and we can drop the E^2 term since it will be $\mathcal{O}(\alpha^4)$.

I do not think that it is possible to solve this equation analytically with the cutoffs in place, and with the light-front kinematic constraints $-1 \le 1 - 2y_i \le 1$. In order to determine the binding energy to leading order, we need to evaluate the regions of phase space removed by the cutoffs.

If we want to find a cutoff for which the ground state is dominated by the electron-positron component of the wave function, we need the first cutoff to remove the important part of the electron-positron-photon phase space. Using the 'guess' that $|\kappa| = \mathcal{O}(\alpha m)$ and $1 - 2y = \mathcal{O}(\alpha)$, this requires

$$\Lambda^2 < \alpha m^2 \,. \tag{146}$$

On the other hand, we cannot allow the cutoff to remove the region of the electron-positron phase space from which the wave function receives most of its strength. This requires

$$\Lambda^2 > \alpha^2 m^2 \,. \tag{147}$$

While it is not necessary, the most elegant way to proceed is to introduce 'new' variables,

$$\kappa_i = k_{\perp i} \,, \tag{148}$$

$$y_i = \frac{1}{2} + \frac{k_z}{2\sqrt{\mathbf{k}_\perp^2 + k_z^2 + m^2}} \,. \tag{149}$$

This change of variables can be 'discovered' in a number of ways, but they basically take us back to equal time coordinates, in which both boost and rotational symmetries are kinematic after a nonrelativistic reduction.

For cutoffs that satisfy $\alpha m^2 > \Lambda^2 > \alpha^2 m^2$, Eq. (145) simplifies tremendously when all terms of higher order than α^2 are dropped. Using the scaling behavior of the momenta, and the fact that we will find E^2 is $\mathcal{O}(\alpha^4)$, Eq. (145) reduces to:

$$-E\phi(\mathbf{k}_1) = \frac{\mathbf{k}_1^2}{m}\phi(\mathbf{k}_1) - \alpha \int \frac{d^3k_2}{(2\pi)^3}\frac{1}{(\mathbf{k}_1^2 - \mathbf{k}_2^2)}\phi(\mathbf{k}_2) \,. \tag{150}$$

The step function cutoffs drop out to leading order, leaving us with the familiar non-relativistic Schrödinger equation for positronium in momentum space. The solution is

$$\phi(\mathbf{k}) = \frac{\mathcal{N}}{(\mathbf{k}^2 + mE)^2} \,, \tag{151}$$

$$E = \frac{1}{4}\alpha^2 m \,. \tag{152}$$

N is a normalization constant.

This is the Bohr energy for the ground state of positronium, and it is obvious that the entire nonrelativistic spectrum is reproduced to leading order.

Beyond this leading order result the calculations become much more interesting, and in any Hamiltonian formulation they rapidly become complicated. There is no analytic expansion of the binding energy in powers of α, since negative mass-squared states appear to signal vacuum instability when α is negative ; but our simple strategy for performing field theory calculations can be improved by taking advantage of the weak coupling expansion.[1]

We can expand the binding energy in powers of α, and as is well known we find that powers of $\log(\alpha)$ appear in the expansion at $\mathcal{O}(\alpha^5)$. We have taken the first step to generate this expansion by expanding the effective Hamiltonian in the explicit powers of α which appear in the renormalization group analysis. The next step is to take advantage of the fact that all bound state momenta are proportional to α, which allows us to expand each of the operators in the Hamiltonian in powers of momenta. The renormalization group analysis justifies an expansion in powers of transverse momenta, and the nonrelativistic reduction leads to an expansion in powers of longitudinal momentum differences. The final step is to regroup terms appearing in bound state perturbation theory. For example, when we compute the first order correction in bound state perturbation theory, we find all powers of α, and these must be grouped order-by-order with terms that appear at higher orders of bound state perturbation theory.

The leading correction to the binding energy is $\mathcal{O}(\alpha^4)$, and producing these corrections is a much more serious test of the renormalization procedure than the calculation shown above. To what order in the coupling must the Hamiltonian be computed to correctly reproduce all masses to $\mathcal{O}(\alpha^4)$? The leading error can be found in the electron mass itself. With the Hamiltonian given above, two-loop effects would show errors in the electron mass that are $\mathcal{O}(\alpha^2\Lambda^2)$. This would appear to present a problem for the calculation of the binding energy to $\mathcal{O}(\alpha^4)$, but remembering that the cutoff must be lowered so that $\alpha m^2 > \Lambda^2 > \alpha^2 m^2$, we see that the error in the electron mass is actually of $\mathcal{O}(\alpha^{3+\delta} m^2)$. This means that to compute masses correctly to $\mathcal{O}(\alpha^4)$ we would have to compute the Hamiltonian to $\mathcal{O}(\alpha^4)$, which requires a fourth-order similarity calculation for QED. Such a calculation has not yet been completed. However, if we compute the splitting between bound state levels instead, errors in the electron mass cancel and we find that the Hamiltonian computed to $\mathcal{O}(\alpha^2)$ issufficient.

In Ref. 10 we have shown that the fine structure of positronium is correctly reproduced when the first- and second-order corrections from bound state perturbation theory are added. This is a formidable calculation, because the exact Coulomb bound and scattering states appear in second-order bound state perturbation theory[¶]

A complete calculation of the Lamb shift in hydrogen would also require a fourth-order similarity calculation of the Hamiltonian; however, the dominant contribution to the Lamb shift that was first computed by Bethe[36] can be computed using a Hamiltonian determined to $\mathcal{O}(\alpha)$.[11] In this calculation a Bloch transformation was used rather than a similarity transformation because the Bloch transformation is simpler and small energy denominator problems can be avoided in analytic QED calculations.

The primary obstacle to using our light-front strategy for precision QED calculations is algebraic complexity. We have successfully used QED as a testing ground for

¶There is a trick which allows this calculation to be performed using only first-order bound state perturbation theory.[35] The trick basically involves using a Melosh rotation.

this strategy, but these calculations can be done much more conveniently using other methods. The theory for which we believe our methods are best suited is QCD.

LIGHT-FRONT QCD

This section relies heavily on the discussion of positronium, because we only require the QCD Hamiltonian determined to $\mathcal{O}(\alpha)$ to discuss a simple confinement mechanism which appears naturally in light-front QCD and to complete reasonable zeroth order calculations for heavy quark bound states. To this order the QCD Hamiltonian in the quark-antiquark sector is almost identical to the QED Hamiltonian in the electron-positron sector. Of course the QCD Hamiltonian differs significantly from the QED Hamiltonian in other sectors, and this is essential for justifying my choice of $\mathcal{H}_0^\Lambda$ for non-perturbative calculations.

The basic strategy for doing a sequence of (hopefully) increasingly accurate QCD bound state calculations is almost identical to the strategy for doing QED calculations. I use coupling coherence to find an expansion for H^Λ in powers of the QCD coupling constant to a finite order. I then divide the Hamiltonian into a non-perturbative part, $\mathcal{H}_0^\Lambda$, and a perturbative part, $\mathcal{V}^\Lambda$., The division is based on the physical argument that adding a parton in an intermediate state should require more energy than indicated by the free Hamiltonian, and that as a result these states will 'freeze out' as the cutoff approaches Λ_{QCD}. When this happens the evolution of the Hamiltonian as the cutoff is lowered further changes qualitatively, and operators that were consistently canceled over an infinite number of scales also freeze, so that their effects in the few parton sectors can be studied directly. A one-body operator and a two-body operator arise in this fashion, and serve to confine both quarks and gluons.

The simple confinement mechanism I outline is certainly not the final story, but it may be the seed for the full confinement mechanism. One of the most serious problems we face when looking for non-perturbative effects such as confinement is that the search itself depends on the effect. A candidate mechanism must be found and then shown to self-consistently produce itself as the cutoff is lowered towards Λ_{QCD}.

Once we find a candidate confinement mechanism, it is possible to study heavy quark bound states with little modification of the QED strategy. Of course the results in QCD will differ from those in QED because of the new choice of $\mathcal{H}_0^\Lambda$, and in higher orders because of the gluon interactions. When we move on to light quark bound states, it becomes essential to introduce a mechanism for chiral symmetry breaking.[7,1] I will discuss this briefly at the end of this section.

When we compute the QCD Hamiltonian to $\mathcal{O}(\alpha)$, several significant new features appear. First are the familiar gluon interactions. In addition to the many gluon interactions found in the canonical Hamiltonian, there are modifications to the instantaneous gluon exchange interactions, just as there were modifications to the electron-positron interaction. For example, a Coulomb interaction will automatically arise at short distances. In addition the gluon self-energy differs drastically from the photon self-energy.

The photon develops a self-energy because it mixes with electron-positron pairs, and this self energy is $\mathcal{O}(\alpha\Lambda^2/P^+)$. When the cutoff is lowered below $4m^2$, this mass term vanishes because it is no longer possible to produce electron-positron pairs. For all cutoffs the small bare photon self-energy is exactly canceled by mixing with pairs below the cutoff. I will not go through the calculation, but because the gluon also mixes with gluon pairs in QCD, the gluon self-energy acquires an infrared divergence, just as the electron did in QED. In QCD both the quark and gluon self-energies are proportional to $\alpha\Lambda^2 \ln(1/\epsilon)/P^+$, where ϵ is the secondary cutoff on parton longitudinal momenta introduced in the last section. This means that even when the primary cutoff Λ^2 is finite, the energy of a single quark or a single gluon is infinite, because we are supposed to let $\epsilon \rightarrow 0$. One can easily argue that this result is meaningless, because the relevant matrix elements of the Hamiltonian are not even gauge invariant; however, since we must live with a variational principle when doing Hamiltonian calculations, this result may be useful.

In QED I argued that the bare electron self-energy was a complete lie, because the bare electron mixes with photons carrying arbitrarily small longitudinal momenta to cancel this bare self-energy and produce a finite mass physical electron. However, in QCD there is no reason to believe that this perturbative mixing continues to arbitrarily small cutoffs. There are *no* massless gluons in the world. In this case, the free QCD Hamiltonian is a complete lie and *cannot be trusted* at low energies.

On the other hand, coupling coherence gives us no choice about the quark and gluon self-energies as computed in perturbation theory. These self-energies appear because of the behavior of the theory at extremely high energies. The question is not whether large self-energies appear in the Hamiltonian. The question is whether these self-energies are canceled by mixing with low energy multi-gluon states. I argue that this cancellation does not occur, and that the infrared divergent quark and gluon self-energies *should be included* in $\mathcal{H}_0^\Lambda$. The transverse scale for these energies is the running scale Λ, and over many orders of magnitude we should see the self-energies canceled by mixing. However, as the cutoff approaches Λ_{QCD}, I speculate that these cancellations cease to occur because perturbation theory breaks down and a mass gap between states with and without extra gluons appears.

But if the quark and gluon self-energies diverge, and the divergences cannot be canceled by mixing between sectors with an increasingly large number of partons, how is it possible to obtain finite mass hadrons? The parton-parton interaction also diverges, and the infrared divergence in the two-body interaction *exactly cancels* the infrared divergence in the one-body operator for color singlet states.

Of course, the cancellation of infrared divergences is not enough to obtain confinement. The cancellation is exact regardless of the relative motion of the partons in a color singlet state, and confinement requires a residual interaction. I will show that the $\mathcal{O}(\alpha)$ QCD Hamiltonian produces a logarithmic potential in both longitudinal and transverse directions. I will not discuss whether a logarithmic confining potential is 'correct,' but to the best of my knowledge there is no rigorous demonstration that the confining interaction is linear, and a logarithmic potential may certainly be of interest phenomenologically for heavy quark bound states.[37, 38] I would certainly be delighted if a better light-front calculation produces a linear potential, but this may not be necessary even for light hadron calculations.

The calculation of how the quark self-energy changes when a similarity transformation lowers the cutoff on energy transfer is almost identical to the electron self-energy calculation. Following the steps in the section on positronium, we find the one-body operator required by coupling coherence,

$$\Sigma_{coh}^{\Lambda}(p) = \frac{g^2 C_F}{8\pi^2 p^+}\left\{2y\Lambda^2 \ln\left(\frac{y^2\Lambda^2}{(y\Lambda^2+m^2)\epsilon}\right) - \frac{3}{2}y\Lambda^2 + \frac{1}{2}\frac{ym^2\Lambda^2}{y\Lambda^2+m^2} + 3m^2\ln\left(\frac{m^2}{y\Lambda^2+m^2}\right)\right\} + \mathcal{O}(\epsilon/y)\,, \tag{153}$$

where $C_F = (N^2-1)/(2N)$ for a SU(N) gauge theory.

The calculation of the quark-antiquark interaction required by coupling coherence is also nearly identical to the QED calculation. Keeping only the infrared singular parts of the interaction, as was done in QED,

$$\tilde{V}^{\Lambda}_{coh} = -4g_\Lambda^2 C_F \sqrt{p_1^+ p_2^+ k_1^+ k_2^+}\, \frac{q_\perp^2}{(q^+)^3} \delta_{\sigma_1\sigma_2}\delta_{\lambda_1\lambda_2}$$
$$\theta(|q^+| - \epsilon P^+)\, \theta\left(\frac{\Lambda^2}{P^+} - |\, p_1^- + k_1^- - p_2^- - k_2^- \,|\right)$$
$$\left[\frac{\theta(|p_1^- - p_2^- - q^-| - \Lambda^2/\mathcal{P}^+)\ \theta(|p_1^- - p_2^- - q^-| - |k_2^- - k_1^- - q^-|)}{p_1^- - p_2^- - q^-}\right.$$
$$\left. + \frac{\theta(|k_2^- - k_1^- - q^-| - \Lambda^2/\mathcal{P}^+)\ \theta(|k_2^- - k_1^- - q^-| - |p_1^- - p_2^- - q^-|)}{k_2^- - k_1^- - q^-}\right]. \tag{154}$$

The instantaneous gluon exchange interaction is

$$V^{\Lambda}_{instant} = -4g_\Lambda^2 C_F \sqrt{p_1^+ p_2^+ k_1^+ k_2^+} \left(\frac{1}{q^+}\right)^2 \delta_{\sigma_1\sigma_2}\delta_{\lambda_1\lambda_2}$$
$$\times\, \theta(|q^+| - \epsilon P^+)\, \theta\left(\frac{\Lambda^2}{P^+} - |\, p_1^- + k_1^- - p_2^- - k_2^- \,|\right). \tag{155}$$

Just as in QED the coupling coherent interaction induced by gluon exchange above the cutoff partially cancels instantaneous gluon exchange. For the discussion of confinement the part of V_{coh} that remains is not important, because it produces the short range part of the Coulomb interaction. However, the part of the instantaneous interaction that is not canceled is

$$\tilde{V}^{\Lambda}_{instant} = -4g_\Lambda^2 C_F \sqrt{p_1^+ p_2^+ k_1^+ k_2^+} \left(\frac{1}{q^+}\right)^2 \delta_{\sigma_1\sigma_2}\delta_{\lambda_1\lambda_2}$$
$$\times \theta\left(\frac{\Lambda^2}{P^+} - |\, p_1^- + k_1^- - p_2^- - k_2^- \,|\right)\theta(|p_1^+ - p_2^+| - \epsilon P^+)$$
$$\times \left[\theta(\frac{\Lambda^2}{P^+} - |p_1^- - p_2^- - q^-|)\ \theta(|p_1^- - p_2^- - q^-| - |k_2^- - k_1^- - q^-|) + \right.$$
$$\left. \theta(\frac{\Lambda^2}{P^+} - |k_2^- - k_1^- - q^-|)\ \theta(|k_2^- - k_1^- - q^-| - |p_1^- - p_2^- - q^-|)\right]. \tag{156}$$

Note that this interaction contains a cutoff that projects onto exchange energies below the cutoff, because the interaction has been screened by gluon exchange above the cutoffs. This interaction can become important at long distances, if parton exchange below the cutoff is dynamically suppressed. In QED I argued that this singular long range interaction is exactly canceled by photon exchange below the cutoff, because such exchange is not suppressed no matter how low the cutoff becomes. Photons are massless and experience no significant interactions, so they are exchanged to arbitrarily low energies as effectively free photons. This cannot be the case for gluons.

For the discussion of confinement, I will place only the most singular parts of the quark self-energy and the quark-antiquark interaction in $\mathcal{H}_0^\Lambda$. To see that all infrared divergences cancel and that the residual long range interaction is logarithmic, we can study the matrix element of these operators for a quark-antiquark state,

$$|\Psi(P)\rangle = \sum_{\sigma\lambda}\sum_{rs}\int \frac{dp^+ d^2p_\perp}{16\pi^3 p^+}\frac{dk^+ d^2k_\perp}{16\pi^3 k^+}\sqrt{p^+k^+}16\pi^3\delta^3(P-p-k)$$
$$\phi(p,\sigma,r;k,\lambda,s)\, b^{r\dagger}(p,\sigma)\, d^{s\dagger}(k,\lambda)|0\rangle\,, \tag{157}$$

where r and s are color indices and I will choose ϕ to be a color singlet and drop color indices. The cancellations we find do not occur for the color octet configuration. The matrix element is,

$$
\begin{aligned}
\langle\Psi(P)|\mathcal{H}_0^\Lambda|\Psi(P')\rangle &= 16\pi^3 P^+\delta^3(P-P') \times \\
&\Bigg\{ \int \frac{dy d^2\kappa}{16\pi^3}\left[\frac{g_\Lambda^2 C_F\Lambda^2}{2\pi^2 P^+}\ \ln(\frac{1}{\epsilon})\right]|\phi(\kappa,y)|^2 \\
&- \frac{4g_\Lambda^2 C_F}{P^+}\int \frac{dy_1 d^2\kappa_1}{16\pi^3}\frac{dy_2 d^2\kappa_2}{16\pi^3}\theta\left(\Lambda^2 - \left|\frac{\kappa_1^2+m^2}{y_1(1-y_1)} - \frac{\kappa_2^2+m^2}{y_2(1-y_2)}\right|\right) \\
&\times\Bigg[\theta\left(\Lambda^2 - \left|\frac{\kappa_1^2+m^2}{y_1} - \frac{\kappa_2^2+m^2}{y_2} - \frac{(\kappa_1-\kappa_2)^2}{|y_1-y_2|}\right|\right) \\
&\times\theta\left(\left|\frac{\kappa_1^2+m^2}{y_1} - \frac{\kappa_2^2+m^2}{y_2} - \frac{(\kappa_1-\kappa_2)^2}{|y_1-y_2|}\right| - \right. \\
&\qquad\left.\left|\frac{\kappa_2^2+m^2}{1-y_2} - \frac{\kappa_1^2+m^2}{1-y_1} - \frac{(\kappa_1-\kappa_2)^2}{|y_1-y_2|}\right|\right) \\
&+\theta\left(\Lambda^2 - \left|\frac{\kappa_2^2+m^2}{1-y_2} - \frac{\kappa_1^2+m^2}{1-y_1} - \frac{(\kappa_1-\kappa_2)^2}{|y_1-y_2|}\right|\right) \\
&\times\theta\left(\left|\frac{\kappa_2^2+m^2}{1-y_2} - \frac{\kappa_1^2+m^2}{1-y_1} - \frac{(\kappa_1-\kappa_2)^2}{|y_1-y_2|}\right| - \right. \\
&\qquad\left.\left|\frac{\kappa_1^2+m^2}{y_1} - \frac{\kappa_2^2+m^2}{y_2} - \frac{(\kappa_1-\kappa_2)^2}{|y_1-y_2|}\right|\right)\Bigg] \\
&\times\theta\Big(|\,y_1-y_2\,|-\epsilon\Big)\left(\frac{1}{y_1-y_2}\right)^2\phi^*(\kappa_2,y_2)\phi(\kappa_1,y_1)\Bigg\}.
\end{aligned}
\tag{158}
$$

Here I have chosen a frame in which the center-of-mass transverse momentum is zero, assumed that the longitudinal momentum scale introduced by the cutoffs is that of the bound state, and used Jacobi coordinates,

$$
p_i^+ = y_i P^+\ ,\quad p_{i\perp} = \kappa_i\ ;\quad k_i^+ = (1-y_i)P^+\ ,\quad k_{i\perp} = -\kappa_i\ . \tag{159}
$$

The first thing I want to do is show that the last term is divergent and the divergence exactly cancels the first term. My demonstration is not elegant, but it is straightforward. The divergence results from the region $y_1 \sim y_2$. In this region the second and third cutoffs restrict $(\kappa_1-\kappa_2)^2$ to be small compared to Λ^2, so we should change variables,

$$
Q = \frac{\kappa_1+\kappa_2}{2}\ ,\quad Y = \frac{y_1+y_2}{2}\ ;\quad q = \kappa_1-\kappa_2\ ,\quad y = y_1-y_2\ . \tag{160}
$$

Using these variables we can approximate the above interaction near $q = 0$ and $y = 0$. The double integral becomes

$$
\begin{aligned}
&\frac{-4g_\Lambda^2 C_F}{P^+}\int \frac{dY d^2Q}{16\pi^3}\frac{dy d^2q}{16\pi^3}\theta(1-Y)\theta(Y)|\phi(Q,Y)|^2 \\
&\qquad\theta\left(\Lambda^2 - \frac{q^2}{|y|}\right)\theta(|y|-\epsilon)\theta(\eta-|y|)\left(\frac{1}{y}\right)^2,
\end{aligned}
\tag{161}
$$

where η is an arbitrary constant that restricts $|y|$ from becoming large. Completing the q and y integration we get

$$
-\frac{g_\Lambda^2 C_F\Lambda^2}{2\pi^2 P^+}\ln\left(\frac{1}{\epsilon}\right)\int\frac{dY d^2Q}{16\pi^3}\theta(1-Y)\theta(Y)|\phi(Q,Y)|^2\ . \tag{162}
$$

The divergent part of this exactly cancels the first term on the right-hand side of Eq. (122). This cancellation occurs for any state, and this cancellation is unusual because it is between the expectation value of a one-body operator and the expectation value of a two-body operator. The cancellation resembles what happens in the Schwinger model and is easily understood. It results from the fact that a color singlet has no color monopole moment. If the state is a color octet the divergences are both positive and cannot cancel. Since the cancellation occurs in the matrix element, we can let $\epsilon \rightarrow 0$ before diagonalizing $\mathcal{H}_0^\Lambda$.

The fact that the divergences cancel exactly does not indicate that confinement occurs. This requires the residual interactions to diverge at large distances, which means small momentum transfer. Equivalently, we need the color dipole self-energy to diverge if the color dipole moment diverges because the partons separate to large distance.

My analysis of the residual interaction is neither elegant nor complete. For a more complete analysis see Ref.12. I show that the interaction is logarithmic in the longitudinal direction at zero transverse separation and logarithmic in the transverse direction at zero longitudinal separation, and I present the full angle-averaged interaction without derivation.

In order to avoid the infrared divergence, which is canceled, I compute spatial derivatives of the potential. First consider the potential in the longitudinal direction. Given a momentum space expression, set $x_\perp = 0$ and the Fourier transform of the longitudinal interaction requires the transverse momentum integral of the potential,

$$\frac{\partial}{\partial z}V(z) = \int \frac{d^3q}{(2\pi)^3} iq_z V(q_\perp, q_z) e^{iq_z z} \,. \tag{163}$$

We are interested only in the long range potential, so we can assume that q_z is arbitrarily small during the analysis and approximate the step functions accordingly.

For our interaction this leads to

$$\frac{\partial}{\partial x^-}V(x^-) = -4g_\Lambda^2 C_F P^+ \int \frac{dq^+ d^2q_\perp}{16\pi^3}\theta(P^+ - |q^+|)$$
$$\theta\Big(\frac{\Lambda^2}{P^+} - \frac{q_\perp^2}{|q^+|}\Big)\Big(\frac{1}{q^+}\Big)^2 (iq^+) e^{iq^+x^-} \,. \tag{164}$$

Completing the $q_\perp$ integration we have

$$\begin{aligned}
\frac{\partial}{\partial x^-}V(x^-) &= -\frac{ig_\Lambda^2 C_F \Lambda^2}{4\pi^2}\int dq^+ \theta(P^+ - |q^+|)\frac{q^+}{|q^+|} e^{iq^+x^-} \\
&= \frac{g_\Lambda^2 C_F \Lambda^2}{2\pi^2}\int_0^{P^+} dq^+ \sin(q^+x^-) \\
&= \frac{g_\Lambda^2 C_F \Lambda^2}{2\pi^2}\Big(\frac{1}{x^-} - \frac{\cos(P^+x^-)}{x^-}\Big) \\
&= \frac{g_\Lambda^2 C_F \Lambda^2}{2\pi^2}\frac{\partial}{\partial x^-}\ln(|x^-|) + \text{short range} \,.
\end{aligned} \tag{165}$$

To see that the term involving a cosine in the next-to-last line produces a short range potential, simply integrate it. At large $|x^-|$, which is the only place we can trust our approximations for the original integrand, this yields a logarithmic potential, as promised.

Next consider the potential in the transverse direction. Here we can set $x^- = 0$ and get

$$\frac{\partial}{\partial x_\perp^i} V(x_\perp) = -4g_\Lambda^2 C_F P^+ \int \frac{dq^+ d^2q_\perp}{16\pi^3} \theta(P^+ - |q^+|)$$
$$\theta\left(\frac{\Lambda^2}{P^+} - \frac{q_\perp^2}{|q^+|}\right)\left(\frac{1}{q^+}\right)^2 (iq_\perp^i) e^{i\mathbf{q}_\perp \cdot \mathbf{x}_\perp} . \qquad (166)$$

Here I have used the fact that the integration is dominated by small q^+ to simplify the integrand again. Completing the q^+ integration this becomes

$$\begin{aligned}\frac{\partial}{\partial x_\perp^i} V(x_\perp) &= -\frac{ig_\Lambda^2 C_F \Lambda^2}{2\pi^3} \int d^2q_\perp \frac{q_\perp^i}{q_\perp^2} e^{i\mathbf{q}_\perp \cdot \mathbf{x}_\perp} + \text{short range} \\ &= \frac{g_\Lambda^2 C_F \Lambda^2}{\pi^2} \frac{x_\perp^i}{x_\perp^2} \\ &= \frac{g_\Lambda^2 C_F \Lambda^2}{\pi^2} \frac{\partial}{\partial x_\perp^i} \ln(|\mathbf{x}_\perp|) . \end{aligned} \qquad (167)$$

Once again, this is the derivative of a logarithmic potential, as promised. The strength of the long-range logarithmic potential is not spherically symmetrical in these coordinates, with the potential being larger in the transverse than in the longitudinal direction. Of course, there is no reason to demand that the potential is rotationally symmetric in these coordinates.

The angle-averaged potential for two heavy quarks with mass M is[12]

$$\frac{g_\Lambda^2 C_F \Lambda^2}{\pi^2}\left(\ln(\mathcal{R}) - \mathrm{Ci}(\mathcal{R}) + 2\frac{\mathrm{Si}(\mathcal{R})}{\mathcal{R}} - \frac{(1-\cos(\mathcal{R}))}{\mathcal{R}^2} + \frac{\sin(\mathcal{R})}{\mathcal{R}} - \frac{5}{2} + \gamma_E\right) , \qquad (168)$$

where,

$$\mathcal{R} = \frac{\Lambda^2}{2M} r . \qquad (169)$$

I have assumed the longitudinal momentum scale in the cutoff equals the longitudinal momentum of the state.

The full potential is not naively rotationally invariant. Of course, the two body interaction in QED is also not rotationally invariant. The leading term in an expansion in powers of momenta yields the rotationally invariant Coulomb interaction, but higher order terms are not rotationally invariant. These higher order terms do not spoil rotational invariance of physical results because they must be combined with higher order interactions in the effective Hamiltonian, and with corrections involving photons at higher orders in bound state perturbation theory. There is no reason to expect, and no real need for a potential that is rotationally invariant; however, to proceed we need to decide which part of the potential must be treated non-perturbatively. The simplest reasonable choice is the angle-averaged potential, which is what we have used in heavy quark calculations.[12, 13]

Had we computed the quark-gluon or gluon-gluon interaction, we would find essentially the same residual long range two-body interaction in every Fock space sector, although the strengths would differ because different color operators appear. In QCD gluons have a divergent self-energy and experience divergent long range interactions with other partons if we use coupling coherence. In this sense, the assumption that gluon exchange below some cutoff is suppressed is consistent with the Hamiltonian that results from this assumption. To show that gluon exchange is suppressed when

$\Lambda \rightarrow \Lambda_{QCD}$, rather than some other scale (*i.e.*, zero as in QED), a non-perturbative calculation of gluon exchange is required. This is exactly the calculation bound state perturbation theory produces, and bound state perturbation theory suggests how the perturbative renormalization group calculation might be modified to generate these confining interactions self-consistently.

If perturbation theory, which produced this potential, continues to be reasonable, the long range potential will be exactly canceled in QCD just as it is in QED. We need this exact cancellation of new forces to occur at short distances and turn off at long distances, if we want asymptotic freedom to give way to a simple constituent confinement mechanism. At short distances the divergent self-energies and two-body interactions cannot be ignored, but they should exactly cancel pairwise if these divergences appear only when gluons are emitted and absorbed in immediately successive vertices, as I have speculated. The residual interaction must be analyzed more carefully at short distances, but in any case a logarithmic potential is less singular than a Coulomb interaction, which does not disturb asymptotic freedom. This is the easy part of the problem. The hard part is discovering how the potential can survive at any scale.

A perturbative renormalization group will not solve this problem. The key I have suggested is that interactions between quarks and gluons in the intermediate states required for cancellation of the potential will eventually produce a non-negligible energy gap. I am unable to detail this mechanism without an explicit calculation, but let me sketch a naive picture of how this might happen.

Focus on the quark-antiquark-gluon intermediate state, which mixes with the quark-antiquark state to screen the long range potential. The free energy of this intermediate state is always higher than that of the quark-antiquark free energy, as is shown using a simple kinematic argument.[1] However, if the gluon is massless, the energy gap between such states can be made arbitrarily small. As we use a similarity transformation to run a vertex cutoff on energy transfer, mixing persists to arbitrarily small cutoffs since the gap can be made arbitrarily small. Wilson has suggested using a gluon mass to produce a gap that will prevent this mixing from persisting as the cutoff approaches[1] Λ_{QCD}, and I am suggesting a slightly different mechanism.

If we allow two-body interactions to act in both sectors to all orders, even the Coulomb interaction can produce quark-antiquark and quark-antiquark-gluon bound states. In this respect QCD again differs qualitatively from QED because the photon cannot be bound and the energy gap is always arbitrarily small even when the electron-positron interaction produces bound states. If we assume that a fixed energy gap arises between the quark-antiquark bound states and the quark-antiquark-gluon bound states, and that this gap establishes the important scale for non-perturbative QCD, these states must cease to mix as the cutoff goes below the gap.

An important ingredient for any calculation that tries to establish confinement self-consistently is a *seed mechanism*, because it is possible that it is the confining interaction itself which alters the evolution of the Hamiltonian so that confinement can arise. I have proposed a simple seed mechanism whose perturbative origin is appealing because this allows the non-perturbative evolution induced by confinement to be matched on to the perturbative evolution required by asymptotic freedom.

I provide only a brief summary of our heavy quark bound state calculations,[12, 13] and refer the reader to the original articles for details. We follow the strategy that has been successfully applied to QED, with modifications suggested by the fact that gluons experience a confining interaction. We keep only the angle-averaged two-body interaction in $\mathcal{H}_0^\Lambda$, so the zeroth order calculation only requires the Hamiltonian matrix elements in the quark-antiquark sector to $\mathcal{O}(\alpha)$. All of the matrix elements are given above.

For heavy quark bound states[13] we can simplify the Hamiltonian by making a nonrelativistic reduction and solving a Schrödinger equation, using the potential in Eq. (168). We must then choose values for Λ, α, and M. These should be chosen differently for bottomonium and charmonium. The cutoff for which the constituent approximation works well depends on the constituent mass, as in QED where it is obviously different for positronium and muonium. In order to fit the ground state and first two excited states of charmonium, we use $\Lambda = 2.5GeV$, $\alpha = 0.53$, $M_c = 1.6GeV$. In order to fit these states in bottomonium we use $\Lambda = 4.9GeV$, $\alpha = 0.4$, and $M_b = 4.8GeV$.|| Violations of rotational invariance from the remaining parts of the potential are only about 10%, and we expect corrections from higher Fock state components to be at least of this magnitude for the couplings we use.

These calculations show that the approach is reasonable, but they are not yet very convincing. There are a host of additional calculations that must be done before the success of this approach can be judged, and most of them await the solution of several technical problems.

In order to test the constituent approximation and see that full rotational invariance is emerging as higher Fock states enter the calculation, we must be able to include gluons. If gluons are massless (which need not be true since the gluon mass is a relevant operator that must in principle be tuned), we cannot continue to employ a nonrelativistic reduction. In any case, we are primarily interested in light hadrons and must learn how to perform relativistic calculations to study them. The primary difficulty is that evaluation of matrix elements of the interactions given above involves high dimensional integrals which display little symmetry in the presence of cutoffs. Worse than this is the fact that the confinement mechanism requires cancellation of infrared divergences which have prevented us from using Monte Carlo methods to date.[31] These difficulties are avoided when a nonrelativistic reduction is made, but there is little more that we can do for which such a reduction is valid.

I conclude this section with a few comments on chiral symmetry breaking. While quark-quark, quark-gluon, and gluon-gluon confining interactions appear in the hamiltonian at $O(\alpha)$, chiral symmetry is not broken at any finite order in the perturbative expansion; so symmetry-breaking operators must be allowed to appear *ab initio*. Light-front chiral symmetry differs from equal-time chiral symmetry in several interesting and important aspects.[1,7,9] For example, quark masses in the kinetic energy operator do not violate light-front chiral symmetry; and the only operator in the canonical hamiltonian that violates this symmetry is a quark-gluon coupling linear in the quark mass. Again, the primary technical difficulty is the need for relativistic bound state calculations, and real progress on chiral symmetry cannot be made before we are able to perform relativistic calculations with constituent gluons. If transverse locality is maintained, simple renormalization group arguments indicate that chiral symmetry should be broken by a relevant operator in order to preserve well-established perturbative results at high energy. The only relevant operator that breaks chiral symmetry involves gluon emission and absorption, leading to the conclusion that the pion contains significant constituent glue. This situation is much simpler than what we originally envisioned,[1] because it does not require the addition of any operators that cannot be found in the canonical QCD hamiltonian; and we have long known that relevant operators must be tuned to restore symmetries.

|| There are several minor errors in Ref. 13, which are discussed in Ref.14. I have also chosen $\mathcal{P}^+ = P^+$, and in those papers $\mathcal{P}^+ = P^+/2$; and I have absorbed this change in a redefinition of Λ.

Acknowledgments

I would like to thank the staff of the Newton Institute for their wonderful hospitality, and the Institute itself for the support that made this exceptional school possible. I benefitted from many discussions at the school, but I would like to single out Pierre van Baal both for his patient assistance with all problems and for the many insights into QCD he has given me. I would also like to thank the many theorists who have helped me understand light-front QCD, including Edsel Ammons, Matthias Burkardt, Stan Glazek, Avaroth Harindranath, Tim Walhout, Wei-Min Zhang, and Ken Wilson. I especially want to thank Brent Allen, Martina Brisudová, Billy Jones, and Sergio Szpigel; who have been responsible for most of the recent progress in this program. This work has been partially supported by National Science Foundation grants PHY-9409042 and PHY-9511923.

REFERENCES

1. K. G. Wilson, T. S. Walhout, A. Harindranath, W.-M. Zhang, R. J. Perry and St. D. Glazek, Phys. Rev. **D49**, 6720 (1994).
2. P.A.M. Dirac, Rev. Mod. Phys. **21**, 392 (1949).
3. R. J. Perry and K. G. Wilson, Nucl. Phys. **B403**, 587 (1993).
4. R. J. Perry, Ann. Phys. **232**, 116 (1994).
5. R. J. Perry, Phys. Lett. **300B**, 8 (1993).
6. W. M. Zhang and A. Harindranath, Phys. Rev. **D48**, 4868; 4881; 4903 (1993).
7. D. Mustaki, *Chiral symmetry and the constituent quark model: A null-plane point of view,* preprint hep-ph/9404206.
8. R. J. Perry, *Hamiltonian Light-Front Field Theory and Quantum Chromodynamics.* Proceedings of *Hadrons '94,* V. Herscovitz and C. Vasconcellos, eds. (World Scientific, Singapore, 1995), and revised version hep-th/9411037.
9. K. G. Wilson and M. Brisudová, *Chiral Symmetry Breaking and Light-Front QCD.* Proceedings of the International Workshop on *Light-cone QCD,* S. Glazek, ed. (World Scientific, Singapore, 1995).
10. B. D. Jones, R. J. Perry and St. D. Glazek, Phys. Rev. **D55**, 6561 (1997).
11. B. D. Jones and R. J. Perry, Phys. Rev. **D55**, 7715 (1997).
12. M. Brisudová and R. J. Perry, Phys. Rev. **D 54**, 1831 (1996).
13. M. Brisudová, R. J. Perry and K. G. Wilson, Phys. Rev. Lett. **78**, 1227 (1997).
14. M. Brisudová, S. Szpigel and R. J. Perry, "Effects of Massive Gluons on Quarkonia in Light-Front QCD." Preprint hep-ph/9709479.
15. K. G. Wilson, *Light-Front QCD,* OSU internal report (1990).
16. A. Harindranath, *An Introduction to Light-Front Dynamics for Pedestrians,* Lectures given at the International School on Light-Front Quantization and Non-perturbative QCD, Ames, Iowa, 1996; preprint hep-ph/9612244.
17. St. D. Glazek and K.G. Wilson, Phys. Rev. **D48**, 5863 (1993).
18. St. D. Glazek and K.G. Wilson, Phys. Rev. **D49**, 4214 (1994).
19. St. D. Glazek and K. G. Wilson, *Asymptotic Freedom and Bound States in Hamiltonian Dynamics.* Preprint hep-th/9707028.
20. M. Burkardt, Phys. Rev. **D44**, 4628 (1993).
21. M. Burkardt, *Much Ado About Nothing: Vacuum and Re-normalization on the Light-Front,* Nuclear Summer School NUSS 97, preprint hep-ph/9709421.
22. S.J. Chang, R.G. Root, and T.M. Yan, Phys. Rev. **D7**, 1133 (1973).
23. K. G. Wilson, Rev. Mod. Phys. **47**, 773 (1975).
24. J. Schwinger, Phys. Rev. **125**, 397 (1962), Phys. Rev. **128**, 2425 (1962).
25. J. Lowenstein and A. Swieca, Ann. Phys. (N.Y.) **68**, 172 (1971).
26. H. Bergknoff, Nucl. Phys. **B122**, 215 (1977).
27. K. G. Wilson, Phys. Rev. **B445**, 140 (1965).
28. K. G. Wilson, Phys. Rev. **D2**, 1438 (1970).
29. K. G. Wilson and J. B. Kogut, Phys. Rep. **12C**, 75 (1974).

30. F.J. Wegner, Phys. Rev. **B5**, 4529 (1972); **B6**, 1891 (1972); F.J. Wegner, in *Phase Transitions and Critical Phenomena, C.* Domb and M.S. Green, Eds., Vol. 6 (Academic Press, London, 1976).
31. B. Allen, unpublished work.
32. F. Wegner, Ann. Physik **3**, 77 (1994).
33. R. Oehme, K. Sibold, and W. Zimmerman, Phys. Lett. **B147**, 115 (1984); R. Oehme and W. Zimmerman, Comm. Math. Phys. **97**, 569 (1985); W. Zimmerman, Comm. Math. Phys. **97**, 211 (1985); J. Kubo, K. Sibold, and W. Zimmerman, Nuc. Phys. **B259**, 331 (1985); R. Oehme, Prog. Theor. Phys. Supp. **86**, 215 (1986).
34. S. J. Brodsky and G. P. Lepage, in *Perturbative Quantum Chromodynamics,* A. H. Mueller, Ed. (World Scientific, Singapore, 1989).
35. M. Brisudová and R. J. Perry, Phys. Rev. **D54**, 6453 (1996).
36. H. A. Bethe, Phys. Rev. **72**, 339 (1947).
37. C. Quigg and J.L. Rosner, Phys. Lett. **71B**, 153 (1977).
38. Chris Quigg, *Realizing the Potential of Quarkonium.* Preprint hep-ph/9707493.

INSTANTONS IN QCD AND RELATED THEORIES

E. Shuryak[1,2]

[1]Isaac Newton Institute for Mathematical Sciences
20 Clarkson Road, Cambridge CB3 0EH, UK
[2]Physics Department*, State University of New York
Stony Brook, NY 11794, USA

INTRODUCTION

Overview

QCD is now more than 20 years old, and naturally it has reached a certain level of maturity. Perturbative QCD has been developed in great detail, with most hard processes calculated beyond leading order, compared to data and compiled in reviews and textbooks. However, the development of non-perturbative QCD has proven to be a much more difficult task.

This is hardly surprising. While perturbative QCD could build on the methods developed in the context of QED, strategies for dealing with the non-perturbative aspects of field theories first had to be developed. The gap between hadronic phenomenology on the one side and exactly (or partly) solvable field theories is still huge. While some fascinating discoveries (instantons among them) have been made, and important insights emerged from lattice simulations and hadronic phenomenology, a lot of work remains to be done in order to unite these approaches and truly understand the phenomena involved.

In order to develop a meaningful strategy we have to first see if the problem can be split into few separate steps: as usual in physics, this may be possible if some hierarchy of scales exist. And indeed, it was found that looking at the problem from several angles that it seems to be the case. Historically the first idea about the scales of the non-perturbative physics was suggested already in the early 60's. Nambu and Jona-Lasinio[1] (NJL) have suggested a model, inspired by an analogy to superconductivity. A hypothetical four-fermion interaction was introduced in order to explain the chiral symmetry breaking, creation of pions as Goldstone bosons, etc. The scale at which all this happens enters their model as a cut-off parameter, $\Lambda_{\chi SB} \sim 1$ GeV, and this effective theory (containing pions and constituent quarks) works in the gap *between* this scale and some lower hadronic scale, presumably related to confinement. Although the progress in understanding confinement is still very slow, the fundamental mechanism of

*Permanent address.

deviations from perturbative QCD and of the chiral symmetry breaking is now clarified in significant details.

The recent comprehensive review[2] should be consulted for details (only general ideas and some results will be described below). A qualitative picture of the vacuum suggested, is as follows: the non-perturbative gluon fields are concentrated mostly in *small-size* ($\rho \sim 1/3$ fm) topological fluctuations, the instantons. These have rather strong gauge fields, thus explaining why the parameter $\Lambda_{\chi SB} \sim 1$ GeV is so surprisingly large, why the glueballs are so heavy[3], why the scalar glueball is so small in size, why ordinary hadrons may have noticeable "intrinsic charm", etc. In the physics of light quarks it is more important that instantons play a role similar to that of a phonon exchange in a superconductor, providing the fundamental mechanism for the four-fermion attractive interaction in scalar and pseudoscalar channels. It breaks the chiral symmetry and creates the quark condensate, generates the quark "constituent mass" of about 400 MeV, explains the pion properties, etc.[4, 5]. In short, it does all what the NJL interaction is supposed to do.

Instantons also do this job better, because they generate vertices with the particular non-locality related to the instanton size: so many nasty questions related with the non-renormalizable NJL model have been naturally solved. Furthermore, a numerical but practical approach was discovered, allowing to include *all* orders in the instanton-induced 't Hooft effective interaction: this is something which in NJL-based calculations has never even attempted. Furthermore, this instanton-induced interaction is *not* identical to the NJL one. It drives the famous chiral anomaly and violates the chiral $U_A(1)$ symmetry[6], *quantitatively* explaining large η' mass.

These were the main conclusions of the applications of the simplest instanton-based vacuum model, the so called Random Instanton Liquid Model[7] (RILM), in the 1980's. It is a simple ad hoc model, an ensemble with random space-time positions and orientations, with two parameters: the mean size and spacing between instantons, $\rho \sim 1/3$ fm, $R \sim 1$ fm.

At the beginning of the 90's larger-scale studies of the RILM were made, and they have shown that it works far better than can be expected. The actual calculations are based on numerical evaluation of the quark propagator in a multi-instanton gauge field, which is then plugged in the correlator and averaged over the ensemble. The correlators were calculated in the region up to distances about 1.5 fm, where they are falling by 3-5 decades. Compared to phenomenology[10] and/or lattice results[11] whenever possible, the RILM *quantitatively* describes dozens of different hadronic correlation functions, including all main mesonic and baryonic channels. In fact, agreement is not only reached at small distances (where one can use a single-instanton approximation), but also at large ones, since the low-lying hadronic states (like pion, rho, nucleon, etc.) appear to be bound in the RILM and have reasonable masses and other parameters. In particular, it was found that instanton-induced effective interactions generate spin dependent forces between quarks: producing deeply bound nucleons (octet baryons), weakly bound[8] Δ (and decuplet members) and describing spin and orbital splittings in heavy-light hadrons[9] (D,B etc.). Although the RILM has no confinement, and thus in principle should have unphysical multi-quark cuts in its spectral density, in all major correlators it has so far been practically impossible to detect any sign of this problem. Thus, in practical terms, even the simplest instanton-based model, the RILM, outperformed other approaches (such as QCD sum rules) by a huge margin. Further studies of many other problems of hadronic physics in its framework seems justified now. Those may include various form-factors, wave and structure functions, etc.

The next step was development of a self-consistent Interacting Instanton Liquid

Model (IILM). It is a statistical model with a partition function similar to QCD, with gluonic fields represented by a superposition of instantons and the fermionic determinant calculated in the subspace of instanton zero modes. (It means that all orders in the 't Hooft effective interaction are included.) It is simple enough to be solved numerically, producing very good results. Furthermore, being self-consistent it satisfy the relevant theorems (such as chiral perturbation theory and other low energy theorems) exactly. One can also use it for studies of new phases at high temperature/density or for studies of other QCD-like theories.

Let us now mention some open problems. Why the large-size instantons are not there? How to treat consistently close instanton-antiinstanton pairs? What is the relation between instantons and the abelian-projected monopoles, and what is the reason for strong correlation between them found on the lattice?

Finally, to complete the overview, let me mention that recent progress in supersymmetric gauge theories with $N = 1$ due to Seiberg[58] and $N = 2$ due to Seiberg and Witten[59] had revealed that in this theory instantons seem to be the *only* dynamics there is! When the exact solution is expanded in inverse powers of Higgs VEV a at large a, one can see that (apart of a single perturbative one-loop log) all the power terms are just multi-instanton contributions. The first two orders have been calculated[60, 61, 62] and found to be exactly right, and all higher orders will probably be calculated soon. The challenge now is to develop the "instanton liquid" for $N = 1$ and $N = 2$ SUSY theories and bridge the gap between our phenomenological knowledge in QCD and recent theoretical advances in SUSY theories.

Scales of non-perturbative QCD

QCD is now firmly established as the correct theory of the strong interactions, but most calculations are performed in the perturbative domain - hard reactions with large momentum transfer. The limits of PQCD is a fundamental input we now consider. Although the value of the QCD scale parameter entering perturbative logs (and Particle Data Tables) is $\Lambda_{QCD} \sim 0.2$ GeV ~ 1 fm^{-1} (the exact value depending on the definition), the applicability of the parton model to hard processes is limited to reactions involving a scale of at least $Q \sim 1$ GeV. Where does this scale actually come from?

Another determination of the scale of non-perturbative effects in QCD is possible from below, using effective theories. The first estimate of this kind goes back to the Nambu and Jona-Lasinio model[1], inspired in the early 60's by the analogy between chiral symmetry breaking and superconductivity. It postulates a four-fermion interaction which, if it exceeds a certain strength, leads to quark condensation, the appearance of pions as Goldstone bosons, etc. The scale at which this interactions should be cut off (in order to reproduced known properties of the pion etc.) enters the model as an explicit *upper* cut-off on the momenta, $\Lambda_{\chi SB} \sim 1$ GeV.

It was further argued that the scales for chiral symmetry breaking and confinement are very different[12]: $\Lambda_{\chi SB} \gg \Lambda_{conf} \sim \Lambda_{QCD}$. In particular, it implies that constituent quarks (and pions) should have sizes smaller than those of typical hadrons, explaining the success of the non-relativistic quark model. This idea was developed by Georgi and Manohar[13] in a systematic fashion: they have considered the ratio of these two scales as the natural expansion parameter in chiral effective Lagrangians. They also argued that an effective theory of pions and constituent quarks is the natural description in the intermediate regime $\Lambda_{conf} < Q < \Lambda_{\chi SB}$, at which models of hadronic structure operate.

Although the progress in understanding confinement is still very slow, the fundamental mechanism of chiral symmetry breaking has been clarified in significant detail.

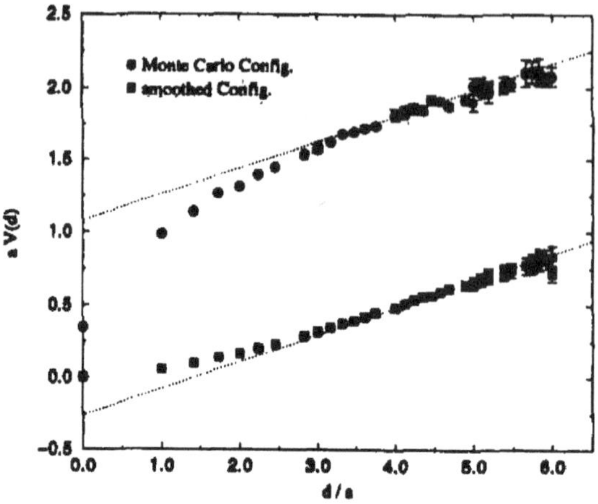

Figure 1. The $\bar{q}q$ potential before (above) and after (below) smoothening of the gauge fields, from Ref.[14].

It is the purpose of this review to explain the main ideas and results. For a more detailed exposition and technical details, the reader is invited to consult the more comprehensive work[2]. The picture of the vacuum that has emerged over the years has been termed the instanton liquid model[7].

The bottom line: it is the multiplicity, and especially the small size of instantons which explains why the strong interaction scale in QCD, $\Lambda_{\chi SB} \sim 1$ GeV, is so large as compared to Λ_{pert}, the place where perturbation theory blows up.

Hadronic structure

Understanding the structure of hadrons and the regularities in the spectrum of hadrons is an old problem. Indeed, the first two attempts to understand hadronic structure, the non-relativistic quark model and current algebra, predate the development of QCD. The former still provides a very simple and phenomenologically successful scheme to describe the spectrum and the properties of hadrons, based on the idea that hadrons are loosely bound composites of massive ($\sim$ 300 MeV) constituent quarks. On the other hand, current algebra led to the conclusion that the current quark masses that appear as symmetry breaking terms in the Lagrangian are tiny, on the order of a few MeV.

To reconcile these seemingly conflicting results was one of the major challenges to model builders. Another challenging task is to understand the relative importance of the various forces acting between quarks. Hadronic models incorporate (i) perturbative one-gluon-exchange interactions, (ii) confining (string) potentials, (iii) pion or other collective exchanges and (iv) quasi-local instanton-induced interactions (which can be either two or three body forces). To understand the interplay of these interactions on the basis of hadronic spectroscopy alone is a challenging task; below we will argue that the systematic study of hadronic correlation functions is a much more appropriate tool.

Here, we only want to make a few comments on the role of these forces. First, from the study of hard processes and the analysis of correlation functions (determined from experiment or on the lattice), it has become clear that the perturbative treatment of gluon fields at the hadronic scale ($1\ \mathrm{fm} = 5\ \mathrm{GeV}^{-1}$) is impossible*.

*Since gluons propagate perturbatively only at distances $\ll 1\ \mathrm{GeV}^{-1}$, they effectively decouple from low energy hadronic physics.

Second, one may argue that the numerical role of confinement for the masses of hadrons made of light quarks appears to be small. It has been known for a long time that the effective confining potential that provides an optimal description of low-lying hadrons in the constituent quark model is weaker than the one deduced from heavy quark ($\bar{c}c, \bar{b}b$) states. It is tempting to attribute this difference to the extended nature of constituent quarks, in contrast to the point-like c or b quark. Constituent quarks have form-factors and only interact with sufficiently soft gluonic modes. This can be simulated on the lattice by measuring the string tension after smoothing the gauge fields. An example is shown in Fig. 1. Although the string tension (the slope at large distances) is the same for both potentials, the smoothed potential is much smaller for $r \lesssim 3a \sim 1$ fm. This suggests that the string potential only affects the tail of hadronic wave functions. This can be seen more explicitly by comparing hadronic correlators and wave functions in full and cooled† configurations, see[15].

There is another indirect hint that confinement effects are not dominant in light hadrons. The MIT bag model[16] assumes that confinement leads to the creation of a bubble of the perturbative phase that contains valence quarks (and gluons) surrounded by the non-perturbative vacuum in which quarks are confined. Hadronic spectroscopy then determines the difference in vacuum energy between the two phases, the so-called bag pressure. It turns out that the MIT bag constant is more than an order of magnitude smaller than the phenomenologically determined non-perturbative energy density[17]. This implies that the non-perturbative vacuum fields (e.g. instantons) are not switched off inside hadrons, and can only be slightly modified.

Over the years many hadronic models have been developed. These models cure many of the defects of the simple MIT bag model, in particularly they preserve chiral symmetry. However, relation between these models and the underlying field theory remains unclear. This is hardly surprising: hadrons are collective excitations of the vacuum, like phonons in solids, so any model that attempts to reproduce their properties without addressing the structure of the ground state is meaningless.

The strategy we wish to follow here (motivated by the hierarchy of scales discussed above) is to treat the phenomenon of chiral symmetry breaking first. Starting from a satisfactory theory of constituent quarks and pions, one can go to the next scale and try to build a quantitative theory of hadrons. We will argue that the physics of chiral symmetry breaking is quite well understood, while the physical nature of confinement remains unclear. However, given the arguments above, this should provide a reasonable approximation for hadrons made of light quarks.

The importance of instantons in the context of chiral symmetry breaking is related to the fact that the Dirac operator has a chiral zero mode in the field of an instantons. These zero modes correspond to localized quark states that can become collective if many instantons and anti-instantons interact. The resulting de-localized state corresponds to the wave function of the quark condensate. In addition to that, instanton zero modes generate an effective four fermion interaction.

This brings us back to one of the oldest approaches to hadronic structure, the already mentioned NJL model. Nambu and Jona-Lasinio showed that a short-range attractive force between fermions, if strong enough, can rearrange the vacuum and the ground state becomes superconducting. There is a gap in the fermionic spectrum, corresponding to the constituent quark mass, and a non-zero quark condensate. The role of instantons in QCD (similar to that of phonon exchange in a superconductor) is to provide the fundamental mechanism for the four-fermion interaction. Instantons fix the strength of the interaction and its spin-isospin dependence. In addition to that,

†Cooling is a specific smoothing method designed to isolate instanton contributions.

asymptotic freedom and the finite size of instantons provide a natural cutoff (as opposed to the ad hoc cutoff in the NJL model).

How to calculate?

In this lectures there is no place for technical details, which may be found in the original papers. Rather we try to explain the logical structure and possible alternatives of the theory. For that reason, discussion of possible methods to calculate something with instantons is put into the introduction, with only results of the calculation reported below.

In the QCD partition function there are two types of fields, gluons and quarks, and so the first question one addresses is which integral to take first.

(i) One way is to eliminate *gluonic* degrees of freedom first. Physical motivation for that may be that gluonic states are heavy and an effective fermionic theory should be better suited to derive an effective low-energy theory. In the instanton framework, this approach was started by 't Hooft who discovered that instantons lead to new effective interactions between light quarks. We will present the explicit form of such four-fermion interaction for two-flavor QCD below.

It is a well-trotted path and one can follow it along the development for a similar four-fermion theory, the NJL model. One can do simple mean field or random field approximation (RPA) diagrams, and find the mean condensate and properties of the mesons. Unfortunately, it is difficult to do more. For example, baryons are states with three quarks, and using quasi-local four-fermion Lagrangians for the three body problem is technically a very difficult (although solvable) quantum mechanical problem. There were no attempts to sum more complicated diagrams.

(ii) The opposite strategy can be to do fermion integral first. It is a simple step, because they only enter quadratically, leading to a fermionic determinant. This is the way lattice people proceed. In the instanton approximation, it leads to the Interacting Instanton Liquid Model, defined by the following partition function:

$$Z = \sum_{N_+, N_-} \frac{1}{N_+! N_-!} \int \prod_i^{N_+ + N_-} [d\Omega_i \, d(\rho_i)] \exp(-S_{\text{int}}) \prod_f^{N_f} \det(\hat{D} + m_f) , \qquad (1)$$

describing a system of pseudo-particles interacting via the bosonic action and the fermionic determinant. Here $d\Omega_i = dU_i d^4 z_i d\rho_i$ is the measure in color orientation, position and size, associated with single instantons and $d(\rho)$ is the single instanton density $d(\rho) = dn_{I,\bar{I}}/d\rho d^4 z$ (see (21) furtheron).

The gauge interaction between instantons is approximated by a sum of pure two-body interaction $S_{\text{int}} = \frac{1}{2}\sum_{I \neq J} S_{\text{int}}(\Omega_{IJ})$. Genuine three body effects in the instanton interaction are not important as long as the ensemble is reasonably dilute. Implementation of this part of the interaction (quenched simulation) is quite analogous to usual statistical ensembles made of atoms.

As already mentioned, quark exchanges between instantons are included in the fermionic determinant. Finding a diagonal set of fermionic eigenstates of the Dirac operator is similar to what people are doing, e.g., in quantum chemistry when electron states for molecules are calculated. The difficulty of our problem is however much higher, because this set of fermionic states should be determined for *all* configurations which appear during the Monte-Carlo process.

If the set of fermionic states is however limited to the subspace of instanton zero modes, the problem becomes tractable numerically. Typical calculations in the IILM involved up to N~ 100 instantons/anti-instantons: it means that the determinants of

$N \times N$ matrices are involved. Such determinants can be evaluated by the ordinary workstation (and even PC these days) so quickly, that straightforward Monte Carlo simulation of IILM is possible in a matter of minutes. On the other hand, expanding the determinant in a sum of products of matrix elements, one can easily identify the sum of all closed loop diagrams up to order N in the 't Hooft interaction. Thus, in this way we take care of (practically) all orders!

Now that we know how to proceed, what should we actually calculate?

Correlators as a bridge between PQCD and hadronic physics

In a (relativistic) field theory, correlation functions of gauge invariant local operators are the proper tool to study the spectrum of the theory. The correlation functions can be calculated either from the physical states (mesons, baryons, glueballs) or in terms of the fundamental fields (quarks and gluons) of the theory. In the latter case, we have a variety of techniques at our disposal, ranging from perturbative QCD, the operator product expansion (OPE), to models of QCD and ultimately to lattice simulations. For this reason, correlation functions provide a bridge between hadronic phenomenology on the one side and the underlying structure of the QCD vacuum on the other side.

Loosely speaking, hadronic correlation functions play the same role for understanding the forces between quarks as the *NN* scattering phase did in the case of nuclear forces. In the case of quarks, however, confinement implies that we cannot define scattering amplitudes in the usual way. Instead, one has to focus on the behavior of gauge invariant correlation functions at short and intermediate distance scales. The available theoretical and phenomenological information about these functions was recently reviewed in[10].

Euclidean point-to-point correlation functions are defined as

$$\Pi_h(x) = \langle 0|j_h(x)j_h(0)|0\rangle, \tag{2}$$

where $j_h(x)$ is a local operator with the quantum numbers of a hadronic state h. We will concentrate on mesonic and baryonic currents of the type

$$j_{\rm mes}(x) = \delta^{ab}\bar{\psi}^a(x)\Gamma\psi^b(x), \tag{3}$$

$$j_{\rm bar}(x) = \epsilon^{abc}(\psi^{aT}(x)C\Gamma\psi^b(x))\Gamma'\psi^c(x). \tag{4}$$

Here, a, b, c are color indices and Γ, Γ' are isospin and Dirac matrices. In the following, we will only consider space-like separations $\tau = \sqrt{-x^2}$. In this case, correlation functions are exponentially suppressed rather than oscillatory at large distance.

Hadronic correlation functions can be written in terms of the spectrum and the coupling constants of the physical excitations with the quantum numbers of the current j_h. This connection is based on the standard dispersion relation

$$\Pi(Q^2) = \frac{1}{\pi}\int ds\,\frac{{\rm Im}\Pi(s)}{s+Q^2} + a_0 + a_1 Q^2 + \ldots, \tag{5}$$

(where $Q^2 = -q^2$ is the Euclidean momentum transfer and we have indicated possible subtraction constants a_i) and the spectral decomposition $(\rho(s) \equiv \frac{1}{\pi}{\rm Im}\Pi(s))$

$$\rho(s=-q^2) = (2\pi)^3\sum_n \delta^4(q-q_n)\langle 0|j_h(0)|n\rangle\langle n|j_h^\dagger(0)|0\rangle. \tag{6}$$

A spectral representation of the coordinate space correlation function is obtained by Fourier transforming (5),

$$\Pi(\tau) = \int ds\,\rho(s)D(\sqrt{s},\tau), \tag{7}$$

where $D(m,\tau) = mK_1(m\tau)/(4\pi^2\tau)$ is the Euclidean propagator of a scalar particle with mass m. Note that for large arguments the correlation function decays exponentially, $\Pi(\tau) \sim \exp(-m\tau)$, where the decay is governed by the lowest pole in the spectral function.

Correlation functions that involve quarks fields only (like the meson and baryon currents introduced above) can be expressed in terms of the full quark propagator. For an isovector meson current $j_{I=1} = \bar{u}\Gamma d$ (where Γ is only a Dirac matrix), the correlator only has a "one-loop" contribution ($S^{ab}(x,y)$ is the quark propagator)

$$\Pi_{I=1}(x) = \langle \mathrm{Tr}\left[S^{ab}(0,x)\Gamma S^{ba}(x,0)\Gamma\right]\rangle. \tag{8}$$

The averaging is performed over all gauge configurations, with the weight function $\det(i\not{D} + im)\exp(-S)$. Correlators of isosinglet meson currents $j_{I=0} = \frac{1}{\sqrt{2}}(\bar{u}\Gamma u + \bar{d}\Gamma d)$ receive an additional two-loop, or disconnected, contribution

$$\Pi_{I=0}(\tau) = \langle \mathrm{Tr}\left[S^{ab}(0,x)\Gamma S^{ba}(x,0)\Gamma\right]\rangle - 2\langle \mathrm{Tr}\left[S^{aa}(0,0)\Gamma\right] \mathrm{Tr}\left[S^{bb}(x,x)\Gamma\right]\rangle. \tag{9}$$

Analogously, baryon correlators can be expressed as vacuum averages of three quark propagators.

At short distance, asymptotic freedom implies that the correlation functions are determined by free quark propagation, $S^{ab}(x,y) = \delta_{ab}S_0(x-y)$. Using the free quark propagator $S_0(x) \sim (\gamma\cdot x)/x^4$, we conclude that mesonic and baryonic correlation functions at short distance behave as $\Pi_{mes} \sim 1/x^6$ and $\Pi_{bar} \sim 1/x^9$, respectively. Deviations from asymptotic freedom at intermediate distances can be studied using the operator product expansion (OPE). The OPE systematically accounts for the interaction with non-perturbative quark and gluon condensates. Historically, QCD sum rules based on the OPE played an important role in establishing the connection between hadronic phenomenology and the structure of the QCD vacuum.

There are a number of sources for phenomenological information about hadronic correlation functions[10]. The ideal situation is that the spectral function is determined from the optical theorem and an experimentally accessible cross section, as in the case of the vector-isovector (rho meson) channel from $\sigma(e^+e^- \to (I = 1 \text{ hadrons}))$ and in the a_1 channel from hadronic decays of the τ lepton, $\Gamma(\tau \to \nu_\tau + \text{hadrons})$. In most cases, the information is much more limited and only the contribution of a few resonances is known. The high energy behavior, of course, can always be extracted from perturbation theory. Ultimately, the best source of information about hadronic correlation functions comes from the lattice. At present, most lattice calculations use complicated nonlocal sources, but some studies of correlation functions of local sources have been reported[18, 11].

Concluding this section, we would like to emphasize that any model of the QCD vacuum or of hadronic structure should be compared to the available information on hadronic correlation functions. Only in this way can the structure of hadrons and the effective forces between quarks be connected to the structure the QCD vacuum.

INSTANTONS

Instantons and tunneling

In this section we remind the reader of a few basic facts about instantons. All of this material can be found in reviews or text books, see for example[19, 20, 2]. Instantons are solutions of the classical equations of motion in imaginary (or Euclidean) time. This means that, unlike solitons, instantons are not physical objects in Minkowski space, but tunneling paths that connect different vacua of the theory.

"Tunneling" phenomena in quantum mechanics were discovered by George Gamow in the late 20's in the context of alpha decay of nuclei. He explained one of the *mysteries of radioactivity*: why the alpha clusters, formed inside the nuclei, should hit its walls each 10^{-22} sec or so, for a time as long as sometimes billions of years, in order to get their freedom. He emphasized in his multiple lectures and popular books how unusual quantum tunneling is, being so orthogonal to classical mechanics. And still, it is possible to use *classical mechanics* for the description of classically forbidden phenomena!

An instanton is precisely a *classical path through a barrier, so* one can understandably raise the question of how it can work. In quantum mechanics, as in classical mechanics, energy is conserved and the Schrödinger equation can be understood as

$$E = \frac{\langle p^2 \rangle}{2m} + \langle V(x) \rangle \tag{10}$$

In the classically allowed region, $p^2 > 0$ and the wave function is a wave $\psi \sim \exp(ipx)$ with real p. However, if we are in the classically forbidden region $E < V$, so we must have negative kinetic energy or *imaginary p*.

In quantum mechanics this simply means that the wave function decays with distance, $\psi \sim \exp(-|p|x)$. If so, one is able to understand why tunneling is such a rare event. Now, if p is imaginary, why not try to interpret it as *motion in imaginary time*? Changing t to $\tau = it$, we have the new classical equation of motion, with an opposite sign for the force:

$$m\frac{d^2x}{d\tau^2} = -F = -\frac{d(-V)}{dx}. \tag{11}$$

This is the same as flipping the potential upside down! Thus classical paths describing tunneling do exist, though in imaginary time.

In quantum field theory instantons also appear as a sub-barrier path, being extremas of the (euclidean) QCD partition function

$$Z = \int DA_\mu \exp(-S) \prod_f^{N_f} \det\left(i\not{D} + im_f\right), \tag{12}$$

$$S = \frac{1}{4g^2}\int d^4x\, G^a_{\mu\nu}G^a_{\mu\nu}. \tag{13}$$

Here, S is the gauge field action and the determinant of the Dirac operator $i\not{D} = i\gamma_\mu(\partial_\mu - iA_\mu)$ accounts for the contribution of fermions. In the semi-classical approximation, we look for saddle points of the functional integral (12), i.e. configurations that minimize the classical action S. This means that saddle point configurations are solutions of the classical equations of motion.

These solutions can be found using the identity

$$S = \frac{1}{4g^2}\int d^4x \left[\pm G^a_{\mu\nu}\tilde{G}^a_{\mu\nu} + \frac{1}{2}\left(G^a_{\mu\nu} \mp \tilde{G}^a_{\mu\nu}\right)^2\right], \tag{14}$$

where $\tilde{G}_{\mu\nu} = \frac{1}{2}\epsilon_{\mu\nu\rho\sigma}G_{\rho\sigma}$ is the dual field strength tensor (the field strength tensor in which the roles of electric and magnetic fields are reversed). Since the last term is always positive, it is clear that the action is minimal if the field is (anti) self-dual

$$G^a_{\mu\nu} = \pm\tilde{G}^a_{\mu\nu}. \tag{15}$$

The action of a self-dual field configuration is determined by its topological charge

$$Q = \frac{1}{32\pi^2}\int d^4x\, G^a_{\mu\nu}\tilde{G}^a_{\mu\nu}. \tag{16}$$

From (14) we have $S = (8\pi^2|Q|)/g^2$. For finite action configurations, Q has to be an integer. The instanton is a solution with[21] $Q = 1$

$$A^a_\mu(x) = \frac{2\eta_{a\mu\nu}x_\nu}{x^2+\rho^2}, \tag{17}$$

where the 't Hooft symbol $\eta_{a\mu\nu}$ is defined by

$$\eta_{a\mu\nu} = \begin{cases} \epsilon_{a\mu\nu} & \mu,\nu = 1,2,3, \\ \delta_{a\mu} & \nu = 4, \\ -\delta_{a\nu} & \mu = 4. \end{cases} \tag{18}$$

and ρ is an arbitrary parameter characterizing the size of the instanton. The classical instanton solution has a number of degrees of freedom, known as collective coordinates. In addition to the size, the solution is characterized by the instanton position z_u and the color orientation matrix R^{ab} (corresponding to color rotations $A^a_u \to R^{ab}A^b_\mu$). A solution with topological charge $Q = -1$ can be constructed by replacing $\eta_{a\mu\nu} \to \bar{\eta}_{a\mu\nu}$, where $\bar{\eta}_{a\mu\nu}$ is defined by changing the sign of the last two equations in (18).

The physical meaning of the instanton solution becomes clear if we consider the classical Yang-Mills Hamiltonian (in the temporal guage, $A_0 = 0$)

$$H = \frac{1}{2g^2}\int d^3x\,(E_i^2 + B_i^2), \tag{19}$$

where E_i^2 is the kinetic and B_i^2 the potential energy term. The classical vacua corresponds to configurations with zero field strength. For non-abelian gauge fields this limits the gauge fields to be "pure gauge" $A_i = iU(\vec{x})\partial_i U(\vec{x})^\dagger$. Such configurations are characterized by a topological winding number n_W which distinguishes between gauge transformations U that are not continuously connected to the identity.

This means that there is an infinite set of classical vacua enumerated by an integer n. Instantons are tunneling solutions that connect the different vacua. They have potential energy $B^2 > 0$ and kinetic energy $E^2 < 0$, their sum being zero at any moment in time. Since the instanton action is finite, the barrier between the topological vacua can be penetrated, and the true vacuum is a linear combination $|\theta\rangle = \sum_n e^{in\theta}|n\rangle$, called the theta vacuum. In QCD, the value of θ is an external parameter. If $\theta \neq 0$ the QCD vacuum breaks CP invariance. Experimental limits on CP violation require * $\theta < 10^{-9}$.

The *rate* of tunneling between different topological vacua will from now on be called the "instanton density" n (in space-time). It is determined by the semi-classical (WKB) method. From the single instanton action one expects

$$P_{\text{tunneling}} \sim \exp(-8\pi^2/g^2). \tag{20}$$

The factor in front of the exponent can be determined by taking into account fluctuations $A_\mu = A^{cl}_\mu + \delta A_\mu$ around the classical instanton solution. This calculation was performed by 't Hooft in a classic paper[22]. The result is

$$dn_I = \frac{0.47\exp(-1.68N_c)}{(N_c-1)!(N_c-2)!}\left(\frac{8\pi^2}{g^2}\right)^{2N_c}\exp\left(-\frac{8\pi^2}{g^2(\rho)}\right)\frac{d^4z\,d\rho}{\rho^5}, \tag{21}$$

*The question why θ happens to be so small is known as the "strong CP problem". Most likely, the resolution of the strong CP problem requires physics outside QCD and we will not discuss it any further.

where $g^2(\rho)$ is the running coupling constant at the scale of the instanton size. Taking into account quantum fluctuations, the effective action depends on the instanton size. Using the one-loop beta function the result can be written as $dn_I/(d^4z) \sim d\rho\, \rho^{-5}(\rho\Lambda)^b$ where $b = (11N_c/3) = 11$ is the first coefficient of the beta function. Since b is a large number, small size instantons are strongly suppressed. On the other hand, there appears to be a divergence at large ρ (where the perturbative analysis based on the one loop beta function is not applicable).

The instanton liquid

Above we concentrated on the effects of a single instanton, but in order to understand the structure of the QCD vacuum we have to consider ensembles with a finite density (N/V) of instantons and anti-instantons. The repulsive interaction between them provides us with a method to deal with large size instantons. The instanton ensemble is described by the partition function in the space of collective coordinates (1). It is a typical statistical mechanics system, described by four-dimensional pseudo-particles interacting via the bosonic interaction $S_{\rm int}$ and the (non-local) fermion determinant $\det(i\not{D} + im_f)$. This means that we can evaluate the partition function using standard techniques of statistical mechanics.

The simplest method is the mean field approximation. If we ignore all correlations among instantons, the partition function can be represented in terms of a single instanton distribution[23] $\mu(\rho)$

$$Z \simeq \frac{1}{N_+!N_-!}\left(V\int d\rho\,\mu(\rho)\right)^{N_++N_-} . \tag{22}$$

For small ρ, the instanton distribution is given by the single instanton result (21) while for larger ρ interactions are important and the size distribution is modified. If the average effective interaction between instantons is repulsive, large instantons are suppressed and the distribution is peaked at some average size $\bar{\rho}$. Precisely how to define the interaction between very close (or very large) instantons is not very well understood.

Let us mention a few simple consequences of the instanton liquid model. First, we would like to consider the gluon condensate. The volume integral of $G^2_{\mu\nu}$ in the field of a single instanton is $32\pi^2$, independent of the size of the instanton. If we can ignore interactions between instantons, we get

$$\langle G^2_{\mu\nu}\rangle = (32\pi^2)\left(\frac{N}{V}\right). \tag{23}$$

Due to tunneling, the energy density of the QCD vacuum is less than that of the perturbative vacuum (which we take to be zero). The classical energy density inside an instanton is zero everywhere (this is true for any self-dual field configuration), but quantum mechanically this result is modified

$$\epsilon = -\frac{b}{4}\left(\frac{N}{V}\right). \tag{24}$$

where b is the first coefficient of the beta function. This means that the energy density of the non-perturbative vacuum is about 1 $\mathrm{GeV/fm^3}$ lower than in the perturbative one. (Note that it is about 20 time larger than the original MIT bag constant!)

The topological susceptibility is related to fluctuations of the topological charge. The average topological charge $\langle Q\rangle = \langle (N_+ - N_-)\rangle$ vanishes. In pure gauge theory, the instanton ensemble is fairly random and local fluctuations of the number of pseudo-particles are Poissonian. This means that $(N_+ - N_-) \sim \sqrt{N}$ and

$$\chi_{top} = \lim_{V\to\infty} \frac{\langle Q^2 \rangle}{V} = \lim_{V\to\infty} \frac{\langle (N_+ - N_-)^2 \rangle}{V} \simeq \left(\frac{N}{V}\right). \tag{25}$$

In the presence of light quarks, the situation is more complicated. Light quarks lead to correlations between instantons and anti-instantons and the topological charge is screened, $\chi_{top} \simeq m\langle \bar{q}q \rangle$, and vanishes in the chiral limit.

Phenomenology of the instanton liquid

The first attempt to estimate the instanton density phenomenologically was made in[24]. These authors used the value of the gluon condensate $\langle 0|(G^a_{\mu\nu})^2|0\rangle \simeq 0.5\,\text{GeV}^4$ obtained from a QCD sum rule analysis of the charmonium spectrum[25]. If the gluon condensate is dominated by (weakly interacting) instantons, then we can estimate their density from (23). This estimate provides an *upper limit* for the instanton density

$$\left(\frac{N}{V}\right) \leq \frac{1}{32\pi^2} \langle (G^a_{\mu\nu})^2 \rangle \simeq 1\,\text{fm}^{-4}. \tag{26}$$

A similar estimate can be obtained from the (pure gauge) topological susceptibility. The value of χ_{top} in quenched QCD can be estimated from the Witten-Veneziano relation[26, 27]

$$\chi_{top} = 2 f_\pi^2 (m_\eta^2 + m_{\eta'}^2 - 2m_K^2) = (180\,\text{MeV})^4. \tag{27}$$

Using the result (25), we again get $(N/V) \simeq 1\ \text{fm}^{-4}$.

The other important parameter characterizing the instanton ensemble is the typical instanton *size*. If the total tunneling rate can be calculated from the semi-classical 't Hooft formula, then we can estimate the critical size by determining the maximum size up to which the rate has to be integrated in order to reproduce the phenomenological instanton density

$$\int_0^{\rho_{max}} d\rho\, n_{scl}(\rho) = n_{phen} \tag{28}$$

Using $n_{phen} = 1\ \text{fm}^{-4}$, Shifman et al. concluded that $\rho_{max} \simeq 1$ fm. This is a very pessimistic result, because it implies that there are no individual instantons and that their action is not large, so the semi-classical approximation is useless.

However, if instanton interact, the situation may be different. The role of interactions can also be estimated using the value of the gluon condensate

$$dn = dn_{scl} \left[1 + \frac{\pi^4 \rho^4}{2g^4} \langle (G^a_{\mu\nu})^2 \rangle + \ldots \right]. \tag{29}$$

Using the canonical value of the gluon condensate, we see that for $\rho > 0.2$ fm the interaction of instantons with vacuum fields (of whatever origin) is not negligible. Also, we observe that interactions lead to a tunneling rate that grows faster than the semi-classical rate. Based on this result, we[4] suggested that the typical instanton size is significantly smaller,

$$\rho_{max} \sim 1/3\,\text{fm}. \tag{30}$$

If the average instanton is indeed small, we obtain a completely different picture of the QCD vacuum:

1. Since the instanton size is significantly smaller than the typical separation R between instantons, $\rho/R \sim 1/3$, the vacuum is fairly dilute. The fraction of spacetime occupied by strong fields is only a few per cent.

2. The fields inside the instanton are very strong $G_{\mu\nu} \gg \Lambda^2$. This means that the semi-classical approximation is valid, and the typical action is large

$$S_0 = 8\pi^2/g^2(\rho) \sim 10 - 15 \gg 1. \tag{31}$$

Higher order corrections are proportional to $1/S_0$ and presumably small.

3. Instantons retain their individuality and are not destroyed by interactions. From the dipole formula, one can estimate

$$|\delta S_{\text{int}}| \sim (2 - 3) \ll S_0. \tag{32}$$

4. Nevertheless, interactions are important for the structure of the instanton ensemble, since

$$\exp|\delta S_{\text{int}}| \sim 20 \gg 1. \tag{33}$$

This implies that interactions have a significant effect on correlations among instantons, the instanton ensemble in QCD is not a dilute gas, but an interacting liquid.

Improved estimates of the instanton size can be obtained from phenomenological applications of instantons. The average instanton size determines the structure of chiral symmetry breaking, in particular the values of the quark condensate, the pion mass, its decay constant and its form factor. We will discuss these observables in more detail below.

The instanton liquid is "distilled" from the lattice

Instantons start emerging from the "fog" of quantum fluctuations, when lattice configurations were subjected to operations making the gauge field more smooth. The most popular one known as "cooling" rapidly kills the short-wavelength quantum noise and leads to a local minimum, a kind of classical content of the configuration. In Fig. 2 from[15] one can see how it works.

After counting the instantons, and transferring the resulting density in absolute units†, these authors have concluded‡ that the instanton density in *quenched* QCD is about $1.3\text{-}1.6\ \text{fm}^{-3}$.

The average size of an instanton[4] $\rho \sim 1/3$ fm has been confirmed qualitatively many years ago by[30], and now more quantitative measurements[15] give $\rho = .35$ fm. Furthermore, the *shape* of the instanton size distribution function, $d(\rho)$, was recently measured on the lattice[31] (for pure $SU(2)$ theory). Their results are compared in Fig. 3. The agreement is best for large instanton sizes and it is not good for small

†The units are fixed by the measurement of the ρ meson mass and demanding it to be equal to the experimental value.

‡Unfortunately, the cooling method has some systematic errors. If the simplest Wilson action is used, instantons gradually shrink and finally "drop through the lattice". More sophisticated actions prevent that and make lattice instantons stable. Another approach recently developed is called the "inverse blocking": it allows to recover instantons with a radius as small as $.8a$.

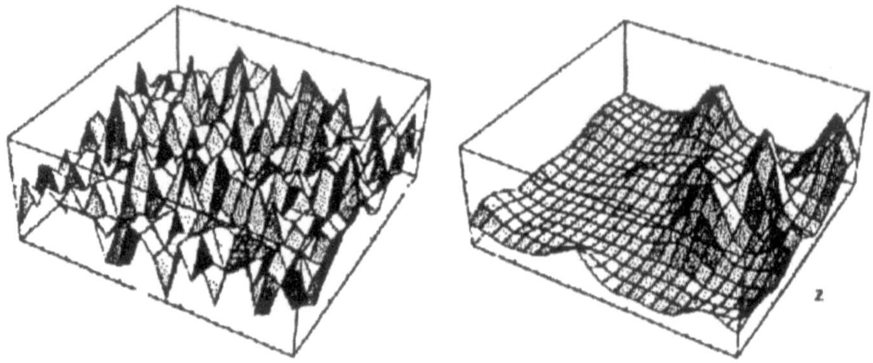

Figure 2. Sample of a gauge configuration before (left) and after (right) cooling. In the latter case instantons are clearly seen. The quantity shown is the action density, and its scale (not shown) is two orders of magnitude larger on the left figure.

ones[§]. That is of course exactly as expected: instantons with the radius $\rho \sim a$ "fall through the lattice" during "cooling". The main fact is a confirmation of a sharp maximum, with strong suppression of large-size[¶] instantons. The open points in Fig. 3 are the "instanton liquid" calculations[29].

Another striking news[15] came from the correlators calculated *after cooling* of lattice configurations (so to say, in the instanton liquid distilled from them). In spite of apparent loss of perturbative effects and (nearly complete) loss of confinement, the correlators stubbornly remained about the same. It is hardly possible to demonstrate more clearly that perturbative phenomena (the Coulomb and spin-spin forces) and confinement are not in fact so important for hadronic properties!

Zero Modes and the $U(1)_A$ anomaly

Instantons interpolate between different topological vacua in QCD. It is then natural to ask if the different vacua can be physically distinguished. This question is answered in the presence of light fermions, because the different vacua have different axial charge. This observation is the key element in understanding the mechanism of "anomalies". Anomalies first appeared in the context of perturbation theory[32, 33]. From the triangle diagram involving an external axial vector current one finds that the flavor singlet current which is conserved on the classical level develops an anomalous divergence on the quantum level

$$\partial_\mu j^5_\mu = \frac{N_f}{16\pi^2} G^a_{\mu\nu} \tilde{G}^a_{\mu\nu}. \tag{34}$$

This anomaly plays an important role in QCD, because it explains the absence of a ninth goldstone boson, the so-called $U(1)_A$ puzzle.

The mechanism of the anomaly is intimately connected with instantons. First, we recognize the integral of the rhs of (34) as $2N_f Q$, where Q is the topological charge. This means that in the background field of an instanton we expect axial charge conservation

[§] In passing, let us mention here one practical aspect of the studies of $d(\rho)$: its measurements at *small* $\rho < 0.2$ fm is potentially the source of by far the most accurate measurements of Λ_{QCD}. As it is well known, here $d(\rho) \sim \Lambda_{QCD}^{b_0}$: thus even relatively poor accuracy in its measurements leads to a Λ_{QCD} value with the accuracy better than measured today by other methods.

[¶] Note that unlike small ones, those are measured reliably.

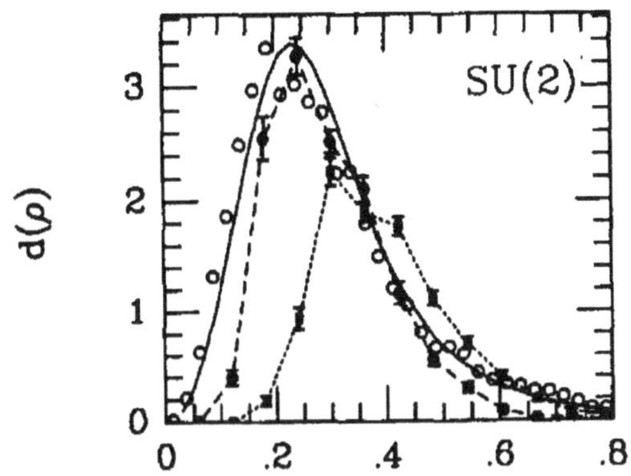

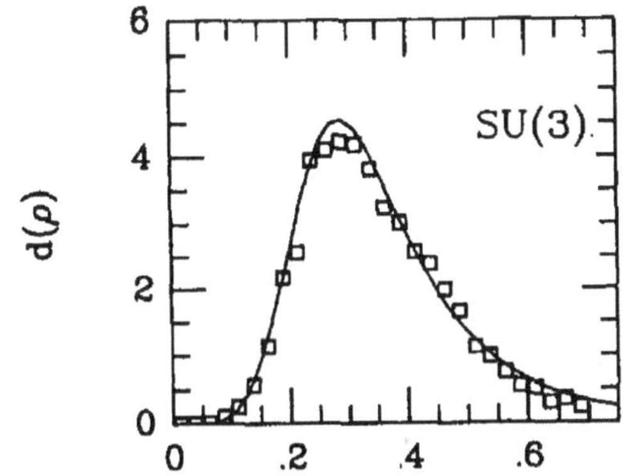

Figure 3. **The instanton size distribution in gauge theories. The open points correspond to interacting instanton liquid", and the closed ones are lattice results by Michael and Spencer, $16^4, 4/g^2 = 2.4$ for squares and $24^4, 4/g^2 = 2.5$ for dots (the dotted and dashed lines just guide the eye). Solid curves correspond to the parameterization discussed in the text. Units are in "femtometers", defined for lattice data by the scalar glueball mass defined as $m_{0^+} = 1.7$ "GeV"; and by $1/\Lambda_{PV}$ for the instanton model.**

to be violated by $2N_f$ units. The crucial property of instantons, originally discovered by 't Hooft, is that the Dirac operator has a zero mode $i\not{D}\psi_0(x) = 0$ in the instanton field. For an instanton in the singular gauge, the zero mode wave function is

$$\psi_0(x) = \frac{\rho}{\pi}\frac{1}{(x^2+\rho^2)^{3/2}}\frac{\gamma\cdot x}{\sqrt{x^2}}\frac{1+\gamma_5}{2}\phi \tag{35}$$

where $\phi^{\alpha m} = \epsilon^{\alpha m}/\sqrt{2}$ is a constant spinor, which couples the color index α to the spin index $m = 1, 2$. Note that the solution is left handed, $\gamma_5\psi_0 = -\psi_0$. Analogously, in the field of an anti-instanton there is a right handed zero mode.

We can now see how axial charge is violated during tunneling[34, 35] For this purpose, let us consider the Dirac Hamiltonian $i\vec{\alpha}\cdot\vec{D}$ in the field of the instanton. The presence of a four dimensional normalizable zero mode implies that there is one left handed state that crosses from positive to negative energy during the tunneling event, while one right handed state crosses the other way. This can be seen as follows: In the adiabatic approximation, solutions of the Dirac equation are given by

$$\psi_i(\vec{x},t) = \psi_i(\vec{x},t=-\infty)\exp\left(-\int_{-\infty}^{t} dt'\,\epsilon(t')\right). \tag{36}$$

The only way we can have a four-dimensional normalizable wave function is if the energy ϵ is positive for $t \to \infty$ and negative for $t \to -\infty$. This explains how axial charge can be violated during tunneling. No fermion ever changes its chirality, all states simply move one level up or down. The axial charge comes, so to say, from the "bottom of the Dirac sea".

The effective Interaction between quarks

Proceeding from pure glue theory to QCD with light quarks, one has to deal with the much more complicated problem of quark-induced interactions. Indeed, on the level of a single instanton we can not even understand the presence of instantons in full QCD. The reason is again related to the existence of zero modes. In the presence of light quarks, the tunneling rate is proportional to the fermion determinant, which is given by the product of the eigenvalues of the Dirac operator. This means that (as $m \to 0$) the fermion zero mode makes the tunneling amplitude vanish and individual instantons cannot exist!

During the tunneling event, the axial charge of the vacuum changes, so instantons have to be accompanied by flipping of the fermion chirality. Consider the fermion propagator in the instanton field

$$S(x,y) = \frac{\psi_0(x)\psi_0^+(y)}{im} + \sum_{\lambda\neq 0}\frac{\psi_\lambda(x)\psi_\lambda^+(y)}{\lambda + im} \tag{37}$$

where $i\not{D}\psi_\lambda = \lambda\psi_\lambda$. For N_f light quark flavors the instanton amplitude is proportional to m_f^N. Instead of the tunneling amplitude, let us calculate a $2N_f$-quark Green's function $\langle\prod_f \bar{\psi}_f(x_f)\Gamma\psi_f(y_f)\rangle$, containing one quark and antiquark of each flavor. Performing the contractions, the amplitude involves N_f fermion propagators (37), so that the zero mode contribution involves a factor m^{N_f} in the denominator.

The result can be written in terms of an effective Lagrangian[22]. It is a non-local $2N_f$-fermion interaction, where the quarks are emitted or absorbed in zero mode wave functions. The result simplifies if we take the long wavelength limit (in fact, the interaction is cut off at momenta $k > \rho^{-1}$) and average over the instanton position and color orientation. For $N_f = 1$ the result is[36,37]

$$\mathcal{L}_{N_f=1} = \int d\rho\, n_0(\rho)\left(m\rho - \frac{4}{3}\pi^2\rho^3\bar{q}_R q_L\right), \tag{38}$$

where $n_0(\rho)$ is the tunneling rate. Note that the zero mode contribution acts like a mass term. For $N_f = 2$, the result is

$$\mathcal{L}_{N_f=2} = \int d\rho\, n_0(\rho)\Big[\prod_f\left(m\rho - \frac{4}{3}\pi^2\rho^3\bar{q}_{f,R}q_{f,L}\right) \tag{39}$$
$$+\frac{3}{32}\left(\frac{4}{3}\pi^2\rho^3\right)^2\left(\bar{u}_R\lambda^a u_L\bar{d}_R\lambda^a d_L - \bar{u}_R\sigma_{\mu\nu}\lambda^a u_L\bar{d}_R\sigma_{\mu\nu}\lambda^a d_L\right)\Big].$$

One can easily check that the interaction is $SU(2)\times SU(2)$ invariant, but $U(1)_A$ is explicitly broken.

INTERACTING INSTANTON LIQUID AND HADRONS

There is no place here to describe the interacting model: let me only mention that it includes the whole zero-mode part of the fermionic determinant, which means that the 't Hooft interaction is included to *all* orders. Methodically it is similar to usual statistical simulations. One new element[38] which was incorporated only recently is a trick which has allowed us to calculate the absolute value of the partition function: as a result the instanton density is now fixed from the standard condition of minimizing the free energy.

The interacting liquid is qualitatively different from the random one in several important aspects. It obeys known low energy theorems (see[38]), screens the topological charge and provides vanishing topological susceptibility[39], etc. Let me only comment that some improvement is definitely seen in vector-axial channels, and they are really dramatic in "repulsive" η, δ channels, in which RILM has produced unphysical negative correlators. One can see that indeed those are due to correlations between instantons and anti-instantons, induced by quark exchanges.

The calculated correlation functions

Correlation functions in the different instanton ensembles were calculated in[40, 8, 38] to which we refer the reader for more details. The results are shown in Figs. 4-5. The pion correlation functions in the different ensembles are qualitatively very similar. The differences are mostly due to different values of the quark condensate (and the physical quark mass) in the different ensembles. Using the Gell-Mann, Oaks, Renner relation, one can extrapolate the pion mass to the physical value of the quark masses. The

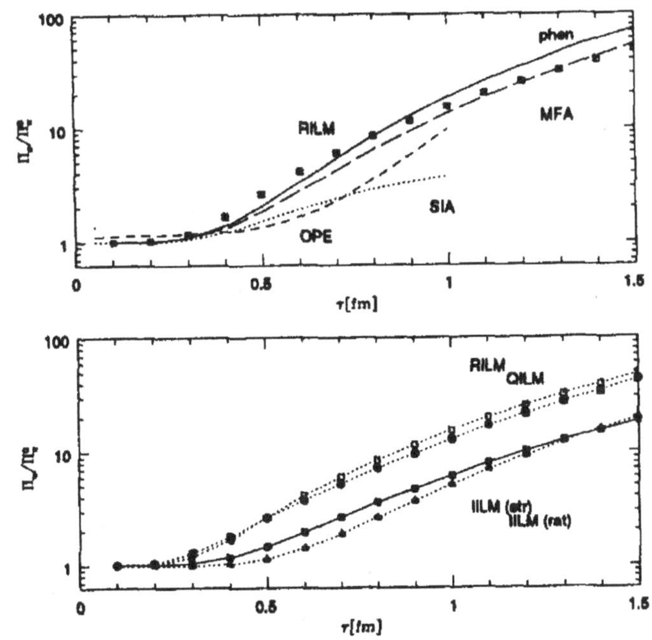

Figure 4. Pion correlation function in various approximations and instanton ensembles. In the top figure we show the phenomenological expectation (solid), the OPE (dashed), the single instanton (dash-dotted) and mean field approximations (dashed) as well as data in the random instanton ensemble. In the bottom figure we compare different instanton ensembles, random (open squares), quenched (circles) and interacting (streamline: solid squares, ratio ansatz solid triangles).

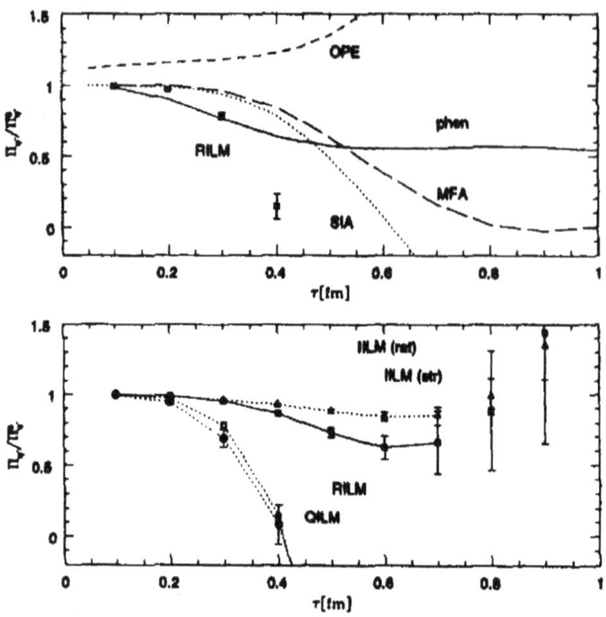

Figure 5. Eta prime meson correlation functions. The various curves and data sets are labeled as in Fig. 4.

results are consistent with the experimental value in the streamline ensemble (both quenched and unquenched), but clearly too small in the ratio ansatz ensemble. This is a reflection of the fact that the ratio ansatz ensemble is not sufficiently dilute.

In Fig. 6 we also show the results in the ρ channel. The ρ meson correlator is not affected by instanton zero modes to first order in the instanton density. The results in

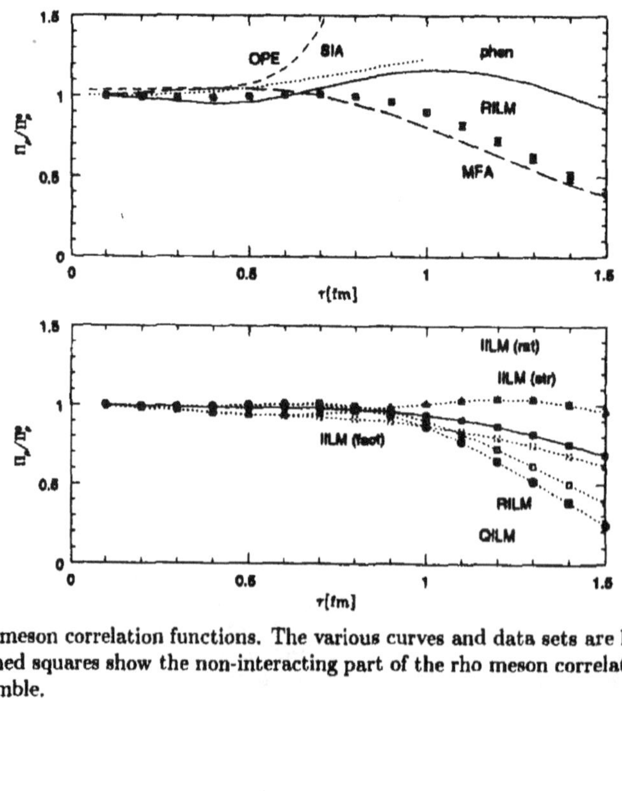

Figure 6. Rho meson correlation functions. The various curves and data sets are labeled as in Fig. 4. The dashed squares show the non-interacting part of the rho meson correlator in the interacting ensemble.

the different ensembles are fairly similar to each other and all fall somewhat short of the phenomenological result at intermediate distances $x \simeq 1$ fm. We have determine the ρ meson mass and coupling constant from a fit. The ρ meson mass is somewhat too heavy in the random and quenched ensembles, but in very good agreement with the experimental value $m_\rho = 770$ MeV in the interacting ensemble.

Since there are no interactions in the ρ meson channel in the RPA, it is important to study whether the instanton model provides any binding at all. In the instanton model, there is no confinement, and m_ρ is close to the two (constituent) quark threshold. In QCD, the ρ meson is also not a true bound state, but a resonance in the 2π continuum. In order to determine whether the continuum contribution in the instanton model is predominantly 2π or 2 quark would require the determination of the corresponding three point functions (which has not been done yet). Instead, we will compare the full correlation function with the non-interacting one, where we use the average (constituent quark) propagator determined in the same ensemble (Fig. 6). As explained above, this comparison provides a measure of the strength of interaction. We observe that there is an attractive interaction generated in the interacting liquid. The interaction is due to correlated instanton-anti-instanton pairs. This is consistent with the fact that the interaction is considerably smaller in the random ensemble. In the random model, the strength of the interaction grows as the ensemble becomes more dense. However, the interaction in the full ensemble is significantly larger than in the random model at the same diluteness. Therefore, most of the interaction comes from dynamically generated pairs.

The situation is drastically different in the η' channel. Among the ~ 40 correlation functions calculated in the random ensemble, only the η' (and the isovector-scalar δ discussed in the next section) are completely unacceptable: The correlation function decreases very rapidly and becomes negative at $x \sim 0.4$ fm. This behavior is incompatible with a normal spectral representation. The interaction in the random ensemble is too repulsive, and the model "over-explains" the $U(1)_A$ anomaly.

The results in the unquenched ensembles (closed and open points) significantly improve the situation. This is related to dynamical correlations between instantons and anti-instantons (topological charge screening). The single instanton contribution is repulsive, but the contribution from pairs is attractive[41]. Only if correlations among instantons and anti-instantons are sufficiently strong, the correlators are prevented from becoming negative. Quantitatively, the δ and η_{ns} masses in the streamline ensemble are still too heavy as compared to their experimental values. In the ratio ansatz, on the other hand, the correlation functions even show an enhancement at distances on the order of 1 fm, and the fitted masses are too light. This shows that the η' channel is very sensitive to the strength of correlations among instantons.

In summary, pion properties are mostly sensitive to global properties of the instanton ensemble, in particular its diluteness. Good phenomenology demands $\bar{\rho}^4 n \simeq 0.03$, as originally suggested in[42]. The properties of the ρ meson are essentially independent of the diluteness, but show sensitivity to $\bar{I}I$ correlations. These correlations become crucial in the η' channel.

After discussing the π, ρ, η' in some detail we only briefly comment on other correlation functions. The remaining scalar states are the isoscalar σ and the isovector δ (the a_0, f_0 in modern PDG notations). The sigma correlator receives a disconnected contribution, which at large distance is dominated by the quark condensate. In order to determine the ground state in this channel, the constant contribution has to be subtracted, which makes it difficult to obtain reliable results. Nevertheless, we find that the instanton liquid favors a (presumably broad) resonance around 500-600 MeV.

The isovector channel is in many ways similar to the η'. In the random ensemble, the interaction is too repulsive and the correlator becomes unphysical. This problem is solved in the interacting ensemble, but the δ is still very heavy, $m_\delta > 1$ GeV.

The remaining non-strange vectors are the a_1, ω and f_1. The a_1 mixes with the pion, which allows a determination of the pion decay constant f_π (as does a direct measurement of the $\pi - a_1$ mixing correlator). In the instanton liquid, disconnected contributions in the vector channels are small. This is consistent with the fact that the ρ, ω and a_1, f_1 are almost degenerate.

Finally, we can also include strange quarks. $SU(3)$ flavor breaking in the 't Hooft interaction nicely accounts for the masses of the K and the η. More difficult is a correct description of $\eta - \eta'$ mixing, which can only be achieved in the full ensemble. The random ensemble also has a problem with the mass splittings among the vectors ρ, K^*, ϕ. This is related to the fact that flavor symmetry breaking in the random ensemble is so strong that the strange and non-strange constituent quark masses are essentially the same. This problem is improved (but not fully solved) in the interacting ensemble.

Baryonic correlation functions

After discussing quark-antiquark systems in the last section, we now proceed to three quark (baryon) channels. As emphasized in[43], existence of a strongly attractive interaction in the pseudoscalar quark-antiquark (pion) channel also implies an attractive interaction in the scalar quark-quark (diquark) channel. This interaction is phenomenologically very desirable, because it immediately explains why the nucleon and lambda are light, while the delta and sigma are heavy.

The proton current can be constructed by coupling a d-quark to a uu-diquark. The diquark has the structure $\epsilon_{abc} u_b C\Gamma u_c$ which requires that the matrix $C\Gamma$ is symmetric. This condition is satisfied for the V and T gamma matrix structures. The two possible currents (with no derivatives and the minimum number of quark fields) with positive parity and spin $1/2$ are given by[44]

$$\eta_1 = \epsilon_{abc}(u^a C\gamma_\mu u^b)\gamma_5\gamma_\mu d^c, \qquad \eta_2 = \epsilon_{abc}(u^a C\sigma_{\mu\nu} u^b)\gamma_5\sigma_{\mu\nu} d^c. \tag{40}$$

It is sometimes useful to rewrite these currents in terms of scalar and pseudoscalar diquarks

$$\eta_{1,2} = (2,4)\left\{\epsilon_{abc}(u^a C d^b)\gamma_5 u^c \mp \epsilon_{abc}(u^a C\gamma_5 d^b)u^c\right\}. \tag{41}$$

Nucleon correlation functions are defined by $\Pi^N_{\alpha\beta}(x) = \langle \eta_\alpha(0)\bar\eta_\beta(x)\rangle$, where α, β are the Dirac indices of the nucleon currents. In total, there are six different nucleon correlators: the diagonal $\eta_1\bar\eta_1$, $\eta_2\bar\eta_2$ and off-diagonal $\eta_1\bar\eta_2$ correlators, each contracted with either the identity or $\gamma\cdot x$. Let us focus on the first two of these correlation functions (for more detail, see[8] and references therein). The OPE predicts[45])

$$\frac{\Pi^N_1}{\Pi^{N0}_2} = \frac{\pi^2}{12}|\langle\bar q q\rangle|\tau^3, \tag{42}$$

$$\frac{\Pi^N_2}{\Pi^{N0}_2} = 1 + \frac{\pi^2}{192}\left\langle\frac{\alpha}{\pi}G^2\right\rangle\tau^4 + \frac{\pi^4}{72}|\langle\bar q q\rangle|^2\tau^6. \tag{43}$$

The vector components of the diagonal correlators receive perturbative quark-loop contributions, which are dominant at small distances. The scalar components of the diagonal correlators as well as the off-diagonal correlation functions are sensitive to chiral symmetry breaking, and the OPE starts at order $\langle \bar{q}q \rangle$ or higher. Single instanton corrections to the correlation functions were calculated in[46, 47 *]. Instantons introduce additional, regular, contributions in the scalar channel and violate the factorization assumption used for the four-quark condensate in the vector channel. Similar to the pion case, both of these effects increase the amount of attraction already seen in the OPE. As shown in[47], they also help to improve the stability of the QCD sum rule predictions for the nucleon mass.

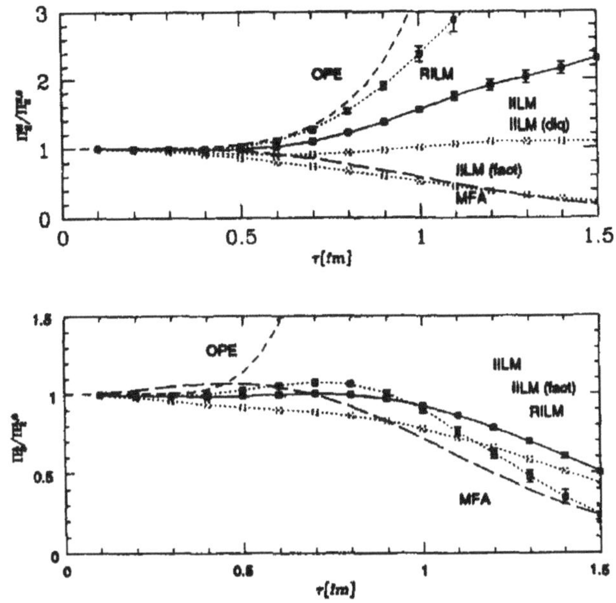

Figure 7. Nucleon and delta correlation functions Π_2^N and Π_2^Δ. Curves labeled as in Figs. 4-6.

The correlation function Π_2^N in the interacting ensemble is shown in Fig. 7. There is a significant enhancement over the perturbative contribution which is nicely described in terms of the nucleon contribution. Numerically, we find[†] $m_N = 1.019$ GeV. In the random ensemble, we have measured the nucleon mass at smaller quark masses and found $m_N = 0.96 \pm 0.03$ GeV. The nucleon mass is fairly insensitive to the instanton ensemble. However, the strength of the correlation function depends on the instanton ensemble. This is reflected by the value of the nucleon coupling constant, which is smaller in the IILM.

The fitted value of the threshold is $E_0 \simeq 1.8$ GeV, indicating that there is little strength in the (unphysical) three quark continuum. The fact that nucleon is bound can also be demonstrated by comparing the full nucleon correlation function with that of three non-interacting quarks (the cube of the average propagator, see Fig. 7). The full correlator is significantly larger than the non-interacting one, indicating the presence of a strongly attractive interaction. At least some of this interaction certainly comes from the attractive scalar diquark correlator which is part of the nucleon correlation function. This raises the question whether the nucleon (in our model) is a strongly bound diquark very loosely coupled to a third quark. In order to check this, we can rewrite the *NN* correlation functions using (41). Treating the nucleon as a non-interacting quark-

*The latter paper corrects some mistakes in the original work by Dorokhov and Kochelev.
[†]Note that this value corresponds to a relatively large current quark mass $m = 30$ MeV.

diquark system, we have

$$\Pi_2^N(\tau) = 4S_V(\tau)\left(\Pi_{sd}(\tau) + \Pi_{pd}(\tau)\right) + 96S_V(\tau)S_S^2(\tau), \tag{44}$$

where $\Pi_{sd,pd}(\tau)$ are the correlation functions of the scalar $j_{sd}^a = \epsilon^{abc}(u^a C\gamma_5 d^b)$ and pseudoscalar diquark $j_{pd}^a = \epsilon^{abc}(u^a C d^b)$ currents. From Fig. 7 we observe that the quark-diquark model explains some of the attraction seen in Π_2^N, but falls short of the numerical results. This means that while diquarks may play some role in making the nucleon bound, there are substantial interactions in the quark-diquark system.

In[8] we studied all six nucleon correlation functions. We showed that all correlation functions can be described with the same nucleon mass and coupling constants. This description is particularly good for the $\Pi_{1,2}^N$ and $\Pi_{4,5}^N$ correlators, but misses some of the contributions at intermediate distances in the $\Pi_{3,6}^N$ channels. This might be an indication that in these channels contributions from higher resonances are important.

Delta correlation functions

In the case of the Δ resonance, there exists only one independent current, given (for the Δ^{++}) by

$$\eta_\mu^\Delta = \epsilon_{abc}(u^a C\gamma_\mu u^b)u^c. \tag{45}$$

However, the spin structure of the correlator $\Pi_{\mu\nu;\alpha\beta}^\Delta(x) = \langle \eta_{\mu\alpha}^\Delta(0)\bar{\eta}_{\nu\beta}^\Delta(x)\rangle$ is much richer. In general, there are ten independent tensor structures, but the Rarita-Schwinger constraint $\gamma^\mu\eta_\mu^\Delta = 0$reduces this number to four.

The mass of the delta resonance is too large in the random model, but closer to experiment in the unquenched ensemble. Note that similar to the nucleon, part of this discrepancy is due to the value of the current mass. Nevertheless, the delta-nucleon mass splitting in the unquenched ensemble is $m_\Delta - m_N = 409$ MeV, still too large as compared to the experimental value 297 MeV. Similar to the ρ meson, there is no binding in the Δ channel on the level of the mean field approximation. Comparing the full correlation function with the non-interacting one (see Fig. 7) we find substantial attraction between the quarks. Again, it would be important to show that the continuum is dominated by $N\pi$ rather than the three quark threshold. In addition to the results presented here, heavy-light baryons were studied in[8, 9].

GLUEBALLS AND INSTANTONS

One of the most interesting problems in hadronic spectroscopy is the question how the glueball states look like. A number of "glueball candidates" have been experimentally observed, but none was unambiguously identified. Lattice simulations provide important qualitative insights, and (although large-scale numerical efforts are still necessary) a few statements appear to be firmly established: (i) The lightest glueball is the scalar, with a mass in the 1.5-1.7 GeV range. (ii) The tensor glueball is significantly heavier $m_{2^{++}}/m_{0^{++}} \simeq 1.4$, with the pseudoscalar one heavier still $m_{0^{-+}}/m_{0^{++}} = 1.5$-$1.8$. (iii) The scalar has a much smaller size than other glueballs. This is seen both from the magnitude of finite size effects[48] and directly from measurements of the wave functions[49, 50]. The size of the scalar glueball (defined through the exponential decay of the wave function) is $r_{0^{++}} \simeq 0.2$ fm, while[50] $r_{2^{++}} \simeq 0.8$ fm*.

*For comparison, a similar measurement for the π and ρ mesons gives[50] 0.32 fm and 0.45 fm, indicating that spin-dependent forces between gluons are stronger than between quarks.

Instantons are expected to give a large contribution to glueball correlation functions because of their very strong classical gluonic field. Roughly speaking, they contribute proportionally to the square of their action $S_0 = 8\pi^2/g^2(\rho) \sim 10 - 15$, so enhancement is about 2 orders of magnitude.

Gluonic currents with the quantum numbers of the lowest glueball states are the field strength squared (0^{++}), the topological charge density (0^{-+}), and the energy momentum tensors (2^{++}): For the scalar channel a *single instanton* contribution is attractive, for the pseudoscalar channel it is repulsive, and for the tensor channel there is no classical contribution since the stress tensor of the self-dual field of an instanton is zero. To study this question in more detail, and in order to determine the effect of light quarks on the glueball correlation functions we have calculated glueball correlation functions numerically in the different instanton ensembles[3]. As usual we parameterize the glueball correlators using a pole plus continuum model for the spectral function. The scalar glueball mass is roughly $m_{0^{++}} = 1.5$ GeV with deviations of up to 0.25 GeV depending on the underlying ensemble. Given the uncertainties in the mass determination (the scalar correlator requires a subtraction), it is not clear whether these differences are significant. The coupling constant is almost independent of the ensemble $\lambda_{0^{++}} = 16 \pm 2\ \mathrm{GeV}^{3\dagger}$. This large coupling, as well as direct study of the wave function have lead to a very small size for a scalar glueball, about .2 fm.

In the pseudoscalar case the classical and one-loop contributions have opposite signs. At distances where they tend to cancel each other our approximation (which neglects the interference between the two) becomes questionable. However, the rapid downturn directly translates into the position of the perturbative threshold, for which we find $\sqrt{s_0} \simeq 2.4$ GeV in the quenched theory. We see no clear evidence for a pseudoscalar glueball state below the continuum threshold. In the unquenched theory we observe the η' signal with $m_{\eta'} \approx 800$ MeV and $\lambda_{\eta'} \sim 7.0\ \mathrm{GeV}^3$. Using the anomaly equation, this corresponds to $f_{\eta'} = 200$ MeV.

The next step is to calculate the glueball decay widths: work is in progress.

INSTANTONS AND CHARM

We have discussed in the introduction a scale of 1 GeV which is related to the field strength at the center of a typical instanton. Note that it is not very small compared to the charm quark mass $m_c^2 \approx 2\ \mathrm{GeV}^2$, and thus one may expect some significant polarization of the charm quark vacuum. It means that hadrons made of light quarks may in fact have some instanton-induced "intrinsic charm", especially those which are most closely related to instantons.

It was found recently[65] that this seems to be the case for η' meson. It started when the CLEO collaboration reported[64] measurements of inclusive and exclusive production of the η' in B-decays:

$$Br(B \to \eta' + X\ ; 2.2\ GeV < E_{\eta'} < 2.7\ GeV) = (7.5 \pm 1.5 \pm 1.1) \cdot 10^{-4}\ , \tag{46}$$

$$Br(B \to \eta' + K) = (7.8^{+2.7}_{-2.2} \pm 1.0) \cdot 10^{-5}\ . \tag{47}$$

Simple estimates[66] show that these data are in severe contradiction with the standard *b*-quark decay into light quarks: Cabbibo suppression V_{ub} leads to decay rates two orders

†This number is significantly larger than the result of QCD sum rule calculations that do not enforce the low energy theorem[51, 52].

of magnitude smaller than the data (both inclusive and exclusive ones). An alternative mechanism, suggested in[66], is based on the Cabbibo favored $b \to c\bar{c}s$ process, followed by a transition of virtual $\bar{c}c$ into the η'. The latter transition may be possible, provided there exists a large intrinsic charm component of the η'. Its quantitative measure can be expressed through the matrix element

$$\langle 0|\bar{c}\gamma_\mu\gamma_5 c|\eta'(q)\rangle \equiv i f^{(c)}_{\eta'} q_\mu \,. \tag{48}$$

and one needs $f^{(c)}_{\eta'} \approx 140$ MeV in order to explain the CLEO data, see[66]. This value is surprisingly large, being only a few times smaller than the analogously normalized residue $\langle 0|\bar{c}\gamma_\mu\gamma_5 c|\eta_c(q)\rangle = i f_{\eta_c} q_\mu$ with $f_{\eta_c} \simeq 400$ MeV known experimentally from the $\eta_c \to \gamma\gamma$ decay. It is also comparable to a similarly defined coupling of η' to the axial current of light quarks: note however that instantons in fact lead to *repulsive* interactions between them, and thus the light quark wave function of η' should be depleted at the origin.

Because the c-quark is heavy, it may only exist in the η' through a virtual loop, and its contribution can be evaluated in terms of gluonic fields. Taking the divergence of the axial current in Eq.(48) one gets

$$f^{(c)}_{\eta'} = \frac{1}{m^2_{\eta'}}\langle 0|2m_c\bar{c}i\gamma_5 c + \frac{\alpha_s}{4\pi}G_{\mu\nu}\tilde{G}_{\mu\nu}|\eta'\rangle \tag{49}$$

which can be further simplified by the operator product expansion in inverse powers of the c–quark mass

$$2m_c\bar{c}i\gamma_5 c = -\frac{\alpha_s}{4\pi}G_{\mu\nu}\tilde{G}_{\mu\nu} - \frac{1}{16\pi^2 m_c^2}g^3 f^{abc}G^a_{\mu\nu}\tilde{G}^b_{\nu\alpha}G^c_{\alpha\mu} + O(G^4/m_c^4) \tag{50}$$

(see the appendix in[66] for a detailed derivation of this result. Further terms in the expansion (50) are neglected in what follows.) Thus the problem is reduced to the matrix element of a particular dimension six pseudo-scalar gluonic operator:

$$f^{(c)}_{\eta'} = -\frac{1}{16\pi^2 m^2_{\eta'}}\frac{1}{m_c^2}\langle 0|g^3 f^{abc}G^a_{\mu\nu}\tilde{G}^b_{\nu\alpha}G^c_{\alpha\mu}|\eta'\rangle \,. \tag{51}$$

The calculation is based on numerical evaluation of the following two-point Euclidean correlation functions

$$K_{22}(x) = \langle 0|g^2 G^a_{\mu\nu}\tilde{G}^a_{\mu\nu}(x) g^2 G^a_{\mu\nu}\tilde{G}^a_{\mu\nu}(0)|0\rangle \tag{52}$$

$$K_{23}(x) = \langle 0|g^2 G^a_{\mu\nu}\tilde{G}^a_{\mu\nu}(x) g^3 f^{abc}G^a_{\mu\nu}\tilde{G}^b_{\nu\lambda}G^c_{\lambda\mu}(0)|0\rangle \tag{53}$$

$$K_{33}(x) = \langle 0|g^3 f^{abc}G^a_{\mu\nu}\tilde{G}^b_{\nu\lambda}G^c_{\lambda\mu}(x)\;\; g^3 f^{abc}G^a_{\mu\nu}\tilde{G}^b_{\nu\lambda}G^c_{\lambda\mu}(0)|0\rangle \tag{54}$$

Studies of $K_{22}(x)$ have been made previously[3], where it was demonstrated that in the "unquenched" ensemble of instantons with dynamical quarks the non-perturbative part changes sign at distances $x > 0.6$ fm, displaying a "Debye cloud" of compensating topological charge. It is identified with the η' contribution, and leads to an estimate

$$\langle 0|g^2 G^a_{\mu\nu}\tilde{G}^a_{\mu\nu}|\eta'\rangle = \frac{16\pi^2}{\sqrt{3}}f_{\eta'}m^2_{\eta'} \approx 7\ \text{GeV}^3 \tag{55}$$

which agrees reasonably well with other estimates in the literature. In this formula we have expressed matrix element (55) in terms of the standard parameter $f_{\eta'} \approx 85$ MeV which is defined as follows

$$\langle 0|\frac{1}{\sqrt{3}}\sum_{i=u,d,s}\bar{q}_i\gamma_\mu\gamma_5 q_i|\eta'\rangle = i f_{\eta'} q_\mu .$$

Using an anomaly in the chiral limit, $m_u = m_d = m_s = 0$, we arrive at (55).

We have calculated by numerical simulation the correlators mentioned. At small x purely perturbative gluons (e.g. $K_{22}^{pert}(x) = 384g^4/\pi^4x^8$) dominate, while the non-perturbative fields can be included via the operator product expansion. At large x we will argue below that (at least with dynamical quarks) the non-perturbative fields dominate.

The quantity $f_{\eta'}^{(c)}$ (51) can be obtained from the correlation functions (52, 53, 54):

$$|\frac{f_{\eta'}^{(c)}\sqrt{3}m_c^2}{f_{\eta'}}| = |\frac{K_{23}(x\to\infty)}{K_{22}(x\to\infty)}| = \sqrt{\frac{K_{33}(x\to\infty)}{K_{22}(x\to\infty)}} \quad (56)$$

It is expected that at large distances the contribution to two other correlators would also be dominated by the non-perturbative field of the instantons. If so, one has a simple estimate for the ratio of matrix elements

$$|\frac{\langle 0|g^3 f^{abc}G^a_{\mu\nu}\tilde{G}^b_{\nu\alpha}G^c_{\alpha\mu}|\eta'\rangle}{\langle 0|g^2 G^a_{\mu\nu}\tilde{G}^a_{\mu\nu}|\eta'\rangle}| = \frac{12}{5}\langle\frac{1}{\rho^2}\rangle \approx (1\sim 1.5)GeV^2 \quad (57)$$

The two numbers given here correspond to averaging over instanton size distribution for two variants of the instanton-anti-instanton interaction, the so-called "streamline" and "ratio-ansatz" ones, and indicate the systematics involved. The latter (giving smaller average size and the larger number above) should be considered preferable, because it better agrees with the size distribution directly obtained from lattice gauge field configurations, see the discussion in[38].

In our measurements of K_{23}, K_{33} both ratios entering (56) were found to stabilize at large enough $x > 0.8$ fm at the same numerical value. We take it as an indication that the η' contribution does in fact dominate, although we were not able to see that all correlators fall off with the right mass. The numerical value of the ratios is about $(1.5 \sim 2.2)$ GeV^2, for the two ensembles mentioned. These numbers are somewhat larger than in (57) because the second operator in the correlator makes it more biased towards smaller instantons.

Proceeding to the final result, we have to consider QCD radiative corrections. The experimental number mentioned above is defined at the scale $\mu_1^2 \approx m_c^2$, which is different from that obtained in the instanton calculation. In the latter case the charge and fields are normalized at $\mu_2^2 \approx gG$ where G is the typical gauge field at the points which contribute most to the correlators. This concludes our derivation of the parameter

$$|\frac{f_{\eta'}^{(c)}}{f_{\eta'}}| \simeq (0.85 \sim 1.22), \quad (58)$$

where the second value is preferable, see above.

The next logical question to ask is whether the connection between strong instanton fields and charm leads to phenomena unrelated to η'. One intriguing direction to study is η_c hadronic decays: their deviations from a simple perturbative pattern (which works well for J/ψ) are well known, see e.g.[67]. Let us also mention the recent intriguing observation by Bjorken[68] that its three leading hadronic decay channels $(\eta'\pi\pi, \eta\pi\pi, KK\pi)$ fit well to a pattern following from the instanton-induced 't Hooft effective Lagrangian.

Another question is whether the "intrinsic charm" (see e.g.[69]) of other hadrons is due to the same mechanism. Especially close to the problem considered is charm contribution to the spin of the nucleon. The relevant matrix element is that of the charm axial current, as above,

$$\langle N|\bar{c}\gamma_\mu\gamma_5 c|N\rangle = g_A^{(c)}\bar{N}\gamma_\mu\gamma_5 N \tag{59}$$

and it could be generated, e.g., by the η' "cloud" of the nucleon. Assuming now the η' dominance in this matrix element one could get the following Goldberger-Treiman type relation $g_A^{(c)} = \frac{1}{2M_N} g_{\eta' NN} f_{\eta'}^{(c)}$. Although the value of $g_{\eta' NN}$ is unknown, and its phenomenological estimates vary significantly $g_{\eta' NN} = 3-7$, one gets from this estimate a surprisingly large contribution

$$\langle N|\bar{c}\gamma_\mu\gamma_5 c|N\rangle = (0.2 \sim 0.5)\bar{N}\gamma_\mu\gamma_5 N \tag{60}$$

comparable to the *sum* (but not each term!) of the light *u, d, s* quarks. Calculation of $g_A^{(c)}$ and $g_{\eta' NN}$ in the instanton model are in progress: their lattice determination would be more than welcomed. Ultimately, the contribution of the charmed quarks in polarized deep-inelastic scattering may be tested experimentally, by tagging the charmed quark jets (e.g. by the COMPASS experiment at CERN).

In summary: it is by now widely known that the Zweig rule is badly broken in all scalar/pseudoscalar channels, and that the (rather large) mass of the η' is in fact due to light-quark-gluon mixing. Furthermore, these phenomena are generally attributed to instantons. We have found that similar phenomena for larger-dimension (multi-gluon) operators are much stronger. The reason for that is the inhomogeneous vacuum, with a very strong field inside small-size instantons. We have found a very significant fraction of charm in η'. Perhaps the same mechanism may help to solve other puzzles, especially related with η_c decays and DIS on the polarized nucleon.

THE CHIRAL PHASE TRANSITIONS

High temperature QCD

Let me first remind that non-zero T is incorporated in quantum field theory in a very simple way: the Euclidean time τ is limited by a period $1/T$, the so called Matsubara time. The instanton solution with periodic boundary conditions, called caloron, is analytically known.

$$A_\mu^a = \bar{\eta}_{\mu\nu}^a \Pi(x)\partial_\nu \Pi^{-1}(x) \tag{61}$$

where

$$\Pi(x) = 1 + \frac{\pi\rho^2}{\beta r}\frac{\sinh(2\pi r/\beta)}{\cosh(2\pi r/\beta) - \cos(2\pi\tau/\beta)} \tag{62}$$

Fermions should be *anti*–periodic, and the corresponding zero mode can also be found.

$$\psi_i^a = \frac{1}{2\sqrt{2}\pi\rho}\sqrt{\Pi(x)}\partial_\mu\left(\frac{\Phi(x)}{\Pi(x)}\right)\left(\frac{1-\gamma_5}{2}\gamma_\mu\right)_{ij}\epsilon_{aj}\,, \tag{63}$$

where $\Phi(x) = (\Pi(x) - 1)\cos(\pi\tau/\beta)/\cosh(\pi r/\beta)$. Note that the zero-mode wave function shows exponential decay $\exp(-\pi r T)$ in the spatial direction, but oscillates in τ.

So if instantons are like atoms with the quark zero mode as a wave function, finite T compresses their special extension and enhances the temporal one. (It looks like atoms in a very strong magnetic field.) That radically change their interactions, which are only strong if instantons are interacting along the time direction. In particular, a pair of such type can be formed, connected to themselves by periodicity.

The main finding in IILM at finite T is that the chiral phase transition is actually driven by a *rearrangement*[53, 41, 38] of the ensemble into a set of instanton anti-instanton "molecules"*. Recently the details of this mechanism were significantly clarified, both by numerical simulations[38] and analytic studies[54].

Let me start with some analytic arguments first. It was shown by Balitsky and Young[55] that in quantum mechanics coupled to fermions (in which the vacuum also is made of $\bar{I}I$ molecules) one can take the integral over the $\bar{I}I$ separation R by the saddle point method. Apart of the contribution from $R = 0$ (the perturbative saddle point) they have also found the non-perturbative contribution from large R. Unfortunately it is complex, unless the parameters are tuned so that the model is supersymmetric†. However, at *sufficiently high T* the non-perturbative saddle point becomes real again. It corresponds to a configuration in which the centers are at the same spatial point, but separated by half a Matsubara box in time $\Delta\tau = 1/(2T)$, the most symmetric orientation of the instanton anti-instanton pair on the Matsubara torus. The effect is largest when the molecule exactly fits onto the torus, i.e. $4\rho \simeq 1/T$., Using the standard size $\rho \simeq 0.33$ fm, one gets $T \simeq 150$ MeV, close to the transition temperature.

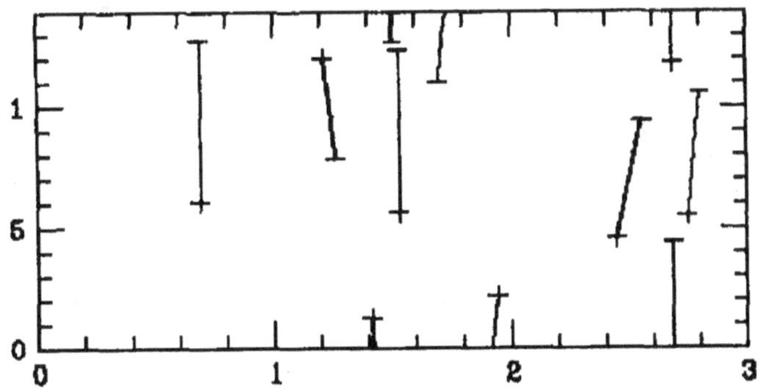

Figure 8. Typical instanton configurations for a temperature $T = 158$ MeV. The plot show projection of a four dimensional box, $(3\Lambda^{-1})^3 \times T^{-1}$, into the 3-4 ($z$ axes-imaginary time) plane. Instanton and anti-instanton positions are indicated by + and − symbols. The lines correspond to the strongest fermionic "bonds".

In a series of recent numerical simulations[3] it was found that this transition indeed goes as expected, see Fig. 8, where molecules are clearly seen. Furthermore, many thermodynamic parameters, the spectra of the Dirac operator, the evolution of the quark condensate and susceptibilities were calculated[38, 54], with results surprisingly consistent with available lattice data.

*Note a similarity to the Kosterlitz-Thouless transition in the $O(2)$ spin model in two dimensions: again one has paired topological objects, vortices in one phase and random liquid in another. The high and low-temperature phase change places, though.

†In this case one can see that the answer they got is in fact correct.

The effect of molecules on the effective interaction between quarks at high temperature can be described by the effective Lagrangian

$$\mathcal{L}_{\mathrm{mol\,sym}} = G\left\{\frac{2}{N_c^2}\left[(\bar\psi\tau^a\psi)^2 - (\bar\psi\tau^a\gamma_5\psi)^2\right]\right. \tag{64}$$

$$\left. - \frac{1}{2N_c^2}\left[(\bar\psi\tau^a\gamma_\mu\psi)^2 + (\bar\psi\tau^a\gamma_\mu\gamma_5\psi)^2\right] + \frac{2}{N_c^2}(\bar\psi\gamma_\mu\gamma_5\psi)^2\right\} + \cdots,$$

with the coupling constant

$$G = \int d\rho_1 d\rho_2 \frac{n(\rho_1,\rho_2)}{8T_{I\bar I}^2}(2\pi\rho_1)^2(2\pi\rho_2)^2. \tag{65}$$

Here, $n(\rho_1,\rho_2)$ is the tunneling probability for the $I\bar I$ pair and $T_{I\bar I}$ is the corresponding overlap matrix element. τ^a is a four-vector with components $(\vec\tau, 1)$.

Completely unexpected results were found in simulations, suggesting that some hadronic states (especially pions and its chiral partner sigma) survive the phase transition as a bound (but not-Goldstone) state. The interaction described above should be responsible for that.

High density QCD

Zero (or low) temperature and high density (or chemical potential μ) is another area in which one expects a phase transition in which chiral symmetry gets restored and quarks are "liberated". We know that normal nuclei consist of what is called nuclear matter, it has a baryon density of about $n_o \approx .16\ \mathrm{fm}^{-3}$, and is dilute enough to consist of individual nucleons. With the appearance of QCD it became apparent in the 70's that very high density matter should be more or less an ideal fermi gas of quarks. A very important feature in this case is Debye screening of color fields, and thus it is referred to as the quark plasma phase. Among other things, this screening cuts off large size instantons[4]

$$dn(\rho,\mu) \approx dn(\rho,\mu=0)\exp(-N_f\rho^2\mu^2) \tag{66}$$

in the same way as the high-temperature quark-gluon plasma does. In general, the properties of high-density quark plasma are rather well understood. However, very little is known about the transition from nuclear matter to the quark plasma. It may proceed via a number of intermediate phases, and many are discussed in literature.

In this subsection we study what can be learned about this transition using the instanton-based approach. One particular instanton-induced effect in vacuum is formation of specific correlations between two quarks, or *diquarks.* It was shown in many phenomenological papers that correlations of such kind seem to exist inside the nucleons, see the recent review[71].

In relation with this, it is very instructive to consider QCD with *two* colors. Discussion of the large-N_c limit of QCD is well known and is now part of text books. We remind only that in this limit the baryons should become very heavy $M_{baryons} \sim N_c^1$, as compared to mesons $M_{mesons} \sim N_c^0$, and all quarks should sit in the same lowest state in a semi-classical common potential well. Quite the opposite picture is the case for $N_c = 2$. In the massless case an additional Pauli-Gursey symmetry (PGSY) exists[72], which incorporates mixing of quarks with anti-quarks. Therefore in this theory baryons should be *degenerate* with appropriate mesons. It reminds us of what we are used to in supersymmetric (SUSY) QCD, however, these degenerate baryons and mesons form different representations of flavor and spin groups, and therefore their numbers do *not* match. So PGSY QCD is a kind of intermediate case between SUSY QCD and ordinary QCD, in which no relation between baryons and mesons exists. As the real world with $N_c = 3$ is in between the two extremes $N_c = 2$ and $N_c = \infty$, the former case deserves to be studied in details. (Furthermore, we will speculate below that the real world is probably even closer to the former rather than the latter extreme.)

In the $N_c = 2$ theory the lowest (scalar) baryon (or diquark) is bound as strongly as the lowest meson, the massless pion. (Furthermore, as we discuss below such strong binding also ensures a very small size of this baryon.) The general pattern of symmetry breaking is $SU(2N_f) \rightarrow Sp(2N_f)$ and the number of Goldstone modes is

$$N_{\text{Goldstone}} = 2N_f^2 - N_f - 1 \tag{67}$$

Let us mention three cases specifically. For $N_f = 1$ there is no Goldstone boson‡. For the most interesting case of $N_f = 2$ the coset (ratio) of the full group over the remaining one is

$$K = SU(4)/Sp(4) = SO(6)/SO(5) = S^5 \tag{68}$$

which means the five dimensional sphere with 5 massless modes: three of those are pions, plus the scalar diquark S and its anti-particle $\bar{S}$. The next case, $N_f = 3$, leads to 14 Goldstone bosons. It is easy to count them: mesons form the usual $SU(3)_F$ octet, plus 3 diquarks and and 3 anti-diquarks belonging to the $\underline{3}$ and 3 representations, because flavor indices are convoluted with ϵ_{ijk}, and so on.

Let us now consider properties of this theory at non-zero density or chemical potential μ. Unlike other QCD-like theories, the two color case allows for straightforward studies on the lattice because the fermionic determinant remains real even for non-zero μ. Nevertheless, very little work was done to take advantage of this fact.

As we have already mentioned in the introduction, the first numerical studies of the instanton liquid model for finite density have been recently performed by T. Schaefer[73]. Among his results, there are those for the $N_c = 2$ theory: it was also found that $< \bar{\psi}\psi >$ starts decreasing with μ only above some critical value

$$\mu_c \approx (mK)^{1/2} \tag{69}$$

Here m is the quark mass and§ $K \approx 1.2$ GeV. (Note that in QCD $K = m_\pi^2/(m_u + m_d) \approx 1.7$ GeV.) Dependence of the mesonic condensate on μ above μ_c was found to be similar to what was seen on the lattice. The diquark condensate and density was not studied in this work.

Let us now review general features expected on theoretical grounds, starting for simplicity with the *massless* $N_c = N_f = 2$ theory. The usual vacuum state has a standard chiral symmetry breaking pattern, with a non-zero value of the "mesonic" condensate $< \bar{q}q >= n_0 \sim \Lambda^3$, like in QCD. There are five massless Goldstone bosons mentioned above. Rotations on the five dimensional sphere along Goldstone directions cost no energy, and by doing so in the direction of the scalar diquark one naturally obtains states with the diquark density from zero to n_0. As it energetically costs nothing, the chemical potential $\mu = \partial E/\partial n_q = 0$, till the maximal density $n_q = 2n_0$ is reached. At this point chiral symmetry gets restored, $< \bar{q}q >= 0$, the five Goldstone bosons are now the pions, sigma and $\bar{S}$. What is very important, the scalar diquark

‡Recall that due to the U(1) anomaly, even the corresponding meson, η', is massive. The diquark, similar to what we need, cannot even exist due to the Pauli principle.

§The units are defined by setting the instanton density to be exactly 1 fm^{-4}.

acquires the role of the former sigma meson, and therefore becomes massive. The reason for this is, as usual, repulsive interaction with particles in the condensate.

If one increases the density further, the system is still a Bose condensate of diquarks. However, the energy starts to grow and the non-zero chemical potential appears. The next qualitative phenomenon[¶] occurs when half of the diquark chemical potential reaches the quark mass, $\mu_S/2 = m_q$: from now on the system is a Bose-Fermi mixture of diquarks and quarks. Equilibrium demands that both components have the same chemical potential.

It is convenient to demonstrate first how it works for non-zero quark mass using the simplest effective Lagrangian, an analog of the linear sigma model. Together with the usual kinetic energy terms, it has a potential

$$V = \lambda(\vec{\pi}^2 + \sigma^2 + S^2 + \tilde{S}^2 - v^2)^2 - A\sigma - \mu^2 S^2 \tag{70}$$

where we have also included the chemical potential for the diquark and and the chiral symmetry breaking mass term. At $\mu = 0$ one obviously gets all the usual results, in particular the Goldstone mass $m_g^2 = A/v$, the sigma mass $m_\sigma^2 = 8\lambda v^2$, etc. For non-zero μ a (mean field) minimization over S produces the equation for its mean value $< S >$

$$4\lambda < S > (\sigma^2 + < S >^2 - v^2) = 2\mu^2 < S > \tag{71}$$

It is therefore either zero (and we are back to the vacuum solution) or

$$< S >^2 = v^2 + \frac{\mu^2}{\lambda} - \frac{A^2}{16\mu^2} \tag{72}$$

Putting it back into the effective energy density one finds for the unusual vacuum

$$\epsilon_{<S>} = -\mu^2 v^2 - \frac{3A^2}{16\mu^2} \tag{73}$$

to be compared to that for the usual vacuum

$$\epsilon_{<\sigma>} = -m_g^2 v^2 + \frac{m_g^4}{16\lambda} \tag{74}$$

It clearly wins if μ is about (but not exactly!) the pion mass. Furthermore, differentiating oureffective **energy**[||] over μ leads to

$$n = 2\mu v^2 - \frac{3m_g^2 v^2}{8\mu^3} \tag{75}$$

(Recall that it can only be used above the transition, at not too small a μ.) The behavior of the mesonic condensate in this phase is

$$< \sigma > = \frac{m_g^2 v}{4\mu^2} \tag{76}$$

which is roughly consistent with lattice data behavior above the critical μ.

Finally, let us look at the *mass* of the diquark. Taking for simplicity the massless $(A = 0 = m_g)$ case, one can easily compute from the effective action that it is given by

[¶]Assuming that deconfinement has also happened at about the same conditions as chiral restoration.

[||]It is worth recalling that it is actually $\epsilon_{eff} = \epsilon - \mu n$ where ϵ is the true energy and n is the density. Therefore true energy is positive and grows with μ.

$m_S = (8\lambda v^2 + 10\mu^2)^{1/2}$, growing from the value which the sigma meson has at $\mu = 0$. As we commented above, after it reaches twice the quark mass the diquarks become unstable against decay and one should populate both diquark Bose gas and quark Fermi gas in order to prevent this.

It brings us to the next important issue, properties of *single* quarks propagating in this matter. Generally, it can only be discussed at high enough density, in the deconfined phase. However, most chiral models (including versions of sigma, NJL or instanton-based models, which we will eventually discuss) disregard confinement, producing instead a finite "constituent" quark mass in the vacuum. For the massless theory, symmetry arguments suggest that rotation in our multi-dimensional chiral circle from the usual vacuum with non-zero $< \sigma >$ to the unusual one with non-zero $< S >$ should not change the quark mass value. At the same time, chiral symmetry should be restored in this new vacuum, so the mass should be a "chiral" one. It means that the quark propagator in momentum space should look like

$$S(p) = \frac{p_\mu \gamma_\mu}{p^2 - m_{\text{eff}}^2}. \tag{77}$$

This is indeed the case, as one can see from explicit diagrams: only even number of condensed diquark insertions are allowed in "normal" Green functions.

Let us now return to the real world with $N_c = 3$. The scalar diquark, according to instanton liquid calculations[8] is much lighter than all others, with a mass of about 500 MeV (to be compared with twice the constituent quark mass 700-800 MeV). It is plausible that there exists a "diquark condensate" phase as an intermediate between nuclear matter and the single-quark Fermi gas. What is definitely clear, at *any* density some analog of diquarks exists, now in form of "Cooper pairs" living on the surface of the Fermi sphere: as known from the Bardeen-Cooper-Schrieffer theory of superconductivity, arbitrary small attraction leads to binding. In the instanton ensemble it should look like "polymers" made of alternating instantons and anti-instantons**, local in space and extended indefinitely in the time direction. First indications for existence of these structures have indeed been seen in[73].

Diquarks provide a double advantage over a quark Fermi gas, due to binding $m_{dq} < 2m_q$ and their Bose condensation. However, it would be wrong to assume that any number of diquarks can be put into the $p = 0$ state: diquarks are composite objects and, like nucleons, they should have a repulsive "core". We have accounted for this effect by introducing a scattering length $a \approx .3\,\text{fm}$ into the expression for the energy of the diquark Bose gas (counted per *quark*)

$$\frac{\epsilon^B}{n_q} = \frac{4\pi a n_{dq}}{m}[1 + \frac{128}{15\sqrt{\pi}}(a^3 n_{dq})^{1/2}] \tag{78}$$

where n_{dq} is the diquark density. The first term is the mean field result from condensed diquarks, like included in our sigma-model discussion above, and the second term (calculated by Lee and Yang in the 50's) comes from non-condensate diquarks. Repulsion makes the diquark Bose gas less favorable than some optimal Bose-Fermi mixture. Results of our calculations are shown in Fig. 9, in which matter consists of (i) diquark Bose gas, being in chemical equilibrium with (ii) blue-green quarks, and neutralized in color and electric charge by the appropriate amount of (iii) red quarks. For definiteness, we use a quark mass of 400 MeV and a diquark mass of 500 MeV.

In addition we have introduced the energy of color strings. In order to explain how we have done it, we recall an old puzzle: for $\bar{c}c, \bar{b}b$ quarkonia we have an excellent description with a standard **Coulomb+linear** potential, while for hadrons made

**Note the clear correlation with "molecules" we discussed above for high T phase.

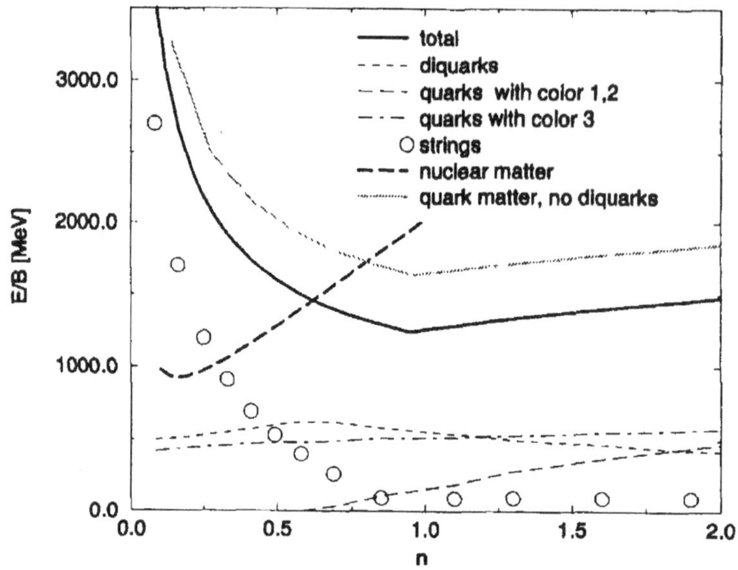

Figure 9. Energy per baryon (in MeV) versus baryon charge density n_b (fm^{-3}). Curves explained in the text.

of light quarks one uses a much weaker confining potential. (Many models like NJL and IILM even obtain correct hadronic masses without it: most of it just comes from the constituent quark mass.) The puzzle seems to be solved as follows: due to the extended nature of constituent quarks (as opposed to the point-like c or b ones), they have form-factors and interact only with sufficiently soft gluonic modes. On the lattice this effect was studied by measuring the string tension after smoothing the gauge fields: the resulting potential looks as

$$V = K|R - R_0|\theta(R - R_0) \tag{79}$$

with the standard string tension $K \approx 1$ GeV/fm and $R_0 \approx .7$ fm (see the example in[14]). In other words, a string seems to be formed only if constituent quarks do not overlap. We evaluate the average energy of a string schematically, substituting $R = n_{tot}^{-1/3}$ where n_{tot} is the total density of all quarks and diquarks. As seen from the figure, it prevents nuclear matter from dissolution into a gas of diquarks and quarks. The latter obtains a shallow (meta-stable) minimum at a density around six times nuclear density. (Most probably, at such densities confinement forces disappear due to deconfinement phenomenon anyway.) Existence of the minimum may be an artifact of the crude model: the bottom-line is that diquarks allow for significant overall energy gain relative to the quark matter (upper grey curve), and they also make quark matter very "soft".

What is not taken into account in this model is the effect of diquark condensation on the vacuum ($< \bar{q}q >$) and on the dispersion relation of the "blue-green" quarks. (The latter may be justified by the fact that at $n_b \sim 1$ fm^{-3} their role is not yet large.)

Phases of QCD with more quark flavors

In this section we discuss further the role of quarks in QCD, adding more flavors to it. If we add too much of them, namely $N_f > 33/2$ (here and below we imply 3

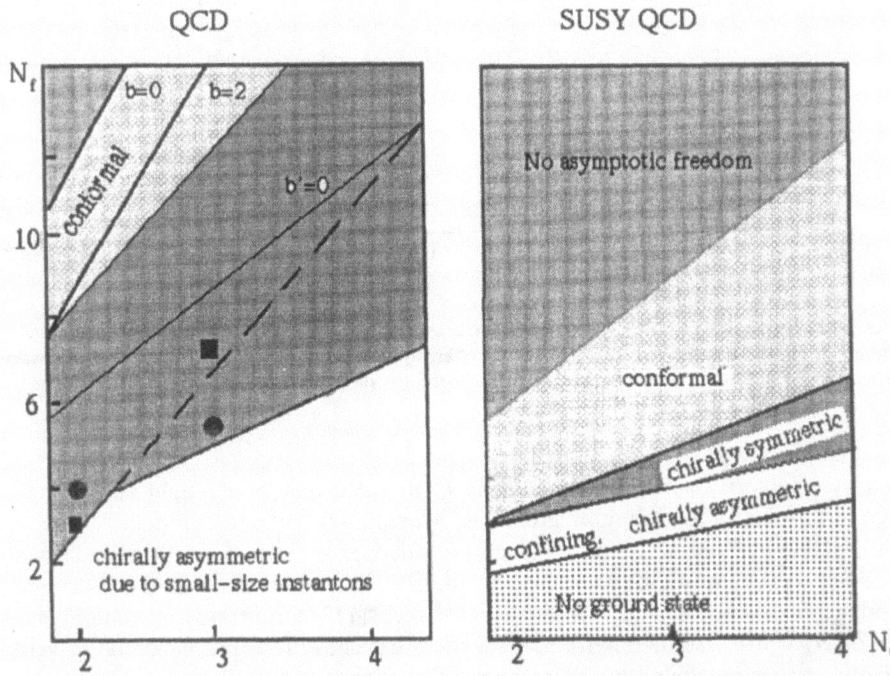

Figure 10. Schematic phase diagram of zero T QCD (a) and supersymmetric QCD (b) as a function of the number of colors N_c and the number of flavors N_f. Squares show where lattice calculations have found the infrared fixed point, dots are where the instanton ensemble is purely "molecular", with the unbroken chiral symmetry.

colors), the asymptotic freedom is lost and we get an *uninteresting* field theory with a charge growing at small distances, basically a theory as bad as QED! So our "most flavored" QCD is with 16 flavors, and we work our way down from there (see Fig. 10). The phase we are in at that point is actually a rather simple one, known as the Banks-Zaks conformal domain[56]. It has an infrared fixed point at small coupling $g_*^2/16\pi^2 = -b/b' << 1$ (b, b' are the one and two-loops coefficients of the beta function). It happens in the perturbative domain, so the charge is small both at small and large distances. There are no particles in this phase, and all correlators decay as powers of the distance. In this phase the non-perturbative phenomena like instantons are exponentially suppressed, $\exp(-\mathrm{const}/g_*^2)$. However, as one lowers the fermion number, the fixed point g_*^2 moves to larger values and eventually disappears. Lattice simulations of multi-flavor QCD were recently reported in[57]. These authors studied QCD with up to 240 flavors. Studying the sign of the beta function in the weak and strong coupling domains, they confirmed the existence of an infrared fixed point as low as at $N_f = 7$.

It is not known which phase we find next at $N_f = 5 - 6$, most probably it is the so called Coulomb phase (basically QGP).

The results of the interacting instanton model are summarized by Fig. 11, which shows how one singular point at $N_f = 2$develops into the discontinuity line for $N_f = 3$. The value of T_c goes down with increasing N_f and at $N_f = 5$[††] one finds that the

[††]Note that the $N_f = 4$ case is missing. It is because I have found the condensate to be small and comparable to finite-size effects. In order to separate those one should do calculations in different boxes, which is time consuming.

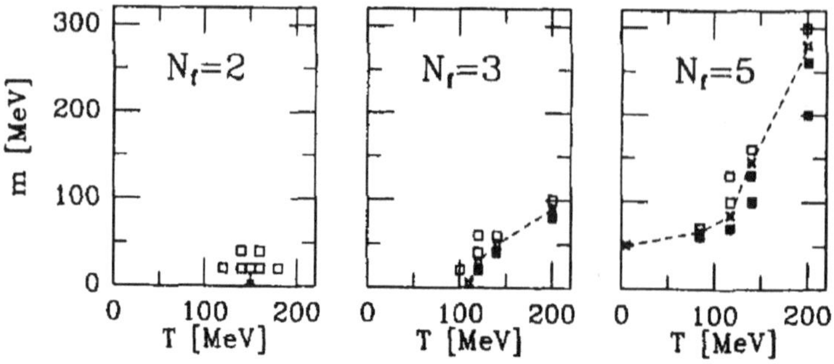

Figure 11. **Phase diagram (temperature-quark mass plane) of the instanton liquid for different numbers of quark flavors, N_f=2, 3 and 5. The open squares indicate a non-zero chiral condensate, while solid ones indicate that it is zero. The dashed lines show the approximate location of the discontinuity line.**

chiral symmetry is restored even at $T = 0$, provided quarks are light enough. New lattice results from the Columbia group (R. Mawhinney, private communication) deal with QCD with $N_f = 4$. When the measured masses are extrapolated to small quark masses, $m \to 0$, it was found that a dramatic and significant drop in chiral symmetry breaking effects occurs: the $\pi - \sigma, \rho - a_1, N - N^*(1/2^-)$ splittings are much smaller than for $N_f = 0-3$, studied before. It suggests that chiral restoration is indeed nearby, similarly to what was found in the instanton calculations.

Now we return to Figure 10, and explain its rhs, showing a similar phase diagram for the $N = 1$ SUSY QCD based on[58]. Not going into details, let me only mention that the vacuum is now definitely dominated by instanton-antiinstanton molecules, and their contribution can be calculated without problems (in QCD there are large perturbative contributions). There are two phases which are impossible in QCD: a case without a ground state (molecules force the system toward infinite Higgs VEV) and also a funny situation with unbroken chiral symmetry and confinement (hadrons exist but are degenerate in parity). It is amusing that chiral symmetry is restored at $N_f = N_c + 1$, similar to what was found in QCD in the instanton model. Also note, that in this case the existence of the Coulomb phase is a proven fact.

Finally, let me comment that nowhere the role of instantons is more clearly seen than in the supersymmetric theories. With the perturbative contributions basically cancelled out, there is no longer a problem to separate instantons from the perturbation series. For example, for $N = 2$ extended supersymmetric QCD general constraints are so powerful that Witten and Seiberg have been able to determine not just the effective potential, but the complete low energy effective action[59]. Its expansion shows powers of the instanton densities only, suggesting those to be the sole dynamical features. Furthermore, if the Higgs VEV a is large $(a \gg \Lambda)$, the semi-classical description is valid, instantons are small $\rho \sim a^{-1}$ and the instanton ensemble is dilute $n\rho^4 \sim \Lambda^4/a^4 << 1$. One can then calculate the first and the second order effects in density[60, 61, 62], and the results have explicitly reproduced the Witten-Seiberg expansion. Not only individual diagrams are all well convergent, but the instanton-based series seem to be convergent as well, explaining where and why the monopole mass vanishes. Furthermore, one can work out[63] a correct large N_c limit of the $N = 2$ theory, finding that (contrary to naive expectations) the instantons are not suppressed. It would be interesting to see what the instanton contribution is to the $N = 2$ theory for other terms in the effective Lagrangian (with more derivatives than in the Seiberg-Witten result), or for $N = 1$ theories, bridging the gap between $N = 0$ QCD and the $N = 2$ theory.

Acknowledgments

These lectures are based on works done with my collaborators, especially with T. Schaefer and J. Verbaarschot, who contributed immensely to progress reported in them. I am very grateful to Pierre van Baal, the major organizer of this school, for the wonderful opportunity to learn more about new developments in field theory.

REFERENCES

1. Y. Nambu and G. Jona-Lasinio, *Phys. Rev.*, 122:345, 1961.
2. T. Schäfer and E. V. Shuryak, 1996, Instantons in QCD, hep-ph/9610451, Rev.Mod.Phys, in press.
3. T. Schäfer and E. V. Shuryak, *Phys. Rev. Lett.*, 75:1707, 1995.
4. E. V. Shuryak, *Nucl. Phys.,*, B198:83, 1982.
5. D. I. Diakonov and V. Yu. Petrov, *Nucl. Phys.*, B272:457, 1986.
6. G. 't Hooft, *Phys. Rep.*, 142:357, 1986.
7. E. V. Shuryak, *Nucl. Phys.*, B203:93, 1982.
8. T. Schäfer, E. V. Shuryak, and J. J. M. Verbaarschot, *Nucl. Phys.*, B412:143, 1994.
9. S. Chernyshev, M. A. Nowak, and I. Zahed, hep-ph/9510326.
10. E. V. Shuryak, *Rev. Mod. Phys.*, 65:1, 1993.
11. M. C. Chu, J. M. Grandy, S. Huang, and J. W. Negele, *Phys. Rev.*, D 48:3340, 1993.
12. E. Shuryak, *Phys. Lett.*, 107B:103, 1981.
13. H. Georgi and A. Manohar, *Nucl. Phys.*, B234:189, 1984.
14. M. Feurstein, E. M. Ilgenfritz, M. Muller-Preussker and S. Thurner, hep-lat/9611024.
15. M. C. Chu, J. M. Grandy, S. Huang, and J. W. Negele, *Phys. Rev.*, D49:6039, 1994.
16. T. DeGrand, R.L. Jaffe, K. Johnson, and J. Kiskis, *Phys. Rev.*, D12:2060, 1975.
17. E. V. Shuryak, *Phys. Lett.*, B79:135, 1978.
18. M. C. Chu, J. M. Grandy, S. Huang, and J. W. Negele, *Phys. Rev. Lett.*, 70:225, 1993.
19. R. Rajaraman, *Solitons and Instantons,* North Holland, Amsterdam, 1982.
20. M. Shifman, *Instantons in Gauge Theories,* World Scientific, Singapore, 1994.
21. A. A. Belavin, A. M. Polyakov, A. A. Schwartz, and Yu. S. Tyupkin, *Phys. Lett.*, 59B:85, 1975.
22. G. 't Hooft, *Phys. Rev.*, D14:3432, 1976.
23. D. I. Diakonov and V. Yu. Petrov, *Phys. Rev.*, D50:266, 1994.
24. M. A. Shifman, A. I. Vaishtein, and V. I. Zakharov, *Phys.Lett.*, B76:971, 1978.
25. M. A. Shifman, A. I. Vainshtein, and V. I. Zakharov, *Nucl. Phys.*, B147:385, 448, 1979.
26. E. Witten, *Nucl.Phys.*, B156:269, 1979.
27. G. Veneziano, *Nucl.Phys.*, B159:213, 1979.
28. C. Michael and P. S. Spencer, *Phys. Rev.*, D52:4691, 1995.
29. E. V. Shuryak, *Phys. Rev.*, D52:5370, 1995.
30. M. I. Polikarpov and A. I. Veselov, *Nucl. Phys.*, B297:34, 1988.
31. C. Michael and P. S. Spencer, *Nucl. Phys. (Proc. Suppl.)*, B42:261, 1995.
32. S. L. Adler, *Phys. Rev.*, 177:2426, 1969.
33. J. S. Bell and R. Jackiw, *Nuovo. Cim.,* A60:47, 1969.
34. V. N. Gribov, 1981, KFKI-1981-66 preprint, Budapest.
35. M. A. Shifman, Sow. *Phys. Usp.*, 32:289, 1989.
36. G. 't Hooft, *Phys. Rev. Lett.*, 37:8, 1976.
37. M.A. Shifman, A.I. Vainshtein, and V.I. Zakharov, *Nucl.Phys.*, B163:46, 1980.
38. T. Schäfer and E. V. Shuryak, *Phys. Rev*, D53:6522, 1996.
39. E. V. Shuryak and J. J. M. Verbaarschot, *Phys. Rev.*, D52:295, 1995.
40. E. V. Shuryak and J. J. M. Verbaarschot, *Nucl. Phys.*, B410:55, 1993.
41. T. Schäfer, E. V. Shuryak, and J. J. M. Verbaarschot, *Phys. Rev.*, D51:1267, 1995.
42. E. V. Shuryak, *Nucl. Phys.*, B203:116, 1982.
43. E. V. Shuryak and J. L. Rosner, *Phys. Lett.*, B218:72, 1989.
44. B. L. Ioffe, *Nucl. Phys.,*, B188:317, 1981.
45. V. M. Belyaev and B. L. Ioffe, *Sov. Phys. JETP,, 83:976,* 1982.
46. A. E. Dorokhov and N. I. Kochelev, *Z. Phys., C* 46:281, 1990.
47. H. Forkel and M. K. Banerjee, *Phys.Rev.Lett.*, 71:484, 1993.
48. P. van Baal and A. S. Kronfeld, *Nucl. Phys. B (Proc. Suppl.)*, 9:227, 1989.

49. K. Ishikawa, G. Schierholz, H. Schneider, and M. Teper, *Phys.Lett.*, 128B:309, 1983.
50. F. de Forcrand and K.-F. Liu, *Phys. Rev. Lett.*, 69:245, 1992.
51. S. Narison, *Z. Phys.*, C26:209, 1984.
52. E. Bagan and S. Steele, *Phys. Lett.*, B243:413, 1990.
53. E.-M. Ilgenfritz and E. V. Shuryak, *Phys. Lett.*, B325:263, 1994.
54. E. Shuryak and M. Velkovaky, Mean field approach to instanton effects close to the phase transition, hep-ph/9603234.
55. I.I. Balitsky and A. V. Yung, *Nucl. Phys.*, B274:475, 1986.
56. T. Banks and A. Zaks, *Nucl. Phys.*, B82:196, 1982.
57. Y. Iwasaki, K. Kanaya, S. Kaya, S. Sakai, and T. Yoshie, *Z. Phys.*, C71:343, 1996.
58. N. Seiberg, *Phys. Rev.*, D49:6857, 1994.
59. N. Seiberg and E. Witten, *Nucl. Phys.*, B426:19, 1994.
60. D. Finnel and P. Pouliot, *Nucl. Phys.*, B453:225, 1995.
61. K. Ito and N. Sasakura, *Phys. Lett.*, B382:95, 1996.
62. N. Dorey, V.V. Khoze, and M.P. Mattis, *Phys. Rev.*, D54:2921, 1996.
63. M.R. Douglas, S.H. Shenker, Nucl.Phys.B447:271-296,1995; hep-th/9503163.
64. P. Kim (CLEO), Talk at FCNC 1997, Santa Monica, CA (Feb 1997).
65. E. Shuryak and A. Zhitnitsky, The gluon/charm content of the η' meson and instantons, hep-ph/9706316.
66. I. Halperin and A. Zhitnitsky, $B \to \eta' + K$ decay as a unique probe of η meson, hep-ph/9704412.
67. M.Shifman, Z.Phys.C4 (1980) 345.
68. J.D. Bjorken, CP and B Physics: Progress and Prospects, hep-ph/9706524.
69. S.J. Brodsky, P. Hoyer, A.H. Mueller and W.-K. Tang, Nucl.Phys. B369 (1992) 519; S. J. Brodsky, Beyond standard QCD, hep-ph/9503391.
70. R. Rapp, E. Shuryak and M. Velkovsky, High density matter and instantons, NTG-SUNY preprint, 1997.
71. M. Anselmino et al, Rev.Mod.Phys. 65 (1993) 1199.
72. W. Pauli, Nuovo Cim. 6 (1957) 205; F. Gursey, ibid 7 (1958) 411.
73. T. Schaefer, Instantons and the chiral phase transition at non-zero baryon density, hep-ph/9708256.

UNIVERSAL BEHAVIOR IN DIRAC SPECTRA

Jacobus Verbaarschot

Department of Physics and Astronomy
University at Stony Brook, SUNY
Stony Brook, NY 11794, USA

INTRODUCTION

In the chiral limit the QCD Lagrangian has an important global chiral symmetry: Specifically, for QCD with three colors and N_f massless flavors, the invariance group acting on the quark fields is $SU(N_f) \times SU(N_f) \times U_A(1)$. In spite of the lightness of the up and down quark, no signatures of this symmetry have been observed in nature. On the contrary, the fact that the pion mass is much lighter than the other hadron masses indicates that $SU(N_f) \times SU(N_f)$ is spontaneous broken with the pions as Goldstone bosons. The breaking of $U_A(1)$ follows from the absence of parity doublets. For example, the mass of the δ is very different from the mass of the pion. The description of the broken phase requires the introduction of an order parameter which is zero in the restored phase. For the chiral phase transition the order parameter is the chiral condensate $\langle\bar{\psi}\psi\rangle$. Large scale lattice QCD simulations performed during the past decade show that for two light flavors the value of the chiral condensate is approximately $-(220\,MeV)^3$ at zero temperature and vanishes above a critical temperature of about 140 MeV (see reviews by DeTar[1], Ukawa[2] and Smilga[3] for recent results on this topic). As was in particular pointed out by Shuryak[4], not necessarily both symmetries are restored at the same temperature[5]. This may lead to interesting observable consequences.

According the Banks-Casher formula[6], the chiral condensate is directly related to the spectrum of the Euclidean Dirac operator near zero virtuality (here and below, we use virtuality for the value of the Euclidean Dirac eigenvalues). However, the eigenvalues *fluctuate* about their average position as the gauge fields vary over the ensemble of gauge field configurations. The main question that will be addressed in these two lectures is to what extent such fluctuations are universal. Inspired by the study of spectra of complex systems[7], we will conjecture that spectral fluctuations on the scale of a typical eigenvalue spacing are universal. If this conjecture is true, such spectral fluctuations can be obtained from a broad class of theories with global symmetries of QCD as common ingredient. We will derive them from the simplest models in this class: chiral Random Matrix Theory (chRMT).

A first argument in favor of universality in Dirac spectra came from the analysis of the finite volume QCD partition function[8]. As has been shown by Gasser and Leutwyler[9], for box size L in the range

$$1/\Lambda \ll L \ll 1/m_\pi, \tag{1}$$

(Λ is a typical hadronic scale and m_π is the pion mass) the mass dependence of the QCD partition function is completely determined by its global symmetries. As a consequence, fluctuations of Dirac eigenvalues near zero virtuality are constrained by, but not determined by, an infinite family of sum rules[8] (also called Leutwyler-Smilga sum rules). For example, the simplest Leutwyler-Smilga sum rule can be obtained from the microscopic spectral density[10] (the spectral density near zero virtuality on a scale of a typical eigenvalue spacing). On the other hand, the infinite family of Leutwyler-Smilga sum rules is not sufficient to determine the microscopic spectral density. The additional ingredient is universality. A priori there is no reason that fluctuations of Dirac eigenvalues are in the same universality class as chRMT. Whether or not QCD is inside this class is a dynamical question that can only be answered by full scale lattice QCD simulations. However, the confidence in an affirmative answer to this question has been greatly enhanced by universality studies within chiral Random Matrix Theory. The aim of such studies is to show that spectral fluctuations do not depend on the details of the probability distribution. Recently, it has been shown that the microscopic spectral density is universal for a wide class of probability distributions[11, 12, 13, 14, 15, 16]. We will give an extensive review of these important new results.

Because of the $U_A(1)$ symmetry of the Dirac operator two types of spectral fluctuations can be distinguished. Spectral fluctuations near zero virtuality and spectral fluctuations in the bulk of the spectrum. The fluctuations of Dirac eigenvalues near zero virtuality are directly related to the approach to the thermodynamic limit of the chiral condensate. In particular, knowledge of the microscopic spectral density provides us with a quantitative explanation[17] of finite size corrections to the valence quark mass of dependence of the chiral condensate[18].

Recently, it has become possible to obtain *all* eigenvalues of the lattice QCD Dirac operator on reasonably large lattices[19, 20], making a direct verification of the above conjecture possible. This is one of the main objectives of these lectures. This is easiest for correlations in the bulk of the spectrum. Under the assumption of spectral ergodicity[21], eigenvalue correlations can be studied by spectral averaging instead of ensemble averaging[22, 23]. On the other hand, in order to study the microscopic spectral density, a very large number of independent gauge field configurations is required. First lattice results confirming the universality of the microscopic spectral density have been obtained recently[20].

At this point I wish to stress that there are two different types of applications of Random Matrix Theory. In the first type, fluctuations of an observable are related to its average. Because of universality it is possible to obtain exact results. In general, the average of an observable is not given by Random Matrix Theory. There are many examples of this type of universal fluctuations ranging from atomic physics to quantum field theory (a recent comprehensive review was written by Guhr, Müller-Groeling and Weidenmüller[24]). Most examples are related to fluctuations of eigenvalues. Typical examples are nuclear spectra[25], acoustic spectra[26], resonances in resonance cavities[27], S-matrix fluctuations[28, 29] and universal conductance fluctuations[30]. In these lectures we will discuss correlations in the bulk of Dirac spectra and the microscopic spectral density. The second type of application of Random Matrix Theory is as a schematic model of disorder. In this way one obtains *qualitative* results which may be helpful in understanding some physical phenomena. There are numerous examples in this category. We only mention the Anderson model of localization[31], neural networks[32], the Gross-Witten model of QCD[33] and quantum gravity[34] . In these lectures we will discuss chiral random matrix models at nonzero temperature and chemical potential. In particular, we will review recent work by Stephanov[35] on the quenched approximation at nonzero chemical potential.

In the first lecture we will review some general properties of Dirac spectra including the Banks-Casher formula. From the zeros of the partition function we will show that there is an intimate relation between chiral symmetry breaking and correlations of Dirac eigenvalues. Starting from Leutwyler-Smilga sum-rules the microscopic spectral density will be introduced. We will discuss the statistical analysis of quantum spectra. It will be argued that spectral correlations of 'complex' systems are given by Random Matrix Theory. We will end the first lecture with the introduction of chiral Random Matrix Theory.

In the second lecture we will compare the chiral random matrix model with QCD and discuss some of its properties. We will review recent results showing that the microscopic spectral density and eigenvalue correlations near zero virtuality are strongly universal. Lattice QCD results for the microscopic spectral density and correlations in the bulk of the spectrum will be discussed in detail. We will end the second lecture with a review of chiral Random Matrix Theory at nonzero chemical potential. Novel features of spectral universality in nonhermitean matrices will be discussed.

THE DIRAC SPECTRUM

Introduction

The Euclidean QCD partition function is given by

$$Z_{\rm QCD}(m,\theta) = \int dA \det(\gamma D + m) e^{-S_{\rm YM}/\hbar + i\theta\nu}, \tag{2}$$

where $\gamma D = \gamma_\mu(\partial_\mu + iA_\mu)$ is the anti-Hermitean Dirac operator and $S_{\rm YM}$ is the Yang-Mills action. The integral over field configurations includes a sum over all topological sectors with topological charge ν. Each sector is weighted by $\exp(i\theta\nu)$. Phenomenologically the value of the vacuum θ-angle is consistent with zero. We use the convention that the Euclidean gamma matrices are Hermitean with $\{\gamma_\mu, \gamma_\nu\} = 2\delta_{\mu\nu}$. The integral is over all gauge field configurations, and for definiteness, we assume a lattice regularization of the partition function.

Our main object of interest is the spectrum of the Dirac operator. The eigenvalues λ_k are defined by

$$\gamma D \phi_k = i\lambda_k \phi_k. \tag{3}$$

The spectral density is given by

$$\rho(\lambda) = \sum_k \delta(\lambda - \lambda_k). \tag{4}$$

Correlations of the eigenvalues can be expressed in terms of the two-point correlation function

$$\rho_2(\lambda, \lambda') = \langle \rho(\lambda)\rho(\lambda') \rangle, \tag{5}$$

where $\langle \cdots \rangle$ denotes averaging with respect to the QCD partition function (2). The connected two-point correlation function is obtained by subtraction of the product of the average spectral densities

$$\rho_c(\lambda, \lambda') = \rho_2(\lambda, \lambda') - \langle \rho(\lambda) \rangle \langle \rho(\lambda') \rangle. \tag{6}$$

Because of the $U_A(1)$ symmetry

$$\{\gamma_5, \gamma D\} = 0, \tag{7}$$

the eigenvalues occur in pairs $\pm\lambda$ or are zero. The eigenfunctions are given by ϕ_k and $\gamma_5 \phi_k$, respectively. If $\gamma_5 \phi_k = \pm \phi_k$, then necessarily $\lambda_k = 0$. This happens for a solution of the Dirac operator in the field of an instanton. In a sector with topological charge ν the Dirac operator has ν exact zero modes with positive chirality. In order to represent the low energy sector of the Dirac operator for field configurations with topological charge ν, it is natural to choose a chiral basis with n right-handed states and $m \equiv n + \nu$ left-handed states. Then the Dirac matrix has the block structure

$$\begin{pmatrix} 0 & W \\ W^\dagger & 0 \end{pmatrix}, \tag{8}$$

where W is an $n \times m$ matrix. For $m = 2$ and $n = 1$, one can easily convince oneself that the Dirac matrix has exactly one zero eigenvalue. We leave it as an exercise to the reader to show that in general the Dirac matrix has $|m - n|$ zero eigenvalues.

In terms of the eigenvalues of the Dirac operator, the QCD partition function can be rewritten as

$$Z_{\rm QCD}(m, \theta) = \sum_\nu e^{i\nu\theta} \prod_f m_f^{|\nu|} \int_\nu dA \prod_f \prod_k (\lambda_k^2 + m_f^2) e^{-S_{YM}/\hbar} \tag{9}$$

where $\int_\nu dA$ denotes the integral over field configurations with topological charge ν, and $\prod_f$ is the product over N_f flavors with mass m_f. The partition function in the sector of topological charge ν is obtained by Fourier inversion

$$Z_\nu(m) = \frac{1}{2\pi} \int_0^{2\pi} d\theta e^{-i\nu\theta} Z_{\rm QCD}(m, \theta). \tag{10}$$

The fluctuations of the eigenvalues of the QCD Dirac operator are induced by the fluctuations of the gauge fields. Formally, one can think of integrating out all gauge field configurations for fixed values of the Dirac eigenvalues. The transformation of integration variables from the fields, A, to the eigenvalues, λ_k, leads to a nontrivial "Jacobian". Universality in Dirac spectra has its origin in this "Jacobian".

The free Dirac spectrum can be obtained immediately from the square of the Dirac operator. For a box of volume $L_1 \times L_2 \times L_3 \times L_4$ one finds

$$\lambda_{\vec{n}} = 2\pi \left((\frac{n_1}{L_1})^2 + (\frac{n_2}{L_2})^2 + (\frac{n_3}{L_3})^2 + (\frac{n_4 + \frac{1}{2}}{L_4})^2 \right)^{1/2}, \tag{11}$$

where we have used periodic boundary conditions in the spatial directions and anti-periodic boundary conditions in the time direction. The spectral density is obtained by counting the total number of eigenvalues in a shell of radius $\lambda L/2\pi$. The result is

$$\rho^{\rm free}(\lambda) \sim V\lambda^3. \tag{12}$$

For future reference, we note that in the generic case, when the sides of the hypercube are related by an irrational number, asymptotically, the eigenvalues are uncorrelated, i.e.

$$\rho_2(\lambda, \lambda') = \langle \rho(\lambda) \rangle \langle \rho(\lambda') \rangle. \tag{13}$$

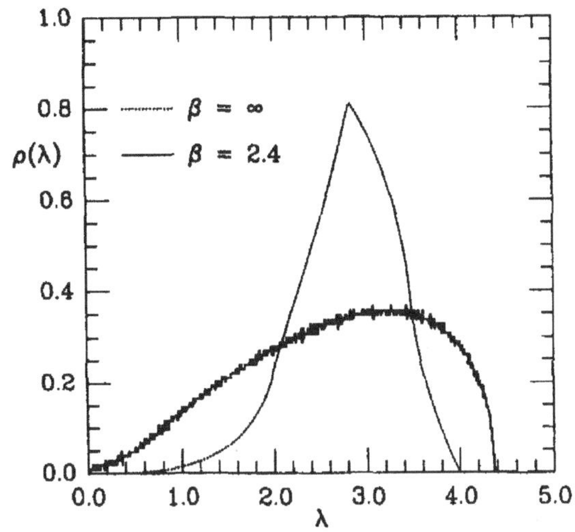

Figure 1. The free Dirac spectral density (dotted curve) and the spectral density of the Kogut-Susskind Dirac operator for a gauge field configuration generated with $\beta = 2.4$ (histogram). Both spectral densities are on a 12^4 lattice and are normalized to unit area.

Two examples of Dirac spectra are shown in Fig. 1. The dotted curve represents the free Kogut-Susskind Dirac spectrum on a 12^4 lattice with periodic boundary conditions in the spatial directions anti-periodic boundary conditions in the time direction. For an $N_1 \times N_2 \times N_3 \times N_4$ lattice this spectrum is given by

$$\lambda_{\vec{n}} = 2\left(\sin^2\left(\frac{\pi n_1}{N_1}\right) + \sin^2\left(\frac{\pi n_2}{N_2}\right) + \sin^2\left(\frac{\pi n_3}{N_3}\right) + \sin^2\left(\frac{\pi(n_4+0.5)}{N_4}\right)\right)^{1/2}. \quad (14)$$

Here, $n_i = 0, 1, \cdots, [N_i/2]$ $(i = 1, 2, 3)$ and $n_4 = 0, 1, \cdots, [(N_4+1)/2]$. The Kogut-Susskind Dirac spectrum for an $SU(2)$ gauge field configuration with $\beta = 2.4$ on the same size lattice is shown by the histogram in the same figure (full curve). We clearly observe an accumulation of small eigenvalues.

The Banks-Casher Relation

The order parameter of the chiral phase transition, $\langle\bar{\psi}\psi\rangle$, is nonzero only below the critical temperature. As was shown by Banks and Casher[6], $\langle\bar{\psi}\psi\rangle$ is directly related to the eigenvalue density of the QCD Dirac operator per unit four-volume

$$\Sigma \equiv |\langle\bar{\psi}\psi\rangle| = \lim \frac{\pi\langle\rho(0)\rangle}{V}. \quad (15)$$

It is elementary to derive this relation. The chiral condensate follows from the partition function (9) (all quark masses are chosen equal),

$$\langle\bar{\psi}\psi\rangle = -\lim \frac{1}{VN_f}\partial_m \log Z(m) = -\lim \frac{1}{V}\langle\sum_k \frac{2m}{\lambda_k^2 + m^2}\rangle. \quad (16)$$

If we express the sum as an integral over the average spectral density, and take the thermodynamic limit before the chiral limit, so that many eigenvalues are less than m, we recover (15). The order of the limits in (15) is important. First we take the thermodynamic limit, next the chiral limit and, finally, the field theory limit. As can be observed from (16) the sign of $\langle\bar{\psi}\psi\rangle$ changes if m crosses the imaginary axis.

An important consequence of the Bank-Casher formula (15) is that the eigenvalues near zero virtuality are spaced as

$$\Delta\lambda = 1/\rho(0) = \pi/\Sigma V. \tag{17}$$

This should be contrasted with the eigenvalue spectrum of the non-interacting Dirac operator. Then eq. (12) results in an eigenvalue spacing equal to $\Delta\lambda \sim 1/V^{1/4}$. Clearly, the presence of gauge fields leads to a strong modification of the spectrum near zero virtuality. Strong interactions result in the coupling of many degrees of freedom leading to extended states and correlated eigenvalues. On the other hand, for uncorrelated eigenvalues, the eigenvalue distribution factorizes, and for $\lambda \neq 0$,we have $\rho(\lambda) \sim \lambda^{2N_f+1}$ in the chiral limit, i.e. no breaking of chiral symmetry. One consequence of the interactions is level repulsion of neighboring eigenvalues. Therefore, the two smallest eigenvalues of the Dirac operator, $\pm\lambda_{\rm min}$ repel each other, and the Dirac spectrum will have a gap at $\lambda = 0$ with a width of the order of $1/\Sigma V$.

Spectral Correlations and Zeros of the Partition Function

The study of zeros of the partition function has been a fruitful tool in statistical mechanics[37, 38]. In QCD, both zeros in the complex fugacity plane and the complex mass plane have been studied[39, 40]. Since the QCD partition function is a polynomial in m it can be factorized as (all quark masses are taken to be equal to m)

$$Z_{\rm QCD}(m,\theta) = \prod_k (m - m_k). \tag{18}$$

Because configurations of opposite topological charge occur with the same probability, the coefficients of this polynomial are real, and the zeros occur in complex conjugate pairs. For an even number of flavors the zeros occur in pairs $\pm m_k$. In a sector with topological charge ν, this is also the case for even $N_f \times \nu$. The chiral condensate is given by

$$\Sigma(m) = -\lim \frac{1}{VN_f}\partial_m \log Z_{\rm QCD}(m,\theta) = -\lim \frac{1}{VN_f}\sum_k \frac{1}{m - m_k}. \tag{19}$$

For an even number of flavors, $\Sigma(m)$ is an odd function of m. In order to have a discontinuity at $m = 0$, the zeros in this region have to coalesce into a cut along the imaginary axis in the thermodynamic limit.

In the hypothetical case that the eigenvalues of the Dirac operator do not fluctuate the zeros are located at $m_k = \pm i\lambda_k$. In the opposite case, of uncorrelated eigenvalues, the eigenvalue distribution factorizes and all zeros are located at $\pm i\sigma$, where $\sigma^2 = \langle\lambda_k^2\rangle$. As a result, the chiral condensate does not show a discontinuity across the imaginary axis and is equal to zero.

We hope to convince the reader that the presence of a discontinuity is intimately related to correlations of eigenvalues[41]. Let us study the effect of pair correlations for one flavor in the sector of zero topological charge. The fermion determinant can be written as

$$\langle\prod_k(m^2+\lambda_k^2)\rangle = \sum_k \binom{N}{k} m^{2(N-k)}\langle\lambda_1^2\cdots\lambda_k^2\rangle. \tag{20}$$

There are

$$\binom{k}{2l}(2l-1)!!$$

ways of selecting l pairs from $\lambda_1^2\cdots\lambda_k^2$. The average of each pair of different eigenvalues is given by

$$\langle\lambda_m^2\lambda_n^2\rangle = \sigma^4 + C_2, \tag{21}$$

where σ^2 is the expectation value of λ_k^2 and C_2 is the connected correlator

$$C_2 = \langle\lambda_m^2\lambda_n^2\rangle - \langle\lambda_m^2\rangle\langle\lambda_n^2\rangle, \qquad m \neq n. \tag{22}$$

This results in the partition function

$$Z(m) = \sum_{k=0}^{N}\sum_{l=0}^{[\frac{k}{2}]} m^{2(N-k)}\binom{N}{k}\binom{k}{2l}(2l-1)!!C_2^l\sigma^{2(k-2l)}. \tag{23}$$

After interchanging the two sums, one can easily show that $Z(m)$ can be expressed as a multiple of a Hermite polynomial

$$Z(m) = (-\frac{C_2}{2})^{N/2}\mathrm{H}_N((\sigma^2+m^2)/\sqrt{-2C_2}). \tag{24}$$

In terms of the zeros of the Hermite polynomials, z_k, the zeros of the partition function are located at

$$m_k^2 = z_k\sqrt{-2C_2} - \sigma^2. \tag{25}$$

Asymptotically, the zeros of the Hermite polynomials are given by $z_k \approx \pi k/2\sqrt{N}$ (with integer $\pm k$). In order for the zeros to join into a cut in the thermodynamic limit, they have to be spaced as $\sim 1/N$. This requires that

$$C_2 \sim -\frac{1}{N}. \tag{26}$$

The density of zeros is then given by

$$\frac{dk}{dm} \sim Nm. \tag{27}$$

We conclude that pair correlations are sufficient to generate a cut of $Z(m)$ in the complex m-plane, but the chiral symmetry remains unbroken. Pair correlations alone cannot suppress the effect of the fermion determinant.

Leutwyler-Smilga Sum Rules

We have shown that pair-correlations are not sufficient to generate a discontinuity in the chiral condensate. In this subsection we start from the assumption that chiral symmetry is broken spontaneously, and look for consistency conditions that are imposed on the Dirac spectrum. As has been argued by Gasser and Leutwyler[9] and Leutwyler and Smilga[8], in the mesoscopic range (1), the mass dependence of the QCD partition function is given by (for simplicity, all quark masses have been taken equal)

$$Z^{\rm eff}(m,\theta) \sim \int_{U\in G/H} dU e^{mV\Sigma {\rm Re\,Tr}\, U e^{i\theta/N_f}}. \qquad (28)$$

The integral is over the Goldstone manifold associated with chiral symmetry breaking from G to H. For three or more colors with fundamental fermions $G/H = SU(N_f) \times SU(N_f)/SU(N_f)$. The finite volume partition function in the sector of topological charge ν follows by Fourier inversion according to (10). The partition function for $\nu = 0$ is thus given by (28) with the integration over $SU(N_f)$ replaced by an integral over $U(N_f)$. The case of $N_f = 1$ isparticularlysimple.Thenonlya $U(1)$ integration remains, and the partition function is given by[8] $Z^{\rm eff}_{\nu=0}(m) = I_0(mV\Sigma)$. Its zeros are regularly spaced along the imaginary axis in the complex m-plane, and, in the thermodynamic limit, they coalesce into a cut.

The Leutwyler-Smilga sum-rules are obtained by expanding the partition function $Z_\nu(m)$ in powers of m before and after averaging over the gauge field configurations and equating the coefficients. This corresponds to an expansion in powers of m of both the QCD partition function (2) and the finite volume partition function (28) in the sector of topological charge ν. As an example, we consider the coefficients of m^2 in the sector with $\nu = 0$. This results in the sum-rule

$$\langle \sum{}' \frac{1}{\lambda_k^2} \rangle = \frac{\Sigma^2 V^2}{4N_f}, \qquad (29)$$

where the prime indicates that the sum is restricted to nonzero positive eigenvalues.

The next order sum rules are obtained by equating the coefficients of order m^4. They can be combined into

$$\langle \sum_{k,l}{}' \frac{1}{\lambda_k^2 \lambda_l^2} \rangle - \langle \sum_k{}' \frac{1}{\lambda_k^2} \rangle \langle \sum_l{}' \frac{1}{\lambda_l^2} \rangle = \frac{\Sigma^4 V^4}{16 N_f^2 (N_f^2 - 1)}. \qquad (30)$$

We conclude that chiral symmetry breaking leads to correlations of the inverse eigenvalues. However, if one performs an analysis similar to the one in previous section, it can be shown easily that pair correlations given by (30) do not result in a cut in the complex m-plane. Apparently, chiral symmetry breaking requires a subtle interplay of all types of correlations.

For two colors with fundamental fermions or for adjoint fermions the pattern of chiral symmetry breaking is different. Sum rules for the inverse eigenvalues can be derived along the same lines. The general expression for the simplest sum-rule can be summarized as[42, 43]

$$\langle \sum{}' \frac{1}{\lambda_k^2} \rangle = \frac{\Sigma^2 V^2}{4(|\nu| + (\dim(G/H)+1)/N_f)}, \qquad (31)$$

where $\dim(G/H)$ is the number of generators of the coset manifold in (28).

The Leutwyler-Smilga sum-rules can be expressed as an integral over the average spectral density and spectral correlation functions. For the sum rule (29) this results in

$$\frac{1}{V^2\Sigma^2} \int \frac{\langle \rho(\lambda) \rangle d\lambda}{\lambda^2} = \frac{1}{4N_f}. \qquad (32)$$

If we introduce the microscopic variable

$$u = \lambda V \Sigma, \qquad (33)$$

this integral can be rewritten as

$$\int \frac{1}{V\Sigma}\langle\rho(\frac{u}{V\Sigma})\rangle\frac{du}{u^2} = \frac{1}{4N_f}. \tag{34}$$

The thermodynamic limit of the combination that enters in the integrand,

$$\rho_S(u) = \lim_{V\to\infty}\frac{1}{V\Sigma}\langle\rho(\frac{u}{V\Sigma})\rangle, \tag{35}$$

will be called the *microscopic spectral density*[10]. This limit exists if chiral symmetry is broken. Our conjecture is that $\rho_S(u)$ is a universal function that only depends on the global symmetries of the QCD partition function. Because of universality it can be derived from the simplest theory with the global symmetries of the QCD partition function. Such theory is a chiral Random Matrix Theory which will be introduced later in these lectures.

We emphasize again that the $U_A(1)$ symmetry of the QCD Dirac spectrum leads to two different types of eigenvalue correlations: spectral correlations in the bulk of the spectrum and spectral correlations near zero virtuality. The simplest example of correlations of the latter type is the microscopic spectral density defined in (35).

We close this subsection with two unrelated side remarks. First, the QCD Dirac operator is only determined up to a constant matrix. We can exploit this freedom to obtain a Dirac operator that is maximally symmetric. For example, the Wilson lattice QCD Dirac operator, D^W, is neither Hermitean nor anti-Hermitean, but $\gamma_5 D^W$ is Hermitean.

Second, the QCD partition function can be expanded in powers of m^2 before or after averaging over the gauge field configurations. In the latter case one obtains sum rules for the inverse zeros of the partition function. As an example we quote,

$$\sum \frac{1}{m_k^2}\Big|_{\nu=0} = \frac{\Sigma^2V^2}{4}, \tag{36}$$

where we have averaged over field configurations with zero topological charge.

SPECTRAL CORRELATIONS IN COMPLEX SYSTEMS

Statistical Analysis of Spectra

Spectra for a wide range of complex quantum systems have been studied both experimentally and numerically (a excellent recent review has been given by Guhr, Miiller-Groeling and Weidenmüller[24]). One basic observation is that the scale of variations of the average spectral density and the scale of the spectral fluctuations separate. This allows us to unfold the spectrum, i.e. we rescale the spectrum in units of the local average level spacing. Specifically, the unfolded spectrum is given by

$$\lambda_k^{\rm unf} = \int_{-\infty}^{\lambda_k}\langle\rho(\lambda')\rangle d\lambda', \tag{37}$$

with unfolded spectral density

$$\rho_{\rm unf}(\lambda) = \sum_k \delta(\lambda - \lambda_k^{\rm unf}). \tag{38}$$

The fluctuations of the unfolded spectrum can be measured by suitable statistics. We will consider the nearest neighbor spacing distribution, *P(S),* and moments of the number of levels in an interval containing n levels on average. In particular, we will consider the number variance, $\Sigma_2(n)$, and the first two cumulants, $\gamma_1(n)$ and $\gamma_2(n)$. Another useful statistic is the $\Delta_3(n)$-statistic introduced by Dyson and Mehta[44]. It is related to $\Sigma_2(n)$ via a smoothening kernel. The advantage of this statistic is that its fluctuations as a function of n are greatly reduced. Both $\Sigma_2(n)$ and $\Delta_3(n)$ can be obtained from the pair correlation function defined as

$$Y_2(\lambda, \lambda') = -\langle \rho_{\rm unf}(\lambda)\rho_{\rm unf}(\lambda')\rangle + \langle \rho_{\rm unf}(\lambda)\rangle\langle \rho_{\rm unf}(\lambda')\rangle. \tag{39}$$

Analytical expressions for the above statistics can be obtained for the eigenvalues of the invariant random matrix ensembles. They are defined as ensembles of Hermitean matrices with Gaussian independently distributed matrix elements, i.e. with probability distribution given by

$$P(H) \sim e^{-\frac{N\beta}{2}{\rm Tr} H^\dagger H}. \tag{40}$$

Depending on the anti-unitary symmetry, the matrix elements are real, complex or quaternion real. They are called the Gaussian Orthogonal Ensemble (GOE), the Gaussian Unitary Ensemble (GUE) and the Gaussian Symplectic Ensemble (GSE), respectively. Each ensemble is characterized by its Dyson index β which is defined as the number of independent variables per matrix element. For the GOE, GUE and the GSE we thus have $\beta = 1$, 2 and 4, respectively.

Independent of the value of β, the average spectral density is the semicircle,

$$\langle \rho(\lambda)\rangle = \frac{N}{\pi}\sqrt{2-\lambda^2}. \tag{41}$$

Analytical results for all spectral correlation functions have been derived for each of the three ensembles[45] via the orthogonal polynomial method. We only quote the most important results. The nearest neighbor spacing distribution, which is known exactly in terms of a power series, is well approximated by

$$P(S) \sim S^\beta \exp(-a_\beta S^2), \tag{42}$$

where a_β is a constant of order one. The asymptotic behaviour of the pair correlation function is given by[45]

$$Y_2(\lambda, \lambda') \sim \frac{1}{\pi^2(\lambda-\lambda')^2} \quad \text{for} \quad \beta = 1, \tag{43}$$

$$Y_2(\lambda, \lambda') \sim \frac{\sin^2 \pi(\lambda-\lambda')}{\pi^2(\lambda-\lambda')^2} \quad \text{for} \quad \beta = 2, \tag{44}$$

$$Y_2(\lambda, \lambda') \sim -\frac{\cos 2\pi(\lambda-\lambda')}{4(\lambda-\lambda')} + \frac{1+(\pi/2)\sin 2\pi(\lambda-\lambda')}{4\pi^2(\lambda-\lambda')^2} \quad \text{for} \quad \beta = 4. \tag{45}$$

The $1/(\lambda-\lambda')^2$ tail of the pair correlation function results in a logarithmic dependence of the asymptotic behavior of $\Sigma_2(n)$ and $\Delta_3(n)$,

$$\Sigma_2(n) \sim (2/\pi^2\beta)\log n \quad \text{and} \quad \Delta_3(n) \sim \beta\Sigma_2(n)/2. \tag{46}$$

Characteristic features of random matrix correlations are level repulsion at short distances and a strong suppression of fluctuations at large distances.

For uncorrelated eigenvalues the level repulsion is absent and one finds

$$P(S) = \exp(-S), \tag{47}$$

and

$$\Sigma_2(n) = n \quad \text{and} \quad \Delta_3(n) = n/15. \tag{48}$$

Spectral Universality

The main conclusion of numerous studies of eigenvalue spectra of complex systems is that spectral correlations of classically chaotic systems are given by RMT[24]. As illustration of this so called Bohigas conjecture, we mention three examples from completely different fields. The first example is the nuclear data ensemble in which the above statistics are evaluated by superimposing level spectra of many different nuclei[25]. The second example concerns correlations of acoustic resonances in irregular quartz blocks[26]. In both cases the statistics that were considered are, within experimental accuracy, in complete agreement with the GOE statistics. The third example pertains to the zeros of Riemann's zeta function. Extensive numerical calculations[46] have shown that asymptotically, for large imaginary part, the correlations between the zeros are given by the GUE.

The Gaussian random matrix ensembles introduced above can be obtained[45] from two assumptions: i) The probability distribution is invariant under unitary transformations; ii) The matrix elements are statistically independent. If the invariance assumption is dropped it can be shown with the theory of free random variables[47] that the average spectral density is still given by a semicircle if the variance of the probability distribution is finite. For example, if the matrix elements are distributed according to a rectangular distribution, the average spectral density is a semicircle. On the other hand, if the independence assumption is released the average spectral density is typically not a semicircle. For example, this is the case if the quadratic potential in the probability distribution is replaced by a more complicated polynomial potential $V(H)$. Using the supersymmetric method for Random Matrix Theory, it was shown by Hackenbroich and Weidenmüller[48] that the same supersymmetric nonlinear σ-model is obtained for a wide range of potentials $V(H)$. This implies that spectral correlations of the unfolded eigenvalues are independent of the potential. Remarkably, this result could be proved for all three values of the Dyson index.

Several examples have been considered where both the invariance assumption and the independence assumption are relaxed. We mention $H \to H + A$, where A is an arbitrary fixed matrix, and the probability distribution of H is given by a polynomial $V(H)$. It was shown by P. Zinn-Justin[49] that also in this case the spectral correlations are given by the invariant random matrix ensembles. For a Gaussian probability distribution the proof was given by Brézin and Hikami[50].

The domain of universality has been extended in the direction of real physical systems by means of the Gaussian embedded ensembles[51, 52]. The simplest example is the ensemble of matrix elements of n-particle Slater determinants of a two-body operator with random two-particle matrix elements. It can be shown analytically that the average spectral density is a Gaussian. However, according to substantial numerical evidence, the spectral correlations are in complete agreement with the invariant random matrix ensembles[51].

A large number of examples have been found that fall into one of universality classes of the invariant random matrix ensembles. This calls out for a more general approach. Naturally, one thinks in terms of the renormalization group. This approach was pioneered by Brézin and Zinn-Justin[56]. The idea is to integrate out rows and columns of a random matrix and to show that the Gaussian ensembles are a stable fixed point. This was made more explicit in a paper by Higuchi *et al.*[57]. However, much more work is required to arrive at a natural proof of spectral universality.

Although the above mentioned universality studies provide support for the validity of the Bohigas conjecture, the ultimate goal is to derive it directly from the underlying classical dynamics. An important first step in this direction was made by Berry[55]. He showed that the asymptotics of the two-point correlation function is related to sum-rules for isolated classical trajectories. Another interesting approach was introduced by Andreev *et al*[58] who were able to obtain a supersymmetric nonlinear sigma model from spectral averaging. In this context we also mention the work of Altland and Zirnbauer[59] who showed that the kicked rotor can be mapped onto a supersymmetric sigma model.

CHIRAL RANDOM MATRIX THEORY

Introduction of the Model

In this section we will introduce an instanton liquid[60, 61] inspired chiral RMT for the QCD partition function. In the spirit of the invariant random matrix ensembles we construct a model for the Dirac operator with the global symmetries of the QCD partition function as input, but otherwise Gaussian random matrix elements. The chRMTthatobeystheseconditionsisdefinedby[10,62,63,64]where

$$Z^{\beta}_{N_f,\nu}(m_1,\cdots,m_{N_f}) = \int DW \prod_{f=1}^{N_f} \det(\mathcal{D}+m_f)e^{-\frac{N\beta}{4}{\rm Tr}V(W^\dagger W)}, \tag{49}$$

where

$$\mathcal{D} = \begin{pmatrix} 0 & iW \\ iW^\dagger & 0 \end{pmatrix}, \tag{50}$$

and W is a $n \times m$ matrix with $\nu = |n-m|$ and $N = n+m$. As is the case in QCD, we assume that ν does not exceed $\sqrt{N}$, so that, to a goodapproximation, $n = N/2$. The parameter N is identified as the dimensionless volume of space time. The potential V is defined by

$$V(\phi) = \sum_{k\geq 1} \frac{a_k}{k}\phi^k. \tag{51}$$

The simplest case is the Gaussian case when $V(\phi) = \Sigma^2\phi$. Below it will be shown that the microscopic spectral density is independent of the higher order terms in this potential provided that the average spectral density near zero remains nonzero. The matrix elements of W are either real ($\beta = 1$, chiral Gaussian Orthogonal Ensemble (chGOE)), complex ($\beta = 2$, chiral Gaussian Unitary Ensemble (chGUE)), or quaternion real ($\beta = 4$, chiral Gaussian Symplectic Ensemble (chGSE)). This partition function is invariant under

$$W \to U^\dagger W V \tag{52}$$

where the $n \times n$ matrix U and the $m \times m$ matrix V are orthogonal matrices for $\beta = 1$, unitary matrices for $\beta = 2$, and symplectic matrices for $\beta = 4$. This invariance makes it possible to express the partition function in terms of eigenvalues of W defined by

$$W = U^\dagger \Lambda V. \tag{53}$$

Here, Λ is a diagonal matrix with diagonal matrix elements $\lambda_k \geq 0$. In terms of the eigenvalues the partition function (49) is given by

$$Z^{\beta}_{N_f,\nu}(m_1,\cdots,m_{N_f}) = \int d\lambda |\Delta(\lambda_k^2)|^\beta \prod_k \lambda_k^{2N_f+\beta\nu+\beta-1} \prod_f m_f^\nu \prod_{f,k}(\lambda_k^2+m_f^2)e^{-\frac{N\beta}{4}\sum_k V(\lambda_k^2)}, \tag{54}$$

where the Vandermonde determinant is defined by

$$\Delta(\lambda_k^2) = \prod_{k<l}(\lambda_k^2 - \lambda_l^2). \tag{55}$$

This model reproduces the following symmetries of the QCD partition function:

- The $U_A(1)$ symmetry. All nonzero eigenvalues of the random matrix Dirac operator occur in pairs $\pm\lambda$ or are zero.
- The topological structure of the QCD partition function. The Dirac matrix has exactly $|\nu| \equiv |n - m|$ zero eigenvalues. Thisidentifies ν as the topological sector of the model.
- The flavor symmetry is the same as in QCD. For $\beta = 2$ it is $SU(N_f) \times SU(N_f)$, for $\beta = 1$ it is $SU(2N_f)$ and for $\beta = 4$ it is $SU(N_f)$.
- The chiral symmetry is broken spontaneously with chiral condensate given by

  $$\Sigma = \lim_{N\to\infty} \pi\rho(0)/N. \tag{56}$$

  (N is interpreted as the (dimensionless) volume of space time.) The symmetry breaking pattern is[42] $SU(N_f) \times SU(N_f)/SU(N_f)$, $SU(2N_f)/Sp(N_f)$ and $SU(N_f)/O(N_f)$ for $\beta = 2$, 1 and 4, respectively, the same as in QCD[65].
- The anti-unitary symmetries. For three or more colors with fundamental fermions the Dirac operator has no anti-unitary symmetries, and generically, the matrix elements of the Dirac operator are complex. The matrix elements W_{kl} of the corresponding random matrix ensemble are chosen arbitrary complex as well ($\beta =$ 2). For $N_c = 2$, the Dirac operator in the fundamental representation satisfies

  $$[C\tau_2 K, i\gamma D] = 0, \tag{57}$$

  where C is the charge conjugation matrix and K is the complex conjugation operator. Because, $(C\tau_2 K)^2 = 1$, the matrix elements of the Dirac operator can always be chosen real, and the corresponding random matrix ensemble is defined with real matrix elements ($\beta = 1$). For two or more colors with gauge fields in the adjoint representation the anti-unitary symmetry of the Dirac operator is given by

  $$[CK, i\gamma D] = 0. \tag{58}$$

  Because $(CK)^2 = -1$, it is possible to rearrange the matrix elements of the Dirac operator into real quaternions. The matrix elements W_{kl} of the corresponding random matrix ensemble are chosen quaternion real ($\beta = 4$).

Together with the invariant random matrix ensembles, the chiral ensembles are part of a larger classification scheme. Apart from the random matrix ensembles discussed in this review, this classification also includes random matrix models for disordered super-conductors[66]. As pointed out by Zirnbauer[67], all known universality classes of Hermitean random matrices are tangent to the large classes of symmetric spaces in the classification given by Cartan. There is a one-to-one correspondence between this classification and the classification of the large families of Riemannian symmetric superspaces[67].

Calculation of the Microscopic Spectral Density

The joint eigenvalue distribution of the nonzero eigenvalues for zero masses follows immediately from the partition function (54). For N_f flavors and topological charge ν the result for arbitrary potential is given by[62]

$$\rho_\beta(\lambda_1, \cdots, \lambda_n) = C_{\beta,n} \prod_{k<l} |\lambda_k^2 - \lambda_l^2|^\beta \prod_k \lambda_k^{2N_f+\beta\nu+\beta-1} \exp\left(-\frac{n\beta}{2} \sum_k V(\lambda_k^2)\right), \quad (59)$$

where $C_{\beta,n}$ is a normalization constant. For $\beta = 2$ the average spectral density and the spectral correlation functions can be derived from (59) with the help of the orthogonal polynomial method[45]. The orthogonal polynomials in the variable $x = \lambda^2$ are denned by

$$\int_0^\infty dx x^a e^{-nV(x)} P_n^a(x) P_m^a(x) = h_n^a \delta_{mn}. \quad (60)$$

where

$$a = N_f + |\nu| \quad \text{for} \quad \beta = 2. \quad (61)$$

In the Gaussian case, $V(\lambda) = \Sigma^2\lambda$, the associated polynomials are the generalized Laguerre polynomials. That is why this ensemble is also known as the Laguerre ensemble[68, 69]. The Vandermonde determinant can be rewritten as $\Delta(x_i) = \det_{kl}[x_k^{l-l}]$. By the addition of linear combinations of rows this determinant can be expressed in terms of the orthogonal polynomials (60), $\Delta(x_i) \sim \det_{kl}[P_{l-1}^a(x_k)]$. By multiplying the two determinants one obtains

$$\rho_\beta(x_1, \cdots, x_n) = \frac{1}{n!} \det K^a(x_k, x_l), \quad (62)$$

where the kernel is defined by

$$K^a(x,y) = \sum_{k=0}^{n-1} \frac{1}{h_k^a} P_k^a(x) P_k^a(y) (xy)^{a/2} e^{-n(V(x)+V(y))/2}. \quad (63)$$

The average spectral density is obtained by integrating over $x_2, \cdots, x_n$,

$$\rho(x) = n \int \prod_{k=2}^n dx_k \rho(x, x_2, \cdots, x_n) = K^a(x,x). \quad (64)$$

In terms of the original variables, $\lambda = \sqrt{x}$, the spectral density is given by

$$\rho(\lambda) = 2\lambda K^a(\lambda^2, \lambda^2). \quad (65)$$

In the Gaussian case the spectral density is thus given by

$$\rho(\lambda) = 2\lambda^{2a+1}\Sigma^{2a+2}n^{a+1}e^{-n\Sigma^2\lambda^2} \sum_{k=0}^{n-1} \frac{1}{h_k^a} L_k^a(\lambda^2\Sigma^2 n) L_k^a(\lambda^2\Sigma^2 n), \quad (66)$$

where the L_k^a are the generalized Laguerre polynomials. With the help of the Christoffel-Darboux formula the sum can be expressed into the nth order Laguerre polynomial and its derivative (with $L_n^{a\prime}(x) = -L_{n-1}^{a+1}(x)$),

$$\rho(\lambda) = \frac{2\Sigma n^{-a} n!}{\Gamma(a+n)} \left(\frac{z}{2}\right)^{2a+1} e^{-z^2/4n} \left(L_{n-1}^a\left(\frac{z^2}{4n}\right) L_{n-1}^{a+1}\left(\frac{z^2}{4n}\right) - L_n^a\left(\frac{z^2}{4n}\right) L_{n-2}^{a+1}\left(\frac{z^2}{4n}\right) \right). \quad (67)$$

Here, we introduced the microscopic variable

$$z = N\Sigma\lambda = 2n\Sigma\lambda, \tag{68}$$

and used the explicit expression for the normalization of the Laguerre polynomials, $h_k^a = \Gamma(a + k + 1)/k!$. The microscopic limit is defined by

$$\lim_{N\to\infty} \frac{1}{N\Sigma}\rho(\frac{\lambda}{N\Sigma}) \tag{69}$$

and can be evaluated with the help of the asymptotic relation

$$\lim_{n\to\infty} \frac{1}{n^\alpha} L_n^\alpha(\frac{x}{n}) = x^{-\frac{\alpha}{2}} J_\alpha(2\sqrt{x}), \tag{70}$$

where J_α is the ordinary Bessel function of degree α. In order to take the this limit most conveniently, we substitute the recursion relations

$$\begin{aligned} L_{n-1}^{a+1}(z) &= L_{n-1}^a(z) + L_{n-2}^{a+1}(z), \\ L_n^a(z) &= L_{n-1}^a(z) + L_n^{a-1}(z). \end{aligned} \tag{71}$$

This results in the microscopic spectral density

$$\rho_S(z) = \frac{z}{2}(J_a^2(z) - J_{a+1}(z)J_{a-1}(z)). \tag{72}$$

From the asymptotic relation for the Bessel function

$$J_\nu(z) \sim \left(\frac{2}{\pi z}\right)^{1/2} \cos(z - \frac{\pi}{2}\nu - \frac{\pi}{4}) \quad \text{for} \quad z \to \infty \tag{73}$$

we find that

$$\lim_{x\to\infty} \rho_S(x) = \frac{1}{\pi}. \tag{74}$$

The result for the average spectral density follows from the asymptotic properties of the Laguerre polynomials. It given by the semicircular distribution

$$\rho(\lambda) = (n\Sigma^2/\pi)\sqrt{4/\Sigma^2 - \lambda^2}. \tag{75}$$

Notice that the microscopic limit of the average spectral density coincides with the asymptotic limit (74) of the microscopic spectral density.

The two-point correlation function is obtained by integrating the joint spectral density over all eigenvalues except two. The microscopic limit can again be expressed in terms of Bessel functions[63].

The spectral density and the two-point correlation function were also derived within the framework of the supersymmetric method of Random Matrix Theory[70].

The calculation of the average spectral density and the spectral correlations functions for $\beta = 1$ and $\beta = 4$ is much more complicated. However, with the help of skew-orthogonal polynomials[71,72,73] exact analytical results for finite N can be obtained as well.

The microscopic spectral density for $SU(2)$ with fundamental fermions ($\beta = 1$) is given by[74]

$$\rho_S(z) = \frac{1}{4}J_{2a+1}(z) + \frac{1}{2}\int_0^\infty dw(zw)^{2a+1}\epsilon(z-w)\left(\frac{1}{w}\frac{d}{dw} - \frac{1}{z}\frac{d}{dz}\right) \times \frac{wJ_{2a}(z)J_{2a-1}(w) - zJ_{2a-1}(z)J_{2a}(w)}{(zw)^{2a}(z^2 - w^2)}, \tag{76}$$

where a is the combination

$$a = N_f - \frac{1}{2} + \frac{|\nu|}{2}. \tag{77}$$

The microscopic spectral density in the symmetry class with $\beta = 4$ was first calculated by Nagao and Forrester[75]. It is given by

$$\rho_S(z) = 2z^2 \int_0^1 du u^2 \int_0^1 dv \ \ [J_{4a-1}(2uvz)J_{4a}(2uz) - vJ_{4a-1}(2uz)J_{4a}(2uvz)] \tag{78}$$

with $4a = N_f + 2|\nu| + 1$.

The spectral correlations in the bulk of the spectrum are given by the invariant random matrix ensemble with the same value of β. For $\beta = 2$ this was already shown three decades ago by Fox and Kahn[68]. For $\beta = 1$ and $\beta = 4$ this was only proved recently[73].

Duality between Flavor and Topology

As one can observe from the joint eigenvalue distribution, for $\beta = 2$ the dependence on N_f and ν enters only through the combination $N_f + |\nu|$. This allows for the possibility of trading topology for flavors. In this section we will work out this duality for the finite volume partition function. This relation completes the proof of the conjectured expression[76] for the finite volume partition function for different quark masses and topological charge ν.

For $\beta = 2$ the partition function (54) obeys the relation

$$\frac{Z_{N_f,\nu}(m_1, \cdots, m_{N_f})}{\prod_f m_f^\nu} \sim Z_{N_f+\nu,0}(m_1, \cdots, m_{N_f}, 0, \cdots, 0), \tag{79}$$

where the argument of the last factor has ν zeros.

As an example, the simplest nontrivial identity of this type is given by

$$Z_{1,1}(m) \sim m Z_{2,0}(m, 0). \tag{80}$$

Let us prove this identity without relying on Random Matrix Theory. According to the definition (28) we have

$$Z_{1,1}(m) \sim \int d\theta e^{i\theta} e^{mV\Sigma \cos\theta}, \tag{81}$$

and

$$Z_{2,0}(m, 0) \sim \int_{U \in U(2)} dU e^{V\Sigma \mathrm{Re\,Tr}(MU)}, \tag{82}$$

where M is a diagonal matrix with diagonal elements m and 0. The integral over U can be performed by diagonalizing U according to $U = U_1 e^{i\phi_k} U_1^{-1}$, and choosing U_1 and ϕ_k as new integration variables. The Jacobian of this transformation is

$$J \sim \Delta^2(e^{i\phi_k}). \tag{83}$$

The integral over U_1 can be performed using the Itzykson-Zuber formula. This results in

$$Z_{2,0} \sim \int d\phi_1 d\phi_2 \frac{|e^{-i\phi_1} - e^{-i\phi_2}|^2}{m(\cos\phi_1 - \cos\phi_2)} (e^{mV\Sigma\cos\phi_1} - e^{mV\Sigma\cos\phi_2}). \tag{84}$$

Both terms in the last factor result in the same contribution to the integral. Let us consider only the first term $\sim \exp(mV\Sigma\cos\phi_1)$. Then the integral over ϕ_2 has to be defined as a principal value integral. If we use the identity

$$\frac{|e^{-i\phi_1} - e^{-i\phi_2}|^2}{(\cos\phi_1 - \cos\phi_2)} = 2\left(\cos\phi_1 - \frac{\sin\phi_1}{\tan((\phi_1+\phi_2)/2)}\right), \tag{85}$$

the ϕ_2-integral of the term proportional to $\sin\phi_1$ gives zero because of the principal value prescription. The term proportional to $\cos\phi_1$ trivially results in $Z_{1,1}$. We leave it as an exercise to the reader to generalize this proof to arbitrary N_f and ν.

The group integrals in finite volume partition function (28) were evaluated by Leutwyler and Smilga[8] for *equal* quark masses. An expression for *different* quark masses was obtained by Jackson *et al.*[76]. The expression could only be proved for $\nu = 0$. The above duality can be used to relate a partition function at arbitrary ν to a partition function at $\nu = 0$. This completes the proof of the conjectured expression for arbitrary topological charge.

UNIVERSALITY IN CHIRAL RANDOM MATRIX THEORY

In the chiral ensembles, two types of universality studies can be performed. First, the universality of correlations in the bulk of the spectrum. As discussed above, they are given by the invariant random matrix ensembles. Second, the universality of the microscopic spectral density and the eigenvalue correlations near zero virtuality. The aim of such studies is to show that microscopic correlations are stable against deformations of the chiral ensemble away from the Gaussian probability distribution. Recently, a number of universality studies on microscopic correlations have appeared. They will be reviewed in this section.

We wish to emphasize that all universality studies for the chiral ensembles have been performed for $\beta = 2$. The reason is that $\beta = 1$ and $\beta = 4$ are mathematically much more involved. It certainly would be of interest to extend such studies to these cases as well.

In addition to the analytical studies to be discussed below the universality of the microscopic spectral density also follows from numerical studies of models with the symmetries of the QCD partition function. In particular, we mention strong support in favor of universality from a different branch of physics, namely from the theory of universal conductance fluctuations. In that context, the microscopic spectral density of the eigenvalues of the transmission matrix was calculated for the Hofstadter[81] model, and, to a high degree of accuracy, it agrees with the random matrix prediction[82]. Other studies deal with a class random matrix models with matrix elements with a diverging variance. Also in this case the microscopic spectral density is given by the universal expressions[83] (76) and (72).

The conclusion that emerges from all numerical and analytical work on modified chiral random matrix models is that the microscopic spectral density and the correlations near zero virtuality exhibit a strong universality that is comparable to the stability of microscopic correlations in the bulk of the spectrum.

Of course, QCD is much richer than chiral Random Matrix Theory. One question that should be asked is at what scale (in virtuality) QCD spectral correlations deviate from RMT. This question has been studied by means of instanton liquid simulations. Indeed, at macroscopic scales, it was found that the number variance shows a linear dependence instead of the logarithmic dependence observed at microscopic scales[84]

More work is needed to determine the point where the crossover between these two regimes takes place.

Invariant Deformations of the Gaussian Random Matrix Ensembles

In a first class of universality studies one considers probability distributions that maintain unitary invariance. In this case the joint probability distribution is given by (59). In this section we consider the simplest nontrivial case with only a_1 and a_2 different from zero and present the proof of Akemann, Damgaard, Magnea and Nishigaki[11,12] for this case. We wish to point out that the general case only leads to minor complications.

This case was first studied by Brézin, Hikami and Zee[13]. They showed that the microscopic spectral density is independent of a_2. A general proof valid for arbitrary potential was given by Akemann, Damgaard, Magnea and Nishigaki[11, 12]. The essence of the proof is a remarkable generalization of the identity for the Laguerre polynomials,

$$\lim_{n\to\infty} L_n\left(\frac{x}{n}\right) = J_0(2\sqrt{x}) \,, \tag{86}$$

to orthogonal polynomials determined by an arbitrary potential V. This relation was proved by deriving a differential equation from the continuum limit of the recursion relation for orthogonal polynomials. This proof has been extended to the microscopic correlation functions of all chiral ensembles in a recent work by Akemann, Damgaard, Magnea and Nishigaki[12].

In the normalization $P_k(0) = 1$, the orthogonal polynomials (60) satisfy the recursion relation

$$xP_k(x) = -r_k\left[P_{k+1}(x) - P_k(x) - Q_k(P_k(x) - P_{k-1}(x))\right], \tag{87}$$

where

$$r_k = -\frac{p_k}{p_{k+1}} \quad \text{and} \quad Q_k = \frac{h_k r_{k-1}}{r_k h_{k-1}}. \tag{88}$$

Here, the coefficient of x^k in $P_k(x)$ is denoted by p_k. The fractions r_k and the normalizations h_k can be determined from the relations

$$\int_0^\infty dx\; nV'(x)e^{-nV(x)}P_k(x)P_k(x) = -\int_0^\infty dx\frac{d}{dx}\left[e^{-nV(x)}P_k(x)P_k(x)\right] = 1, \tag{89}$$

$$\int_0^\infty dx\; nV'(x)e^{-nV(x)}xP_k(x)P_k(x) - h_k - 2kh_k = -\int_0^\infty dx\frac{d}{dx}\left[e^{-nV(x)}xP_k(x)P_k(x)\right] = 0. \tag{90}$$

For the potential

$$V(x) = a_1 x + \frac{a_2}{2}x^2, \tag{91}$$

these relations reduce to

$$na_1 h_k + na_2 r_k h_k(1+Q_k) = 1, \tag{92}$$

$$na_1 r_k h_k(1+Q_k) + na_2 r_k^2\left(h_{k+1} + h_k(1+Q_k)^2 + h_{k-1}Q_k^2\right) - h_k - 2kh_k = 0. \tag{93}$$

In order to proceed we take the continuum limit of these relations, i.e. $k \to \infty$ and $n \to \infty$ at fixed $k/n = t$. With

$$
\begin{aligned}
r(t) &= r(k/n) + \mathcal{O}(1/n), \\
h(t) &= \frac{1}{n} h(k/n) + \mathcal{O}(1/n^2),
\end{aligned}
\tag{94}
$$

to leading order in $1/n$, these recursion relations can be rewritten as

$$
a_1 + 2a_2 r(t) = \frac{1}{h(t)}, \tag{95}
$$

$$
a_1 r(t) + 3a_2 r^2(t) = t. \tag{96}
$$

The functions $r(t)$ and $h(t)$ are given by the solution of these equations. However, we do not need the explicit solution. A more useful property is that they satisfy a differential equation that does not depend on a_1 and a_2:

$$
r'(t)h(t) - 2h'(t)r(t) = h^2(t). \tag{97}
$$

Remarkably, as was shown by Akemann, Damgaard, Magnea and Nishigaki[11,12], this relation is valid for any polynomial potential. This relation is the essential ingredient in reducing the continuum limit of the recursion relation (87) to a Bessel equation.

To leading order in $1/n$ the r.h.s. of the recursion relation (87) is zero. We have to collect the terms of subleading order. In the continuum limit we may write

$$
\begin{aligned}
r_{k-1} &= r(t) - \frac{1}{n} r'(t), \\
h_{k-1} &= \frac{1}{n} h(t) - \frac{1}{n^2} h'(t), \\
P_{k-1}(x) &= P(t,x) - \frac{1}{n}\frac{d}{dt} P(t,x), \quad \text{etc..}
\end{aligned}
\tag{98}
$$

This results in the differential equation

$$
n^2 x P(t,x) = -h(t) \frac{d}{dt} \frac{r(t)}{h(t)} \frac{d}{dt} P(t,x). \tag{99}
$$

We observe that the continuum limit of the recursion relation exists if we take at the same time the microscopic limit with fixed $n\sqrt{x}$. If we introduce the new variables

$$
\begin{aligned}
v &= n\lambda = n\sqrt{x}, \\
u(t) &= \frac{2\sqrt{r(t)}}{h(t)} = \int_0^t \frac{dt'}{\sqrt{r(t')}}
\end{aligned}
\tag{100}
$$

the differential equation reduces to the Bessel equation

$$
\frac{1}{u}\frac{d}{du} u \frac{d}{du} P(u, v^2) + v^2 P(u, v^2) = 0. \tag{101}
$$

Here, and below we omit the argument t of $u(t)$. The general solution of this Bessel equation is given by

$$
P(u, v^2) = A J_0(uv) + B N_0(uv) \tag{102}
$$

The constants are determined by the boundary condition that the orthogonal polynomials are normalized as $P_n(0) = 1$. This results in $A = 1$ and $B = 0$.

The average spectral density follows from (63) and (64). For $a = 0$ (we omit the superscript a) we obtain

$$\rho(\lambda) = 2\lambda e^{-nV(\lambda^2)} \sum_{k=0}^{n-1} \frac{1}{h_k} P_k(\lambda^2) P_k(\lambda^2). \tag{103}$$

With the help of the Christoffel-Darboux formula the sum over the polynomials can be expressed in $P_{n-1}(\lambda^2)$ and its derivative. This results in

$$\rho(\lambda) = 2\lambda e^{-nV(\lambda^2)} \frac{r_{n-1}}{h_{n-1}} \left(P'_{n-1}(\lambda^2) P_n(\lambda^2) - P'_n(\lambda^2) P_{n-1}(\lambda^2) \right), \tag{104}$$

where the prime denotes differentiation with respect to λ^2. The leading order terms contributing to the continuum limit of this equation cancel. The subleading terms follow from the third equation in (98). In terms of the new variables we have,

$$P_{n-1}(\lambda^2) \sim P_n(\lambda^2) - \frac{1}{n}\frac{d}{dt} P(u, \lambda^2) = P_n(\lambda^2) - \frac{1}{n}\frac{d}{dt} J_0(uv), \tag{105}$$

$$P'_n(\lambda^2) = \frac{d}{d\lambda^2} P_n(\lambda^2) \sim n^2 \frac{d}{dv^2} J_0(uv). \tag{106}$$

Taking both the continuum limit and the microscopic limit, eq. (104) reduces to

$$\begin{aligned} \rho(\lambda) &= 2\lambda n^2 \frac{r(t)}{h(t)} \left(-\frac{1}{2v} J_0(uv) \frac{d}{dv}\frac{d}{dt} J_0(uv) + \frac{1}{2v}\frac{d}{dv} J_0(uv) \frac{d}{dt} J_0(uv) \right) \\ &= \lambda n^2 \frac{r(t) u u'}{h(t)} \left(-\frac{1}{vu} J_0'(uv) J_0(uv) - J_0''(uv) J_0(uv) + J_0'(uv) J_0'(uv) \right). \end{aligned} \tag{107}$$

All polynomials that enter in the precursor to this equation are of the n'th order. Therefore, we can put $t = 1$. If we also use the relations $J_0'(z) = -J_1(z)$ and $J_1' = -J_1(z)/z + J_0(z)$, and the definition of u (see eq. (100)), we obtain

$$\rho(\lambda) = \frac{1}{2} \lambda n^2 u^2(1) \left(J_0(u(1)v) J_0(u(1)v) + J_1(u(1)v) J_1(u(1)v) \right). \tag{108}$$

For $\lambda \to \infty$ the microscopic spectral density should coincide with $\rho(0)$. This allows us to identify $u(1)$ as

$$u(1) = \frac{\pi \rho(0)}{n}. \tag{109}$$

For the microscopic spectral density defined by (with $z = N\Sigma\lambda = \pi\rho(0)\lambda$, notice that $N = 2n$)

$$\rho_S(z) = \lim_{N\to\infty} \frac{1}{N\Sigma} \rho(\frac{z}{N\Sigma}) = \lim_{n\to\infty} \frac{1}{\pi\rho(0)} \rho(\frac{z}{\pi\rho(0)}), \tag{110}$$

we obtain

$$\rho_S(z) = \frac{z}{2} (J_0^2(z) + J_1^2(z)), \tag{111}$$

in agreement with the result (72) derived for the Gaussian chiral ensemble. This result is independent of the specific shape of the random matrix potential. This proves its universality.

Noninvariant Deformations of the Chiral Gaussian Random Matrix Ensembles

In a second class of universality studies one considers deformations of the Gaussian random matrix ensemble that violate unitary invariance. In particular, one has considered the case where the matrix W in (50) is replaced by

$$W \to W + A, \tag{112}$$

whereas the probability distribution of W is Gaussian. Because of the unitary invariance, the matrix A can always be chosen diagonal. The simplest case with $A = \pi T$ times the identity was considered by Jackson *et al.*[15]. This model provides a schematic model of the chiral phase transition. In this section we will give a detailed derivation of the average spectral density. The aim is to shown that the spectrum changes dramatically with variations of the temperature parameter. Nevertheless, it could be shown[15] that the microscopic spectral density is temperature independent in the broken phase.

We will show that for large matrices, the average resolvent defined by

$$G(z) = \frac{1}{2n}\mathrm{Tr}\left\langle \frac{1}{z + i\epsilon - \mathcal{D}} \right\rangle \tag{113}$$

obeys the cubic equation[77, 78, 15] (the parameter $\Sigma = 1$ in (49))

$$G^3 - 2zG^2 + G(z^2 - \pi^2T^2 + 1) - z = 0. \tag{114}$$

Here, $\mathcal{D}$ is the ensemble of random matrices

$$\mathcal{D} = \begin{pmatrix} 0 & W + \pi T \\ W^\dagger + \pi T & 0 \end{pmatrix}, \tag{115}$$

with probability distribution given by

$$P(W) = \exp(-n\mathrm{Tr}WW^\dagger). \tag{116}$$

The average over $P(W)$ is denoted by the brackets $\langle\cdots\rangle$. Because the operator (115) has only a finite support, it is possible to expand the resolvent in a geometric series in $1/(z - K)$ for z sufficiently large. Here, K is the matrix

$$K = \begin{pmatrix} 0 & \pi T \\ \pi T & 0 \end{pmatrix}. \tag{117}$$

One finds by inspection that $G(z)$ satisfies

$$G(z) = \mathrm{Tr}\frac{1}{z-K} + \mathrm{Tr}\frac{1}{z-K}\left\langle \begin{pmatrix} 0 & W \\ W^\dagger & 0 \end{pmatrix} \mathcal{G} \begin{pmatrix} 0 & W \\ W^\dagger & 0 \end{pmatrix} \mathcal{G} \right\rangle \tag{118}$$

where $\mathcal{G}$ is the matrix

$$\mathcal{G} = \left\langle \frac{1}{z - H} \right\rangle. \tag{119}$$

It should be clear that $\mathcal{G}$ is block diagonal with the block structure

$$\mathcal{G} = \begin{pmatrix} g\mathbf{1}_n & h\mathbf{1}_n \\ h\mathbf{1}_n & g\mathbf{1}_n \end{pmatrix}, \tag{120}$$

where $\mathbf{1_n}$ is the $n \times n$ identity matrix. Therefore, we find that $G(z) = g$. The average over W can be carried out immediately to give

$$\left\langle \begin{pmatrix} 0 & W \\ W^\dagger & 0 \end{pmatrix} \mathcal{G} \begin{pmatrix} 0 & W \\ W^\dagger & 0 \end{pmatrix} \right\rangle = \frac{1}{n} \begin{pmatrix} g\mathbf{1}_n & 0 \\ 0 & g\mathbf{1}_n \end{pmatrix} . \tag{121}$$

This yields the following matrix equation for g and h:

$$\begin{pmatrix} z & -\pi T \\ -\pi T & z \end{pmatrix} \begin{pmatrix} g & h \\ h & g \end{pmatrix} = \mathbf{1} + \begin{pmatrix} g & 0 \\ 0 & g \end{pmatrix} \begin{pmatrix} g & h \\ h & g \end{pmatrix} , \tag{122}$$

which leads to the two independent equations

$$\begin{aligned} zg - \pi T h &= 1 + g^2 , \\ zh - \pi T g &= gh . \end{aligned} \tag{123}$$

Elimination of h yields the equation

$$zg - \frac{\pi^2 T^2 g}{z - g} = 1 + g^2 . \tag{124}$$

Evidently, it can be rewritten as a cubic equation for g. By taking the trace of $\mathcal{G}$ one obtains the announced cubic equation (114) for $G(z)$

The average spectral density given by

$$\rho(\lambda) = -\frac{1}{\pi} \mathrm{Im} G(z = \lambda) \tag{125}$$

is a semicircle at $\pi T = 0$ and splits into two arcs at $\pi T = 1$. For the spectral density at zero one obtains $\rho(0) = \sqrt{1 - \pi^2 T^2}$, and therefore chiral symmetry is broken for $\pi T < 1$, and is restored above this temperature. In spite of this drastic change in average spectral density, it could be shown[15] with the help of a supersymmetric formulation of Random Matrix Theory that the microscopic spectral density does not depend on T.

The super-symmetric method in the first paper by Jackson *et al.*[15] is not easily generalizable to higher order correlation functions. A natural way to proceed is to employ the super-symmetric method introduced by Guhr[79]. In the case of $\beta = 2$this method results in an analytical expression for the kernel determining all correlation functions. This approach was followed in two papers, one by Guhr and Wettig[14] and one by Jackson *et al.*[16]. The latter authors studied microscopic correlation functions for A in (112) proportional to the identity, whereas Guhr and Wettig considered an arbitrary diagonal matrix A. It was shown that independent of the matrix A, the correlations are given by the Bessel kernel[80]. Of course, a necessary condition on the matrix A is that chiral symmetry is broken. Guhr and Wettig also showed that correlations in the bulk of the spectrum are insensitive to A.

The deformation $W \to W + A$ with the probability distribution for W given by an arbitrary invariant potential has not yet been considered. We have no doubt that universality proofs along the lines of methods developed by P. Zinn-Justin[49] can be given.

LATTICE QCD RESULTS

In this section we will focus ourselves on the spectral correlations of the lattice QCD Dirac operator. Both correlations in the bulk of the spectrum and the microscopic spectral density will be studied. Consistent with universality arguments presented above, we find that spectral correlations are in complete agreement with chiral Random Matrix Theory.

Correlations in the Bulk of the Spectrum

Recently, Kalkreuter[19] calculated *all* eigenvalues of the $N_c = 2$ lattice Dirac operator both for Kogut-Susskind (KS) fermions and Wilson fermions for lattices as large as 12^4. For the Kogut-Susskind Dirac operator, D^{KS}, we use the convention that it is anti-Hermitean. Because of the Wilson-term, the Wilson Dirac operator, D^W, is neither Hermitean nor anti-Hermitean. Its Hermiticity relation is given by $D^{W\dagger} = \gamma_5 D^W \gamma_5$. Therefore, the operator $\gamma_5 D^W$ is Hermitean. However, it does not anti-commute with γ_5, and its eigenvalues do not occur in pairs $\pm\lambda_k$.

In the case of $SU(2)$, the anti-unitary symmetry of the Kogut-Susskind and the Wilson Dirac operator is given by[85,22],

$$[D^{KS}, \tau_2 K] = 0, \quad \text{and} \quad [\gamma_5 D^W, \gamma_5 C K \tau_2] = 0. \tag{126}$$

Because

$$(\tau_2 K)^2 = -1, \quad \text{and} \quad (\gamma_5 C K \tau_2)^2 = 1, \tag{127}$$

the matrix elements of the KS Dirac operator can be arranged into real quaternions, whereas the Wilson Dirac operator can be expressed into real matrix elements. Therefore, we expect that eigenvalue correlations in the bulk of the spectrum are described by the GSE and the GOE, respectively[22]. The microscopic correlations for KS fermions are described by the chGSE. However, the microscopic correlations for Wilson fermions are not described by the chGOE but rather by the GOE. Because of the anti-unitary symmetry, the eigenvalues of the KS Dirac operator are subject to the Kramers degeneracy, i.e. they are double degenerate.

In both cases, the Dirac matrix is tri-diagonalized by Cullum's and Willoughby's Lanczos procedure[86] and diagonalized with a standard QL algorithm. This improved algorithm makes it possible to obtain *all* eigenvalues. This allows us to test the accuracy of the eigenvalues by means of sum-rules for the sum of the squares of the eigenvalues of the lattice Dirac operator. Typically, the numerical error in the sum rule is of order 10^{-8}.

As an example, in Fig. 1 we show a histogram of the overall Dirac spectrum for KS fermions at $\beta = 2.4$. Results for the spectral correlations are shown in Figs. 2, 3 and 4. The results for KS fermions are for 4 dynamical flavors with $ma = 0.05$ on a 12^4 lattice. The results for Wilson fermions were obtained for two dynamical flavors on a $8^3 \times 12$ lattice. For the values of β and κ we refer to the labels of the figure. For $\beta > 2.4$, with our lattice parameters for KS fermions, the Dirac spectrum near zero virtuality develops a gap. Of course, this is an expected feature of the weak coupling domain. For small enough values of κ the Wilson Dirac spectrum shows a gap at $\lambda = 0$ as well. In the scaling domain the value of κ is just above the critical value of κ.

The eigenvalue spectrum is unfolded by fitting a second order polynomial to the integrated spectral density of a stretch of 500-1000 eigenvalues. The results for $\Sigma_2(n)$, $\Delta_3(n)$ and $P(S)$ in Fig. 2 show an impressive agreement with RMT predictions. The fluctuations in $\Sigma_2(n)$ are as expected from RMT. The advantage of Δ_3-statistic is well illustrated by this figure. We also investigated[23] the n dependence of the first two cumulants of the number of levels in a stretch of length n. Results presented in Fig. 4 show a perfect agreement with RMT. Spectra for different values of β have been analyzed as well. It is probably no surprise that random matrix correlations are found at stronger couplings. What is surprising, however, is that even in the weak-coupling domain ($\beta = 2.8$) the eigenvalue correlations are in complete agreement with Random

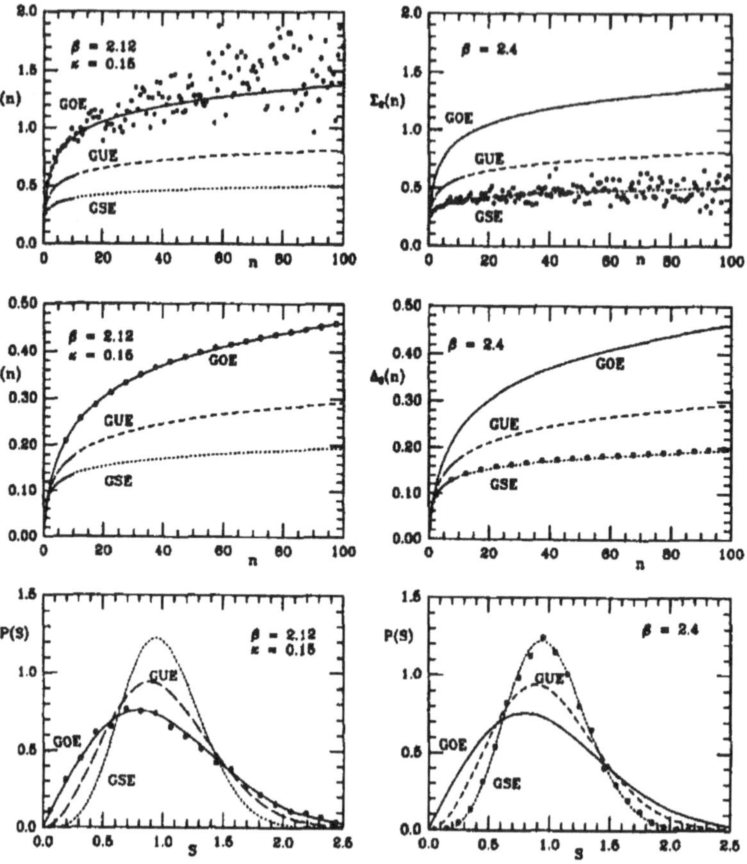

Figure 2. Spectral correlations of Dirac eigenvalues for Wilson fermions (left) and KS-fermions (right). Results are shown for the number variance, $\Sigma_2(n)$, the Δ_3−statistic and the nearest neighbor spacing distribution, $P(S)$. The solid, dashed and dotted curves represent the analytical result for the GOE, GUE and GSE, respectively.

Matrix Theory. Finally, we have studied the stationarity of the ensemble by analyzing level sequences of about 200 eigenvalues (with relatively low statistics). No deviations from random matrix correlations were observed all over the spectrum, including the region near $\lambda = 0$. This justifies the spectral averaging which results in the good statistics in Figs. 2 and 3.

In the case of three or more colors with fundamental fermions, both the Wilson and Kogut-Susskind Dirac operator do not possess any anti-unitary symmetries. Therefore, our conjecture is that in this case the spectral correlations in the bulk of the spectrum of both types of fermions can be described by the GUE. In the case of two fundamental colors the continuum theory and Wilson fermions are in the same universality class. It is an interesting question of how spectral correlations of KS fermions evolve in the approach to the continuum limit. Certainly, the Kramers degeneracy of the eigenvalues remains. However, since Kogut-Susskind fermions represent 4 degenerate flavors in the continuum limit, the Dirac eigenvalues should obtain an additional two-fold degeneracy. We are looking forward to more work in this direction.

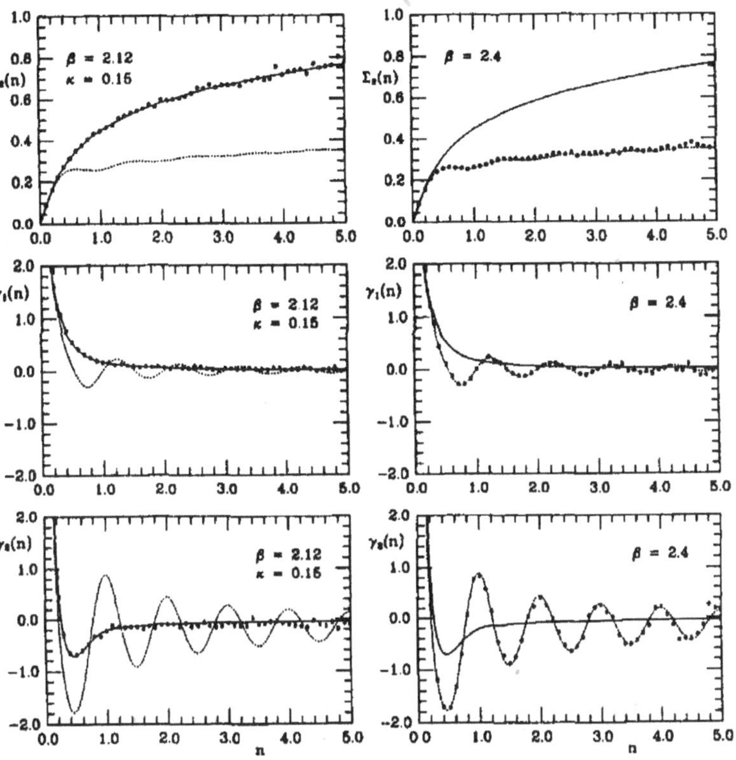

Figure 3. The number variance, $\Sigma_2(n)$ and the first two cumulants, $\gamma_1(n)$ and $\gamma_2(n)$ as a function of n for eigenvalues of the Wilson Dirac operator (left) and the Kogut-Susskind Dirac operator (right). The full and dotted curves represent the analytical result for the GOE and the GSE, respectively.

The Microscopic Spectral Density

The advantage of studying spectral correlations in the bulk of the spectrum is that one can perform spectral averages instead of ensemble averages requiring only a relatively small number of equilibrated configurations. This so called spectral ergodicity cannot be exploited in the study of the microscopic spectral density. In order to gather sufficient statistics for the microscopic spectral density of the lattice Dirac operator a large number of independent configurations is needed. One way to proceed is to generate instanton-liquid configurations which can be obtained much more cheaply than lattice QCD configurations. Results of such analysis[87] show that for $N_c = 2$ with fundamental fermions the microscopic spectral density is given by the chGOE. For $N_c = 3$ it is given by the chGUE. One could argue that instanton-liquid configurations can be viewed as smoothened lattice QCD configurations. Roughening such configurations will only improve the agreement with Random Matrix Theory.

Of course, the ultimate goal is to test the conjecture of microscopic universality for realistic lattice QCD configurations. In order to obtain a very large number of independent gauge field configurations one is necessarily restricted to relatively small lattices. The first study in this direction was reported recently[20, 88]. In this work, the quenched $SU(2)$ Kogut-Susskind Dirac operator was diagonalized for lattices with linear dimension of 4, 6, 8 and 10, and a total number of configurations of 9978, 9953, 3896 and 1416, respectively. The results were compared with predictions from the chGSE.

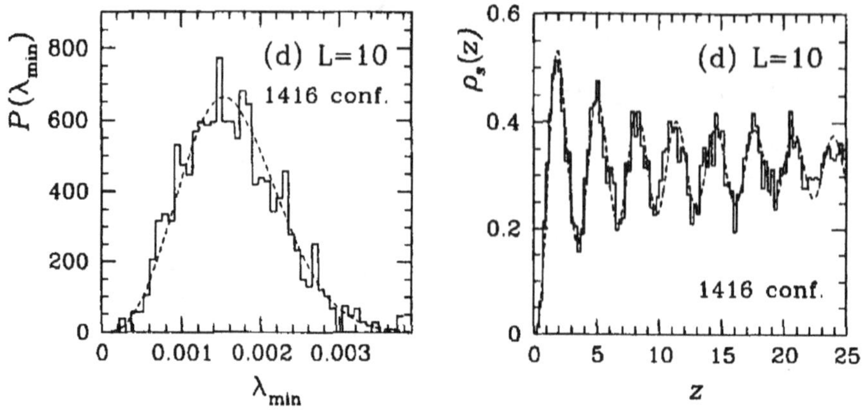

Figure 4. The distribution of the smallest eigenvalue (left) and the microscopic spectral density (right) of the Kogut-Susskind Dirac operator for two colors and $\beta = 2.0$. Lattice results are represented by the histogram, and the analytical results for the chGSE are given by the dashed curves.

We only show results for the largest lattice. For more detailed results, including results for the two-point correlation function, we refer to the original work. In Fig. 4 we show the distribution of the smallest eigenvalue (left) and the microscopic spectral density (right). The lattice results are given by the full line. The dashed curve represents the random matrix results. The distribution of the smallest eigenvalue was derived by Forrester[89] and is given by

$$P(\lambda_{\min}) = \alpha\sqrt{\frac{\pi}{2}}(\alpha\lambda_{\min})^{3/2} I_{3/2}(\alpha\lambda_{\min}) e^{-\frac{1}{2}(\alpha\lambda_{\min})^2}, \tag{128}$$

where $\alpha = V\Sigma$. The random matrix result for the microscopic spectral density is given in eq. (78). We emphasize that the theoretical curves have been obtained without any fitting of parameters. The input parameter, the chiral condensate, is derived from the same lattice calculations. The above simulations were performed at a relatively strong coupling of $\beta = 2$. Recently, the same analysis[90] was performed for $\beta = 2.2$ and for $\beta = 2.5$ on a 16^4 lattice. In both cases agreement with the random matrix predictions was found[90].

An alternative way to probe the Dirac spectrum is via the valence quark mass dependence of the chiral condensate[18] defined as

$$\Sigma(m) = \frac{1}{N}\int d\lambda \langle\rho(\lambda)\rangle \frac{2m}{\lambda^2 + m^2}. \tag{129}$$

The average spectral density is obtained for a fixed sea quark mass. For masses well beyond the smallest eigenvalue, $\Sigma(m)$ shows a plateau approaching the value of the chiral condensate Σ. In the mesoscopic range (1), we can introduce $u = \lambda m N$ and $x = mN\Sigma$ as new variables. Then the microscopic spectral density enters in $\Sigma(m)$. For three fundamental colors the microscopic spectral density for $\beta = 2$ (eq. (72)) applies and the integral over λ in (129) can be performed analytically. The result is given by[17],

$$\frac{\Sigma(x)}{\Sigma} = x(I_a(x)K_a(x) + I_{a+1}(x)K_{a-1}(x)), \tag{130}$$

where $a = N_f + |\nu|$ and I_a and K_a are modified Bessel functions. In Fig. 2 we plot this ratio as a function of x (the 'volume' V is equal to the total number of Dirac eigenvalues) for lattice data of two dynamical flavors with mass $ma = 0.01$ and $N_c = 3$ on a $16^3 \times 4$ lattice. We observe that the lattice data for different values of β fall on a single curve. Moreover, in the mesoscopic range this curve coincides with the random matrix prediction for $N_f = \nu = 0$.

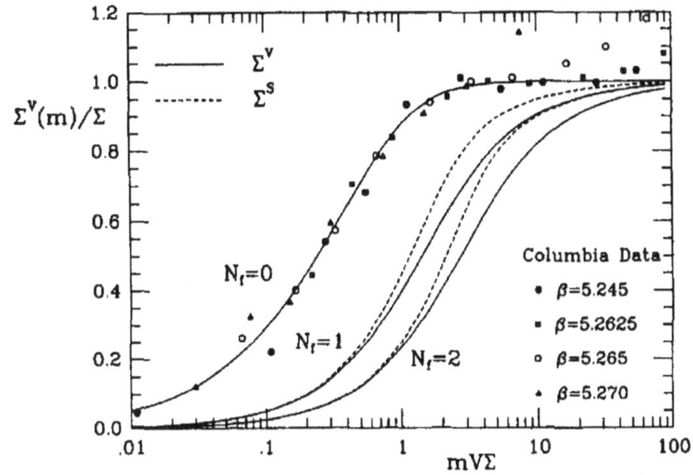

Figure 5. The valence quark mass dependence of the chiral condensate $\Sigma^V(m)$ plotted as $\Sigma^V(m)/\Sigma$ versus $mV\Sigma$. The dots and squares represent lattice results by the Columbia group[18] for values of β as indicated in the label of the figure.

Apparently, the zero modes are completely mixed with the much larger number of nonzero modes. For eigenvalues much smaller than the sea quark mass, one expects quenched ($N_f = 0$) eigenvalue correlations. In the same figure the dashed curves represent results for the quark mass dependence of the chiral condensate (i.e. the mass dependence for equal valence and sea quark masses). In the sector of zero topological charge one finds[91, 9, 8]

$$\frac{\Sigma^S(u)}{\Sigma} = \frac{I_1(u)}{I_0(u)} \quad \text{for} \quad N_f = 1, \tag{131}$$

and

$$\frac{\Sigma^S(u)}{\Sigma} = \frac{I_1^2(u)}{u(I_0^2(u) - I_1^2(u))} \quad \text{for} \quad N_f = 2. \tag{132}$$

We observe that both expressions do not fit the data. Also notice that, according to Göckeler *et al.*[92], eq. (131) describes the valence mass dependence of the chiral condensate for non-compact QED with quenched Kogut-Susskind fermions. However, we were not able to derive their result (no derivation is given in the paper).

CHIRAL RANDOM MATRIX THEORY AT $\mu \neq 0$

Generalities

At nonzero temperature T and chemical potential μ a *schematic* random matrix model of the QCD partition function is obtained by replacing the Dirac operator in

(49) by[77, 93, 78, 35]

$$\mathcal{D} = \begin{pmatrix} 0 & iW + i\Omega_T + \mu \\ iW^\dagger + i\Omega_T + \mu & 0 \end{pmatrix}. \tag{133}$$

Here, $\Omega_T = T \otimes_n (2n+1)\pi\mathbf{1}$ are the matrix elements of $i\gamma_0\partial_0$ in a plane wave basis with anti-periodic boundary conditions in the time direction. Below, we will discuss a model with Ω_T absorbed in the random matrix and $\mu \neq 0$. The aim of this model is to explore the effects of the non-Hermiticity of the Dirac operator. For example, the random matrix partition function (49) with the Dirac matrix (133) is well suited for the study of zeros of this partition function in the complex mass plane and in the complex chemical potential plane. Numerical results for the location of zeros could be explained analytically[94].

The term $\mu\gamma_0$ does not affect the anti-unitary symmetries of the Dirac operator. This is also the case in lattice QCD where the color matrices in the forward time direction are replaced by $U \to e^{\mu}U$ and in the backward time direction by $U^\dagger \to e^{-\mu}U^\dagger$. For this reason the universality classes are the same as at zero chemical potential.

The Dirac operator that will be discussed in this section is thus given by

$$\mathcal{D}(\mu) = \begin{pmatrix} 0 & iW + \mu \\ iW^\dagger + \mu & 0 \end{pmatrix}, \tag{134}$$

where the matrix elements of the $n \times n$ matrix W are either real ($\beta = 1$), complex ($\beta = 2$) or quaternion real ($\beta = 4$). For all three values of β the eigenvalues of $\mathcal{D}(\mu)$ are scattered in the complex plane.

Since many standard random matrix methods rely on convergence properties based on the Hermiticity of the random matrix, direct application of most methods is not possible. The simplest way out is the Hermitization[95] of the problem, i.e we consider the Hermitean operator

$$D^H(z, z^*) = \begin{pmatrix} \kappa & z - \mathcal{D}(\mu) \\ z^* - \mathcal{D}^\dagger(\mu) & \kappa \end{pmatrix}. \tag{135}$$

For example, the generating function in the supersymmetric method of Random Matrix Theory[96, 97] is then given by[98, 99, 100, 101]

$$Z(J, J^*) = \left\langle \frac{\det(D^H(z+J, z^*+J^*)}{\det(D^H(z, z^*))} \right\rangle. \tag{136}$$

The determinants can be rewritten as fermionic and bosonic integrals. Convergence is assured by the Hermiticity and by the infinitesimal increment κ. The resolvent follows from the generating function by differentiation with respect to the source terms

$$G(z, z^*) = \mathrm{Tr}\left\langle \frac{1}{z - \mathcal{D}(\mu)} \right\rangle = \frac{\partial}{\partial J} Z(J, J^*)\bigg|_{J=J^*=0}. \tag{137}$$

Notice that, after averaging over the random matrix, the partition function depends in a non-trivial way on both z and z^*. The spectral density is then given by

$$\rho(z) = \frac{1}{\pi}\frac{\partial}{\partial z^*} G(z, z^*). \tag{138}$$

Therefore, outside the domain of the eigenvalues $G(z, z^*)$ is a function of z only, whereas for z inside the domain of eigenvalues, $G(z, z^*)$ should depend on z^* as well.

Alternatively, one can use the replica trick[102, 35] with generating function given by

$$Z(z, z^*, \mu) = \langle \det^{N_f} D_H(z, z^*) \rangle. \tag{139}$$

The resolvent is then given by

$$G(z, z^*) = \lim_{N_f \to 0} \frac{1}{2nN_f} \frac{\partial}{\partial z} \log Z(z, z^*, \mu). \tag{140}$$

The idea is to perform the calculation for integer values of N_f and perform the limit $N_f \to 0$ at the end of the calculation. Although the replica limit, $N_f \to 0$, fails in general[104], it is expected to work for det $D_H(z, z^*)$ because it is positive definite (or zero). Then the partition function is a smooth function of N_f. For other techniques addressing nonhermitean matrices we refer to the recent papers by Feinberg and Zee[95] and Nowak and co-workers[103]. One recent method that does not rely on the Hermiticity of the random matrices is the method of complex orthogonal polynomials[36]. This method was used by Fyodorov *et al.*[36] to calculate the number variance and the nearest neighbor spacing distribution in the regime of weakly nonhermitean matrices. As surprising new result, they found an $S^{5/2}$ repulsion law.

In the physically relevant case of QCD with three colors, the fermion determinant is complex for nonzero chemical potential. Its phase prevents the convergence of fully unquenched Monte-Carlo simulations (see Kogut *et al.*[105] for the latest progress in this direction). However, it is possible to perform quenched simulations. In such calculations it was found that the critical chemical potential $\mu_c \sim \sqrt{m}$, instead of a third of the nucleon mass[106]. This phenomenon was explained analytically by Stephanov[35] with the help of the above random matrix model. He could show that for small μ the eigenvalues are distributed along the imaginary axis in a band of width $\sim \mu^2$ leading to a critical chemical potential of $\mu_c \sim \sqrt{m}$. A detailed derivation of his solution will be given in the next section. As has been argued above, the quenched limit is necessarily obtained from a partition function in which the fermion determinant appears as

$$\lim_{N_f \to 0} |\det \mathcal{D}(\mu)|^{N_f}, \tag{141}$$

instead of the same expression without the absolute value signs. The partition function with the absolute value of the determinant can be interpreted as a partition function of an equal number of fermions and conjugate fermions. The critical value of the chemical potential, equal to half the pion mass, is due to Goldstone bosons with a net baryon number consisting of a quark and conjugate anti-quark. The reason that the quenched limit does not correspond to the standard QCD partition function is closely related to the failure of the replica trick in the case of a determinant with a nontrivial phase.

The Stephanov Solution

In this section we study the quenched limit of the partition function (139). We give a detailed derivation of the results originally obtained by Stephanov[35]. The determinants in (139) can be written as Grassmann integrals. Since Grassmann integrals are always convergent the infinitesimal increment κ can be put equal to zero. This results in the partition function

$$Z(z, z^*, \mu) = \int \mathcal{D}W \mathcal{D}\psi^* \mathcal{D}\psi \mathcal{D}\phi^* \mathcal{D}\phi \exp[-n\Sigma^2 \operatorname{Tr} WW^\dagger]$$
$$\times \exp\left[-\sum_{k=1}^{N_f} \psi^{k*} \begin{pmatrix} z & -iW + \mu \\ -iW^\dagger + \mu & z \end{pmatrix} \psi^k - \sum_{k=1}^{N_f} \phi^{k*} \begin{pmatrix} z^* & iW + \mu \\ iW^\dagger + \mu & z^* \end{pmatrix} \phi^k \right], \tag{142}$$

where W is an arbitrary complex $n \times n$ matrix. The Gaussian integrals over W can be performed trivially,

$$
\begin{aligned}
Z(z,z^*,\mu) = \int \mathcal{D}\psi^*\mathcal{D}\psi\mathcal{D}\phi^*\mathcal{D}\phi \exp\Big[& z(\psi^{f*}_{Ri}\psi^f_{Ri} + \psi^{f*}_{Lk}\psi^f_{Lk}) + z^*(\phi^{f*}_{Ri}\phi^f_{Ri} + \phi^{f*}_{Lk}\phi^f_{Lk}) \\
& + \mu(\psi^{f*}_{Ri}\psi^f_{Li} + \psi^{f*}_{Lk}\psi^f_{Rk} + \phi^{f*}_{Ri}\phi^f_{Li} + \phi^{f*}_{Lk}\phi^f_{Rk}) \\
- \frac{1}{n\Sigma^2}\Big(\psi^{f*}_{Lk}\psi^f_{Ri}\psi^{g*}_{Ri}\psi^g_{Lk} & - \psi^{f*}_{Lk}\psi^f_{Ri}\phi^{g*}_{Ri}\phi^g_{Lk} - \phi^{f*}_{Lk}\phi^f_{Ri}\psi^{g*}_{Ri}\psi^g_{Lk} + \phi^{f*}_{Lk}\phi^f_{Ri}\phi^{g*}_{Ri}\phi^g_{Lk}\Big)\Big].
\end{aligned}
\tag{143}
$$

The four-fermion terms can be written as the difference of two squares. Each square can be linearized by the Hubbard-Stratonovitch transformation according to

$$\exp(-AQ^2) \sim \int d\sigma \exp(-\frac{\sigma^2}{4A} - iQ\sigma) \ . \tag{144}$$

Using this, the fermionic integrals can be performed, and the partition function can be written as an integral over the complex $2N_f \times 2N_f$ matrices, σ and $\tilde{\sigma}$,

$$Z(z,z^*,\mu) = \int \mathcal{D}\tilde{\sigma}\mathcal{D}\sigma \exp[-n\Sigma^2 \mathrm{Tr}(\sigma - \zeta)k(\tilde{\sigma} - \zeta)k] \det{}^n\left(\sigma\tilde{\sigma} - \mu^2\right). \tag{145}$$

In a $N_f \times N_f$ block structure notation, the matrices σ, $\tilde{\sigma}$, k and ζ are given by

$$
\begin{aligned}
\sigma &= \begin{pmatrix} a & ib \\ ic & d \end{pmatrix}, & \tilde{\sigma} &= \begin{pmatrix} a^\dagger & ic^\dagger \\ ib^\dagger & d^\dagger \end{pmatrix}, \\
k &= \begin{pmatrix} 1 & 0 \\ 0 & -1 \end{pmatrix}, & \zeta &= \begin{pmatrix} z & 0 \\ 0 & z^* \end{pmatrix},
\end{aligned}
\tag{146}
$$

where a, b, c and d are arbitrary complex $N_f \times N_f$ matrices. The resolvent is obtained by differentiation with respect to z according to (140) with the averaged partition function given by (145). This results in

$$G = \frac{1}{2N_f}\mathrm{Tr}(2z + a + a^\dagger). \tag{147}$$

In the thermodynamic limit the integrals in (145) can be evaluated by a saddle-point method. Notice that a variable and its complex conjugate have to be considered as independent integration variables. The saddle point equations are given by

$$\frac{1}{\tilde{\sigma}\sigma - \mu^2}\tilde{\sigma} = k(\tilde{\sigma} + \zeta)k, \tag{148}$$

$$\frac{1}{\sigma\tilde{\sigma} - \mu^2}\sigma = k(\sigma + \zeta)k. \tag{149}$$

In general the solution of the saddle point equations is not unique. However, a unique solution is obtained from the requirement that $\partial_z^* G(z)$ is positive definite. The saddle point equations have two obvious solutions

$$\sigma = \tilde{\sigma} \quad \text{or} \quad \sigma = k\tilde{\sigma}k, \tag{150}$$

for which the matrix equations (148) and (149) coincide.

It can be shown that the first possibility results in an unphysical solution. As usual in applications of the replica trick, we assume that the replica (or flavor) symmetry remains unbroken. This implies that, at the saddle point, each block in (146) is diagonal. With (150) the solution of our saddle point equations is reduced to a 2×2 matrix problem

$$\sigma = (\sigma k \sigma k - \mu^2)(k \sigma k + \zeta). \tag{151}$$

Consistent with the second solution in (150) we use the parametrization

$$\sigma = \begin{pmatrix} a & ib \\ -ib^* & d \end{pmatrix}. \tag{152}$$

One solution of (151) can be written down immediately, namely $b = b^* = 0$. Then the equations for a and d reduce to the same cubic equation. This solution results in a partition function which factorizes in a product of a z dependent part and in a z^* dependent part. This resolvent is therefore an analytic function of z valid outside the domain of the eigenvalues. The cubic equation for the resolvent in this domain can be obtained from the cubic equation (114) by the replacement $T \to i\mu$.

Let us now focus on the solution inside the domain of eigenvalues with $b \neq 0$. From the off-diagonal elements of (151) we obtain the equations

$$\begin{aligned} (a+d)(d-z^*) - (a^2 - bb^* - \mu^2) &= 1, \\ (a+d)(a-z) - (d^2 - bb^* - \mu^2) &= 1. \end{aligned} \tag{153}$$

The difference of these equation results in

$$2(a+d)(d-a) + (a+d)(z-z^*) = 0, \tag{154}$$

with solution given by

$$a - d = \frac{1}{2}(z - z^*). \tag{155}$$

From the sum of the two equations one obtains

$$bb^* = 1 + \frac{1}{2}(a+d)(z+z^*) - ad - \mu^2. \tag{156}$$

The boundary of the domain of the eigenvalues is given by the set of points where this solution merges with the factorized solution, i.e. where $b = 0$. With (155) the condition $b = 0$ can be expressed as an equation for a, or by (147) ($G = a - z$ for diagonal a) an equation for the resolvent. On the boundary of the domain the equation for the resolvent is given by

$$(2G + 2x + iy)x - (G + x + iy)(G + x) + 1 - \mu^2 = 0. \tag{157}$$

For $b = 0$ the Stephanov solution and the solution of the cubic equation merge into each other. We thus have a second order phase transition, and on the boundary, the second derivative of the free energy corresponding to (145) vanishes.

From the diagonal elements of (151) one obtains

$$\begin{aligned} a &= (a-z)(a^2 - bb^* - \mu^2) + bb^*(a+d), \\ d &= (d-z^*)(d^2 - bb^* - \mu^2) + bb^*(a+d). \end{aligned} \tag{158}$$

In the first equation we substitute the first equation of (153) in the first term of the r.h.s. and the expression (156) for bb^* in the second term of the r.h.s.. This results in

$$d - a + z + (a+d)(\frac{1}{2}(z - z^*)(a-d) - \mu^2 + zz^*) = 0. \tag{159}$$

Together with (155) this results in an equation for a with solution

$$a = \frac{x}{\mu^2 - x^2} + \frac{i}{2}y. \tag{160}$$

For the resolvent one obtains

$$G = a - z = \frac{x}{\mu^2 - x^2} - x - \frac{i}{2}y. \tag{161}$$

This results in the spectral density

$$\rho(x, y) = \frac{1}{4\pi}\left(\frac{x^2 + \mu^2}{(\mu^2 - x^2)^2} - 1\right) \tag{162}$$

inside the boundary given by

$$y^2 = 4 - 4\mu^2 + \frac{4x^2}{\mu^2 - x^2} - \frac{x^2}{(\mu^2 - x^2)^2}. \tag{163}$$

The latter equation has obtained by the substitution of (161) in (157). Both (161) and (163) were first derived by Stephanov[35,108]. The closed curves in Fig. 6 show the boundary of the domain of eigenvalues given by (163). Numerical results for the eigenvalues are represented by the dots in the same figure.

For small μ, the width of the domain of eigenvalues is $\sim \mu^2$. This explains that the critical value of the chemical potential is given $\mu_c \sim \sqrt{m}$., This result explains the quenched results found in lattice QCD at nonzero chemical potential.

Random Matrix Triality at Nonzero Chemical Potential

In this section, we study the Dirac operator (134) for all three values of β. Both for $\beta = 1$ and $\beta = 4$ the fermion determinant, $\det(\mathcal{D}(\mu) + m)$, is real. This is obvious for $\beta = 1$. For $\beta = 4$ the reality follows from the identity $q^* = \sigma_2 q \sigma_2$ for a quaternion real element q, and the invariance of a determinant under transposition. We thus conclude that quenching works for an even number of flavors. Consequently, chiral symmetry will be restored for arbitrarily small nonzero μ, whereas a condensate of a quark and a conjugate anti-quark develops. Indeed, this phenomenon has been observed in the strong coupling limit of lattice QCD with two colors[109].

In the quenched approximation, the spectral properties of the random matrix ensemble (134) can be easily studied numerically by simply diagonalizing a set of matrices with probability distribution (49). In Fig. 6 we show numerical results[110] for the eigenvalues of a few 100×100 matrices for $\mu = 0.15$ and $\mu = 0.5$. The dots represent the eigenvalues in the complex plane. The solid line is the analytical result[35] for the boundary of the eigenvalues which is given by the algebraic curve (163). This result was first derived by Stephanov for $\beta = 2$ (see previous section). However, the method that was used can be extended[110] to $\beta = 1$ and $\beta = 4$. Although the effective partition function is much more complicated, it can be shown without too much effort that the solutions of the saddle point equations are the same if the variance of the probability distribution is scaled as $1/\beta$. In particular, the boundary of the domain of eigenvalues is the same in each of the three cases. However, as one observes from Fig. 6, for $\beta = 1$ and $\beta = 4$ the spectral density deviates significantly from the saddle-point result. For $\beta = 1$ we find an accumulation of eigenvalues on the imaginary axis, whereas for $\beta = 4$ we find a depletion of eigenvalues in this domain. This depletion can be understood as

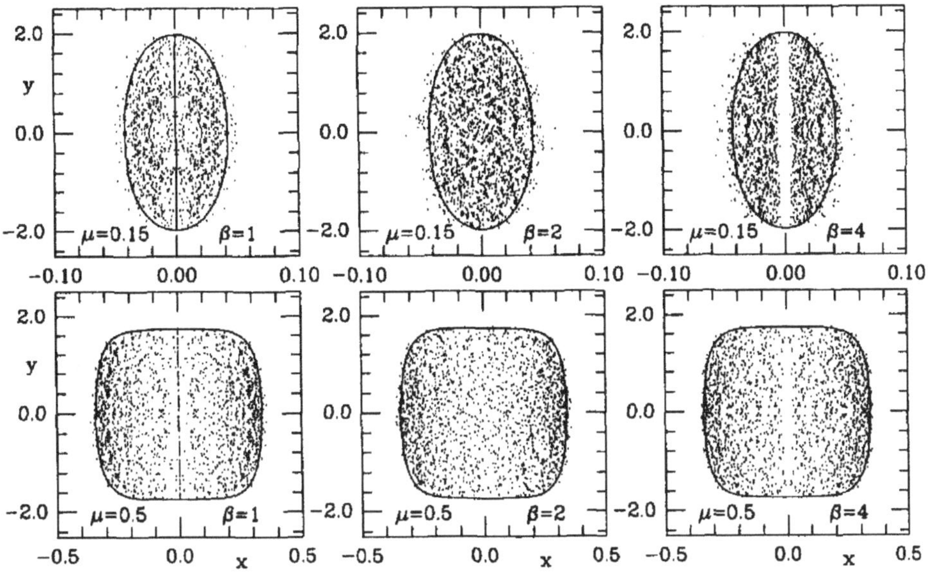

Figure 6. Scatter plot of the real (x), and the imaginary parts (y) of the eigenvalues of the random matrix Dirac operator. The values of β and μ are given in the labels of the figure. The full curve shows the analytical result for the boundary.

follows. For $\mu = 0$ all eigenvalues are doubly degenerate. This degeneracy is broken at $\mu \neq 0$ which produces the observed repulsion of the eigenvalues.

The number of purely imaginary eigenvalues for $\beta = 1$ appears to scale as $\sqrt{N}$. This explains that this effect is not visible in a leading order saddle point analysis. From a perturbative analysis of (139) one obtains a power series in $1/N$. Clearly, the $\sqrt{N}$ dependence requires a truly nonperturbative analysis of the partition function (49) with the Dirac operator (134). Such a $\sqrt{N}$ scaling behavior is typical for the regime of weak non-hermiticity first identified by Fyodorov *et al.*[100]. Using the supersymmetric method for the generating function (136) the $\sqrt{N}$ dependence was obtained analytically by Efetov[101].

A similar cut below a cloud of eigenvalues was found in instanton liquid simulations[111] for $N_c = 2$ at $\mu \neq 0$ and in a random matrix model of arbitrary real matrices[99]. The depletion of the eigenvalues along the imaginary axis was observed earlier in lattice QCD simulations with staggered fermions[112]. Obviously, more work has to be done in order to arrive at a complete characterization of universal features[36] in the spectrum of nonhermitean matrices.

CONCLUSIONS

We have argued that there is an intimate relation between correlations of Dirac eigenvalues and the breaking of chiral symmetry. In the chiral limit, the fermion determinant suppresses gauge field configurations with small Dirac eigenvalues. Correlations counteract this suppression, and are a necessary ingredient of chiral symmetry breaking. From the study of eigenvalue correlations in strongly interacting systems, we have concluded that they are described naturally by a Random Matrix Theory with the global symmetries of the physical system. In QCD, this led to the introduction of chiral Random Matrix Theories. They provided us with an analytical understanding of the statistical properties of the eigenvalues on the scale of a typical level spacing. It could be shown analytically that the microscopic spectral density is strongly universal. These results constitute the foundation of the impressive agreement between lattice QCD and chiral Random Matrix Theory for the microscopic spectral density and for spectral correlations in the bulk of the spectrum.

An extension of this model to nonzero chemical potential provided us with a complete analytical understanding of the failure of the quenched approximation observed in lattice QCD simulations at finite density. Some intriguing properties of previously obtained lattice QCD Dirac spectra and instanton liquid Dirac spectra at finite density could be explained as well.

Acknowledgements

This work was partially supported by the US DOE grant DE-FG-88ER40388. NATO is acknowledged for financial support. In particular, we wish to thank Pierre van Baal for organizing this wonderful summer school. We benefitted from discussions with M. Stephanov and T. Wettig. M.K. **Şener** and M. Stephanov are thanked for a critical reading of the manuscript. Finally, I thank all my collaborators on whose work this review is based.

REFERENCES

1. C. DeTar, *Quark-gluon plasma in numerical simulations of QCD*, in *Quark gluon plasma 2*, R, Hwa ed., World Scientific 1995.
2. A. Ukawa, Nucl. Phys. Proc. Suppl. **53** (1997) 106.
3. A. Smilga, *Physics of thermal QCD*, hep-ph/9612347.
4. E. Shuryak, Comments Nucl. Part. Phys. **21** (1994) 235.
5. C. Bernard, T. Blum, C. DeTar, S. Gottlieb, U. Heller, J. Hetrick, K. Rummukainen, R. Sugar, D. Toussaint and M. Wingate, Phys. Rev. Lett. **78** (1997) 598.
6. T. Banks and A. Casher, Nucl. Phys. **B169** (1980) 103.
7. O. Bohigas, M. Giannoni, Lecture notes in Physics **209** (1984) 1; O. Bohigas, M. Giannoni and C. Schmit, Phys. Rev. Lett. **52** (1984) 1.
8. H. Leutwyler and A. Smilga, Phys. Rev. **D46** (1992) 5607.
9. J. Gasser and H. Leutwyler, Phys. Lett. **188B**(1987) 477.
10. E.V. Shuryak and J.J.M. Verbaarschot, Nucl. Phys. **A560** (1993) 306.
11. S. Nishigaki, Phys. Lett. 387 B (1996) 707.
12. G. Akemann, P. Damgaard, U. Magnea and S. Nishigaki, Nucl. Phys. **B 487[FS]** (1997) 721.
13. E. Brézin, S. Hikami and A. Zee, Nucl. Phys. **B464** (1996) 411.
14. T. Guhr and T. Wettig, *Universal spectral correlations of the Dirac operator at finite temperature*, hep-th/9704055, Nucl. Phys. **B** (in press); T. Guhr and T. Wettig, J. Math. Phys. **37** (1996) 6395.
15. A.D. Jackson, M.K. Sener and J.J.M. Verbaarschot, Nucl. Phys. **B479** (1996) 707.
16. A.D. Jackson, M.K. Sener and J.J.M. Verbaarschot, *Universality of correlation functions in random matrix models of QCD*, hep-th/9704056, Nucl. Phys. **B** (in press).
17. J.J.M. Verbaarschot, Phys. Lett. **B368** (1996) 137.
18. S. Chandrasekharan, Nucl. Phys. Proc. Suppl. **42** (1995) 475; S. Chandrasekharan and N. Christ, Nucl. Phys. Proc. Suppl. **42** (1996) 527; N. Christ, Lattice 1996.
19. T. Kalkreuter, Phys. Lett. **B276** (1992) 485; Phys. Rev. **D48** (1993) 1; Comp. Phys. Comm. **95** (1996) 1.
20. M.E. Berbenni-Bitsch, S. Meyer, A. Schäfer, J.J.M. Verbaarschot, and T. Wettig, *Microscopic Universality in the spectrum of the lattice Dirac operator*, hep-lat/9704018.
21. A. Pandey, Ann. Phys. **134** (1981) 119.
22. M.A. Halasz and J.J.M. Verbaarschot, Phys. Rev. Lett. **74** (1995) 3920.
23. M.A. Halasz, T. Kalkreuter and J.J.M. Verbaarschot, Nucl. Phys. Proc. Suppl. **53** (1997) 266.
24. T. Guhr, A. Müller-Groeling and H.A. Weidenmüller, *Random Matrix Theories in quantum physics: Common concepts*, cond-mat/9707301, Phys. Rep. (in press).

25. R. Haq, A. Pandey and O. Bohigas, Phys. Rev. Lett. **48** (1982) 1086.
26. C. Ellegaard, T. Guhr, K. Lindemann, H.Q. Lorensen, J. Nygard and M. Oxborrow,Phys. Rev. Lett. **75** (1995) 1546.
27. S. Deus, P. Koch and L. Sirko, Phys. Rev. **E 52** (1995) 1146; H. Gräf, H. Harney, H. Lengeler, C. Lewenkopf, C. Rangacharyulu, A. Richter, P. Schardt and H.A. Weidenmüller, Phys. Rev. Lett. **69** (1992) 1296.
28. T. Ericson, Phys. Rev. Lett. **5** (1960) 430.
29. H.A. Weidenmüller, Ann. Phys. (N.Y.) **158** (1984) 78; in *Proceedings of T. Ericson's 60th birthday.*
30. Y. Imry, Europhysics Lett. **1** (1986) 249; B.L. Altshuler, P.A. Lee and R.A. Webb (eds.), *Mesoscopic Phenomena in Solids*, North-Holland, New York, 1991; S. Iida, H.A. Weidenmüller and J. Zuk, Phys. Rev. Lett. **64** (1990) 583; Ann. Phys. (N.Y.) **200** (1990), 219; C.W.J. Beenakker, Rev. Mod. Phys. **69** (1997) 731.
31. P.W. Anderson, Phys. Rev. **109** (1958) 1492.
32. H. Sommers, A. Crisanti, H. Sompolinsky and Y. Stein, Phys. Rev. Lett. **60** (1988) 1895.
33. D. Gross and E. Witten, Phys. Rev. **D21** (1980) 446; S. Chandrasekharan, Phys. Lett. **B395** (1997) 83.
34. P. Di Francesco, P. Ginsparg, and J. Zinn-Justin, Phys. Rep. **254** (1995) 1.
35. M. Stephanov, Phys. Rev. Lett. **76** (1996) 4472.
36. Y. Fyodorov, B. Khoruzhenko and H. Sommers, Phys. Rev. Lett. **79** (1997) 557.
37. C.N. Yang and T.D. Lee, Phys. Rev. **87** (1952) 104, 410.
38. V. Matteev and R. Shrock, J. Phys. A: Math. Gen. **28** (1995) 5235.
39. J. Vink, Nucl. Phys. **B323** (1989) 399.
40. I. Barbour, A. Bell, M. Bernaschi, G. Salina and A. Vladikas, Nucl. Phys. B386 (1992) 683.
41. J. Osborn and J.J.M. Verbaarschot, in preparation.
42. A. Smilga and J.J.M. Verbaarschot, Phys. Rev. **D51** (1995) 829.
43. M.A. Halasz and J.J.M. Verbaarschot, Phys. Rev. **D52** (1995) 2563.
44. F. Dyson and M. Mehta, J. Math. Phys. **4** (1963) 701.
45. M. Mehta, *Random Matrices*, Academic Press, San Diego, 1991.
46. A.M. Odlyzko, Math. Comput. **48** (1987) 273.
47. D. Voiculescu, K. Dykema and A. Nica, *Free Random Variables*, Am. Math. Soc., Providence RI, 1992.
48. G. Hackenbroich and H.A. Weidenmüller, Phys. Rev. Lett. **74** (1995) 4118.
49. P. Zinn-Justin, Nucl. Phys. **B497** (1997) 725.
50. E. Brézin and S. Hikami, *An extension of level spacing universality*, cond-mat/9702213.
51. T.A. Brody, J. Flores, J.B. French, P.A. Mello, A. Pandey and S.S.M. Wong, Rev. Mod. Phys. **53** (1981) 385.
52. J.J.M. Verbaarschot and M.R. Zirnbauer, Ann. Phys. (N.Y.) **158** (1984) 78.
53. J.J.M. Verbaarschot, H.A. Weidenmüller and M.R. Zirnbauer, Ann. Phys. (N.Y.) **153** (1984) 367.
54. E. Brézin and A. Zee, Nucl. Phys. **B453** (1995) 531.
55. M. Berry, Proc. Roy. Soc. London **A 400** (1985) 229.
56. E. Brézin and J. Zinn-Justin, Phys. lett. **B288** (1992) 54.
57. S. Higuchi, C.Itoi, S.M. Nishigaki and N. Sakai, Phys. Lett. **B398** (1997) 123.
58. A.V. Andreev, O. Again, B.D. Simons and B.L. Altshuler, Nucl. Phys. **B482** (1996) 536.
59. A. Altland and M. Zirnbauer, Phys. Rev. Lett. **77** (1996) 4536.
60. T. Schäfer and E. Shuryak, Instantons in QCD, hep-ph/9610451, Rev. Mod. Phys. (1997).
61. D.I. Diakonov and V.Yu. Petrov, Nucl. Phys. **B272** (1986) 457.
62. J.J.M. Verbaarschot, Phys. Rev. Lett. **72** (1994) 2531; Phys. Lett. **B329** (1994) 351.
63. J.J.M. Verbaarschot and I. Zahed, Phys. Rev. Lett. **70** (1993) 3852.
64. J.J.M. Verbaarschot, Nucl. Phys. Proc. Suppl. **53** (1997) 88.
65. M. Peskin, Nucl. Phys. **B175** (1980) 197; S. Dimopoulos, Nucl. Phys. **B168** (1980) 69; M. Vysotskii, Y. Kogan and M. Shifman, Sov. J. Nucl. Phys. **42** (1985) 318; D.I. Diakonov and V.Yu. Petrov, Lecture notes in physics, **417**, Springer 1993.
66. A. Altland, M.R. Zirnbauer, Phys. Rev. Lett. **76** (1996) 3420; *Novel Symmetry Classes in Mesoscopic Normal-Superconducting Hybrid Structures*, cond-mat/9602137.
67. M.R. Zirnbauer, J. Math. Phys. **37** (1996) 4986; F.J. Dyson, Comm. Math. Phys. **19** (1970) 235.
68. D. Fox and P.Kahn, Phys. Rev. **134** (1964) B1152; (1965) 228.
69. B. Bronk, J. Math. Phys. **6** (1965) 228.

70. A.V. Andreev, B.D. Simons, and N. Taniguchi, Nucl. Phys **B432 [FS]** (1994) 487.
71. F. Dyson, J. Math. Phys. **13** (1972) 90.
72. G. Mahoux and M. Mehta, J. Phys. I Prance **I** (1991) 1093.
73. T. Nagao and M. Wadati, J. Phys. Soc. Japan **60** (1991) 2998; J. Phys. Soc. Jap. **60** (1991) 3298; J. Phys. Soc. Jap. bf 61 (1992) 78; J. Phys. Soc. Jap. bf 61 (1992) 1910. **61** (1992) 78, 1910.
74. J.J.M. Verbaarschot, Nucl. Phys. B426 (1994) 559.
75. T. Nagao and P.J. Forrester, Nucl. Phys. **B435** (1995) 401.
76. A.D. Jackson, M.K. Sener and J.J.M. Verbaarschot, Phys. Lett. **B 387** (1996) 355.
77. A.D. Jackson and J.J.M. Verbaarschot, Phys. Rev. **D53** (1996) 7223.
78. M. Stephanov, Phys. Lett. **B275** (1996) 249; Nucl. Phys. Proc. Suppl. **53** (1997) 469.
79. T. Guhr, J. Math. Phys. **32** (1991) 336.
80. C. Tracy and H. Widom, Comm. Math. Phys. **161** (1994) 289.
81. D. Hofstadter, Phys. Rev. **B14** (1976) 2239.
82. K. Slevin and T. Nagao, Phys. Rev. Lett. **70** (1993) 635.
83. J. Kelner and J.J.M. Verbaarschot, in preparation.
84. J. Osborn and J.J.M. Verbaarschot, in progress.
85. S. Hands and M. Teper, Nucl. Phys. **B347** (1990) 819.
86. J. Cullum and R.A. Willoughby, J. Comp. Phys. **44** (1981) 329.
87. J.J.M. Verbaarschot, Nucl. Phys. **B427** (1994) 534.
88. T. Wettig, T. Guhr, A. Schäfer and H. Weidenmüller, *The chiral phase transition, random matrix models, and lattice data*, hep-ph/9701387.
89. P. Forrester, Nucl. Phys. **B[FS]402** (1993) 709.
90. T. Wettig, private communication 1997.
91. T. Jolicoeur and A. Morel, Nucl. Phys. **B262** (1985) 627.
92. M. Göckeler, R. Horsley, E. Laermann, P. Rakow, G. Schierholtz, R. Sommer and U.-J. Wiese, Nucl. Phys. **B334** (1990) 527.
93. T. Wettig, A. Schäfer and H. Weidenmüller, Phys. Lett. **B367** (1996) 28.
94. M.A. Halasz, A.D. Jackson and J.J.M. Verbaarschot, Phys. Lett. **B395** (1997) 293; *Fermion determinants in matrix models of QCD at nonzero chemical potential*, hep-lat/9703006, Phys. Rev. D (in press).
95. J. Feinberg and A. Zee, *Non-Hermitean Random Matrix Theory: method of hermitization*, cond-mat/9703118; *Nongaussian nonhermitean random matrix theory: phase transition and addition formalism*, cond-mat/9704191; *Nonhermitean random matrix theory: method of hermitean reduction*, cond-mat/9703087.
96. K. Efetov, Adv. Phys. **32**, 53 (1983).
97. J.J.M. Verbaarschot, H.A. Weidenmüller, and M.R. Zirnbauer, Phys. Rep. **129**, 367 (1985).
98. Y. Fyodorov and H. Sommers, JETP Lett. **63** (1996), 1026.
99. B. Khoruzhenko, J. Phys. A: Math. Gen. **29**, L165 (1996).
100. Y. Fyodorov, B. Khoruzhenko and H. Sommers, Phys. Lett. **A 226**, 46 (1997).
101. K. Efetov, Phys. Rev. Lett. **79** (1997) 491; *Quantum disordered systems with a direction*, cond-mat/9706055.
102. V.L. Girko, *Theory of random determinants*, Kluwer Academic Publishers, Dordrecht, 1990.
103. M.A. Nowak, *New developments in nonhermitean random matrix models*, hep-ph/9708418.
104. J.J.M. Verbaarschot and M.R. Zirnbauer, J. Phys. **A17** (1985) 1093.
105. I. M. Barbour, S. E. Morrison, E. G. Klepfish, J. B. Kogut and M.-P. Lombardo, *Results on Finite Density QCD*, hep-lat/9705042.
106. I. Barbour, N. Behihil, E. Dagotto, F. Karsch, A. Moreo, M. Stone and H. Wyld, Nucl. Phys. **B275** (1986) 296; M.-P. Lombardo, J.B. Kogut and D.K. Sinclair, Phys. Rev. **D54** (1996) 2303.
107. I.M. Barbour, S. Morrison and J. Kogut, *Lattice gauge theory simulation at nonzero chemical potential in the chiral limit*, hep-lat/9612012.
108. M. Stephanov, private communication.
109. E. Dagotto, F. Karsch and A. Moreo, Phys. Lett. **169 B**, 421 (1986).
110. M.A. Halasz, J. Osborn and J.J.M. Verbaarschot, *Random matrix triality at nonzero chemical potential*, hep-lat/9704007, Phys. Rev. **D** (in press).
111. Th. Schäfer, *Instantons and the Chiral Phase Transition at non-zero Baryon Density*, hep-ph/9708256.
112. C. Baillie, K.C. Bowler, P.E. Gibbs, I.M. Berbour and M. Rafique, Phys. Lett. **197B**, 195 (1987).

DUALITY AND OBLIQUE CONFINEMENT

G. 't Hooft*

Institute for Theoretical Physics
University of Utrecht, P.O.Box 80 006
3508 TA Utrecht, the Netherlands

ABSTRACT

By construction, renormalized non-Abelian gauge theories do not allow for point-like magnetic charges. However, a procedure exists called 'the Abelian projection', that transforms these theories into apparently non-renormalizable Abelian theories with magnetic point charges. The small-distance treatment of magnetic charges in gauge theories differs completely from that of electric charges. Yet the effects of these charges, in particular in the infrared region, are very similar. This is expressed in the notion called 'duality', and allows one to predict the possibility of exotic confinement modes, 'oblique confinement'. An exact duality relation is shown to exist which applies to electric and magnetic fluxes in a box.

1. THE ABELIAN PROJECTION

Consider an *SU*(*N*) Yang-Mills gauge theory. Fermions and/or scalar fields may be there, but will not be looked at. In the early '70s, an important issue in these theories was the understanding of their infrared behaviour, in particular quark confinement, which cannot be understood in terms of perturbation theory. For the renormalization of the theory it was considered mandatory to fix the gauge freedom, by adding a gauge fixing term to the Lagrangian[1]. Gauge-fixing gives rise to ghost fields, and although we have learned how to handle these ghost in perturbation theory, and how to renormalize the theory in the presence of these ghosts, we now understand that these same ghosts obscure our view on the physical degrees of freedom, in particular at large distances[2].

Giving a gauge fixing condition at one point in space-time, may affect the field values at some other point, and since this change is a pure gauge transformation, this effect produces an unphysical action at a distance. This is why ghosts arise. Therefore, ghosts can be avoided if we choose a gauge fixing procedure that does not require gauge transformations elsewhere: a 'local' gauge fixing. Let us try to do this[3].

*e-mail: `g.thooft@fys.ruu.nl`

• *Step 1.*
Pick a field $X(\mathbf{x}, t)$ (elementary or composite) in the adjoint representation. We choose the adjoint representation because, in the absence of fermions, this is the simplest non-trivial representation available. X must be a single field (a scalar or one component of a vector). Examples:

$$X^{ij} = G^{ij}_{12} \qquad \text{or} \qquad X^{ij} = (D_\mu G_{\alpha\beta} D_\mu G_{\alpha\beta})^{ij} \ . \tag{1}$$

X is a self-adjoint $N \times N$ matrix: $X = X^\dagger$. It transforms as

$$X'(\mathbf{x},t) = \Omega(\mathbf{x},t) X(\mathbf{x},t) \Omega^{-1}(\mathbf{x},t) \, , \tag{2}$$

which is completely local (i.e., no $\partial_\mu \Omega$ is involved.)

• *Step 2.*
Pick the gauge Ω in which this field X is diagonal:

$$X = \begin{pmatrix} \lambda_1 & & & \\ & \lambda_2 & & \emptyset \\ \emptyset & & \ddots & \\ & & & \lambda_N \end{pmatrix} , \qquad \lambda_1 \geq \lambda_2 \geq \cdots \geq \lambda_N \, . \tag{3}$$

Note that this can always be done, but it does not fix Ω entirely.

One may, for instance, choose as a gauge-fixing term

$$\mathcal{L}^{\text{gauge}} = \sum_{i>j} \alpha_{ij} X_{ij} \, , \tag{4}$$

where α_{ij} are Lagrange multiplier fields.

• *Step 3.*
Observe that this leaves a local gauge group,

$$\frac{U(1)^N}{U(1)_{\text{diag}}} = U(1)^{N-1} \tag{5}$$

as an invariance group:

$$X = \Omega X \Omega^{-1} \quad \text{iff} \quad [X, \Omega] = 0 \quad \rightarrow$$
$$\Omega = \begin{pmatrix} \omega_1 & & \emptyset \\ & \ddots & \\ \emptyset & & \omega_N \end{pmatrix} = \exp i \begin{pmatrix} \Lambda_1 & & \emptyset \\ & \ddots & \\ \emptyset & & \Lambda_N \end{pmatrix} , \tag{6}$$
$$\text{with} \quad \sum_i \Lambda_i = 0 \, .$$

• *Step 4.*
Fix the residual (Abelian) gauge just as one is accustomed to do in QED. For instance:

$$\mathcal{L}^{\text{gauge, Abelian}} = \sum_i \beta_i \partial_\mu A^\mu_{ii} \, , \qquad (\sum_i \beta_i \equiv 0) \, . \tag{7}$$

This is a *non-local* gauge condition, but there is not much that can be done about that; it is not very harmful either, since this refers to an Abelian local symmetry, which is as well understood as the standard theory of quantum electrodynamics.

Adding (4) with (7) we find the total gauge fixing term:

$$\mathcal{L}^{\text{gauge}} = \sum_{i>j} \alpha_{ij} X_{ij} + \sum_i \beta_i \partial_\mu A_\mu^{ii} \,. \tag{8}$$

- *Step 5.*

Formally, we still need the corresponding ghost Lagrangian:

$$\mathcal{L}^{\text{ghost}}(\eta, \bar{\eta}) = \sum_{i>j} \bar{\eta}_{ij} [\eta, X]_{ij} + \sum_i \bar{\eta}_i \partial_\mu D_\mu \eta^{ii} =$$
$$\sum_{i>j} \bar{\eta}_{ij} \eta_{ij} (\lambda_j - \lambda_i) + \sum_i \bar{\eta}_i \partial^2 \eta_{ii} + ig \sum_i \bar{\eta}_i \partial_\mu \left\{ A_\mu^{ik} \eta_{ki} - A_\mu^{ki} \eta_{ik} \right\} . \tag{9}$$

It is not so difficult to convince oneself however that this ghost does not contribute to the physical amplitudes. The interaction term in Eq. (9) turns the diagonal components of the η field into off-diagonal components, but there is no way back. So, even though the diagonal gauge fixing required a propagating ghost (the propagator is the second term in (9)), this ghost is as harmless as in ordinary Abelian theories. Thus, we see that the ghost can be ignored.

- *Step 6.*

Observe that, in this gauge, we have all characteristics of a $U(1)^{N-1}$ Abelian gauge theory. All fields carry $U(1)^{N-1}$ charges q_e^i (with $\sum_i q_e^i = 0$). For instance, the off-diagonal parts of the gauge field A_μ^{ij} has $q_e^i = 1$ and $q_e^j = -1$. The diagonal components, A_μ^{ii}, behave exactly as $N-1$ massless photons, but the off-diagonal components are no longer obviously protected from obtaining a mass. Once the gauge has been fixed as in (8), all three polarizations of these vector fields are physical, as in massive vector particles.

- *Step 7.*

But this gauge fixing procedure may lead to singularities, whenever two eigenvalues λ_i of the field X coincide. Since we ordered the eigenvalues (see Eq. (3)), this may happen only for two consecutive eigenvalues: $\lambda_i = \lambda_{i+1}$. Prior to gauge fixing, the field X near such a point may take values of the form

$$X(\mathbf{x}, t) = \begin{pmatrix} \lambda_1 & & & & \\ & \ddots & & & \\ & & \lambda & 0 & \\ & & 0 & \lambda & \\ & & & & \ddots \end{pmatrix} + \sum_{k=1}^{3} a_k(\mathbf{x}, t) \begin{pmatrix} \ddots & & \\ & (\sigma_k) & \\ & & \ddots \end{pmatrix} . \tag{10}$$

Here, a_k are three space-time dependent functions that vanish at the point of the singularity. If X were to be taken to be a C^∞ function of space-time, the loci of points where all three a_k vanish are isolated points in 3-space, forming trajectories in time, and therefore, they are particle-like. Indeed, they have all characteristics of magnetic monopoles. They are magnetically charged with respect to the Abelian subgroup $U(1)_i \otimes U(1)_{i+1}^{-1}$, so we lable this charge as

$$g_i = (0, \cdots, 0, 1, -1, 0, \cdots, 0) \,. \tag{11}$$

In conclusion: the original non-Abelian $SU(N)$ theory is formally equivalent to an Abelian $U(1)^{N-1}$ theory containing electric charges as described under Step 6 and magnetic charges as described by (11). All these charges are bare, pointlike objects.

2. OBLIQUE CONFINEMENT

In a **Θ-vacuum,** the magnetically charged objects receive an electric charge, in addition to their magnetic charges. This is seen as follows[4]. Let us write the Abelian parts of the Lagrangian, after fixing the non-Abelian part of the gauge (Abelian projection) as

$$\begin{aligned}\mathcal{L} &= -\tfrac{1}{2}\mathrm{Tr}\,F_{\mu\nu}F_{\mu\nu} + \tfrac{1}{2}i\Theta\mathrm{Tr}\,g^2F_{\mu\nu}\tilde{F}_{\mu\nu} + \cdots \\ &= \mathrm{Tr}\,(\dot{\mathbf{A}}^2 - \mathbf{B}^2) - g^2\Theta\cdot 2\mathrm{Tr}\,\dot{\mathbf{A}}\mathbf{B} + \cdots \\ &= \mathrm{Tr}\,(\dot{\mathbf{A}} - g\Theta\mathbf{B})^2 - (1+g^2\Theta^2)\mathrm{Tr}\,\mathbf{B}^2 + \cdots .\end{aligned} \tag{12}$$

Here, **B** describes the magnetic fields of the monopoles, whose values are fixed by topological constraints. This is exactly the Lagrangian of a theory without **Θ** term but with a background **E** field of strength $-g\Theta\mathbf{B}$.

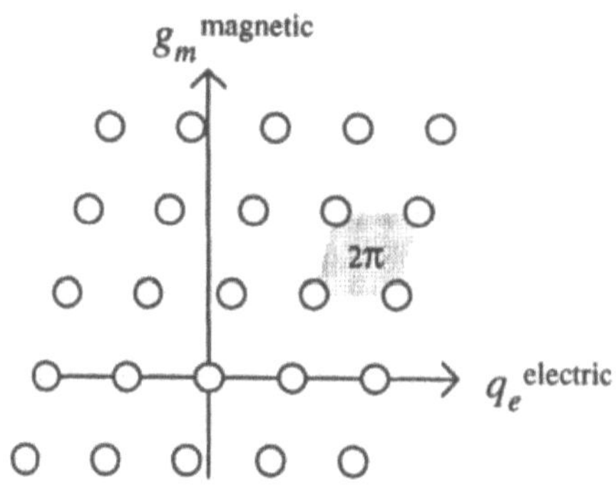

Figure 1. The oblique lattice of electric and magnetic charges.

The fact that these monopoles cany fractional electric charge is not in contradiction with the Dirac quantization condition[5] for electric and magnetic charges, because, given two particles 1 and 2 in a $U(1)$ theory, with electric charges q_e^1 and q_e^2 and magnetic charges g_m^1 and g_m^2, this condition reads

$$q_e^1 g_m^2 - q_e^2 g_m^1 = 2\pi n\,, \tag{13}$$

where n is an integer. If we plot the possible electric and magnetic charges as in Fig. 1, this condition amounts to the requirement that the area of the shaded box in Fig. 1 be a multiple of 2π. In our theory the Abelian gauge group is $U(1)^{N-1}$,in which case the Dirac condition reads

$$\sum_i (q_e^{1i} g_m^{2i} - q_e^{2i} g_m^{1i}) = 2\pi n\,. \tag{14}$$

If the *Higgs mechanism* is activated (for instance if a scalar charged field develops a vacuum expectation value), then all electrically charged particles only carry short range electric fields (since the A_μ^{ii} bosons obtain masses); they can roam about freely. Magnetic monopoles in this case will be *confined* by Nielsen-Olesen vortex tubes[6] (the

Figure 2. The Abrikosov vortex connecting magnetic monopoles in a super conductor.

Abrikosov vortex in a super conductor, see Fig. 2). In a plane that dissects the vortex, the Higgs field makes a full rotation; the vortex stays located near the zero of the Higgs field.

At large distance scales, this vortex behaves just as a string. It displays string degrees of freedom, whereas all other effects due to the exchanges of quantum fields will only be of short range (there are no massless particle types).

It is important to note that the *Higgs phase* is an aggregation phase in which the color-electrically charged particles have undergone Bose condensation (super conductivity). *Confinement* is a different aggregation phase of the system. Instead of the electrically charged particles, we expect the magnetically charged ones to undergo Bose condensation. In terms of the long distance degrees of freedom, this phenomenon could occur equally well. Because there is a complete dual symmetry between the electric fields and the magnetic fields, and since both electrically charged and magnetically charged particles were present, magnetic superconductivity is the dual counterpart of the Higgs mechanism. The particles that are bose condensed may roam about freely in the vacuum. They screen all other particles that carry the same combination of electric and magnetic charges. All other particles, however, undergo the same fate as what happens with the magnetic monopoles in the Higgs phase: they are confined. This is how we can understand the confinement phenomenon in theories such as QCD. Confinement is the dual analogue of the Higgs mechanism.

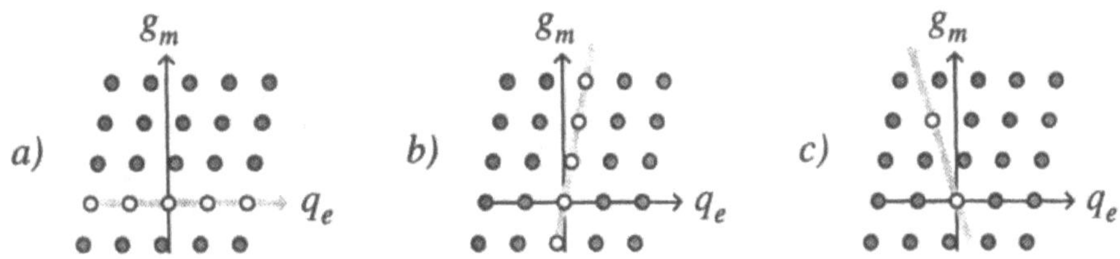

Figure 3. The light circles are liberated; the dark ones are confined.
a) The Higgs phase, b) The confinement phase, c) Oblique confinement.

It is also possible that *no* Bose condensation takes place at all. In this case, al particles carry long-range Abelian force fields. We call this the Coulomb phase. The Coulomb phase is self-dual, apart from the fact that the electric charge units q_e^i need not have the same values as the magnetic charge units g_m^i. They must obey Eq. (14).

As it was argued at the beginning of this section, the tilt angle in Fig. 1 and Fig. 3 is determined by the Θ angle. If we could vary Θ continuously from 0 to 2π, we would see the pattern of Fig. 1 shifting continuously, such that the configuration at $\Theta = 2\pi$ coincides with the one for $\Theta = 0$. Since the physics of the system must be periodic in Θ, we conclude that there is no fundamental distinction between magnetic monopoles and dyons (particles with both magnetic and electric charge).

If confinement takes place (Fig. 3b), we see however that one series of charge combination Bose condenses, and particles with the same charge ratios are liberated, the others confined. But, symmetry arguments tell us that, if Θ comes close to 2π, the series of charges that Bose condenses must be a different one. There must be intermediate values for Θ, at which the system jumps from one confinement mode to the other. Alternatively, it could be that at some Θ values, the system jumps back into the Higgs phase or the Coulomb phase. Thus, we see that QCD must exhibit phase transitions as Θ is varied from 0 to 2π. There must be at least one phase transition, which would then occur at $\Theta = \pi$.

A more exotic aggregation phase is conceivable. It could be that the objects that Bose condense carry more exotic charge combinations, for instance a dyon with $g_m = 2q_e$, (that is, a combination that would be impossible for monopoles with just one unit of magnetic charge. See Fig. 3c. In particular when Θ is close to π, such confinement modes are conceivable. It is this phase that we call oblique confinement.

3. FLUX IN A BOX

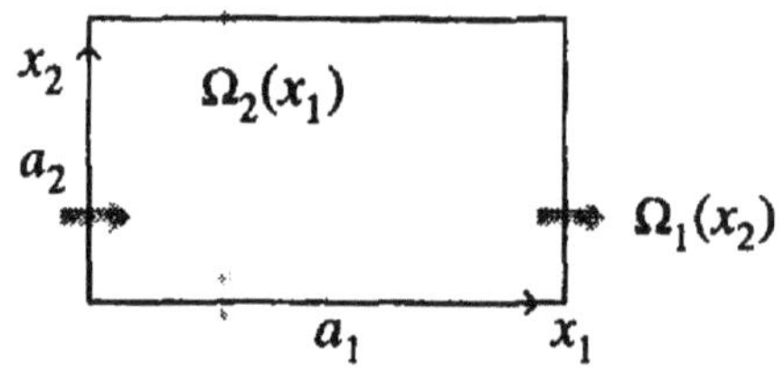

Figure 4. Boundary conditions in two-dimensional plane.

Again consider a pure $SU(N)$ gauge theory inside a 4 dimensional box with periodic boundary conditions. The lengths of the sides of the box are $a_1, \ldots, a_4$. We choose the gauge fields to be periodic *modulo* gauge transformations. Introducing the shorthand notation

$$A'_\mu = \Omega A_\mu\,, \qquad \text{or} \qquad A'_\mu(x') = \Omega(x_\perp)A_\mu(x)$$

$$\text{to mean} \qquad A'_\mu(x') = \Omega(x_\perp)\big(A_\mu(x) + \frac{1}{gi}\partial_\mu\big)\Omega^{-1}(x_\perp)\,, \tag{15}$$

we first formulate our boundary conditions in a two-dimensional plane: (see Fig. 4)

$$A_\mu(a_1,\, x_2) = \Omega_1(x_2)A_\mu(0,\, x_2)\,; \tag{16}$$

$$A_\mu(x_1,\, a_2) = \Omega_2(x_1)A_\mu(x_1, 0)\,. \tag{17}$$

This gives *two* conditions relating $A_\mu(a_1, a_2)$ to $A_\mu(0,0)$, which clearly should not conflict with each other:

$$A_\mu(a_1,\, a_2) = \Omega_1(a_2)\Omega_2(0)A_\mu(0,\, 0) = \Omega_2(a_1)\Omega_1(0)A_\mu(0,\, 0)\,. \tag{18}$$

From this we deduce the requirement that

$$\Omega_1(a_2)\Omega_2(0) = \Omega_2(a_1)\Omega_1(0)\, Z_{12}\,, \tag{19}$$

where Z_{12} is an element of the center on the gauge group (in this case $SU(N)$):

$$Z_{12} = e^{\frac{2\pi i}{N} n_{12}}\,, \tag{20}$$

where N is the number of 'colors' in the gauge group $SU(N)$, and n_{12} is an integer defined *modulo N.* Because of continuity these integers cannot depend on any of the (other) coordinates, but they do depend on the topological orientations of the plane. There are 6 orientations (in 4 dimensions), so we have 6 numbers $n_{\mu\nu}$ with $\mu\nu =$ 12, …, 34. For future reference we introduce the relabeling

$$\begin{matrix} n_{12} = m_3 & & n_{14} = n_1 \\ n_{23} = m_1 & \text{and} & n_{24} = n_2\,. \\ n_{31} = m_2 & & n_{34} = n_3 \end{matrix} \tag{21}$$

Besides these six numbers, there is one more topological index that can take any (positive or negative) integral value, the instanton winding number ν. This number is defined by multiplying[†] the gauge transformation matrices $\Omega(x_\perp)$ mentioned above by another matrix $\Omega(x_\perp)$ with a winding number ν from the mapping of the boundary $S(3)$ of the box onto $SU(N)$.

One can show that

$$\int \tfrac{1}{2}\mathrm{Tr}\, F_{\mu\nu}\tilde{F}_{\mu\nu} = \frac{8\pi}{g^2}\Big(\nu + \frac{\mathbf{n}\cdot\mathbf{m}}{N}\Big) . \tag{22}$$

This is proven, for instance, by constructing some simple field configurations, and then arguing that all other cases are obtained by continuous deformations. One may write (see Eqs. 21)

$$\frac{\mathbf{n}\cdot\mathbf{m}}{N} = \frac{n_{\mu\nu}\,\tilde{n}_{\mu\nu}}{8N} . \tag{23}$$

Now we continue by demanding that the box is in Euclidean space, and we define the amplitudes

$$W(\mathbf{n},\,\mathbf{m},\,\nu,\,a_1,\,a_2,\,a_3,\,a_4) \equiv \int \mathcal{D}A\; e^{iS}\Big|_{\mathbf{n},\mathbf{m},\nu,a_\mu} . \tag{24}$$

The interpretation of these amplitudes is elaborated in Ref[7]. The integers m_1, m_2, m_3 are the *magnetic* fluxes in the three spatial directions of a three-dimensional box. The length a_4 in the Euclidean time direction is the inverse temperature β. If we wish to know the free energy F of a gauge field configuration in the box with *electric* fluxes e_1, e_2, e_3 in the same box, then one can show that

$$e^{-\beta F(\mathbf{e},\mathbf{m},\Theta)} = C\sum_{n_i=0}^{N-1}\sum_{\nu=-\infty}^{\infty} e^{i\Theta(\nu + \frac{\mathbf{n}\cdot\mathbf{m}}{N}) - \frac{2\pi i}{N}(\mathbf{n}\cdot\mathbf{e})}\; W(\mathbf{n},\,\mathbf{m},\,\nu) . \tag{25}$$

We see that this expression depends on $\mathbf{e}$ and $\mathbf{m}$ only via the combination

$$\mathbf{e} - \frac{\Theta}{2\pi}\mathbf{m} , \tag{26}$$

which again shows that any object emitting magnetic flux behaves as if carrying (fractional) electric flux as well.

4. DUALITY

A striking feature of this expression is that is obeys an exact duality relation[7]. One can relate the free energy of magnetic fluxes with that of electric ones. The relation is obtained by means of a rotation over 90° in Euclidean space. Let us interchange the axes 1 and 2, and also the axes 3 and 4, using the $SO(4)$ rotation matrix

$$\mathcal{O} = \begin{pmatrix} 0 & -1 & & \\ 1 & 0 & \multicolumn{2}{c}{\emptyset} \\ & & 0 & 1 \\ \multicolumn{2}{c}{\emptyset} & -1 & 0 \end{pmatrix} . \tag{27}$$

Defining the following notation for the transverse components of a vector,

$$\tilde{x} \equiv \begin{pmatrix} x_1 \\ x_2 \end{pmatrix} , \qquad \hat{x} \equiv \begin{pmatrix} x_2 \\ x_1 \end{pmatrix} , \tag{28}$$

[†]To some extent, the result of this operation depends on how the equations (18) are implemented in the construction of the matrices Ω_i, but Eq. (22) defines the index ν uniquely.

we find that this rotation produces the substitutions:

$$\tilde{n} \leftrightarrow \tilde{m}\,, \qquad a_1 \leftrightarrow a_2\,, \qquad a_3 \leftrightarrow \beta\,;$$
$$n_3 \leftrightarrow n_3\,, \qquad m_3 \leftrightarrow m_3\,, \tag{29}$$

and the final result is

$$e^{-\beta F(\tilde{e}, e_3, \tilde{m}, m_3, \tilde{a}, a_3, \beta)} =$$
$$\frac{1}{N^2}\sum_{\tilde{n},\tilde{\ell}} e^{-\frac{i}{N}\tilde{n}(2\pi\tilde{e}-\Theta\tilde{m}) + \frac{i}{N}\tilde{m}(2\pi\tilde{\ell}-\Theta\tilde{n})} \cdot e^{-a_3F(\tilde{\ell}, e_3, \tilde{n}, m_3, \hat{a}, \beta, a_3)}$$
$$= \frac{1}{N^2}\sum_{\tilde{n},\tilde{\ell}} e^{\frac{2\pi i}{N}(\tilde{m}\cdot\tilde{\ell} - \tilde{n}\cdot\tilde{e}) - a_3F(\tilde{\ell}, e_3, \tilde{n}, m_3, \hat{a}, \beta, a_3)}\,. \tag{30}$$

Thus, the Θ dependence drops out.

From this duality relation, one can derive several interesting features concerning the behaviour of electric and magnetic fluxes. If, for instance, there is quark confinement, then the electric flux lines will carry energy proportional to their lengths, which can be written as

$$E_{e1} \approx \varrho a_1\,, \tag{31}$$

(for the flux in the 1-direction), where ϱ is the string constant for the electric flux. One may derive[7] that then the energy E_{m1} of a magnetic flux in the 1-direction drops as

$$E_{m1} \approx C\, a_1\, e^{-\varrho a_2 a_3}\,. \tag{32}$$

If the Coulomb phase is realized, the duality relation agrees with the usual expressions for the energy of electric and magnetic Abelian fields.

REFERENCES

1. G. 't Hooft and M. Veltman, Nucl.Phys. **B50** (1972) 318.
2. V. Gribov, Nucl. Phys. **B139** (1978) 1.
3. G. 't Hooft, Nucl. Phys. **B190** [FS3] (1981) 455; Phys. Scripta **25** (1982) 133.
4. E. Witten, Phys. Lett, **B86** (1979) 283.
5. P.A.M. Dirac, Proc. Roy. Soc. **A133** (1934) 60; Phys. Rev. **74** (1948) 817.
6. H.B. Nielsen and P. Olesen, Phys. Lett. **32B** (1970) 203.
7. G. 't Hooft, Acta Phys. Austr., Suppl.**22** (1980) 531.

ABELIAN PROJECTIONS AND MONOPOLES

M.N. Chernodub and M.I. Polikarpov*

Institute of Theoretical and Experimental Physics
B. Cheremushkinskaya 25, Moscow, 117259, Russia

INTRODUCTION

In these lectures we give an introduction to the theory of the confinement of color in lattice gauge theories. For the sake of self-consistency, we explain all definitions. The reader is supposed to be familiar with basic notions of lattice gauge theory.

Lattice field theories were originally formulated[1] in order to explain the confinement of color in nonabelian gauge theories. The leading term of the strong coupling $(1/g^2)$ expansion in lattice gauge theories yields the area law for the Wilson loop. The D dimensional theory is reduced to the 2–dimensional one, the field strength tensors are independent on different plaquettes and the confinement has a stochastic nature. The expectation value of the Wilson loop is simply $< P >^S$, where $< P >$ is the expectation value of the plaquette, S is the area of the minimal surface spanned on the loop. Therefore, the string tension is $\sigma = -\ln < P >$.

The numerical study of lattice gauge theories initiated by Creutz[2] shows that the strong coupling expansion of lattice gauge theory has no relation to continuum physics: the effects of lattice regularization are very strong, the results are not rotationally invariant, and there is no scaling for hadron masses[†].

In the weak coupling region (where the existence of the continuum limit was found in numerical calculations), the confinement mechanism has no stochastic origin. There are several approaches to explain color confinement. We will describe the most traditional one, namely, confinement caused by the dual Meissner effect. The linear confining potential can be explained by the formation of a string (flux tube) connecting a quark and an anti-quark. A well-known example of the string–like solution of the classical equations of motion is given in Ref.[3]. If we have a medium of condensed electric charges (a superconductor), then between the monopole and anti-monopole an Abrikosov string is formed, see Figure 1(a). To explain the confinement of electric charges, we need a condensate of magnetic monopoles, Figure 1(b). This simple qualitative idea was suggested by 't Hooft and Mandelstam[4]. There is a long way from this picture to real QCD. First, we have to explain how the abelian gauge field with monopoles is obtained from

*Lectures presented by Mikhail Polikarpov

[†]Here, scaling means independence of the ratios of the hadron masses m_k/m_i on the parameters of the theory such as the bare coupling and the cutoff parameter.

the non-abelian gauge field. Secondly, we have to explain why a dual superconductor is involved here. We discuss these two questions in these lectures.

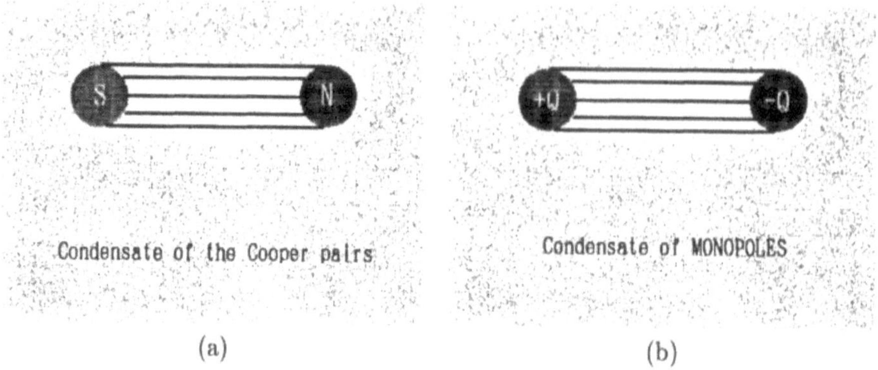

Figure 1. The Abrikosov string between the monopoles in the superconductor (a) and an analogue of the Abrikosov string between the electrically charged particles in the dual superconductor (b).

At present, we have no analytic proof of the existence of the condensate of abelian magnetic monopoles in gluodynamics and in chromodynamics. However, in all theories allowing for an analytical proof of confinement, the latter is due to the condensation of monopoles. These analytical examples are: compact electrodynamics[5], the Georgi–Glashow model[6], and super-symmetric Yang–Mills theory[7].

On the other hand, many numerical facts (some of these are discussed in these lectures) suggest that the vacuum in $SU(2)$ and $SU(3)$ lattice gauge theories behaves as a dual superconductor. As an illustration, we give two figures obtained by numerical calculations in $SU(2)$ lattice gluodynamics.

In Figure 2, taken from Ref.[8], the action density (vertical axis) of the $SU(2)$ fields is shown. The two peaks correspond to the quark–anti-quark pair, the formation of the flux tube is clearly seen. In Figure 3, taken from Ref.[9], the abelian monopole currents near the center of the flux tube formed by the quark–anti-quark pair are shown. It is seen that the monopoles wind around the center of the flux tube just as the Cooper pairs wind around the center of the Abrikosov string.

We first explain how to get the abelian fields and monopoles from the non-abelian fields. Then we present the results of numerical studies of the confinement mechanism in lattice gluodynamics.

All technical details such as the formalism of differential forms on the lattice are given in the Appendices.

ABELIAN MONOPOLES FROM NON-ABELIAN GAUGE FIELDS

In this Section we discuss the question how to obtain the abelian monopoles from non-abelian gauge fields.

The Method of Abelian Projection

The abelian monopoles arise from non-abelian gauge fields as a result of the abelian projection suggested by 't Hooft[10]. The abelian projection is a partial gauge fixing under which the abelian degrees of freedom remain unfixed. For example, the abelian

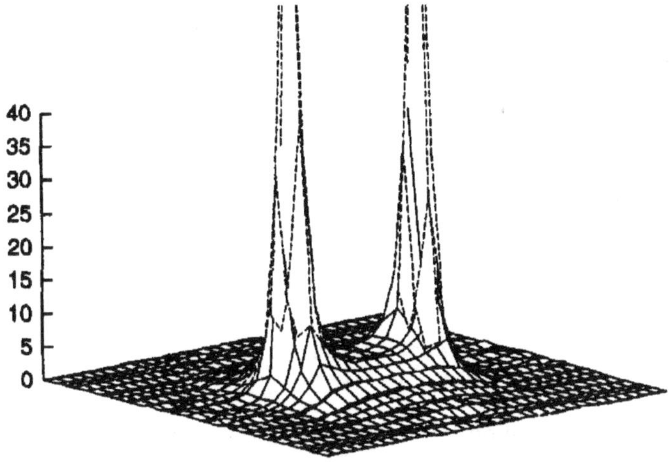

Figure 2. The action density distribution[8] at $\beta = 2.635$, the distance between quark and anti-quark sources is $\approx 1.35 fm$. Two horizontal axes correspond to two spatial lattice axes.

projection of a theory with $SU(N)$ gauge symmetry leads to a theory with $[U(1)]^{N-1}$ gauge symmetry.

Since the original $SU(N)$ gauge symmetry group is compact, the remaining abelian gauge group is also compact. But the abelian gauge theories with compact gauge symmetry group possess abelian monopoles. Therefore $SU(N)$ gauge theory in the abelian gauge has abelian monopoles.

First, consider the simplest example of $\boldsymbol{F_{12}}$ *abelian projection* for $SU(2)$ gauge theory. This gauge is defined by the following condition:

$$\hat{F}_{12}(x) = \text{diagonal matrix}. \qquad (1)$$

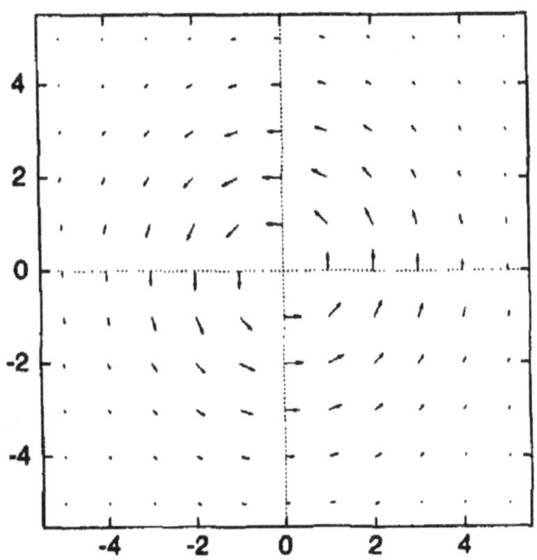

Figure 3. The monopole currents around the string tube which is formed between the static quark and anti-quark in gluodynamics, Ref.[9].

It is easy to fix to this gauge, since under gauge transformations the field strength tensor $\hat{F}_{12}$ transforms as

$$\hat{F}_{12}(x) \to \hat{F}'_{12}(x) = \Omega^{+}(x)\hat{F}_{12}(x)\Omega(x). \tag{2}$$

If we fix to the F_{12} gauge, then the field strength tensor $\hat{F}'_{12}(x)$ is invariant under $U(1)$ gauge transformations:

$$\hat{F}'_{12}(x) = \Omega^{+}_{U(1)}(x)\hat{F}'_{12}(x)\Omega_{U(1)}(x), \tag{3}$$

where

$$\Omega_{U(1)}(x) = \begin{pmatrix} e^{i\alpha(x)} & 0 \\ 0 & e^{-i\alpha(x)} \end{pmatrix}, \quad \alpha \in [0, 2\pi). \tag{4}$$

Therefore, the gauge condition (1) fixes the $SU(2)$ gauge group up to the diagonal $U(1)$ subgroup.

The $SU(2)$ gauge field

$$\hat{A}_{\mu} = \begin{pmatrix} A^{3}_{\mu} & A^{+}_{\mu} \\ A^{-}_{\mu} & -A^{3}_{\mu} \end{pmatrix} \tag{5}$$

transforms under the gauge transformations as

$$\hat{A}_{\mu} \to \Omega^{+}\hat{A}_{\mu}\Omega - \frac{i}{g}\Omega^{+}\partial_{\mu}\Omega. \tag{6}$$

If we fix to the F_{12} gauge, then under the remaining $U(1)$ gauge transformation (4) the components of the nonabelian gauge field A transform as

$$a_{\mu} \to a_{\mu} - \frac{1}{g}\partial_{\mu}\alpha, \quad a_{\mu} = A^{3}_{\mu}; \qquad A^{+}_{\mu} \to e^{2i\alpha}A^{+}_{\mu}, \quad A^{+}_{\mu} = A^{1}_{\mu} + iA^{2}_{\mu}. \tag{7}$$

Thus, in the abelian gauge the field a_{μ} plays the role of the abelian gauge field and the field A^{+}_{μ} is a charge 2 abelian vector matter field.

We thus obtain abelian fields from non-abelian ones. It occurs that abelian projection is also responsible for the appearance of abelian monopoles. If a is a regular abelian field, then $\mathrm{div}\vec{H} = 0$, since $\vec{H} = \mathrm{curl}\,\vec{a}$. But the abelian gauge field may be non-regular, since the matrix of the gauge transformation Ω may contain singularities. The nonabelian field strength tensor $\hat{F}_{\mu\nu}$ is *not invariant* under singular gauge transformations:

$$\hat{F}_{\mu\nu}[A] \to \hat{F}_{\mu\nu}[A^{(\Omega)}] = \Omega^{+}\hat{F}_{\mu\nu}[A]\Omega + \hat{F}^{sing}_{\mu\nu}[\Omega], \tag{8}$$

where

$$\hat{F}^{sing}_{\mu\nu}[\Omega] = -\frac{i}{g}\Omega^{+}(x)[\partial_{\mu}\partial_{\nu} - \partial_{\nu}\partial_{\mu}]\Omega(x). \tag{9}$$

Therefore, if we fix to the abelian gauge (1), using the singular gauge rotation matrices, the *abelian* field strength tensor

$$f_{\mu\nu} = \partial_{\mu}a_{\nu} - \partial_{\nu}a_{\mu} \tag{10}$$

may contain singularities (the Dirac strings)

$$f_{\mu\nu} = f^{r}_{\mu\nu} + f^{s}_{\mu\nu}, \qquad \varepsilon_{\mu\nu\alpha\beta}\partial_{\nu}f^{r}_{\alpha\beta} = 0 \quad \text{and} \quad \varepsilon_{\mu\nu\alpha\beta}\partial_{\nu}f^{s}_{\alpha\beta} \neq 0, \tag{11}$$

and therefore, $\mathrm{div}\vec{H} \neq 0$.

The charge of the monopole can be calculated[11] by means of the Gauss law. Let us choose an abelian monopole in a certain time slice and surround it by an infinitesimally small sphere S. The monopole charge is

$$m = \frac{1}{4\pi}\oint \vec{H}\, d\vec{\sigma} = \frac{1}{8\pi}\int d\sigma_{\mu\nu}\, \varepsilon_{\mu\nu\alpha\beta} f_{\alpha\beta} = \frac{i}{4\pi g}\int d\sigma_{\mu\nu}\, \varepsilon_{\mu\nu\alpha\beta}\mathrm{Tr}[\Omega^+\partial_\alpha\Omega\Omega^+\partial_\beta\Omega]$$
$$= -\frac{i}{4\pi g}\int d\sigma_{\mu\nu}\, \varepsilon_{\mu\nu\alpha\beta}\partial_\alpha\mathrm{Tr}[\Omega^+\partial_\beta\Omega] = 0, \pm\frac{1}{2g}, \pm\frac{1}{g}, \dots \quad (12)$$

The quantization of the abelian monopole charge is due to topological reasons: the surface integral in eq.(12) is equal to the winding number of *SU(2)* over the sphere *S* surrounding the monopole. The physics of quantization is simple: the electric charge is fixed by the gauge transformation (7), and magnetic charge should obey the Dirac quantization condition. The last integral can be seen as the magnetic flux through the Dirac string that arises as a consequence of the gauge singularity.

Various Abelian Projections

There is an infinite number of abelian projections. In the previous section we have considered the $\hat{F}_{12}$ abelian gauge. Instead of the diagonalization of the tensor component $\hat{F}_{12}$ by the gauge transformation, we can diagonalize any operator X which transforms under the gauge rotation as follows: $X \to \Omega^+ X \Omega$. Each operator X defines an abelian projection. At finite temperature one can consider the so-called *Polyakov abelian gauge* which is defined by the diagonalization of the Polyakov line.

The most interesting numerical results are those obtained in the Maximal Abelian (MaA) gauge. This gauge is defined by the maximization of the functional

$$\max_{\Omega} R[\hat{A}^\Omega], \qquad R[\hat{A}] = -\int d^4x\, [(A^1_\mu)^2 + (A^2_\mu)^2], \quad (13)$$

The condition of a local extremum of the functional R is

$$(\partial_\mu \pm igA^3_\mu)A^\pm = 0. \quad (14)$$

Clearly, this condition (as well as the functional R[A]) is invariant under the U(1) gauge transformations (7). The meaning of the MaA gauge is simple: by gauge transformations we make the field $\hat{A}_\mu$ as diagonal as possible.

Abelian Projection on the Lattice

The SU(2) gauge fields U_l on the lattice are defined by SU(2) matrices attached to the links l. These lattice fields are related to the continuum SU(2) fields $\hat{A}$: $U_{x,\mu} = e^{iag\hat{A}_\mu(x)}$, here a is the lattice spacing. Under the gauge transformation, the field U_l transforms as $U^\Omega_{x,\mu} = \Omega^+_x U_{x,\mu}\Omega_{x+\hat{\mu}}$, the matrices of the gauge transformation are attached to the sites x of the lattice.

The Maximal Abelian gauge is defined on the lattice by the following **condition**[11]:

$$\max_{\Omega} R[\hat{U}^\Omega_l], \qquad R[U_l] = \sum_l Tr[\sigma_3 U^+_l \sigma_3 U_l], \quad l = \{x, \mu\}. \quad (15)$$

This gauge condition corresponds to an abelian gauge, since R is invariant under the gauge transformations defined by the matrices (4).

Let us parametrize the link matrix U in the standard way

$$U_l = \begin{pmatrix} \cos\varphi_l e^{i\theta_l} & \sin\varphi_l e^{i\chi_l} \\ -\sin\varphi_l e^{-i\chi_l} & \cos\varphi_l e^{-i\theta_l} \end{pmatrix}, \tag{16}$$

where $\theta, \chi \in [-\pi, +\pi)$ and $\varphi \in [0, \pi)$. In this parameterization,

$$R[U_l] = \sum_l \cos 2\varphi_l . \tag{17}$$

Thus, the maximization of R corresponds to the maximization of the diagonal elements of the link matrix (16).

Under the $U(1)$ gauge transformations, the components of the gauge field (16) are transformed as

$$\theta_{x,\mu} \to \theta_{x,\mu} + \alpha_x - \alpha_{x+\hat{\mu}}, \quad \chi_{x,\mu} \to \chi_{x,\mu} + \alpha_x + \alpha_{x+\hat{\mu}}, \quad \varphi_{x,\mu} \to \varphi_{x,\mu} \tag{18}$$

Therefore, in the MaA the gauge, the field θ is the $U(1)$ gauge field, the field χ is the abelian charge 2 vector matter field, the field φ is the non-charged vector matter field.

Monopoles on the Lattice

A configuration of abelian gauge fields θ_l can contain monopoles. The position of the monopoles is defined by the lattice analogue of the Gauss theorem. Consider the elementary three–dimensional cube C (Figure 4(a)) on the lattice.

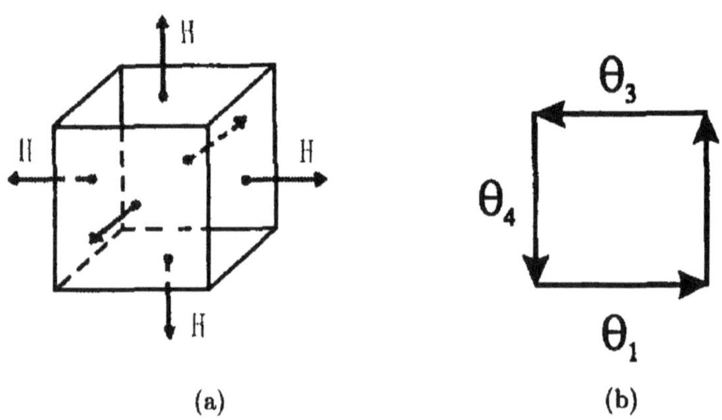

Figure 4. The magnetic flux through the boundary of the cube C (a) and lattice field strength tensor (b).

The abelian magnetic flux $\vec{H}$ through the surface of the cube C is given by the formula

$$m = \frac{1}{2\pi} \sum_{P \in \partial C} \bar{\theta}_P , \tag{19}$$

where $\bar{\theta}_P$ is the magnetic field defined as follows. Consider the plaquette angle $\theta_P = \theta_1 + \theta_2 - \theta_3 - \theta_4 \equiv d\theta$, the θ_i's are attached to the links i which form the boundary of the plaquette P, Figure 4(b). The definition of $\bar{\theta}_P$ is $\bar{\theta}_P = \theta_P + 2\pi k$, where the integer k is such that $-\pi < \bar{\theta}_P \leq \pi$. The restriction of $\bar{\theta}_P$ to the interval $(-\pi, \pi]$ is natural since (as we point out in Appendix A) the abelian action for the compact fields θ_l is a periodic function of $\bar{\theta}_P$. Equation (19) is the lattice analogue of the continuum formula $m = \oint \vec{H}\, d\vec{S}$. Due to the compactness of the lattice field θ $(-\pi < \theta \leq \pi)$ there exist singularities (Dirac strings), and therefore, $\mathrm{div}\vec{H} \neq 0$.

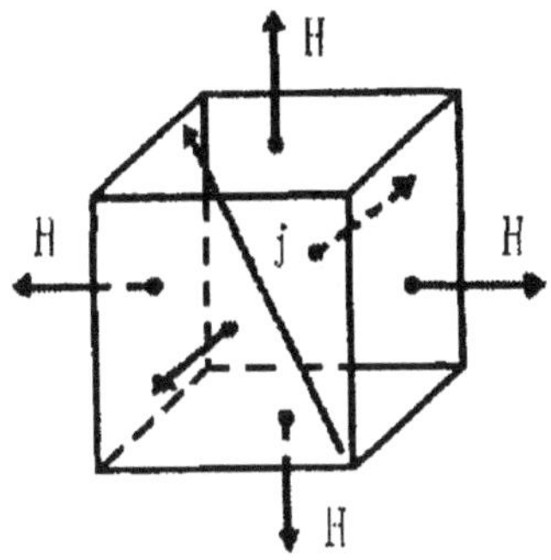

Figure 5. The definition of the magnetic monopole current.

The magnetic charge m defined by eq. (19) has the following properties:

1. m is quantized: $m = 0, \pm 1, \pm 2$;
2. If $m \neq 0$, then there exists a magnetic current j. This current is attached to the link dual to cube C, Figure 5.
3. Monopole currents j are conserved: $\delta j = 0$, the currents form closed loops on the 4D lattice. The proof is given in Example 3 of Appendix A.

VACCUM OF GLUODYNAMICS AS A (DUAL) SUPERCONDUCTOR

As we have just shown, the non-abelian gauge field in the abelian projection is reduced to abelian fields and abelian monopoles. The static quark-anti-quark pair in the abelian projection becomes the electric charge–anti-charge pair, and if the monopoles are condensed, then the quark and anti-quark are connected to each other by an analogue of the Abrikosov string. A more detailed discussion of the confinement in the abelian projection is given in Refs.[10]. Thus, in order to justify the monopole confinement mechanism we have to prove the existence of the monopole condensate. For lattice gluodynamics we have a lot of numerical facts which confirm the monopole confinement mechanism. We discuss these at the end of this section. But first we give an analytical example which shows how the monopole condensate appears in compact electrodynamics.

Compact $U(1)$ Lattice Gauge Theory

We show that lattice compact electrodynamics can be represented as a dual Abelian Higgs model, with the Higgs particles being monopoles. The partition function of the compact $U(1)$ gauge theory can be written as

$$\mathcal{Z} = \int_{-\pi}^{+\pi} \mathcal{D}\theta_l \exp\{-S(d\theta)\}\,, \tag{20}$$

where, as below, $\int\limits_{-\pi}^{+\pi} \mathcal{D}\theta_l = \prod\limits_l \int\limits_{-\pi}^{+\pi} d\theta_l$ is the integral over all link variables. In the continuum limit $(a \to 0)$, $S(\theta_P) \propto f_{\mu\nu}^2 a^4$.

By means of the duality transformation (see Appendix B), the theory (20) can be rewritten in the form

$$\mathcal{Z} = \text{const.} \cdot \sum_{\substack{{}^*n({}^*C_1)\in \mathbb{Z} \\ \delta^* n=0}} \exp\{-S^d(d^*n)\}, \tag{21}$$

where $d^*n = {}^*n_1 + {}^*n_2 - {}^*n_3 - {}^*n_4$ is the plaquette constructed from the integers n (see Figure 6). The dual action is defined by eq.(B.3).

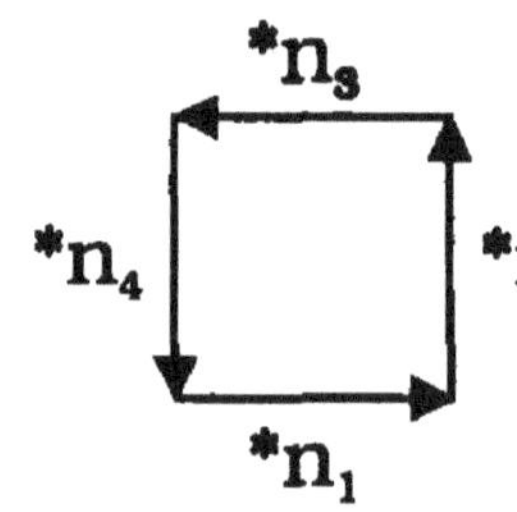

Figure 6. Field strength tensor on the dual lattice.

The integer valued variable $n_{x,\mu}$ is dual to the field $\theta_{x,\mu}$. In the continuum limit, $\theta_{x,\mu} \propto A_\mu(x)$ is the ordinary $U(1)$ gauge field and $n_{x,\mu} \propto B_\mu(x)$ is the dual $U(1)$ gauge field[‡], $\partial_\mu A_\nu - \partial_\nu A_\mu = \frac{1}{2}\varepsilon_{\mu\nu\alpha\beta}\partial_\alpha B_\beta$.

The partition function (21) can be represented as that of the (dual) Abelian Higgs model.

$$\mathcal{Z} = \text{const.} \cdot \lim_{\gamma\to\infty} \lim_{\lambda\to\infty} \int\limits_{-\infty}^{+\infty} \mathcal{D}^*B \int \mathcal{D}^*\Phi \exp\{-S^{AH}({}^*B, {}^*\Phi)\}, \tag{22}$$

where the action of the (dual) Abelian Higgs model is

$$S^{AH}({}^*B, {}^*\Phi) = \sum_P S^d(d^*B) + \frac{1}{2}\sum_x \sum_{\mu=1}^{4} |{}^*\Phi_x - e^{i^*B_{x,\mu}}{}^*\Phi_{x+\hat\mu}|^2 + \lambda \sum (|{}^*\Phi_x|^2 - \gamma)^2 \tag{23}$$

Here d^*B is the plaquette variable constructed from the link variable ${}^*B_l$, where ${}^*B_l$ is the dual gauge field. In the continuum limit, we have $S(d^*B) \to \beta(\partial_\mu B_\nu - \partial_\nu B_\mu)^2$. The second term in eq.(23) has the following continuum limit: $|\Phi_x - e^{i^*B_{x,\mu}}{}^*\Phi_{x+\hat\mu}|^2 \to |(\partial_\mu - iB_\mu)\Phi|^2$. The action S^{AH} is invariant under the following gauge transformations: ${}^*B \to {}^*B - d^*\alpha$, ${}^*\Phi \to {}^*\Phi e^{i^*\alpha}$.

In the London limit, $\lambda \to \infty$, the mass of the Higgs particle tends to infinity. The radius of the Higgs field $|{}^*\Phi|$ is fixed and only the phase ${}^*\varphi$ of the Higgs field ${}^*\Phi = |{}^*\Phi| e^{i^*\varphi}$ is a physical degree of freedom. In the unitary gauge ${}^*\varphi = 0$, the action of the Abelian Higgs model (23) is

$$S^{AH} = \sum_P S^d(d^*B) + \frac{\gamma}{2}\sum_x \sum_{\mu=4}^{4} |1 - e^{i^*B_{x,\mu}}|^2 . \tag{24}$$

[‡]In the continuum notations we do not use the symbol "*" to denote the dual fields.

In the limit $\gamma \to \infty$, the photon mass $m^2_{ph} = \gamma/\beta$ becomes infinite and the field $B_{x,\mu}$ is equal to zero modulo 2π: ${}^*B_{x,\mu} = 2\pi {}^*n_{x,\mu}$. In this limit, the partition function (24) reduces to

$$\mathcal{Z} = \text{const.} \cdot \sum_{\substack{{}^*n({}^*C_1)\in\mathbb{Z} \\ \delta^*n=0}} e^{-S^d(\mathrm{d}^*n)}\,, \qquad S(\mathrm{d}^*n) = 4\pi^2\beta\,(\mathrm{d}^*n)^2\,. \tag{25}$$

Therefore, the compact $U(1)$ gauge theory is equivalent to the dual abelian Higgs model in the double limit

$$\textbf{Compact}\, U(1)\, \textbf{gauge theory} = \lim_{\gamma\to\infty}\lim_{\lambda\to\infty}\,(\textbf{Abelian Higgs model}) \tag{26}$$

The gauge fields B_l in eq.(23) are dual to the original gauge fields θ_l, and these interact via the covariant derivative with the Higgs field Φ. Therefore, the field Φ carries the magnetic charge, and due to the Higgs potential in eq.(23), these monopoles are condensed at the classical level. It is well known that in the quantum $4D$ compact electrodynamics there exists a confinement–deconfinement phase transition. It can be shown by numerical calculations[12, 13] that in the confinement phase the monopoles are condensed and that in the deconfinement phase the monopoles are not condensed.

What is the Theory Dual to Gluodynamics?

In any abelian projection, lattice gluodynamics corresponds to some abelian gauge theory. This abelian theory, in general, is very complicated and non-local. Nevertheless, in the Maximal Abelian projection, numerical calculations show that the abelian monopoles are important degrees of freedom, and that they are responsible for the confinement. Moreover, as we show in the next section (see also Refs.[14, 15]), the distribution of monopole currents indicates that, at large distances, gluodynamics is equivalent to the dual Abelian Higgs model, the Higgs particles are abelian monopoles and these are condensed in the confinement phase.

NUMERICAL FACTS IN $4D$ $SU(2)$ GLUODYNAMICS

The standard scheme of numerical calculations can be described as follows.

1. Generate the lattice Yang–Mills fields U_l using the standard Monte-Carlo method. Thus we obtain the configurations of fields distributed according to the Boltzmann factor $e^{-S^{YM}}$;

2. Perform the abelian gauge fixing and extract abelian gauge fields from the non-abelian ones. In the case of the MaA projection, the abelian gauge fixing is a rather time-consuming problem.

3. Extract abelian monopole currents from the abelian gauge fields. As mentioned above, the monopole currents form closed paths on the dual lattice. In Figure 7 we show the abelian monopole currents for the confinement (a) and the deconfinement (b) phases. It is seen that in the confinement phase the monopoles form a dense cluster, and there is a number of small mutually disjoint clusters. In the deconfinement phase the monopole currents are dilute.

4. Calculate expectation values of various operators using the monopole currents. Below we discuss a number of numerical facts which show that abelian monopoles in the MaA gauge are the appropriate degrees of freedom to describe confinement.

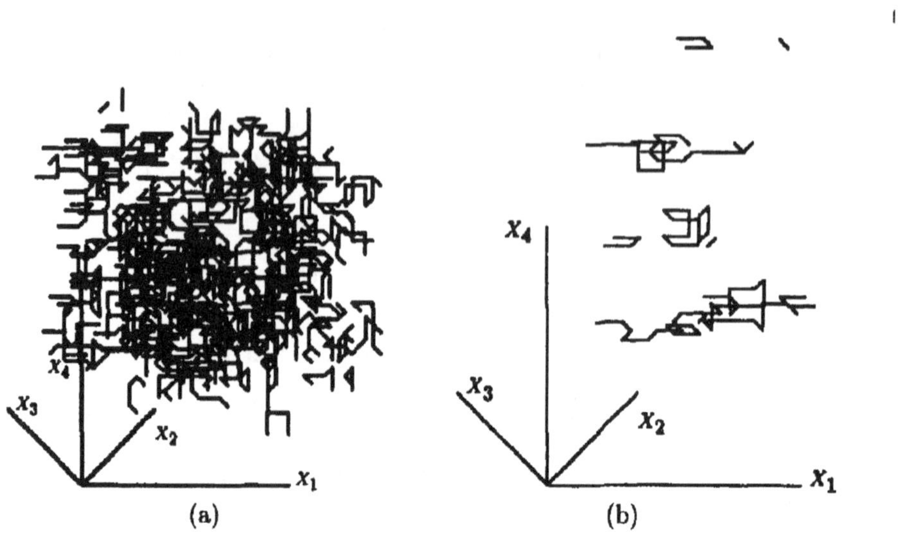

Figure 7. The abelian monopole currents for the confinement (a) ($\beta = 2.4$, 10^4 lattice) and the deconfinement (b) phases ($\beta = 2.8$, $12^3 \cdot 4$ lattice).

Fact 1: Abelian and Monopole Dominance

The notion of the "abelian dominance" introduced in Ref.[16] means that the expectation value of a physical quantity $< \mathcal{X} >$ in the nonabelian theory coincides with (or is very close to) the expectation value of the corresponding abelian operator in the abelian theory obtained by abelian projection. Monopole dominance means that the same quantity can be calculated in terms of the monopole currents extracted from the abelian fields. If we have N configurations of nonabelian fields on the lattice, the abelian and monopole dominance means that

$$\frac{1}{N}\sum_{conf}\mathcal{X}(\hat{U}_{nonabelian}) = \frac{1}{N}\sum_{conf}\mathcal{X}'(U_{abelian}) = \frac{1}{N}\sum_{conf}\mathcal{X}''(j). \tag{27}$$

Here each sum is taken over all configurations; $U_{abelian} = e^{i\theta_l}$ is the abelian part of the nonabelian field $\hat{U}_{nonabelian}$, j is the monopole current extracted from $U_{abelian}$. It is clear that $\frac{1}{N}\sum_{conf}\mathcal{X}(\hat{U}_{nonabelian})$ is a gauge invariant quantity, while the abelian and the monopole contributions depend on the type of abelian projection. In numerical calculations the equalities (27) can be satisfied only approximately.

Among the well-studied problems is that of the abelian and the monopole dominance for the string tension [16, 17, 18, 19]. In this case, $\mathcal{X}(\hat{U}_{nonabelian}) = \sigma_{SU(2)}$, $\mathcal{X}(U_{abelian}) = \sigma_{U(1)}$ and the string tension $\sigma_{SU(2)}$ ($\sigma_{U(1)}$) is calculated by means of the nonabelian (abelian) Wilson loops, $Tr\prod_{l\in C}\hat{U}_l$ ($\prod_{l\in C} e^{i\theta_l}$). An accurate numerical study of the MA projection of $SU(2)$ gluodynamics on the 32^4 lattice at $\beta = 2.5115$ is performed in Ref.[19].

The corresponding string tension can be taken from the potential between a heavy quark and antiquark: $V(r) = V_0 + \sigma r - \frac{e}{r}$, where r is the distance between the quark and antiquark. The abelian and the nonabelian potentials are shown in Figure 8. The contribution of the photon and the monopole parts to the abelian potential is shown in Figure 9.

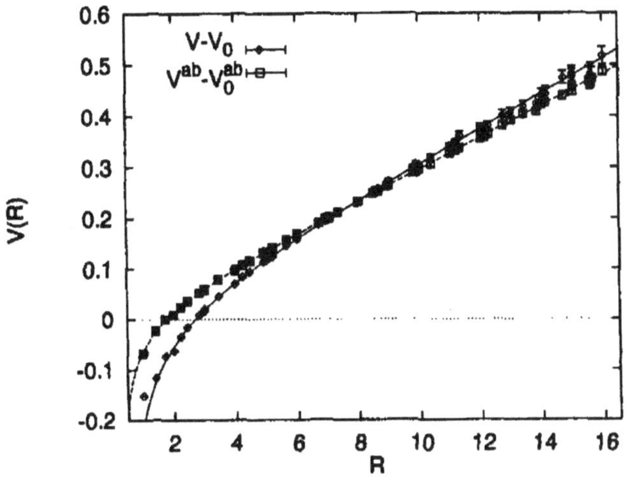

Figure 8. Abelian and nonabelian potentials (with the self energy V_0 subtracted), Ref.[19].

The differences in the slopes of the linear part of the potentials in Figure 8 and Figure 9 yield the following relations: $\sigma_{U(1)} \approx 92\% \sigma_{SU(2)}$, $\sigma_j \approx 95\% \sigma_{U(1)}$, where σ_j is the monopole current contribution to the string tension. It is important to study a widely discussed idea that in the continuum limit ($\beta \to \infty$), the abelian and the monopole dominance is exact (27): $\sigma_{SU(2)} = \sigma_{U(1)} = \sigma_j$.

There are many examples of abelian and monopole dominance in the MA projection. The monopole dominance for the string tension has been found for the $SU(2)$ positive plaquette model in which Z_2 monopoles are suppressed[20], and also for the $SU(2)$ string tension at finite temperature[21], and for the string tension in $SU(3)$ gluodynamics[22]. Abelian and monopole dominance for $SU(2)$ gluodynamics has been found in[23, 24] for the Polyakov line and for critical exponents for the Polyakov line, for the value of the quark condensate, for the topological susceptibility and also for the hadron masses in quenched $SU(3)$ QCD with Wilson fermions[25].

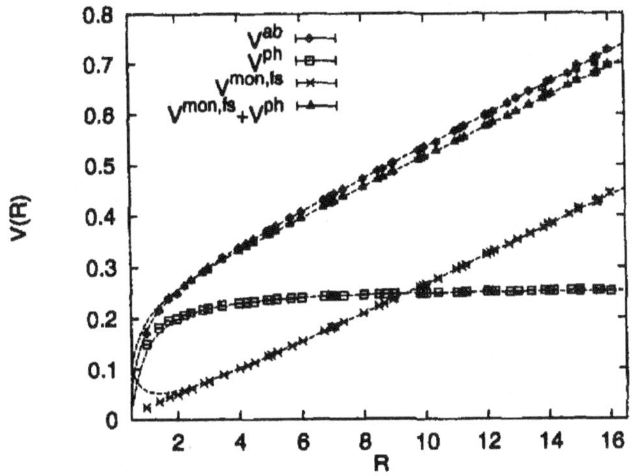

Figure 9. The abelian potential (diamonds) in comparison with the photon contribution (squares), the monopole contribution (crosses) and the sum of these two parts (triangles), Ref.[19].

Fact 2: London Equation for Monopole Currents

In the ordinary superconductor the current of the Cooper pairs satisfies the London equation. If the vacuum of gluodynamics behaves as a dual superconductor, then it is natural to assume that the abelian monopole currents should satisfy the dual London equation in the presence of the dual string

$$\vec{E} - \delta^2 \operatorname{curl} \vec{j}_m = \Phi_m \delta(x_x)\delta(x_y)\vec{n}_z\,, \tag{28}$$

where $\vec{E}$ is the abelian electric field, $\delta^{-1} = m_{ph}$ is the dual photon mass, $\Phi_m = \frac{2\pi}{g_m}$ is the electric flux of the dual string and g_m is the magnetic charge of the monopole. The string is placed along the *oz* axis, the vector $\vec{n}_z$ is parallel to the direction of the string and we assume for simplicity that the core of the string is a delta-function.

Indeed, as shown numerically in Ref.[26], the dual London equation is satisfied in the MaA projection of $SU(2)$ gluodynamics. A recent detailed investigation[27, 9] of the electric field profiles and the distribution of monopole currents around the string shows that the structure of the chromo-electric string in the MaA projection is very similar to that of the Abrikosov string in the superconductor.

The following physical question is relevant: what kind of superconductor do we have. If $m_{ph} < m_{mon}$, then the superconductor is of the second type, the Abrikosov vortices are attracted to each other. If $m_{ph} > m_{mon}$, then the superconductor is of the first type and the vortices repel each other. The computation[26] of the dual photon mass from the dual London equation shows that $m_{ph} \approx m_{mon}$. This means that the vacuum of gluodynamics is close to the border between type-I and type-II dual superconductors. The same conclusion is also obtained in Ref.[28]. A recent detailed study of the abelian flux tube[9] shows that the dual photon mass is definitely smaller then the mass of the monopole, and therefore the vacuum of $SU(2)$ is the type-II dual superconductor (see also Ref.[14]).

Fact 3: Monopole Condensate

If the vacuum of $SU(2)$ gluodynamics in the abelian projection is similar to the dual superconductor, then the value of the monopole condensate should depend on the temperature as a disorder parameter: at low temperatures it should be nonzero, and it should vanish above the deconfinement phase transition.

The behaviour of the monopole condensate can be studied with the help of the monopole creation operator. In gluodynamics this operator can be derived by means of the Fröhlich and Marchetti construction [29] for the compact electrodynamics. For $SU(2)$ lattice gluodynamics in the MA projection, it is convenient to study the effective constraint potential for the monopole creation operator (similar calculations were performed for compact electrodynamics in Ref.[13]):

$$V_{eff}(\Phi) = -\ln(< \delta(\Phi - \frac{1}{V}\sum_x \Phi_{mon}(x)) >) \tag{29}$$

where φ is the monopole field. If this potential has a Higgs-like form

$$V_{eff}(\Phi) \propto \lambda(|\Phi|^2 - \Phi_c^2)^2\,, \qquad \lambda, \Phi_c^2 > 0\,, \tag{30}$$

then a monopole condensate exists. The dual superconductor picture predicts this behaviour of the effective constraint potential in the confinement phase. In the deconfinement phase the following form of the potential is expected:

$$V_{eff}(\Phi) \propto m^2|\Phi|^2 + \lambda|\Phi|^4 + \dots\,, \qquad m^2, \lambda > 0\,; \tag{31}$$

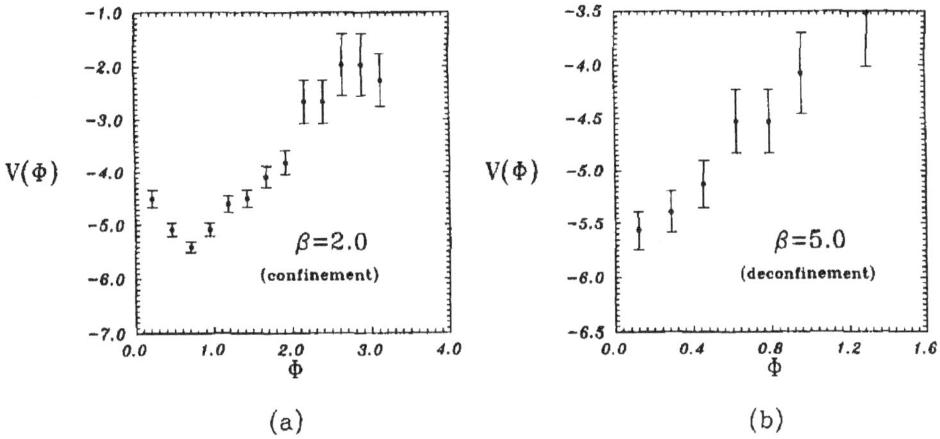

Figure 10. $V(\Phi)$ for the confinement (a) and the deconfinement (b) phases, Ref.[31].

(no monopole condensate).

The numerical calculation of the effective constraint potential (29) is very time-consuming, and therefore, in Refs.[30, 31] the following quantity $V(\Phi)$ has been studied:

$$V(\Phi) = -\ln(< \delta(\Phi - \Phi_{mon}(x)) >). \tag{32}$$

Numerical calculations of this quantity were performed on lattices of size $4 \cdot L^3$, for L = 8, 10, 12, 14, 16 with anti–periodic boundary conditions in space directions for the abelian fields. Periodic boundary conditions are forbidden, since a *single* magnetic charge cannot physically exist in a closed volume with periodic boundary conditions due to the Gauss law. The results on finite lattices have been extrapolated to the infinite volume, since near the phase transition finite volume effects are very strong.

In Figures 10 (a,b) the (right-hand side of the) effective potential (32) is shown for the confinement and the deconfinement phases, the calculations being performed on a $4 \cdot 12^3$ lattice.

In the confinement phase (Figure 10 (a)), the minimum of $V(\Phi)$ is shifted from zero, while in the deconfinement phase, the minimum is at the zero value of the monopole field Φ. The value Φ_c of the monopole field at which the potential has a minimum is equal to the value of the monopole condensate.

The potential shown in Figure 10 (b) corresponds to a trivial potential with a minimum at a zero value of the field: $\Phi_c = 0$. The dependence of Φ_c, the minimum of the potential, on the spatial size L of the lattice is shown in Figure 11 (a).

We fit the data for Φ_c by the formula $\Phi_c = AL^\alpha + \Phi_c^{inf}$, where A, α and Φ_c^{inf} are the fitting parameters. It occurs that $\alpha = -1$ within statistical errors. Figure 11(b) shows the dependence on β of the value of the monopole condensate extrapolated to the infinite spatial volume Φ_c^{inf}. It is clearly seen that Φ_c^{inf} vanishes at the point of the phase transition and it plays the role of the order parameter.

The monopole creation operator[32] in the monopole current representation is studied in Ref.[33]. First the monopole action is reconstructed from the monopole currents in the MA projection, and after that the expectation value of the monopole creation operator is calculated in the quantum theory of monopole currents. Again, the monopole creation operator depends on the temperature as the disorder parameter. A slightly different monopole creation operator was studied in Refs.[34, 35].

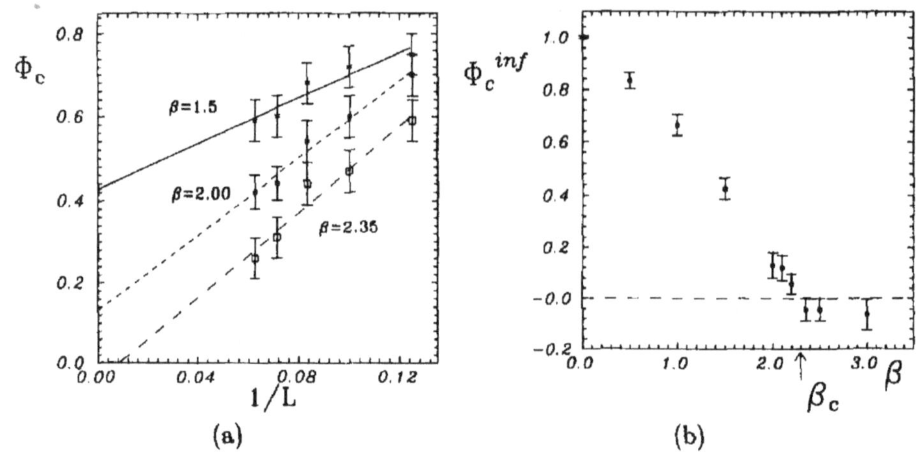

Figure 11. The dependence of (a) Φ_c on the spatial size of the lattice for three values of β and (b) of Φ_c^{inf} on β, Ref.[31].

Abelian Monopole Action: Analytical examples

In the next section we explain how to study the action of the monopoles extracted from $SU(2)$ fields in the MaA projection. Now we give several examples of monopole actions corresponding to abelian gauge theories.

Example 1: $4D$ compact abelian electrodynamics

We have already discussed the duality transformation of $4D$ compact electrodynamics. There is another exact transformation which represents the partition function as a sum over monopole currents. This transformation was initially performed by Berezinsky [36] and by Kosterlitz and Thouless[37] for the $2D$ XY model. Accordingly, we call it the BKT transformation. For compact electrodynamics with the Villain action this transformation was found by Banks, Myerson and Kogut[38]. The partition function for compact electrodynamics with the Villain action is

$$\mathcal{Z}^{Villain} = \int_{-\pi}^{+\pi} \mathcal{D}\theta \sum_{m(C_2)\in\mathbb{Z}} \exp\left\{-||d\theta + 2\pi m||^2\right\}. \qquad (33)$$

As explained in Appendix C, the BKT transformation of this partition function has the form

$$\mathcal{Z} = [\int_{-\infty}^{+\infty} \mathcal{D}A \exp\{-\beta||dA||^2\}] \times [\sum_{\substack{{}^*j({}^*C_1)\in\mathbb{Z} \\ \delta^*j=0}} \exp\{-4\pi^2\beta\,({}^*j, \Delta^{-1}\,{}^*j)\}]. \qquad (34)$$

Here the partition function for the non–compact gauge field A is inessential because it is Gaussian. All the dynamics is in the partition function for the monopole currents ${}^*j$ which are lying on the dual lattice and form closed loops $(\delta^*j = 0)$.

It is possible to find the BKT transformation for compact electrodynamics with a general form for the action[31]. In this general case the monopole action is non-local (see Appendix C).

Example 2: Abelian Higgs theory in the London limit

The partition function for the Abelian Higgs model is (cf., eqs.(22,23[§]):

$$\mathcal{Z}^{AH} = \int_{-\infty}^{+\infty} \mathcal{D}B \int_{-\pi}^{+\pi} \mathcal{D}\varphi \sum_{m(C_2)\in \mathbb{Z}} \exp\Big\{-\beta||dB||^2 - \gamma||d\varphi + B + 2\pi m||^2\Big\}, \qquad (35)$$

where B is the non-compact gauge field and φ is the phase of the Higgs field (the radial part of the Higgs field is frozen, since we consider the London limit).

After the BKT transformation [39, 40], the partition function takes the form

$$\mathcal{Z}^{AH} = \text{const.} \cdot \sum_{\substack{j(C_1)\in \mathbb{Z} \\ \delta j = 0}} \exp\Big\{-S^{AH}_{mon}(j)\Big\}, \quad S^{AH}_{mon}(j) = \frac{1}{4\beta}(j, \Delta^{-1} j) + \frac{1}{4\gamma}||j||^2, \qquad (36)$$

where Δ^{-1} is the inverse lattice Laplacian and j is the current of the Higgs particles.

Example 3: Abelian Higgs theory near the London limit

As pointed out by T. Suzuki[14], the numerical data show that in the MaA projection the currents in lattice $SU(2)$ gluodynamics behave as the currents of the Higgs particles in the Abelian Higgs model near the London limit. This means that the coefficient λ of the Higgs potential $\lambda(|\Phi|^2 - 1)^2$, eq.(23), is large but finite. In this case it is impossible to get an explicit expression for the monopole action, but there exists a $1/\lambda$ expansion of this action

$$S_{mon}(^*j) = \frac{1}{4\beta}(^*j, \Delta^{-1} {}^*j) + \Big(\frac{1}{4\gamma} + a_1\Big)||^*j||^2 + a_2 \sum_x \Big(\sum_{\mu=-4}^{4} {}^*j^2_{x,\mu}\Big)^2$$
$$+ a_3 \sum_x \sum_{\mu=1}^{4} {}^*j^4_{x,\mu} + \dots, \qquad (37)$$

where

$$a_k = \sum_{n=1}^{\infty} \frac{f_k^{(n)}}{\lambda^n},$$

$f_k^{(n)}$ are the coefficients which can be calculated[14]. In the limit $\lambda \to +\infty$, this action is reduced to $S^{AH}_{mon}(^*j)$, eq.(36).

Monopole Action from "Inverse Monte–Carlo"

The $1/\lambda$ expansion of the monopole action has an important application in lattice gluodynamics. It is possible to show that the monopole currents in the MaA projection of $SU(2)$ gluodynamics are in a sense equivalent to the currents generated by the theory with the action (37). This means that (at least at large distances) gluodynamics is equivalent to the Abelian Higgs model. The details can be found in the lecture of T. Suzuki [14]. Below we briefly describe some of these results.

It occurs that from a given distribution of currents on the $4D$ lattice it is possible to find the action of the currents. This can be done by the Swendsen ("inverse Monte–Carlo") [41, 42] method. By the usual Monte–Carlo method we get an ensemble of the

[§]We assume the Villain form for the interaction of the gauge and Higgs fields.

fields $\{\Phi\}$ with the probability distribution $e^{-S(\Phi)}$. The "inverse Monte–Carlo" allows us to reconstruct the action $S(\Phi)$ from the distribution of the fields $\{\Phi\}$.

The procedure of the reconstruction of the monopole action for lattice gluodynamics is the following. First the $SU(2)$ gauge fields are generated by the usual Monte–Carlo method. Then, the MaA gauge fixing is performed and the abelian monopole currents are extracted from the abelian gauge fields. Finally, the inverse Monte–Carlo method is applied to the ensemble of these currents and the coefficients $\beta, \gamma, a_1, a_2, a_3, \ldots$ of the action (37) are determined. It occurs that at large distances the distribution of the currents is well described by the action (37). In that sense, the effective action of gluodynamics is the Abelian Higgs model, the monopoles play the role of the Higgs particles. In contrast to compact electrodynamics, the monopole potential is not infinitely deep, see Figure 12(a,b).

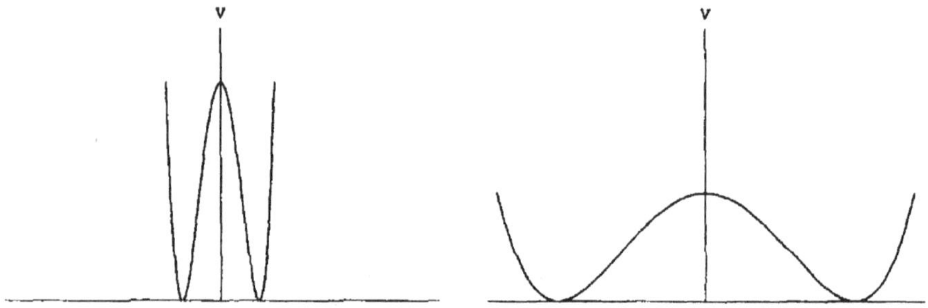

Figure 12. Schematic representation of the monopole potential $\lambda(|\Phi|^2 - 1)^2$ for compact electrodynamics (a), $\lambda \to +\infty$; and for lattice $SU(2)$ gluodynamics (b), λ is finite.

Various Representations for the Partition Function

There are several equivalent representations of the partition function:

$$\mathcal{Z}^j = \sum_{\delta j=0} \exp\{-S_{mon}(j)\}, \tag{38}$$

where S_{mon} is given by (37).

Representation 1 as the dual abelian Higgs model:

As we have already said, the model (38) is related to the (dual) Abelian Higgs theory by the inverse BKT transformation

$$\mathcal{Z}^j \propto \mathcal{Z}^{AH} = \int \mathcal{D}B\,\mathcal{D}\Phi \exp\{-S^{AH}(B,\Phi)\}, \tag{39}$$

where S^{AH} is given by eq.(23).

Representation 2 in terms of the Nielsen-Olesen strings:

Using the BKT transformation for the partition function (39), one can show that model (38) is equivalent[43] to the string theory:

$$\mathcal{Z}^j \propto \mathcal{Z}^{str} = \sum_{\substack{\sigma(C_2)\in \mathbb{Z} \\ \delta\sigma=0}} e^{-S^{AH}_{string}(\sigma)} \tag{40}$$

$$S^{AH}_{string}(\sigma) = 4\pi\gamma(\sigma, (\Delta + m^2)^{-1}\sigma) + O(\frac{1}{\lambda}), \quad m^2 = \frac{\gamma}{\beta}. \tag{41}$$

Here σ is the closed world sheet of the Nielsen-Olesen string. In the continuum limit[44] we have

$$S^{AH}_{string} \propto \int d\sigma_{\mu\nu}(\tilde{x})\mathcal{D}^{(4)}_m(\tilde{x} - \tilde{x}')d\sigma_{\mu\nu}(\tilde{x}'), \tag{42}$$

where $\mathcal{D}^{(4)}_m$ is the free scalar propagator, $(\Delta + m^2)\mathcal{D}^{(4)}_m(x) = \delta^{(4)}(x)$.

Representation 3 in terms of the gauge field θ:

Using the inverse dual transformation for the partition function (39), one can get the representation of the monopole partition function in terms of the *compact* gauge field θ which is dual to the gauge field *B*, (23), (39):

$$\mathcal{Z}^j \propto \mathcal{Z}^\theta = \int_{-\pi}^{+\pi} \sum_{n(C_2)\in\mathbb{Z}} \mathcal{D}\theta e^{-S^o(\theta,n)}; \tag{43}$$

$$S^o = \frac{1}{16\pi^2\beta}||\mathrm{d}\theta + 2\pi n||^2 - \frac{1}{16\pi^2\gamma}(\mathrm{d}\theta + 2\pi n, \Delta(\mathrm{d}\theta + 2\pi n)) + O(\frac{1}{\lambda}). \tag{44}$$

Representation 4 in terms of hypergauge fields:

Applying the duality transformation to the partition function (43), we get the representation in terms of the hypergauge (Kalb–Ramond) field

$$\mathcal{Z}^j \propto \mathcal{Z}^{HG} = \sum_{h(C_2)\in\mathbb{Z}} \int \mathcal{D}A \exp\{-S(A,h)\}$$

$$S(A,h) = \frac{1}{4\gamma}||dh||^2 + \frac{1}{4\beta}||dA + h||^2 + O(\frac{1}{\lambda}), \tag{45}$$

where *h* is the hypergauge field $(h_{\mu\nu}(x))$ interacting with the gauge field A $(A_\mu(x))$;the field *h* is dual to the monopole field Φ, and the field *A* is dual to the field *B*.

The action *S*(*A*, *h*) is invariant under gauge transformations: $A \to A + \mathrm{d}\alpha$, $h \to h$, and under *hypergauge* transformations $A \to A - \gamma$, $h \to h + \mathrm{d}\gamma$.

Representation 5 by fourier transformation of the string world sheets:

This representation is intermediate between the representations (40) and (39):

$$\mathcal{Z} \propto \mathcal{Z}^h = \int \mathcal{D}\xi \int \mathcal{D}\zeta \sum_{\sigma(C_2)\in\mathbb{Z}} \exp\left\{2\pi i(\xi,\sigma) - \tilde{S}[\xi,\zeta] + O(\frac{1}{\lambda})\right\}, \tag{46}$$

$$S[\xi,\zeta] = \frac{1}{4\beta}||\xi + \mathrm{d}\zeta||^2 + \frac{1}{4\gamma}||\mathrm{d}\xi||^2, \tag{47}$$

where σ are the world sheets of the Abrikosov-Nielsen-Olesen strings. Again, there exists a hypergauge symmetry $(\xi \to \xi + \mathrm{d}\gamma,\ \zeta \to \zeta - \gamma)$, which reflects the fact that the string worlds sheets are closed (or equivalently, conservation of the magnetic flux).

It is possible to derive a lot of physical consequences from these representations[14]. One of the most interesting is that the classical string tension which is calculated from the string action (41) coincides, within statistical errors, with the string tension in $SU(2)$ lattice gauge theory. The coefficients β and γ in eq.(41) are found by the inverse Monte–Carlo method, as we have just explained.

Abelian Monopoles as Physical Objects

The abelian monopoles arise in the continuum theory[10] from singular gauge transformation (8)-(12) and it is not clear whether these monopoles are "real" objects. A physical object is something which carries action and below we only discuss the question if there are any correlations between abelian monopole currents and the $SU(2)$ action. In Ref.[42] it was found that the total action of the $SU(2)$ fields is correlated with the total length of the monopole currents, so there exists a global correlation. We now discuss the local correlations between the action density and the monopole currents[45].

In lattice calculations, the monopole current $j_\mu(x)$ lies on the dual lattice and it is natural to consider the correlator of the current and the dual action density[¶]:

$$C_1 =< \frac{1}{2}Tr\left(j_\mu(x)\frac{1}{2}\varepsilon_{\nu\mu\alpha\beta}F_{\alpha\beta}(x)\right)^2 > - < j_\mu^2(x) >< \frac{1}{2}TrF_{\alpha\beta}^2(x) > . \quad (48)$$

The density of $j_\mu(x)$ strongly depends on $\beta = \frac{4}{g^2}$, therefore it is convenient to normalize C_1:

$$C_2 = \frac{< \frac{1}{2}\mathrm{Tr}\left(j_\mu(x)\frac{1}{2}\varepsilon_{\nu\mu\alpha\beta}F_{\alpha\beta}(x)\right)^2 >}{< j_\mu^2(x) >< \frac{1}{2}\mathrm{Tr}F_{\alpha\beta}^2(x) >} - 1 . \quad (49)$$

For the static monopole we have $j_0(x) \neq 0,\ j_i(x) = 0,\ i = 1,2,3,$ and the fact that $C_2 \neq 0$ means that the magnetic part S_m of the $SU(2)$ action is correlated with $j_\mu(x)$.

The correlator C_2 is related to the relative excess of the action carried by the monopole current. The expectation value of S_m on the monopole current is

$$S_m =< \frac{1}{24}\mathrm{Tr}\left(n_\mu(x)\frac{1}{2}\varepsilon_{\nu\mu\alpha\beta}F_{\alpha\beta}(x)\right)^2 >, \quad (50)$$

where $n_\mu(x) = j_\mu(x)/|j_\mu(x)|$. The relative excess of the action on the monopole current is

$$\eta = \frac{S_m - S}{S}, \quad S =< \frac{1}{24}TrF_{\alpha\beta}^2(x) > . \quad (51)$$

where $S =< \frac{1}{24}TrF_{\alpha\beta}^2(x) >$ is the expectation value of the action, the coefficient $\frac{1}{24}$ corresponds to the lattice definition of the "plaquette action" $< S >$. If $j_\mu(x) = 0, \pm 1$, then $\eta = C_2$. Since in lattice calculations at sufficiently large values of β the probability of $j_\mu(x) = \pm 2$ is small in the MaA projection, we have $\eta \approx C_2$ at large values of β. From numerical calculations we have found that $\eta = C_2$ with 5% accuracy for $\beta > 1.5$ on lattices of sizes 10^4 and $12^3 \cdot 4$. The lattice definition of S_m is

$$S_m =< \sum_{\nu=1}^{4} n_\nu^2(x) \cdot \frac{1}{6} \sum_{P\in\partial C_\nu(x)} \left(1 - \frac{1}{2}Tr\, U_P\right) >, \quad (52)$$

where the summation is over the plaquettes P which are the faces of the cube $C_\nu(x)$; the cube $C_\nu(x)$ is dual to $n_\nu(x)$; U_P is the plaquette matrix. Thus, S_m is the average action on the plaquettes closest to the magnetic current (see Figure 5(b)). The lattice definition of S is standard: $S =< \left(1 - \frac{1}{2}Tr\, U_P\right) >$. We use these definitions of S and S_m in our lattice calculations.

It is interesting to study the quantity (51) in various gauges. In Figure 13(a) the relative excess of the magnetic action density near the monopole current η is shown for

[¶]Here and below, formulae correspond to the continuum limit implied by the lattice regularization.

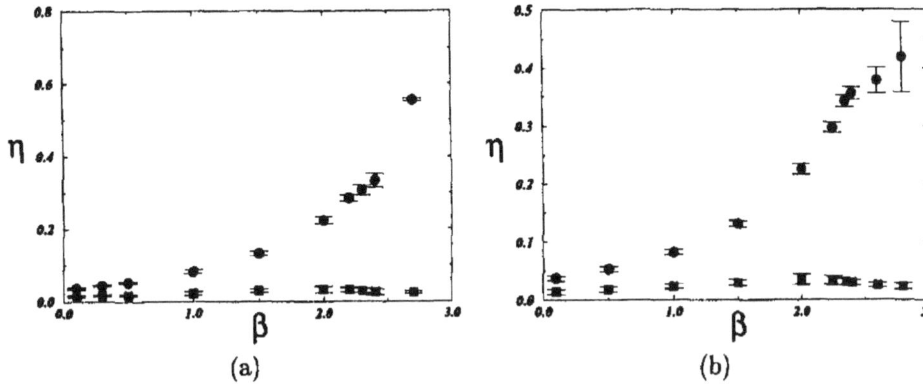

Figure 13. The relative excess η, eq.(51), for 10^4 lattice (a) and for $12^3 \cdot 4$ lattice (b) in the MA (circles) and the Polyakov abelian (squares) gauges, Ref.[45].

a 10^4 lattice. Circles correspond to the MaA projection, and squares to the Polyakov gauge. The data for the F_{12} projection coincide, within statistical errors, with those for the Polyakov gauge.

The quantity (51) at finite temperatures has also been studied. In Figure 13(b) the relative excess of the magnetic action density near the monopole current η is shown for $12^3 \cdot 4$ lattice. Again, the circles correspond to the MaA projection and the squares to the Polyakov gauge.

Thus we have shown that in the MaA projection the *abelian* monopole currents are surrounded by regions with a high *nonabelian* action. This fact presumably means that the monopoles in the MaA projection are physical objects. It does not mean that they have to be real objects in the Minkowsky space. What we have found is that these currents carry $SU(2)$ action in Euclidean space. It is important to understand what is the general class of configurations of $SU(2)$ fields which generate the monopole currents. Some specific examples are known. These are instantons[46] and the BPS–monopoles (periodic instantons)[40].

Monopoles are Dyons

A dyon is an object which has both electric and magnetic charge. In the field of a single instanton the monopole currents in the MaA projection are accompanied by electric currents[47]. The qualitative explanation of this fact is simple. Consider the (anti)self-dual configuration

$$F_{\mu\nu}(A) = \pm {}^*F_{\mu\nu}(A) \,. \tag{53}$$

The MaA projection is defined[11] by the minimization of the off-diagonal components of the non-abelian gauge field, so that in the MaA gauge one can expect the abelian component of the commutator term $1/2Tr(\sigma^3[A_\mu, A_\nu]) = \varepsilon^{3ab} A^a_\mu A^b_\nu$ to be small compared with the abelian field-strength $f_{\mu\nu}(A) = \partial_{[\mu} A^3_{\nu]}$. Therefore, in the MaA projection eq.(53) yields

$$f_{\mu\nu}(A) \approx \pm {}^*f_{\mu\nu}(A) \,. \tag{54}$$

Due to eq.(54), the monopole currents have to be correlated with electric ones, since

$$j^e_\mu = \partial_\nu f_{\mu\nu}(A) \approx \pm \partial_\nu {}^*f_{\mu\nu}(A) = j^m_\mu \,. \tag{55}$$

It is known that the (anti-) instantons produce abelian monopole currents[46]. The abelian monopole trajectories may go through the center of the instanton (Figure 14(a)) or form a circle around it (Figure 14(b)).

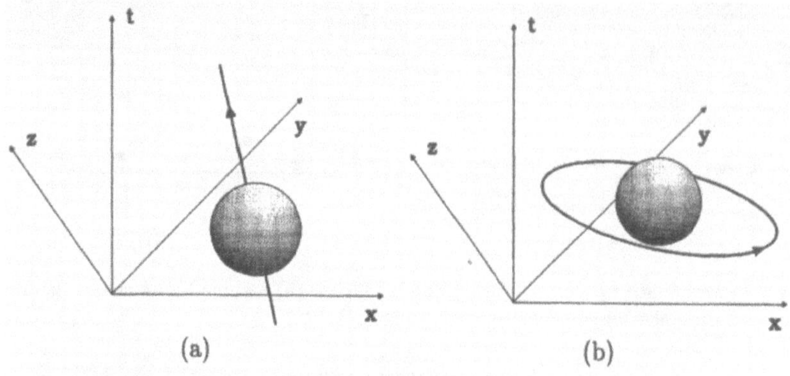

Figure 14. The instantons (spheres) inducing monopoles (thin lines).

Due to eq.(55), the monopole currents are accompanied by electric currents in the self-dual fields. Therefore, the monopoles are dyons for an instanton background. However, the real vacuum is not an ensemble of instantons, and below we discuss the correlation of j^e and j^m in the vacuum of lattice gluodynamics.

There is a lot of vacuum models: the instanton gas, instanton liquid, toron liquid, *etc.* Suppose that the vacuum is a "topological medium", and there are self-dual and anti-self-dual regions (domains). In this vacuum the electric and the magnetic currents should correlate with each other. But the sign of the correlator depends on the topological charge of the domain.

We define the electric current as[47]

$$J_\mu^e(x) = \frac{1}{2\pi}\sum_\nu [\bar{\theta}_{\mu\nu}(x) - \bar{\theta}_{\mu\nu}(x-\hat{\nu})]. \tag{56}$$

In the continuum limit, this definition corresponds to the usual one: $J_\mu^e = \partial_\nu f_{\mu\nu}$. The electric currents are conserved ($\partial_\mu J_\mu^e = 0$) and are attached to the links of the original lattice. Electric currents are not quantized.

In order to calculate the correlators of the type $<J_\mu^e(x)J_\mu^m(x)...>$,one has to define the electric current on the dual lattice or the magnetic current on the original lattice. We define the electric current on the dual lattice in the following way:

$$J_\mu^e(y) = \frac{1}{16}\sum_{x\in {}^*C(y,\mu)} \left[J_\mu^e(x) + J_\mu^e(x-\hat{\mu})\right]. \tag{57}$$

Here the summation on the r.h.s. is over eight vertices x of the 3-dimensional cube ${}^*C(y,\mu)$ to which the current $J_\mu^e(y)$ is dual. The point y lies on the dual lattice and the points x on the original one.

For the topological charge density operator we use the simplest definition:

$$Q(x) = \frac{1}{2^9\pi^2}\sum_{\mu_1,\dots,\mu_4=-4}^{4} \varepsilon^{\mu_1,\dots,\mu_4} Tr[U_{\mu_1,\mu_2}(x)U_{\mu_3,\mu_4}(x)], \tag{58}$$

where U_{μ_1,μ_2} is the plaquette matrix. On the dual lattice the topological charge density corresponding to the monopole current $J_\mu^m(y)$ is defined by taking the average over the eight sites nearest to the current $J_\mu^m(y)$: $Q(y) = \frac{1}{8}\sum_x Q(x)$.

The simplest (connected) correlator of electric and magnetic currents is given by $< J^m_\mu J^e_\mu >_c \equiv < J^m_\mu J^e_\mu > - < J^m_\mu >< J^e_\mu >=< J^m_\mu J^e_\mu >$, here we used the fact that $< J^m_\mu >=< J^e_\mu >= 0$ due to the Lorentz invariance. The connected correlator $< J^m_\mu J^e_\mu >_c$ is equal to zero, since J^m_μ and J^e_μ have opposite parities. A scalar quantity can be constructed if we multiply $J^m_\mu J^e_\mu$ with the topological charge density. The corresponding connected correlator

$$< J^m_\mu J^e_\mu Q >_c =< J^m_\mu J^e_\mu Q > \tag{59}$$

is nonzero for the vacuum consisting of (anti-)self-dual domains (*cf.* eq. (55)). Note that we derived eq.(59) using the equalities $< J^m_\mu >=< J^e_\mu >=< Q >= 0$.

We consider $SU(2)$ lattice gauge theory on a 8^4 lattice with the Wilson action. We calculate[48] the correlator $<J^m_\mu(y)J^e_\mu(y)Q(y)>$, using 100 statistically independent configurations at each value of β.

This correlator strongly depends on β and it is convenient to normalize it by dividing by $\rho^m\rho^e$. Here ρ^m and ρ^e are the monopole and the electric current densities:

$$\rho_{m(e)} = \frac{1}{4V}\sum_l |J_l^{m(e)}|, \tag{60}$$

V is the lattice volume (the total number of sites).

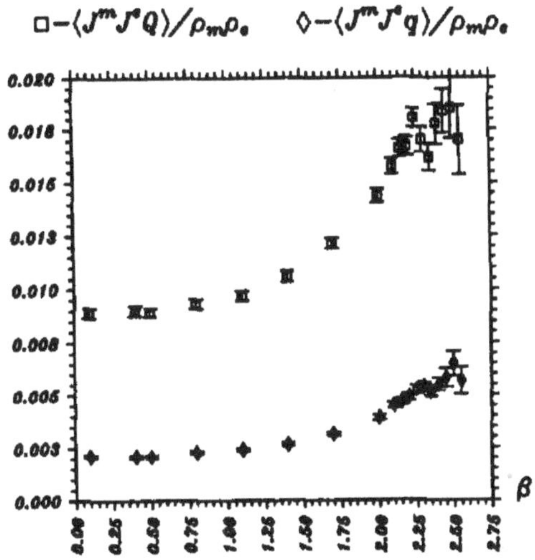

Figure 15. Correlators $<J^m J^e Q>/\rho_m\rho_e$ and $<J^m J^e q>/\rho_m\rho_e$ as functions of β, Ref.[48].

The correlators $<J^m J^e Q>/\rho_m\rho_e$ and $<J^m J^e q>/\rho_m\rho_e$ $(q(y) = Q(y)/|Q(y)|)$ are represented in Figure 15. As one can see from Figure 15, the product of the electric and the magnetic currents is correlated with the topological charge.

Thus our results show that in the vacuum of lattice gluodynamics the magnetic current is correlated with the electric current, the abelian monopoles have an electric charge. The sign of the electric charge depends on the sign of the topological charge density.

CONCLUSIONS

Now we briefly summarize the properties of the abelian monopole currents in the MaA projection of lattice $SU(2)$ gluodynamics:

- Monopoles are responsible for $\approx$ 90% of the $SU(2)$ string tension.
- Monopole currents satisfy the London equation for a superconductor.
- Monopoles are condensed in the confinement phase.
- The effective monopole Lagrangian is similar to the Lagrangian of the dual Abelian–Higgs model.
- Monopoles carry the $SU(2)$ action.
- Monopoles are dyons.

The main conclusion which can be obtained from these facts is that the vacuum of lattice gluodynamics behaves as a dual superconductor: the monopole currents are condensed and they are responsible for confinement of color.

We have to note that the approach discussed above is not unique and there are several other approaches to the confinement problem in non-abelian gauge theories. We cannot discuss all these here, but mention the description of confinement in terms of Z_2 vortices[49], the description of the QCD vacuum in terms of the non-abelian dual lagrangians[50], and the study of QCD by means of the cumulant expansion[51].

ACKNOWLEDGMENTS

Authors are grateful to E.T. Akhmedov, P. van Baal, F.V. Gubarev, T.L. Ivanenko, Yu.A. Simonov and T. Suzuki for useful discussions. M.N.Ch and M.I.P. acknowledge the kind hospitality of the Theoretical Department of the Kanazawa University. This work has been supported by the JSPS Program on Japan – FSU scientists collaboration, and also by the Grants: INTAS-94-0840, INTAS-94-2851, INTAS-RFBR-95-0681, and Grant No. 96-02- 17230a of the Russian Foundation for Fundamental Sciences.

Appendix A: Differential Forms on the Lattice

Here we briefly summarize the main notions from the theory of differential forms on the lattice. The calculus of differential forms was developed for field theories on the lattice in ref.[62]. Since all lattice formulas have a direct continuum interpretation, the standard formalism of differential geometry can have wide applications in lattice theories. The advantage of the calculus of differential forms consists in the general character of the obtained expressions. Most of the transformations depend neither on the space-time dimension nor on the rank of the fields. With minor modifications, the transformations are valid for lattices of any form (triangular, hypercubic, random, etc).

A differential form of rank k on the lattice is a skew-symmetrical function ϕ_k defined on k-dimensional cells C_k of the lattice. The scalar lattice field is a 0-form. The $U(1)$ gauge field is a 1-form. The exterior differential operator d is defined as follows:

$$(\mathrm{d}\phi)(C_{k+1}) = \sum_{C'_k \in \partial C_{k+1}} \phi(C'_k) \tag{A.1}$$

Here ∂C_k is the boundary of the k-cell C_k. Thus, the operator d increases the rank of the form by one. For instance, in Wilson's $U(1)$ theory with the dynamic variables $U = e^{i\theta}$, the action $\cos(d\theta)$ depends on the plaquette $d\theta(C_2)$ constructed from the links $\theta(C_1)$.

The dual lattice is defined as follows. The sites of the dual lattice are placed at the centers of the D–dimensional cells of the original lattice. The object dual to an oriented k-cell (C_k) is an oriented$(D-k)$–cell $({}^*C_k)$, which lies on the dual lattice and has an intersection with C_k. The dual of the cubic lattice is also a cubic lattice obtained by shifting the lattice along each axis by 1/2 of the lattice spacing. Every k-form on the lattice corresponds to a $(D-k)$-form on the dual lattice: ${}^*\phi({}^*C_k) = \phi(C_k)$. he co-derivative is defined as follows:

$$\delta = {}^*\mathrm{d}^*. \tag{A.2}$$

This operator decreases the rank of the form

$$(\delta\phi)(C_{k-1}) = \sum_{all\ C_k : C_{k-1} \in \partial C_k} \phi(C_k). \tag{A.3}$$

The square of d and δ is equal to zero:

$$\begin{aligned} \mathrm{d}^2 &= 0, \\ \delta^2 &= 0. \end{aligned} \tag{A.4}$$

The first equality is a consequence of the well-known geometrical fact: "the boundary of the boundary is the empty set"; the second equality follows from the first one and the definition of δ (A.2).

Now we discuss the lattice version of the Laplace operator, which is defined as

$$\Delta = \delta\mathrm{d} + \mathrm{d}\delta. \tag{A.5}$$

This operator acts on 0–forms (scalar fields) in the same way as the usual finite-difference version of the continuous Laplacian. For the forms of non-zero rank the relation (A.5) is the generalization of the usual Laplacian. The obvious properties of δ are

$$\begin{aligned} \delta\Delta &= \Delta\delta \\ \mathrm{d}\Delta &= \Delta\mathrm{d} \\ 1 &= \delta\Delta^{-1}\mathrm{d} + \mathrm{d}\Delta^{-1}\delta \end{aligned} \tag{A.6}$$

The last relation, called the Hodge identity, implies the widely used decomposition formula for an arbitrary k–form:

$$\phi = \delta\Delta^{-1}(\mathrm{d}\phi) + \mathrm{d}\Delta^{-1}(\delta\phi) + \phi_h, \tag{A.7}$$

where ϕ_h is a harmonic form, $\delta\phi_h = \mathrm{d}\phi_h = 0$. The number of the harmonic forms depends on the topology of the space-time; for instance, the number of the harmonic 1–forms is equal to 0,1 and D respectively, for the space–time topology R^D, $R^{D-1} \otimes S^1$ and S^D.

For two forms of the same rank the scalar product is introduced in a natural way:

$$(\phi, \psi) = \sum_{all\ C_k} \phi(C_k)\psi(C_k) \tag{A.8}$$

This scalar product agrees with the definitions of the d and δ operators in the sense that the following formula of "integration by parts" is valid:

$$(\mathrm{d}\phi, \psi) = (\phi, \delta\psi). \tag{A.9}$$

The norm of a k-form is defined as usual by

$$\|\phi\|^2 = \sum_{C_k} \phi(C_k)^2 = (\phi, \phi) \tag{A.10}$$

We illustrate the above definitions by four simple examples.

Example 1. Let us show that the general action for the compact gauge field is a periodic function of the angle $\bar{\theta}_P$ corresponding to the plaquette. By definition, $\bar{\theta}_P = \mathrm{d}\theta + 2\pi k$, where the integer valued 2–form k is defined in such a way that $-\pi < \bar{\theta}_P \leq \pi$ Under a gauge transformation we have $\theta' = \theta + \mathrm{d}\chi$ and $\bar{\theta}'_P = \mathrm{d}\theta + \mathrm{d}^2\chi + 2\pi k' = \mathrm{d}\theta + 2\pi k'$, where k' is an integer chosen in such a way that $-\pi < \theta' \leq \pi$. Therefore, the gauge invariance requires the periodicity condition on the action: $S(\bar{\theta}_P) = S(\bar{\theta}_P + 2\pi k)$.

Example 2. Note, that the definition of the norm (A.10) allows us to write in concise form the action of the lattice theory, e.g., $S = \sum_{C_2}[\mathrm{d}(A(C_1))]^2 = \|\mathrm{d}A\|^2$ for noncompact electrodynamics. We often use a similar notation for the Villain action.

Example 3. Let us show that in standard compact lattice electrodynamics there exists a conservation law. As we have mentioned before, the compact character of the fields implies the existence of monopoles. The monopole charge inside an elementary 3-dimensional cube is defined by $m = \frac{1}{4\pi}\mathrm{d}\bar{\theta}_P = \frac{1}{2}\mathrm{d}k$, where $\bar{\theta}_P = \mathrm{d}\theta + 2\pi k$, as in Example 1. In the continuum limit, the above definition corresponds to the Gauss law: $m = \frac{1}{4\pi}\oint_S \vec{H}\mathrm{d}\vec{S}$, where S is the surface of the elementary cube. If $j = {}^*m$, then $\delta j = {}^*\mathrm{d}^{**}m = \frac{1}{2}{}^*\mathrm{d}^2 k = 0$. Since m is a rank 3 form, it follows that j is a rank 1 form and the equation

$$\delta j = 0 \tag{A.11}$$

means that for each site the incoming current j is equal to the outgoing one. Therefore, the monopole current j is conserved. It is easy to prove that j is gauge invariant. This result is well known for 4-dimensional lattice QED[38]. Note that all transformations considered above are valid for any space-time dimension D and for the field θ of any rank k; the current j in this case is a $(D-k-2)$–form. For the XY model in $D = 3$ we have $k = 0$ and the conservation law (A.11) for the 1-form j implies that the excitations, called vortices, form closed loops. In the 4-dimensional XY model the conserved quantity j is represented by a 2-form, which means that there exist excitations forming closed surfaces. These objects are related to "global" cosmic strings, forming closed surfaces[53] in 4-dimensional space–time.

Example 4. To perform the BKT transformation (see Appendix C) we have to solve the so-called cohomological equation

$$\mathrm{d}l = n, \tag{A.12}$$

where n is a given fc-form and l is a $(k-1)$-form to be found. Using the Hodge decomposition, we can easily show that $l(n) = \Delta^{-1}\delta n$ is a particular solution which, being added to the general solution of the homogeneous equation, yields the general solution of (A.12): $l = \Delta^{-1}\delta n + \mathrm{d}m$ (m is an arbitrary $(k-2)$-form).

If n is an integer valued form, then the solution l of (A.12) can also be found in terms of integer numbers. We explicitly construct such a solution for $k = 2$, The generalization for arbitrary values of k is obvious. Let us assign the value 0 to the links which belong to a given maximal tree on the lattice (a maximal tree is a maximal set of links which do not contain closed loops). To each link that does not belong to the tree we attach the value equal to the value of the plaquette formed by the link and the tree (there exists one and only one such plaquette). It is easy to see that we have thus constructed the required particular solution $l(n)$. The general solution has the form $l = l(n) + \mathrm{d}m$, $m \in \mathbb{Z}$. For simplicity, we have neglected finite volume effects (harmonic forms) in the above construction.

Appendix B: Duality transformation

In this Appendix we perform the duality transformation for $U(1)$ compact gauge theory with an arbitrary form of the action. Initially the duality transformation was discussed by Kramers and Wannier[54] for the 2–dimensional Ising model.

Let us consider the $U(1)$ gauge theory with an arbitrary periodic action $S[\mathrm{d}\theta]$, $S[\ldots, X_P + 2\pi, \ldots] = S[\ldots, X_P, \ldots]$, where θ_l is the compact $U(1)$ link field and P denotes any plaquette of the original lattice. We use the formalism of differential forms on the lattice (see Appendix A). We start from the partition function of the compact $U(1)$ theory:

$$\mathcal{Z} = \int_{-\pi}^{+\pi} \mathcal{D}\theta \exp\left\{-S[\mathrm{d}\theta]\right\}. \tag{B.1}$$

The Fourier expansion of the function $e^{-S(X)}$ yields

$$\mathcal{Z} = \mathrm{const.} \cdot \int_{-\pi}^{+\pi} \mathcal{D}\theta \sum_{k(C_2)\in\mathbb{Z}} e^{-S^d(k)} e^{i(k,\mathrm{d}\theta)}, \tag{B.2}$$

where k *is* an integer valued two-form, and the action S^d is

$$e^{-S^d(k)} = \mathrm{const.} \cdot \int_{-\pi}^{+\pi} \mathcal{D}X \exp\left\{-S[X] - i(k, X)\right\}. \tag{B.3}$$

The integration over the field θ in eq.(B.2) gives the constraint $\delta k = 0$, which can be resolved by introducing the new 3-form n: $k = \delta n$. For simplicity, it is assumed here that we are dealing with a lattice having trivial topology (e.g. R^4). In the case of a lattice with a nontrivial topology, arbitrary harmonic forms must be added to the r.h.s. of this equation. Finally, changing the summation $\sum_{k(C_2)\in\mathbb{Z}} \to \sum_{{}^*n({}^*C_1)\in\mathbb{Z}}$, we get the dual representation of compact $U(1)$ gauge theory:

$$\mathcal{Z} \propto \mathcal{Z}^d = \sum_{{}^*n({}^*C_1)\in\mathbb{Z}} e^{-S^d({}^*\mathrm{d}n)}. \tag{B.4}$$

Consider now compact $U(1)$ gauge theory with the action in the Villain form eq.(33). Now the integral is Gaussian and the dual action is $S^d({}^*\mathrm{d}n) = ||{}^*\mathrm{d}n||^2/4\beta$.

Appendix C: The BKT transformation

In this Appendix we show how to perform the BKT transformation for compact $U(1)$ gauge theory with an arbitrary form for the action (B.1). Let us insert the unity

$$1 = \int_{-\infty}^{+\infty} \mathcal{D}G\, \delta(G - k)$$

(here G is a real-valued two-form) into the sum (B.2)

$$\mathcal{Z} = \int_{-\pi}^{+\pi} \mathcal{D}\theta \exp\{-S_P(\mathrm{d}\theta)\} = \text{const.} \cdot \int_{-\pi}^{+\pi} \mathcal{D}\theta \int_{-\infty}^{+\infty} \mathcal{D}G \sum_{k(C_2)\in \mathbb{Z}} \delta(G-k) e^{-S^d(G)} e^{i(G,\mathrm{d}\theta)} \quad (C.1)$$

Using the Poisson summation formula

$$\sum_{k(C_2)\in \mathbb{Z}} \delta(G-k) = \sum_{k(C_2)\in \mathbb{Z}} e^{2\pi i(G,k)}, \quad (C.2)$$

we get

$$\mathcal{Z} = \text{const.} \cdot \int_{-\pi}^{+\pi} \mathcal{D}\theta \int_{-\infty}^{+\infty} \mathcal{D}G \sum_{k(C_2)\in \mathbb{Z}} e^{-S^d(G)} e^{i(\mathrm{d}\theta + 2\pi k, G)}. \quad (C.3)$$

Now we perform the BKT transformation with respect to the integer valued 2-form k:

$$k = m[j] + \mathrm{d}q, \quad \mathrm{d}m[j] = j, \quad \mathrm{d}j = 0, \quad (C.4)$$

where q and j are one- and three-forms, respectively. First, we change the summation variable, $\sum_{k(C_2)\in \mathbb{Z}} = \sum_{\substack{j(C_3)\in \mathbb{Z} \\ \mathrm{d}j=0}} \sum_{q(C_1)\in \mathbb{Z}}$. Using the Hodge–de–Rahm decomposition, we adsorb the d–closed part of the 2–form k into the compact variable θ:

$$\mathrm{d}\theta + 2\pi k = \mathrm{d}\theta_{n.c.} + 2\pi\delta\Delta^{-1}j, \quad \theta_{n.c.} = \theta + 2\pi\Delta^{-1}\delta m[j] + 2\pi q. \quad (C.5)$$

Substituting eq.(C.5) in eq.(C.3) and integrating over the noncompact field $\theta_{n.c.}$, we get the following representation of the partition function:

$$\mathcal{Z} = \text{const.} \cdot \int_{-\infty}^{+\infty} \mathcal{D}G \sum_{\substack{{}^*j({}^*C_1)\in \mathbb{Z} \\ \delta^* j=0}} e^{-S^d(G)} \exp\{2\pi i(G, \delta\Delta^{-1}j)\}\, \delta(\delta G). \quad (C.6)$$

The constraint $\delta G = 0$ can be solved by $G = \delta H$, where H is a real valued 3–form. Substituting this solution into eq.(C.6) we obtain the final expression for the BKT-transformed action on the dual lattice:

$$\mathcal{Z} = \text{const.} \cdot \sum_{\substack{{}^*j({}^*C_1)\in \mathbb{Z} \\ \delta^* j=0}} e^{-S_{mon}({}^*j)}. \quad (C.7)$$

where

$$S_{mon}({}^*j) = -\ln\left(\int_{-\infty}^{+\infty} \mathcal{D}H e^{-S^d(\delta H)} \exp\{2\pi i({}^*H, {}^*j)\} \right). \quad (C.8)$$

We have used the relation $\mathrm{d}\delta\Delta^{-1}j \equiv j,\ \forall j : \mathrm{d}j = 0$. Therefore, for the general $U(1)$ action $S[\mathrm{d}\theta]$ the monopole action (C.8) is nonlocal and it is expressed through the integral over the entire lattice ($\int_{-\infty}^{+\infty} \mathcal{D}H$)

As an example we consider the Villain form of the $U(1)$ action (33). Repeating all the steps we get the following monopole action:

$$S_{mon}({}^*j) = 4\pi^2\beta({}^*j, \Delta^{-1}{}^*j). \quad (C.9)$$

REFERENCES

1. K. Wilson, *Phys. Rev.* **D10** (1974) 2445.
2. Creutz, *Phys. Rev. Lett.* **43** (1979) 553; *Phys. Rev.* **D21** (1980) 1006, 2308; *Phys. Rev. Lett.* **45** (1980) 313;
3. A.A. Abrikosov, *JETP* **32** (1957) 1442.
4. S. Mandelstam, Phys. Rep., 23C (1976) 245; G. 't Hooft, "High Energy Physics", Zichichi, Editrice Compositori, Bolognia, 1976.
5. A.M. Polyakov, *Phys.Lett.* **59B** (1975) 82.
6. A.M. Polyakov *Nucl.Phys.* **B120** (1977) 429.
7. N. Sieberg and E. Witten: *Nucl.Phys.* **B341** (1994) 484.
8. G.S. Bali K. Schilling and Ch. Schlichter, *Phys.Rev.* **D51** (1995) 5165.
9. Ch. Schlichter, G.S. Bali and K. Schilling, to be published.
10. G. 't Hooft, Nucl. Phys., **B190** [FS3] (1981) 455; G. 't Hooft, these proceedings.
11. A.S. Kronfeld, M.L. Laursen, G. Schierholz and U.J. Wiese, *Phys.Lett.* **198B** (1987) 516; A.S. Kronfeld, G. Schierholz and U.J. Wiese, *Nucl.Phys.* **B293** (1987) 461.
12. L. Polley and U.J. Wiese, *Nucl.Phys.* **B356** (1991) 629.
13. M.I. Polikarpov, L. Polley and U.J. Wiese, *Phys.Lett.* **B253** (1991) 212.
14. T. Suzuki *et al.*, these proceedings.
15. S. Kato *et al.*, preprint KANAZAWA 97-15, hep-lat/9709092.
16. T. Suzuki and I. Yotsuyanagi, Phys. Rev.**D42** (1990) 4257.
17. S. Hioki *et. al., Phys.Lett.* **B272** (1991) 326; *ERRATUM-ibid.* **B281** (1992) 416.
18. J.D. Stack, S.D. Neiman and R.J. Wensley, *Phys.Rev.* **D50** (1994) 3399.
19. G. Bali, V. G. Bornyakov, M. Müller-Preussker and K. Schilling, *Phys. Rev.* **D54** (1996) 2863.
20. J.D. Stack, S.D. Neiman, *Phys.Lett.* **B377** (1996) 113.
21. J.D. Stack, R.J. Wensley and S.D. Neiman, *Phys.Lett.* **B385** (1996) 261; J.D. Stack, *Nucl.Phys.Proc.Suppl.* **53** (1997) 524, hep-lat/9607064.
22. K. Yee, *Nucl.Phys.* **B** *(Proc. Suppl.)* **34** (1994) 189.
23. S. Kitahara, *preprint KANAZAWA-96-12.*
24. S. Ejiri, *Nucl.Phys.Proc.Suppl.* **53** (1997) 491, hep-lat/9608001
25. T. Suzuki, S. Kitahara, T. Okude, F. Shoji, K. Moroda and O. Miyamura, *Nucl. Phys.* **B** *(Proc. Suppl.)* **47** (1996) 374, hep-lat/9509016; T. Suzuki, *preprint KANAZAWA-96-08*, hep-lat/9606015.
26. V. Singh, R.W. Haymaker and D.A. Browne *Phys.Rev.* **D47** (1993) 1715.
27. Ch. Schlichter, G.S. Bali and K. Schilling, hep-lat/9709114.
28. Y. Matsubara, S. Ejiri and T. Suzuki *Nucl.Phys.* **B** *Proc.Suppl.* **34** (1994) 176, hep-lat/9311061.
29. J. Fröhlich and P.A. Marchetti, *Europhys. Lett.* **2** (1986) 933.
30. M.N. Chernodub, M.I. Polikarpov and A.I. Veselov, *Nuclear Physics* **B** *(Proc. Suppl.)* 49 (1996) 307; hep-lat/9512030.
31. M.N. Chernodub, M.I. Polikarpov and A.I. Veselov, *Phys.Lett.* **B399** (1997) 267.
32. T. Kennedy and C. King, *Comm.Math.Phys.* **104** (1986) 327.
33. N. Nakamura, V. Bornyakov, S. Ejiri, S. Kitahara, Y. Matsubara and T. Suzuki, *preprint KANAZAWA-96-16*, hep-lat/9608004.
34. A. Di Giacomo, these proceedings
35. L. Del Debbio, A. Di Giacomo, G. Paffuti and P. Pieri, *Phys.Lett.* **B355** (1995) 255; also talk by A. Di Giacomo, these proceedings.
36. V.L. Berezinskii, *Sov. Journal JETP,* **32** (1971) 493.
37. J.M. Kosterlitz, D.J. Thouless, *J.Phys.* **C6** (1973) 1181.
38. T. Banks, R. Myerson and J. Kogut, *Nucl.Phys.* **B129** (1977) 493.
39. T.L. Ivanenko and M.I. Polikarpov, preprint *ITEP 49-91* (1991).
40. J. Smit and A. van der Sijs, *Nucl.Phys.* **B355** (1991) 603.
41. R.H. Swendsen, *Phys. Rev. Lett.* **52** (1984) 1165; *Phys. Rev.* **D30** (1984) 3866,3875.
42. H. Shiba and T. Suzuki, *Phys. Lett.* **B343** (1995) 315; *Phys. Lett. B351* (1995) 519.
43. M.I. Polikarpov, U.-J. Wiese and M.A. Zubkov, *Phys. Lett.,* **272B** (1991) 326.
44. P. Orland, *Nucl.Phys.* **B428** (1994) 221; E.T. Akhmedov *et al., Phys.Rev.* **D53** (1996) 2087.
45. B.L.G. Bakker, M.N. Chernodub and M.I. Polikarpov, *preprint KANAZAWA-97-10,* hep-lat/9706007, preprint *ITEP-TH-43-97*, hep-lat/9709038.
46. O.Miyamura, S.Origuchi, *'QCD monopoles and Chiral Symmetry Breaking in SU(2) Lattice Gauge Theory'*, RCNP Confinement 1995, Osaka, Japan, Mar 22-26, 1995, p.137;

M.N.Chernodub, F.V.Gubarev, *JETP Lett.* 62 (1995) 100; R.C.Brower, K.N.Orginos, Chung-I Tan, *Phys. Rev*.**D55** (1997) 6313; A.Hart, M.Teper, *Phys. Lett.* **371** (1996) 261; M. Feurstein, H. Markum and S. Thurner, *Phys.Lett.* **B396** (1997) 203; M.Fukushima *et al.*, *Phys. Lett.* **B399** (1997) 141.
47. V. Bornyakov and G. Schierholz, *Phys.Lett.* **B384** (1996) 190.
48. M.N. Chernodub, F.V. Gubarev and M.I. Polikarpov, *preprint ITEP-TH-44-97*, hep-lat/9709039.
49. M. Faber, J. Greensite, S. Olejnik, hep-lat/9710039; L. Del Debbio, M. Faber, J. Greensite, S. Olejnik, hep-lat/9709032; hep-lat/9708023.
50. M. Baker, J.S. Ball, N. Brambilla, G.M. Prosperi and F. Zachariasen, *Phys.Rev.* **D54** (1996) 2829; *Erratum-ibid.* **D56** (1997) 2475; M. Baker, J.S. Ball and F. Zachariasen, *J.Mod.Phys.* **A11** (1996) 343.
51. Yu.A. Simonov, *Phys.Usp.* **39** (1996) 313; hep-ph 9709344; H.G. Dosh and Yu.A. Simonov *Phys.Lett.* **205B** (1988) 339.
52. P. Becher and H. Joos, *Z.Phys.* **C15** (1982) 343; A.H. Guth, *Phys. Rev.* **D21** (1980) 2291
53. M.I. Polikarpov and U.J. Wiese, *preprint HLRZ-90-78* (1990); A.K. Bukenov, M.I. Polikarpov, A.V. Pochinskii and U.J. Wiese, *Phys.Atom.Nucl.* **56** (1993) :122.
54. H.A. Kramers and G.H. Wannier, *Phys.Rev.* **60** (1941) 252.

THE DUAL SUPERCONDUCTOR PICTURE FOR CONFINEMENT

Adriano Di Giacomo

Dip. Fisica Università and INFN,
Piazza Torricelli 2, 56100 Pisa, Italy

INTRODUCTION

At short distances QCD vacuum mimics the Fock space ground state of perturbation theory: deep inelastic scattering experiments, jet production in high energy reactions, QCD sum rules provide empirical evidence for that.

In fact colour is confined, and quarks and gluons never appear as free particles in asymptotic states: the ground state is very different from the perturbative vacuum. Its exact structure is not known. Many models have been attempted to give an approximate description of it: some of them are based on "mechanisms", i.e. they assume that the degrees of freedom relevant for large distance physics behave as other well understood physical systems.

The most attractive mechanism for colour confinement is *Dual superconductivity of type II of QCD vacuum*[1, 2]. Dual means interchange of electric with magnetic with respect to ordinary superconductors. The idea is that the chromoelectric field in the region of space between a $Q\bar{Q}$ pair is constrained by the dual Meissner effect into Abrikosov[3] flux tubes, with constant energy per unit length. The energy is then proportional to the distance

$$E = \sigma R \tag{1}$$

and this means confinement.

The lattice is the ideal tool to study (at least numerically) large distance phenomena from first principles. There is indeed evidence from lattice simulations that:

1) The string tension exists: large Wilson loops $W(R,T)$ describing a pair of static quarks at a distance $\boldsymbol{R}$ propagating for a time T, obey the area law[4]

$$W(R,T) \simeq \exp(-\sigma RT) \tag{2}$$

Since in general $W(R,T) \simeq \exp(-V(R)T)$, the observed behaviour (Eq.(2)) confirms confinement as defined by Eq.(1).

2) Chromoelectric flux tubes have been observed, joining a $Q\bar{Q}$ pair propagating in Wilson loops[5,6]. Their transverse size is ~ 0.5 fin.

3) String like modes of these flux tubes have been detected[7].

4) Particles belonging to higher representations than quarks also experience a string tension at intermediate distances, which, for $SU(2)$, depends on the colour spin J as[8]

$$\sigma_J = kJ(J+1)$$

or

$$\frac{\sigma_J}{\sigma_{1/2}} = \frac{3}{4}J(J+1) \tag{3}$$

Observations 1)-3) support the idea of dual superconductivity. We will discuss in detail the implications of 4) in the following.

The problem we address in these lectures is: can dual superconductivity of QCD vacuum be directly tested?

The plan of the lectures is as follows. We will recall basic superconductivity and its order parameter, in order to clarify what we are looking for. We will then define dual superconductivity and its disorder parameter. We will construct the disorder parameter for $U(1)$ pure gauge theory.

We will then check the construction with the $X-Y$ 3d model (liquid He4). Subsequently we will revisit the abelian projection, which reduces the problem of dual superconductivity in QCD to a $U(1)$ problem. The Heisenberg ferromagnet will prove a useful laboratory to check this procedure. We will finally show how dual superconductivity can be directly detected in $SU(2)$ and $SU(3)$ gauge theories.

A discussion of the results and of their physical consequences is contained in the final section.

BASIC SUPERCONDUCTIVITY: THE ORDER PARAMETER[9]

A relativistic version of a superconductor is the abelian Higgs model

$$\mathcal{L} = -\frac{1}{4}F_{\mu\nu}F_{\mu\nu} + (D_\mu\Phi)^\dagger(D_\mu\Phi) - V(\Phi) \tag{4}$$

$F_{\mu\nu}$ is the electromagnetic field strength, D_μ is the covariant derivative

$$D_\mu\Phi = (\partial_\mu - iqA_\mu)\Phi \tag{5}$$

and $V(\Phi)$ the potential of the scalar field

$$V(\Phi) = \frac{1}{4}\left(\Phi^\dagger\Phi - \mu^2\right)^2 \tag{6}$$

If μ^2 is positive the field Φ has a nonzero vacuum expectation value. Since Φ is a charged field this is nothing but a spontaneous breaking of the $U(1)$ symmetry related to charge conservation. The ground state is a superposition of states with different electric charges, a phenomenon which is usually called "condensation" of charges.

A convenient parametrization of Φ is

$$\Phi = \rho e^{i\theta q} \qquad \rho = \rho^\dagger > 0 \tag{7}$$

Under gauge transformations

$$A_\mu \to A_\mu - \partial_\mu \alpha \qquad \theta \to \theta + \alpha \tag{8}$$

The covariant derivative (5) reads in this notation

$$D_\mu \Phi = e^{i\theta q} \left[\partial_\mu - iq(A_\mu - \partial_\mu \theta)\right] \rho \tag{9}$$

The quantity $\tilde{A}_\mu = A_\mu - \partial_\mu \theta$ is gauge invariant. Moreover

$$F_{\mu\nu} = \partial_\mu A_\nu - \partial_\nu A_\mu = \partial_\mu \tilde{A}_\nu - \partial_\nu \tilde{A}_\mu \tag{10}$$

The equation of motion reads, neglecting loop corrections (or looking at $\mathcal{L}$ as an effective lagrangian)

$$\partial_\mu F^{\mu\nu} + \frac{\tilde{m}^2}{2} \tilde{A}_\nu = 0 \qquad \tilde{m} = \sqrt{2} q \langle \Phi \rangle \tag{11}$$

In the gauge $A_0 = 0$ a static configuration has $\partial_0 \vec{A} = 0$, $\partial_0 \Phi = 0$ so that $E_i = F_{0i} = 0$ Eq.(11) implies that

$$\vec{\nabla} \wedge \vec{H} + \frac{\tilde{m}^2}{2} \vec{\tilde{A}} = 0 \tag{12}$$

The term $\tilde{m}^2/2\, \vec{\tilde{A}}$ in Eq.(12) is a consequence of spontaneous symmetry breaking and is an stationary electric current (London current). A persistent current with $\vec{E} = 0$, means $\rho = 0$ since $\rho \vec{j} = \vec{E}$ and hence superconductivity.

The curl of Eq.(12), reads

$$\nabla^2 \vec{H} - \frac{\tilde{m}^2}{2} \vec{H} = 0 \tag{13}$$

The magnetic field has a finite penetration depth $1/\tilde{m}$, and this is nothing but the Meissner effect. The key parameter is $\langle \Phi \rangle$, which is the order parameter for superconductivity: it signals spontaneous breaking of charge conservation.

Besides $\lambda_A = 1/\tilde{m}$ there is another parameter with dimension of a length, $\lambda_\Phi = \mu^{-1}$. If $\lambda_A \geq \sqrt{2}\lambda_\Phi$ the superconductor is called of the second kind otherwise it is of the first kind. For a superconductor of the first kind there is a Meissner effect for an external magnetic field $H < H_c$ (critical field); for $H > H_c$ the field penetrates the bulk and superconductivity is destroyed. For a superconductor of the second kind instead a penetration by Abrikosov flux tubes of transverse size $1/\tilde{m}$ is energetically favoured. Flux tubes repel each other. By increasing the external field the number of flux tubes increases. When they touch each other the field penetrates the bulk and superconductivity is destroyed.

In a dual superconductor the role of the electric and magnetic field is interchanged. The $U(1)$ symmetry related to magnetic charge conservation is spontaneously broken, i.e. monopoles condense in the vacuum. An order parameter for dual superconductivity will then be the vacuum expectation value of a field carrying nonzero magnetic charge.

Monopoles

The equations of motion for the electromagnetic field in the presence of an electric current j_μ and of a magnetic current j^M_μ are

$$\begin{aligned} \partial^\mu F_{\mu\nu} &= j_\nu \\ \partial^\mu F^*_{\mu\nu} &= j^M_\nu \end{aligned} \tag{14}$$

If both j_μ and j^M_μ are zero (no charges, no monopoles) photons are free, and the equations of motion are invariant under the transformation

$$F_{\mu\nu} \to \cos\theta\, F_{\mu\nu} + \sin\theta\, F^*_{\mu\nu}$$
$$F^*_{\mu\nu} \to \cos\theta\, F^*_{\mu\nu} - \sin\theta\, F_{\mu\nu} \tag{15}$$

for any θ. In particular if $\theta = \pi/2$, Eq.'s (15) give $\vec{E} \to \vec{H}$, $\vec{H} \to -\vec{E}$ which is known as a duality transformation. In nature $j^M_\mu = 0$. The general solution of Eq.(14) is then written in terms of vector potential A_μ

$$F_{\mu\nu} = \partial_\mu A_\nu - \partial_\nu A_\mu \tag{16}$$

and $\partial^\mu F^*_{\mu\nu} = 0$ is identically satisfied (Bianchi identities).

If a monopole exists Bianchi identities are violated. However they can be preserved, and with them the description in terms of A_μ, by considering the monopole as the end point of a thin solenoid (Dirac string) connecting it to infinity: the flux of the Coulomb like magnetic field, $\vec{H} = \frac{M}{4\pi}\frac{\vec{r}}{r^3}$, $\Phi(H) = M$ is conveyed to infinity by the string[10].

The string is invisible if the parallel transport of any electric charge around it is trivial:

$$e^{iq\oint \vec{A}d\vec{x}} = e^{iq\Phi(H)} = e^{i2\pi n} = 1$$

or

$$qM = 2\pi n \tag{17}$$

Eq.(17) is known as the Dirac quantization condition, and constrains the $U(1)$ group to be compact. If one insists to describe the system in terms of A_μ, monopoles are non-local objects with nontrivial topology. One could introduce dual vector potential B_μ, such that $F^*_{\mu\nu} = \partial_\mu B_\nu - \partial_\nu B_\mu$. Dual Bianchi identities would read $\partial^\mu F_{\mu\nu} = 0$, monopoles would be pointlike but electric charges could only exist if dual strings were attached to them.

There is another manifestation of duality, which originates from statistical mechanics. The prototype example is the 2d Ising model. The model is defined on a square lattice, by associating to each site i a field $\sigma(i)$ wich can assume 2 values, say ± 1. The action can be written

$$\mathcal{L} = -\beta \sum_{i,j} \sigma(i)\sigma(j) \tag{18}$$

the sum running on nearest neighbours. The partition function $K[\beta]$ is known exactly in the thermodynamical limit. At high β (low temperatures) the system is magnetized $\langle\sigma(i)\rangle \neq 0$; at low β it is disordered. A dual description can be given of the same system, by associating to each link (dual lattice site) a variable σ' with value -1 if the values of σ in the sites connected by the link are the same, $+1$ otherwise.

It can be rigorously proven that the partition function in terms of the new variables has the same form as the original one $K[\beta]$

$$K^*[\beta] = K[\beta^*] \tag{19}$$

with the only change

$$\beta \to \beta^* \qquad \beta^* = \frac{1}{2}\text{arcsinh}\left(\frac{1}{\sinh 2\beta}\right) \simeq \frac{1}{\beta} \tag{20}$$

A relation like Eq.(19) is called a duality relation. It maps high temperature (strong coupling) regime of K^* to the low temperature (weak coupling) of K[11]. Similar relations

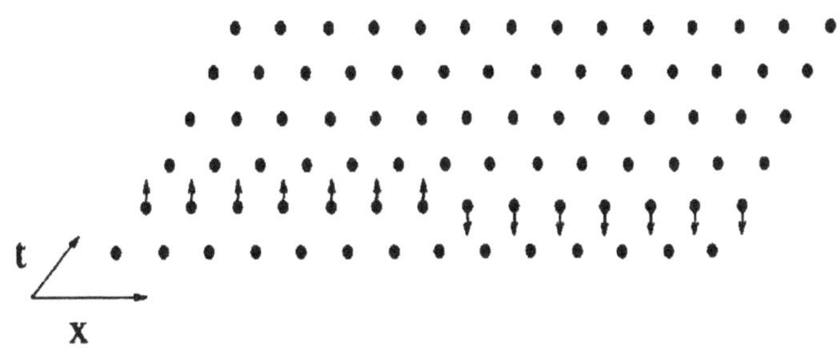

Figure 1. A kink in the Ising model.

have been recently discovered in SUSY QCD with $N = 2$ supersymmetry, and more generally in models of string theory[12].

If we look at the Ising model as the euclidean version of some 1+1 dimensional field theory, a configuration at fixed t, with $\sigma(i) = -1,\ i < i_0\ \sigma(i) = +1,\ i \geq i_0$ will appear on the dual lattice as a single spin up. This configuration has topology, it is a kink, see fig.1. The excitations of the dual lattice are kinks. At low temperature the system is magnetized, $\langle\sigma\rangle \neq 0$ and very few kinks are present. At high temperature $\langle\sigma\rangle = 0$, but using the duality relation $\langle\sigma^*\rangle \neq 0$. $\langle\sigma^*\rangle$ is called a disorder parameter, as opposed to $\langle\sigma\rangle$, which is the order parameter. The relation

$$\langle\sigma\rangle \cdot \langle\sigma^*\rangle = 0 \tag{21}$$

can be proven in the thermodynamical limit. $\langle\sigma^*\rangle$ signals the condensation of kinks.

We shall next address the study of monopole condensation in $U(1)$ compact gauge theory. We shall define a disorder parameter for this system, which describes the condensation of monopoles in the vacuum at high temperature (low β). The parameter will be the v.e.v. of an operator with nonzero monopole charge, and will thus signal dual superconductivity. The same construction will then be used for non-abelian gauge theories after abelian projection.

MONOPOLE CONDENSTATION IN COMPACT $U(1)$: A DISORDER PARAMETER[13]

Like any other gauge theory, compact $U(1)$ is defined in terms of parallel transport along the links joining nearest neighbours on the lattice

$$U_\mu(n) = \exp(ieaA_\mu(n)) \tag{22}$$

a being the lattice spacing.

In the following we shall denote $eaA_\mu(n)$ as $\theta_\mu(n)$. The action is written in terms of the parallel transport $\Pi_{\mu\nu}$ around the elementary square of the lattice in the plane μ, ν

$$\begin{aligned}\Pi_{\mu\nu} &= \exp \mathrm{i}\,[\theta_\mu(n) + \theta_\nu(n+\hat\mu) - \theta_\mu(n+\hat\mu) - \theta_\nu(n)] \equiv \exp(-\mathrm{i}\theta_{\mu\nu}(n)) \\ \theta_{\mu\nu}(n)) &= \Delta_\mu\theta_\nu(n) - \Delta_\nu\theta_\mu(n) \underset{a\to 0}{\simeq} a^2 eF_{\mu\nu}\end{aligned} \tag{23}$$

$$S = \beta \sum_{n,\mu<\nu} (1 - \cos\theta_{\mu\nu}) \tag{24}$$

As $a \to 0$

$$S \simeq \beta \sum_{n,\mu\neq\nu} \frac{\theta^2_{\mu\nu}(n)}{4} \tag{25}$$

and Eq.(24) describes photons if $\beta = 1/e^2$.

The generating functional of the theory (partition function) is

$$Z(\beta) = \int \prod_{n,\mu} \left[\frac{d\theta_\mu(n)}{2\pi}\right] \exp(-S) \tag{26}$$

The theory is compact, since S depends on the cos of the angular variables, and is invariant under change of variables

$$\theta_\mu(n) \to \theta_\mu(n) + f_\mu(n) \tag{27}$$

with arbitrary $f_\mu(n)$. A special case of Eq.(27) are gauge transformations.

A critical β, $\beta_c \simeq 1.011$ exists such that for $\beta > \beta_c$ the theory describes free photons. For $\beta < \beta_c$ electric charge is confined: Wilson loops obey the area law[14] Eq.(2) and flux tubes are observed[15].

A variant of the theory is provided by the Villain action

$$\exp(-S) = \sum_m \exp\left[-\frac{\beta}{2} \sum_{n,\mu<\nu} |\theta_{\mu\nu} - 2\pi m|^2\right] \tag{28}$$

For this variant, condensation of monopoles has rigorously been proven as a mechanism of confinement[16]. Recently the proof has been extended to more general forms of the action, including[17] the Wilson action Eq.(23). Monopoles are identified and counted by the following procedure[18].

Since by construction $\pi \leq \theta_\mu(n) \leq \pi$, it follows from Eq.(23) that

$$-4\pi \leq \theta_{\mu\nu}(n) \leq 4\pi$$

$\theta_{\mu\nu}$ can be redefined modulo an integer multiple of 2π, $n_{\mu\nu}$, as

$$\theta_{\mu\nu} = \bar{\theta}_{\mu\nu} + 2\pi n_{\mu\nu} \qquad -\pi < \bar{\theta}_{\mu\nu} \leq \pi \tag{29}$$

and the monopole current as

$$\rho_\mu^M = \frac{1}{6}\varepsilon_{\mu\nu\rho\sigma}\Delta_\nu n_{\rho\sigma} \equiv \partial_\nu F^*_{\mu\nu} \tag{30}$$

The total number of monopoles is

$$N^M = \sum_n \rho_0^M(n) \tag{31}$$

N^M is large in the confined phase, and drops to zero in the deconfined phase. It has sometimes been identified with the disorder parameter for monopole condensation. Of course N^M commutes with the monopole charge, and therefore cannot signal by any means spontaneous breaking of magnetic $U(1)$.

The disorder parameter

The basic idea of the construction[11, 19] of a disorder parameter is the simple formula for translations

$$e^{ipa}|x\rangle = |x + a\rangle$$

If we identify in our field theory

$$\begin{aligned} x &\to \vec{A}(\vec{x}, t) \\ p &\to \vec{E}(\vec{x}, t) = -i\frac{\delta}{\delta\vec{A}(\vec{x}, t)} \end{aligned} \tag{32}$$

then the operator

$$\mu(\vec{y},t) = \exp\left[\frac{i}{e}\int d^3x\, \vec{E}(\vec{x},t)\vec{b}(\vec{x}-\vec{y})\right] \tag{33}$$

operating on field states in the Schrödinger representation will give

$$\mu|\vec{A}(\vec{x},t)\rangle = |\vec{A}(\vec{x},t) + \frac{1}{e}\vec{b}(\vec{x}-\vec{y})\rangle$$

i.e. it will add a monopole to any field configuration provided that $\vec{b}$ is the vector potential describing the field produced by the monopole

$$\frac{1}{e}\vec{b}(\vec{r}) = \frac{1}{e}\frac{m}{2\pi}\frac{\vec{r}\wedge\vec{n}}{r(r-\vec{r}\cdot\vec{n})} \tag{34}$$

The gauge has been chosen to have the Dirac string in the direction $\vec{n}$. μ does not dependent on the choice of the gauge for $\vec{b}$ if $\vec{E}$ obeys Gauss' law.

On a lattice[20], after Wick rotation, and with the identification

$$a^2E^i = \frac{1}{e}\mathrm{Im}\Pi^{0i} + \mathcal{O}(a^4) \simeq \frac{1}{e}\sin\theta^{0i} \tag{35}$$

$$\mu(\vec{y},n_0) = \exp\left[-\beta\sum_{\vec{n}} b^i(\vec{n}-\vec{y})\sin\theta^{0i}(\vec{n},n_0)\right] \tag{36}$$

Here $b_i(\vec{n})$ is the discretized transcription of Eq.(34) and the β in front comes from the normalization 1/e in Eq.(22) times the $1/e$ appearing in the monopole charge.

A better definition (compactified) which shifts the angle θ_{0i} and not $\sin\theta_{0i}$ is[13]

$$\mu(\vec{y},n_0) = \exp\beta\sum_{n}\left[S(\theta^{0i}(\vec{n},n_0) + b^i(\vec{n}-\vec{y})) - S(\theta^{0i}(\vec{n},n_0))\right] \tag{37}$$

which reduces to Eq.(36) at first order in $\vec{b}$. We denote by 5 in Eq.(37) the density of the action. The action can be any of form, e.g. Eq.(24) or (28), provided Eq.(25) is satisfied.

If an arbitrary number of monopoles or antimonopoles is created at time n_0, then $b^i(\vec{n}-\vec{y})$ has to be replaced by the corresponding field configuration. To compute correlation functions of operators at different times, the rule is

$$\langle\mu(\vec{y}_1,n_1^0),\ldots\mu(\vec{y}_k,n_k^0)\rangle = \frac{1}{Z}\int\prod\left[\frac{d\theta_\mu(n)}{2\pi}\right]\exp\left(\beta(S+\Delta S)\right) \tag{38}$$

where $S+\Delta S$ is obtained by replacing in the action the plaquettes $\Pi^{0i}(\vec{n},n_\ell^0) = 1 - \cos(\theta^{0i}(\vec{n},n_\ell^0))$ by $1-\cos\,(\theta^{0i}(\vec{n},n_\ell^0) + b^i(\vec{n}-\vec{y}_\ell))$. $(1\le\ell\le k)$

In particular we will study the correlator

$$\mathcal{D}(x_0) = \langle\mu(\vec{0},x_0)\mu(\vec{0},0)\rangle \tag{39}$$

At large enough x_0

$$\mathcal{D}(x_0) \underset{|x_0|\to\infty}{\simeq} A\exp(-M|x_0|) + \langle\mu\rangle^2 \tag{40}$$

The last equality follows from the cluster property, translation invariance and C invariance of the vacuum. Our aim will be to extract M and $\langle\mu\rangle$ from numerical determinations of $\mathcal{D}(x_0)$. $\langle\mu\rangle$ is the disorder parameter: a nonzero value of it in the thermodynamic limit signals dual superconductivity. M is the mass of the lightest excitation with the quantum numbers of a monopole and is a lower limit to the mass of the effective Higgs field which produces superconductivity.

As explained above

$$\mathcal{D}(x_0) = \frac{1}{Z[S]} Z[S + \Delta S] \tag{41}$$

where $S + \Delta S$ is obtained from S by the change

$$\theta^{i0}(\vec{n}, 0) \to \theta^{i0}(\vec{n}, 0) + b^i(\vec{n}) \tag{42}$$

$$\theta^{i0}(\vec{n}, x_0) \to \theta^{i0}(\vec{n}, x_0) - b^i(\vec{n}) \tag{43}$$

and $b^i(\vec{n})$ is defined by Eq.(34). Since

$$\theta^{i0}(\vec{n}, 0) = -\theta^i(\vec{n}, 1) + \theta^i(\vec{n}, 0) + \theta^0(\vec{n} + \hat{\imath}, 0) - \theta^0(\vec{n}, 0)$$

the replacement, Eq.(42), amounts to the change

$$\theta^i(\vec{n}, 1) \to \theta^i(\vec{n}, 1) - b_i(\vec{n}) \equiv \bar{\theta}^i(\vec{n}, 1) \tag{44}$$

The change, Eq.(44), can be reabsorbed in a redefinition of variables of the Feynman integral defining $Z[S + \Delta S]$ which leaves the measure invariant [Eq.(27)]. As a consequence

$$\theta^{ij}(\vec{n}, 1) \to \theta^{ij}(\vec{n}, 1) + \Delta^i b^j - \Delta^j b^i$$

meaning that a monopole is added at $x_0 = 1$. Moreover

$$\theta^{0i}(\vec{n}, 1) \to \theta^{0i}(\vec{n}, 1) + b^i(\vec{n})$$

Again this change can be reabsorbed by a change of variables

$$\theta^i(\vec{n}, 2) \to \theta^i(\vec{n}, 2) - b_i(\vec{n})$$

after which

$$\theta^{ij}(\vec{n}, 2) \to \theta^{ij}(\vec{n}, 2) + \Delta^i b^j - \Delta^j b^i$$

and

$$\theta^{i0}(\vec{n}, 2) \to \theta^{i0}(\vec{n}, 2) + b^i(\vec{n})$$

The construction can be repeated till $n_0 = x_0$, when the addition of $b_i(\vec{n})$ to $\theta^{i0}(\vec{n}, n_0)$ cancels with the term $-b_i(\vec{n})$ in Eq.(43). The change $Z[S + \Delta S]$ amounts to add a monopole on the site $\vec{n} = \vec{y}$, propagating from time 0 to time x_0.

Measuring $\mathcal{D}(x_0)$, to extract M and $\langle\mu\rangle$, is nontrivial due to large fluctuations, ΔS is the change of the action on a spatial volume V: it fluctuates roughly as $\sqrt{V}$,which means a fluctuation $\sim \exp(\sqrt{V})$ for $\langle\mu\rangle$. A way out of this difficulty is to measure, instead of $\mathcal{D}(x_0)$, the quantity

$$\rho(x_0) = \frac{d}{d\beta} \ln \mathcal{D}(n_0) = \langle S\rangle_S - \langle S + \Delta S\rangle_{S+\Delta S} \tag{45}$$

The last equality trivially follows from the definition of Z. $\langle S\rangle_S$ means average of S computed by weighting with the action S. Since $\mathcal{D}(x_0)_{\beta=0} = 1$

$$\mathcal{D}(x_0) = \exp\left(\int_0^\beta \rho(\beta) d\beta\right) \tag{46}$$

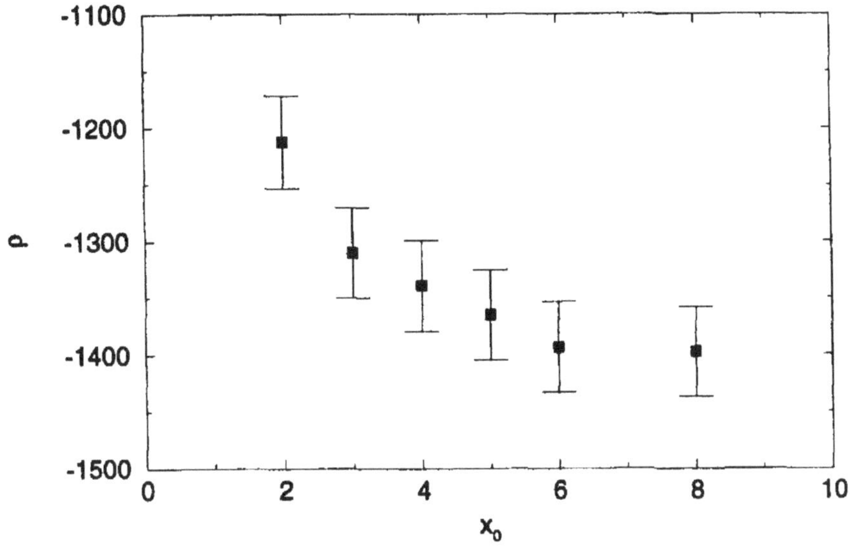

Figure 2. Monopole antimonopole correlation. (Lattice $8^3 \times 16$.)

As $x_0 \to \infty$

$$\rho \underset{x_0 \to \infty}{\simeq} 2\frac{d}{d\beta}\ln\langle\mu\rangle + C\exp(-Mx_0) \tag{47}$$

Fig.2 shows the behaviour of ρ as a function of x_0 consistent with Eq.(47). In fig.3 $\rho(\infty) \equiv 2\frac{d}{d\beta}\ln\langle\mu\rangle$ is shown as a function of β. A huge negative peak appears at the phase transition, which, according to the definitions Eq.(45), Eq.(46) reflects a sharp decrease of $\langle\mu\rangle$. This can be appreciated from fig.4 where a direct measurement of $\langle\mu\rangle$ is shown, even with large errors.

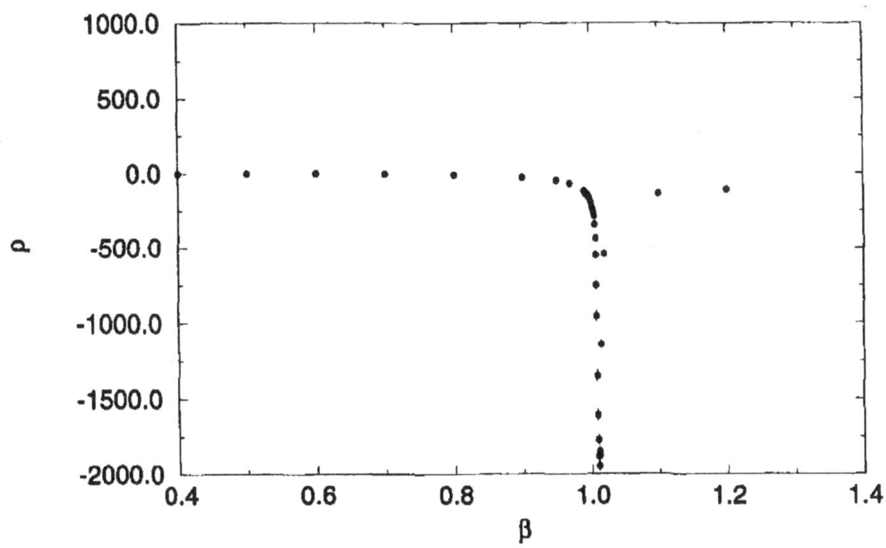

Figure 3. ρ_∞ as a function of β. The negative peak signals the phase transition. (Lattice $8^3 \times 16$.)

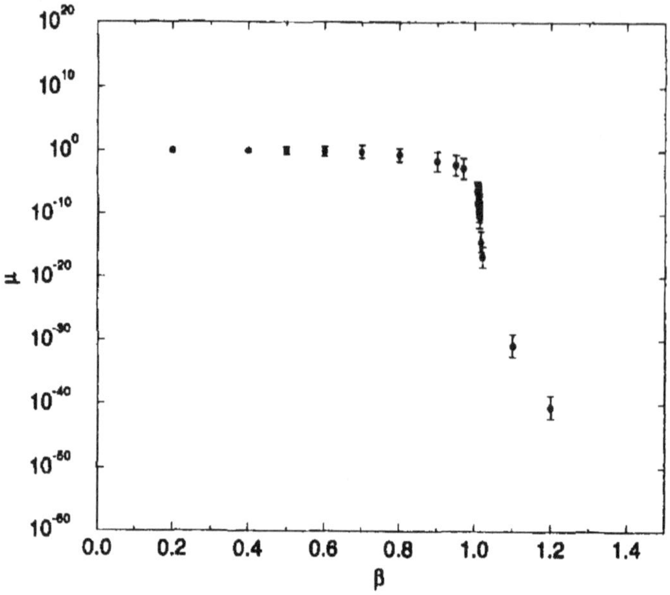

Figure 4. $\langle\mu\rangle$ v.s. β. (Lattice $10^3 \times 20$.)

At $\beta > \beta_c$ the system describes free photons: μ and $\rho(\infty)$ can be computed in perturbation theory by a gaussian integration. The numerical result for a lattice $L^3 \times 2L$ is

$$\rho_\infty = -10.1 \cdot L + 9.542 \qquad (\beta > \beta_c) \tag{48}$$

ρ_∞ tends to $-\infty$ in the thermodynamical limit $L \to \infty$, or $\langle\mu\rangle \to 0$.. In fact $\langle\mu\rangle$ is an analytic function of β at finite volume, and cannot be exactly zero for $\beta > \beta_c$, since it would be identically zero everywhere. Only as $V \to \infty$,, Lee - Yang singularities develop and $\langle\mu\rangle = 0$ for $\beta > \beta_c$ (48), as a respectable order parameter. For $\beta < \beta_c$, ρ_∞ tends to a finite value as $V \to \infty$. This can be seen from fig.5 but it is also a theorem proven

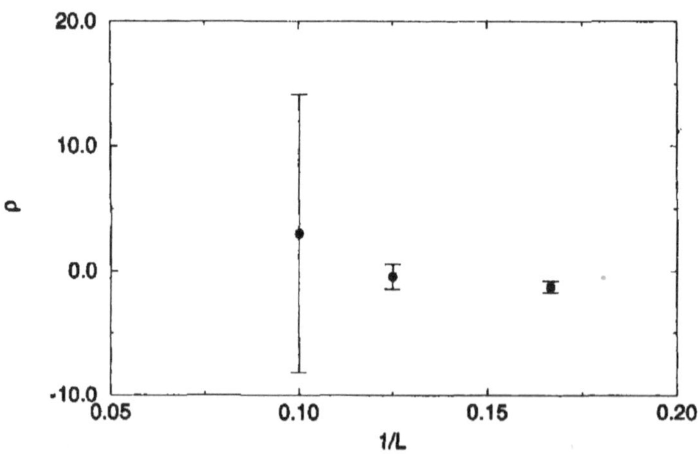

Figure 5. ρ_∞ versus $1/L$ for $\beta = 1.009$.

in ref.(16), which generalizes the result of ref.(15) for the Villain action.

For $\beta \sim \beta_c$ there is a phase transition, which is weak first order or second order. In any case the correlation length will grow as $\beta \to \beta_c$ in a certain interval of $(\beta_c - \beta)$ with some effective critical index ν:

$$\xi \simeq (\beta_c - \beta)^{-\nu} \tag{49}$$

A finite size scaling analysis can be done as follow. In general, by dimensional reasons

$$\mu = \mu\left(\frac{L}{\xi}, \frac{a}{\xi}\right) \tag{50}$$

As $\beta \to \beta_c$, $a/\xi \to 0$ and $\mu \simeq \mu(L/\xi, 0)$ or

$$\mu = f\left(L^{1/\nu}(\beta_c - \beta)\right) \tag{51}$$

which implies in turn that

$$\rho = \frac{d}{d\beta}\ln\langle\mu\rangle = -L^{1/\nu}\frac{f'(L^{1/\nu}(\beta_c - \beta))}{f(L^{1/\nu}(\beta_c - \beta))} \tag{52}$$

or $\rho L^{-1/\nu}$ is a universal function of $L^{1/\nu}(\beta_c - \beta)$

Data from lattices of different size will lay on the same universal curve only for appropriate values of β_c and ν. The best values can be then determined. We obtain

$$\beta_c = 1.01160(5) \qquad \nu = 0.29(2) \tag{53}$$

$\mu \simeq (\beta_c - \beta)^\delta$ as $\beta \to \beta_c$, and

$$\frac{\rho}{L^{1/\nu}} \simeq -\frac{\delta}{L^{1/\nu}(\beta_c - \beta)}$$

For δ we obtain $\delta = 1.1 \pm 0.2$.

Our result is then that $\langle\mu\rangle \neq 0$ for $\beta < \beta_c$ in the thermodynamic limit, i.e. that the system is a dual superconductor for $\beta < \beta_c$.

We have also measured the penetration depth of the electric field, i.e. the mass of the photon, m, by the method of ref.(18). m properly scales as $\beta \to \beta_c$ with index ν, and the indication is that it is substantially smaller than M, fig.6, or that the superconductor is type II.

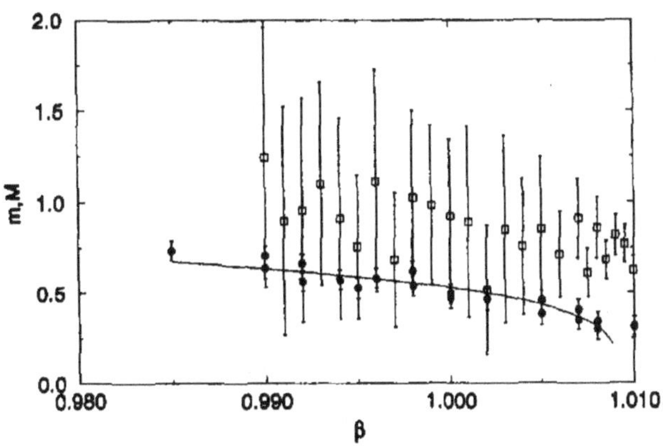

Figure 6. Mass of the monopole M (squares), and mass of the dual photon m (circles) vs. β.

An alternative method to demonstrate dual superconductivity is to detect the London current in the flux tube configurations[15] between $Q\bar{Q}$. For a detailed comparison of our approach with ref.16 we refer to ref.13.

As for a direct determination of $\langle\mu\rangle$ it can be shown that $\langle\mu\rangle$ has a gaussian distribution in the sense of the central limit theorem[13]. However, due to the exponential dependence on ΔS the expectation value of $\langle\mu\rangle$ is not centered at the minimum of ΔS. Let $\Pi(y)$ be the distribution of $y = \beta\Delta S - \beta\langle\Delta S\rangle$, the distribution probability for μ is

$$P(\mu) = \Pi\left(\ln(\frac{\mu}{\bar{\mu}})\right) d\ln(\frac{\mu}{\bar{\mu}})$$

If $\Pi(y)$ is gaussian with width σ_y,

$$\langle\mu\rangle = \bar{\mu}\exp(\frac{\sigma_y^2}{2}) \qquad \sigma_\mu = \bar{\mu}\exp(\sigma_y^2)$$

The fluctuations σ_μ are larger than $\langle\mu\rangle$

A careful analysis requires to account for higher cumulants. The displacement of the maximum of $P(\mu)$ with respect to $\langle\mu\rangle$ should be kept into account when computing the so called constrained effective potential.

As a final comment it can be shown that our construction gives the same result as that of ref.16 in the case of the Villain action. We have thus a disorder parameter which is a reliable tool to detect dual superconductivity.

3d $X - Y$ MODEL[21] (LIQUID He_4)

We have successfully repeated the construction for the $X - Y$ model in 3d, where vortices condense to produce the phase transition. The result can be checked against experiment (liquid He_4). The model is denned on a 3d cubic lattice. An angle $\theta(i)$ is defined on each site i. The action reads

$$S = \beta\sum_i\sum_{\mu=0}^{2}[1 - \cos(\Delta_\mu\theta(i))] \tag{54}$$

$\Delta_\mu\theta(i) = \theta(i + \hat{\mu}) - \theta(i)$. The partition function is

$$Z = \int\prod_i \frac{d\theta(i)}{2\pi}\exp(-S) \tag{55}$$

Z is a periodic functional of $\theta(i)$ with period 2π (compactness). Z is invariant under the change of variables

$$\theta(i) \to \theta(i) + f(i) \tag{56}$$

with arbitrary $f(i)$ and so is any correlator of compact fields, in spite of the fact that the transformation (56) is not a symmetry of the action. In running the indices in Eq.(56) from 0 to 2 we anticipate that we shall consider the theory as the euclidean version of a 2+1 dimensional field theory.

As $\beta \to \infty$

$$S \simeq \frac{\beta}{2}\sum_{\mu,i}[\Delta_\mu(\theta_i)]^2 \tag{57}$$

and the theory describes a massless scalar field. At $\beta = \beta_c = 0.454$ a 2nd order phase transition takes place. Below β_c vortices are expected to condense in the vacuum. Like for monopoles, condensation has always been demonstrated by the drop of the density of vortices when β rises through β_c. We will show instead that a spontaneous symmetry breaking of the $U(1)$ symmetry related to the conservation of vortex number takes place and that a legitimate disorder parameter can be defined.

We define

$$A_\mu = \partial_\mu \theta \tag{58}$$

Under the transformation, Eq.(56), A_μ undergoes a gauge transformation

$$A_\mu \to A_\mu + \partial_\mu f \tag{59}$$

The invariance under Eq.(56) means gauge invariance if the theory is phrased in terms of A_μ. From Eq.(58) it follows

$$F_{\mu\nu} = \partial_\mu A_\nu - \partial_\nu A_\mu = 0 \tag{60}$$

In fact Eq.(60) is valid apart from singularities. In terms of A_μ

$$\theta(x) = \exp(\mathrm{i} \int_C^x A_\mu dx^\mu) \tag{61}$$

and, if Eq.(60) holds, the choice of the path C used in Eq.(61) is irrelevant. A current j_μ can be defined as the dual of $F_{\mu\nu}$:

$$j_\mu = \varepsilon_{\mu\alpha\beta} \partial^\alpha A^\beta \tag{62}$$

and

$$\partial^\mu j_\mu = 0 \tag{63}$$

is the analog of the Bianchi identities. The conserved quantity associated to Eq.(63) is the vorticity

$$Q = \int d^2x \; j^0(\vec{x}, t) = \int d^2x (\vec{\nabla} \wedge \vec{A})_0 = \oint \vec{A} \cdot d\vec{x} \tag{64}$$

Since $\vec{A} = \vec{\nabla}\theta$, single–valuedness of the action implies that $Q = 2\pi n$. If there are no singularities it follows from Eq.(58) that $\vec{\nabla} \wedge \vec{A} = 0$, or from Eq.(64) that $Q = 0$.

There exist however configurations with nontrivial vorticity. An example is

$$\bar{\theta}_{(q)}(\vec{x} - \vec{y}) = q \arctan \frac{(x_2 - y_2)}{(x_1 - y_1)} \tag{65}$$

For this configuration

$$A_0 = 0 \qquad \vec{A} = \frac{q}{r} \vec{v}_\theta \tag{66}$$

$\vec{v}_\theta$ being the unit vector in the polar coordinate direction θ. If $\vec{A}$ is the field of velocities, $\bar{\theta}_{(q)}$ describes a vortex. For thisconfiguration $j^0 = 2\pi \delta^{(2)}(\vec{x} - \vec{y})$ and

$$\oint_C \vec{A} \cdot d\vec{x} = 2\pi q \tag{67}$$

if the path C encloses $\vec{y}$ once.

A disorder parameter describing condensation of vortices can be defined. In the continuum this is nothing but addition to the field $\theta(x)$ of any configuration, the field $\bar{\theta}$ describing a vortex. The conjugate momentum to θ being $\beta \sin \partial_0 \theta = \delta \mathcal{L} / \delta(\partial_0 \theta)$

$$\mu(\vec{y}, t) = \exp \left[\mathrm{i}\beta \int d^2x \; \sin(\partial_0 \theta(\vec{x}, t)) \bar{\theta}_{(q)}(\vec{x} - \vec{y}) \right] \tag{68}$$

analogous to Eq.(38).

After Wick rotation the compactified version of Eq.(68) becomes (see Eq.(37))

$$\mu(\vec{y},t) = \exp\left\{-\beta\sum_{\vec{n}}\left[\cos(\Delta_0\theta(\vec{n},t) - \bar{\theta}_{(q)}(\vec{n}-\vec{y})) - \cos(\Delta_0\theta(\vec{n},t)\right]\right\} \tag{69}$$

When computing correlation functions of $\mu(\vec{y},t)$,, Eq.(69) produces a change of the term in the action containing $\Delta_0\theta$ at time t from $\Delta_0\theta(\vec{n},t)$ to $\Delta_0\theta(\vec{n},t) - \bar{\theta}_{(q)}(\vec{n}-\vec{y})$. The correlator

$$\mathcal{D}(t) = \langle\mu(\vec{0},t)\mu(\vec{0},0)\rangle \tag{70}$$

will behave as

$$\mathcal{D}(t) \underset{t\to\infty}{\simeq} \langle\mu\rangle^2 + Ae^{-Mt} \tag{71}$$

and, from the definition of μ

$$\mathcal{D}(t) = \frac{Z[S+\Delta S]}{Z[S]} \tag{72}$$

The $S+\Delta S$ is obtained from S by the modification

$$\Delta_0\theta(\vec{n},0) \to \Delta_0\theta(\vec{n},0) - \bar{\theta}_{(q)}(\vec{n}-\vec{y}) \tag{73}$$
$$\Delta_0\theta(\vec{n},t) \to \Delta_0\theta(\vec{n},t) + \bar{\theta}_{(q)}(\vec{n}-\vec{y}) \tag{74}$$

A change of variables $\theta(\vec{n},1) \to \theta(\vec{n},1) - \bar{\theta}_{(q)}$ reabsorbs the modification in $\Delta_0\theta(\vec{n},0)$, but shifts $\Delta_i\theta(\vec{n},1) = A_i(\vec{n},1)$ by a vortex $\Delta_i\bar{\theta}_{(q)} = A_i^{(q)}$ and sends

$$\Delta_0\theta(\vec{n},1) \to \Delta_0\theta(\vec{n},1) - \bar{\theta}_{(q)}(\vec{n}-\vec{y})$$

Similarly to what was done with $U(1)$ monopoles the construction can be repeated till $n_0 = t$ is reached and $-\bar{\theta}_{(q)}$ cancels $\bar{\theta}_{(q)}$ of Eq.(74).

$D(t)$ describes a vortex at $\vec{y} = 0$ propagating in time from 0 to t. Instead of $D(t)$, $\rho(t) = \frac{d}{d\beta}\ln\mathcal{D}(t)$ can be studied. At large t

$$\rho(t) \simeq \rho + Ae^{-Mt} \tag{75}$$
$$\rho = 2\frac{d}{d\beta}\ln\langle\mu\rangle \qquad \langle\mu\rangle = \exp\left[\frac{1}{2}\int_0^\beta \rho_\infty(\beta)\,d\beta\right] \tag{76}$$

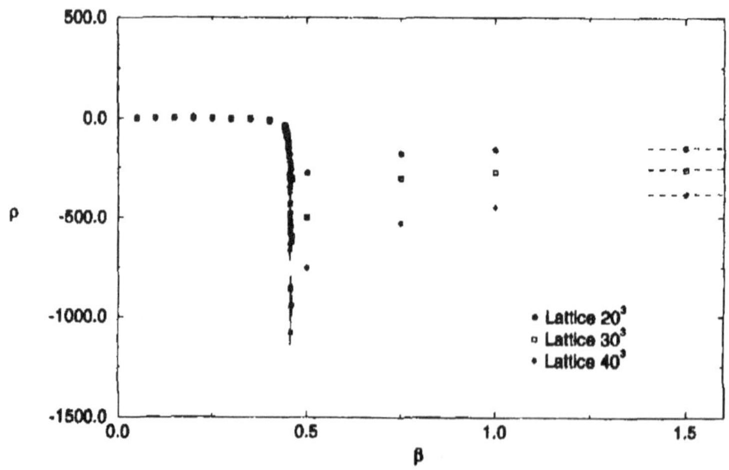

Figure 7. ρ vs β for $X-Y$ model. The peak is at the transition to a superfluid.

From Eq. (76) the disorder parameter $\langle\mu\rangle$ can be determined. The typical behaviour of ρ is shown in fig.6. In the thermodynamical limit, $\langle\mu\rangle \neq 0$ signals spontaneous symmetry breaking of the symmetry Eq.(63), and hence condensation of vortices. At large β the theory is free and ρ can be computed by a gaussian integration. The result for a cubic lattice of size L is

$$\rho = -11.33 \cdot L + 72.669 \tag{77}$$

As $L \to \infty$ this implies $\langle\mu\rangle = 0$. For $\beta < \beta_c$, ρ tends to a finite value, as $L \to \infty$.

Around β_c a finite size scaling analysis can be done as in the $U(1)$ model giving

$$\nu = 0.669 \pm 0.065 \qquad [.670(7)] \tag{78}$$
$$\beta_c = 0.4538 \pm 0.0003 \qquad [.45419(2)] \tag{79}$$
$$\delta = 0.740 \pm 0.029 \tag{80}$$

δ is the index of μ. The numbers in parenthesis on the right are the accepted determinations by other methods[22]. The agreement is good.

MONOPOLES IN QCD. REVISITING THE ABELIAN PROJECTION[23]

At the classical level monopoles in non-abelian gauge theories can be defined by the usual multipole expansion[24, 25]. At large distances the magnetic monopole field obeys abelian equations. By a suitable choice of gauge the direction of the Dirac string can be chosen, and magnetic charges are identified by a diagonal matrix in the fundamental representation with positive or negative integer eigenvalues. For $SU(N)$, $N-1$ magnetic charges exist, which correspond to a group $U(1)^{N-1}$.

This classification coincides with the so called abelian projection. Let $\vec{\Phi} \cdot \vec{\sigma}$ be any field belonging to the adjoint representation: in what follows we shall refer to $SU(2)$ for the sake of simplicity. For $SU(N)$ the procedure is analogous with some formal complication. A gauge transformation $U(x)$ which diagonalizes $\vec{\Phi} \cdot \vec{\sigma}$ is called an abelian projection. $U(x)$ will be singular at the locations where $\vec{\Phi} = 0$. These locations are world lines of $U(1)$ Dirac monopoles. $U(1)$ is the residual invariance under rotation around the axis of $\vec{\Phi} \cdot \vec{\sigma} = \Phi\sigma_3$ after diagonalization.

These monopoles are supposed to condense and produce dual superconductivity. There is a large arbitrariness in the choice of $\vec{\Phi} \cdot \vec{\sigma}$, i.e. in the identification of monopoles. We will discuss this point in detail below.

The meaning of the abelian projection can be better understood[26] in the Georgi Glashow model[27]. This is an $SO(3)$ gauge theory coupled to a triplet Higgs

$$\mathcal{L} = -\frac{1}{4}\vec{G}_{\mu\nu}\vec{G}_{\mu\nu} + (D_\mu\vec{\Phi})^\dagger(D_\mu\vec{\Phi}) - V(\vec{\Phi}) \tag{81}$$

with

$$D_\mu\vec{\Phi} = (\partial_\mu - g\vec{W}_\mu\wedge)\vec{\Phi}$$

and

$$V(\vec{\Phi}) = \frac{\lambda}{4}\left[(\vec{\Phi}^2)^2 - \mu^2\vec{\Phi}^2\right]$$

The model admits monopoles as soliton solutions in the spontaneous broken phase where $\langle\vec{\Phi}\rangle = \vec{\Phi}^0 \neq 0$.

Usually a fixed (space independent) frame of reference is used in colour space, e.g. $\vec{\xi}_i^0$ $(i = 1, 2, 3)$., the unit vectors of an orthogonal cartesian frame. $\vec{\xi}_i^0 \wedge \vec{\xi}_j^0 = \varepsilon_{ijk}\vec{\xi}_k^0$. One can, however, define a Body Fixed Frame (BFF) by 3 orthogonal unit vectors $\vec{\xi}_i(x)$

$$\vec{\xi}_i \wedge \vec{\xi}_j = \varepsilon_{ijk}\vec{\xi}_k \tag{82}$$

such that $\vec{\xi}_3(x) = \hat{\Phi}(x)$ is parallel to the direction $\hat{\Phi}(x)$ of the $\vec{\Phi}$ field. This system is defined up to a rotation around $\vec{\xi}_3$. A gauge group element $R(x)$ exists such that

$$\vec{\xi}_i(x) = R^{-1}(x)\vec{\xi}_i^0 \tag{83}$$

By construction $R(x)$ is the gauge transformation which implements the abelian projection, since $R(x)\vec{\xi}_3 = \vec{\xi}_3^0$. Since $\vec{\xi}_i^2 = 1$, $\partial_\mu\vec{\xi}_i = \vec{\omega}_\mu \wedge \vec{\xi}_i$ or

$$D_\mu(\omega)\vec{\xi}_i \equiv (\partial_\mu - \vec{\omega}_\mu\wedge)\vec{\xi}_i = 0 \tag{84}$$

Eq.(84) implies $[D_\mu, D_\nu] = G_{\mu\nu}(\vec{\omega}) = 0$. The field $\vec{\omega}_\mu$ is a pure gauge. The last equality reads explicitly

$$\partial_\mu\vec{\omega}_\nu - \partial_\nu\vec{\omega}_\mu + \vec{\omega}_\mu \wedge \vec{\omega}_\nu = 0 \tag{85}$$

Eq.(85) is true apart from singularities. As a consequence of Eq.(85)

$$\vec{\Phi}(x) = P\exp\left[i\int_C^x \vec{\omega}_\mu \cdot \vec{T}\, dx^\mu\right]\vec{\Phi}^0 \tag{86}$$

Due to Eq.(85) the integral in Eq.(86) is independent of the path C. This is true apart from singularities which can give a nontrivial connection to space time.

A 't Hooft[28]-Polyakov[29] monopole configuration has a zero of $\vec{\Phi}(x)$ at the location of the monopole, and in that point the abelian projection $R(x)$ has a singularity.

The singularities of $\vec{\omega}_\mu$ can be studied by expressing $\vec{\xi}(x)$ in terms of polar coordinates $(\theta(x), \psi(x))$, with polar axis 3 in colour space. The singularities come from the fact that $\psi(x)$ is not defined at the sites where $\theta(x) = 0, \pi$. In terms of $\vec{G}_{\mu\nu}(\omega)$ the potentially singular term at $\theta(x) = 0, \pi$ is[26]

$$F^3_{\mu\nu}(x) = -\cos\theta(x)(\partial_\mu\partial_\nu - \partial_\nu\partial_\mu)\psi(x) \tag{87}$$

The singularity exists where $\theta = 0, \pi$ or at the sites where $\hat{\Phi}$ is in the direction 3, and the field is parallel to $\hat{\Phi}$ in colour space and abelian. The singularity is a string with flux $\pm 2n\pi$ $(n = 0, 1\ldots)$ The abelian field is related to the residual $U(1)$ symmetry along the 3d axis. The field strength is the abelian part of $\vec{F}_{\mu\nu}$, or

$$\mathcal{F}_{\mu\nu} = \hat{\Phi}\cdot\vec{G}_{\mu\nu} - \frac{1}{g}\hat{\Phi}\cdot(D_\mu\hat{\Phi} \wedge D_\nu\hat{\Phi}) \tag{88}$$

Indeed, in the abelian projected frame $D_\mu\hat{\Phi} = g\vec{A}_\mu \wedge \vec{\xi}_3^0$ and

$$\frac{1}{g}\hat{\Phi}\cdot(D_\mu\vec{\Phi} \wedge D_\nu\vec{\Phi}) = g(\vec{W}_\mu \wedge \vec{W}_\nu)\cdot\vec{\Phi}$$

which cancels the non-abelian term of $\hat{\Phi}\cdot\vec{G}_{\mu\nu}$ in Eq.(88). $\mathcal{F}_{\mu\nu}$ is nothing but a covariant expression for the $U(1)$ field identified by the abelian projection. It is a gauge invariant quantity. For a 't Hooft-Polyakov monopole configuration $\mathcal{F}_{\mu\nu}$ is the field of a pointlike Dirac monopole, located at the zero of $\vec{\Phi}(x)$.

In QCD there are no fundamental Higgs fields. The idea is that monopoles could be defined by any composite field $\vec{\Phi}(x)$ in the adjoint representation[23]. No unique criterion is known for the choice of the operator $\vec{\Phi}$ in QCD. Popular choices are[23]:

1) $\hat{\Phi}(x)$ is the Polyakov line

$$P(\vec{x}, x_0) = \oint_{C,x_0}^{x_0} P\exp(\mathrm{i}\vec{A}_\mu \cdot \vec{T}\, dx^\mu) \tag{89}$$

The path C being the line $\vec{x}$ = constant along the time axis closing at infinity by periodic boundary conditions.

2) Any component $\vec{F}_{\mu\nu}$ of the field strength tensor.

3) The operator implicitly defined by the maximization of

$$\sum_{\mu,n} \mathrm{Tr}\left\{\sigma_3 U_\mu(n)\sigma_3 U_\mu^\dagger(n)\right\} \tag{90}$$

This choice is known as maximal abelian projection.

4) For $SU(3)$ $(F_{\mu\nu})^2$ since, contrary to the case of $SU(2)$, it is not a singlet, due to the d algebra.

For each of these choices a $U(1)$ gauge field (actually 2 for $SU(3)$) can be identified, which couples to monopoles, and a creation operator of monopoles can be constructed on the same lines as for compact $U(1)$.

A strategy to answer the question whether QCD vacuum is a dual superconductor is to detect the condensation of the monopoles defined by different abelian projections in the confined phase and across the transition to the deconfined phase. As shown in the previous sections a reliable tool (disorder parameter) exists for that, which has been successfully tested in systems which are well understood. Before proceeding to that we will test the ideas of this section on a simple system: the Heisenberg ferromagnet.

The Heisenberg ferromagnet[30]

The action is

$$\mathcal{L} = \frac{1}{2}\sum_{\mu=0}^{2}\sum_{x} \Delta_\mu\vec{n}(x)\cdot\Delta_\mu\vec{n}(x) = \sum_{\mu=0}^{2}\sum_{x}(1 - \vec{n}(x+\hat{\mu})\cdot\vec{n}(x)) \tag{91}$$

x runs on a cubic 3d lattice and $\vec{n}$ is a vector of unit length, $\vec{n}^2 = 1$.

We shall look at the model as a 2+1 dimensional field theory. The Feynman functional

$$Z = \int \prod_x d\Omega(x)\, \exp(-\beta S) \tag{92}$$

has a much bigger symmetry than the lagrangian. Any local rotation $\vec{n}(x) \to R(x)\vec{n}(x)$ even if it does not leave $\mathcal{L}$ invariant, is reabsorbed in a change of variables in the Feynman integral, leaving the measure invariant. Assuming constant boundary condition at infinity, $\vec{n}_0$, we can write

$$\vec{n}(x) = R^{-1}(x)\vec{n}_0 \tag{93}$$

R^{-1} is determined up to a rotation along $\vec{n}$. As in the Georgi Glashow model a gauge field $\vec{\omega}_\mu$ can be defined by introducing a Body Fixed Frame $\vec{\xi}_i$, with $\vec{\xi}_3 = \vec{n}$. Then

$$\partial_\mu\vec{\xi}_i = \vec{\omega}_\mu \wedge \vec{\xi}_i$$

or

$$D_\mu\vec{\xi}_i = (\partial_\mu - \mathrm{i}\vec{\omega}\wedge)\vec{\xi}_i = 0$$

which implies $F_{\mu\nu}(\omega) = 0$, apart from singularities and

$$\vec{n}(x) = P \exp\left[i \int_{\infty, C}^{x} \vec{\omega}_\mu \cdot \vec{T}\, dx^\mu \right] \vec{n}_0 \tag{94}$$

independent of the path C, $\vec{\omega}_\mu$ being a pure gauge. Again this is true apart from singularities. A conserved current exists,

$$\vec{j}_\mu = \frac{1}{8\pi} \varepsilon_{\mu\alpha\beta} \partial_\alpha \vec{n} \wedge \partial_\beta \vec{n} \tag{95}$$

$\vec{j}_\mu$ is parallel to $\vec{n}$, since both $\partial_\alpha \vec{n}$ and $\partial_\beta \vec{n}$ are orthogonal to it. The corresponding conserved quantity is

$$Q = \int d^2x\, j^0(x) = \frac{1}{4\pi} \int d^2x\, \vec{n} \cdot (\partial_1 \vec{n} \wedge \partial_2 \vec{n}) \tag{96}$$

which is nothing but the topological charge of the 2 dimensional version of the model. Instantons of the 2 dimensional model look as solitons of the 2+1 dimensional one.

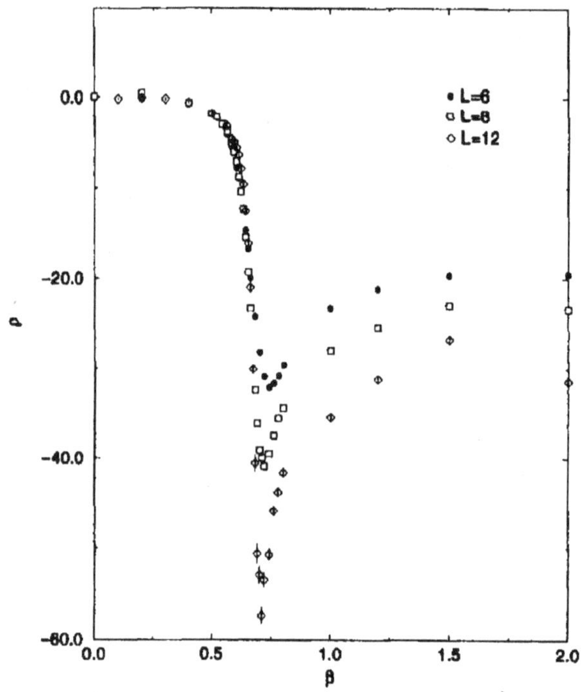

Figure 8. ρ v.s. β. Heisenberg model.

$\vec{j}_\mu$ can also be written as

$$\vec{j}_\mu = \frac{\vec{n}}{8\pi} \left[\varepsilon_{\mu\alpha\beta} \left(\vec{\omega}_\alpha \wedge \vec{\omega}_\beta \right) \cdot \vec{n} \right] \tag{97}$$

Except for the singularities corresponding to the locations of the instantons $\vec{\omega}_1 \wedge \vec{\omega}_2 = \partial_1 \vec{\omega}_2 - \partial_2 \vec{\omega}_1$ and

$$Q = \frac{1}{4\pi} \oint \vec{\omega}_\alpha \cdot \vec{n}\, dx_\alpha \tag{98}$$

A nontrivial connection is created by the presence of solitons. A direct calculation shows that the field $\vec{\omega}_\mu$ has Dirac string singularities propagating in time from the center of the soliton. The transition from a magnetized phase to a disordered phase can be seen as a condensation of solitons.

The system undergoes a second order phase transition at $\beta = \beta_c \simeq 0.7$ from the magnetized phase to a disordered phase. We have investigated if the solitons described above condense in the disordered phase. A disorder parameter can be constructed, on the same line as in the compact $U(1)$ and in 3d $X - Y$ model, as the vev of the creation operator of a soliton, as follows

$$\langle \mu(x) \rangle = \frac{Z[S + \Delta S]}{Z[S]} \tag{99}$$

where $S + \Delta S$ is obtained from S by the change

$$(\Delta_0 \vec{n}(\vec{x}, x_0))^2 \equiv (\vec{n}(\vec{x}, x_0 + 1) - \vec{n}(\vec{x}, x_0))^2 \to (R_q^{-1} \vec{n}(\vec{x}, x_0 + 1) - \vec{n}(\vec{x}, x_0))^2 \tag{100}$$

R_q is a time independent transformation which adds a soliton of charge q to a configuration.

Numerical simulations show that in the thermodynamical limit $\langle \mu(x) \rangle$ vanishes in the ordered (magnetized) phase, is different from zero in the disordered phase, and at the phase transition obeys a finite size scaling law from which the critical indices and the transition temperature can be extracted. A typical form of $\rho = d \ln\langle\mu\rangle / d\beta$ is shown in fig.8.

The results are still preliminary but agree with the values known from other methods[31]. We get

$$\beta_c = 0.69 \pm 0.01 \tag{101}$$

and

$$\nu = 0.7 \pm 0.2 \tag{102}$$

This shows that the phase transition to disorder in the Heisenberg magnet can be viewed as condensation of solitons.

The string structure of the singularities of the field $\vec{\omega}_\mu$ is similar to Georgi Glashow model, and the field strength tensor is generated by the topology of the field $\vec{n}$, even in the absence of gauge fields. Indeed going back to the previous section, the monopoles only depend on the Higgs field. Monopoles exposed by the abelian projection are not lattice artifacts, but reflect the dynamics of the field $\vec{\Phi}$.

DUAL SUPERCONDUCTIVITY IN QCD

I will present the first results of a systematic exploration of monopole condensation below and above the deconfining transition in different abelian projections[33, 32].

Besides the disorder parameter we also measure, at $T = 0$, the monopole anti-monopole correlation function to extract a lower limit to the effective Higgs mass, as well as the penetration depth of the electric field, i.e. the mass of the photon which produces the (dual) Meissner effect. We use for that APE QUADRIX machines.

The creation operator for a monopole is constructed in analogy with the $U(1)$ operator as follows: the Π^{0i} plaquettes in the action at, say the time $n_0 = 0$,, when the monopole is created are modified as follows

$$\begin{aligned} \Pi^{0i}(\vec{n}, n_0) &= \mathrm{Tr}\left\{U_i(\vec{n},0)U_0(\vec{n}+\hat{\imath},0)U_i^\dagger(\vec{n},1)U_0^\dagger(\vec{n},0)\right\} \\ &\to \mathrm{Tr}\left\{U_i'(\vec{n},0)U_0(\vec{n}+\hat{\imath},0)U_i^\dagger(\vec{n},1)U_0^\dagger(\vec{n},0)\right\} \end{aligned} \tag{103}$$

In Eq.(103)

$$U_i'(\vec{n},0) = \mathrm{e}^{\mathrm{i}\Lambda(\vec{n})} \exp(\mathrm{i}\hat{\Phi}\cdot\vec{\sigma}\frac{b_i^{\perp}(\vec{n})}{2}) U_i(\vec{n},0) \exp(\mathrm{i}\hat{\Phi}\cdot\vec{\sigma}\frac{b_i^{\perp}(\vec{n})}{2}) \mathrm{e}^{-\mathrm{i}\Lambda(\vec{n}+\hat{i})} \tag{104}$$

We call $S + \Delta S$ the resulting action. Then

$$\langle\mu\rangle = \frac{Z[S+\Delta S]}{Z[S]} \tag{105}$$

The vector potential describing the field of the monopole b_i has been split in a transverse part $b_i^{\perp}$ with $\partial_i b_i = 0$ and a gauge part $\partial_i \Lambda$. The gauge dependence in Eq.(105) can be reabsorbed in a redefinition of the temporal links, which leaves the measure invariant.

A redefinition of U_i to $\tilde{U}_i = \mathrm{e}^{\mathrm{i}\sigma_3 \frac{b_i^{\perp}(n)}{2}} U_i \mathrm{e}^{\mathrm{i}\sigma_3 \frac{b_i^{\perp}(n)}{2}}$ in the abelian projected gauge can be reabsorbed in a change of variables which leaves the measure invariant. The space plaquettes however acquire in each link factors

$$\mathrm{e}^{\mathrm{i}\vec{\sigma}\cdot\hat{\Phi}\frac{b_i^{\perp}(n)}{2}} \tilde{U}_i \mathrm{e}^{\mathrm{i}\vec{\sigma}\cdot\hat{\Phi}\frac{b_i^{\perp}(n)}{2}} \tag{106}$$

In the abelian projected gauge the generic $\tilde{U}_i$ can be written as

$$\tilde{U}_i = \mathrm{e}^{\mathrm{i}\sigma_3\alpha_i}\mathrm{e}^{\mathrm{i}\sigma_2\gamma_i}\mathrm{e}^{\mathrm{i}\sigma_3\beta_i} = \mathrm{e}^{\mathrm{i}\sigma_3\alpha_i}\mathrm{e}^{\mathrm{i}\sigma_2\gamma_i}\mathrm{e}^{-\mathrm{i}\sigma_3\alpha_i}\mathrm{e}^{\mathrm{i}\sigma_3(\alpha_i+\beta_i)} = V\mathrm{e}^{\mathrm{i}\sigma_3\theta_i} \tag{107}$$

with $\exp(\mathrm{i}\sigma_3\theta_i)$ the abelian link and $\theta_i = \alpha_i + \beta_i$. The transformation Eq.(104) adds b_i to θ_i or adds the magnetic field of the monopole to the abelian magnetic field

$$\Delta_i\theta_j - \Delta_j\theta_i \to \Delta_i\theta_j - \Delta_j\theta_i + \Delta_i b_j - \Delta_j b_i \tag{108}$$

An abelian projected monopole has been created. $\langle\mu\rangle$, or better $\rho = d(\ln\langle\mu\rangle)/d\beta$, is measured across the deconfining phase transition. Typical behaviours are shown in fig.9 and fig. 10 for the monopole defined by the Polyakov loop.

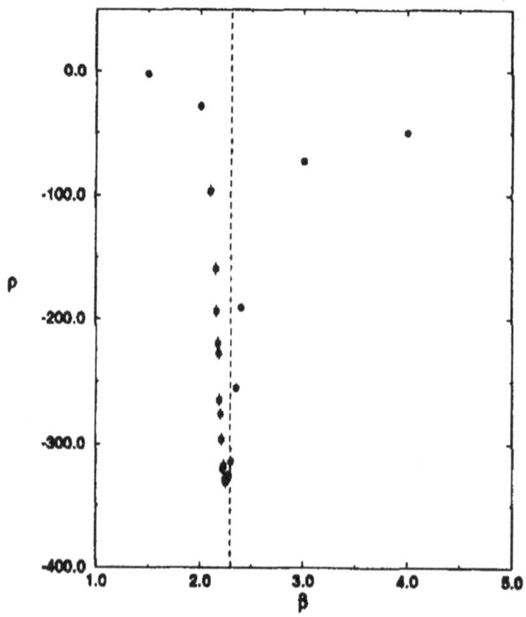

Figure 9. ρ v.s. β $SU(2)$ gauge theory. (Lattice $12^3 \times 4$.)

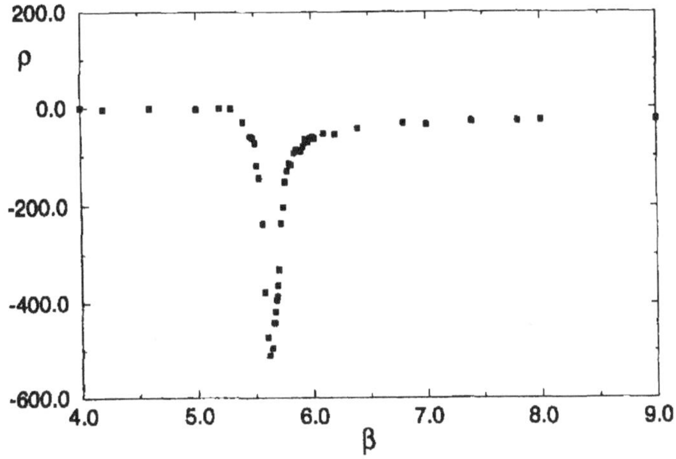

Figure 10. ρ v.s. β $SU(3)$ gauge theory. (Lattice $12^3 \times 4$.)

A careful analysis of the thermodynamical limit produces evidence of dual superconductivity below the deconfining temperature T_c for different choices of $\hat{\Phi}$: Polyakov line, $F_{\mu\nu}$, maximal abelian.

We need more work and more statistics to determine the critical indices and the type of superconductor.

DISCUSSION

Most of the success of the dual superconductivity mechanism for confinement is presently based on the abelian dominance and on the monopole dominance, numerically observed in the maximal abelian projection[34,35,36] In $SU(2)$ gauge theory after maximal abelian projection, which amounts to maximizing the quantity

$$A = \sum_{n,\mu} \mathrm{Tr}\left\{\sigma_3 U_\mu(n) \sigma_3 U_\mu^\dagger(n)\right\}$$

with respect to gauge transformations, all the links are on the average parallel to the 3 axis within 80 – 90%. The string tension computed from the abelian part of the Wilson loops accounts for 80 – 90% of the full string tension. Similar behaviour is found for many other quantities. This is called abelian dominance. In addition, the abelian plaquettes can be split as in Eq.(29) in a monopole part $n_{\mu\nu}$ and the residual angle $\tilde{\theta}_{\mu\nu}$, usually called the Coulomb field. The likewise empirical observation is that the contribution of the Coulomb part to the abelian quantities is a small factor, so that for example the string tension is dominated in fact by the contribution of the abelian monopoles (monopole dominance).

Apparently such dominance is not observed in other abelian projections, except maybe in the Polyakov line projection where, if the string tension is measured from the correlation between Polyakov lines, it is 100% abelian dominated by construction.

The more or less explicitly expressed idea is then that some abelian projection (the maximal abelian) is better than others and is "the abelian projection", identifying the degrees of freedom relevant for confinement. On the one hand the fact that in this projection links are in the abelian direction at 80 – 90% implies dominance as a kinematical fact: on the other hand maybe it is nontrivial that such a gauge exists.

The attitude presented in this paper is somewhat complementary: dual superconductivity is related to symmetry, and the way to detect it is to look for symmetry. Prom this point of view, independent of possible abelian dominances, what we find is more similar to the idea of 't Hooft that all abelian projections are physically equivalent. We find condensation of max abelian, Polyakov line, field strength monopoles.

There are a few aspects of the mechanism which need further understanding.

1) Any abelian projection implies that the gluons corresponding to the residual $U(1)$'s have no electric charge with respect to them, and hence are not confined. This contrasts with the observation about the string tension in the adjoint representation reported in the introduction.

2) If the mechanism of confinement were superconductivity produced by condensation of the monopoles defined by some abelian projection then the electromagnetic field in the flux tubes joining $q - \bar{q}$ pairs should belong to the projected $U(1)$. Lattice exploration show that it is isotropically distributed in color space[6, 37].

3) There are infinitely many abelian projections which can be obtained from each other by continuous transformations, e.g. by shifting the zero of $\vec{\Phi}(x)$. If one of them were privileged, it is hard to understand how others, which differ by small continuous changes, could be so different.

Most probably a more complicated and new mechanism is at work, a kind of non-abelian dual superconductivity, which manifests itself as abelian superconductivity in different abelian projected gauges. Our aim is to understand this.

Acknowledgements

It is a special pleasure to thank all the collaborators who contributed to this research, in particular Giampiero Paffuti, Manu Mathur and Luigi Del Debbio.

REFERENCES

1. G. 't Hooft, in "High Energy Physics", EPS International Conference, Palermo 1975, ed. A. Zichichi.
2. S. Mandelstam, *Phys. Rep.* 23C:245 (1976).
3. A.B. Abrikosov, *JETP* 5:1174 (1957).
4. M. Creutz, *Phys. Rev.* D21:2308 (1980).
5. R.W. Haymaker, J. Wosiek, *Phys. Rev.* D36:3297 (1987).
6. A. Di Giacomo, M. Maggiore and Š.Olejník: *Phys. Lett.* B236:199 (1990); *Nucl. Phys.* B347:441 (1990)
7. M.Caselle, R. Fiore, F. Gliozzi, M. Hasenbush, P. Provero, *Nucl. Phys.* B486:245 (1997).
8. J. Ambjorn, P. Olesen, C.Peterson, *Nucl.Phys.* B240:189 (1984).
9. S. Weinberg, *Progr. of Theor. Phys. Suppl.* 86:43 (1986).
10. P.A.M. Dirac: *Proc. Roy. Soc.* (London), Ser. A, 133:60 (1931).
11. L.P. Kadanoff, H. Ceva, *Phys. Rev.* B3:3918 (1971).
12. N. Sieberg, E. Witten: *Nucl. Phys.* B341:484 (1994).
13. A. Di Giacomo, G. Paffuti, *A disorder parameter for dual superconductivity in gauge theories,* hep-lat/9707003, to appear in Phys. Rev. D.
14. R.J. Wensley, J.D. Stack *Phys. Rev. Lett.* 63:1764 (1989).
15. V. Singh, R.W. Haymaker, D.A. Brown, *Phys. Rev.* D47:1715 (1993).
16. J. Fröhlich, P.A. Marchetti, *Commun. Math. Phys.* 112:343 (1987).
17. V. Cirigliano, G.Paffuti, *Magnetic Monopoles in U(1) lattice gauge theory,* hep-lat/9707219.
18. T. De Grand, D. Toussaint, *Phys. Rev.* D22:2478 (1980).
19. B.C. Marino, B. Schror, J.A. Swieca, *Nucl. Phys.* B200:473 (1982).

20. L. Del Debbio, A. Di Giacomo, G. Paffuti, *Phys. Lett.* B349:513 (1995).
21. G. Di Cecio, A. Di Giacomo, G. Paffuti, M. Trigiante, *Nucl. Phys.* B489:739 (1997).
22. A.P. Gottlob, M. Hasenbush, Cern preprint, CERN-TH 6885-93.
23. G. 't Hooft, *Nucl Phys.* B190:455 (1981).
24. S. Coleman, *Erice Lectures 1981,* Plenum Press, Ed. A. Zichichi.
25. P. Goddard, J. Nuyts, D. Olive, *Nucl. Phys.* B125:1 (1977).
26. A. Di Giacomo, M. Mathur, *Phys. Lett.* B400:129 (1997).
27. H. Georgi, S. Glashow, *Phys. Rev. Lett.* 28:1494 (1972).
28. G. 't Hooft, *Nucl Phys.* B79:276 (1974).
29. A.M. Polyakov, *JEPT Lett.* 20:894 (1974).
30. A. Di Giacomo, D. Martelli, G. Paffuti, *in preparation.*
31. C. Helm, W. Jank, *Phys. Rev.* B48:936 (1993).
32. L. Del Debbio, A. Di Giacomo, G. Paffuti and P. Pieri, *Phys. Lett.* B355:255 (1995).
33. A. Di Giacomo, B. Lucini, L. Montesi, G. Paffuti, hep-lat/9709005.
34. T. Suzuki, *Nucl Phys. (Proc. Suppl.)* B30:176 (1993).
35. A. Di Giacomo, *Nucl Phys. (Proc. Suppl)* B47:136 (1996).
36. M.I. Polikarpov, *Nucl Phys.* (Proc. Suppl.) B53:134 (1997).
37. J. Grensite, J. Winchester, *Phys. Rev.* D40:4167 (1989).

DUAL LATTICE BLOCKSPIN TRANSFORMATION AND PERFECT MONOPOLE ACTION FOR SU(2) GAUGE THEORY

Tsuneo Suzuki[1*], Maxim N. Chernodub[2], Seikou Kato[1], Shun-Ichi Kitahara[3], Naoki Nakamura[1], Mikhail I. Polikarpov[2]

[1]Department of Physics, Kanazawa University, Kanazawa 920-11, Japan
[2]Institute of Theoretical and Experimental Physics, Moscow 117259 Russia
[3]Jumonji University, Niiza, Saitama 352, Japan

INTRODUCTION

It is a very important problem to understand low-energy hadron physics starting directly from quantum chromodynamics (QCD). Numerical Monte-Carlo simulations of lattice QCD have been done extensively. To obtain the continuum limit reliably, we have to consider much larger lattices than presently available ones.

Recently, considerable efforts have been made to improve a lattice action in various ways in order to get more reliable data on available lattices[1,2]. Theoretically the best is a perfect action which could be obtained as the renormalized trajectory after renormalization transformations like a block-spin one. However, a block-spin transformation using nonabelian link variables can not be carried out practically in QCD unless we are restricted to too simple a truncated form of lattice action[3].

The fixed point action (the classical perfect action) which is fixed analytically is interesting[4]. It is a perfect action in the ultraviolet limit. However, it is not clear how far the fixed point action is different from the perfect action in the interesting infrared region.

Is it possible to fix a perfect lattice action which is valid in the infrared region? Note that there are various ways of renormalization transformations. It seems important to pay attention to a dynamical variable which plays a dominant role in the infrared region. The usual nonabelian link variables do not look suitable to describe long-range behaviors of QCD.

In this respect, the old conjecture which was proposed by 't Hooft is very interesting[5]. It says that QCD is reduced to abelian $U(1) \times U(1)$ theory with electric and magnetic charges after performing a partial gauge fixing called abelian projection. Quark confinement could be understood by the dual Meissner effect due to condensation of the magnetic monopoles.

*Seminar presented by Tsuneo Suzuki

Actually there have been many numerical data supporting the above conjecture when we perform the abelian projection in the maximally abelian (MA) gauge, using so-called abelian and monopole dominance[6,7,8].

However, why the MA gauge is so beautiful is not known yet, these facts indicate that the abelian (monopole) component after abelian projection is dominant in the infrared QCD. It is interesting to perform a block-spin transformation paying attention to these abelian components.

The preliminary studies have been done by Shiba and Suzuki[9,10]. They get a $U(1)$ effective action in terms of monopole currents in $T = 0$ $SU(2)$ gauge theory after performing a dual transformation numerically. The effective action is fixed also for extended monopoles[11]. Considering an extended monopole corresponds to making a block spin transformation on the dual lattice. They restricted the forms of monopole current interactions only to quadratic ones. This action seems to show scaling behavior, that is, it depends only on a physical scale $b = na(\beta)$.

However there are some insufficiencies in their works. The biggest problem is that the obtained action violates discrete rotational invariance. Since the renormalized trajectory leads us to the continuum theory, such a violation of the lattice rotational invariance is unnatural. This suggests that the restriction of the action to quadratic forms alone was too naive.

We report here on our recent results in which a discrete rotational invariant action is successfully obtained analyzing longer-range region on larger lattices.

STEPS OF CALCULATIONS

The process of our study is the following (see Fig.1):

1. We first generate the thermalized nonabelian link fields $\{U(s,\mu)\}$ using a simple Wilson action for pure $SU(2)$ gauge theory. The lattices considered are 16^4, 24^4, 32^4 and 48^4 for $\beta = 2.0 \sim 2.8$

2. Next we perform abelian projection in the MA gauge and then separate abelian link variables $\{u_\mu(s)\}$ from gauge-fixed $SU(2)$ link fields $\{U'(s,\mu)\}$. The MA gauge fixing condition is to maximize the following operator:

$$R = Tr \sum_{s,\mu} \left[U(s,\mu)\sigma_3 U^\dagger(s,\mu)\sigma_3 \right]. \tag{1}$$

Then

$$X(s) = \sum_{\mu} \left[U(s,\mu)\sigma_3 U^\dagger(s,\mu) + U^\dagger(s-\hat{\mu},\mu)\sigma_3 U(s-\hat{\mu},\mu) \right] \tag{2}$$

is diagonalized. After the gauge-fixing, the $SU(2)$ link fields are expressed as

$$U'(s,\mu) = \begin{pmatrix} \sqrt{1-|c_\mu(s)|^2} & -c_\mu^*(s) \\ c_\mu(s) & \sqrt{1-|c_\mu(s)|^2} \end{pmatrix} \begin{pmatrix} u_\mu(s) & 0 \\ 0 & u_\mu^*(s) \end{pmatrix}, \tag{3}$$

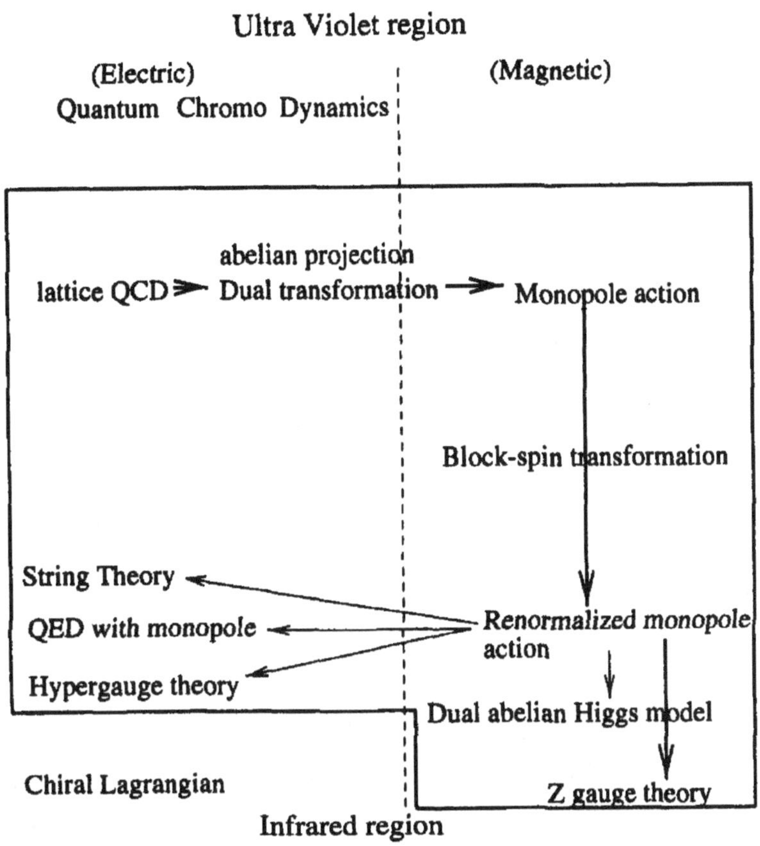

Figure 1. The flowchart of the study. The electric/magnetic means the original/dual lattice. In this report, only the dual abelian Higgs model and string model are discussed explicitly among various equivalent effective theories on the lattice.

where

$$u_\mu(s) = e^{i\theta_\mu(s)}, \quad (-\pi \leq \theta_\mu(s) < \pi). \tag{4}$$

Here $u_\mu(s)$ transforms as the gauge field of the residual $U(1)$ symmetry and $c_\mu(s)$ as the charged matter field.

3. The abelian dominance suggests that operators $\mathcal{O}(u_\mu(s))$ alone are enough to describe the infrared behavior of QCD. Then a sufficiently local abelian effective action is expected to exist:

$$Z = \int Du[\int Dce^{-S(u,c)}\delta(X^{\pm})\Delta_F(U)] \tag{5}$$

$$= \int Due^{-S_{eff}(u)}, \tag{6}$$

where $X^{\pm}$ is the off-diagonal part of (2) and $\Delta_F(U)$ is the Faddeev-Popov determinant.

4. Monopole currents can be defined from the abelian plaquette variables $\theta_{\mu\nu}(s)$ following DeGrand and Toussaint[12]. The abelian plaquette variables are

$$\theta_{\mu\nu}(s) \equiv \theta_\mu(s) + \theta_\nu(s+\hat{\mu}) - \theta_\mu(s+\hat{\nu}) - \theta_\nu(s), \quad (-4\pi < \theta_{\mu\nu}(s) < 4\pi). \quad (7)$$

It is decomposed into two terms:

$$\theta_{\mu\nu}(s) \equiv \bar{\theta}_{\mu\nu}(s) + 2\pi n_{\mu\nu}(s), \qquad (-\pi \leq \bar{\theta}_{\mu\nu}(s) < \pi). \qquad (8)$$

Here, $\bar{\theta}_{\mu\nu}(s)$ is interpreted as the electro-magnetic flux through the plaquette and the integer $n_{\mu\nu}(s)$ corresponds to the number of Dirac strings penetrating the plaquette. One can define the quantized conserved monopole currents $k_\mu(s)$:

$$k_\mu(s) = \frac{1}{2}\epsilon_{\mu\nu\rho\sigma}\partial_\nu n_{\rho\sigma}(s+\hat{\mu}). \qquad (9)$$

5. The monopole dominance suggests that operators $\mathcal{O}(k_\mu(s))$ alone describe the infrared behavior of QCD. An effective monopole action is expected to exist:

$$\begin{aligned} Z &= (\prod\sum)\int Du\delta(k,u)e^{-S_{eff}(u)} \\ &= (\prod_{s,\mu}\sum_{k_\mu(s)=-\infty}^{\infty})(\prod_s \delta_{\partial'_\mu k_\mu(s),0})e^{-S[k]}, \end{aligned} \qquad (10)$$

where $\delta(k,u)$ stands for the condition (9). Since the monopole currents are attached to the dual lattice, Eq.(10) is a kind of dual transformation.

6. Now we perform a block-spin transformation in terms of the monopole currents on the dual lattice to investigate the perfect action in the infrared region. We study $n = 1,2,3,4,6,8,12$ extended monopole currents[11] as an n blocked operator:

$$k_\mu^{(n)}(s) = \sum_{i,j,l=0}^{n-1} k_\mu(ns + (n-1)\hat{\mu} + i\hat{\nu} + j\hat{\rho} + l\hat{\sigma}). \qquad (11)$$

Actually the definition includes the usual block-spin transformation. For example,

$$k_\mu^{(2)}(s) = \sum_{i,j,l=0}^{1} k_\mu(2s + \hat{\mu} + i\hat{\nu} + j\hat{\rho} + l\hat{\sigma}), \qquad (12)$$

$$\begin{aligned} k_\mu^{(4)}(s) &= \sum_{i,j,l=0}^{3} k_\mu(4s + 3\hat{\mu} + i\hat{\nu} + j\hat{\rho} + l\hat{\sigma}) \\ &= \sum_{i,j,l=0}^{1} k_\mu^{(2)}(2s + \hat{\mu} + i\hat{\nu} + j\hat{\rho} + l\hat{\sigma}) \end{aligned} \qquad (13)$$

The renormalized lattice spacing is $b = na(\beta)$ and the continuum limit is taken as the limit $n \to \infty$ for a fixed physical length b. The renormalization flow is determined by the relation

$$e^{-S^{(n)}[k^{(n)}]} = (\prod_{s,\mu}\sum_{k_\mu(s)=-\infty}^{\infty})(\prod_s \delta_{\partial'_\mu k_\mu(s),0})\delta(k^{(n)},k)e^{-S[k]}, \qquad (14)$$

where $\delta(k^{(n)},k)$ stands for the definition (11).

Table 1. The physical scale $a(\beta)$

β	$a(\beta)\sqrt{\kappa}$	3σ
2.0000	.69516	.01504
2.1000	.60641	.01060
2.2000	.48037	.00520
2.3000	.35887	.00361
2.4000	.26784	.00828
2.5000	.18551	.00315
2.6000	.13455	.00386
2.7000	.10155	.00212
2.8000	.07771	.00167

7. Since we study small β regions, we have to fix the physical scale $a(\beta)$ without use of the theoretical asymptotic beta function. In pure $SU(2)$ gauge theory, we have fixed it from evaluating the string tension κ_L from the monopole part of abelian Wilson loops. The lattice spacing $a(\beta)$ is given by the relation $a(\beta) = \sqrt{\kappa/\kappa_L}$, where κ is the physical string tension. Since the monopole contribution to the Wilson loop is known to have only linear and constant terms (no Coulomb term)[13], it is the best operator to fix the string tension for such small β region where Wilson loops are small. To eliminate the finite-size effects, we have used the data of $n = 1$ monopoles on 16^4 lattice for $\beta = 2.0 - 2.3$, $n = 2$ monopoles on 24^4 lattice for $\beta = 2.4 - 2.6$ and $n = 4$ monopoles on 48^4 lattice for $\beta = 2.7 - 2.8$. Although the statistical errors of the original data are small, we adopt 3σ as the errors considering the differences of the string tensions determined by the different extended monopoles. There may be other systematic errors coming from the possible difference between string tensions from nonabelian and abelian Wilson loops. The physical scale in units of $1/\sqrt{\kappa}$ is given in Table 1.

INVERSE MONTE-CARLO METHOD

The renormalization flows are obtained numerically from the ensemble of thermalized (extended) monopole currents $\{k_\mu^{(n)}(s)\}$ with the aid of an inverse Monte-Carlo method first developed by Swendsen[14] and extended to closed monopole currents by Shiba and Suzuki[9, 10].

Consider the expectation value of some operator $\mathcal{O}_i[k]$ assuming the monopole action $S[k] = \sum_i g_i S_i[k]$:

$$\langle \mathcal{O}_i[k] \rangle = \frac{\prod_{s,\mu} \sum_{k_\mu(s)=-\infty}^{\infty} \mathcal{O}_i[k] \exp\left(-\sum_i g_i S_i[k]\right)}{\prod_{s,\mu} \sum_{k_\mu(s)=-\infty}^{\infty} \exp\left(-\sum_i g_i S_i[k]\right)}. \quad (15)$$

Define $\tilde{S}[k]$ a a part of $S[k]$ which contains currents along a special plaquette $(s', \hat{\mu}', \hat{\nu}')$, i.e., $\{k_{\mu'}(s'), k_{\nu'}(s' + \hat{\mu}'), k_{\mu'}(s' + \hat{\nu}'), k_{\nu'}(s')\}$. Using $\tilde{S}[k]$, Eq.(15) is rewritten in the following form:

$$\langle \mathcal{O}_i[k] \rangle = \frac{\prod_{s,\mu} \sum_{k_\mu(s)=-\infty}^{\infty} \overline{\mathcal{O}}_i[k] \exp\left(-\sum_i g_i S_i[k]\right)}{\prod_{s,\mu} \sum_{k_\mu(s)=-\infty}^{\infty} \exp\left(-\sum_i g_i S_i[k]\right)} \quad (16)$$

$$= \langle \overline{\mathcal{O}}_i[k] \rangle, \quad (17)$$

where

$$\overline{O}_i[k] = \frac{\sum_{M=-\infty}^{\infty} O_i[k'] \exp\left(-\sum_i g_i \tilde{S}_i[k']\right)}{\sum_{M=-\infty}^{\infty} \exp\left(-\sum_i g_i \tilde{S}_i[k']\right)}, \tag{18}$$

$$k'_\mu(s) \equiv k_\mu(s) + M(\delta_{s,s'}\delta_{\mu,\mu'} + \delta_{s,s'+\hat{\mu}'}\delta_{\mu,\nu'} - \delta_{s,s'+\hat{\nu}'}\delta_{\mu,\mu'} - \delta_{s,s'}\delta_{\mu,\nu'}). \tag{19}$$

The identity (17) can be used to fix the monopole action $S[k]$ iteratively. Define $\overline{O}_i[k,\tilde{g}]$ which is equal to Eq.(18) with trial coupling constants $\{\tilde{g}_i\}$ instead of real ones $\{g_i\}$. When g_i is not equal to $\tilde{g}_i$ for all i, we get the following expansion up to the first order in $g_i - \tilde{g}_i$:

$$\langle O_i - \overline{O}_i \rangle = \langle \overline{O_i S_j} - \overline{O}_i \overline{S}_j \rangle (g_j - \tilde{g}_j). \tag{20}$$

For details, see Ref.9.

Practically, we first need to restrict the type of interactions. It is natural that short-distant interactions and two-point interactions are more important. Here we adopt discrete rotational invariant terms composed of two-point self and Coulomb interactions and 4- and 6-point interactions as a dominant part. To test the validity of the dominant form, we add general two-point interactions up to lattice distance three. The contribution of the additional terms means insufficiencies of the dominant form of the action adopted.

$$S[k] = \frac{g_m^2}{2} \sum_{s,s'} \sum_{\mu=1}^{D} k_\mu(s) \Delta^{-1}_{s,s'} k_\mu(s') + p \sum_s \sum_{\mu=1}^{D} k_\mu^2(s) - q \sum_s \left(\sum_{\mu=-D}^{D} k_\mu^2(s) \right)^2$$
$$+ r \sum_s \left(3 \left(\sum_{\mu=-D}^{D} k_\mu^2(s) \right)^3 + \sum_{\mu,\nu,\rho=-D}^{D} k_\mu^2(s) k_\nu^2(s) k_\rho^2(s+\hat{\mu}) \right) + S_2[k]. \tag{21}$$

Here $\Delta^{-1}_{s,s'}$ is the lattice Coulomb propagator and $S_2[k] = \sum_i f_i S_2^{(i)}[k]$. $S_2^{(i)}[k]$'s are additional two-point interactions. Since the additional term $S_2[k]$ is found to be very small, we neglect it from now on. Other higher-order interactions are assumed to be small. For details, see Ref.15.

RESULTS

The results of our analyses are the following:

1. The inverse Monte-Carlo method works very well and the coupling constants are fixed beautifully in a discrete rotational invariant way, although we need to start with good initial conditions. Since the 4- and the 6-point interactions receive large contriburions from higher charged currents, the choice of the initial conditions is very important to get convergence. This is different from the old approach[9, 10] with 2-point interactions, where the convergence was achieved easily.

2. The first four terms, i.e., self + Coulomb + 4-point + 6-point interactions are found to be enough for the long-range region $b > 1/\sqrt{\kappa}$.

3. Both the limits $V \to \infty$ and $a \to 0$ must be taken to derive the continuum results from our finite-lattice analyses. To get the infinite volume limit, we have determined the actions from the different lattice sizes for each β and monopole size. Especially, the results for 4^3 monopoles show that the volume dependence is

weak (Fig.2). Hence we have used the χ^2 fit assuming the constant behaviors and have fixed the values of the infinite-volume limit for each monopole size. We also have tried the linear fit and the difference of the fits is included in the expected error values of the infinite-volume limit.

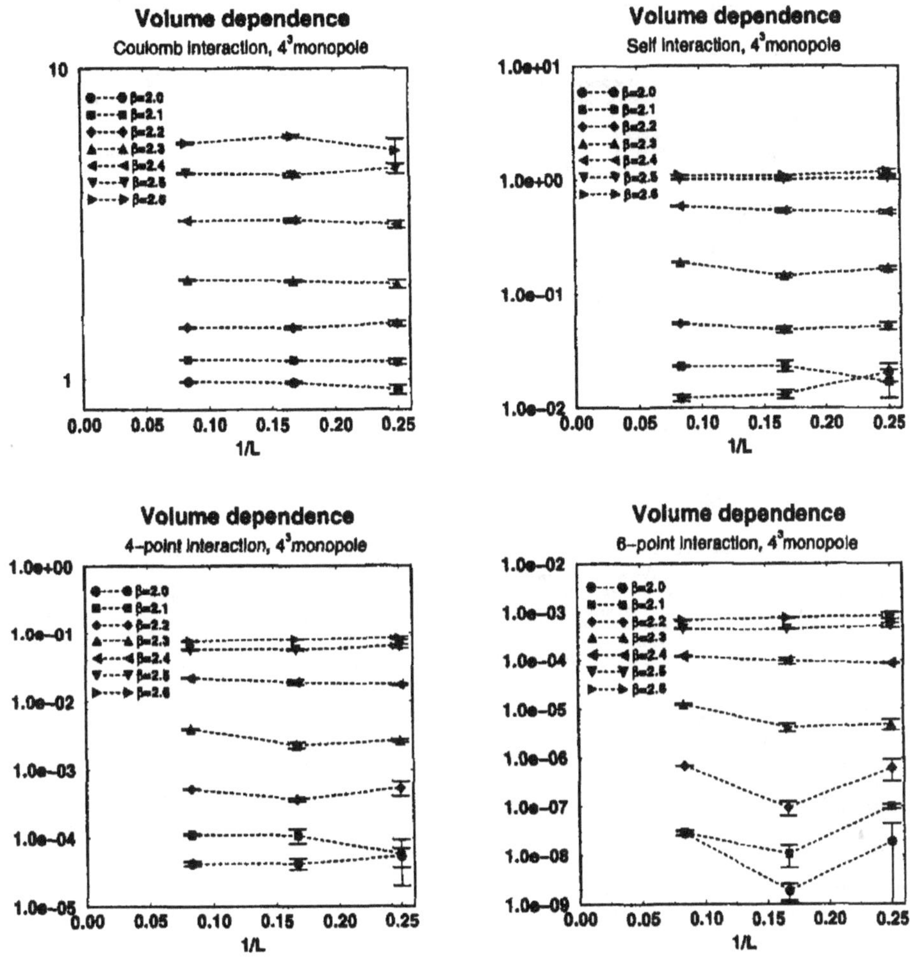

Figure 2. Volume dependence of four-dominant coupling constants.

4. The physical scale is fixed as shown in Table 1. The coupling constants are plotted versus $b = na(\beta)$ for each n extended monopole as shown in Fig.3. For small $b < 1/\sqrt{\kappa}$, all n data seem to be the same. When we go to higher b, the couplings of smaller n extended monopoles begin to deviate from the universal curve. All data with higher extended monopoles $(n \geq 4)$ seem to scale in the coupling constants except that of the Coulomb interaction.

5. To study the $n \to \infty$ limit for each b, we have tried to fix the b-dependence for each n data using the χ^2 fit. In this step, we adopt a simple power function of b.

6. The dominant four couplings are plotted versus $1/n$ for each fixed b in Fig.4. As discussed earlier, there seems no n dependence for the small b region. The

Coulomb coupling seems to have small $1/n$ dependence for the large values of b. Other three couplings show almost constant behavior for at least $n \geq 6$ even in the large b region. The χ^2 fit has been performed assuming constant or linear relations in each case and the continuum limit of these couplings have been deduced.

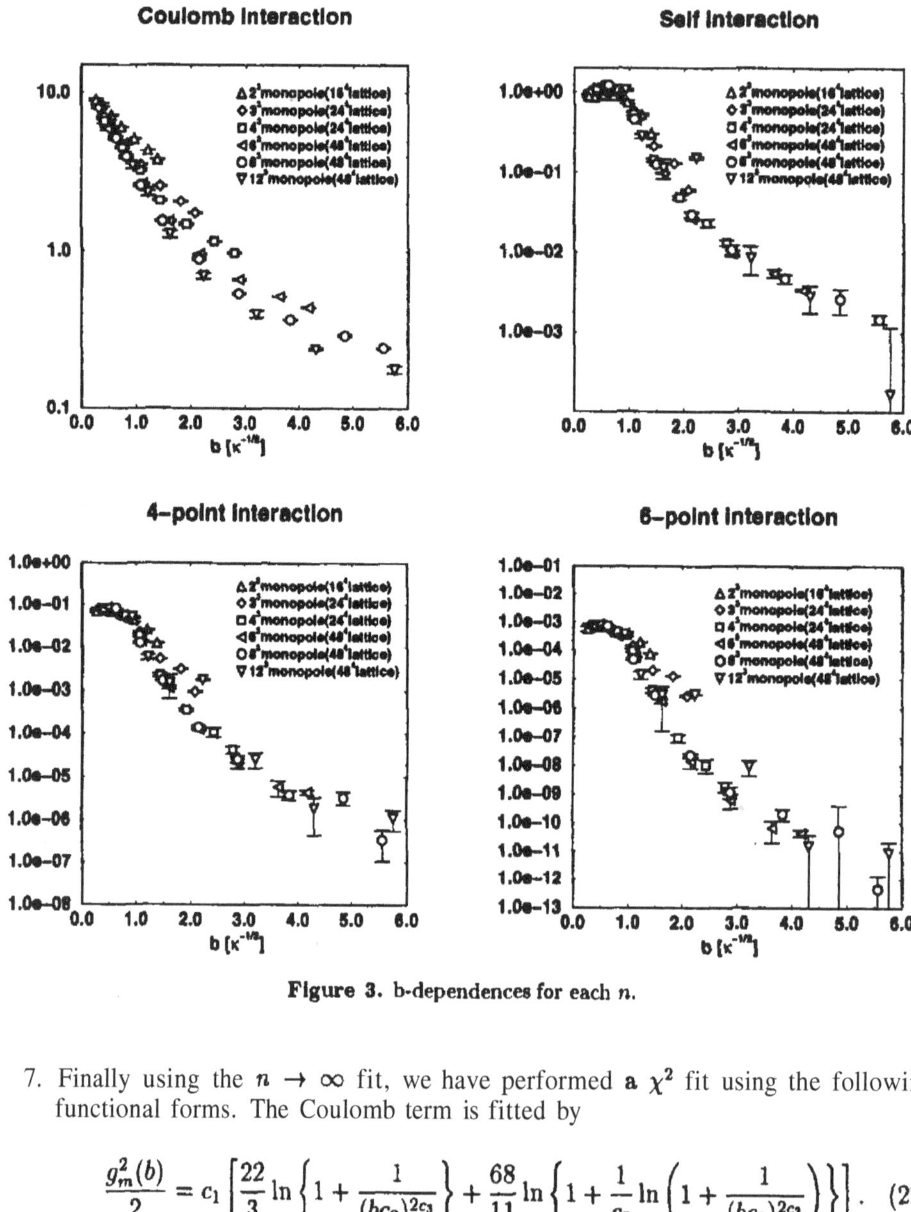

Figure 3. b-dependences for each n.

7. Finally using the $n \to \infty$ fit, we have performed a χ^2 fit using the following functional forms. The Coulomb term is fitted by

$$\frac{g_m^2(b)}{2} = c_1 \left[\frac{22}{3} \ln \left\{ 1 + \frac{1}{(bc_2)^{2c_3}} \right\} + \frac{68}{11} \ln \left\{ 1 + \frac{1}{c_3} \ln \left(1 + \frac{1}{(bc_2)^{2c_3}} \right) \right\} \right]. \quad (22)$$

This function has power behavior in the large b region and takes the form expected from the 2-loop perturbative β-function of pure $SU(2)$ gauge theory in the small b region. The coupling constants of the other three dominant terms p, q and r are fitted by a simple power function of b. Thus the almost perfect monopole action is obtained. The explicit results are shown in Fig.5.

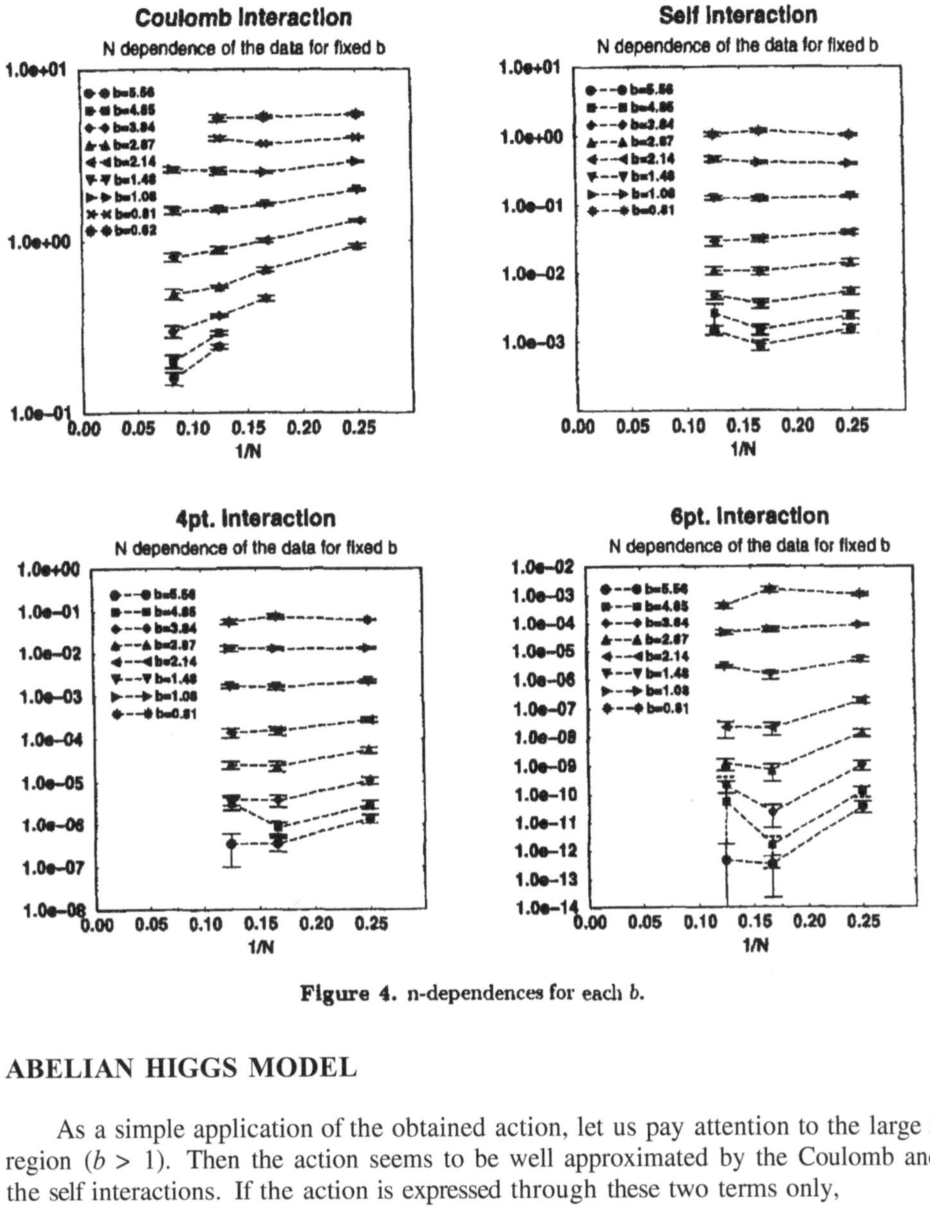

Figure 4. n-dependences for each b.

ABELIAN HIGGS MODEL

As a simple application of the obtained action, let us pay attention to the large b region ($b > 1$). Then the action seems to be well approximated by the Coulomb and the self interactions. If the action is expressed through these two terms only,

$$Z = \sum_{{}^*k\in Z,\delta {}^*k=0} \exp\left\{-\frac{g_m^2}{2}({}^*k,\Delta^{-1}{}^*k) + p\|{}^*k\|^2\right\}, \tag{23}$$

the partition function can be exactly transformed to other representations[16] (here and below we use the lattice differential form formalism).

Using an auxiliary field ${}^*C({}^*c_1)$, the partition function can be written as

$$Z \propto \int_{-\infty}^{\infty} D{}^*C({}^*c_1) \sum_{{}^*k({}^*c_1)\in Z,\delta {}^*k=0} \exp\left\{-\frac{1}{2g_m^2}\|d{}^*C\|^2 - i({}^*C,{}^*k) - p\|{}^*k\|^2\right\}, \tag{24}$$

where ${}^*c_i$ is the i-th dual cell on the lattice. Introducing the phase ${}^*\phi({}^*c_0)$, the current

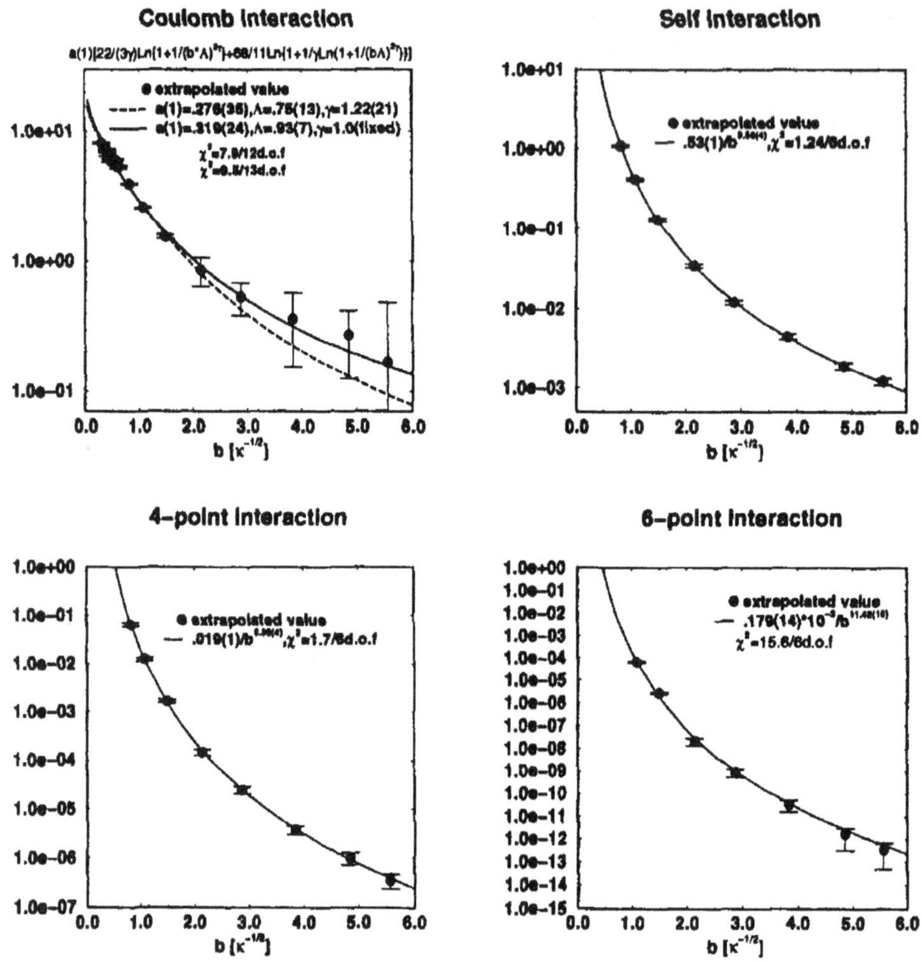

Figure 5. b-dependences of the perfect action. In the case of the Coulomb interactions, $(a(1), \Lambda, \gamma)$ are used instead of (c_1, c_2, c_3) in Eq.22.

conservation law can be rewritten in the following form:

$$\delta(\delta\,{}^*k) = const. \int_{-\pi}^{\pi} D\,{}^*\phi \exp[i({}^*\phi, \delta\,{}^*k)]. \tag{25}$$

Substituting Eq.(25) into Eq.(24), we get

$$Z \propto \int_{-\infty}^{\infty} D\,{}^*C({}^*c_1) \int_{-\pi}^{\pi} D\,{}^*\phi({}^*c_0) \sum_{{}^*k({}^*c_1)\in Z}$$
$$\exp\left\{-\frac{1}{2g_m^2}\|d\,{}^*C\|^2 + i(d\,{}^*\phi - {}^*C, {}^*k) - p\|\,{}^*k\|^2\right\}. \tag{26}$$

Using the Poisson summation formula

$$\int DF \sum_{l\in Z} \exp[2\pi i(F, l)] f(F) = \sum_{k\in Z} f(k), \tag{27}$$

we get the expression of the partition function:

$$Z \propto \int_{-\infty}^{\infty} D\,{}^*C({}^*c_1) \int_{-\infty}^{\infty} D\,{}^*F({}^*c_1) \int_{-\pi}^{\pi} D\,{}^*\phi({}^*c_0) \sum_{{}^*l({}^*c_1)\in Z}$$

$$\exp\left\{-\frac{1}{2g_m^2}\|d{}^*C\|^2 + i(d{}^*\phi + 2\pi{}^*l - {}^*C, {}^*F) - p\|{}^*F\|^2\right\}. \tag{28}$$

The Gaussian integral with respect to the auxiliary field ${}^*F({}^*c_1)$ leads to the abelian-Higgs model on the dual lattice (the radius of the Higgs field is fixed):

$$Z^{AH} = \int_{-\infty}^{\infty} D{}^*C({}^*c_1) \int_{-\pi}^{\pi} D{}^*\phi({}^*c_0) \sum_{{}^*l({}^*c_1)\in Z}$$
$$\exp\left\{-\frac{1}{2g_m^2}\|d{}^*C\|^2 - \frac{1}{4p}\|d{}^*\phi + 2\pi{}^*l - {}^*C\|^2\right\} \tag{29}$$

When we take into account the 4- and the 6-point interactions of the monopole current, we get the dual abelian-Higgs model with the unfixed radius of the Higgs field. Let us start from the partition function of the dual abelian-Higgs model:

$$\mathcal{Z}^{A.H.} = \int_{-\infty}^{+\infty} \mathcal{D}\theta \int_{-\infty}^{+\infty} \mathcal{D}\phi \exp\left\{-\mathcal{S}^{A.H.}\right\}, \tag{30}$$

$$\mathcal{S}^{A.H.}[\theta,\phi] = \sum_P \frac{\beta}{2}\theta_p^2 + \lambda\sum_n (\phi^*(n)\phi(n) - 1)^2 + \sum_n \phi^*(n)\phi(n) \tag{31}$$

$$- \gamma\sum_{n,\mu}(\phi^*(n)U_\mu(n)\phi(n+\mu) + h.c.), \tag{32}$$

where $U_\mu = \exp(i\theta_\mu)$ is the dual gauge field, θ_P is the field strength tensor and $\phi(n) = \rho_n \exp(i\varphi_n)$ is the complex Higgs field.

One can rewrite the above integral as the sum over the closed monopole currents $*k$ using the analogue of the BKT transformation[17]. The monopole action calculated in the saddle-point approximation up to $O(\lambda^{-2})$ terms has the form:

$$\mathcal{S}^{mon}[{}^*k] = \frac{1}{4\beta}\left({}^*k, \Delta^{-1}{}^*k\right) + \left\{\frac{1}{4\gamma} + \frac{1}{\lambda}\left(\frac{15}{32\gamma} - 1\right) + \frac{1}{\lambda^2}\left(\frac{865}{2^9\gamma} - \frac{439}{2^6} + 4\gamma\right)\right\}\|{}^*k\|^2$$
$$+ \left\{-\frac{1}{2^8\gamma^2\lambda} + \frac{1}{\lambda^2}\left(\frac{7}{2^9\gamma} - \frac{157}{2^{12}\gamma^2}\right)\right\}\sum_x\left(\sum_{\mu=-D}^{D} {}^*k_{x,\mu}^2\right)^2$$
$$- \frac{1}{2^{11}\gamma^2\lambda^2}\sum_x\sum_{\mu=1}^{D} {}^*k_{x,\mu}^4$$
$$+ \frac{1}{2^{13}\gamma^3\lambda^2}\sum_x\sum_{\mu,\nu,\delta=-D}^{D} {}^*k_{x,\mu}^2{}^*k_{x,\nu}^2\left(3{}^*k_{x,\delta}^2 + {}^*k_{x+\nu,\delta}^2\right) + O\left(\frac{1}{\lambda^3}\right). \tag{33}$$

When we consider the terms up to $O(\lambda^{-1})$ and determine the b dependence of the parameters β,γ,λ from the monopole action obtained numerically, we can estimate the type of the superconductivity of the QCD vacuum from the Ginzburg-Landau parameter κ_{GL} defined by

$$2\kappa_{GL}^2 = \frac{m_H^2}{m_A^2} = \frac{4\beta\lambda}{\gamma^2}. \tag{34}$$

It is found that the QCD vacuum is a type-II superconductor for $b \geq 1/\sqrt{\kappa}$ (see Fig.6 where the fit with $c_3 = 1$ is used in the Coulomb coupling).

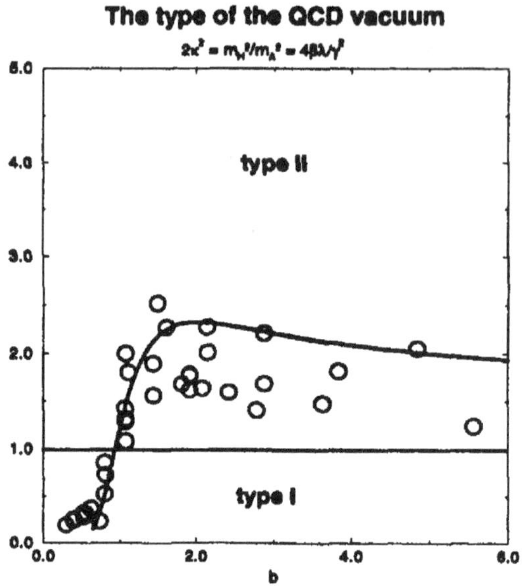

Figure 6. The b dependence of the Ginzburg-Landau parameter for the QCD vacuum. The straight line denotes the critical line between type-I and type-II superconductivity.

STRING TENSION

When the 4- and the 6-point interactions are neglected, we can also get the representation of the string model[16]. Let us perform the BKT transformation for ${}^*l({}^*c_1)$in the abelian-Higgs model ${}^*l = {}^*s + d\,{}^*r$, $d\,{}^*s = {}^*\sigma$, ${}^*\phi_{nc} = {}^*\phi + 2\pi(\delta\Delta^{-1}\,{}^*s + {}^*r)$, we get the partition function

$$Z \propto \int_{-\infty}^{\infty} D\,{}^*C({}^*c_1) \int_{-\infty}^{\infty} D\,{}^*\phi_{nc}({}^*c_0) \sum_{{}^*\sigma({}^*c_2)\in Z, d\,{}^*\sigma=0}$$

$$\exp\left\{-\frac{1}{2g_m^2}\|d\,{}^*C\|^2 - \frac{1}{4p}\|d\,{}^*\phi_{nc} + 2\pi\delta\Delta^{-1}\,{}^*\sigma - {}^*C\|^2\right\} \tag{35}$$

Choosing the gauge ${}^*\phi_{nc} = 0$ and integrating over ${}^*C({}^*c_1)$, we get the string model on the lattice:

$$Z^{string} = \sum_{\sigma(c_2)\in Z, \delta\sigma=0} \exp\left\{-\frac{\pi^2}{p}\left(\sigma, \left(\Delta + \frac{g_m^2}{2p}\right)^{-1}\sigma\right)\right\}. \tag{36}$$

The condition $\delta\sigma = 0$ means that the world sheets form closed surfaces.

Using such representations, one can evaluate the string tension analytically. The Wilson loop is estimated by the following action of the hadronic string model on the lattice.

$$\langle W\rangle_m = \frac{1}{Z} \sum_{\sigma(c_2)\in Z, \delta\sigma=J} \exp\left[-\frac{\pi^2}{p(b)}\left(\sigma, \left(\Delta + M^2(b)\right)^{-1}\sigma\right)\right].$$

where

$$M^2 = \frac{g_m^2}{2p}. \tag{37}$$

Here we have written only the contribution from monopoles.

Our numerical results show that M^2 is large in the large b region. In this case, a classical picture may be reliable in the string model and the string tension is approximated by the self coupling of the world sheets:

$$\kappa_L = \frac{\pi^2}{p(b)(\Delta + M^2(b))}(0). \tag{38}$$

The physical string tension is obtained as follows:

$$\kappa = \frac{\kappa_L}{b^2} = \frac{\pi^2}{b^2 p(b)(\Delta + M^2(b))}(0), \tag{39}$$

where $p(b) = 0.53(1)/b^{3.56(4)}$ and $g_m^2(b)/2$ is given by Eq.(22).

Since M^2 is large, it can be evaluated by the $1/M^2$ expansion. The results shown in Fig.7 (the fit with $c_3 = 1$ is used) show that for $b > 1/\sqrt{\kappa}$, the obtained action reproduces the experimental string tension rather well. It is very interesting that we can almost reproduce the physical string tension analytically with the action obtained. This means that the monopole action is very near to the perfect action.

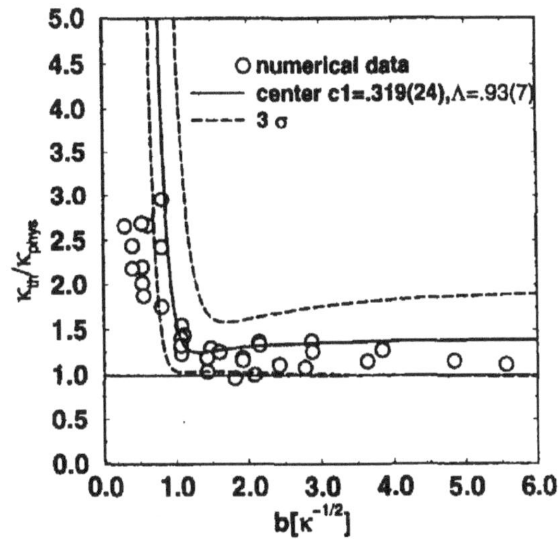

Figure 7. Theoretical string tension.

Acknowledgements

M.I.P. and M.N.Ch. feel much obliged for the kind reception given to them by the staff of Department of Physics of Kanazawa University. This work is supported by the Supercomputer Project (No.97-17) of High Energy Accelerator Research Organization (KEK) and the Supercomputer Project of the Institute of Physical and Chemical Research (RIKEN). T.S. is financially supported by JSPS Grant-in Aid for Exploratory Research (No.09874060). M.I.P. and M.N.Ch. were supported by the JSPS Program on Japan – FSU scientists collaboration, by the grants INTAS-94-0840, INTAS-94-2851, INTAS-RFBR-95-0681 and RFBR-96-02-17230a.

REFERENCES

1. P. Weisz, Contribution to this Proceedings and references therein.
2. P. Lepage, Contribution to this Proceedings and references therein.
3. Ph. de Forcrand *et al.*, *Nucl. Phys.* B(Proc. Suppl.) 53:938 (1997).
4. P. Hasenfratz and F. Niedermayer, *Nucl. Phys.*B414:785 (1994); P. Hasenfratz, Contribution to this Proceedings and references therein.
5. G. 't Hooft, *Nucl. Phys.* B190:455 (1981).
6. A.S. Kronfeld *et al.*, *Phys. Lett.* B 198:516 (1987); A.S. Kronfeld *et al., Nucl. Phys.* B 293:461 (1987).
7. T. Suzuki and I. Yotsuyanagi, *Phys. Rev.* D42:4257 (1990).
8. S. Kitahara *et al.*, *Prog. Theor. Phys.* 93:1 (1995) and references therein.
9. H. Shiba and T. Suzuki, *Phys. Lett.* B 343:315 (1995).
10. H. Shiba and T. Suzuki, *Phys. Lett.* B 351:519 (1995) and references therein.
11. T.L. Ivanenko *et al, Phys. Lett.* B 252:631 (1990).
12. T.A. DeGrand and D. Toussaint, *Phys. Rev.* D22:2478 (1980).
13. H. Shiba and T. Suzuki, *Phys. Lett.* B 333:461 (1994).
14. R.H. Swendsen, *Phys. Rev. Lett.* 52:1165 (1984); *Phys. Rev.* D30:3866,3875 (1984).
15. S. Kato et al, Kanazawa Univ. Preprint KANAZAWA 97-17.
16. M.I. Polikarpov *et al.*, *Phys. Lett.* B309:133 (1993).
17. T. Banks, R. Myerson and J. Kogut, *Nucl. Phys.* B129:493 (1977).

INTRODUCTION TO RIGID SUPERSYMMETRIC THEORIES

P.C. West

Department of Mathematics
King's College, London, UK

PREFACE

In these lectures we discuss the supersymmetry algebra and its irreducible representations. We construct the theories of extended rigid supersymmetry and give their superspace formulations. The perturbative quantum properties of the extended supersymmetric theories are derived, including the superconformal invariance of a large class of these theories. The superconformal transformations in four dimensional superspace are derived and encoded into one superconformal Killing superfield. It is shown that the anomalous dimensions of chiral operators in a superconformal quantum field are related to their R weight.

Some of this material follows the book[1] by the author. Certain chapters of this book are reproduced here, however, in other sections the reader is referred to the relevant parts of Ref.[1].

In this review the section on superconformal theories and two subsections on flat directions and non-holomorphicity are new material.

The aim of the lectures is to provide the reader with the material required to understand more recent developments in the non-perturbative properties of quantum extended supersymmetric theories.

THE SUPERSYMMETRY ALGEBRA

This section is identical to chapter 2 of Ref.[1]. The equation numbers are kept the same as in this book. I thank World Scientific Publishing for their permission to reproduce this material.

In the 1960's, with the growing awareness of the significance of internal symmetries such as $SU(2)$ and larger groups, physicists attempted to find a symmetry which would combine in a non-trivial way the space-time Poincaré group with an internal symmetry group. After much effort it was shown that such an attempt was impossible within the context of a Lie group. Coleman and Mandula[2] showed on very general assumptions that any Lie group which contained the Poincaré group P, whose generators $\boldsymbol{P}_a$ and

J_{ab} satisfy the relations

$$\begin{aligned}
[P_a, P_b] &= 0 \\
[P_a, J_{bc}] &= (\eta_{ab}P_c - \eta_{ac}P_b) \\
[J_{ab}, J_{cd}] &= -(\eta_{ac}J_{bd} + \eta_{bd}J_{ac} - \eta_{ad}J_{bc} - \eta_{bc}J_{ad})
\end{aligned} \tag{2.1}$$

and an internal symmetry group G with generators T_s such that

$$[T_r, T_s] = f_{rst}T_t \tag{2.2}$$

must be a direct product of P and G; or in other words

$$[P_a, T_s] = 0 = [J_{ab}, T_s] \tag{2.3}$$

They also showed that G must be of the form of a semisimple group with additional $U(1)$ groups.

It is worthwhile to make some remarks concerning the status of this no-go theorem. Clearly there are Lie groups that contain the Poincaré group and internal symmetry groups in a non-trivial manner; however, the theorem states that these groups lead to trivial physics. Consider, for example, two-body scattering; once we have imposed conservation of angular momentum and momentum the scattering angle is the only unknown quantity. If there were a Lie group that had a non-trivial mixing with the Poincaré group then there would be further gnerators associated with space-time. The resulting conservation laws will further constrain, for example, two-body scattering, and so the scattering angle can only take on discrete values. However, the scattering process is expected to be analytic in the scattering angle, θ, and hence we must conclude that the process does not depend on θ at all.

Essentially the theorem shows that if one used a Lie group that contained an internal group which mixed in a non-trivial manner with the Poincaré group then the S-matrix for all processes would be zero. The theorem assumes among other things, that the S-matrix exists and is non-trivial, the vacuum is non-degenerate and that there are no massless particles. It is important to realise that the theorem only applies to symmetries that act on S-matrix elements and not on all the other many symmetries that occur in quantum field theory. Indeed it is not uncommon to find examples of the latter symmetries. Of course, no-go theorems are only as strong as the assumptions required to prove them.

In a remarkable paper Gelfand and Likhtman[3] showed that provided one generalised the concept of a Lie group one could indeed find a symmetry that included the Poincaré group and an internal symmetry group in a non-trivial way. In this section we will discuss this approach to the supersymmetry group; having adopted a more general notion of a group, we will show that one is led, with the aid of the Coleman-Mandula theorem, and a few assumptions, to the known supersymmetry group. Since the structure of a Lie group, at least in some local region of the identity, is determined entirely by its Lie algebra it is necessary to adopt a more general notion than a Lie algebra. The vital step in discovering the supersymmetry algebra is to introduce generators Q^i_α, which satisfy anti-commutation relations, i.e.

$$\begin{aligned}
\{Q^i_\alpha, Q^j_\beta\} &= Q^i_\alpha Q^j_\beta + Q^j_\beta Q^i_\alpha \\
&= \text{some other generator}
\end{aligned} \tag{2.4}$$

The significance of the i and a indices will become apparent shortly. Let us therefore assume that the supersymmetry group involves generators P_a, J_{ab}, T_s and possibly some other generators which satisfy commutation relations, as well as the generators Q^i_α $(i = 1, 2, \ldots, N)$. We will call the former generators which satisfy Eqs. (2.1), (2.2) and (2.3) to be even and those satisfying Eq. (2.4) to be odd generators.

Having let the genie out of the bottle we promptly replace the stopper and demand that the supersymmetry algebras have a Z_2 graded structure. This simply means that the even and odd generators must satisfy the rules:

$$\begin{aligned} [\text{even}, \text{even}] &= \text{even} \\ \{\text{odd}, \text{odd}\} &= \text{even} \\ [\text{even}, \text{odd}] &= \text{odd} \end{aligned} \tag{2.5}$$

We must still have the relations

$$[P_a, T_s] = 0 = [J_{ab}, T_s] \tag{2.6}$$

since the even (bosonic) subgroup must obey the Coleman-Mandula theorem.

Let us now investigate the commutator between J_{ab} and Q^i_α. As a result of Eq. (2.5) it must be of the form

$$[Q^i_\alpha, J_{ab}] = (b_{ab})_\alpha^\beta Q^i_\beta \tag{2.7}$$

since by definition the Q^i_α are the only odd generators. We take the α indices to be those rotated by J_{ab}. As in a Lie algebra we have some generalised Jacobi identities. If we denote an even generator by B and an odd generator by F we find that

$$\begin{aligned} [[B_1, B_2], B_3] + [[B_3, B_1], B_2] + [[B_2, B_3], B_1] &= 0 \\ [[B_1, B_2], F_3] + [[F_3, B_1], B_2] + [[B_2, F_3], B_1] &= 0 \\ \{[B_1, F_2], F_3\} + \{[B_1, F_3], F_2\} + [\{F_2, F_3\}, B_1] &= 0 \\ [\{F_1, F_2\}, F_3] + [\{F_1, F_3\}, F_2] + [\{F_2, F_3\}, F_1] &= 0 \end{aligned} \tag{2.8}$$

The reader may verify, by expanding each bracket, that these relations are indeed identically true.

The identity

$$[[J_{ab}, J_{cd}], Q^i_\alpha] + [[Q^i_\alpha, J_{ab}], J_{cd}] + [[J_{cd}, Q^i_\alpha], J_{ab}] = 0 \tag{2.9}$$

upon use of Eq. (2.7) implies that

$$[b_{ab}, b_{cd}]_\alpha^\beta = -\eta_{ac}(b_{bd})_\alpha^\beta - \eta_{bd}(b_{ac})_\alpha^\beta + \eta_{ad}(b_{bc})_\alpha^\beta + \eta_{bc}(b_{ad})_\alpha^\beta \tag{2.10}$$

This means that the $(b_{cd})_\alpha^\beta$ form a representation of the Lorentz algebra or in other words the Q^i_α carry a representation of the Lorentz group. We will select Q^i_α to be in the $(0, \frac{1}{2}) \oplus (\frac{1}{2}, 0)$ representation of the Lorentz group, i.e.

$$[Q^i_\alpha, J_{ab}] = \frac{1}{2}(\sigma_{ab})_\alpha^\beta Q^i_\beta \tag{2.11}$$

We can choose Q^i_α to be a Majorana spinor, i.e.

$$Q^i_\alpha = C_{\alpha\beta}\bar{Q}^{\beta i} \tag{2.12}$$

where $C_{\alpha\beta} = -C_{\beta\alpha}$ is the charge conjugation matrix (see Appendix A of Ref.[1]). This does not represent a loss of generality since, if the algebra admits complex conjugation as an involution we can always redefine the supercharges so as to satisfy (2.12) (see Note 1 at the end of this section).

The above calculation reflects the more general result that the Q^i_α must belong to a realization of the even (bosonic) subalgebras of the supersymmetry group. This is a simple consequence of demanding that the algebra be Z_2 graded. The commutator of any even generator B_1, with Q^i_α is of the form

$$[Q^i_\alpha, B_1] = (h_1)^{i\beta}_{\alpha j} Q^j_\beta \tag{2.13}$$

The generalised Jacobi identity

$$[[Q^i_\alpha, B_1], B_2] + [[B_1, B_2], Q^i_\alpha] + [[B_2, Q^i_\alpha], B_1] = 0 \tag{2.14}$$

implies that

$$[h_1, h_2]^{i\beta}_{\alpha j} Q^j_\beta = [Q^i_\alpha [B_1, B_2]] \tag{2.15}$$

or in other words the matrices h represent the Lie algebra of the even generators.

The above remarks imply that

$$[Q^i_\alpha, T_r] = (l_r)^i_j Q^j_\alpha + (t_r)^i_j (i\gamma_5)^\beta_\alpha Q^j_\beta \tag{2.16}$$

where $(l_r)^i_j + i\gamma_5 (t_r)^i_j$ represent the Lie algebra of the internal symmetry group. This results from the fact that δ^α_β and $(\gamma_5)^\beta_\alpha$ are the only invariant tensors which are scalar and pseudoscalar.

The remaining odd-even commutator is $[Q^i_\alpha, P_a]$. A possibility that is allowed by the generalised Jacobi identities that involve the internal symmetry group and the Lorentz group is

$$[Q^i_\alpha, P_a] = c(\gamma_a)^\beta_\alpha Q^i_\beta \tag{2.17}$$

However, the $[[Q^i_\alpha, P_a], P_b] + \ldots$ identity implies that the constant $c = 0$, i.e.

$$[Q^i_\alpha, P_a] = 0 \tag{2.18}$$

More generally we could have considered $(c\gamma_a + d\gamma_a\gamma_5)Q$, on the right-hand side of (2.17), however, then the above Jacobi identity and the Majorana condition imply that $c = d = 0$. (See Note 2 at the end of this section). Let us finally consider the $\{Q^i_\alpha, Q^j_\beta\}$• anticommutator. This object must be composed of even generators and must be symmetric under interchange of $\alpha \leftrightarrow \beta$ and $i \leftrightarrow j$. The even generators are those of the Poincaré group, the internal symmetry group and other even generators which, from the Coleman-Mandula theorem, commute with the Poincaré group, i.e. they are scalar and pseudoscalar. Hence the most general possibility is of the form

$$\{Q^i_\alpha, Q^j_\beta\} = r(\gamma^a C)_{\alpha\beta} P_a \delta^{ij} + s(\sigma^{ab} C)_{\alpha\beta} J_{ab} \delta^{ij} + C_{\alpha\beta} U^{ij} + (\gamma_5 C)_{\alpha\beta} V^{ij} \tag{2.19}$$

We have not included a $(\gamma^b \gamma_5 C)_{\alpha\beta} L^{ij}_b$ term as the (Q, Q, J_{ab}) Jacobi identity implies that L^{ij}_b mixes nontrivially with the Poincaré group and so is excluded by the no-go theorem.

The fact that we have only used numerically invariant tensors under the Poincaré group is a consequence of the generalised Jacobi identities between two odd and one even generators.

To illustrate the argument more clearly, let us temporarily specialise to the case $N = 1$ where there is only one supercharge Q_α. Equation (2.19) then reads

$$\{Q_\alpha, Q_\beta\} = r(\gamma^a C)_{\alpha\beta} P_a + s(\sigma^{ab} C)_{\alpha\beta} J_{ab}.$$

Using the Jacobi identity

$$\{[P_a, Q_\alpha], Q_\beta\} + \{[P_a, Q_\beta], Q_\alpha\} + [\{Q_\alpha, Q_\beta\}, P_a] = 0,$$

we find that

$$0 = s(\sigma^{cd}C)_{\alpha\beta}[J_{cd}, P_a] = s(\sigma^{cd}C)_{\alpha\beta}(-\eta_{ac}P_d + \eta_{ad}P_c),$$

and, consequently, $s = 0$. We are free to scale the generator P_a in order to bring $r = 2$.

Let us now consider the commutator of the generator of the internal group and the supercharge. For only one supercharge, Eq. (2.16) reduces to

$$[Q_\alpha, T_r] = l_r Q_\alpha + i(\gamma_5)_\alpha^\beta t_r Q_\beta.$$

Taking the adjoint of this equation, multiplying by $(i\gamma^0)$ and using the definition of the Dirac conjugate given in Appendix A of Ref.[1], we find that

$$[\bar{Q}^\alpha, T_r] = l_r^* \bar{Q}^\alpha + \bar{Q}^\beta (it_r^*)(\gamma_5)_\beta^\alpha.$$

Multiplying by $C_{\gamma\alpha}$ and using Eq. (2.12), we arrive at the equation

$$[Q_\alpha, T_r] = l_r^* Q_\alpha + it_r^*(\gamma_5)_\alpha^\beta Q_\beta.$$

Comparing this equation with the one we started from, we therefore conclude that

$$l_r^* = l_r, \qquad t_r^* = t_r.$$

The Jacobi identity

$$[\{Q_\alpha, Q_\beta\}, T_r] + [[T_r, Q_\alpha], Q_\beta] + [[T_r, Q_\beta], Q_\alpha] = 0$$

results in the equation

$$\begin{aligned}&[0 + (l_r\delta_\alpha^\gamma + it_r(\gamma_5)\alpha^\gamma)2(\gamma_a C)_{\gamma\beta}P_a] + (\alpha \leftrightarrow \beta) = \\ &\quad 2P_a\{l_r(\gamma_a C)_{\alpha\beta} + it_r(\gamma_5\gamma_a C)_{\alpha\beta}\} + (\alpha \leftrightarrow \beta) = 0.\end{aligned}$$

Since $(\gamma_a C)_{\alpha\beta}$ and $(\gamma_5\gamma_a C)_{\alpha\beta}$ are symmetric and antisymmetric in $\alpha\beta$ respectively, we conclude that $l_r = 0$ but t_r has no constant placed on it. Consequently, we find that we have only one internal generator R and we may scale it such that

$$[Q_\alpha, R] = i(\gamma_5)_\alpha^\beta Q_\beta.$$

The $N = 1$ supersymmetry algebra is summarised in Eq. (2.27).

Let us now return to the extended supersymmetry algebra. The even generators $U^{ij} = -U^{ji}$ and $V^{ij} = -V^{ji}$ are called central charges[4] and are often also denoted by Z. It is a consequence of the generalised Jacobi identities $((Q, Q, Q)$ and $(Q, Q, Z))$ that they commute with all other generators including themselves, i.e.

$$[U^{ij}, \text{anything}] = 0 = [V^{ij}, \text{anything}] \tag{2.20}$$

We note that the Coleman-Mandula theorem allowed a semi-simple group plus $U(1)$ factors. The details of the calculation are given in note 5 at the end of the section. Their role in supersymmetric theories will emerge in later sections.

In general, we should write, on the right-hand side of (2.19), $(\gamma^a C)_{\alpha\beta}\omega^{ij}P_a + \ldots$, where ω^{ij} is an arbitrary real symmetric matrix. However, one can show that it is possible to redefine (rotate and rescale) the supercharges, whilst preserving the Majorana condition, in such a way as to bring ω^{ij} to the form $\omega^{ij} = r\delta^{ij}$ (see Note 3 at the end of this section). The $[P_a, \{Q_\alpha^i, Q_\beta^j\}] + \ldots = 0$ identity implies that $s = 0$ and we can normalise P_a by setting $r = 2$ yielding the final result

$$\{Q_\alpha^i, Q_\beta^j\} = 2(\gamma_a C)_{\alpha\beta}\delta^{ij}P^a + C_{\alpha\beta}U^{ij} + (\gamma_5 C)_{\alpha\beta}V^{ij} \tag{2.21}$$

In any case r and s have different dimensions and so it would require the introduction of a dimensional parameter in order that they were both non-zero.

Had we chosen another irreducible Lorentz representation for Q^i_α, one other then $(j+\frac{1}{2},j)\oplus(j,j+\frac{1}{2})$, we would not have been able to put P_a, i.e. a $(\frac{1}{2},\frac{1}{2})$ representation, on the right-hand side of Eq. (2.21). The simplest choice is $(0,\frac{1}{2})\oplus(\frac{1}{2},0)$. In fact this is the only possible choice (see Note 4).

Finally, we must discuss the constraints placed on the internal symmetry group by the generalised Jacobi identity. This discussion is complicated by the particular way the Majorana constraint of Eq. (2.12) is written. A two-component version of this constraint is

$$\bar{Q}_{\dot{A}i} = (Q^i_A)^*; \quad A,\dot{A} = 1,2 \tag{2.22}$$

(see Appendix A of Ref.[1] for two-component notation). Equation (2.19) and (2.16) then become

$$\begin{aligned} \{Q^i_A, \bar{Q}_{\dot{B}j}\} &= -2i(\sigma^a)_{A\dot{B}}\delta^i_j P_a \\ \{Q^i_A, Q^j_B\} &= \varepsilon_{AB}(U^{ij} + iV^{ij}) \\ [Q^i_A, J_{ab}] &= +\frac{1}{2}(\sigma_{ab})_A{}^B Q^i_B \end{aligned} \tag{2.23}$$

and

$$[Q^i_A, T_r] = (l_r + it_r)^i_j Q^j_A \tag{2.24}$$

Taking the complex conjugate of the last equation and using the Majorana condition we find that

$$[Q_{\dot{A}l}, T_r] = Q_{\dot{A}k}(U^\dagger_r)^k_l \tag{2.25}$$

where $(U_r)^i_j = (l_r + it_r)^i_j$. The $(Q,\bar{Q},T)$ Jacobi identity then implies that δ^i_j be an invariant tensor of G, i.e.

$$U_r + U^\dagger_r = 0 \tag{2.26}$$

Hence U_r is an antihermitian matrix and so represents the generators of the unitary group $U(N)$. However, taking account of the central charge terms in the (Q, Q, T) Jacobi identity one finds that there is for every central charge an invariant antisymmetric tensor of the internal group and so the possible internal symmetry group is further reduced. If there is only one central charge, the internal group is $Sp(N)$ while if there are no central charges it is $U(N)$.

To summarise, once we have adopted the rule that the algebra be Z_2 graded and contain the Poincaré group and an internal symmetry group then the generalised Jacobi identities place very strong constraints on any possible algebra. In fact, once one makes the further assumption that Q^i_α are spinors under the Lorentz group then the algebra is determined to be of the form of equations (2.1), (2.6), (2.11), (2.16), (2.18) and (2.21).

The simplest algebra is for $N = 1$ and takes the form

$$\begin{aligned} \{Q_\alpha, Q_\beta\} &= 2(\gamma_a C)_{\alpha\beta}P^a \\ [Q_\alpha, P_a] &= 0 \\ [Q_\alpha, J_{cd}] &= \frac{1}{2}(\sigma_{cd})_\alpha{}^\beta Q_\beta \\ [Q_\alpha, R] &= i(\gamma_5)_\alpha{}^\beta Q_\beta \end{aligned} \tag{2.27}$$

as well as the commutation relations of the Poincaré group. We note that there are no central charges (i.e. $U^{11} = V^{11} = 0$), and the internal symmetry group becomes just a chiral rotation with generator R.

We now wish to prove three of the statements above. This is done here rather than in the above text, in order that the main line of argument should not become obscured by technical points. These points are best clarified in two-component notation.

Note 1:

Suppose we have an algebra that admits a complex conjugation as an involution; for the supercharges this means that

$$(Q^i_A)^* = b_i{}^j Q_{\dot{A}j}; \quad (Q_{\dot{A}j})^* = d^j{}_k Q^k_A$$

There is no mixing of the Lorentz indices since $(Q^i_A)^*$ transforms like $Q_{\dot{A}i}$, namely in the $(0, \frac{1}{2})$ representation of the Lorentz group, and not like Q^i_A which is in the $(\frac{1}{2}, 0)$ representation. The lowering of the i index under * is at this point purely a notational device. Two successive * operations yield the unit operation and this implies that

$$(b_i{}^j)^* d^j{}_k = \delta^i_k \tag{2.28}$$

and in particular that $b_i{}^j$ is an invertible matrix. We now make the redefinitions

$$\begin{aligned} Q'^i{}_A &= Q^i_A \\ Q'_{\dot{A}i} &= b_i{}^j Q_{\dot{A}j} \end{aligned} \tag{2.29}$$

Taking the complex conjugate of Q'_{Ai}, we find

$$(Q'^i{}_A)^* = (Q^i_A)^* = b_i{}^j Q_{\dot{A}j} = Q'_{\dot{A}j}$$

while

$$(Q'_{\dot{A}i})^* = (b_i{}^j)^*(Q_{\dot{A}j})^* = (b_i{}^j)^* d^j{}_k Q^k_A = Q^i_A \tag{2.30}$$

using Eq. (2.28).

Thus the $Q'^i{}_A$ satisfy the Majorana condition, as required. If the Q's do not initially satisfy the Majorana condition, we may simply redefine them so that they do.

Note 2:

Suppose the $[Q_A, P_a]$ commutator were of the form

$$[Q_A, P_a] = e(\sigma_a)_{A\dot{B}} Q^{\dot{B}} \tag{2.31}$$

where e is a complex number and for simplicity we have suppressed the i index. Taking the complex conjugate (see Appendix A of Ref.[1]), we find that

$$[Q_{\dot{A}}, P_a] = -e^*(\sigma_a)_{B\dot{A}} Q^B \tag{2.32}$$

Consideration of the $[[Q_A, P_a], P_b] + \ldots = 0$ Jacobi identity yields the result

$$-|e|^2(\sigma_a)_{A\dot{B}}(\sigma^b)^{C\dot{B}} - (a \leftrightarrow b) = 0 \tag{2.33}$$

Consequently $e = 0$ and we recover the result

$$[Q_A, P_a] = 0 \tag{2.34}$$

Note 3:

The most general form of the $Q^{Ai}, Q^{\dot{B}}_j$ anticommutator is

$$\{Q^{Ai}, Q^{\dot{B}}_j\} = -2iU^i_j(\sigma^m)^{A\dot{B}} P_m + \text{terms involving other Dirac matrices} \tag{2.35}$$

Taking the complex conjugate of this equation and comparing it with itself, we find that U is a Hermitian matrix

$$(U^i_j)^* = U^j_i \tag{2.36}$$

We now make a field redefinition of the supercharge

$$Q'^{Ai} = B^i_j Q^{Aj} \tag{2.37}$$

and its complex conjugate

$$Q'^{\dot{A}}_i = (B^i_j)^* Q^{\dot{A}}_j \tag{2.38}$$

Upon making this redefinition in Eq. (2.35), the U matrix becomes replaced by

$$U'^i_j = B^i_k U^k_l (B^j_l)^* \quad \text{or} \quad U' = BUB^\dagger \tag{2.39}$$

Since U is a Hermitian matrix, we may diagonalise it in the form $c_i\delta^i_j$ using a unitarity matrix B. We note that this preserves the Majorana condition on Q^{Ai}. Finally, we may scale $Q^i \to (1/\sqrt{c^i})Q^i$ to bring U to the form $U = d_i\delta^i_j$, where $d_i = \pm 1$. In fact, taking $A = B = 1$ and $i = j = k$, we realise that the right-hand side of Eq. (2.35) is a positive definite operator and since the energy $-iP_0$ is assumed positive definite, we can only find $d_i = +1$. The final result is

$$\{Q^{Ai}, Q^{\dot{B}}_j\} = -2i\delta^i_j(\sigma^m)^{A\dot{B}}P_m \tag{2.40}$$

Note 4:

Let us suppose that the supercharge Q contains an irreducible representation of the Lorentz group other than $(0,\frac{1}{2})\oplus(\frac{1}{2},0)$, say, the representation $Q_{A_1\ldots A_n,\dot{B}_1\ldots\dot{B}_m}$ where the A and B indices are understood to be separately symmetrised and $n+m$ is odd in order that Q is odd and $n+m > 1$. By projecting the $\{Q,Q^\dagger\}$ anticommutator we may find the anti-commutator involving $Q_{A_1\ldots A_n,\dot{B}_1\ldots\dot{B}_m}$ and its hermitian conjugate. Let us consider in particular the anticommutator involving $Q = Q_{11\ldots1,\dot{1}\dot{1}\ldots\dot{1}}$, this must result in an object of spin $n+m > 1$. However, by the Coleman-Mandula no-go theorem no such generator can occur in the algebra and so the anticommutator must vanish, i.e. $QQ^\dagger + Q^\dagger Q = 0$.

Assuming the space on which Q acts has a positive definite norm, one such example being the space of on-shell states, we must conclude that Q vanishes. However if $Q_{11\ldots1,\dot{1}\dot{1}\ldots\dot{1}}$ vanishes, so must $Q_{A_1\ldots A_n,\dot{B}_1\ldots\dot{B}_m}$ by its Lorentz properties, and we are left only with the $(0,\frac{1}{2})\oplus(\frac{1}{2},0)$ representation.

Note 5:

We now return to the proof of equation (2.20). Using the (Q,Q,Z) Jacobi identity it is straightforward to show that the supercharges Q commutes with the central charges Z. The (Q,Q,U) Jacobi identity then implies that the central charges commute with themselves. Finally, one considers the (Q,Q,T_r) Jacobi identity, this relation shows that the commutator of T_r and Z takes the generic form $[T_r, Z] = \ldots Z$. However, the generators T_r and Z form the internal symmetry group of the supersymmetry algebra and from the no-go theorem we know that this group must be a semisimple Lie group times $U(1)$ factors. We recall that a semisimple Lie group is one that has no normal Abelian subgoups other that the group itself and the identity element. As such, we must conclude that T_r and Z commute, and hence our final result that the central charges commute with all generator, that is they really are central.

Although the above discussion started with the Poincaré group, one could equally well have started with the conformal or (anti-)de Sitter groups and obtained the superconformal and super (anti-)de Sitter algebras. For completeness, we now list these algebras. The superconformal algebra which has the generators $P_n, J_{mn}, D, K_n, A, Q^{\alpha i}$, $S^{\alpha i}$ and the internal symmetry generators T_r and A is given by the Lorentz group plus:

$$
\begin{aligned}
[J_{mn}, P_k] &= \eta_{nk}P_m - \eta_{mk}P_n \\
[J_{mn}, K_k] &= \eta_{nk}K_m - \eta_{mk}K_n \\
[D, P_K] &= -P_K \quad [D, K_K] = +K_K \\
[P_m, K_n] &= -2J_{mn} + 2\eta_{mn}D \quad [K_n, K_m] = 0 \quad [P_n, P_m] = 0 \\
[Q^i_\alpha, J_{mn}] &= \frac{1}{2}(\gamma_{mn})_\alpha^{\ \beta} Q^i_\beta \quad [S^i_\alpha, J_{mn}] = \frac{1}{2}(\gamma_{mn})_\alpha^{\ \beta} S^{\beta i} \\
\{Q^i_\alpha, Q^j_\beta\} &= -2(\gamma^n C^{-1})_{\alpha\beta} P_n \delta^{ij} \\
\{S^i_\alpha, S^j_\beta\} &= +2(\gamma^n C^{-1})_{\alpha\beta} K_n \delta^{ij} \\
[Q^i_\alpha, D] &= \frac{1}{2}Q^i_\alpha \quad [S^i_\alpha, D] = -\frac{1}{2}S^i_\alpha \\
[Q^i_\alpha, K_n] &= -(\gamma_n)_\alpha^{\ \beta} S^i_\beta \quad [S^i_\alpha, P_n] = (\gamma_n)_\alpha^{\ \beta} Q^i_\beta \\
[Q^i_\alpha, T_r] &= (\delta_\alpha^\beta (\tau_{r_1})^i_{\ j} + (\gamma_5)_\alpha^{\ \beta}(\tau_{r_2})^i_{\ j}) Q^j_\beta \\
[S^i_\alpha, T_r] &= (\delta_\alpha^\beta (\tau_{r_1})^i_{\ j} - (\gamma_5)_\alpha^{\ \beta}(\tau_{r_2})^i_{\ j}) Q^j_\beta \\
[Q^i_\alpha, A] &= -i(\gamma_5)_\alpha^{\ \beta} Q^i_\beta \left(\frac{4-N}{4N}\right) \\
[S^i_\alpha, A] &= \frac{4-N}{4N} i(\gamma_5)_\alpha^{\ \beta} S^i_\beta \\
\{Q^i_\alpha, S^j_\beta\} &= -2(C^{-1})_{\alpha\beta} D\delta^{ij} + (\gamma^{mn}C^{-1})_{\alpha\beta} J_{mn}\delta^{ij} + 4i(\gamma_5 C^{-1})_{\alpha\beta} A\delta^{ij} \\
&\quad -2((\tau_{r_1})^{ij}(C^{-1})_{\alpha\beta} + (\tau_{r_2})^{ij}(\gamma_5 C^{-1})_{\alpha\beta}) T_r
\end{aligned}
\tag{2.41}
$$

The T_r and A generate $U(N)$ and $\tau_1 + \gamma_5\tau_2$ are in the fundamental representation of $SU(N)$.

The case of $N = 4$ is singular and one can have either

$$[Q^i_\alpha, A] = 0 \quad \text{or} \quad [Q^i_\alpha, A] = -i(\gamma_5)_\alpha^{\ \beta} Q^i_\beta$$

and similarly for S^i_α and A. One may verify that both possibilities are allowed by the $N = 4$ Jacobi identities and so form acceptable superalgebras.

The anti-de Sitter superalgebra has generators M_{mn}, $T_{ij} = -T_{ji}$ and $Q^{\alpha i}$, and is given by

$$
\begin{aligned}
[M_{mn}, M_{pq}] &= \eta_{np}M_{mq} + 3 \text{ terms} \\
[M_{mn}, T_{ij}] &= 0 \quad [Q^i_\alpha, M_{mn}] = \frac{1}{2}(\gamma_{mn})_\alpha^{\ \beta} Q^i_\beta \\
[Q^i_\alpha, T^{jk}] &= -2i(\delta^{ij}Q^k_\alpha - \delta^{ik}Q^j_\alpha) \\
\{Q^i_\alpha, Q^j_\beta\} &= \delta^{ij}(\gamma_{mn}C^{-1})_{\alpha\beta} i M_{mn} + (C^{-1})_{\alpha\beta} T^{ij} \\
[T^{ij}, T^{kl}] &= -2i(\delta^{jk}T^{il} + 3 \text{ terms})
\end{aligned}
\tag{2.42}
$$

MODELS OF RIGID SUPERSYMMETTY

The Wess-Zumino Model

This section is identical to chapter 5 of Ref.[1]. The equation numbers are kept the same as in this book. I thank World Scientific Publishing for their permission to reproduce this material.

The first four-dimensional model in which supersymmetry was linearly realised was found by Wess and Zumino[5] by studying two-dimensional dual models[6]. In this section we discuss the Wess-Zumino model which is the simplest model of $N = 1$ supersymmetry.

Let us assume that the simplest model possesses one fermion χ_α which is a Majorana spinor, i.e.

$$\chi_\alpha = C_{\alpha\beta}\bar{\chi}^\beta \tag{5.1}$$

On shell, that is, when

$$\partial\!\!\!/\chi = \text{ interaction} \tag{5.2}$$

χ_α has two degrees of freedom or two helicity states. Applying the rule concerning equal numbers of fermionic and bosonic degrees of freedom to the on-shell states we find that we must add two bosonic degrees of freedom to χ_α in order to form a realization of supersymmetry. These could either be two spin-zero particles or one massless vector particle which also has two helicity states on-shell. We will consider the former possibility in this section and the latter possibility, which is the $N = 1$ Yang-Mills theory, further on.

An irreducible representation of $N = 1$ supersymmetry can be carried either by one parity even spin-zero state, one parity odd spin-zero state and one Majorana spin - $\frac{1}{2}$, or by one massless spin-one and one Majorana spin -$\frac{1}{2}$. Taking the former possibility we have a Majorana spinor χ_α and two spin-zero states which we will assume to be represented by a scalar field A and pseudoscalar field B. For simplicity we will begin by constructing the free theory; the fields A, B, χ_α are then subject to

$$\partial^2 A = \partial^2 B = \partial\!\!\!/\chi = 0 \tag{5.3}$$

We now wish to construct the supersymmetry transformations that are carried by this irreducible realization of supersymmetry. Since $\bar{\varepsilon}^\alpha Q_\alpha$ is dimensionless and Q_α has mass dimension $+\frac{1}{2}$, the parameter $\bar{\varepsilon}^\alpha$ must have dimension $-\frac{1}{2}$. On grounds of linearity, dimension, Lorentz invariance and parity we may write down the following set of transformations:

$$\begin{aligned} \delta A &= \bar{\varepsilon}QA = \bar{\varepsilon}\chi, \qquad \delta B = i\bar{\varepsilon}\gamma_5\chi \\ \delta\chi &= \partial\!\!\!/(\alpha A + \beta i\gamma_5 B)\varepsilon \end{aligned} \tag{5.4}$$

where a and β are undetermined parameters.

The variation of A is straightforward; however, the appearance of a derivative in $\delta\chi$ is the only way to match dimensions once the transformations are assumed to be linear. The reader will find no trouble verifying that these transformations do leave the set of field equations of Eq.(5.3) intact.

We can now test whether the $N = 1$ supersymmetry algebra of the previous section is represented by these transformations. The commutator of two supersymmetries on A is given by

$$[\delta_1, \delta_2]A = [\bar{\varepsilon}_1 Q, \bar{\varepsilon}_2 Q]A \tag{5.5}$$

which, using Eq.(2.27), becomes

$$\begin{aligned} [\delta_1, \delta_2]A &= 2\bar{\varepsilon}_2\gamma^a\varepsilon_1 P_a A \\ &= 2\bar{\varepsilon}_2\gamma^a\varepsilon_1 \partial_a A \\ \text{since} \qquad P_a &= \partial_a \end{aligned} \tag{5.6}$$

On the other hand the transformation laws of Eq.(5.4) imply that

$$\begin{aligned} [\delta_1, \delta_2]A &= \bar{\varepsilon}_2\partial\!\!\!/(\alpha A + i\gamma_5\beta B)\varepsilon_1 - (1 \leftrightarrow 2) \\ &= 2\alpha\bar{\varepsilon}_2\partial\!\!\!/\varepsilon_1 A \end{aligned} \tag{5.7}$$

The term involving B drops out because of the properties of Majorana spinors (see Appendix A of Ref.[1]). Provided $\alpha = +1$ this is indeed the 4-translation required by the algebra. We therefore set $\alpha = +1$. The calculation for B is similar and yields $\beta = +1$. For the field χ_α the commutator of two supersymmetries gives the result

$$\begin{aligned}
[\delta_{\varepsilon_1}, \delta_{\varepsilon_2}]\chi &= \partial\!\!\!/[\bar{\varepsilon}_1\chi + i\gamma_5\bar{\varepsilon}_1 i\gamma_5\chi]\varepsilon_2 - (1 \leftrightarrow 2) \\
&= -\frac{1}{4}\bar{\varepsilon}_1\gamma^R\varepsilon_2\partial\!\!\!/[\gamma_R + i\gamma_5\gamma_R i\gamma_5]\chi - (1 \leftrightarrow 2) \\
&= +\frac{1}{2}\bar{\varepsilon}_2\gamma^a\varepsilon_1 2\partial\!\!\!/\gamma_a\chi \\
&= 2\bar{\varepsilon}_2\partial\!\!\!/\varepsilon_1\chi - \bar{\varepsilon}_2\gamma^a\varepsilon_1\gamma_a\partial\!\!\!/\chi
\end{aligned} \tag{5.8}$$

The above calculation makes use of a Fierz rearrangement (see Appendix A of Ref.[1]) as well as the properties of Majorana spinors. However, χ_α is subject to its equation of motion, i.e. $\partial\!\!\!/\chi = 0$, implying the final result

$$[\delta_1, \delta_2]\chi = 2\bar{\varepsilon}_2\partial\!\!\!/\varepsilon_1\chi \tag{5.9}$$

which is the consequence dictated by the supersymmetry algebra. The reader will have no difficulty verifying that the fields A, B and χ_α and the transformations

$$\begin{aligned}
\delta A &= \bar{\varepsilon}\chi, \quad \delta B = i\bar{\varepsilon}\gamma_5\chi \\
\delta\chi &= \partial\!\!\!/(A + i\gamma_5 B)\varepsilon
\end{aligned} \tag{5.10}$$

form a representation of the whole of the supersymmetry algebra *provided A, B* and χ_α are on-shell (i.e. $\partial^2 A = \partial^2 B = \partial\!\!\!/\chi = 0$).

We now wish to consider the fields A, B and χ_α when they are no longer subject to their field equations. The Lagrangian from which the above field equations follow is

$$L = -\frac{1}{2}(\partial_\mu A)^2 - \frac{1}{2}(\partial_\mu B)^2 - \frac{1}{2}\bar{\chi}\partial\!\!\!/\chi \tag{5.11}$$

It is easy to prove that the action $\int d^4x L$ is indeed invariant under the transformation of Eq.(5.10). This invariance is achieved *without the use of the field equations.* The trouble with this formulation is that the fields A, B and χ_α do not form a realization of the supersymmetry algebra when they are no longer subject to their field equations, as the last term in Eq.(5.8) demonstrates. It will prove useful to introduce the following terminology. We shall refer to an irreducible representation of supersymmetry carried by fields which are subject to their equations of motion as an *on-shell representation.* We shall also refer to a Lagrangian as being *algebraically on-shell* when it is formed from fields which carry an on-shell representation, that is, do not carry a representation of supersymmetry off-shell, and the Lagrangian is invariant under these on-shell transformations. The Lagrangian of Eq.(5.11) is then an algebraically on-shell Lagrangian.

That A, B and χ_α cannot carry a representation of supersymmetry off-shell can be seen without any calculation, since these fields do not satisfy the rule of equal numbers of fermions and bosons. Off-shell, A and B have two degrees of freedom, but χ_α has four degrees of freedom. Clearly, the representations of supersymmetry must change radically when enlarged from on-shell to off-shell.

A possible way out of this dilemma would be to add two bosonic fields F and G which would restore the fermion-boson balance. However, these additional fields would have to occur in the Lagrangian so as to give rise to no on-shell states. As such, they must occur in the Lagrangian in the form $+\frac{1}{2}F^2 + \frac{1}{2}G^2$ assuming the free action to be only bilinear in the fields and consequently be of mass dimension two. On dimensional grounds their supersymmetry transformations must be of the form

$$\delta F = \bar{\varepsilon}\partial\!\!\!/\chi \quad \delta G = i\bar{\varepsilon}\gamma_5\partial\!\!\!/\chi \tag{5.12}$$

where we have tacitly assumed that F and G are scalar and pseudoscalar respectively. The fields F and G cannot occur in δA on dimensional grounds, but can occur in $\delta\chi_\alpha$ in the form

$$\delta\chi = [(\mu F + i\tau\gamma_5 G) + \partial\!\!\!/(A + i\gamma_5 B)]\varepsilon \tag{5.13}$$

where μ and τ are undetermined parameters.

We note that we can only modify transformation laws in such a way that on-shell (i.e., when $F = G = \partial\!\!\!/\chi = \partial^2 A = \partial^2 B = 0$) we regain the on-shell transformation laws of Eq.(5.10).

We must now test if these new transformations do form a realization of the supersymmetry algebra. In fact, straightforward calculation shows they do, provided $\mu = \tau = +1$. This representation of supersymmetry involving the fields A, B, χ_α, F and G was found by Wess and Zumino[5] and we now summarize their result:

$$\begin{aligned} &\delta A = \bar{\varepsilon}\chi \qquad \delta B = i\bar{\varepsilon}\gamma_5\chi \\ &\delta\chi = [F + i\gamma_5 G + \partial\!\!\!/(A + i\gamma_5 B)]\varepsilon \\ &\delta F = \bar{\varepsilon}\partial\!\!\!/\chi \qquad \delta G = i\bar{\varepsilon}\gamma_5\partial\!\!\!/\chi \end{aligned} \tag{5.14}$$

The action which is invariant under these transformations, is given by the Lagrangian

$$A = \int d^4x \left\{ -\frac{1}{2}(\partial_\mu A)^2 - \frac{1}{2}(\partial_\mu B)^2 - \frac{1}{2}\bar{\chi}\partial\!\!\!/\chi + \frac{1}{2}F^2 + \frac{1}{2}G^2 \right\} \tag{5.15}$$

As expected the F and G fields occur as squares without derivatives and so lead to no on-shell states.

The above construction of the Wess-Zumino model is typical of that for a general free supersymmetric theory. We begin with the on-shell states, and construct the on-shell transformation laws. We can then find the Lagrangian which is invariant without use of the equations of motion, but contains no auxiliary fields. One then tries to find a set of auxiliary fields that give an off-shell algebra. Once this is done one can find a corresponding off-shell action. How one finds the nonlinear theory from the free theory is discussed in the later chapters of Ref.[1].

The first of these two steps is always possible; however, there is no sure way of finding auxiliary fields that are required in all models, except with a few rare exceptions. This fact is easily seen to be a consequence of our rule for equal numbers of fermi and bose degrees of freedom in any representation of supersymmetry. It is only spin 0's, when represented by scalars, that have the same number of field components off-shell as they have on-shell states. For example, a Majorana spin-$\frac{1}{2}$ when represented by a spinor χ_α has a jump of 2 degrees of freedom between on and off-shell and a massless spin-1 boson when represented by a vector A_μ has a jump of 1 degree of freedom. In the latter case it is important to subtract the one gauge degree of freedom from A_μ thus leaving 3 field components off-shell (see chapter 6 of Ref.[1]). Since the increase in the number of degrees of freedom from an on-shell state to the off-shell field representing it changes by different amounts for fermions and bosons, the fermionic-boson balance which holds on-shell will not hold off-shell if we only introduce the fields that describe the on-shell states. The discrepancy must be made up by fields, like F and G, that lead to no on-shell states. These latter type of fields are called *auxiliary fields*. The whole problem of finding representations of supersymmetry amounts to finding the auxiliary fields.

Unfortunately, it is not at all easy to find the auxiliary fields. Although the fermi-bose counting rule gives a guide to the number of auxiliary fields it does not actually tell you what they are, or how they transform. In fact, the auxiliary fields are only known for almost all $N = 1$ and 2 supersymmetry theories and for a very few $N = 4$ theories and not for the higher N theories. In particular, they are not known for the $N = 8$ supergravity theory.

Theories for which the auxiliary fields are not known can still be described by a Lagrangian in the same way as the Wess-Zumino theory can be described without the use of F and G, namely, by the so called algebraically on-shell Lagrangian formulation, which for the Wess-Zumino theory was given in Eq. (5.11). Such 'algebraically on-shell Lagrangians' are not too difficult to find at least at the linearized level. As explained in chapter 8 of Ref.[1] we can easily find the relevant on-shell states of the theory. The algebraically on-shell Lagrangian then consists of writing down the known kinetic terms for each spin.

Of course, we are really interested in the interacting theories. The form of the interactions is however often governed by symmetry principles such as gauge invariance in the above example or general coordinate invariance in the case of gravity theories. When the form of the interactions is dictated by a local symmetry there is a straightforward, although maybe very lengthy way of finding the nonlinear theory from the linear theory. This method, called Noether coupling, is described in chapter 7 of Ref.[1]. In one guise or another this technique has been used to construct nonlinear 'algebraically on-shell Lagrangians' for all supersymmetric theories.

The reader will now ask himself whether algebraically on-shell Lagrangians may be good enough. Do we really need the auxiliary fields? The following example is a warning against over-estimating the importance of a Lagrangian that is invariant under a set of transformations that mix fermi-bose fields, but do not obey any particular algebra.

Consider the Lagrangian

$$L = -\frac{1}{2}(\partial_\mu A)^2 - \frac{1}{2}\bar{\chi}\not{\partial}\chi \tag{5.16}$$

whose corresponding action is invariant under the transformations

$$\partial A = \bar{\varepsilon}\chi \quad \partial\chi = \not{\partial}A\varepsilon \tag{5.17}$$

However, this theory has nothing to do with supersymmetry. The algebra of transformations of Eq. (5.17) does not close on or off-shell without generating transformations which, although invariances of the free theory, can never be generalized to be invariances of an interacting theory. In fact, the on-shell states do not even have the correct fermi-bose balance required to form an irreducible representation of supersymmetry. This example illustrates the fact that the 'algebraically on-shell Lagrangians' rely for their validity, as supersymmetric theories, on their on-shell algebra.

As a final remark in this section it is worth pointing out that the problem of finding the representations of any group is a mathematical question not dependent on any dynamical considerations for its resolution. Thus the questions of which are physical fields and which are auxiliary fields is a model-dependent statement.

The $N = 1$ Yang-Mills Theory

The construction of the $N = 1$ Yang-Mills theory in x-space, presented during the lectures, follows closely chapter 6 of Ref.[1].

The Extended Theories

The $N = 2$ Yang-Mills theory and $N = 2$ matter, as well as their most general renormalizable coupling, were constructed along the lines of chapter 12 of Ref.[1].

THE IRREDUCIBLE REPRESENTATIONS OF SUPERSYMMETRY

Irreducible representations of supersymmetry were constructed using the method of induced representations. This provides a complete list of the possible supersymmetric theories in four dimensions. The material can be found in chapter 8 of Ref.[1].

SUPERSPACE

Construction of Superspace

Superspace was constructed as the coset space of the super-Poincare group divided by the Lorentz group. Details can be found in chapter 14 of Ref.[1].

Superspace Formulations of Rigid Supersymmetric Theories

The formulation of the Wess-Zumino model and $N = 1, 2$ Yang-Mills theories in superspace can be found in chapter 15 of Ref.[1].

QUANTUM PROPERTIES OF SUPERSYMMETRIC MODELS

Super-Feynman Rules and the Non-renormalisation Theorem

The super-Feynman rules of the Wess-Zumino model and $N = 1$ Yang-Mills theory were derived and the non-renormalisation theorem was proved. For details see chapter 17 of Ref.[1].

Flat Directions

The potential in a supersymmetric theory is given by the squares of the auxiliary fields. In this section we consider an $N = 1$ supersymmetric model which contains Wess-Zumino multiplets coupled to the $N = 1$ Yang-Mills multiplet with gauge group G. Let us denote the auxiliary fields of the Wess-Zumino multiplets by the complex field $\mathcal{F}^i$ where the index i labels the Wess-Zumino multiplets and those of the $N = 1$ Yang-Mills multiplet by D^a where $a = 1, \ldots,$ dimension of G. Then the classical potential is given by

$$V = |\mathcal{F}^i|^2 + \frac{1}{2} \sum_a (D^a)^2 \tag{5.2.1}$$

For a general $N = 1$ renormalizable theory the auxiliary fields are given by

$$\mathcal{F}^i = \frac{\partial W(z^j)}{\partial z^i} \tag{5.2.2}$$

and

$$D^a = -g\bar{z}_i(T^a)^i_j z^j + \zeta^a \tag{5.2.3}$$

In equation (5.2.2) W is the superpotential which we recall occurs in the superspace formulation of the theory as $(\int d^4x d^2\theta W + c.c)$] and z^i are the scalars of the Wess-Zumino multiplet. For a renormlizable theory, the superpotential has the form $W(z) = \frac{1}{3!} d_{ijk} z^i z^j z^k + \frac{1}{2!} m_{ij} z^i z^j + e_i z^i$. In equation (5.2.3) g is the gauge coupling constant and

$(T^a)^i_j$ are the generators of the group G to which these scalars z^i belong. The terms in the auxiliary fields which are independent of z^i can only occur when we have $U(1)$ factors for D^a and auxiliary fields $\mathcal{F}^i$ that transform trivial under G. The resulting ζ^a and e_i are constants.

Clearly, the potential is positive definite. Another remarkable feature of the potential is that it generically has flat directions. This means that minimizing the potential does not specify a unique field configuration. In other words there exists a vacuum degeneracy. The simplest example is for a Wess-Zumino model in the adjoint representation coupled to a $N = 1$ Yang-Mills multiplet. Taking the superpotential for this theory to vanish the potential is given by

$$V = \frac{1}{2} \sum_a (-g f_{abc} \bar{z}^b z^c)^2 \tag{5.2.4}$$

Clearly, the minimum is given by field configurations whose only non-zero vacuum expectation values are $< z^a > H^a$ where H^a are the Cartan generators of the algebra. This theory is precisely the $N = 2$ supersymmetric Yang-Mills theory when written in terms of $N = 1$ supermuliplets.

In a general quantum field theory such a vacuum degeneracy would be removed by quantum corrections to the potential. However, things are different in supersymmetric theories. In fact, if supersymmetry is not broken the potential does not receive any perturbative quantum corrections[7]. It obviously follows that if supersymmetry is not broken then the vacuum degeneracy is not removed by perturbative quantum corrections[7]. This result was first proved before the advent of the non-renormalisation theorem as formulated in Ref.[8], but it is particularly obvious given this theorem. For the effective potential we are interested in field configurations where the spinors vanish and the space-time derivatives of all fields are set to zero. For such configurations, the gauge invariant superfields do not contain any θ dependence as only their first component is non-zero. Quantum corrections, however, contain an integral over all of superspace and to be non-zero requires a $\theta^2\bar{\theta}^2$ factor in the integrand. For the field configurations of interest to us such an integral over the full superspace must vanish and as a result we find that there are no quantum corrections to the effective potential if supersymmetry is not broken.

Finally, we recall why the expectation values of the auxiliary fields vanish if supersymmetry is preserved. In this case the expectation value of the supersymmetry transformations of the spinors must vanish. The transformation of the spinors contain auxiliary fields which occur without space-time derivatives and the bosonic fields which correspond to the dynamical degrees of freedom of the theory. The latter occur with space-time derivative, as they have mass dimension one and ϵ has dimension $-\frac{1}{2}$. Consequently, if the expectation values of supersymmetry transformations of the spinors vanish so do the expectation values of all the auxiliary fields. By examining the supersymmetry transformations of the spinors given earlier the reader may verify that there are no loop holes in this argument.

Clearly, the rigid $N = 2$ and $N = 4$ theories can be written in terms of $N = 1$ supermultiplets and so the flat directions that occur in these theories are also not removed by quantum corrections. Although this might be viewed as a problem in these theories it has been turned to advantage in the work of Seiberg and Witten. These authors realized that the dependence of these theories on the expectation values of the scalar fields, or the moduli, obeyed interesting properties that can be exploited to solve for part of the effective action of these theories.

Non-holomorphicity

The non-renormalisation theorem states that perturbative quantum corrections to the effective action are of the form

$$\int d^4x_1 \ldots \int d^4x_n \int d^4\theta G(x_1, \ldots x_n) f(\varphi(x_1,\theta), \ldots, V(x_1,\theta), \ldots, D^A\varphi(x_1,\theta), \ldots) \tag{5.3.1}$$

where φ and V are the superfields that contain the Wess-Zumino and $N = 1$ Yang-Mills fields respectively.

The most significant aspect of this result is that the corrections arise from a single superspace integral over **all** of superspace, that is they contain a $d^4\theta = d^2\theta d^2\bar{\theta}$ integral and not a sub-integral of the form $d^2\theta$ or $d^2\bar{\theta}$, Such sub-integrals play an important role in supersymmetric theories. For example, the superpotential in the superspace formulation of the Wess-Zumino model has the form $\int d^4x d^2\theta W + c.c.$

While their is no question that this formulation of the non-renormalisation theorem is correct, with the passing of time, it was taken by many workers to mean that their could never be any quantum corrections which were sub-integrals i.e. that is of the form

$$\int d^4x d^2\theta \hat{W}(\varphi) \tag{5.3.2}$$

In particular, it was often said that there could be no quantum corrections to the superpotential.

Consider, however, the expression

$$\int d^4x d^4\theta [(-\frac{D^2}{4\partial^2})\varphi^n] = \int d^4x d^2\theta (-\frac{\bar{D}^2}{4})(-\frac{D^2}{4\partial^2})\varphi^n = \int d^4x d^2\theta \varphi^n \tag{5.3.3}$$

where we have used the relation $\bar{D}^2 D^2 \psi = 16\partial^2\psi$ where ψ is any chiral superfield. This maneuver illustrates the important point that although an expression can be written as a full superspace integral, it can also be expressible as a local integral over only a subspace of superspace. The above expression when written in terms of the full superspace integral is non-local, however, any effective action contains many non-local contributions. The occurrence of the $\frac{1}{\partial^2}$ is the signal of a massless particle. For a massive particle one would instead find a factor of $\frac{1}{(\partial^2+m^2)}$ which can not be rewritten as a sub-integral. Hence, only when massless particles circulate in the quantum loops can we find a contribution to the effective action which can be written as a sub-superspace integral.

The first example of such a correction to the superpotential was found in Ref.[9]. In Ref.[10] it was shown that all the proofs of the non-renormalisation theorem allowed contributions to the effective action which were integrals over a subspace of superspace if massless particles were present. It was also shown[10] that such corrections were not some pathological exception, but that they generically occurred whenever massless particles were present. This lecture follows the first part of Ref.[10] and the reader is referred there for a much more complete discussion and several examples. In the Wess-Zumino model such corrections first occur at two loops and were calculated in[11], while in the Wess-Zumino model coupled to $N = 1$ Yang-Mills theory the corrections occur even at one loop[12]. An alternative way of looking at such corrections was given in Refs.[13, 14].

The reader may wonder what such corrections have to do with non-holomorphicity. The answer is that the corrections we have been considering are non-holomorphic in the coupling constants. The situation is most easily illustrated in the context of the massless Wess-Zumino model where the superpotential is of the form $\int d^4x d^2\theta \lambda\varphi^3 + c.c.$ Since the propagator connects φ to $\bar{\varphi}$ we get no corrections at all if we do not include terms that contain both λ and $\bar{\lambda}$. Consequently, the corrections we find to the superpotential must contain λ and $\bar{\lambda}$ and so is non-holomorphic in λ.

We can of course prevent the occurrence of such terms if we give masses to all the particles or we do not integrate over the infra-red region of the loop momentum integration for the massless particles. Such is the case if we calculate the Wilsonian effective action. However, if the terms considered here affect the physics in an important way one will necessarily miss such effects and they will only become apparent when one carries out the integrations that one had previously excluded.

Perturbative Quantum Properties of Extended Theories of Supersymmetry

Many of the perturbative properties of the extended theories of supersymmetry were derived. These include the finiteness, or superconformal invariance, of the $N = 4$ Yang-Mills theory, the demonstration that $N = 2$ Yang-Mills theory coupled to $N = 2$ matter has a perturbative beta-function that only has one-loop contributions and the existence of a large class of superconformally invariant quantum $N = 2$ theories. Details can be found in chapter 18 of Ref.[1].

SUPERCONFORMAL THEORIES

The Geometry of Superconformal Transformations

The superspace that we used was defined as the coset space of the super-Poincare group divided by the Lorentz group and internal symmetry group, see chapter 14 of Ref.[1]. This superspace is called Minkowski superspace. For the case of the $N = 1$ super-Poincare group, the superspace is parameterised by the coordinates $Z^M = (x^\mu, \theta^A, \theta^{\dot{A}})$ corresponding to the generators P^μ and $Q^A, Q^{\dot{A}}$ which generate transformations that are not contained in the isotropy subgroup. We can construct on superspace a set of preferred frames with supervierbeins $E_M{}^\pi$. The covariant derivatives are given by $D_M = E_M^\pi \frac{\partial}{\partial Z^\pi}$. Their precise form being

$$D_A = \frac{\partial}{\partial \theta^A} - i(\sigma^m)_{A\dot{B}}\theta^{\dot{B}}\partial_m, \; D_{\dot{A}} = \frac{\partial}{\partial \theta^{\dot{A}}} - i(\sigma^m)_{B\dot{A}}\theta^B\partial_m, \tag{6.1}$$

and

$$D_m = \frac{\partial}{\partial x^m} \equiv \partial_m \tag{6.2}$$

We can read off the components of the inverse supervierbien from these equations.

For superconformal theories it is more natural to consider a superspace which is constructed from the coset space found by dividing the superconformal group by the subgroup which is generated by Lorentz transformations $J_{\mu\nu}$, dilations D, special translations K_μ and special supersymmetry transformations $S_{Ai}, S_{\dot{A}i}, i = 1, \ldots, N$ and the internal symmetry generators. The internal group for the superconformal algebra contains the group $U(N) = SU(N) \times U(1)$ although in the case of $N = 4$ the $U(1)$ factor does not act on the supercharges.

This coset construction leads to the same Minkowski superspace with the same transformations for the super-Poincare group, but it has the advantage that it automatically encodes the action of the superconformal transformations on the superspace. These transformations were first calculated by Martin Sohnius in Ref.[15].

The purpose of this section is to give an alternative method of calculating the superconformal transformations in four dimensions which will enable us to give a compact superspace form for the superconformal transformations. In particular, all the parameters of the transformations will be encode in one superfield which we can think of as the superspace equivalent of a conformal Killing vector. This formulation was first given by B. Conlong and P. West and can be found in Ref.[16]. One reason for reviewing this work here is that there is still not a readable account readily available in the literature. This section was written in collaboration with B. Conlong. Some reviews on this subject can be found in[17].

Conformal transformations in Minkowski space are defined to be those transformations which preserve the Minkowski metric up to scale (see chapter 25 of Ref.[1] for a review). However, superspace does not have a natural metric since the tangent space group contains the Lorentz group which does not relate the bosonic to the fermionic sectors of the tangent space. The treatment we now give follows that given in chapter 25 of Ref.[1] for the case of two dimensional superconformal transformations.

There are two methods to define a superconformal transformation.

1. We can demand that it is a superdiffeomorphism which preserves part of the bosonic part of the supersymmetric line element

   $$dL^m dL^n \eta_{mn} \tag{6.3}$$

   where

   $$dL^m \equiv dZ^M E_M^{\ m} = dx^m + i\bar{\theta}\gamma^m d\theta \tag{6.4}$$

   up to an arbitrary local scale factor.

2. We can alternatively demand that it is a superdiffeomorphism which preserves the spinor components of the superspace covariant derivatives up to an arbitrary local scale factor. More precisely, a superconformal transformation is one such that

   $$D_A \mapsto D'_A = f_A{}^B(z) D_B, \quad \bar{D}_{\dot{A}} \mapsto \bar{D}'_{\dot{A}} = \bar{f}_{\dot{A}}{}^{\dot{B}}(z)\bar{D}_{\dot{B}} \tag{6.5}$$

   We note that the transformation must preserve each chirality spinor derivative separately. In fact, these two definitions are equivalent and we will work with only the second definition.

Carrying out a super-reparameterisation $z^M = (x^\mu, \theta^A, \bar{\theta}^{\dot{A}}) \mapsto z^{M\prime} = (x^{\mu\prime}, \theta^{A\prime}, \bar{\theta}^{\dot{A}\prime})$ upon the spinorial covariant derivatives we find that a finite superconformal transformation obeys the constraints

$$D_A \bar{\theta}'^{\dot{B}} = 0, \quad D_A x'^\mu + i\sigma^\mu_{B\dot{B}} \bar{\theta}'^{\dot{B}} (D_A \theta'^B) = 0 \tag{6.6}$$

and

$$\bar{D}_{\dot{A}} \theta'^B = 0, \quad \bar{D}_{\dot{A}} x'^\mu + i\sigma^\mu_{B\dot{B}} \theta'^B (\bar{D}_{\dot{A}} \bar{\theta}'^{\dot{B}}) = 0 \tag{6.7}$$

The corresponding transformation of the covariant derivatives being

$$D_A = (D_A \theta'^B) D'_B, \quad \bar{D}_{\dot{A}} = (\bar{D}_{\dot{A}} \bar{\theta}'^{\dot{B}}) \bar{D}'_{\dot{B}} \tag{6.8}$$

We now consider an infinitesimal transformation

$$z^\pi \mapsto z'^\pi = z^\pi + G^\pi \tag{6.9}$$

where $G^\pi = (G^\mu, G^A, G^{\dot{A}})$ is a set of infinitesimal superfields. Equations (6.6) and (6.7) then become

$$D_A G^{\dot{A}} = 0, \quad D_A G^\mu + i\sigma^\mu_{A\dot{B}} G^{\dot{B}} + i\sigma^\mu_{B\dot{B}} \bar{\theta}^{\dot{B}} D_A G^B = 0 \tag{6.10}$$

and

$$\bar{D}_{\dot{A}}\bar{G}^{B} = 0, \quad \bar{D}_{\dot{A}}G^{\mu} + i\sigma^{\mu}_{A\dot{A}}G^{A} + i\sigma^{\mu}_{B\dot{B}}\theta^{B}\bar{D}_{\dot{A}}\bar{G}^{\dot{B}} = 0 \tag{6.11}$$

The vector field corresponding to such an infinitesimal transformation is given by $V = G^{\pi}\partial_{\pi}$. However, this can also be written as $V = F^{M}D_{M}$ where the change of basis corresponds to the relation $F^{M} = G^{\pi}E_{\pi}^{\ M}$., In terms of components this change is given by

$$F^{\mu} = G^{\mu} - i\sigma^{\mu}_{A\dot{A}}\bar{\theta}^{\dot{A}}G^{A} - i\sigma^{\mu}_{A\dot{A}}\theta^{A}G^{\dot{A}} \tag{6.12}$$

as well as

$$F^{A} = G^{A}, \ F^{\dot{A}} = G^{\dot{A}} \tag{6.13}$$

We shall denote the vector component by $F^{\mu}, \ldots$ or $F^{n}, \ldots$ even though it should strictly speaking carry the latter m, n,... indices. It is straightforward to verify that equations (6.10) and (6.11) now take the neater form

$$D_{A}F^{\mu} = -2i\sigma^{\mu}_{A\dot{A}}\bar{F}^{\dot{A}}, \ \bar{D}_{\dot{A}}F^{\mu} = -2i\sigma^{\mu}_{A\dot{A}}F^{A} \tag{6.14}$$

and

$$D_{\dot{A}}F^{B} = 0 = D_{A}F^{\dot{B}}. \tag{6.15}$$

A somewhat quicker derivation of this result can be given by first writing the infinitesimal change in the covariant derivatives under an infinitesimal superdiffeomorphism in the form $D_{M} \to D_{M} + [V, D_{M}]$. Using the form for V given above which contains the covariant derivatives and then using the fact that the only non-zero commutator or anti-commutator, where appropriate, of the covariant derivatives is

$$\{D_{A}, D_{\dot{B}}\} = -2i(\sigma^{n})_{A\dot{B}}\partial_{n} \tag{6.16}$$

we recover equations (6.14) and (6.15).

Equation (6.14) can be rewritten as

$$F^{\dot{A}} = -\frac{1}{8i}\sigma_{\mu}^{A\dot{A}}D_{A}F^{\mu}, \ F^{A} = -\frac{1}{8i}\sigma_{\mu}^{A\dot{A}}\bar{D}_{\dot{A}}F^{\mu} \tag{6.17}$$

from which it is apparent that all transformations may be expressed in terms of F^{μ} alone. Using equations (6.12) and (6.13) we find that the explicit transformations of the coordinates are given by

$$\begin{aligned} x'^{\mu} &= x^{\mu} + (F^{\mu} - \frac{1}{8}\theta^{A}D_{A}F^{\mu} + \frac{1}{8}\theta^{A}(\sigma^{\mu}{}_{\nu})_{AC}D^{C}F^{\nu} - \frac{1}{8}\bar{\theta}^{\dot{A}}\bar{D}_{\dot{A}}F^{\mu} + \frac{1}{8}\bar{\theta}^{\dot{A}}(\bar{\sigma}^{\mu}{}_{\nu})_{\dot{A}\dot{C}}\bar{D}^{\dot{C}}F^{\nu}) \\ \theta'^{A} &= \theta^{A} - \frac{1}{8i}\sigma_{\mu}^{A\dot{A}}\bar{D}_{\dot{A}}F^{\mu} \\ \bar{\theta}'^{\dot{A}} &= \bar{\theta}^{\dot{A}} - \frac{1}{8i}\sigma_{\mu}^{A\dot{A}}D_{A}F^{\mu} \end{aligned} \tag{6.18}$$

Let us define $F_{B\dot{B}} \equiv \sigma^{\mu}_{B\dot{B}}F_{\mu}$, whereupon equation (6.14) becomes

$$D_{A}F_{B\dot{B}} = -4i\varepsilon_{AB}F_{\dot{B}}, \ \ \bar{D}_{\dot{A}}F_{B\dot{B}} = -4i\varepsilon_{\dot{A}\dot{B}}F_{B} \tag{6.19}$$

from which we may deduce the constraint

$$D_{(A}F_{B)\dot{B}} = 0, \ \ \bar{D}_{(\dot{A}}F_{|B|\dot{B})} = 0 \tag{6.20}$$

Acting with D_{C} on $D_{A}F_{B\dot{B}}$ and using equation (6.19) we conclude that

$$D_{C}D_{A}F_{B\dot{B}} = -4i\varepsilon_{AB}D_{C}F_{\dot{B}} = -D_{A}D_{C}F_{B\dot{B}} = -4i\varepsilon_{CB}D_{A}F_{\dot{B}} = 0. \tag{6.21}$$

The last step follows by tracing with ε^{AB}. Consequently, we find that equation (6.15) follows from equation (6.14) or equivalently equation (6.20) and so the superconformal transformations are encoded in $F_{B\dot{B}}$ subject to equation (6.20). We shall refer to equation (6.20) as the superconformal Killing equation, and the field $F_{B\dot{B}}$ as the superconformal Killing vector, these being the natural analogues of the conformal Killing equation and the usual Killing vector in Minkowski space.

To find the consequences for the x-space component fields within $F_{B\dot{B}}$ we expand the superfield $F_{B\dot{B}}$ as a Taylor series in θ and solve the superconformal Killing equation order by order in θ. Writing $F_{B\dot{B}}$ as

$$\begin{aligned} F_{B\dot{B}}(x,\theta,\bar{\theta}) &= \zeta_{B\dot{B}}(x) + \theta^A \chi_{AB\dot{B}}(x) + \bar{\theta}^{\dot{A}} \bar{\chi}_{\dot{A}B\dot{B}}(x) + \theta^A \bar{\theta}^{\dot{A}} A_{AB\dot{B}\dot{A}}(x) \\ &+ \frac{1}{2}\theta^2 f_{B\dot{B}}(x) + \frac{1}{2}\bar{\theta}^2 g_{B\dot{B}}(x) \\ &+ \frac{1}{2}\theta^2 \bar{\theta}^{\dot{A}} \lambda_{B\dot{B}\dot{A}}(x) + \frac{1}{2}\bar{\theta}^2 \theta^A \mu_{BA\dot{B}}(x) + \frac{1}{4}\theta^2\bar{\theta}^2 \kappa_{B\dot{B}}(x) \end{aligned} \tag{6.22}$$

and substituting this expression into the superconformal Killing equation we find that the resulting constraints are solved by the solution

$$\begin{aligned} F_{B\dot{B}} &= \zeta_{B\dot{B}} + \theta_B \bar{\chi}_{\dot{B}} - \bar{\theta}_{\dot{B}} \chi_B + i\theta^A \bar{\theta}^{\dot{A}} \sigma^\mu_{B\dot{A}} \partial_\mu \zeta_{A\dot{B}} \\ &+ i\theta_B \bar{\theta}_{\dot{B}} \alpha - \frac{i}{2}\theta^2 \bar{\theta}^{\dot{A}} \sigma^\mu_{B\dot{B}} \partial_\mu \bar{\chi}_{\dot{B}} \\ &+ \frac{i}{2}\bar{\theta}^2 \theta^A \sigma^\mu_{A\dot{B}} \partial_\mu \chi_B \end{aligned} \tag{6.23}$$

In this equation a is constant, $\zeta_{B\dot{B}}$ is a conformal Killing vector which satisfies

$$\partial_{(A}{}^{(\dot{A}} \zeta_{B)}{}^{\dot{B})} = 0 \tag{6.24}$$

and χ_B and $\bar{\chi}_{\dot{B}}$ are conformal spinors which obey the relation

$$\partial_{(A}{}^{\dot{A}} \chi_{B)} = 0, \quad \partial_{(\dot{A}}{}^{A} \bar{\chi}_{\dot{B})} = 0 \tag{6.25}$$

The solutions to equations (6.24) and (6.25) are given by

$$\begin{aligned} \zeta_\mu &= a_\mu + \lambda x_\mu + \Lambda_\mu{}^\nu x_\nu + k_\mu x^2 - 2(k\cdot x)x_\mu \\ \chi_B &= 4(i\varepsilon_B + \bar{\eta}^{\dot{B}} \sigma^\mu_{B\dot{B}} x_\mu) \\ \bar{\chi}_{\dot{B}} &= -4(i\bar{\varepsilon}_{\dot{B}} + \eta^B \sigma^\mu_{B\dot{B}} x_\mu) \end{aligned} \tag{6.26}$$

where a_μ, λ, $\Lambda_{\mu\nu}$, ε_B and η_B are constant parameters.

Combining equations (6.26) and (6.23) it is clear that the parameters a_μ, λ, $\Lambda_{\mu\nu}$, k_μ, α, ε_B and η_B are translations, dilations, Lorentz rotations, special conformal transformations, chiral transformations, chiral rotations, supersymmetry transformations and special supersymmetry transformations respectively.

Having found the superconformal transformations on superspace we now turn our attention to the transformations of superfields under a superconformal transformation. If φ is a general superfield, which may carry Lorentz indices, then, its transformation is of the form

$$\delta\varphi(z) = \delta z^\pi \partial_\pi \varphi(z) + J\varphi(z) \tag{6.27}$$

where J is a superfield which arises from the non-trivial action of generators from the isotropy group acting on φ at the origin of the superspace. This factor is most pedagogically worked out by considering the superfields as induced representations.

However, here we content ourselves with the final result which for a general superfield φ is given by

$$\begin{aligned}\delta\varphi = &(F\cdot\partial)\varphi - \frac{1}{8i}\sigma_\mu^{A\dot{A}}(\bar{D}_{\dot{A}}F^\mu D_A\varphi + D_A F^\mu \bar{D}_{\dot{A}}\varphi)\\ &+\left\{\frac{1}{4}(\partial\cdot F)\Delta + \frac{1}{96i}\sigma^{\mu A\dot{A}}([D_A,\bar{D}_{\dot{A}}]F_\mu)\mathcal{A} - 2(\partial^{[\mu}F^{\nu]})\sigma_{\mu\nu}\right.\\ &\left.+\frac{1}{8}\partial^\mu(D^A F_\mu)S_A + \frac{1}{8}\partial^\mu(\bar{D}^{\dot{A}}F_\mu)\bar{S}_{\dot{A}} - \frac{1}{8}\partial^\mu(\partial\cdot F)\kappa_\mu\right\}\varphi\end{aligned} \tag{6.28}$$

In this equation the symbols $\{\Delta, \Sigma_{\mu\nu}, \mathcal{A}, S_A, \bar{S}_{\dot{A}}, \kappa_\mu\}$ are constants that are the values of the corresponding generators of the isotropy group acting on the superfield when it is taken to be at the origin of superspace. For almost all known situations, only the parameters Δ, $\Sigma_{\mu\nu}$ and $\mathcal{A}$, which correspond to the dilation, Lorentz and $U(1)$ transformations respectively, are non-zero. The first part of the result is just the shift in the coordinates which is given by

$$\delta z^\pi \partial_\pi\varphi = (F\cdot\partial)\varphi - \frac{1}{8i}\sigma_\mu^{A\dot{A}}(\bar{D}_{\dot{A}}F^\mu D_A\varphi + D_A F^\mu\bar{D}_{\dot{A}}\varphi) \tag{6.29}$$

while J is given by

$$\begin{aligned}J = &\frac{1}{4}(\partial\cdot F)\Delta + \frac{1}{96i}\sigma^{\mu A\dot{A}}[D_A,\bar{D}_{\dot{A}}]F_\mu\mathcal{A} - 2\partial^{[\mu}F^{\nu]}\sigma_{\mu\nu}\\ &+\frac{1}{8}\partial^\mu(D^A F_\mu)S_A + \frac{1}{8}\partial^\mu(\bar{D}^{\dot{A}}F_\mu)\bar{S}_{\dot{A}} - \frac{1}{8}\partial^\mu(\partial\cdot F)\kappa_\mu\end{aligned} \tag{6.30}$$

We can verify that equation (6.28) reproduces some of the known results. Let us consider dilations which are generated by taking $F_\mu = \lambda x_\mu$. For this case, equation (6.28) becomes

$$\delta\varphi = \lambda(x^\mu\partial_\mu + \frac{1}{2}\theta^A\partial_A + \frac{1}{2}\bar{\theta}^{\dot{A}}\bar{\partial}_{\dot{A}})\varphi + \lambda\Delta\varphi \tag{6.31}$$

which we recognise as the well known result. In fact, by writing J as the most general form possible which is linear in $F_{B\dot{B}}$, contains covariant derivatives and is consistent with dimensional analysis, evaluating the result for particular transformations we can also arrive at the correct J.

We can apply equation (6.28) to the case of a chiral and anti-chiral superfield. For simplicity, let us consider a lorentz scalar chiral superfield whose $S_A, \bar{S}_{\dot{A}}, \kappa_\mu$ values also vanish. The result is

$$\delta\varphi = (F\cdot\partial)\varphi - \frac{1}{8i}\sigma_\mu^{A\dot{A}}\bar{D}_{\dot{A}}F^\mu D_A\varphi + \Delta(\frac{1}{4}(\partial\cdot F)\varphi + \frac{1}{48i}\sigma^{\mu A\dot{A}}([D_A,\bar{D}_{\dot{A}}]F_\mu)\varphi) \tag{6.32}$$

and

$$\delta\bar{\varphi} = (F\cdot\partial)\bar{\varphi} - \frac{1}{8i}\sigma_\mu^{A\dot{A}}D_A F^\mu\bar{D}_{\dot{A}}\bar{\varphi} + \Delta(\frac{1}{4}(\partial\cdot F)\bar{\varphi} - \frac{1}{48i}\sigma^{\mu A\dot{A}}([D_A,\bar{D}_{\dot{A}}]F_\mu)\bar{\varphi}) \tag{6.33}$$

The reader will observe that the dilation and A weights of the chiral superfield are tied together, a fact that can be established by taking the straightforward reduction of equation (6.28) and making sure the transformed superfield is still chiral or anti-chiral as appropriate. We will discuss this result from a more general perspective in the next section.

Anomalous Dimensions of Chiral Operators at a Fixed Point

Let us consider a supersymmetric theory at a fixed point of the renormalisation group, i.e. $\beta = 0$. Such a theory should be invariant under superconformal transformations. As in all supersymmetric theories some of the observables are given by chiral operators which by definition obey the equation

$$D_{\dot{A}}\varphi = 0 \tag{6.34}$$

where φ denotes the chiral operator involved. It follows that this equation must itself be invariant under any superconformal transformation i.e. $D_{\dot{A}}\delta\varphi = 0$. Choosing a special supersymmetry transformation we conclude that

$$\{D_{\dot{A}}, S_{\dot{B}}\}\varphi = 0 \tag{6.35}$$

In this equation we can swop the covariant derivative for the generator of supersymmetry transformations using the equation

$$D_{\dot{A}} = \frac{\partial}{\partial\theta^{\dot{A}}} - i(\sigma^m)_{A\dot{A}}\theta^A\partial_m = Q_{\dot{A}} - 2i(\sigma^m)_{A\dot{A}}\theta^A\partial_m \tag{6.36}$$

We then conclude that

$$\{Q_{\dot{A}}, S_{\dot{B}}\}\varphi = 0 \tag{6.37}$$

plus terms that contain space-time derivatives. However, in this equation the condition must hold separately on the parts of the equation containing space-time derivatives and those that do not. The advantage of writing the equation in this form is that the anti-commutator is one of the defining relations of the superconformal algebra, namely

$$\{Q_{\dot{A}}, S_{\dot{B}}\} = \epsilon_{\dot{A}\dot{B}}(2D - 4iA) + (\bar{\sigma}^{\mu\nu})_{\dot{A}\dot{B}}J_{\mu\nu} \tag{6.38}$$

where D and A are the generators of the dilations and $U(1)$ transformations in the superconformal algebra which we gave in the first part of these lectures. If we restrict the superconformal algebra to just its super-Poincare subgroup then the A generator is identified with the generator of R transformations. The latter satisfies the relation $[Q_A, R] = iQ_A$ comparing this with the equivalent commutator in the superconformal group (i.e. $[Q_A, A] = -i\frac{3}{4}Q_A$) we thus find that the generators are related by $A = -\frac{3}{4}R$.

Consequently for a Lorentz invariant chiral operator we conclude that $D = -i\frac{3}{2}R$. One can also find this result by substituting the explicit expressions for $D_{\dot{A}}$ and $S_{\dot{A}}$ in equation (6.35) and setting $\theta = 0$. We summarise the result in the theorem

Theorem[18]

Any Lorentz invariant operator in a four dimensional supersymmetric theory at a fixed point has its anomalous dimension Δ and chiral R weight, related by the equation

$$\Delta = -i\frac{3}{2}R. \tag{6.39}$$

In any conformal theory we can determine the two and three point Green's functions using conformal invariance alone. However, one can not normally use this symmetry alone to fix the anomalous weights of any operators. Since non-trivial fixed points are outside the range of usual perturbation theory, these must be calculated using techniques such as the **ϵ-expansion**. The result so obtained are approximations and in some case one can not reliably calculate the anomalous dimensions at all. However, in supersymmetric theories at a fixed point one can determine the anomalous dimensions of chiral operators in superconformal theories exactly in terms of their R weight. However, in many situations one does know the R weight of the chiral operators of interest and we so can indeed exploit the above theorem to find their anomalous dimensions exactly[18]. We shall shortly demonstrate this procedure with some examples.

We must first fix the normalisation of the dilation and R weights that is implied by the superconformal algebra. The relation $[P_\mu, D] = P_\mu$ implies that P_μ has dilation weight one. On the other hand, the relationship $[Q_A, R] = iQ_A$ implies that Q_A has R weight 1. Consequently, θ^A has R weight -1 meaning that it transforms as $\theta^A \to e^{-ia}\theta^A$ where a is the parameter of R transformations.

As our first example, let us consider the Wess-Zumino model in four dimensions and suppose that it had a non-trivial fixed point at which the interaction was of the usual form;

$$\int d^4x d^2\theta \varphi^3 \tag{6.40}$$

Using the above scaling of θ we find that φ transforms as $\varphi \to e^{i\frac{2}{3}a}\varphi$ and as a result φ has R weight $\frac{2}{3}$. Using our theorem we find that φ had dilation weight one. This is the canonical dilation weight of φ, that is, the weight it would have in the free theory. It can be argued that if φ has its canonical weight then the theory must be free and so such a non-trivial fixed point can not exist[19]. It can also be argued that this result implies the the Wess-Zumino model is a trivial field theory meaning that the only consistent value of the coupling constant as we remove the cutoff is zero[20].

Now let us consider the Wess-Zumino model in three dimensions and suppose it has a non-trivial fixed point at which the interaction is given by

$$\int d^3x d^2\theta \varphi^3. \tag{6.41}$$

This is the supersymmetric generalisation of the Ising model. Running through the same argument as above, but taking into account the modified form of the three dimensional superconformal algebra, we find that φ has anomalous dimension $\frac{1}{6}$. However, in this case the canonical weight of φ is $1/2$. Clearly, the theory can not be free with such a dilation weight. Such a non-trivial fixed point is known to exist by using the epsilon expansion which also gives an anomalous dimension in agreement with this result[9].

The theorem in this section can also be used to fix the anomalous dimensions for the chiral operators in the two dimensional $N = 2$ supersymmetric Landau-Ginsburg models whose superpotential at the fixed point take the form

$$\int d^2x d^2\theta \varphi^n + c.c \tag{6.42}$$

The anomalous dimensions agree with the correspondence between these models at their fixed points and the $N = 2$ minimal series of superconformal models. This result was first conjectured in[21] and shown by using the epsilon expansion in Ref.[22].

The theorem can also be applied to four dimensional gauge invariant operators composed form the $N = 1$ Yang-Mills field strength W_A. Such a connection was used to argue that $N = 1$ super QED is trivial[18] and has been used extensively by Seiberg in recent work on dualities between certain $N = 1$ supersymmetric theories.

REFERENCES

1. P. West, "Introduction to Supersymmetry and Supergravity", (1990), Extended and Revised Second Edition, World Scientific Publishing, Singapore.

2. S. Coleman and J. Mandula, *Phys. Rev.* **159**, 1251 (1967).
3. Y.A. Golfand and E.S. Likhtman, *JETP Lett.* **13,** 323 (1971).
4. R. Hagg, J. Lopuszanaki and M. Sohnius, *Nucl. Phys.* **B88**, 61 (1975).
5. J. Wess and B. Zumino, *Nucl. Phys.* **B70**, 139 (1974).
6. P. Ramond, *Phys. Rev.* **D3**, 2415 (1971); A. Neveu and J.H. Schwarz, *Nucl Phys.* **B31**, 86 (1971); *Phys. Rev.* **D4**, 1109 (1971); J.-L. Gervais and B. Sakita, *Nucl. Phys.* **B34**, 477, 632 (1971); F. Gliozzi, J. Scherk and D.I. Olive, *Nucl. Phys.* **B122**, 253 (1977).
7. P. West, *Nucl. Phys.* **B106**, 219 (1976).
8. M. Grisaru, M. Rocek and W. Siegel, *Nucl. Phys.* **B159**, 429 (1979).
9. P. Howe and P. West, "Chiral Correlators in Landau-Ginsburg Theories and $N = 2$ Super-conformal models", *Phys. Lett.* **B227**, 397 (1989).
10. P. West, "A Comment on the Non-Renomalization Theorem in Supersymmetric Theories", *Phys. Lett.* **B258**, 369 (1991).
11. I. Jack and T. Jones and P. West, "Not the no-renomalization Theorem" , *Phys. Lett.* **B258**, 375 (1991).
12. P. West, "Quantum Corrections to the Supersymmetric Effective Superpotential and Resulting Modifications of Patterns of Symmetry Breaking", *Phys. Lett.* **B261**, 396 (1991).
13. M. Shifman and A. Vainshtein, *Nucl. Phys.* **B359**, 571 (1991).
14. L. Dixon, V. Kaplunovsky and J. Louis, *Nucl. Phys.* **B355**, 649 (1991).
15. M. Sohnius, PhD thesis, University of Karlsruhe, (1976); *Phys. Rep.* **128**, 39 (1985).
16. B. P. Conlong and P. West, *in*: B. P. Conlong, Ph. D. Thesis, University of London (1993).
17. I. Buchbinder and S. Kuenko, " Ideas and Methods of Supersymmetry and Supergravity", (1995), Institute of Physics; John Park, "$N = 1$ Superconformal Symmetry in 4 Dimensions", hep-th/9703191.
18. B. P. Conlong and P. C. West, *J. Phys.* **A26**, 3325 (1993).
19. S. Ferrara, J. Iliopoulos and B. Zumino,*Nucl. Phys.* **B77**, 413 (1974).
20. J. Verbaarschot and P. West, "Renormalons in Supersymmetric Theories", *Int. J. Mod. Phys.* **A6**, 2361 (1991).
21. D. Kastor, E. Martinec and S. Shenker, "RG flow in $N = 1$ Discrete Series", (1988), EFI preprint 88-31.
22. P. Howe and P. West, "$N = 2$ Superconformal Models, Landau-Ginsburg Hamiltonians and the epsilon-expansion", *Phys. Lett.* **B223**, 377 (1989).

NON-PERTURBATIVE GAUGE DYNAMICS IN SUPERSYMMETRIC THEORIES. A PRIMER

M. Shifman

Theoretical Physics Institute, Univ. of Minnesota
Minneapolis, MN 55455, USA

ABSTRACT

I give an introductory review of recent, fascinating developments in supersymmetric gauge theories. I explain pedagogically the miraculous properties of supersymmetric gauge dynamics allowing one to obtain exact solutions in many instances. Various dynamical regimes emerging in supersymmetric Quantum Chromodynamics and its generalizations are discussed. I emphasize those features that have a chance of survival in QCD and those which are drastically different in supersymmetric and non-supersymmetric gauge theories. Unlike most of the recent reviews focusing almost entirely on the progress in extended supersymmetries (the Seiberg-Witten solution of $N = 2$ models), these lectures are mainly devoted to $N = 1$ theories. Developments "after Seiberg" (domain walls in supersymmetric gluodynamics) are briefly discussed.

PREFACE

All fundamental interactions established in nature are described by non-Abelian gauge theories. The standard model of the electroweak interactions belongs to this class. In this model, the coupling constant is weak, and its dynamics is fully controlled (with the possible exception of a few, rather exotic problems, like baryon number violation at high energies).

Another important example of non-Abelian gauge theories is Quantum Chromodynamics (QCD). This theory has been under intense scrutiny for over two decades, yet remains mysterious. Interaction in QCD becomes strong at large distances. What is even worse, the degrees of freedom appearing in the Lagrangian (microscopic variables – colored quarks and gluons in the case at hand) are not those degrees of freedom that show up as physical asymptotic states (macroscopic degrees of freedom – colorless hadrons). Color is permanently confined. What are the dynamical reasons of this phenomenon?

Color confinement is believed to take place even in pure gluodynamics, i.e. with no dynamical quarks. Adding massless quarks produces another surprise. The chiral symmetry of the quark sector, present at the Lagrangian level, is spontaneously broken

Confinement, Duality, and Nonperturbative Aspects of QCD
Edited by Pierre van Baal, Plenum Press. New York, 1998

(realized nonlinearly) in the physical amplitudes. Massless pions are the remnants of the spontaneously broken chiral symmetry. What can be said, theoretically, about the pattern of the spontaneous breaking of the chiral symmetry?

Color confinement and the spontaneous breaking of the chiral symmetry are the two most sacred questions of strong non-Abelian dynamics; and the progress of understanding them is painfully slow. At the end of the 1970's Polyakov showed that in 3-dimensional compact electrodynamics (the so called Georgi-Glashow model, a primitive relative of QCD) color confinement does indeed take place[1]. Approximately at the same time a qualitative picture of how this phenomenon could actually happen in 4-dimensional QCD was suggested by Mandelstam[2] and 't Hooft[3]. Some insights, though quite limited, were provided by models of the various degree of fundamentality, and by numerical studies on lattices. This is, basically, all we had before 1994, when a significant breakthrough was achieved in understanding both issues in supersymmetric (SUSY) gauge theories.

Unlike the Georgi-Glashow model in three dimensions mentioned above, which is quite a distant relative of QCD, four-dimensional supersymmetric gluodynamics and supersymmetric gauge theories with matter come much closer to genuine QCD. Moreover, the dynamics of these theories is rich and interesting by itself, which accounts for the attention they have attracted in the last three years when we have been witnessing an unprecedented progress. It turns out that supersymmetry helps unravel several intriguing and extremely elegant properties which shed light on subtle aspects of the gauge theories in general. These developments can, eventually, lead us to a breakthrough in QCD. Today we are definitely one step closer to the dream of every QCD practitioner: understanding of two most salient properties of QCD, color confinement and spontaneous breaking of the chiral symmetry, in analytic terms.

Strongly coupled supersymmetric gauge theories is the topic of this lecture course. My task, as I see it, is educational rather than providing a comprehensive coverage. I will try to be as pedagogical as possible focusing on basic ideas and approaches and avoiding, whenever possible, more technical and involved aspects.

BASICS OF SUPERSYMMETRIC GAUGE THEORIES

In this Lecture we will start our excursion in supersymmetric gauge theories. There is a long way to go before we will be able to discuss a variety of fascinating results obtained in this field recently. As a first step let me briefly review some basic features of these theories and key elements of the formalism we will need below.

Introducing supersymmetry

Supersymmetry relates bosonic and fermionic degrees of freedom[4,5]. A necessary condition for any theory to be supersymmetric is the balance between the number of the bosonic and fermionic degrees of freedom, having the same mass and the same "external" quantum numbers, e.g. color. Let us consider several simplest examples of practical importance.

A scalar complex field ϕ has two degrees of freedom (a particle plus antiparticle). Correspondingly, its fermion superpartner is the Weyl (two-component) spinor, which also has two degrees of freedom – say, the left-handed particle and the right-handed antiparticle. Alternatively, instead of working with the complex fields, one can introduce real fields, with the same physical content: two real scalar fields ϕ_1 and ϕ_2 describing two "neutral" sin-0 particles, plus the Majorana (real four-component) spinor describing a "neutral" spin-1/2 particle with two polarizations. (By neutral I mean that the corresponding antiparticles are identical to their particles). This family has a balanced number of the degrees of freedom both in the massless and massive cases. Below we will see that in the superfield formalism it is described, in a concise form, by one *chiral superfield.*

When we speak of the quark flavors in QCD we count the Dirac spinors. Each Dirac spinor is equivalent to two Weyl spinors. Therefore, in supersymmetric QCD (SQCD) each flavor requires two chiral superfields. Sometimes, the superfields from this chiral pair are referred to as subflavors. Two subflavors comprise one flavor.

Another important example is vector particles, gauge bosons (gluons in QCD, W bosons in the Higgs phase). Each gauge boson carries two physical degrees of freedom (two transverse polarizations). The appropriate superpartner is the Majorana spinor. Unlike the previous example the balance is achieved only for massless particles, since the massive vector boson has three, not two, physical degrees of freedom. The superpartner to the massless gauge boson is called gaugino. Notice that the mass still can be introduced through the (super) Higgs mechanism. We will discuss the Higgs mechanism in supersymmetric gauge theories later on.

In counting the degrees of freedom above, the external quantum numbers were left aside. Certainly, they should be the same for each member of the superfamily. For instance, if the gauge group is $SU(2)$, the gauge bosons are "color" triplets, and so are gauginos. In other words, the Majorana fields describing gauginos are provided by the "color" index a taking three different values, $a = 1, 2, 3$.

If we consider the free field theory with the balanced number of degrees of freedom, the vacuum energy vanishes. Indeed, the vacuum energy is the sum of the zero-point oscillation frequencies for each mode of the theory,

$$E_{\rm vac} = \sum_{\vec{k}} \left\{ \omega_{\vec{k}}^{\rm boson} - \omega_{\vec{k}}^{\rm ferm} \right\} . \tag{1.1}$$

I remind that the modes are labeled by the three-momentum $\vec{k}$; say, for massive particles

$$\omega_{\vec{k}} = \sqrt{m^2 + \vec{k}^2} .$$

It is important that the boson and fermion terms enter with the opposite signs and cancel each other, term by term. This observation, which can be considered as a precursor to supersymmetry, was made by Pauli[6] in 1950! If interactions are introduced in such a way that supersymmetry remains unbroken, the vanishing of the vacuum energy is preserved in dynamically nontrivial theories.

Balancing the number of degrees of freedom is the necessary but not sufficient condition for supersymmetry in dynamically nontrivial theories, of course. All vertices must be supersymmetric too. This means that each line can be substituted by that of a superpartner. Let us consider, for instance, QED, the simplest gauge theory. We start from the electron-electron-photon coupling (Fig. 1*a*). Now, in SQED the electron is accompanied by two selectrons (two, because the electron is described by the four-component Dirac spinor rather than the Weyl spinor). Thus, supersymmetry requires the selectron-selectron-photon vertices, (Fig. 1*b*), with the same coupling constant. Moreover, the photon can be substituted by its superpartner, photino, which generates the electron-selectron-photino vertex (Fig. 1*c*), with the same coupling. In the old-fashioned language of the pre-SUSY era we would call this vertex the Yukawa coupling. In the supersymmetric language this is the gauge interaction since it generalizes the gauge interaction coupling of the photon to the electron.

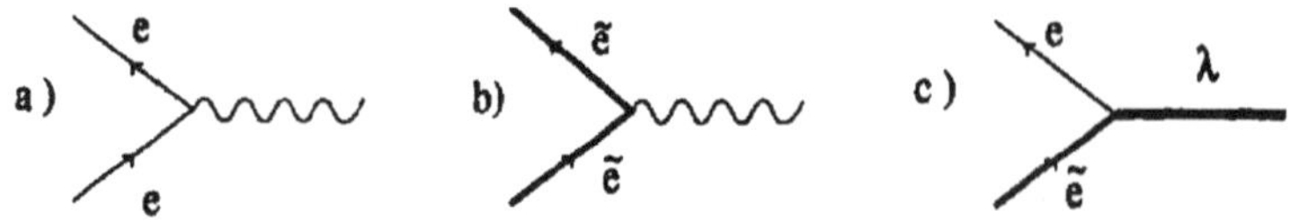

Figure 1. Interaction vertices in QED and its supergeneralization, SQED. (*a*) $\bar{e}e\gamma$ vertex; (*b*) selectron coupling to photon; (*c*) electron-selectron-photino vertex. All vertices have the same coupling constant. The quartic self-interaction of selectrons is also present, but not shown.

With the above set of vertices one can show that the theory is supersymmetric at the level of trilinear interactions, provided that the electrons and the selectrons are degenerate in mass, while the photon and photino fields are both massless. To make it fully supersymmetric one should also add some quartic terms, describing self-interactions of the selectron fields, as we will see shortly.

Now, the theory is dynamically nontrivial, the particles - bosons and fermions – are not free and still $E_{vac} = 0$. This is the *first miracle* of supersymmetry.

The above pedestrian (or step-by-step) approach to supersymmetrizing the gauge theories is quite possible, in principle. Moreover, historically the first supersymmetric model derived by Golfand and Likhtman, SQED, was obtained in this way[4]. This is a painfully slow method, however, which is totally out of use at the present stage of the theoretical development. The modern efficient approach is based on the superfield formalism, introduced in 1974 by Salam and Strathdee[7] who replaced conventional four-dimensional space by superspace.

Superfield formalism: bird's eye view

I will be unable to explain this formalism, even briefly. The reader is referred to the text-books and numerous excellent reviews, see the list of recommended literature at the end. Below some elements are listed mostly with the purpose of introducing relevant notations, to be used throughout the entire lecture course. (A summary of our notation and conventions is given in the Appendix.)

If conventional space-time is parametrized by the coordinate four-vector x_μ, superspace is parametrized by x_μ and two Grassmann variables, θ and $\bar{\theta}$. The Grassmann numbers obey all standard rules of arithmetic except that they anticommute rather than commute with each other. In particular, the product of a Grassmann number with itself is zero, for this reason.

With respect to the Lorentz properties, θ and $\bar{\theta}$ are spinors. As well known, the four-dimensional Lorenz group is equivalent to $SU(2) \times SU(2)$ and, therefore, there exist two types of spinors, left-handed and right-handed, denoted by undotted and dotted indices, respectively; θ_α is the left-handed spinor while $\bar{\theta}_{\dot{\alpha}}$ is the right-handed one $(\alpha, \dot{\alpha} = 1, 2)$. The indices of the right-handed spinors are supplied by dots to emphasize the fact that their transformation law does not coincide with that of the left-handed spinors.

The Lorentz scalars can be formed as a convolution of two dotted or two undotted spinors, $\theta^\alpha\theta_\alpha$ or $\bar{\theta}_{\dot{\alpha}}\bar{\theta}^{\dot{\alpha}}$, with one lower and one upper index. Raising and lowering of indices is realized by virtue of the antisymmetric (Levi-Civita) symbol,

$$\bar{\theta}_{\dot{\alpha}} = \epsilon_{\dot{\alpha}\dot{\beta}}\bar{\theta}^{\dot{\beta}}\,, \quad \theta_\alpha = \epsilon_{\alpha\beta}\theta^\beta\,,$$

where

$$\epsilon^{12} = \epsilon^{\dot{1}\dot{2}} = 1\,; \quad \epsilon_{12} = \epsilon_{\dot{1}\dot{2}} = -1$$

so that $\epsilon^{\alpha\beta}\epsilon_{\beta\gamma} = \delta^\alpha_\gamma$. When one raises or lowers the index of θ the ϵ symbol must be placed to the left of θ.

A shorthand notation when the indices of the spinors are implicit is widely used, for instance,

$$\theta^2 \equiv \theta^\alpha\theta_\alpha$$

and

$$\bar\theta^2 \equiv \bar\theta_{\dot\alpha}\bar\theta^{\dot\alpha} .$$

Notice that in convoluting the undotted indices one writes first the spinor with the upper index while for the dotted indices the first spinor has the lower index. The ordering is important since the elements of the spinors are anticommuting Grassmann numbers.

It remains to be added that the vector quantities can be obtained from two spinors – one dotted and one undotted. Thus, $\theta_\alpha\bar\theta_{\dot\alpha}$ transforms as a Lorentz vector.

Now, we can introduce the notion of supertranslations in the superspace $\{x,\theta,\bar\theta\}$. The generic (infinitesimal) supertransformation has the form

$$\theta \to \theta + \varepsilon\,, \quad \bar\theta \to \bar\theta + \bar\varepsilon\,,$$

$$x_{\alpha\dot\beta} \to x_{\alpha\dot\beta} + 2i\varepsilon_\alpha\bar\theta_{\dot\beta} - 2i\theta_\alpha\bar\varepsilon_{\dot\beta}\,. \tag{1.2}$$

The supertranslations generalize conventional translations in ordinary space.

One can also consider the so called chiral and antichiral superspaces (chiral realizations of the supergroup); the first one does not explicitly contain $\bar\theta$ while the second does not contain θ. It is not difficult to see that a point from the chiral superspace is parametrized by $\{x_L,\theta\}$, and that from the antichiral superspace is parametrized by $\{x_R,\bar\theta\}$. Here

$$(x_L)_{\alpha\dot\alpha} = x_{\alpha\dot\alpha} - 2i\theta_\alpha\bar\theta_{\dot\alpha}\,, \quad (x_R)_{\alpha\dot\alpha} = x_{\alpha\dot\alpha} + 2i\theta_\alpha\bar\theta_{\dot\alpha}\,. \tag{1.3}$$

Under this definition the supertransformations corresponding to the shifts in θ and $\bar\theta$, respectively, leave us inside the corresponding superspace. Indeed, if $\theta \to \theta + \varepsilon$ and $\bar\theta \to \bar\theta + \bar\varepsilon$, then

$$(x_L)_{\alpha\dot\beta} \to (x_L)_{\alpha\dot\beta} - 4i\theta_\alpha\bar\varepsilon_{\dot\beta} \text{ and } (x_R)_{\alpha\dot\beta} \to (x_R)_{\alpha\dot\beta} + 4i\varepsilon_\alpha\bar\theta_{\dot\beta}\,. \tag{1.4}$$

Superfields provide a very concise description of supersymmetry representations. They are very natural generalizations of conventional fields. Say, the scalar field $\phi(x)$ in the $\lambda\phi^4$ theory is a function of x. Correspondingly, superfields are functions of x and θ's. For instance, the chiral superfield $\Phi(x_L,\theta)$ depends on θ and x_L (and has no explicit $\bar\theta$ dependence). If we Taylor-expand it in powers of θ we get the following formula:

$$\Phi(x_L,\theta) = \phi(x_L) + \sqrt{2}\psi(x_L)\theta + \theta^2F(x)\,. \tag{1.5}$$

There are no higher-order terms in the expansion since higher powers of θ vanish due to the Grassmannian nature of this parameter. For the same reason the argument of the last component of the chiral superfield, F, is set equal to x. The distinction between x and x_L is not important in this term. The last component of the chiral superfield is always called F. F terms of the chiral superfields are non-dynamical, they appear in the Lagrangian without derivatives. We will see later that F terms play a distinguished role.

The lowest component of the chiral superfield is a complex scalar field ϕ, and the middle component is a Weyl spinor ψ. Each of these fields describes two degrees of freedom, so the appropriate balance is achieved automatically. Thus, we see that the superfield is a concise form of representing a set of components. The transformation law of the components follows immediately from Eq. (1.4), for instance, $\delta\phi(x) = \sqrt{2}\psi(x)\varepsilon$, and so on.

The antichiral superfields depend on x_R and $\bar\theta$. The chiral and antichiral superfields describe the matter sectors of the theories to be studied below. The gauge field appears from the so called vector superfield V which depends on both, θ and $\bar\theta$ and satisfies the condition $V = V^\dagger$. The component expansion of the vector superfield has the form

$$V(x,\theta,\bar\theta) = C(x) + i\theta\chi(x) - i\bar\theta\bar\chi(x) + \frac{i}{2}\theta^2[M(x)+iN(x)] - \frac{i}{2}\bar\theta^2[M(x)-iN(x)] -$$

$$2\theta_\alpha\bar\theta_{\dot\beta}v^{\alpha\dot\beta}(x) + 2\left\{i\theta^2\bar\theta_{\dot\beta}\left[\bar\lambda^{\dot\beta} - \frac{i}{4}\partial^{\alpha\dot\beta}\chi_\alpha\right] + \text{h.c.}\right\} + \theta^2\bar\theta^2\left[D(x) - \frac{1}{4}\Box C(x)\right]. \quad (1.6)$$

The components C, D, M, N and $v^{\alpha\dot\beta}$ must be real to satisfy the condition $V = V^\dagger$. The vector field $v^{\alpha\beta}$ gives its name to the entire superfield.

The last component of the vector superfield, apart from a full derivative, is called the " D term". D terms also play a special role.

Let me say a few words about the gauge transformations. For simplicity I will consider the case of the Abelian (U(1)) gauge group. In the non-Abelian case the corresponding formulae become more bulky, but the essence stays the same.

As is well known, in nonsupersymmetric gauge theories the matter fields transform under the gauge transformations as

$$\phi(x) \to e^{i\alpha(x)}\phi(x)\,, \quad \phi(x)^\dagger \to e^{-i\alpha(x)}\phi(x)^\dagger\,, \quad (1.7)$$

while the gauge field

$$v_\mu(x) \to v_\mu(x) + \partial_\mu\alpha(x)\,, \quad (1.8)$$

where $\alpha(x)$ is an arbitrary function of x. Equations (1.7) and (1.8) prompt the supersymmetric version of the gauge transformations,

$$\Phi(x_L,\theta) \to e^{i\Lambda}\Phi(x_L,\theta)\,, \quad \bar\Phi(x_R,\bar\theta) \to e^{-i\bar\Lambda}\bar\Phi(x_R,\bar\theta) \quad (1.9)$$

and

$$V \to V - i(\Lambda - \bar\Lambda) \quad (1.10)$$

where Λ is an arbitrary chiral superfield, $\bar\Lambda$ is its antichiral partner. $\bar\Phi e^V\Phi$ is then a gauge invariant combination playing the same role as $\mathcal{D}_\mu\phi^\dagger\mathcal{D}_\mu\phi$ in non-supersymmetric theories. Let me parenthetically note that supersymmetrization of the gauge transformations, Eqs. (1.9), (1.10), was the path which led Wess and Zumino[5] to the discovery of the supersymmetric theories (independently of Golfand and Likhtman).

In components

$$C \to C - i(\phi - \phi^\dagger)\,, \quad \chi \to \chi - \sqrt{2}\psi\,, \quad M + iN \to M + iN + 2F$$

$$v_{\alpha\dot\beta} \to v_{\alpha\dot\beta} + \frac{1}{2}\partial_{\alpha\dot\beta}(\phi + \phi^\dagger)\,, \quad \lambda \to \lambda\,, \quad D \to D\,. \quad (1.11)$$

We see that the C, χ, M and N components of the vector superfield can be gauged away. This is what is routinely done when the component formalism is used. This gauge bears the name of its inventors – it is called the *Wess-Zumino* gauge. Imposing the Wess-Zumino gauge condition in supersymmetric theory one actually does not fix the gauge completely. The component Lagrangian one arrives at in the Wess-Zumino gauge still possesses the gauge freedom with respect to non-supersymmetric (old-fashioned) gauge transformations.

It remains to introduce spinorial derivatives. They will be denoted by capital D and $\bar{D}$,

$$D_\alpha = \frac{\partial}{\partial\theta^\alpha} - i\partial_{\alpha\dot{\beta}}\bar{\theta}^{\dot{\beta}}\,, \quad \bar{D}_{\dot{\alpha}} = -\frac{\partial}{\partial\bar{\theta}^{\dot{\alpha}}} + i\partial_{\beta\dot{\alpha}}\theta^\beta\,. \tag{1.12}$$

The relative signs in Eq. (1.12) are fixed by the requirements $D_\alpha(x_R)_{\beta\dot{\gamma}} = 0$ and $\bar{D}_{\dot{\alpha}}(x_L)_{\beta\dot{\gamma}} = 0$.

To make the spinorial derivatives distinct from the regular covariant derivative the latter will be denoted by the script $\mathcal{D}$. The supergeneralization of the field strength tensor of the gauge field has the form

$$W_\alpha(x_L,\theta) = \frac{1}{8}\bar{D}^2(e^{-V}D_\alpha e^V) =$$

$$i\lambda_\alpha(x_L) - \theta_\alpha D(x_L) - i\theta^\beta G_{\alpha\beta}(x_L) + \theta^2\mathcal{D}_{\alpha\dot{\alpha}}\bar{\lambda}^{\dot{\alpha}}(x_L)\,, \tag{1.13}$$

where $G_{\alpha\beta}$ is the gauge field strength tensor in the spinorial form.

This brief excursion in the formalism, however boring it might seem, is necessary for understanding physical results to be discussed below. I will try to limit such excursions to absolute minimum, but we will not be able to avoid them completely. Now, the stage is set, and we are ready to submerge in the intricacies of the supersymmetric gauge dynamics.

Simplest supersymmetric models

In this section we will discuss some simple models. Our basic task is to reveal general features playing the key role in various unusual dynamical scenarios realized in supersymmetric gauge theories. One should keep in mind that all theories with matter can be divided in two distinct classes: chiral and non-chiral matter. The second class includes supersymmetric generalization of QCD, and all other models where each matter multiplet is accompanied by the corresponding conjugate representation. In other words, a mass term is possible for all matter fields. Even if the massless limit is considered, the very possibility of adding the mass term is very important for dynamics. In particular, dynamical SUSY breaking cannot happen in the non-chiral models.

Models with chiral matter are those where the mass term is impossible. The matter sector in such models is severely constrained by the absence of internal anomalies in the theory. The most well-known example of this type is the *SU*(5) model with equal number of chiral quintets and (anti)decuplets. Each quintet and anti-decuplet, together, is called a *generation*; when the number of generations is three this is nothing but the most popular grand unified theory of electroweak interactions. The chiral models are singled out by the fact that dynamical SUSY breaking is possible, in principle, only in this class. In the present lecture course dynamical SUSY breaking is not our prime concern. Rather, we will focus on various non-trivial dynamical regimes. Most of the regimes to be discussed below manifest themselves in the non-chiral models, which are simpler. Therefore, the emphasis will be put on the non-chiral models, digression to the chiral models will be made occasionally.

Supersymmetric gluodynamics

To begin with we will consider supersymmetric generalization of pure gluodynamics – i.e. the theory of gluons and gluinos. The Lagrangian has the form[8]

$$\mathcal{L} = -\frac{1}{4g^2}G^a_{\mu\nu}G^a_{\mu\nu} + \frac{\vartheta}{32\pi^2}G^a_{\mu\nu}\tilde{G}^a_{\mu\nu} + \frac{i}{2g^2}\bar{\lambda}^a\mathcal{D}_\mu\gamma^\mu\lambda^a \tag{1.14}$$

where $G^a_{\mu\nu}$ is the gluon field strength tensor, $\tilde{G}^a_{\mu\nu}$ is the dual tensor, g is the gauge coupling constant, ϑ is the vacuum angle, and $\mathcal{D}_\mu$ is the covariant derivative. Moreover, λ^a is the gluino field, which can be described either by four-component Majorana (real) fields or two-component Weyl (complex) fields.

In terms of superfields

$$\mathcal{L} = \mathrm{Re}\left\{\frac{1}{2g_0^2}\mathrm{Tr}\int d^2\theta W^2\right\}, \tag{1.15}$$

where the superfield W is a color matrix,

$$W = W^a T^a,$$

T^a are the generators of the gauge group (in the fundamental representation), $\mathrm{Tr}(T^aT^b) = (1/2)\delta^{ab}$. It is very important that the gauge constant $1/g_0^2$ in Eq. (1.15) can be treated as a complex parameter. The subscript 0 emphasizes the fact that the gauge couplings in Eqs. (1.15) and (1.14) are different,

$$\frac{1}{g_0^2} = \frac{1}{g^2} - i\frac{\vartheta}{8\pi^2}, \tag{1.16}$$

its real part is the conventional gauge coupling while the imaginary part is proportional to the vacuum angle. Thus, the gauge coupling becomes complexified in SUSY theories. This fact has far-reaching consequences.

Equivalence between Eqs. (1.15) and (1.14) is clear from Eq. (1.13). The F component of W^2 includes the kinetic term of the gaugino field (or gluino, I will use these terms indiscriminately),

$$\bar{\lambda}\mathcal{D}_\mu\gamma^\mu\lambda,$$

and that of the gauge field,

$$G^a_{\mu\nu}G^a_{\mu\nu} + iG^a_{\mu\nu}\tilde{G}^a_{\mu\nu}.$$

Superficially the model looks very similar to conventional QCD; the only difference is that the quark fields belonging to the fundamental representation of the gauge group in QCD are replaced by the gluino field belonging to the adjoint representation in supersymmetric gluodynamics. Like QCD, supersymmetric gluodynamics is a strong coupling non-Abelian theory. Therefore, it is usually believed that

- only colorless asymptotic states exist;
- the Wilson loop (in the *fundamental* representation) is subject to the area law (confinement);
- a mass gap is dynamically generated; all particles in the spectrum are massive.

I would like to stress the word "believe" since the above features are hypothetical. Although the theory does indeed look pretty similar to QCD, supersymmetry brings in remarkable distinctions – some quantities turn out to be exactly calculable. Namely, we know that the gluino condensate develops,

$$\langle\lambda^{\alpha a}\lambda^a_\alpha\rangle = \mathrm{const.}\times\Lambda^3 e^{2\pi i k/N_c} \tag{1.17}$$

where N_c is the number of colors (an $SU(N_c)$ gauge group is assumed and the vacuum angle ϑ is set equal to zero), Λ is the scale parameter of supersymmetric gluodynamics, k is an integer ($k = 0, 1, ..., N_c - 1$), and the constant in Eq. (1.17) is *exactly* calculable[9, 10]. A discrete Z_{2N_c} symmetry of the model, a remnant of the anomalous

$U(1)$, is spontaneously broken by the gluino condensate* down to Z_2. Correspondingly, there are N_c degenerate vacua, counted by the integer parameter k. Supersymmetry is unbroken – all vacua have vanishing energy density.

Moreover, the Gell-Mann–Low function of the model, governing the running of the gauge coupling constant, is also exactly calculable[13],

$$\beta(\alpha) = -\frac{3N_c\alpha^2}{2\pi}\frac{1}{1-N_c\alpha/(2\pi)}\,. \tag{1.18}$$

By "exactly" I mean that all orders of perturbation theory are known, and one can additionally show that in the case at hand there are no nonperturbative contributions.

Equations (1.17) and (1.18) historically were the first examples of non-trivial (i.e. non-vanishing) quantities exactly calculated in four-dimensional field theories in the strong coupling regime. These examples, alone, show that the supersymmetric gauge dynamics is full of hidden miracles. We will encounter many more examples in what follows. Eventually, after learning more about supersymmetric theories, you will be able to understand how Eqs. (1.17) and (1.18) are derived. But this will take some time. Here I would like only to add an explanatory remark regarding the vacuum degeneracy in supersymmetric gluodynamics. At the classical level Lagrangian (1.14) has a $U(1)$ symmetry corresponding to the phase rotations of the gluino fields,

$$\lambda \to e^{i\alpha}\lambda\,. \tag{1.19}$$

The corresponding current is sometimes called the R_0 current; it is a superpartner of the energy-momentum tensor and the supercurrent. The R_0 current exists in any supersymmetric theory. Moreover, in conformally invariant theories – and supersymmetric gluodynamics is conformally invariant at the classical level – it is conserved[14]. In the spinor notation the R_0 current has the form $J_{\alpha\dot\alpha} = (1/g^2)\bar\lambda_{\dot\alpha}\lambda_\alpha$, while in the Majorana notation the very same current takes the form $J_\mu = (1/g^2)\bar\lambda\gamma_\mu\gamma_5\lambda$. (Let me parenthetically note that the vector current of the Majorana gluino identically vanishes. The proof of this fact is left as an exercise.) The conservation of the axial current above is broken by the triangle anomaly,

$$\partial^\mu J_\mu = \frac{N_c}{16\pi^2}G^a_{\mu\nu}\tilde G^a_{\mu\nu}$$

So, there is no continuous $U(1)$ symmetry in the model. By the same token, the conformal invariance is ruined by the anomaly in the trace of the energy-momentum tensor. As a matter of fact, the divergence of the R_0 current and the trace of the energy-momentum tensor can be combined in one superfield[15] (see the Appendix).

However, a remnant of the would-be symmetry remains, in the form of the discrete phase transformations of the type (1.19) with $\alpha = \pi k/N_c$. The gluino condensate further breaks this symmetry to Z_2 corresponding to $\lambda \to -\lambda$. The number of degenerate vacuum states, N_c, coincides with Witten's index for the $SU(N_c)$ theory[16], an invariant

*I hasten to add that it was argued recently[11] that supersymmetric gluodynamics actually has two phases: one with the spontaneously broken Z_{2N_c} invariance, and another, unconventional, phase where the chiral Z_{2N_c} symmetry is unbroken and the gluino condensate does not develop. Dynamics of the chirally symmetric phase is drastically different from what we got used to in QCD. In particular, although no invariance is spontaneously broken, massless particles appear, and no mass gap is generated. This development is too fresh, however, to be included in this lecture course.
The existence of the gluino condensate was anticipated[12] from the analysis of the so called Veneziano-Yankielowicz effective Lagrangian, even prior to the first dynamical calculation[9]. The Veneziano-Yankielowicz Lagrangian, very useful for orientation, is not a genuinely Wilsonian construction, and one must deal with it extremely cautiously in extracting consequences. For a recent discussion see Ref.[11].

which counts the number of the boson zero energy states minus the number of the fermion zero energy states. If Witten's index is non-vanishing supersymmetry cannot be spontaneously broken, of course.

An interesting aspect, related to the discrete degeneracy of the vacuum states, is the ϑ dependence. What happens with the vacua if $\vartheta \neq 0$? The question was answered in Ref.[10]. The ϑ dependence of the gluino condensate is

$$\langle\lambda\lambda\rangle_\vartheta = \langle\lambda\lambda\rangle_{\vartheta=0} e^{\frac{i\vartheta}{N_c}} . \tag{1.20}$$

This shows that the N_c vacua are intertwined as far as the ϑ evolution is concerned. When ϑ changes continuously from 0 to 2π the first vacuum becomes second, the second becomes third, and so on, in a cyclic way.

SU(2) SQCD with one flavor

As the next step on a long road leading us to understanding of supersymmetric gauge dynamics we will consider SUSY generalization of $SU(2)$ QCD with the matter sector consisting of one flavor. This model will serve us as a reference point in all further constructions.

Since the gauge group is $SU(2)$ we have three gluons and three superpartners - gluinos. As far as the matter sector is concerned, let us remember that one quark flavor in QCD is described by a Dirac field, a doublet with respect to the gauge group. One Dirac field is equivalent to two chiral fields: a left-handed and a right-handed, both transforming according to the fundamental representation of $SU(2)$. Moreover, the right-handed doublet is equivalent to the left-handed anti-doublet, which in turn is equivalent to a doublet. The latter fact is specific to the $SU(2)$ group, all whose representations are (pseudo)real. Thus, the Dirac quark reduces to two left-handed Weyl doublet fields.

Correspondingly, in SQCD each of them will acquire a scalar partner. Thus, the matter sector will be built from two superfields, S_1 and S_2. In what follows we will use the notation S_f^α where $\alpha = 1,2$ is the color index, and $f = 1,2$ is a "subflavor" index. Two subflavors comprise one flavor. The chiral superfield has the usual form, see Eq. (1.5).

In the superfield language the Lagrangian of the model can be represented in a very concise form

$$\mathcal{L} = \frac{1}{2g_0^2}\mathrm{Tr}\int d^2\theta W^2 + \frac{1}{4}\int d^2\theta d^2\bar\theta \bar S_f e^V S_f + \left(\frac{m_0}{4}\int d^2\theta S^{\alpha f} S_{\alpha f} + \mathrm{H.c.}\right) , \tag{1.21}$$

where the superfields V and W_α are matrices in the color space, for instance, $V \equiv V^a\tau^a/2$, with τ^a denoting the Pauli matrices. The subscript 0 indicates that the mass parameter and the gauge coupling constant are bare parameters, defined at the ultra-violet cut off. In what follows we will omit this subscript to ease the notation in several instances where it is unimportant.

If we take into account the rules of integration over the Grassmann numbers we immediately see that the integral over $d^2\theta$ singles out the θ^2 component of the chiral superfields W^2 and S^2, i.e. the F terms. Moreover, the integral over $d^2\theta d^2\bar\theta$ singles out the $\theta^2\bar\theta^2$ component of the real superfield $\bar S e^V S$, i.e. the D term.

Note that the $SU(2)$ model under consideration, with *one flavor* possesses a global $SU(2)$ ("subflavor") invariance allowing one to freely rotate the superfields $S_1 \leftrightarrow S_2$. This symmetry holds even in the presence of the mass term, see Eq. (1.21), and is specific for $SU(2)$ gauge group, with its pseudoreal representations. All indices corresponding to the $SU(2)$ groups (gauge, Lorentz and subflavor) can be lowered and raised by means of the ε symbol, according to the general rules.

The Lagrangian presented in Eq. (1.21) is not generic. Renormalizable models with a richer matter sector usually allow for one more type of F terms, namely

$$\int d^2\theta S^3 .$$

These terms are called the Yukawa interactions, since one of the vertices they include corresponds to a coupling of two spinors to a scalar. Strictly speaking, they should be called the super-Yukawa terms, since spinor-spinor-scalar vertices arise also in the (super)gauge parts of the Lagrangian. This jargon is widely spread, however; eventually you will get used to it and learn how to avoid confusion. The combination of the F terms $S^2 + S^3$ is generically referred to as superpotential. The conventional potential of self-interaction of the scalar fields stemming from the given superpotential is referred to as *scalar potential.*

It is instructive to pass from the superfield notations to components. We will do this exercise now in some detail, putting emphasis on those features which are instrumental in the solutions to be discussed below. Once the experience is accumulated the need for the component notation will subside.

Let us start from W^2. The corresponding F term was already discussed below Eq. (1.16). There is one new important point, however. We omitted the square of the D term present in $W^2|_F$, see Eq. (1.13),

$$\Delta \mathcal{L}_D^{(W)} = \frac{1}{2g^2} D^a D^a . \tag{1.22}$$

If the matter sector of the theory is empty, this term is unimportant. Indeed, the D field enters with no derivatives, and, hence, can be eliminated from the Lagrangian by virtue of the equations of motion. With no matter fields $D = 0$. In the presence of the matter fields, however, eliminating D we get a non-trivial term constructed from the scalar fields, which is of a paramount importance. This point will be discussed later; here let me only note that the sign of D^2 in the Lagrangian, Eq. (1.22), is unusual, it is positive.

The next term to be considered is $\int d^2\theta d^2\bar\theta \bar S e^V S$. Calculation of the D component of $\bar S e^V S$ is a more time-consuming exercise since we must take into account the fact that 5 depends on x_L while $\bar S$ depends on x_R; both arguments differ from a;. Therefore, one has to expand in this difference. The factor e^V sandwiched between $\bar S$ and 5 covariantizes all derivatives. Needless to say that the field V is treated in the Wess-Zumino gauge. It is not difficult to check that

$$\frac{1}{4}\int d^2\theta d^2\bar\theta \bar S_f e^V S_f = \bar\psi_f i \not{D} \psi_f - \phi_f^\dagger D^2 \phi_f +$$

$$\left[(\psi_f \lambda)\phi_f^\dagger + \text{H.c.}\right] + D^a \phi_f^\dagger T^a \phi_f \tag{1.23}$$

where T^a are the matrices of the color generators. In the $SU(2)$ theory $T^a = \tau^a/2$. Now we see why the D^2 term is so important in the presence of matter; D^a does not vanish anymore. Moreover, using the equation of motion we can express D^a in terms of the squark fields, generating in this way a quartic self-interaction of the scalar fields,

$$V_D = \frac{1}{2g^2} D^a D^a , \quad D^a = -\frac{g^2}{2}\left(\phi_1^\dagger \tau^a \phi_1 + \phi_2^\dagger \tau^a \phi_2\right) . \tag{1.24}$$

In the old-fashioned language of the pre-SUSY era one would call the term $(\psi_f\lambda)\phi_f^\dagger$ from Eq. (1.23) the Yukawa interaction. The SUSY practitioner would refer to this term as to the gauge coupling since it is merely a supersymmetric generalization of the quark-quark-gluon coupling. I mention these terms here because later on their analysis will help us establish the form of the conserved R currents.

Vacuum valleys

Let us examine the D potential V_D more carefully, neglecting for the time being F terms altogether. As is well-known, the energy of any state in any supersymmetric theory is positive-definite. The minimal energy state, the vacuum, has energy exactly at zero. Thus, in determining the classical vacuum we must find all field configurations corresponding to vanishing energy. From Eq. (1.24) it is clear that in the Wess-Zumino gauge the *classical space of vacua* (sometimes called the moduli space of vacua) is defined by the *D-flatness condition*

$$D^a = 0 \text{ for all } a\,. \tag{1.25}$$

More exactly, Eq. (1.25) is called the Wess-Zumino gauge D flatness condition. Since this gauge is always implied, we will omit the reference to the Wess-Zumino gauge.

The D potential V_D represents a quartic self-interaction of the scalar fields, of a very peculiar form. Typically in the ϕ^4 theory the potential has one – at most several – minima. In other words, the space of the vacuum fields corresponding to minimal energy, is a set of isolated points. The only example with a continuous manifold of points of minimal energy which was well studied previously is the spontaneous breaking of a global continuous symmetry, say, $U(1)$. In this case all points belonging to this vacuum manifold are physically equivalent. The D potential (1.24) has a specific structure – minimal (zero) energy is achieved along entire directions corresponding to the solution of Eq. (1.25). It is instructive to think of the potential as of a mountain ridge; the D flat directions then present the flat bottom of the valleys. Sometimes, for transparency, I will call the D flat directions *the* vacuum *valleys*. Their existence was first noted in Ref.[17]. As we will see, different points belonging to the bottom of the valleys are physically inequivalent. This is a remarkable feature of the supersymmetric gauge theories.

In the case of the $SU(2)$ theory with one flavor it is not difficult to find the D flat direction explicitly. Indeed, consider the scalar fields of the form

$$\phi_1 = v\begin{pmatrix}1\\0\end{pmatrix}, \quad \phi_2 = v\begin{pmatrix}0\\1\end{pmatrix}, \tag{1.26}$$

where v is an arbitrary complex constant. It is obvious that for any value of v all D^a's vanish. D^1 and D^2 vanish because $\tau^{1,2}$ are off-diagonal matrices; D^3 vanishes after summation over two subflavors.

It is quite obvious that if $v \neq 0$ the original gauge symmetry $SU(2)$ is totally spontaneously broken. Indeed, under the condition (1.26) all three gauge bosons acquire masses $\sim gv$. Thus, we deal here with the supersymmetric generalization of the Higgs phenomenon. Needless to say that supersymmetry is not broken. It is instructive to trace the reshuffling of the degrees of freedom before and after the Higgs phenomenon. In the unbroken phase, corresponding to $v = 0$, we have three massless gauge bosons (6 degrees of freedom), three massless gaugino (6 degrees of freedom), four matter fermions (the Weyl fermions, 8 degrees of freedom), and four matter scalars (complex scalars, 8 degrees of freedom). In the broken phase three matter fermions combine with the gauginos to form three massive Dirac fermions (12 degrees of freedom). Moreover, three matter scalars combine with the gauge fields to form three *massive* vector fields

(9 degrees of freedom) plus three massive (real) scalars. What remains massless? One complex scalar field, corresponding to the motion along the bottom of the valley, v, and its fermion superpartner, one Weyl fermion. The balance between the fermion and boson degrees of freedom is explicit.

A gauge invariant description of the system of the vacuum valleys was suggested in Refs.[18,19] (see also[17]). In these works it was noted that the set of proper coordinates parametrizing the space of the classical vacua is nothing else but the set of all independent (local) products of the chiral matter fields existing in the theory. Since this point is very important let me stress once more that the variables to be included in the set are polynomials built from the fields of one and the same chirality *only*. These variables are clearly gauge invariant.

At the intuitive level this assertion is almost obvious. Indeed, if there is a D flat direction, the motion along the degenerate bottom of the valley must be described by some effective (i.e. composite) chiral superfield, which is gauge invariant and has no superpotential. The opposite is also true. If we are able to build some chiral gauge invariant (i.e. colorless) superfield Φ, as a local product of the chiral matter superfields of the theory at hand, then the energy is guaranteed to vanish, since (in the absence of the F terms) all terms which might appear in the effective Lagrangian for ϕ necessarily contain derivatives. Here ϕ is the lowest component of the above superfield Φ. In other words, then, changing the value of ϕ we will be moving along the bottom of the valley.

A formal proof of the fact that the classical vacua are fully described by the set of local (gauge invariant) products of the chiral fields comprising the matter sector is given in the recent work[20] which combines and extends results scattered in the literature[17, 18, 21, 22].

The approach based on the chiral polynomials is very convenient for establishing the fact of the existence (non-existence) of the moduli space of the classical vacua, and in counting the dimensionality of this space. For instance., in the $SU(2)$ model with one flavor there exists only one invariant, $S^2 \equiv S_{\alpha f} S_{\beta g} \epsilon^{\alpha\beta} \epsilon^{fg} = 2v^2$. Correspondingly, there is only one vacuum valley – a one-dimensional complex manifold. The remaining three (out of four) complex scalar fields are eaten up in the super-Higgs phenomenon by the vector fields, which immediately tells us that a generic point from the bottom of the valley corresponds to fully broken gauge symmetry.

In other cases we will have a richer structure of the moduli space of the classical vacua. In some instances no chiral invariants can be built at all. Then the D flat directions are absent.

If the D flat directions exist, and the gauge symmetry is spontaneously broken, then the constraints of the type of Eq. (1.26) can be viewed as a gauge fixing condition. This is nothing else but the unitary gauge in SUSY. Those components of the matter superfields which are set equal to zero are actually eaten up by the vector particles which acquire the longitudinal components through the super-Higgs mechanism.

Although constructing the set of the chiral invariants is helpful in the studies of the general properties of the space of the classical vacua, sometimes it is still necessary to explicitly parametrize the vacuum valleys, just in the same way as it is done in Eq. (1.26). As we have seen, this problem is trivially solvable in the $SU(2)$ model with one flavor. For higher groups and representations the general situation is much more complicated, and the generic solution is not found. Many useful tricks for finding explicit parametrization of the vacuum valleys in particular examples were suggested in Refs.[17,18,19]. A few simplest examples are considered below. More complicated instances are considered in the literature. For instance, a parametrization of the valleys in the $SU(5)$ model with two quintets and two antidecuplets was given in Ref.[23] and in the E(6) model with the 27-plet in Ref.[24]. The correspondence between the explicit parametrization of the D flat directions and the chiral polynomials was discussed recently more than once, see e.g.[25-29]. I would like to single out Ref.[30] where a catalog of the flat directions in the minimal supersymmetric standard model (MSSM) was obtained by analyzing all possible chiral polynomials and eliminating those of them which are redundant.

Once the existence of the D-flat directions is established at the classical level one may be sure that a manifold of the degenerate vacua will survive at the quantum level, provided no F terms appear in the action which might lift the degeneracy. Indeed, in this case the only impact of the quantum corrections is providing an overall Z factor in front of the kinetic term, which certainly does not affect the vanishing of the D terms. The F terms which could lift the degeneracy must be either added in the action by hand (e.g. mass terms), or generated nonperturbatively. A remarkable non-renormalization theorem[31] guarantees that no F terms can be generated perturbatively. We will return to the discussion of this *second miracle* of SUSY further on.

In this respect the supersymmetric theories are fundamentally different from the non-supersymmetric ones. Say, in the good old ϕ^4 theory with the Yukawa interaction

$$\mathcal{L} = |\partial_\mu \phi|^2 + \bar{\psi} i \partial_\mu \gamma^\mu \psi + (g\phi\bar{\psi}\psi + \text{H.c.})$$

we could also assume that that the mass and self-interaction of the scalar field vanish at the classical level. Then, classically, we will have a flat direction – any constant value of ϕ corresponds to the vanishing vacuum energy. However, this vacuum valley does not survive inclusion of the quantum corrections. Already at the one-loop level both the mass term of the scalar field, and its self-interaction, will be generated, and the continuous vacuum degeneracy will inevitably disappear.

In search of the valleys

Although our excursion in the $SU(2)$ model with one flavor is not yet complete, the issue of the D flat directions is so important in this range of problems that we pause here to do, with pedagogical purposes, a few simple exercises. If you choose to skip this section in the first reading it will be necessary to return to it later.

$SU(N_c)$ model with N_f flavors ($N_f < N_c$)[18]

The matter sector includes $2N_f$ subflavors – N_f chiral fields in the fundamental representation of $SU(N_c)$, $S^{\alpha f}$, and N_f chiral fields in the antifundamental representation, $\tilde{S}_{\alpha f}$, where $\alpha = 1, ..., N_c$ and $f = 1, ...N_f$. It is quite obvious that one can form N_f^2 chiral products of the type

$$\tilde{S}^f_\alpha S^{\alpha g}\,, \quad f, g = 1, ..., N_f\,. \tag{1.27}$$

All these chiral invariants are independent. Thus, the moduli space of the classical vacua (the vacuum valley) is a complex manifold of dimensionality N_f^2, parametrized by the coordinates (1.27). A generic point from the vacuum valley corresponds to spontaneous breaking of $SU(N_c) \to SU(N_c - N_f)$ (except for the case when $N_f = N_c - 1$, when the original gauge group is completely broken). The number of the broken generators is $2N_cN_f - N_f^2$; hence, the same amount of the complex scalar fields are eaten up in the super-Higgs mechanism. The original number of the complex scalar fields was $2N_cN_f$. The remaining N_f^2 degrees of freedom are the moduli (1.27) corresponding to the motion along the bottom of the valley.

In this particular problem it is not difficult to indicate a concrete parametrization of the vacuum field configurations. Indeed, consider a set

$$S_1 = \tilde{S}_1^\dagger = v_1 \begin{pmatrix} 1 \\ 0 \\ 0 \\ \dots \\ 0 \end{pmatrix}, S_2 = \tilde{S}_2^\dagger = v_2 \begin{pmatrix} 0 \\ 1 \\ 0 \\ \dots \\ 0 \end{pmatrix}, \dots, S_{N_f} = \tilde{S}_{N_f}^\dagger = v_{N_f} \begin{pmatrix} 0 \\ 0 \\ \dots \\ 1 \\ \dots \\ 0 \end{pmatrix} \qquad (1.28)$$

where the unity in S_{N_f} occupies the N_f-th line and $v_{1,2,\dots,N_f}$ are N_f arbitrary complex numbers. It is rather obvious that for this particular set all D terms vanish. In verifying this assertion it is convenient to consider first those D^a's which lie outside the Cartan subalgebra of $SU(N_c)$. Since the corresponding T^a matrices are off-diagonal each term in the sum $\sum_f (S^\dagger T^a S + \tilde{S}^\dagger T^a \tilde{S})$ vanishes individually. For the generators from the Cartan subalgebra the fundamentals and anti-fundamentals cancel each other.

The point (1.28) is *not* a generic point from the bottom of the valley. This is clear from the fact that it is parametrized by only N_f complex numbers. To get a generic solution one observes that the theory is invariant under the global $SU(N_f) \times SU(N_f)$ flavor rotations (the fundamentals and antifundamentals can be rotated separately). On the other hand, the solution (1.28) is not invariant. Therefore, we can apply a general $SU(N_f) \times SU(N_f)$ rotation to Eq. (1.28) without destroying the condition $D^a = 0$. It is quite obvious that the generators belonging to the Cartan subalgebra of $SU(N_f) \times SU(N_f)$ do not introduce new parameters. The remaining rotations introduce $N_f^2 - N_f$ complex parameters, to be added to $v_1, \dots, v_{N_f}$, altogether N_f^2 parameters, as it was anticipated from counting the number of chiral invariants.

$SU(N_c)$ model with N_f flavors ($N_f = N_c + 1$)

The vacuum valley is parametrized by N_f^2 complex parameters, although the number of the chiral invariants is larger, $N_f^2 + 2N_f$. Not all chiral invariants are independent. For further details see Eq. (2.8) and further.

SU(5) model with one quintet and one (anti)decuplet

This gauge model describes Grand Unification, with one generation of quarks and leptons. This is our first example of non-chiral matter; it is singled out historically – the instanton-induced dynamical supersymmetry breaking was first found in this model[32]. The quintet field is V^α, the (anti)decuplet field is antisymmetric $X_{\alpha\beta}$. It is quite obvious that there are no chiral invariants at all. Indeed, the only candidate, *VVX*, vanishes due to antisymmetricity of $X_{\alpha\beta}$. This means that no D flat directions exist. The same conclusion can be reached by explicitly parametrizing V and X; inspecting then the D-flatness conditions one can conclude that they have no solutions, see e.g. Appendix A in Ref.[33].

SU(5) model with two quintets and two (anti)decuplets and no superpotential

This model (with a small tree-level superpotential term) was the first example of the instanton-induced supersymmetry breaking in the weak coupling regime[34]. It presents another example of the anomaly-free chiral matter sector. Unlike the one-family model (one quintet and one antidecuplet) flat directions do exist (in the absence of a superpotential). The system of the vacuum valleys in the two-family *SU*(5) model was analyzed in Ref.[23]. Generically, the gauge *SU*(5) symmetry is completely broken, so that 24 out of 30 chiral matter superfields are eaten up in the super-Higgs mechanism. Therefore, the vacuum valley should be parametrized by six complex moduli.

Denote the two quintets present in the model as V_f^α $(f = 1,2)$, and the two antidecuplets as $(X_{\bar{g}})_{\alpha\beta}$ where $\bar{g} = 1,2$ and the matrices $X_{\bar{g}}$ are antisymmetric in the color indices α, β. Indices f and $\bar{g}$ reflect the $SU(2)_X \times SU(2)_V$ flavor symmetry of the model. Six independent chiral invariants are

$$M_{\bar{g}} = V_k X_{\bar{g}} V_l \epsilon^{kl},$$

$$B_{\bar{g}f} = X_{\bar{g}} X_{\bar{k}} X_{\bar{l}} V_f \epsilon^{\bar{k}\bar{l}},$$

where the gauge indices in the first line are convoluted in a straightforward manner $V^\alpha X_{\alpha\beta} V^\beta$, while in the second line one uses the ϵ symbol,

$$X_{\bar{g}} X_{\bar{k}} X_{\bar{l}} V_f = \epsilon^{\alpha\beta\gamma\delta\rho}(X_{\bar{g}})_{\alpha\beta}(X_{\bar{k}})_{\gamma\delta}(X_{\bar{l}})_{\rho\kappa}(V_f)^\kappa .$$

The choice of invariants above implies that there are no moduli transforming as {4, 2} under the flavor group (such moduli vanish).

In this model the explicit parametrization of the valley is far from being obvious, to put it mildly. The most convenient strategy for the search is analyzing the five-by-five matrix

$$D^\alpha_\beta = V^\alpha_f \bar{V}^f_\beta - 2(\bar{X}^{\bar{g}})^{\alpha\gamma}(X_{\bar{g}})_{\gamma\beta} \tag{1.29}$$

where $\bar{V} = V^\dagger$ and $\bar{X} = X^\dagger$. If this matrix is proportional to the unit one, the vanishing of the D terms is guaranteed. (Similar strategy based on analyzing analogs of Eq. (1.29) is applicable in other cases as well).

A solution of the D-flatness condition which contains 7 real parameters looks as follows:

$$V_1 = \begin{pmatrix} a_1 \\ 0 \\ 0 \\ a_4 \\ 0 \end{pmatrix}, \quad V_2 = \begin{pmatrix} 0 \\ a_2 \\ 0 \\ 0 \\ 0 \end{pmatrix},$$

$$X_1 = \frac{1}{\sqrt{2}}\begin{pmatrix} 0 & 0 & 0 & s & 0 \\ 0 & 0 & b & 0 & 0 \\ 0 & -b & 0 & 0 & f \\ -s & 0 & 0 & 0 & 0 \\ 0 & 0 & -f & 0 & 0 \end{pmatrix}, \quad X_2 = \frac{1}{\sqrt{2}}\begin{pmatrix} 0 & d & 0 & 0 & g \\ -d & 0 & 0 & 0 & 0 \\ 0 & 0 & 0 & 0 & 0 \\ 0 & 0 & 0 & 0 & h \\ -g & 0 & 0 & -h & 0 \end{pmatrix}, \tag{1.30}$$

where

$$|a_1|^2 = r^2\cos^2\theta, \quad |a_4|^2 = r^2\cos^2\alpha\tan^2\theta, \quad |a_2|^2 = r^2\tan^2\theta\frac{\cos^2\theta - \cos^2\alpha}{\sin^2\theta + \cos^2\alpha},$$

and

$$|s|^2 = \frac{r^2\cos^2\alpha}{\cos^2\theta\sin^2\theta + \cos^2\alpha}, \quad |b|^2 = r^2\sin^2\theta + \cos^2\alpha, \quad |f|^2 = r^2\frac{\cos^2\alpha\cos^2\theta - \cos^2\alpha}{\cos^2\theta\sin^2\theta + \cos^2\alpha},$$

$$|d|^2 = \frac{r^2}{\cos^2\theta}(\cos^2\theta - \cos^2\alpha), \quad |g|^2 = r\cos^2\alpha, \quad |h|^2 = r^2\sin^2\theta. \tag{1.31}$$

Thus, the absolute values of matrix elements are parametrized by three real parameters r, α and θ. Additional 4 parameters appear *via* phases δ of 9 elements a_i, s, b, f, d, g, h. Three phases, out of nine, are related to gauge rotations and are not observable in the gauge singlet sector. Additionally, there are two constraints,

$$\delta_1 - \delta_4 = \delta_h - \delta_g,$$

$$\delta_f - \delta_b = \delta_g - \delta_d,$$

which are readily derived from vanishing of the off-diagonal terms in D^{α}_{β}.

Substituting the above expressions it is easy to check that the invariants

$$B_{\bar{g}_1\bar{g}_2\bar{g}_3 f} = X_{\{\bar{g}_1}X_{\bar{g}_2}X_{\bar{g}_3\}}V_f$$

symmetrized over g_1, g_2, g_3 (i.e. the {4, 2} representation of $SU(2)_X \times SU(2)_V$) do vanish, indeed.

The most general valley parametrization depends on 12 real parameters, while so far we have only 7. The remaining five parameters are provided by $SU(2)_X \times SU(2)_V$ flavor rotations of the configuration (1.30).

Back to the SU(2) model – dynamics of the flat direction

After this rather lengthy digression into the general theory of the vacuum valleys we return to our simplest toy model, *SU*(2) with one flavor. The vacuum valley in this case is parametrized by one complex number, v, which can be chosen at will since for any value of v the vacuum energy vanishes. One can quantize the theory near any value of v. If $v \neq 0$, the theory splits into two sectors – one containing massive particles which form SU(2) triplets, and another sector which includes only one massless Weyl fermion and one massless complex scalar field. These massless particles are singlets with respect to both *SU*(2) groups – color and subflavor.

So far we totally disregarded the mass term in the action, assuming $m = 0$. If $m \neq 0$ the corresponding term in the superpotential lifts the vacuum degeneracy, making the bottom of the valley non-flat. Indeed,

$$F = mv\,,$$

the corresponding contribution in the scalar potential is

$$\Delta V = |mv|^2\,,$$

which makes the theory "slide down" towards the origin of the valley. Since the perturbative corrections do not renormalize the F terms, this type of behavior – sliding down to the origin of the former valley – is preserved to any finite order in perturbation theory.

What happens if one switches on nonperturbative effects?

The non-renormalization theorem[31], forbidding the occurrence of the F terms, does not apply to nonperturbative effects, which, thus, may or may not generate relevant F terms. The possibility of getting a superpotential can be almost completely investigated by analyzing the general properties of the model at hand, with no explicit calculations. Apart from the overall numerical constant, the functional form of the superpotential, if it is generated, turns out to be fixed.

Let me elucidate this point in more detail. First, on what variables can the superpotential depend? The vector fields are massive and are integrated over. Thus, we are left with the matter fields only, and the only chiral invariant is

$$I = S^{\alpha f} S_{\alpha f}\,. \tag{1.32}$$

The superpotential, if it exists, must have the form

$$\mathcal{L}_F = \int d^2\theta\, f(I(x_L, \theta)) + \text{H.c.} \tag{1.33}$$

where f is some function. Notice that the mass term has just this structure, with $f(z) = z$. We will discuss the possible impact of the mass term later, assuming at the beginning that $m = 0$.

Now, our task is to find the function f exploiting the symmetry properties of the model. At the classical level there exist two conserved currents. One of them, the R_0 current, is the superpartner of the energy-momentum tensor and the supercurrent[35]. The divergence of the R_0 current and the trace of the energy-momentum tensor can be combined in one superfield. The R_0 current exists in any supersymmetric theory, and, moreover, in conformally invariant theories it is conserved. Indeed, since the trace of the energy-momentum tensor vanishes in conformally invariant theories the divergence of R_0 vanishes as well. In our present model the R_0 current corresponds to the following rotations of the fields

$$\lambda_\alpha \to e^{i\beta}\lambda_\alpha\,, \quad \psi^f_\alpha \to e^{-(i/3)\beta}\psi^f_\alpha\,, \quad \phi^f_\alpha \to e^{(2i/3)\beta}\phi^f_\alpha\,. \tag{1.34}$$

If we denote this current by R^0_μ, then

$$R^0_\mu = \frac{1}{g^2}\,\bar\lambda\gamma_\mu\gamma_5\lambda - \frac{1}{3}\sum_f\left(\bar\psi^f\gamma_\mu\gamma_5\psi^f - 2\phi^{f\dagger}\overleftrightarrow{D}_\mu\,\phi^f\right). \tag{1.35}$$

The relative phase between ψ and ϕ is established in the following way. Let us try to add the S^3 term to the superpotential. In the model at hand it is actually forbidden by the color gauge invariance, but we ignore this circumstance, since the form of the R_0 current is general, and in other models the S^3 term is perfectly allowed. It violates neither supersymmetry, nor the conformal invariance (at the classical level), which is obvious from the fact that its dimension is 3. The S^3 term in the superpotential produces a $\phi\psi^2$ term in the Lagrangian. In this way we arrive at the relative phase between ψ and ϕ indicated in Eq. (1.34). The relative phase between λ and ψ is fixed by requiring the term $\lambda\psi\phi^\dagger$ in the Lagrangian (the supergeneralization of the gauge coupling) to be invariant under the $U(1)$ transformation at hand. In the superfield language the last two transformations in Eq. (1.34) can be concisely written as

$$S^f \to \exp\left(\frac{2i}{3}\beta\right)S^f\,, \quad \theta_\alpha \to \exp(i\beta)\theta_\alpha\,. \tag{1.36}$$

The second (classically) conserved current is built from the matter fields,

$$J_\mu = \sum_f\left(\bar\psi^f\gamma_\mu\gamma_5\psi^f + \phi^{f\dagger}\overleftrightarrow{D}_\mu\,\phi^f\right). \tag{1.37}$$

The corresponding $U(1)$ transformation is

$$\psi^f_\alpha \to e^{i\gamma}\psi^f_\alpha\,, \quad \phi^f_\alpha \to e^{i\gamma}\phi^f_\alpha\,, \tag{1.38}$$

or, in the superfield notation, $S^f(x_L,\theta) \to \exp(i\gamma)S^f(x_L,\theta)$.

The conservation of both axial currents is destroyed by the quantum anomalies. At one-loop level

$$\partial_\mu R^0_\mu = \frac{5}{3}\frac{\alpha}{4\pi}G^a_{\mu\nu}\tilde G^a_{\mu\nu}\,, \quad \partial_\mu J_\mu = \frac{\alpha}{4\pi}G^a_{\mu\nu}\tilde G^a_{\mu\nu}\,. \tag{1.39}$$

One can form, however, one linear combination of these two currents,

$$R_\mu = R^0_\mu - \frac{5}{3}J_\mu \tag{1.40}$$

which is anomaly free and is conserved even at the quantum level. The occurrence of a strictly conserved axial current, the so called R current, is a characteristic feature of many supersymmetric models. In what follows we will have multiple encounters with the R currents in various models. The one presented in Eq. (1.40) was given in Ref.[18].

Combining both transformations, Eqs. (1.34) and (1.38), in the appropriate proportion, see Eq. (1.40), we conclude that the $SU(2)$ model under consideration is strictly invariant under the following transformation

$$S^f \rightarrow e^{-i\beta} S^f \, , \quad \theta_\alpha \rightarrow e^{i\beta} \theta_\alpha \, . \tag{1.41}$$

This R invariance leaves us with a unique possible choice for the superpotential

$$\mathcal{L}_F = \text{const.} \times \int d^2\theta \frac{1}{I(x_L, \theta)} \rightarrow \text{const.} \times \int d^2\theta \frac{\Lambda^5}{S^{\alpha f} S_{\alpha f}} \tag{1.42}$$

where Λ is a scale parameter of the model, and the factor Λ^5 has been written out on the basis of dimensional arguments.

Whether the superpotential is actually generated, depends on the value of the numerical constant above. In principle, it could have happened that the constant vanished. However, since no general principle forbids the F term (1.42), the vanishing seems highly improbable. And indeed, the direct one-instanton calculation in the weak coupling regime (i.e. $v^2 \gg \Lambda^2$) shows[18, 23] that this term is generated.

The impact of the term (1.42) is obvious. The corresponding extra contribution to the self-interaction energy of the scalar field is

$$\Delta V = |F|^2 = 4 \frac{\Lambda^{10}}{|v|^6} \tag{1.43}$$

where the numerical constant is included in the definition of Λ. Thus, we see that the instanton-generated contribution ruins the indefinite equilibrium along the bottom of the valley, pushing the theory away from the origin. As a matter of fact, in the absence of the mass term, $m = 0$, the theory does not have any stable vacuum at all since the minimal (zero) energy is achieved only at $|v| \rightarrow \infty$. We encounter here an example of the *run-away vacuum* situation.

Switching on the mass term blocks the exits from the valley. Indeed, now

$$F = mv - 2\frac{\Lambda^5}{v^3} \, , \tag{1.44}$$

and the lowest energy state shifts to a finite value of v. It is easy to see that now there are two points at the bottom of the former valley where the energy vanishes, namely

$$v^2 = \sqrt{2} \frac{\Lambda^{5/2}}{m^{1/2}} \quad \text{and} \quad v^2 = -\sqrt{2} \frac{\Lambda^{5/2}}{m^{1/2}} \, . \tag{1.45}$$

In other words, the continuous vacuum degeneracy is lifted, and only two-fold degeneracy survives; the theory has two vacuum states. The number of vacuum states could have been anticipated from a general argument based on Witten's index[16].

I pause here to make a few remarks. First, we observe that the supersymmetric version of QCD dynamically has very little in common with QCD. Indeed, the chiral limit of QCD, when all quark masses are set equal to zero, is non-singular – nothing spectacular happens in this limit except that the pions become strictly massless. At the same time, in the supersymmetric $SU(2)$ model at hand the limit of the massless matter fields results in the run-away vacuum. This situation is quite general, and takes place in many models, although not in all.

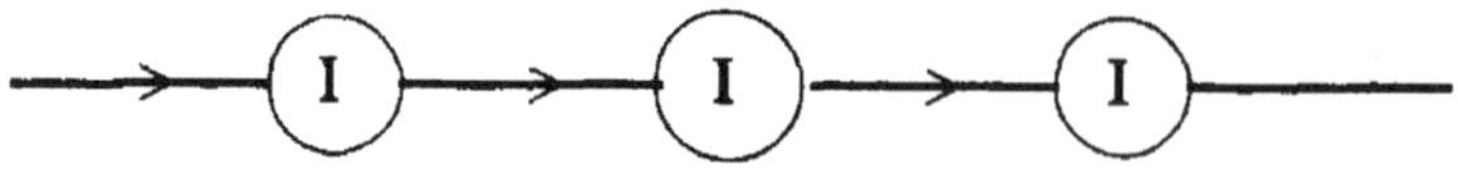

Figure 2. One-particle reducible graphs which might lead to $1/\sigma$ terms. This mechanism is not (and must not be) included in the superpotential.

Second, the analysis of the dynamics of the flat directions presented above is somewhat simplified. Two subtle points deserve mentioning. The general form of the superpotential compatible with the symmetry of the model was established in the massless limit. In this way we arrived at Eq. (1.42). The mass term was then introduced to avoid the run-away vacuum. If $m \neq 0$ the R current is not conserved any more, even at the classical level. To keep the invariance (1.41) alive one must simultaneously rotate the mass parameter,

$$m \to e^{4i\beta} m\,. \tag{1.46}$$

Let us call this invariance, supplemented by the phase rotation of the mass parameter, an *extended R symmetry.*

One could think of m as of a vacuum expectation value of some auxiliary chiral field, to be rotated in a concerted way in order to maintain the R invariance. (We will discuss this trick later on in more detail). It is clear then that multiplying Eq. (1.42) by any function of the dimensionless complex parameter $\sigma = mS^4/\Lambda^5$ is not forbidden by the extended symmetry. Extra arguments are needed to convince oneself that this additional function actually does not appear. Let us assume it does. Then it should be expandable in the Laurent series of the type

$$\sum_n \frac{C_n}{\sigma^n}\,.$$

If negative powers of n were present then the function would grow at large $|\sigma|$, a behavior one can immediately reject on physical grounds. The masses of the heavy particles, which we integrate over to obtain the superpotential, are proportional to gv; large values of $|v|$ imply heavier masses, which implies, in turn, that the impact on the superpotential should be weaker. Thus, all C_n's with negative n must vanish. Positive n are not acceptable as well. If positive powers of n were present then the function would blow up at fixed $|v|$ and $m \to 0$. At fixed $|v|$, however, no dynamically nontrivial singularity develops in the theory. The only mechanism which could provide powers of m in the denominator is a chain of instantons connected by one massless fermion line depicted on Fig. 2. The corresponding contribution, however, is one-particle reducible and should not be included in the effective Lagrangian. This concludes our proof of the fact that Eq. (1.42) is exact.

The second subtle point is related to the discussion of the anomalies in the R_0 and matter axial currents. The consideration presented above assumes that both anomalies are one-loop. Actually, the anomaly in the R_0 current is multiloop[36]. This fact slightly changes the form of the conserved R current. The very fact of existence of the R current remains intact. All expressions for the currents and charges presented above refer to the extreme ultraviolet where the gauge coupling (in asymptotically free theories) tends to zero. The final conclusion that the only superpotential compatible with the symmetry of the model is that of Eq. (1.42) is valid[37].

Thirdly, the consideration above (Eqs. (1.42), (1.45)) strictly speaking, does not tell us what happens at the origin of the valley, $v^2 = 0$, where all expressions become

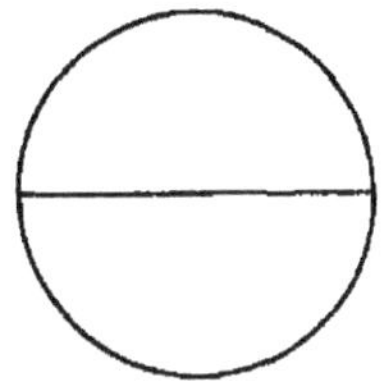

Figure 3. A typical two-loop supergraph for the vacuum energy.

inapplicable. Logically, it is possible to have an extra vacuum state characterized by $S^2 = 0$. This state would correspond to the strong coupling regime and will not be discussed here. The interested reader is referred to Ref.[11].

One last remark before concluding this section. Equation (1.43) illustrates why different points from the vacuum valley are physically inequivalent. In the conventional situation of the pre-SUSY era, the spontaneous breaking of a global symmetry, different vacua differ merely by a phase of v. Since physics depends on the ratio $|\Lambda/v|$, this phase is irrelevant. In supersymmetric theories the vacuum valleys are typically non-compact manifolds. Different points are marked not only by the phase of v, but by its absolute value as well. The dimensionless ratio above is different in different vacua. In particular, if $|\Lambda/v| \ll 1$ we are in the weak coupling regime; if $|\Lambda/v| \sim 1$ we are in the strong coupling regime.

Miracles of supersymmetry

Two of many miraculous dynamical properties of SUSY have been already mentioned – the vanishing of the vacuum energy and the non-renormalization theorem for F terms. It is instructive to see how these features emerge in perturbation theory.

Let us start from the vacuum energy. Consider a typical two-loop (super)graph shown on Fig. 3.

Each line on the graph represents the Green's function of some superfield. We do not even need to know what it is. The crucial point is that (if one works in the coordinate representation) each interaction vertex can be written as an integral over $d^4x d^2\theta d^2\bar{\theta}$. Assume that we substitute explicit expressions for Green's functions and vertices in the integrand, and carry out the integration over the second vertex keeping the first vertex fixed. As a result, we must arrive at an expression of the form

$$\int d^4x d^2\theta d^2\bar{\theta} \times \text{ a function of } x, \theta, \bar{\theta}. \qquad (1.47)$$

Since the superspace is homogeneous (there are no points that are singled out, we can freely make translations, any point in the superspace is equivalent to any other point) the function in Eq. (1.47) can be only constant. If so, the result vanishes because of the integration over the Grassmann variables θ and $\bar{\theta}$.

What remains to be demonstrated is that the one-loop vacuum graphs, not representable in the form given on Fig. 3, also vanish. The one-loop (super)graph, however, is the same as for the free particles, and we know already that for free particles $E_{vac} = 0$, see Eq. (1.1), thanks to the balance between the bosonic and fermionic degrees of freedom.

This concludes the proof of the fact that if the vacuum energy is zero at the classical level it remains there to any finite order - there is no renormalization. What changes if, instead of the vacuum energy, we would consider renormalizations of the F terms?

The proof presented above can be easily modified to include this case as well. Technically, instead of the vacuum loops, we will consider now loop (super)graphs in a background field.

The basic idea is straightforward. In any supersymmetric theory there are several – at least four – supercharge generators. In a generic background all supersymmetries are broken since the background field is generically not invariant under supertransformations. One can select such a background field, however, that leaves a part of the supertransformations as valid symmetries. For this specific background field some terms in the effective action will vanish, others will not. (Typically, F terms do not vanish while D terms do). The nonrenormalization theorems refer to those terms which do not vanish in the background field chosen.

Consider, for definiteness, the Wess-Zumino model[38],

$$S^{WZ} = \frac{1}{4}\int d^4x d^2\theta d^2\bar{\theta}\bar{\phi}\phi + \frac{1}{2}\left[\int d^4x d^2\theta \left(m\phi^2 + g\phi^3\right) + \text{H.c.}\right] . \tag{1.48}$$

An appropriate choice of the background field in this case is

$$\bar{\phi}_0 = 0\,, \quad \phi_0 = C_1 + C_2^\alpha\theta_\alpha + C_3\theta^2\,, \tag{1.49}$$

where $C_{1,2,3}$ are some constants and the subscript 0 marks the background field. This choice assumes that ϕ and $\bar{\phi}$ are treated as independent variables, not connected by the complex conjugation (i.e. we keep in mind a kind of analytic continuation). The x independent chiral field (1.49) is invariant under the action of $\bar{Q}_{\dot{\alpha}}$, i.e. under the transformations

$$\delta\theta_\alpha = 0\,, \quad \delta\bar{\theta}_{\dot{\alpha}} = \bar{\varepsilon}_{\dot{\alpha}}\,, \quad \delta x_{\alpha\dot{\alpha}} = -2i\theta_\alpha\bar{\varepsilon}_{\dot{\alpha}}\,.$$

Now, we proceed in the standard way – decompose the superfields

$$\phi = \phi_0 + \phi_{\text{qu}}\,, \quad \bar{\phi} = \bar{\phi}_0 + \bar{\phi}_{\text{qu}}\,,$$

where the subscript qu denotes the quantum part of the superfield, expand the action in $\phi_{\text{qu}}, \bar{\phi}_{\text{qu}}$, drop the linear terms and treat the remainder as the action for the quantum fields. We then integrate over the quantum fields order by order, keeping the background field fixed. The key element is the fact that in the problem we get for the quantum fields there still exists the exact symmetry under the transformations generated by $\bar{Q}$.

This means that the boson-fermion degeneracy holds, just as in the "empty" vacuum. All lines on the graphs of Fig. 3 have to be treated now as Green's functions in the background field (1.49). After substituting these Green's functions and integrating over all vertices except the first one we come to an expression of the type

$$\int d^4x d^2\bar{\theta} \times \text{ a } \bar{\theta} \text{ independent function} = 0\,.$$

The $\bar{\theta}$ independence follows from the fact that our superspace is homogeneous in the $\bar{\theta}$ direction even in the presence of the background field (1.49). This completes the proof of the non-renormalization theorem for the F terms. Note that the kinetic term *(D* terms) vanishes in the background (1.49), so nothing can be said about its renormalization (and it gets renormalized, of course). The above, somewhat non-standard, proof of the Grisaru-Ro ek-Siegel theorem was suggested in Ref.[36].

A word of caution is in order here. Our consideration tacitly assumes that there are no massless fields which can cause infrared singularities. Infrared singular contributions may lead to the so called *holomorphic anomalies*[39] invalidating the non-renormalization theorem. We will discuss the property of the holomorphy and the corresponding anomalies later, and now will illustrate how infrared singular D term renormalizations can effectively look as F terms. Consider the D term of the form

$$\int d^2\theta d^2\bar{\theta}\frac{D^2}{\Box}F(\phi)\,.$$

It can be rewritten as

$$\int d^2\theta F(\phi)$$

by using the property $[D^2,\bar{D}^2]\propto\Box$ and by integrating by parts in the superspace[21]. It is obvious that the $\Box^{-1}$ singularity can appear only due to massless poles. It was explicitly shown[40] that in the massless Wess-Zumino model such "fake" F terms appear at the two-loop level. The origin of the two-loop and all higher order terms in the Gell-Mann-Low function of supersymmetric gauge theories is the same – they emerge as a "fake" F term which is actually an infrared singular D term[36].

A recent discussion of the "fake" F terms is given in Ref.[41].

Holomorphy

At least some of the miracles of supersymmetry can be traced back to a remarkable property which goes under the name holomorphy. Some parameters in SUSY Lagrangians, usually associated with F terms, are complex rather than real numbers. The mass parameter in the superpotential is an obvious example. Another example mentioned in the section on supersymmetric gluodynamics is the inverse gauge coupling, $1/g_0^2$. Now, it is known for a long time, since the mid-eighties, that appropriately chosen quantities depend on these parameters analytically, with possible singularities in certain well-defined points. It is obvious that the statement that a function (analytically) depends on a complex variable is infinitely stronger than the statement that a function just depends on two real parameters. The power of holomorphy is such that one can obtain a variety of extremely non-trivial results ranging from non-renormalization theorems to exact β functions, the first time ever in dynamically non-trivial four-dimensional theories.

In this section we will outline the basic steps, keeping in mind that the corresponding technology will be of use more than once in what follows. Let us consider, as an example, $SU(2)$ SQCD with one flavor. I have already mentioned that m_0, the complex mass parameter in the action, can be viewed as a vacuum expectation value (of the lowest component) of an auxiliary chiral superfield, let us call it M. It is important that M is singlet with respect to the gauge group and thus, say, fermions from M do not contribute to the triangle anomalies. One can think of the corresponding degrees of freedom as of very heavy particles. Then the only role of M is to develop $\langle M\rangle\neq 0$ which obviously does not violate SUSY and provides the mass term.

The theory is strictly invariant under the following phase transformations

$$W_\alpha\to e^{i\beta}W_\alpha\,,\quad S_f\to e^{-i\beta}S_f\,,\quad M\to e^{4i\beta}M\,,\quad \theta_\alpha\to e^{i\beta}\theta_\alpha\,. \tag{1.50}$$

This is an extended R invariance – extended, because it takes place in the extended theory with the chiral superfield M introduced by hand.

It is rather clear that one chiral superfield can depend only on the expectation value of another superfield of the same chirality – otherwise transformation properties under SUSY would be broken. Thus, the expectation value of W^2 can depend only on that of M; $\bar{M}$ cannot be involved in this relation. Equation (1.50) then tells us that

$$\langle\mathrm{Tr}\,W^2\rangle=\mathrm{const.}(M)^{1/2}. \tag{1.51}$$

In other words, the gluino condensate

$$\langle\lambda\lambda\rangle \propto \sqrt{m_0}\,, \tag{1.52}$$

and this relation is exact as far as the m_0 dependence is concerned. It holds for small $|m_0|$ when the theory is weakly coupled, as well as for large $|m_0|$, when we are in the strong coupling regime.

A similar assertion is valid regarding the vacuum expectation value of S^2. Now, Eq. (1.50) tells us that

$$\langle S^2\rangle = \text{const.}(M)^{-1/2}\,, \tag{1.53}$$

implying, in turn, that

$$\langle\phi^2\rangle \propto 1/\sqrt{m_0}\,. \tag{1.54}$$

It is worth emphasizing that the exact dependence of the condensates on the mass parameter established above refers to the bare mass parameter. If we decided to eliminate the bare mass parameter m_0 in favor of the physical mass of the Higgs field m, we would have to introduce the corresponding Z factor which depends on m in a complicated non-holomorphic way.

Equations (1.52) and (1.54), first derived in Ref.[10] (a similar argument was also given in Ref.[33]), lead to far reaching consequences. Indeed, since the functional dependence of the condensates is fully established, we can calculate the relevant constant at small $|m_0|$, when $\langle\phi^2\rangle$ is large, which ensures weak coupling. I remind that the masses of the gauge bosons in this limit are proportional to

$$M_V^2 \propto |g^2\langle\phi^2\rangle|.$$

The result will still be valid for large $|m_0|$, in the strong coupling regime! This line of reasoning[10], based on holomorphy, lies behind many advances achieved recently in SUSY gauge dynamics.

Let me parenthetically note that a nice consistency check to Eqs. (1.52), (1.54) is provided by the so called Konishi anomaly[42]. In the model at hand the Konishi relation takes the form

$$\bar{D}^2\bar{S}^{\alpha f}e^V S^{\alpha f} = 4m_0 S_{\alpha f}S^{\alpha f} + \frac{1}{2\pi^2}\operatorname{Tr} W^2\,. \tag{1.55}$$

This expression, or more exactly, the second term on the right hand side, is nothing but a supergeneralization of a the triangle anomaly in the divergence of the axial current of the matter fermions, cf. Eq. (1.39). Now, if SUSY is unbroken, the expectation value of the left-hand side must vanish, since the left-hand side i a full superderivative. This fact implies that

$$4m_0\langle S_{\alpha f}S^{\alpha f}\rangle = -\frac{1}{2\pi^2}\langle\operatorname{Tr} W^2\rangle \tag{1.56}$$

which is consistent with Eqs. (1.52), (1.54). Another side remark: the exact proportionality of $\langle\phi^2\rangle$ to $1/\sqrt{m_0}$ presents a somewhat different proof of the fact that the instanton-generated superpotential (1.42) is exact even in the presence of the mass term.

One can take advantage of these observations in many ways. One direction is finding the exact β function of the theory. The idea is as follows. First we assume that $|m_0|$ is small and we are in the weak coupling regime. Then we are able to calculate the gluino condensate – in the weak coupling regime it is saturated by the one-instanton contribution. Since the functional dependence on m_0 is known we can then proceed to the limit of large $|m_0|$, or small ϕ^2. Moreover, the vacuum expectation value of $\lambda\lambda$ is, in principle, a physically measurable quantity. The operator $\lambda\lambda$ has strictly vanishing anomalous dimension since it is the lowest component of the superfield W^2, and the upper component of the same superfield contains the trace of the energy-momentum tensor. This means that if $\langle\lambda\lambda\rangle$ is expressed in terms of the gauge coupling g_0^2 and the ultraviolet cut-off M_0, when one changes the cut off, one should also change g_0^2 in a concerted way, to ensure that $\langle\lambda\lambda\rangle$ stays intact. In this way we obtain a relation between the bare coupling constant and M_0, which is equivalent to the knowledge of the β function.

More concretely, the one-instanton result for the gluino condensate is[43]

$$\langle\lambda\lambda\rangle = \text{const.} \times \frac{M_0^5 e^{-8\pi^2/g_0^2}}{g_0^4 v_0^2} . \tag{1.57}$$

This result is exact; only zero modes in the instanton background contribute in the calculation. It is worth emphasizing that the expectation value of the scalar field appearing in Eq. (1.57) refers to the bare field. The constant on the right-hand side is purely numerical; we will say more about this constant later on, but for the time being its value is inessential.

At the super-symmetric vacuum Eq. (1.56) must hold implying that

$$v_0^2 = \text{const.}\langle\lambda\lambda\rangle m_0^{-1} . \tag{1.58}$$

Combining Eqs. (1.58) and (1.57) we conclude that

$$\langle\lambda\lambda\rangle = \text{const.} M_0^{5/2} m_0^{1/2} e^{-4\pi^2/g_0^2} \frac{1}{g_0^2} . \tag{1.59}$$

When analyzing the response of g_0^2 with respect to the variations of M_0 one should keep in mind that m_0 also depends on M_0, implicitly. Indeed, the physical (low-energy) values of the parameters are kept fixed. This means that we fix the renormalized value of the mass, $m = m_0 Z^{-1}$, where Z is the Z factor renormalizing the kinetic term of the matter fields,

$$(\bar{S}^{\alpha f} e^V S^{\alpha f})_0 \to Z(\bar{S}^{\alpha f} e^V S^{\alpha f})_0 .$$

In this way we arrive at the conclusion that the combination $M_0^{5/2} e^{-4\pi^2/g_0^2} Z^{1/2} g_0^{-2}$ is invariant. Differentiating it with respect to $\ln M_0$ we find the β function,

$$\beta(\alpha) = -\frac{\alpha^2}{2\pi} \frac{5+\gamma}{1-\alpha/\pi} , \tag{1.60}$$

where the β function is defined as

$$\beta(\alpha_0) = d\alpha_0/d\ln M_0 , \quad \alpha_0 = g_0^2/4\pi ,$$

and γ is the anomalous dimension of the matter fields,

$$\gamma = d \ln Z/d \ln M_0 .$$

Note that due to the $SU(2)$ subflavor symmetry of the model at hand both matter fields, S_1 and S_2, have one and the same anomalous dimension.

Equation (1.60) is a particular case of the general β function, sometimes referred to as NSVZ β function,

$$\beta(\alpha) = -\frac{\alpha^2}{2\pi}\left(1 - \frac{T(G)\alpha}{2\pi}\right)^{-1}\left[3T(G) - \sum_i T(R_i)(1-\gamma_i)\right] , \tag{1.61}$$

which can be derived in a similar manner[13]. Here $T(G)$ and $T(R)$ are the so called Dynkin indices defined as follows. Assume that the gauge group is G, and we have a field belonging to the representation R of the gauge group. If T^a is the generator matrix of the group G in the representation R then

$$\mathrm{Tr}\,(T^a T^b) = T(R)\delta^{ab}\,.$$

More exactly, $T(R)$ is one half of the Dynkin index. Moreover, $T(G)$ is $T(R)$ for the adjoint representation. Note that for the fundamental representation of the unitary groups $T(R) = 1/2$. The sum in Eq. (1.61) runs over all subflavors.

As is clear from its derivation, the NSVZ β function implies the Pauli-Villars regularization. It can also be derived purely perturbatively, with no reference to instantons, using only holomorphy properties of the gauge coupling[44]. The relation of this β function to that defined in other, more conventional regularization schemes is investigated in Ref.[48].

In some theories the NSVZ β function is exact – there are no corrections, either perturbative or nonperturbative, as is the case in the SU(2) model with one flavor. In other models it is exact only perturbatively – nonperturbative corrections do modify it. The most important example[46] of the latter kind is $N = 2$ supersymmetry.

The idea of using the analytic properties of chiral quantities for obtaining exact results was adapted for the case of superpotentials in Ref.[47]. As a matter of fact, I have already discussed some elements of the procedure suggested in Ref.[47], in analyzing possible mass dependence of nonperturbatively generated superpotential in the $SU(2)$ model. Let me summarize here the basic stages of the procedure in the general form, and give a few additional examples. Simultaneously, as a byproduct, we will obtain a different proof of some of the non-renormalization theorems considered in the section on miracles of supersymmetry. What is remarkable is that, unlike the proof presented in the section on miracles of supersymmetry, the one given below will be valid both perturbatively and nonperturbatively.

Thus, our task is establishing possible renormalizations of the superpotential $\mathcal{W}$ in a given model. All (complex) coupling constants of the model appearing as coefficients in front of F terms – denote them generically by h_i – are treated as vacuum expectation values of some (auxiliary) chiral superfields. The set of $\{h_i\}$ may include the mass parameters and/or Yukawa constants. Let us assume that if all $h_i = 0$, the model considered possesses a non-anomalous global symmetry group $\mathcal{G}$; however, the couplings $h_i \neq 0$ break this symmetry. Since h_i's are treated now as auxiliary chiral superfields one can always define transformations of these superfields in such a way as to restore the global symmetry $\mathcal{G}$. Then the calculated superpotential depending on the dynamical chiral superfields and on the auxiliary ones, should be invariant under this extended $\mathcal{G}$. This constraint becomes informative if we take into account the fact that the calculated superpotential $\mathcal{W}$ must be a holomorphic function of all chiral superfields. Thus, $\mathcal{W}$ can depend on h_i but cannot depend on h_i^*. A few additional rules apply. The effective superpotential may depend on the dynamically generated scale of the model Λ. It is clear that negative powers of Λ are forbidden since the result should be smooth in the limit when the interaction is switched off. Moreover, if we ensure that the theory is in the weak coupling regime, the possible powers of Λ are only those associated with one, two, three and so on instantons, i.e. $3T(G) - \sum_f T(R_f)$, $2(3T(G) - \sum_f T(R_f))$, and so on, since the instantons are the only source of the nonperturbative parameter Λ in the weak coupling regime. Finally, one more condition comes from analyzing the limit of the small bare couplings h_i. This limit can be often treated perturbatively. Sometimes additional massless fields appear in the limit $h_i = 0$, which are absent for $h_i \neq 0$. When these fields are integrated out and not included in the effective action, W may develop a singularity at $h_i = 0$.

To illustrate the power and elegance of this approach[47] let us turn again to the Wess-Zumino model, Eq. (1.48). If the bare mass and the coupling constant vanish, $m = g = 0$, then the model has two $U(1)$ global invariances – one associated with the rotations of the matter field, $\phi \to e^{i\alpha}\phi$, and another one is the R_0 symmetry, $\phi \to e^{2i\beta/3}\phi$, $\theta \to e^{i\beta}\theta$. To maintain both invariances with m and g switched on we demand that

$$m \to me^{-2i\alpha}, \quad g \to ge^{-3i\alpha}$$

and

$$m \to me^{2i\beta/3}, \quad g \to g$$

under $U(1)_\phi$ and $U(1)_R$, respectively. The most general renormalized superpotential compatible with these symmetries obviously has the form

$$\mathcal{W} = m\phi^2 f\left(\frac{g\phi}{m}\right), \tag{1.62}$$

where f is an arbitrary function. Let us expand it in a power series and consider the coefficient in front of ϕ^n. Prom Eq. (1.62) it is clear that the corresponding term has the form

$$g^{n-2} m^{3-n} \phi^n .$$

The balance of powers of the coupling constant and m is such that this contribution could only be associated with the 1-particle reducible tree graphs, which should not be included in the effective action. Therefore, we conclude that there is no renormalization of the superpotential, $\mathcal{W} = \mathcal{W}_0$.

An example of a more sophisticated situation is provided by the $SU(N)$ theory with N_f flavors and the tree level superpotential[48]

$$\mathcal{W}_0 = hM \sum_f \tilde{Q}_f Q_f + h' M^3 \tag{1.63}$$

where Q_f is a chiral superfield belonging to the fundamental representation of $SU(N)$ while Q_f is a chiral field in the anti-fundamental representation. The color indices are summed over; the additional chiral field M is color-singlet.

Now, if h and h' are set equal to zero, M obviously decouples, and the global symmetries of the model are those of the massless SQCD plus one extra global invariance associated with the rotations of the M field. Massless SQCD is invariant under

$$SU(N_f)_L \times SU(N_f)_R \times U(1)_V \times U(1)_R .$$

The conserved R charge is established from consideration of the anomaly relations analogous to Eq. (1.39). Indeed, it is not difficult to obtain that

$$\partial_\mu R^0_\mu = \left(N - \frac{N_f}{3}\right) \frac{\alpha}{4\pi} G^a_{\mu\nu} \tilde{G}^a_{\mu\nu}, \quad \partial_\mu J_\mu = N_f \frac{\alpha}{4\pi} G^a_{\mu\nu} \tilde{G}^a_{\mu\nu}, \tag{1.64}$$

where the R_0 and J currents are defined in parallel to those in the $SU(2)$ model, see Eqs. (1.35) and (1.37). This means that the conserved R current has the form

$$R_\mu = R^0_\mu - \frac{N - (N_f/3)}{N_f} J_\mu . \tag{1.65}$$

Using this expression it is not difficult to calculate that the R charge of the matter field is $(N_f - N)/N_f$. The R charge of the M field can be set equal to zero.

Now, the superpotential (1.63) explicitly breaks both, $U(1)_R$ and $U(1)_M$. To restore the symmetry we must ascribe to h and h' the following charges

$$h \to \left(-1, 2\frac{N}{N_f}\right), \quad h' \to (-3, 2),$$

where the first charge is with respect to $U(1)_M$ while the second charge is with respect to $U(1)_R$. If $h' = 0$ the model has a rich system of the vacuum valleys. Let us assume that we choose the one for which the expectation value of Q fields vanishes, but the expectation value of $M \neq 0$. Moreover, we will assume that $h\langle M\rangle$ is large. In this "corner" of the valley the matter fields are heavy, and can be integrated over. At very low energies the only surviving (massless) field is M. Our task is to find the effective Lagrangian for the M field. By inspecting the above charge assignments one easily establishes that the most general form of the effective superpotential compatible with all the assignments is

$$\mathcal{W} = h'M^3 f(x), \quad x = \frac{\Lambda^{3-(N_f/N)}(hM)^{N_f/N}}{h'M^3},$$

where Λ is a dynamically generated scale of the (strongly coupled) SU(N) gauge theory. If $\Lambda \to 0$ there should be no singularities. This implies that $f(x)$ *is* expandable in positive powers of x. However, the behavior of the effective superpotential at $h' \to 0$ should also be smooth. These two requirements fix the function f up to a constant,

$$f(x) = 1 + \alpha x, \quad \alpha = \text{const.},$$

and

$$\mathcal{W} = h'M^3 + \alpha\,\Lambda^{3-(N_f/N)}(hM)^{N_f/N}. \tag{1.66}$$

The first term is the same as in the bare superpotential, the second term is generated nonperturbatively[49]. Note that the non-analytical behavior at $h = 0$ is due to the fact that at $h = 0$ there are massless matter fields, and we integrated over them assuming that they are massive. The superpotential (1.66) grows with M. This is natural since the interaction becomes stronger as M increases. The superpotential (1.66) leads to a supersymmetric minimum at $M \neq 0$.

Concluding this section I would like to return to one subtle and very important point for this range of questions, holomorphic anomalies. Consider Eq. (1.59) for the gluino condensate. So far we have studied the analytic dependence of this quantity on m_0. At the same time, however, $1/g_0^2$ is also a coefficient of the F term, which can be viewed as an expectation value of an auxiliary chiral superfield, dilaton/axion. One is tempted to conclude then that the dependence of $\langle\lambda\lambda\rangle$ on $1/g_0^2$ must be holomorphic, and its functional form must follow from consideration of the invariances of the theory. Is this the case?

The answer is yes and no. Let us examine the transformation properties of the SU(2) theory with respect to the matter $U(1)$ rotations, Eq. (1.38), supplemented by the rotation of the mass parameter $m_0 \to m_0 e^{-2i\gamma}$. At the classical level the theory is invariant. The invariance is broken, however, by the triangle anomaly. In order to restore the invariance we must simultaneously shift the vacuum angle ϑ (not to be confused with the supercoordinates θ_α!),

$$\vartheta \to \vartheta - \Delta\text{Arg}(m_0) = \vartheta + 2\gamma. \tag{1.67}$$

Taking into account the fact that $\vartheta = \text{Im}(-8\pi^2/g_0^2)$ we conclude that if Eq. (1.59) contained no pre-exponential factor g_0^{-2} everything would be perfect – $\lambda\lambda$ would be a holomorphic function of g_0^{-2}, precisely the one needed for invariance of $\lambda\lambda$. The pre-exponential factor g_0^{-2} spoils the perfect picture. As a matter of fact, one can see that in the pre-exponential it is $\text{Re}(g_0^{-2})$ which enters, and the holomorphy in g_0^{-2} is absent. The reason is the holomorphic anomaly associated with infrared effects. In Refs.[36, 39] it was first noted that all formal theorems regarding the holomorphic dependences on the gauge coupling constant are valid only provided we define the gauge coupling through the Wilsonian action which, by definition, contains no infrared contributions. What one usually deals with (and refers to as the action) is actually the generator of the one-particle irreducible vertices. In the absence of the infrared singularities these two notions coincide; generally speaking, they are different, however. In particular, the gauge coupling constant in the Wilsonian action, g_W^{-2} is related to g_0^{-2} by the following expression

$$\frac{8\pi^2}{g_W^2} = \frac{8\pi^2}{g_0^2} - T(G)\ln \text{Re} g_0^{-2} \tag{1.68}$$

(in pure gluodynamics, without matter). The gluino condensate is holomorphic with respect to g_W^{-2}. The holomorphic anomaly in the gauge coupling due to massless matter fields was also observed in Ref.[50] in the stringy context, see also[51].

Supersymmetric instanton calculus

As was mentioned more than once, the instanton calculations, combined with specific features of supersymmetry, were instrumental in establishing various exact results in supersymmetric gluodynamics and other theories. We will continue to exploit them in further applications. Needless to say that I will be unable to present supersymmetric instanton calculus to the degree needed for practical uses. The interested reader is referred to Ref.[52]. Here I will limit myself to a few fragmentary remarks.

Technically, the most remarkable feature making the instanton calculations in supersymmetric theories by far more manageable than in non-supersymmetric ones is a residual supersymmetry in the instanton background field. It is clear that picking up a particular external field we typically break (spontaneously) supersymmetry: SUSY generators applied to this field act non-trivially. However, the self-dual (or anti-self-dual) Yang-Mills field, analytically continued to the Euclidean space, to which the instanton belongs, preserves a half of supersymmetry. Depending on the sign of the duality relation either Q_α or $\bar{Q}_{\dot\beta}$ act trivially, i.e. annihilate the background field[53, 54].

The fact that a part of supersymmetry remains unbroken in the instanton background leads to far-reaching consequences. Indeed, the spectrum of fluctuations around this background remains degenerate for bosons and fermions from one and the same superfield, and the form of the modes is in one-to-one correspondence[54], for all modes except the zero modes. An immediate consequence is vanishing of the one-loop quantum correction in the instanton background. Unsurprisingly, a more careful study shows[9] that all higher quantum corrections vanish as well.

Thus, the result of any instanton calculation is essentially determined by the zero modes alone. The problem reduces to quantum mechanics of the zero modes. The structure of the zero modes is governed by a set of relevant symmetries of the theory under consideration[43]. Therefore, all quantities that are saturated by instantons reflect the most general and profound geometrical properties of the theory. One of the examples, the gluino condensate, was already considered above. In the next part we will discuss another example – the instanton-induced modification of the quantum moduli space in SQCD with $N_f = N_c$.

Historically the first application of instantons in supersymmetric gluodynamics was the calculation of the gluino condensate[9] in the strong coupling regime. I mention this result here because although it is 15 years old, there is an intriguing mystery associated with it.

Let us consider for definiteness the $SU(2)$ gluodynamics. In this case there are four gluino zero modes in the instanton field and hence, there is no direct instanton contribution to the gluino condensate $\langle\lambda\lambda\rangle$. At the same time the instanton does contribute to the correlation function

$$\langle\lambda^a_\alpha(x)\lambda^{a\alpha}(x), \lambda^b_\beta(0)\lambda^{b\beta}(0)\rangle\,, \tag{1.69}$$

Here $a, b = 1,2,3$ are the color indices and $\alpha,\beta = 1,2$ are the spinor ones. An explicit instanton calculation shows that the correlation function (1.69) is equal to a non-vanishing constant.

At first sight this result might seem supersymmetry-breaking since the instanton does not generate any bosonic analog of Eq. (1.69). Surprising though it is, supersymmetry does not forbid (1.69) provided that this two-point function is actually an x independent constant. For purposes which will become clear shortly let us sketch here the proof of the above assertion.

Three elements are of importance: (i) the supercharge $\bar{Q}^{\dot\beta}$ acting on the vacuum state annihilates it; (ii) $\bar{Q}^{\dot\beta}$ commuteswith $\lambda\lambda$; (iii) the derivative $\partial_{\alpha\dot\beta}(\lambda\lambda)$ is representable as the anticommutator of $\bar{Q}^{\dot\beta}$ and $\lambda^\beta G_{\beta\alpha}$. (The spinor notations are used.) The second and the third point follow from the fact that $\lambda\lambda$ is the lowest component of the chiral superfield W^2, while $\lambda^\beta G_{\beta\alpha}$ is its middle component.

Now, we differentiate Eq. (1.69), substitute $\partial_{\alpha\dot\beta}(\lambda\lambda)$ by $\{\bar{Q}^{\dot\beta}, \lambda^\beta G_{\beta\alpha}\}$ and obtain zero. Thus, supersymmetry requires the x derivative of (1.69) to vanish[9]. This is exactly what happens if the correlator (1.69) is a constant.

If so, one can compute the result at short distances where it is presumably saturated by small-size instantons, and, then, the very same constant is predicted at large distances, $x \to \infty$. On the other hand, due to the cluster decomposition property which must be valid in any reasonable theory the correlation function (1.69) at $x \to \infty$ reduces to $\langle\lambda\lambda\rangle^2$. Extracting the square root we arrive at a (double-valued) prediction for the gluino condensate.

(The same line of reasoning is applicable in other similar problems, not only for the gluino condensate. The correlation function of the lowest components of any number of superfields of one and the same chirality, if non-vanishing, must be constant. By analyzing the instanton zero modes it is rather easy to catalog all such correlation functions, in which the instanton contribution does not vanish. Thus, for $SU(N)$ gluodynamics one ends up with the N-point function of $\lambda\lambda$. Inclusion of the matter fields, clearly, enriches the list of the instanton-induced "constant" correlators, but not too strongly[55]. The general strategy remains the same as above in all cases.)

Many questions immediately come to one's mind in connection with this argument. First, if the gluino condensate is non-vanishing and shows up in a roundabout instanton calculation through (1.69) why is it not seen in the direct instanton calculation of $\langle\lambda\lambda\rangle$? Second, the constancy of the two-point function (1.69) required by SUSY is ensured in the concrete calculation by the fact that the instanton size ρ turns out to be of order of x. The larger the value of x the larger ρ saturates the instanton contribution. For small x this is alright. At the same time at $x \to \infty$ we do not expect any coherent fields with the size of order x to survive in the vacuum; such coherent fields would contradict our current ideas of the infrared-strong confining theories like SUSY gluodynamics.

If there are no large-size coherent fields in the vacuum how can one guarantee the x independence of (1.69) at all distances?

A tentative answer to the first question might be found in the hypothesis put forward by Amati *et al.*[33]. It was assumed that, instead of providing us with the expectation value of $\lambda\lambda$ in the given vacuum, instantons in the strong coupling regime yield an average value of $\langle\lambda\lambda\rangle$ in all possible vacuum states. If there exist two vacua, with the opposite signs of $\langle\lambda\lambda\rangle$, the conjecture of Amati *et al.* would explain why instantons in the strong coupling regime do not generate $\lambda\lambda$ directly.

When we do the instanton calculation in the weak coupling regime (the Higgs phase) the averaging over distinct vacua does not take place. In the weak coupling regime, we have a marker: a large classical expectation value of the Higgs field tells us in what particular vacuum we do our instanton calculation. In the strong coupling regime, such a marker is absent, so that the recipe of Amati *et al.* seems plausible.

This is not the end of the story, however. One of the instanton computations which was done in the mid-eighties[43] remained a puzzle defying theoretical understanding for years. The result for $\langle\lambda\lambda\rangle$ obtained in the strong coupling regime (i.e. by following the program outlined after Eq. (1.69)) does *not* match $\langle\lambda\lambda\rangle$ calculated in an indirect way, as we did in the section on holomorphy – extending the theory by adding one flavor, doing the calculation in the weakly coupled Higgs phase, and then returning back to SUSY gluodynamics by exploiting the holomorphy of the condensate in the mass parameter. In Ref.[43] it was shown that

$$\langle\lambda\lambda\rangle^2_{scr} = \frac{4}{5}\langle\lambda\lambda\rangle^2_{wcr}\,, \tag{1.70}$$

where the subscripts scr and wcr mark the strong and weak coupling regime calculations.

The hypothesis of Amati *et al.,* by itself, does not explain the discrepancy (1.70). If there are only two vacua characterized by $\langle\lambda\lambda\rangle = \pm\Lambda^3$, the gluino condensate is not affected by the averaging over these two vacuum states, since the contributions of these two vacua to Eq. (1.69) are equal. If, however, there exist an *extra* zero-energy state with $\langle\lambda\lambda\rangle = 0$ involved in the averaging, the final result in the strong coupling regime is naturally different from that obtained in the weak coupling regime in the *given* vacuum. Moreover, the value of the condensate calculated in the strong coupling approach should be smaller, consistently with Eq. (1.70). At the moment there seems to be no other way out of the dilemma[11]. The conclusion of the existence of the extra vacuum with $\langle\lambda\lambda\rangle = 0$ is quite radical, and, perhaps, requires further verification, in particular, in connection with the Witten index counting. What is beyond any doubt, however is that the combination of instanton calculus with holomorphy and other specific features of supersymmetry provides us with the most powerful tool we have ever had in four-dimensional field theories.

It remains to be added that the interest in technical aspects of supersymmetric instanton calculus[43] was revived recently in connection with the Seiberg-Witten solution of the $N = 2$ theory. The solution was obtained[86] from indirect arguments, and it was tempting to verify it by direct instanton calculations[67]. Such calculations require extension of supersymmetric instanton calculus to $N = 2$, which was carried out, in a very elegant way, in Refs.[57,58].

****

Concluding this part of the Lecture let me briefly summarize the main lessons. First, the most remarkable feature of the structure of SUSY gauge theories with matter is the existence of the vacuum valleys – classically flat directions along which the energy vanishes. This degeneracy may or may not be lifted dynamically, at the quantum level.

The $SU(2)$ model with one flavor is an example of the theory where the continuous degeneracy is lifted, and the quantum vacuum has only discrete (two-fold) degeneracy. If this does not happen, the classically flat directions give rise to *quantum moduli space of supersymmetric vacua.* This feature is the key element of the recent developments pioneered by Seiberg.

Second, holomorphic dependences of various chiral quantities enforced by supersymmetry lie behind numerous miracles occurring in SUSY gauge theories – from specific non-renormalization theorems to the exact β functions. This is also an important element of dynamical scenarios to be discussed below. Now the stage is set and we are ready for more adventures and surprises in supersymmetric dynamics.

VARIOUS DYNAMICAL REGIMES IN SUSY GAUGE THEORIES

In the first part I summarized what was known (or assumed) about the intricacies of the gauge dynamics in the eighties. In the following we will discuss the discoveries and exciting results of the recent years. I should say that the current stage of development was opened up by Seiberg, and many ideas and insights to be discussed today I learned from him or extracted from works of his collaborators.

Remarkable facets of the gauge dynamics will be revealed to us. First of all, we will encounter non-conventional patterns of the chiral symmetry breaking. Chiral symmetry breaking is one of the most important phenomena of which very little was known, beyond some empiric facts referring to QCD. In the eighties, when our knowledge of the gauge dynamics was less mature than it is now, it was believed that the massless fermion condensation obeys the so called maximum attraction channel (MAC) hypothesis[89]. In short, one was supposed to consider the one-gluon exchange between fermions, find a channel with such quantum numbers that the attraction was maximal, and then assume the condensation of the fermion pairs in this particular channel. The concrete quantum numbers of the fermion condensates imply a very specific pattern of the chiral symmetry breaking.

In SQCD we will find patterns *contradicting the MAC hypothesis.* This means that the chiral condensates are not governed by the one-gluon exchange, even qualitatively. The basic tool for exploring the chiral condensates is the 't Hooft matching condition. It was exploited for this purpose previously many times, in the context most relevant to us in Ref.[60]. Combining supersymmetry (the fact of the existence of the vacuum manifold) with the matching condition drastically enhances the method.

The second remarkable finding is the observation of conformally invariant theories in four dimensions in the strong coupling regime. The crucial instrument in revealing such theories is Seiberg's "electric-magnetic" duality in the infrared domain, connecting with each other two distinct gauge theories - one of them is strongly coupled while the other is weakly coupled. One can view the gluons and quarks of the weakly coupled theory as bound states of the gluons and quarks of its dual partner. If so, composite gauge bosons can exist! The arguments in favor of the "electric-magnetic" duality are again based on the 't Hooft matching condition (combined with supersymmetry) and some additional indirect consistency checks.

$SU(N_c)$ QCD with N_f flavors – preliminaries

The $SU(2)$ model considered previously is somewhat special since all representations of $SU(2)$ are (pseudo)real. For this reason the flavor sector of this model possesses an enlarged symmetry. Thus, for one flavor we observe the flavor $SU(2)$ symmetry, which is absent if the gauge group is, say, $SU(3)$. Now we will consider a more generic situation. The gauge group is assumed to be $SU(N_c)$ with $N_c > 2$. In accordance with Witten's index, if the matter sector consists of non-chiral matter allowing (at least, in principle) for a mass term for all matter fields, supersymmetry is unbroken.

To describe N_f flavors one has to introduce $2N_f$ chiral superfields, Q^i in the representation N_c and $\tilde{Q}_j$ in the representation $\bar{N}_c$. To distinguish between the fundamental and anti-fundamental representations the flavor indices used are superscripts and subscripts, respectively.

The Lagrangian is very similar to that of the $SU(2)$ model,

$$\mathcal{L} = \frac{1}{2g_0^2} \operatorname{Tr} \int d^2\theta W^2 + \frac{1}{4} \int d^2\theta d^2\bar{\theta} \left(Q^\dagger e^V Q + \tilde{Q}^\dagger e^{-V} \tilde{Q} \right) + \left(\frac{1}{2} \int d^2\theta \, \mathcal{W}(Q, \tilde{Q}) + \text{H.c.} \right), \tag{2.1}$$

where $\mathcal{W}(Q, \tilde{Q})$ is a superpotential which may or may not be present. Then the scalar potential has the form

$$V = \frac{1}{2g^2} D^a D^a + \sum_Q \left| \frac{\partial \mathcal{W}(Q, \tilde{Q})}{\partial Q} \right|^2 + \sum_{\tilde{Q}} \left| \frac{\partial \mathcal{W}(Q, \tilde{Q})}{\partial \tilde{Q}} \right|^2 . \tag{2.2}$$

An example of a possible superpotential is a generalized mass term,

$$\mathcal{W} = m_i^j \tilde{Q}_j Q^i$$

where m_i^j is a mass matrix. Most often we will work under conditions of vanishing superpotential, $\mathcal{W} = 0$.

It is convenient to introduce two $N_f \times N_c$ matrices of the form

$$q = \left\{ Q^1, Q^2, \ldots, Q^{N_f} \right\}, \quad \tilde{q} = \left\{ \tilde{Q}_1, \tilde{Q}_2, \ldots, \tilde{Q}_{N_f} \right\} . \tag{2.3}$$

The rows of these matrices correspond to different values of the color index. Thus, in the first row the color index is 1, in the second row 2, *etc.*, N_c rows altogether. Both matrices can be globally rotated in the color and flavor spaces. Let us assume first that $N_f < N_c$. Then, by applying these rotations one can always reduce the matrix q to the form

$$q = \begin{pmatrix} a_1 & 0 & \ldots & 0 \\ 0 & a_2 & \ldots & 0 \\ \ldots & \ldots & \ldots & \ldots \\ 0 & 0 & \ldots & a_{N_f} \\ 0 & 0 & \ldots & 0 \\ \ldots & \ldots & \ldots & \ldots \end{pmatrix} . \tag{2.4}$$

If $\tilde{q} = q$ we are at the bottom of the vacuum valley – the corresponding energy vanishes. The gauge invariant description is provided by the composite chiral superfield,

$$M_j^i = \tilde{Q}_j Q^i . \tag{2.5}$$

The points belonging to the bottom of the valley are parametrized by the expectation value of M_j^i. Generically, if we are away from the origin the $SU(N_c)$ gauge group is broken down to $SU(N_c - N_f)$. The first group has $N_c^2 - 1$ generators, the second one has $(N_c - N_f)^2 - 1$ generators. Thus, the number of the chiral fields eaten up in the super-Higgs mechanism is $2N_cN_f - N_f^2$. Originally we started from $2N_cN_f$ chiral superfields; N_f^2 remain massless – exactly the number of degrees of freedom in M_j^i. There are exceptional points. When $\det M = 0$ the unbroken gauge subgroup is larger than $SU(N_c - N_f)$ and, correspondingly, we have more than N_f^2 massless particles. At the origin of the vacuum valley the original gauge group $SU(N_c)$ remains unbroken.

The situation changes if the number of flavors is equal to or larger than the number of colors. Indeed, if $N_f > N_c$ the generic form of the matrix q, after an appropriate rotation in the flavor and color space, is

$$q = \begin{pmatrix} a_1 & 0 & \dots & 0 & 0 & \dots & 0 \\ 0 & a_2 & \dots & 0 & 0 & \dots & 0 \\ \dots & \dots & \dots & \dots & \dots & \dots & \dots \\ 0 & 0 & \dots & a_{N_c} & 0 & \dots & 0 \end{pmatrix} . \tag{2.6}$$

The condition defining the bottom of the valley (the vanishing of the energy) is

$$|a_i|^2 - |\tilde{a}_i|^2 = \text{constant independent of } i \tag{2.7}$$

where $i = 1, 2, ..., N_c$. At a generic point of the bottom of the valley the gauge group is completely broken. The gauge invariant chiral variables parametrizing the bottom of the valley (the moduli space) now are

$$M_j^i = \tilde{Q}_j Q^i \,, \quad B = Q^{[i_1} ... Q^{i_{N_c}]} \,, \quad \tilde{B} = \tilde{Q}_{[j_1} ... \tilde{Q}_{j_{N_c}]} \,, \tag{2.8}$$

where the color indices in B and $\tilde{B}$ (they are not written out explicitly) are contracted with the help of the ε symbol; the flavor indices $i_1, ..., i_{N_c}$ and $j_1, ..., j_{N_c}$ then come out automatically antisymmetric, and the square brackets in Eq. (2.8) remind us of this antisymmetrization. *A priori,* the number of the variables B and $\tilde{B}$ is $C_{N_f}^{N_c}$ each, where $C_{N_f}^{N_c}$ are the combinatorial coefficients,

$$C_{N_f}^{N_c} = \frac{N_f!}{N_c!(N_f - N_c)!} \,,$$

since one can pick up N_c flavors out of the total set of N_f in various ways. If we try to calculate now the number of moduli, assuming that all those indicated in Eq. (2.8) are independent, we will see that this number does not match the number of the massless degrees of freedom. Let us consider two examples, $N_f = N_c$ and $N_f = N_c + 1$.

$\mathbf{N_f = N_c}$

The original number of the chiral supefields is $2N_f N_c$; since the gauge symmetry is completely broken the number of the "eaten" superfields is $N_c^2 - 1$; the number of the massless degrees of freedom is thus $N_f^2 + 1 = N_c^2 + 1$. The number of moduli in Eq. (2.8) is $N_f^2 + 2$. One chiral variable is therefore redundant.

$\mathbf{N_f = N_c + 1}$

The number of the chiral supefields is $2N_f N_c$; since the gauge symmetry is completely broken the number of the "eaten" superfields is $N_c^2 - 1$; the number of massless degrees of freedom is N_f^2. The number of moduli in Eq. (2.8) is $N_f^2 + 2N_f$, i.e. $2N_f$ chiral variables are therefore redundant.

In the first case, $N_f = N_c$, the constraint eliminating the redundant chiral variable is

$$\det\{M\} = B\tilde{B} \,, \tag{2.9}$$

while in the second example, $N_f = N_c + 1$, by using merely the definitions of the moduli in Eq. (2.8), it is not difficult to obtain

$$B_i \{M\}_j^i = \{M\}_j^i \tilde{B}^j = 0 \,, \tag{2.10}$$

Table 1. The quantum numbers of the matter fields with respect to the global symmetries (2.13) in SQCD with N_f flavors. The R charges indicated refer to the lowest components of the superfields.

	$SU(N_f)_L$	$SU(N_f)_R$	$U(1)_B$	$U(1)_R$
Q	N_f	1	1	$(N_f - N_c)/N_f$
$\tilde{Q}$	1	$\bar{N}_f$	-1	$(N_f - N_c)/N_f$

and

$$\text{minor}\{M\}_i^j = B_i \tilde{B}^j \,, \tag{2.11}$$

where the left-hand side of the last equation is the minor of the matrix $\{M\}$ (i.e. $(-1)^{i+j}\times$ determinant of the matrix obtained from M by omitting the i-th row and the j-th column. Note that $\det\{M\}$ vanishes in this case. For brevity we will sometimes write Eq. (2.11) in a somewhat sloppy form

$$\det M \left(M^{-1}\right)_i^j = B_i \tilde{B}^j \,. \tag{2.12}$$

At the classical level one could, in principle, eliminate the redundant chiral variables using Eqs. (2.9) or (2.12). One should not hurry with this elimination, however, since at the quantum level the classical moduli fields are replaced by the vacuum expectation values of M_i^j and B, $\tilde{B}$, and although generically the total number of massless degrees of freedom does not change, the quantum version of constraints (2.9) and (2.12) may (and will) be different. Moreover, at some specific points of the valley the number of massless degrees of freedom may increase, as we will see shortly.

SQCD with N_f flavors and no tree-level superpotential has the following global symmetries free from internal anomalies:

$$SU(N_f)_L \times SU(N_f)_R \times U(1)_B \times U(1)_R \tag{2.13}$$

where the conserved R current was introduced in Eq. (1.65), and the quantum numbers of the matter multiplets with respect to these symmetries are collected in Table 1. (For discussion of the subtleties in the R current definition see Ref.[37]. These subtleties, being conceptually important, are irrelevant for our considerations).

The $SU(N_f)_L \times SU(N_f)_R \times U(1)_B$ transformations act only on the matter fields in an obvious way, and do not affect the superspace coordinate θ. As for the extra global symmetry $U(1)_R$ it is defined in such a way that it acts nontrivially on the supercoordinate θ and, therefore, acts differently on the spinor and the scalar or vector components of superfields. The R charges in Table 1 are given for the lowest component of the chiral superfields. If the R charge of the boson component of the given superfield is r then the R charge of the fermion component is, obviously, $r-1$. A part of the above global symmetries is spontaneously broken by the vacuum expectation values of M_j^i and/or $B, \tilde{B}$.

Unlike the $N_c = 2$, $N_f = 1$ model discussed before, instantons do not lift the classical degeneracy, and the bottom of the valley remains flat. The easiest way to see this is to consider a generic point of the bottom of the valley, far away from the origin, where the theory is in the weak coupling regime, and try to write the most general superpotential, compatible with all exact symmetries[18, 19] (it must be symmetric even under those symmetries which may turn out to be spontaneously broken). The symmetry under $SU(N_f)_L \times SU(N_f)_R \times U(1)_B$ is guaranteed if we assume that the superpotential W depends on $\det M$. What about the R symmetry?

For $N_f = N_c$ the R charge of the matter superfield vanishes, as is clear from Table 1. Since the superpotential must have the R charge 2, it is obvious that it cannot be generated. For $N_f > N_c$ the R charge of the matter fields does not vanish, and, in principle, one could have written

$$\mathcal{W} \propto \left(\frac{\Lambda^{3N_c-N_f}}{\det M}\right)^{\frac{1}{N_c-N_f}},$$

an expression which has the right dimension (three) and the correct R charge (two). However, the dimension of Λ does not match the instanton expression which can produce only $\Lambda^{3N_c-N_f}$ (and in the weak coupling regime the instanton is the only relevant nonperturbative contribution). What is even more important, for $N_f > N_c$ the determinant of M vanishes identically. This fact alone shows that no superpotential can be generated, and the flat direction remains flat[61,18,19].

The argument above demonstrates again the power of holomorphy. In non-supersymmetric theories one could built a large number of invariants involving $Q, \tilde{Q}, Q^\dagger$ and $\tilde{Q}^\dagger$. In SUSY theories, as far as the F terms are concerned, one is allowed to use only Q and $\tilde{Q}$ which constraints the possibilities to the extent that nothing is left.

In summary, for $N_f \geq N_c$ the vacuum degeneracy is not lifted. At the origin of the space of moduli, where $B = \tilde{B} = 0$ and M has fewer than $N_c - 1$ non-zero eigenvalues, the gauge symmetry is not fully broken. At this point, the classical moduli space is singular. Far away from the origin, when the expectation values of the squark fields are large, the distinction between the classical and quantum moduli space should be unimportant. In the vicinity of the origin, however, this distinction may be crucial. Our next task is to investigate this distinction. Needless to say that the vicinity of the origin is just the domain of most interesting dynamics. Since the Higgs fields are in the fundamental representation, we are always in the Higgs/confining phase. Far away from the origin the theory is in the weak coupling regime and is fully controllable by well understood methods of weak coupling. In the vicinity of the origin the theory is in the strong coupling regime. The issues to be investigated are the patterns of the spontaneous breaking of the global symmetries and the occurrence of the composite massless degrees of freedom at large distances. Here each non-trivial theoretical result or assertion is a precious asset, a miraculous achievement.

The quantum moduli space

Relations (2.9) and (2.12) are constraints on the classical composite fields. Since in the quantum theory the vacuum valley is parametrized by the *expectation values* of the fields, which may get a contribution from quantum fluctuations, these relations may alter. In other words, the quantum moduli space need not exactly coincide with the classical one. Only in the limit when the vacuum expectation values of the fields parametrizing the vacuum valley become large, much larger than the scale parameter of the underlying theory, we must be able to return to the classical description.

To see that the quantum moduli space does indeed differ from the classical one we will consider here, following Ref.[25], the same two examples, $N_f = N_c$ and $N_f = N_c + 1$. The general strategy used in these explorations is the same as was discussed in detail in connection with $SU(2)$ SQCD[10]: in order to analyze the theory along the classically flat directions one adds the appropriately chosen mass terms (sometimes, other superpotential terms as well), solves the theory in the weak coupling regime, and then analytically continues to the limit where the classical superpotential vanishes.

$\mathbf{N_f = N_c}$

Introduce a mass term for all quark flavors, or more generically, the quark mass matrix

$$m_i^j Q^i \tilde{Q}_j \,, \tag{2.14}$$

(it can always be diagonalized, of course). If we additionally assume that the mass terms for $N_f - 1$ flavors are small and the mass term for one flavor is large then we find ourselves in a situation where an effective low-energy theory is that of $N_c - 1$ flavors. From the first part of this lecture course we know already that in this case the $SU(N_c)$ symmetry is totally broken spontaneously, the theory is in the weak coupling phase, instantons generate a superpotential, and this superpotential, being combined with Eq. (2.14), leads to[18, 19, 43, 33]

$$M_j^i = \langle Q^i \tilde{Q}_j \rangle = \Lambda^2 (\det m)^{\frac{1}{N_c}} \left(\frac{1}{m}\right)_j^i \,,$$

$$B = \langle Q^{i_1} ... Q^{i_{N_c}} \rangle \epsilon_{i_1 ... i_{N_c}} = 0 \,, \quad \tilde{B} = \langle \tilde{Q}_{j_1} ... \tilde{Q}_{j_{N_c}} \rangle \epsilon^{j_1 ... j_{N_c}} = 0 \,. \tag{2.15}$$

Although this result was obtained under a very specific assumption on the values of the mass terms, holomorphy tells us that it is exact. In particular, one can let $m_i^j \to 0$, thus returning to the original massless theory. Equation (2.15) obviously implies that

$$\det M - B\tilde{B} = \Lambda^{2N_c} \,. \tag{2.16}$$

It is instructive to check that this relation stays valid even if $B \neq 0$, $\tilde{B} \neq 0$. To this end one must introduce, additionally, a superpotential $\beta B + \tilde{\beta}\tilde{B}$ where β and $\tilde{\beta}$ are some constants, and redo the instanton calculations. If $\beta \neq 0$ and $\tilde{\beta} \neq 0$, the instanton-induced superpotential changes, non-vanishing values of B and $\tilde{B}$ are generated, the vacuum expectation values M_j^i change as well, but the relation (2.16) stays intact.

Far from the origin, where the semiclassical analysis is applicable, the quantum moduli space (2.16) is close to the classical one. A remarkable phenomenon happens near the origin[25]. In the classical theory where the gluons were massless near the origin, the classical moduli space was singular. Quantum effects eliminated the massless modes by creating a mass gap†. Correspondingly, the singular points with $B = \tilde{B} = 0$ and vanishing eigenvalues of M are eliminated from the moduli space.

In the weak coupling regime dynamics is rather trivial and boring. Let us consider the most interesting domain of the vacuum valley, near the origin, in more detail, "under a microscope". There are several points that are special, they are characterized by an enhanced global symmetry. For instance, if

$$B = \tilde{B} = 0 \quad \text{and} \quad M_j^i = \Lambda^2 \delta_j^i \,, \tag{2.17}$$

the original global $SU(N_f) \times SU(N_f)$ symmetry is spontaneously broken down to the diagonal $SU(N_f)$, while the $U(1)_R$ remains unbroken (the R charges of Q and $\tilde{Q}$ vanish, see Table 1). We are in the vicinity of the origin, where all moduli are either of order of Λ^2 or vanish. Hence, the fundamental gauge dynamics of the quark (squark) matter is strongly coupled. We are in the strong coupling regime.

The spontaneous breaking of the global symmetry implies the existence of the massless Goldstone mesons which, through supersymmetry, entails, in turn, the occurrence of the massless (composite) fermions. These fermions reside in the superfields M, B, and $\tilde{B}$. Their quantum numbers with respect to the unbroken symmetries are indicated in Table 2.

†The latter statement is not quite correct. Massless moduli fields still persist. What is important, however, is that the gluons acquire a dynamical "mass".

Table 2. The quantum numbers of the composite massless fermions with respect to the unbroken global symmetries in SQCD with $N_f = N_c$

	$SU(N_f)$	$U(1)_B$	$U(1)_R$
ψ_M	$N_f^2 - 1$	0	-1
ψ_B	1	N_f	-1
$\psi_{\tilde{B}}$	1	$-N_f$	-1
ψ_Q	N_f	1	-1
$\psi_{\tilde{Q}}$	$\bar{N}_f$	-1	-1
λ	1	0	1

For convenience Table 2 summarizes also the quantum numbers of the fundamental fermions – quarks and gluino. A remark is in order concerning the multiplet of the massless fermions ψ_M. Since M is an $N_f \times N_f$ matrix, naively one might think that the number of these fermions is N_f^2. Actually we must not forget that we are interested in small fluctuations of the moduli fields M, B and $\tilde{B}$ around the expectation values (2.17) subject to the constraint (2.16). It is easy to see that this constraint implies that the matrix of fluctuations $M_j^i - \Lambda^2\delta_j^i$ is traceless, i.e. the fluctuations form the adjoint ($(N_f^2 - 1)$-dimensional) representation of the diagonal $SU(N_f)$.

Massless composite fermions in gauge theories are subject to a very powerful constraint known as the *'t Hooft consistency condition*[62]. As was first noted in[63], the triangle anomalies of the AVV type in the gauge theories with the fermion matter imply the existence of infrared singularities in the matrix elements of the axial currents. (Here A and V stand for the axial and vector currents, respectively). These singularities are unambiguously fixed by the short-distance (fundamental) structure of the theory even if the theory at hand is in the strong coupling regime and cannot be solved in the infrared. The massless composite fermions in the theory, if present, must arrange themselves in such a way as to match these singularities. If they cannot, the corresponding symmetry is spontaneously broken, and the missing infrared singularity is provided by the Goldstone-boson poles coupled to the corresponding broken generators. This device – the 't Hooft consistency condition, or anomaly matching – is widely used in strongly coupled gauge theories: from QCD to technicolor, to supersymmetric models; it allows one to check various conjectures about the massless composite states. (For a pedagogical review see e.g.[64].)

In our case we infer the existence of the massless fermions from the fact that a set of moduli exists, plus supersymmetry. Why do we need to check the matching of the AVV triangles? If we know for sure the pattern of the symmetry breaking – which symmetry is spontaneously broken and which is realized linearly – the matching of the AVV triangles for the unbroken currents must be automatic. The condensates indicated in Eq. (2.17) suggest that the axial $SU(N_f)$ is spontaneously broken while the R current and the baryon current are unbroken. Suggest, but do not prove! For in the strong coupling regime other (non-chiral) condensates might develop too. For instance, on general grounds one cannot exclude the condensate of the type $\langle M_j^i B^\dagger \rangle$ which will spontaneously break the baryon charge conservation. Since this superfield is non-chiral the holomorphy consideration is inapplicable. If the anomalous triangles with the baryon current do match, it will be a strong argument showing that no additional condensates develop, and the pattern of the spontaneous symmetry breaking can be read off from Eq. (2.17). Certainly, this is not a completely rigorous proof, but, rather, a very strong indication.

What is extremely unusual in the pattern implied by Eq. (2.17) is the survival of an unbroken axial current (the axial component of the R current). We must verify that this scheme of symmetry breaking is compatible with the spectrum of the massless composite fermions residing in the superfields M, B, and $\tilde{B}$.

The 't Hooft consistency conditions, to be analyzed in the general case, refer to the so called *external* anomalies of the AVV type. More exactly, one considers those axial currents, corresponding to global symmetries of the theory at hand, which are non-anomalous inside the theory *per se,* but acquire anomalies in weak external backgrounds. For instance, in QCD with several flavors the singlet axial current is internally anomalous – its divergence is proportional to $G\tilde{G}$ where G is the gluon field strength tensor. Thus, it should not be included in the set of the 't Hooft consistency conditions to be checked. The non-singlet currents are non-anomalous in QCD itself, but become anomalous if one includes the photon field, external with respect to QCD. These currents must be checked. The anomaly in the singlet current does not lead to the statement of the infrared singularities in the current while the anomaly in the non-singlet currents does. Those symmetries that are internally anomalous, are non-symmetries.

In our case we first list all those symmetries which are supposedly realized linearly, i.e. unbroken. After listing all relevant currents we then saturate the corresponding triangles. The diagonal $SU(N_f)$ symmetry which remains unbroken is induced by the vector current, not axial. The same is true with regards to $U(1)_B$. The conserved (unbroken) R current has the axial component. Therefore, the list we must consider includes the following triangles

$$U(1)_R^3\,,\ \ U(1)_R SU(N_f)^2\,,\ \ U(1)_R U(1)_B^2\,.$$

One more triangle is of a special nature. One can consider the gravitational field as external, and study the divergence of the R current in this background. This divergence is also anomalous,

$$\partial^\mu(\sqrt{-g}R_\mu) = \frac{1}{192\pi^2}\,[\text{const.}]\,\epsilon_{\mu\nu\lambda\delta}R^{\mu\nu\sigma\rho}R^{\lambda\delta}_{\sigma\rho}\,, \tag{2.18}$$

where g is the metric and $R^{\mu\nu\sigma\rho}$ is the curvature of the gravitational background. The constant in the square brackets depends on the particle content of the theory, and must be matched at the fundamental and composite fermion level. This gravitational anomaly in the R current is routinely referred to as $U(1)_R$. Thus, altogether we have to analyze four triangles. Let us start, for instance, from $U(1)_R SU(N_f)^2$. The relevant quantum numbers of the fundamental-level fermions (ψ_Q, $\psi_{\tilde{Q}}$ and λ), and the composite fermions (ψ_M, ψ_B and $\psi_{\tilde{B}}$) are collected in Table 2. At the fundamental level we have to take into account only ψ_Q and $\psi_{\tilde{Q}}$ since only these fields have both $U(1)_R$ charge and transform non-trivially with respect to $SU(N_f)$. The corresponding triangle is proportional to $-N_f T_{\text{fund}} - N_f T_{\text{anti-fund}} \equiv -N_f$. The factor N_f appears since we have N_f fundamentals and N_f anti-fundamentals. Here T is (one half of) the Dynkin index defined as follows. Assume we have the matrices of the generators of the group G in the representation R. Then

$$\mathrm{Tr}\,(T^a T^b)_R = T_R\,\delta^{ab}\,.$$

For the fundamental representation $T = 1/2$ while for the adjoint representation of $SU(N)$ the index $T = N$. Now, let us calculate the same triangle at the level of the composite fermions. From Table 2 it is obvious that we have to consider only ψ_M, and the corresponding contribution is $-T_{\text{adjoint}} = -N_f$. The match is perfect.

The balance in the $U(1)_R^3$ triangle looks as follows. At the fundamental level we include ψ_Q, $\psi_{\tilde{Q}}$ and λ and get $2N_f^2(-1)^3 + (N_f^2 - 1) = -N_f^2 - 1$. At the composite fermion level we include ψ_M, ψ_B and $\psi_{\tilde{B}}$ and get $(N_f^2 - 1)(-1)^3 - 2 = -N_f^2 - 1$.

By the same token one can check that $U(1)_B^2 U(1)_R$ triangle gives $-2N_f^2$ both at the fundamental and composite levels. The $U(1)_R$ case requires a special comment. The coupling of all fermions to gravity is universal. Therefore, the coefficient in Eq. (2.18) merely counts the number of the fermion degrees of freedom weighed with their R charges. At the fundamental level we, obviously, have $-N_f^2 - N_f^2 + (N_f^2 - 1) = -N_f^2 - 1$, while at the composite level the coefficient is $-(N_f^2 - 1) - 1 - 1 = -N_f^2 - 1$. Again, the match is perfect.

Thus, the massless fermion content of the theory is consistent with the regime implied by Eq. (2.17) – spontaneous breaking of the chiral $SU(N_f)_R \times SU(N_f)_L$ down to vector $SU(N_f)$. The baryon and the $U(1)_R$ currents remain unbroken. This regime is rather similar to what we have in ordinary QCD. The unconventional aspect, as was stressed above, is the presence of the conserved unbroken R current which has the axial component.

This does not mean, however, that all points from the vacuum valley are so reminiscent of QCD. Other points are characterized by different dynamical regimes, with drastic distinctions in the most salient features of the emerging picture. To illustrate this statement let us consider, instead of Eq. (2.17), another point

$$B = -\tilde{B} = \Lambda^{N_f}\,, \quad M_j^i = 0\,. \tag{2.19}$$

This point is characterized by a fully unbroken chiral $SU(N_f) \times SU(N_f)$ symmetry, in addition to the unbroken R symmetry. The only broken generator is that of $U(1)_B$.

This regime is exceptionally unusual from the point of view of the QCD practitioner. As a matter of fact, the emerging picture is directly opposite to what we got used to in QCD: the axial $SU(N_f)$ generators remain unbroken while the vector baryon charge generator is spontaneously broken.

As is well-known, spontaneous breaking of vector symmetries is forbidden in QCD[65]. The no-go theorem of Ref.[65] is based only on very general features of QCD – namely on the vector nature of the quark-gluon vertex. Where does the no-go theorem fail in SQCD?

The answer is quite obvious. The spontaneous breaking of the baryon charge generator in SQCD, apparently defying the no-go theorem of Ref.[65], is due to the fact that in SQCD we have scalar quarks (and the quark-squark-gluino interaction) which invalidates the starting assumptions of the theorem.

Moreover, in QCD general arguments, based on the 't Hooft consistency condition and N_c counting, strongly disfavor[66] the possibility of the linearly realized axial $SU(N_f)$. Although I do not say here that the consideration of Ref.[66] proves the axial $SU(N_f)$ to be spontaneously broken in QCD, there is hardly any space left over for a linear realization. The linear realization is not ruled out only because the argument of Ref.[66] is based on an assumption regarding the N_c dependence (discussed below) which is absolutely natural but still was not derived from first principles. Certain subtleties which I cannot explain now due to time limitations might, in principle, invalidate this assumption. Leaving aside these – quite unnatural – subtleties one can say that the linear realization of the axial $SU(N_f)$ is impossible in QCD.

At the same time, this is exactly what happens in SQCD in the regime specified by Eq. (2.19). Again, the scalar quarks are to blame for the failure of the argument presented in Ref.[66]. In QCD it is difficult to imagine how massless baryons could saturate anomalous triangles since the baryons are composed of N_c quarks; the corresponding contribution naturally tends to be suppressed as $\exp(-N_c)$ at large N_c. In SQCD there exist fermion states built from one quark and one (anti)squark whose contribution to the triangle is not exponentially suppressed.

After this introductory remark it is time to check that the 't Hooft consistency conditions are indeed saturated. The triangles to be analyzed are

$$SU(N_f)^3\,,\quad SU(N_f)^2U(1)_R\,,\quad U(1)_R^3\,,\quad \text{and } U(1)_R\,.$$

The $SU(N_f)$ symmetry is either $SU(N_f)_R$ or $SU(N_f)_L$, but the triangles are the same for both. It is necessary to take into account the fact that the fluctuations around the expectation values (2.19) subject to the constraint (2.16) are slightly different than those indicated in Table 2. Namely, the matrix of fluctuations M_j^i need not be traceless any longer; correspondingly, there are N_f^2 fermions in this matrix transforming as the $(N_f,\,\bar{N}_f)$ representation of $SU(N_f)_R\times SU(N_f)_L$. At the same time the fluctuations of B and $\tilde{B}$ are not independent now, so that $B-\Lambda^{N_f}=\tilde{B}+\Lambda^{N_f}$. One should count only one of them. The $U(1)_R$ quantum numbers remain intact, of course.

With this information in hands, matching of the triangles becomes a straightforward exercise. For instance, the $SU(N_f)^3$ triangle obviously yields $N_f D_{\rm fund}$ both at the quark and composite levels. Here $D_{\rm fund}$ is the cubic Casimir operator for the fundamental representation defined as follows

$$\mathrm{Tr}\,(T^a\{T^b,T^c\})=d^{abc}D\,;$$

the matrices of the generators T^a are taken in the given representation, and the braces denote the anticommutator; d^{abc} stand for the d symbols. The $SU(N_f)^2U(1)_R$ triangle yields $-N_fT_{\rm fund}$ both at the quark and composite levels. Both triangles, $SU(N_f)^3$ and $SU(N_f)^2U(1)_R$, are saturated by ψ_M. Passing to $U(1)_R^3$ we must add the gluino contribution at the fundamental level and that of ψ_B at the composite level. At both levels the coefficient of the $U(1)_R^3$ triangle is $-N_f^2-1$. Finally, $U(1)_R$ counts the number of degrees of freedom weighed with the corresponding R charges. The corresponding coefficient again turns out to be the same, $-N_f^2-1$.

Summarizing, the massless composite fermions residing in the moduli superfields M, B, $\tilde{B}$ saturate all anomalies induced by the symmetries that are supposed to be realized linearly. The conjecture of the unbroken $SU(N_f)_R\times SU(N_f)_L$ and spontaneously broken $U(1)_B$ at the point (2.19) goes through. As a matter of fact, some of these anomaly matching conditions were observed long ago, in Refs.[18, 19].

Sometimes it is convenient to mimic the constraint (2.16) by introducing a Lagrange multiplier superfield X with the superpotential

$$\mathcal{W}=X\left(\det M-B\tilde{B}-\Lambda^{2N_f}\right)\,. \tag{2.20}$$

We could treat, in a similar fashion, any point belonging to the quantum moduli space (2.16). For instance, we could travel from (2.17) to (2.19) observing how the regime continuously changes from the broken axial $SU(N_f)$ to the broken baryon number.

Concluding this part we remind that the case of the gauge group $SU(2)$ is exceptional. Indeed, in this case, the matter sector consisting of 2 fundamentals and 2 anti-fundamentals has $SU(4)$ global flavor symmetry, rather than $SU(2)\times SU(2)$. This is because all representations of $SU(2)$ are (pseudo)real, and fundamentals can be transformed into anti-fundamentals and *vice versa* by applying the $\epsilon^{\alpha\beta}$ symbol. This peculiarity was already discussed in detail in the first part of these lectures. Under these circumstances the pattern of the global symmetry breaking is somewhat different and the saturation of the anomaly triangles must be checked anew. Although this is a relatively simple exercise, we will not do it here. The interested reader is referred to[25].

$\mathbf{N_f = N_c + 1}$

The general strategy is the same as in the previous case. We introduce the mass term (2.14) assuming that two eigenvalues of the mass matrix are large while others are small. Then two heavy flavors can be integrated over, leaving us with the theory with $N_f = N_c - 1$, which can be analyzed in the weak coupling regime. A superpotential is generated on the vacuum valley. Using this superpotential it is not difficult to get the vacuum expectation values of the moduli fields M, $B_i, \tilde{B}^j$. They turn out to be constrained by the following relation[25]:

$$\det M \left(\frac{1}{M}\right)_i^j - B_i \tilde{B}^j = \Lambda^{2N_c-1} m_i^j . \tag{2.21}$$

Note that the vanishing of the determinant, $\det M = 0$, which at the classical level automatically follows from the definition of M_j^i, is gone for the quantum VEV's $\langle Q^i \tilde{Q}_j \rangle$. This is most readily seen if the mass matrix $m_i^j \to m\delta_i^j$. In this case $\langle Q^i \tilde{Q}_j \rangle \sim (\Lambda^{2N_c-1} m)^{1/N_c} \delta_j^i$, $i, j = 1, ..., N_f$.

In the massless limit $m_i^j \to 0$ the quantum constraint (2.21) coincides with the classical one (2.11), or (2.12). Thus, the quantum and classical moduli spaces are identical. Every point from the vacuum valley can be reached by adding appropriate perturbations to the Lagrangian (i.e. mass terms and/or $\beta B + \tilde{\beta}\tilde{B}$).

The only point which deserves special investigation is the origin, which, unlike the situation $N_f = N_c$, remains singular. This is a signature of massless fields. Classically we have massless gluons and massless moduli fields. In the strong coupling regime we expect the gluons to acquire a dynamical mass gap. The classical moduli subject to the constraint (2.12) need not be the only composite massless states, however. Other composite massless states may form too. We will see shortly that they actually appear. At the origin, when $M = B = \tilde{B} = 0$, *all* global symmetries of the Lagrangian are presumably unbroken. In particular, the axial $SU(N_f)$ is realized linearly. Although we have already learned, from the previous example, that such a regime seems to be attainable in SQCD (in sharp contradistinction with QCD), the case $N_f = N_c + 1$ is even more remarkable – we want *all* global symmetries to be realized linearly. (For $N_f = N_c$, in the vacuum where the axial $SU(N_f)$ symmetry was unbroken, the baryon charge generators were spontaneously broken.)

At the origin (and near the origin) the theory is in the strong coupling regime. Let us examine the behavior of the theory in this domain more carefully. When the expectation values of all moduli fields vanish, the global symmetry $SU(N_f) \times SU(N_f) \times U(1)_B \times U(1)_R$ is unbroken provided no other (non-chiral) condensates develop. Is this solution self-consistent?

To answer this question we will try to match all corresponding anomalous *AVV* triangles; in this case we have seven triangles,

$$SU(N_f)^3 ,\ SU(N_f)^2 U(1)_B ,\ SU(N_f)^2 U(1)_R ,$$
$$U(1)_R U(1)_B^2 ,\ U(1)_R^3 ,\ U(1)_R^2 U(1)_B ,\ U(1)_R . \tag{2.22}$$

They must be matched by the composite massless baryons residing in M, B_i and $\tilde{B}^j$. As we will see shortly, to achieve the matching we will need to consider all components of M, B_i and $\tilde{B}^j$ as independent, ignoring the constraint

$$\det M \left(\frac{1}{M}\right)_i^j - B_i \tilde{B}^j = 0 \tag{2.23}$$

Table 3. The quantum numbers of the massless fermions in SQCD with $N_f = N_c + 1$

	$SU(N_f)_L$	$SU(N_f)_R$	$U(1)_B$	$U(1)_R$
ψ_M	N_f	N_f	0	$\frac{2}{N_f} - 1$
ψ_B	$\bar{N}_f$	1	$N_f - 1$	$-\frac{1}{N_f}$
$\psi_{\tilde{B}}$	1	N_f	$-N_f + 1$	$-\frac{1}{N_f}$
ψ_Q	N_f	1	1	$\frac{1}{N_f} - 1$
$\psi_{\tilde{Q}}$	1	$\bar{N}_f$	-1	$\frac{1}{N_f} - 1$
λ	1	1	0	1

defining the vacuum valley both at the classical and the quantum levels. In other words, we will have to deal with a larger number of massless fields than one could infer from the parametrization (2.23) of the vacuum valley. The constraint (2.23) on the vacuum valley will reappear due to the fact that the expanded set of massless fields gets a superpotential $\mathcal{W}(M, B, \tilde{B})$. The requirement of the vanishing of the F term will give us Eq. (2.23).

Thus, our first task is to verify the matching. The quantum numbers of the fundamental quarks and the composite massless fermions can be inferred from Table 1. For convenience we collect them in Table 3.

Since we already have a considerable experience in matching the *AVV* triangles, I will not discuss all triangles from Eq. (2.22). As an exercise let us do just one of them, namely $U(1)_R^3$. In this case, at the fundamental level we have the $\psi_Q, \psi_{\tilde{Q}}$ and λ triangles which yield

$$2N_c N_f \left(\frac{1}{N_f} - 1\right)^3 + N_c^2 - 1\,.$$

At the composite level the $U(1)_R^3$ anomalous triangle is contributed by ψ_M (N_f^2 degrees of freedom), ψ_B and $\psi_{\tilde{B}}$ (each has N_f degrees of freedom). Thus, we get

$$N_f^2\left(\frac{2}{N_f} - 1\right)^3 - 2N_f\left(\frac{1}{N_f}\right)^3 .$$

Both expressions reduce to

$$-N_f^2 + 6N_f - 12 + \frac{8}{N_f} - \frac{2}{N_f^2}\,.$$

Other triangles match too, in a miraculous way. Namely,

$$SU(N_f)^3 \to (N_f - 1)D_{\rm fund}\,,$$

$$SU(N_f)^2 U(1)_R \to -\frac{(N_f - 1)^2}{N_f} T_{\rm fund}\,,$$

$$U(1)_B^2 U(1)_R \to -2(N_f - 1)^2\,,$$

$$SU(N_f)^2 U(1)_B \to (N_f - 1)T_{\rm fund}\,,$$

$$U(1)_R^2 U(1)_B \to 0\,,$$

$$U(1)_R \to -N_f^2 + 2N_f - 2\,.$$

The matching discussed above, was observed many years ago in Ref.[33] where the spectrum of the composite massless particles corresponding to the unconstrained *M, B* and $\tilde{B}$ was conjectured.

Thus, the above spectrum of the composite massless particles appearing at the origin of the vacuum valley in the $N_f = N_c + 1$ theory is self-consistent. We know, however, that the vacuum valley in the model at hand is characterized by Eq. (2.23). The situation seems rather puzzling. How the constraint (2.23) might appear?

The answer to this question was given by Seiberg[25]. If the massless fields, residing in the unconstrained M, B and $\tilde{B}$, acquire a superpotential, then the vacuum values of the moduli fields are obtained through the condition of vanishing F terms. The "right" superpotential will lead to Eq. (2.23) automatically.

So, what is the right superpotential? If it is generated, several requirements are to be met. First, it must be invariant under all global symmetries of the model, including the R symmetry. Second, the vacuum valleys obtained from this superpotential must correspond to Eq. (2.23). Third, away from the origin the only massless fields must be those compatible with the constraint (2.23).

All these requirements are satisfied by the following superpotential[25]:

$$\mathcal{W} = \frac{1}{\Lambda^{2N_f-3}}\left(B_i M^i_j \tilde{B}^j - \det M\right) . \tag{2.24}$$

It is obvious that the condition of vanishing of the F terms corresponding to M^i_j identically coincides with Eq. (2.23); moreover, vanishing of the F terms corresponding to B and $\tilde{B}$ yields two remaining constraints, $B_i M^i_j = \tilde{B}^j M^i_j = 0$. Once we move away from the origin, the moduli M^i_j grow, the fields $B_i, \tilde{B}^j$ acquire masses and can be integrated out. This eliminates $2N_f$ degrees of freedom. This is exactly the amount of the redundant degrees of freedom, see the section on $SU(N_c)$ QCD with N_f flavors - preliminaries. The emerging low-energy theory for the remaining degrees of freedom M^i_j has no superpotential. When $M^i_j \gg \Lambda^2$, the fields $B_i, \tilde{B}^j$ are very heavy, and the low-energy description based on Eq. (2.24) is no longer legitimate. It is interesting to trace the fate of the "baryons" $B_i, \tilde{B}^j$ in the process of this evolution from small to large values of M^i_j. This question has not been addressed in the literature so far.

Let us pause here to summarize the features of the dynamical regime taking place in the $N_f = N_c + 1$ model. The space of vacua is the same at the classical and quantum levels, the origin being singular due to the existence of the massless degrees of freedom. Since we have Higgs fields in the fundamental representation the theory is in the Higgs/confining phase; at the origin and near the origin the theory is strongly coupled and "confines" in the sense that physics is adequately described in terms of gauge invariant composites and their interactions. We think that at the origin all global symmetries of the Lagrangian are unbroken. The number of the massless degrees of freedom here is larger than the dimensionality of the space of vacua. To get the right description of the space of vacua one needs a superpotential, and such a superpotential is generated dynamically. It is a holomorphic function of the massless composites. The vacuum valley for this superpotential coincides with the quantum moduli space of the original theory. As we move along the vacuum valley away from the origin there is no phase transition – the theory smoothly goes into the weak coupling Higgs phase. The "extra" massless fields become massive, and irrelevant for the description of the vacuum valley.

The dynamical regime with the above properties got a special name – now it is referred to as *s- confinement.*

Seiberg's example of the *s*-confining theory was the first, but not the last. Other theories with similar behavior were found, see e.g.[67-73]. The set of *s*-confining models includes even such exotic one as the gauge group G_2 (this is an exceptional group), with five fundamentals[72, 73]. As a matter of fact, it is not difficult to work out a general strategy allowing one to carry out a systematic search of all s-confining theories. This was done in Ref.[74]. Without submerging into excessive technical detail let me outline just one basic point of the procedure suggested in[74].

A necessary condition of the s-confinement is generation of a superpotential at the origin of the moduli space, a holomorphic function of relevant moduli fields. Generically, the form of this superpotential, dictated by the R symmetry plus dimensional arguments is

$$\mathcal{W} \propto \left[\prod_\ell S_\ell^{2T(R_\ell)} \Lambda^{3T(G)-\sum_\ell T(R_\ell)}\right]^{1/(\sum_\ell T(R_\ell)-T(G))} \tag{2.25}$$

The product (sum) runs over all matter fields present in the theory. For instance, in the case of SQCD for each flavor we have to include two subflavors. I remind that $T(R)$ is (one half of) the Dynkin index. Particular combinations of the superfields in the product are not specified; they depend on the particular representations of the matter fields with respect to the gauge group. What is important is only the fact that they all are homogeneous functions of S_ℓ's, of order $2T(R_\ell)$. Note that the combination appearing in Eq. (2.25) is the only one which has correct properties under renormalization, i.e. compatible with the NSVZ β function.

Now, if we want the origin to be analytic (and this is a feature of the s-confinement, by definition), we must ensure that

$$\sum_\ell T(R_\ell) - T(G) = 1 \tag{2.26}$$

(more generically, 1/integer). This severely limits the choice of possible representations since the Dynkin indices are integers. For instance, if the matter sector is vector-like, there exist only two options: (i) Seiberg's model, $SU(N_c)$ color N_c+1 flavors (i.e. N_c+1 fundamentals and N_c+1 anti-fundamentals; (ii) $SU(N_c)$ color with one antisymmetric tensor plus its adjoint plus three flavors.

Not to make a false impression I hasten to add that some models that satisfy Eq. (2.26), are not s-confining. A few simple requirements to be met, which comprise a sufficient condition for s-confinement, are summarized in Ref.[74], which gives also a full list of the s-confining theories.

Conformal window. Duality

Our excursion towards larger values of N_f must be temporarily interrupted here – the methods we used so far fail at $N_f = N_c+2$. One can show that the quantum moduli space coincides with the classical one, just as in the case $N_f = N_c+1$. However, at the origin of the moduli space, description of the large-distance behavior of the theory in terms of the massless fields residing in M, B and $\tilde{B}$ does not go through. These degrees of freedom are irrelevant for this purpose; the dynamical regime of the theory in the infrared is different. To see that M, B and $\tilde{B}$ do not fit suffice to try to saturate the 't Hooft triangles corresponding to the unbroken global symmetries, in the same vein as we did previously for $N_f = N_c+1$. There is no matching! As we will see shortly, the dynamical regime does indeed change in passing from $N_f = N_c+1$ to $N_f = N_c+2$. The correlation functions of the $N_f = N_c+2$ theory at large distances are those of a free theory, like in massless electrodynamics. But the number of free degrees of freedom ("photons" and "photinos") is different from from what one might expect naively. Namely, we will have three "photons" and three "photinos" in the case at hand, in addition to $(N_c+2)\times 2$ free "fermion" fields. These photons and photinos, in a sense, may be considered as the bound states of the original gluons, gluinos, quarks and squarks.

To elucidate this, rather surprising, picture we will have to make a jump in our travel along the N_f axis, leave the domain of N_f close to N_c for a while, and turn to much larger values of N_f. The critical points on the N_f axis are $3N_c$ and $3N_c/2$. That's where a *conformal window* starts and ends. We will return to the $N_f = N_c + 2$ and $N_f = N_c + 1$ theories later on.

At first, let me recall a few well-known facts from ordinary non-supersymmetric QCD. The Gell-Mann-Low function in QCD has the form[75]

$$\beta_{\rm QCD}(\alpha_s) = -\beta_0 \frac{\alpha_s^2}{2\pi} - \beta_1 \frac{\alpha_s^3}{4\pi^2} - \ldots ,$$

$$\beta_0 = 11 - \frac{2}{3}N_f\,, \quad \beta_1 = 51 - \frac{19}{3}N_f\,. \tag{2.27}$$

At small α_s it is negative since the first term always dominates. This is the celebrated asymptotic freedom. With the scale μ decreasing the running gauge coupling constant grows, and the second term becomes important. Generically the second term takes over the first one at $\alpha_s/\pi \sim 1$, when all terms in the α_s expansion are equally important, i.e. in the strong coupling regime. Assume, however, that for some reasons the first coefficient β_0 is abnormally small, and this smallness does not propagate to higher orders. Then the second term catches up with the first one when $\alpha_s/\pi \ll 1$, we are in the weak coupling regime, and higher order terms are inessential. Inspection of Eq. (2.27) shows that this happens when N_f is close to 33/2, say 16 or 15 (N_f has to be less than 33/2 to ensure asymptotic freedom). For these values of N_f the second coefficient β_1 turns out to be negative! This means that the β function develops a zero in the weak coupling regime, at

$$\frac{\alpha_s^*}{2\pi} = \frac{\beta_0}{-\beta_1} \ll 1\,. \tag{2.28}$$

(Say, if $N_f = 15$ the critical value is at 1/44.) This zero is nothing but the infrared fixed point of the theory. At large distances $\alpha_s \to \alpha_s^*$, and $\beta(\alpha_s^*) = 0$, implying that the trace of the energy-momentum vanishes, and the theory is in the conformal regime. There are no localized particle-like states in the spectrum of the theory; rather we deal with massless unconfined interacting quarks and gluons; all correlation functions at large distances exhibit a power-like behavior. In particular, the potential between two heavy static quarks at large distances R will behave as $\sim \alpha_s^*/R$. The situation is not drastically different from conventional QED. The corresponding dynamical regime is, thus, a non-Abelian Coulomb phase. As long as α_s^* is small, the interaction of the massless quarks and gluons in the theory is weak at all distances, short and large, and is amenable to the standard perturbative treatment (renormalization group, etc.). QCD becomes a fully calculable theory.

There is nothing remarkable in the observation that, for a certain choice of N_f, quantum chromodynamics becomes conformal and weakly coupled in the infrared limit. Belavin and Migdal played with this model over 20 years ago[76]. They were quite excited explaining how great it would be if N_f in our world were close to 16, and the theory would be in the infrared conformal regime, with calculable anomalous dimensions. Later on this idea was discussed also by Banks and Zaks[77]. Alas, we do not live in a world with $N_f \approx 16$...

What is much more remarkable is the existence of the infrared conformal regime in SQCD for large couplings, $\alpha_*/\pi \sim 1$. This fact, as many others in the given range of questions, was established by Seiberg[78]. The discovery of the strong coupling conformal regime[78] is based on the so-called *electric-magnetic duality.* Although the term suggests the presence of electromagnetism and the same kind of duality under the substitution $\vec{E} \leftrightarrow \vec{B}$ one sees in the Maxwell theory, actually both elements, "electric-magnetic" and "duality" in the given context are nothing but remote analogies, as we will see shortly.

Analysis starts from consideration of SQCD with N_f slightly smaller than $3N_c$. More exactly, if

$$\varepsilon \equiv 1 - \frac{N_f}{3N_c} \tag{2.29}$$

we assume that $N_c \to \infty$, and $0 < \varepsilon \ll 1$. It is assumed also that we are at the origin of the moduli space – no fields develop VEV's. By examining the NSVZ β function, Eq. (1.61), it is easy to see that in this limit the first coefficient of the β function is abnormally small, and the second coefficient is positive and is of a normal order of magnitude,

$$\beta_0 = 3N_c\varepsilon\,, \quad \beta_1 = -3N_c^2 + \mathcal{O}(\varepsilon)\,. \tag{2.30}$$

To get β_1 I used the fact that

$$\gamma(\alpha) = -\frac{N_c^2 - 1}{2N_c}\frac{\alpha}{\pi} + \mathcal{O}(\alpha^2) \tag{2.31}$$

in the model considered (for a pedagogical review see e.g. the last paper in Ref.[44]). There is a complete parallel with conformal QCD, with 15 or 16 flavors, discussed above. The numerator of the NSVZ β function vanishes at

$$\gamma(\alpha_*) \equiv \gamma_* = 1 - 3\frac{N_c}{N_f} \to -\varepsilon \quad or \quad \frac{N_c\alpha_*}{2\pi} = \varepsilon\,. \tag{2.32}$$

The vanishing of the β function marks the onset of the conformal regime in the infrared domain; the fact that α_* is small means that the theory is weakly coupled in the infrared (it is weakly coupled in the ultraviolet too since it is asymptotically free).

Here comes the breakthrough observation of Seiberg. Compelling arguments can be presented indicating that the original theory with the $SU(N_c)$ gauge group (let us call this theory "electric"), and another theory, with the $SU(N_f - N_c)$ gauge group, the same number offlavors N_f as above, and a specific Yukawa interaction (let us call this theory "magnetic"), flow to one and the same limit in the infrared asymptotics. The corresponding Gell-Mann-Low functions of both theories vanish at their corresponding critical values of the coupling constants. Both theories are in the non-Abelian Coulomb (conformal) phases. By inspecting Eq. (1.61) it is easy to see that, when α_* in the electric theory approaches zero (i.e. $\varepsilon \to 0$), in the magnetic theory $\gamma(\alpha_*)$ approaches –1, i.e. the theory becomes strongly coupled. The opposite is also true. When the magnetic theory becomes weakly coupled, i.e.

$$N_f \to \frac{3}{2}N_c \text{ from above}\,, \quad (\alpha_*)_{\rm magn} \to 0\,, \tag{2.33}$$

in the electric theory $(\gamma_*)_{\rm electr} \to -1$, and the electric theory is strongly coupled in the infrared. This reciprocity relation is, probably, the reason why the correspondence between the two theories is referred to as the electric-magnetic duality. It is worth emphasizing that the correspondence takes place only in the infrared limit. By no means are the above two theories totally equivalent to each other; their ultraviolet behavior is completely different. If $N_f < (3N_c)/2$ the magnetic theory looses asymptotic freedom. Thus, the *conformal window*, where both theories are asymptotically free

Table 4. The quantum numbers of the massless fields of the "electric" theory from the dual pair

	$SU(N_f)_L$	$SU(N_f)_R$	$U(1)_B$	$U(1)_R$
ψ_Q	N_f	0	1	$-N_c/N_f$
$\psi_{\tilde{Q}}$	0	$\bar{N}_f$	-1	$-N_c/N_f$

Table 5. The quantum numbers of the massless fields of the "magnetic" theory from the dual pair

	$SU(N_f)_L$	$SU(N_f)_R$	$U(1)_B$	$U(1)_R$
$\psi_{\mathcal{Q}}$	$\bar{N}_f$	0	$N_c/(N_f - N_c)$	$(N_c - N_f)/N_f$
$\psi_{\tilde{\mathcal{Q}}}$	0	N_f	$-N_c/(N_f - N_c)$	$(N_c - N_f)/N_f$
$\chi = \psi_{\mathcal{M}}$	N_f	$\bar{N}_f$	0	$(N_f - 2N_c)/N_f$

in the ultraviolet and conformally invariant in the infrared extends in the interval $3N_c/2 \leq N_f \leq 3N_c$.

The fact that the conformal window cannot extend below $3N_c/2$ is seen from consideration of the electric theory *per se,* with no reference to the magnetic theory. Indeed, the total (normal + anomalous) dimension of the matter field $Q\tilde{Q}$ in the infrared limit is equal to

$$d = 2 + \gamma_* = \frac{3(N_f - N_c)}{N_f}\,.$$

No physical field can have a dimension less than unity; this is forbidden by the Källèn-Lehmann spectral representation. If $d = 1$ the field is free. The dimension d reaches unity exactly at $3N_c/2$. Decreasing N_f further and assuming that the conformal regime $\beta(\alpha_*) = 0$ is still preserved would violate the requirement $d \geq 1$.

Let us describe the electric and magnetic theories in more detail. I will continue to denote the quark fields of the first theory as Q, while those of the magnetic theory will be denoted by $\mathcal{Q}$. Both have N_f flavors, i.e. $2N_f$ chiral superfields in the matter sector. The same number of flavors is necessary to ensure that global symmetries of the both theories are identical.

The magnetic theory, additionally, has N_f^2 colorless "meson" superfields $\mathcal{M}^i_j$ whose quantum numbers are such as if they were built from a quark and an antiquark. The meson superfields are coupled to the quark ones of the magnetic theory through a superpotential

$$\mathcal{W} = f\mathcal{M}^i_j\mathcal{Q}_i\tilde{\mathcal{Q}}^j\,. \tag{2.34}$$

The quantum numbers of the fields belonging to the matter sectors of the magnetic and electric theories are summarized in Tables 4 and 5.

The quantum numbers of the meson superfield are fixed by the superpotential (2.34). Note a very peculiar relation between the baryon charges of the quarks in the electric and magnetic theories. This relation shows that the quarks of the magnetic theory cannot be expressed, in any polynomial way, through quarks of the electric theory. The connection of one to another is presumably extremely non-local and complicated. The explicit connection between the operators in the dual pairs is known only for a handful of operators which have a symmetry nature[79].

We can now proceed to the arguments establishing the equivalence of these two theories in the infrared limit. The main tool we have at our disposal for establishing the equivalence is again the 't Hooft matching, the same line of reasoning as was used above in verifying various dynamical regimes in $N_f = N_c$ and $N_f = N_c + 1$ models.

Since we are at the origin of the moduli space, all global symmetries are unbroken, and one has to check six highly non-trivial matching conditions corresponding to various triangles with the $SU(N_f)$, $U(1)_R$ and $U(1)_B$ currents at the vertices.

The presence of fermions from the meson multiplet $\mathcal{M}$ is absolutely crucial for this matching. Specifically, one finds for the one-loop anomalies in both theories[78]:

$$\begin{aligned}
SU(N_f)^3 &\to N_c D_{\text{fund}}, \\
SU(N_f)^2 U(1)_R &\to -T_{\text{fund}} N_c^2/N_f, \\
SU(N_f)^2 U(1)_B &\to N_c T_{\text{fund}}, \\
U(1)_B^2 U(1)_R &\to -2N_c^2, \\
U(1)_R^3 &\to N_c^2 - 1 - 2N_c^4/N_f^2, \\
U(1)_R &\to -N_c^2 - 1 .
\end{aligned} \tag{2.35}$$

For example, in the $U(1)_R^3$ anomaly in the electric theory the gluino contribution is proportional to $N_c^2 - 1$ and that of quarks to $-(N_c/N_f)^3 2N_f N_c = -2N_c^4/N_f^2$; altogether $N_c^2 - 1 - 2N_c^4/N_f^2$ as in (2.35). In the dual theory one gets from gluino and quarks another contribution, $(N_f - N_c)^2 - 1 - 2(N_f - N_c)^4/N_f^2$. Then the fermions χ from the meson multiplet $\mathcal{M}$ add an extra $[(N_f - 2N_c)/N_f]^3 N_f^2$, which is precisely the difference.

The last line in Eq. (2.35) corresponds to the anomaly of the R current in the background gravitational field. In the electric $SU(N_c)$ theory the corresponding coefficient is

$$N_c^2 - 1 - 2(N_c/N_f)N_c N_f = -N_c^2 - 1$$

while in the magnetic theory one has $-(N_f - N_c)^2 - 1$. from quarks and gluinos and $[(N_f - 2N_c)/N_f]N_f^2$ from the $\mathcal{M}$ fermions, i.e. the sum is again $-N_c^2 - 1$.

It is not difficult to check the matching of other triangles from Eq. (2.35). The dependence on N_c and N_f is rather sophisticated, and it is hard to imagine that this is an accidental coincidence. The fact that the electric and magnetic theories described above have the same global symmetries is an additional argument in favor of their (infrared) equivalence. Of course, they have different gauge symmetries: $SU(N_c)$ in the first case and $SU(N_f - N_c)$ in the second. The gauge symmetry, however, is not a regular symmetry; in fact, it is not a symmetry at all. Rather, it is a redundancy in the description of the theory. One introduces first more degrees of freedom than actually exist, and then the redundant variables are killed by the gauge freedom. That's why the gauge symmetry has no reflection in the spectrum of the theory. Therefore, distinct gauge groups do not preclude the theories from being dual, generally speaking. On the contrary, the fact that such dual pairs are found is very intriguing; it allows one to look at the gauge dynamics from a new angle.

I have just said that various dual pairs of supersymmetric gauge theories are found. To avoid misunderstanding I hasten to add that although Seiberg's line of reasoning is very compelling it still falls short of *proving* the infrared equivalence. The theory in the strong coupling regime is not directly solved, and we are hardly any closer now to the solution than we were a decade ago. The infrared equivalence has the status of a good solid conjecture substantiated by a number of various indirect arguments we have at our disposal (see the next section).

If we accept this conjecture we can make a remarkable step forward compared to the conformal limit of QCD studied in the weak coupling regime in the 70's and 80's. Indeed, if N_f is close to $3N_c$ (but slightly lower), i.e. we are near the right edge of the conformal window, the weakly coupled electric theory is in the conformal regime. Since it is equivalent (in the infrared) to the magnetic theory, which is strongly coupled at these values of N_f we, thus, establish the existence of a *strongly coupled* superconformal gauge theory. Moreover, when N_f is slightly higher than $3N_c/2$, i.e. near the left edge of the conformal window, the magnetic theory is weakly coupled and in the conformal regime. Its dual, the electric theory, which is strongly coupled near the left edge of the conformal window, must then be in the conformal regime too. In the middle of the conformal window, when both theories are strongly coupled, strictly speaking we do not know whether or not they stay superconformal. In principle, it is possible that they both leave the conformal regime. This could happen, for instance, if α_*, the solution of the equation $\gamma(\alpha_*) = 1 - 3N_cN_f^{-1}$, (temporary) becomes larger than $2\pi/T(G)$, the position of the zero of the *denominator* of the NSVZ β function, as we go further away from the point $N_f = 3N_c$ in the direction of $N_f = 3N_c/2$, and then α_* becomes smaller than $2\pi/T(G)$ again, as we approach $N_f = 3N_c/2$. Such a scenario, although not ruled out, does not seem likely, however.

Traveling along the valleys

So far, the dual pair of theories was considered at the origin of the vacuum valley. Both theories, electric and magnetic, have vacuum valleys and a natural question arises as to what happens if we move away from the origin‡. As a matter of fact, this question is quite crucial, since if the theories are equivalent in the infrared, a certain correspondence between them should persist not only at the origin, but at any other point belonging to the vacuum valley. If a correspondence can be found, it will only strengthen the conjecture of duality.

Thus, let us start from the electric theory and move away from the origin. Consider for simplicity a particular direction in the moduli space, namely,

$$Q = \tilde{Q} = \begin{pmatrix} a_1 \\ 0 \\ \cdots \\ 0 \end{pmatrix}, \tag{2.36}$$

where $Q, \tilde{Q}$ are the superfields comprising, say, the first flavor. Moving along this direction we break the gauge symmetry $SU(N_c)$ down to $SU(N_c - 1)$; $2N_c - 1$ chiral superfields are eaten up in the (super)-Higgs mechanism providing masses to $2N_c - 1$ W bosons. Below the mass scale of these W bosons the effective theory is SQCD with a $SU(N_c-1)$ gauge group and N_f-1 flavors. (Additionally there is one $SU(N_c-1)$ singlet, but it plays no role in the gauge dynamics.) It is not difficult to see that decreasing both N_f and N_c by one unit in the electric theory we move to the right along the axis $N_f/3N_c$. In other words, we move towards the right edge of our conformal window, making the electric theory weaker. From what we already know, we should then expect that the magnetic theory becomes stronger.

Let us have a closer look at the magnetic theory. The vacuum expectation value (2.36) is reflected in the magnetic theory as the expectation value of the (1,1) component of the meson field $\mathcal{M}$. No Higgs phenomenon takes place, but, rather, $\mathcal{M}_1^1 \neq 0$. Then, thanks to the superpotential (2.34), the magnetic quark $\mathcal{Q}_1, \tilde{\mathcal{Q}}_1$ gets a mass, and becomes irrelevant in the infrared limit. The gauge group remains the same, $SU(N_f - N_c)$, but the number of active flavors reduces by one unit (we are left with

‡This question was suggested to me by C. Wetterich. Note that if in the electric theory the vacuum degeneracy manifests itself in arbitrary vacuum expectations of Q and $\tilde{Q}$, in the magnetic theory the expectation values of $\mathcal{Q}$ and $\tilde{\mathcal{Q}}$ vanish. The flat direction corresponds to an arbitrary expectation value of $\mathcal{M}$.

$N_f - 1$ active flavors). This means that the first coefficient of the Gell-Mann-Low function of the magnetic theory becomes more negative and the critical value $\alpha_{*\rm magn}$ increases. The theory becomes coupled stronger, in full accord with our expectations.

Let us now try the other way around. What happens if we introduce the mass term to one of the quarks in the electric theory, say the first flavor? The gauge group remains, of course, the same, $SU(N_c)$. However, in the infrared domain the first flavor decouples, and we are left with $N_f - 1$ active flavors. The first coefficient in the β function of the electric theory becomes more negative; hence, the critical value $\alpha_{*\rm electr}$ increases. We move leftwards, towards the left edge of the conformal window. Correspondingly, the electric theory becomes stronger coupled, and we expect that the magnetic one will be coupled weaker.

What is the effect of the mass term in the magnetic theory? It is rather obvious that the corresponding impact reduces to introducing a mass term in the superpotential (2.34),

$$\mathcal{W}' = f\mathcal{M}^i_j \mathcal{Q}_i \tilde{\mathcal{Q}}^j + m\mathcal{M}^1_1 . \tag{2.37}$$

Extending the superpotential is equivalent to changing the vacuum valley. Indeed, the expectation values of $\mathcal{Q}_1$ and $\tilde{\mathcal{Q}}_1$ do not vanish anymore. Instead, the condition of the vanishing of the F term implies

$$\frac{\partial \mathcal{W}}{\partial \mathcal{M}^1_1} = \mathcal{Q}_1 \tilde{\mathcal{Q}}^1 + m = 0 . \tag{2.38}$$

If m is large, Eq. (2.38) implies, in turn, that the magnetic squarks of the first flavor develop a vacuum expectation value, the magnetic theory turns out to be in the Higgs phase, the gauge group $SU(N_f - N_c)$ is spontaneously broken down to $SU(N_f - N_c - 1)$, and one magnetic flavor is eaten up in the super-Higgs mechanism. We end up with a theory with the gauge group $SU(N_f - N_c - 1)$ and $N_f - 1$ flavors. The $\mathcal{M}^i_1$ and $\mathcal{M}^1_i$ components of the meson field become sterile in the infrared limit. In this theory the first coefficient of the β function is less negative, $\alpha_{*\rm magn}$ is smaller, we are closer to the left edge of the conformal window, as was expected.

Summarizing, we see that Seiberg's conjecture of duality is fully consistent with the vacuum structure of both theories. As a matter of fact, this observation may serve as additional evidence in favor of duality. Simultaneously it makes perfectly clear the fact that, if duality does take place, it can be valid only in the infrared limit; by no means the two theories specified above are fully equivalent.

One may ask what happens if we continue adding mass terms to the electric quarks of the first, second, third, *etc.* flavors. Adding large mass terms we eliminate flavors one by one. In other words, we launch a cascade taking us back to smaller values of N_f. The electric theory becomes stronger and stronger coupled. Simultaneously the dual magnetic theory is coupled weaker and weaker. When the number of active flavors reaches $3N_c/2$, $\alpha_{*\rm magn} = 0$. Eliminating the quark flavors further we leave the conformal window – the magnetic theory looses asymptotic freedom and becomes infrared-trivial, with the interaction switching off at large distances. We find ourselves in the free magnetic phase, or the Landau phase. By duality, the correlation functions in the electric theory (which is superstrongly coupled in this domain of N_f) must have the same trivial behavior at large distances. To be perfectly happy we would need to know the relation between all operators of the electric and magnetic theories, so that given a correlation function in the electric theory we could immediately translate it in the language of essentially free magnetic theory. Alas ... As was already mentioned, this relation is basically unknown. It can be explicitly found only for some operators of a geometric nature. Even though the general relation was not found, the achievement is remarkable. For the first time ever the gauge bosons of the weakly coupled theory (magnetic) are shown to be "bound states" of a strongly coupled theory (electric).

The last but one step in the reduction process is when the number of active flavors is $N_c + 2$. The gauge group of the electric theory is $SU(N_c)$ while that of the magnetic one is $SU(2)$. Under the duality conjecture the large distance behavior of the superstrongly coupled electric theory is determined by the massless modes of the essentially free magnetic theory: three "photons", three "photinos", $2(N_c + 2)$ fields of the type Q and $2(N_c + 2)$ fields of the type $\tilde{Q}$. We can return to the remark made in the very beginning of this section – it is explained now. It becomes clear why all attempts to describe the infrared behavior of the theory in terms of the variables M, B, $\tilde{B}$ failed in the $N_f = N_c + 2$ case: these are not proper massless degrees of freedom.

The last step in the cascade that still can be done is reducing one more flavor, by adding a large mass term in the electric theory, or the corresponding entry of the matrix $\mathcal{M}$ in the superpotential of the magnetic theory. At this last stage the super-Higgs mechanism in the magnetic theory completely breaks the remaining gauge symmetry. We end up with $N_c + 1$ massless flavors interacting with the $(N_c + 1) \times (N_c + 1)$ meson superfield $\mathcal{M}^i_j$ (plus a number of sterile fields inessential for our consideration). This remaining meson superfield $\mathcal{M}^i_j$ is assumed to have no (or small) expectation values, so that we stay near the origin of the vacuum valley. Moreover, it is not difficult to see that the instantons of the broken $SU(2)$ generates the superpotential of the type $\det\mathcal{M}$. This is nothing but a supergeneralization of the 't Hooft interaction[80]. Indeed, the instanton generates fermion zero modes, one mode for each ψ_Q and $\psi_{\tilde{Q}}$. In the absence of the Yukawa coupling $f\mathcal{M}Q\tilde{Q}|_F$, there is no way to contract these zero modes, and their proliferation results in the vanishing of the would-be instanton-induced superpotential. The Yukawa coupling $f\phi_{\mathcal{M}}\psi_Q\psi_{\tilde{Q}}$ lifts the zero modes, much in the same way as the mass term does. (Supersymmetrization of the result is achieved through the vertex $f\psi_{\mathcal{M}}\phi_Q\psi_{\tilde{Q}}$ where the boson field ϕ_Q is induced by the zero mode of ψ_Q and the gluino zero mode). In this way we arrive at the instanton-induced superpotential $\det\mathcal{M}$. The full superpotential of the magnetic theory, describing the interaction of massless (or nearly massless) degrees of freedom which are still left there, is

$$\mathcal{W} = f\mathcal{M}Q\tilde{Q} + \det\mathcal{M}\,. \tag{2.39}$$

Compare it with Eq. (2.24) which was derived for the interaction of the massless degrees of freedom of the $N_f = N_c + 1$ model by exploiting a totally different line of reasoning. Up to a renaming of the fields involved, the coincidence is absolute! Thus, the duality conjecture allows us to rederive the s-confining potential in the "electric" theory, SQCD with the gauge group $SU(N_c)$ and $N_f = N_c + 1$. The fact that two different derivations lead to one and the same result further strengthens the duality conjecture, making one to think that actually it is more than a conjecture: the full infrared equivalence of Seiberg's "electric" and "magnetic" theories does indeed take place.

Patterns of Seiberg's duality in more complicated gauge theories than that discussed above, including non-chiral matter sectors, were studied in a number of publications. The dual pairs proliferate! By now a whole zoo of dual pairs is densely populated. Finding a "magnetic" counterpart to the given "electric" theory remains an art, rather than science - no general algorithm exists which would allow one to generate dual pairs automatically, although there is a collection of some helpful hints and recipes. Systematic searches for dual partners to every given supersymmetric theory is an intriguing and fascinating topic. At the present stage it is too technical, however, to be included in this lecture course. Even a brief discussion of the corresponding advances would lead us far astray. The interested reader is referred to the original literature, see e.g.[81, 82, 83].

DOMAIN WALLS, OR NEW EXACT RESULTS "AFTER SEIBERG"

After 1994 many followers worked on dynamical aspects of non-Abelian SUSY gauge theories. In many instances the development went not in depth but, rather, on the surface. Various "exotic" gauge groups and matter representations were considered, supplementing the list of models considered in the previous section by many new examples with essentially the same dynamical behavior. The corresponding discussion might be interesting to experts but is hardly appropriate here. In this part we will focus on some new findings "after Seiberg" and "Seiberg and Witten". It turns out that little miracles of supersymmetric gauge dynamics are not exhausted by fascinating phenomena considered in the previous sections: the exact β function, the electric-magnetic duality in the conformal window and so on. Here we will discuss a fresh topic: exact results for supersymmetric domain walls. The walls emerge in SUSY gluodynamics, the simplest and the least studied of all supersymmetric gauge theories. In spite of the incredible complexity of this model, and unknown intricacies of its strong dynamics, we will be able to *exactly* calculate the wall energy density (sometimes referred to as the wall tension ε). Again, as in all previous cases, our calculation will be indirect. We will heavily exploit our magic tool kit, presented in the first part of these lectures, supplemented by a couple of new devices. You will need some patience to make yourself familiar with these devices. The effort will be rewarded – eventually we will get an elegant formula expressing ε in terms of the gluino condensate $\langle\lambda\lambda\rangle$.

Domain walls built of supersymmetric glue

Let me remind that SUSY gluodynamics describes gauge interactions of gluons and gluinos. For simplicity we will limit ourselves to the gauge group $SU(N)$. As we learned in the section on supersymmetric gluodynamics several degenerate vacuum states, labeled by the value of the gluino condensate, exist in this theory. There are N vacua in which the discrete chiral Z_{2N} symmetry is spontaneously broken down to Z_2,

$$\langle\lambda\lambda\rangle = \Lambda^3 e^{2\pi i k/N}\,, \tag{3.1}$$

where $k = 0, 1, ..., N-1$ and the vacuum angle ϑ is set equal to zero; Λ is the scale parameter of the theory. Furthermore, it is plausible that an additional chirally symmetric vacuum state exists, with a vanishing value of the gluino condensate, $\langle\lambda\lambda\rangle = 0$ (the so called Kovner-Shifman state[11]). All $N+1$ vacua are supersymmetric, i.e. the vacuum energy density vanishes.

The theories with a discrete set of degenerate vacuum states admit a peculiar class of excitations. Let φ be an order parameter distinguishing between distinct vacua, say, in the first vacuum $\langle\varphi\rangle = v_1$ and in the second vacuum $\langle\varphi\rangle = v_2$ (it is assumed that $v_1 \neq v_2$). Then one can consider a (static) field configuration such that φ depends only on z, the third spatial coordinate, and $\varphi \to v_1$ at $z \to -\infty$ while $\varphi \to v_2$ at $z \to +\infty$. A rapid transition from one asymptotics to another occurs in a thin layer, extending in the xy plane near $z = 0$. Far to the left of $z = 0$ we find ourselves in the first vacuum, $\langle\varphi\rangle = v_1$, far to the right in the second, $\langle\varphi\rangle = v_2$. Thus, far away from the plane, $z = 0$ there is no energy stored in the field φ. However, in the transition layer adjacent to the plane $z = 0$ the field φ has to restructure itself from v_1 to v_2, which costs some energy, both kinetic and potential. The energy density profile is characterized by a sharply peaked energy distribution centered at $z = 0$. This is nothing but a domain wall, a phenomenon familiar to everybody from the schooldays, from the theory of ferromagnets. Integrating over z the volume energy density of the interpolating field configuration, we get the wall tension ε.

The total energy of the domain wall $\mathcal{E}$ is obviously proportional to its area A, so that when the xy extensions of the wall grow to infinity, the total energy becomes infinite too. It is the ratio $\varepsilon = \mathcal{E}/A$ that stays finite. The wall tension ε is a close relative of the string tension σ. The string tension σ in QCD is expected to be $\sim \Lambda^2$. The total energy of the string grows linearly with the string dimension L. The total energy of the wall in supersymmetric gluodynamics grows quadratically with L, and ε is expected to be Λ^3.

The domain wall is not a particle-like configuration, of course; it is an extended object. Nevertheless, the domain walls become an important dynamical feature of any theory where they occur. The walls are topologically stable. After the wall is formed it cannot be removed by any local perturbation. Likewise, one cannot produce a wall by a local source. They appear only as global topological defects. The sectors of the theory with a given number of domain walls (zero, one, two and so on) are totally decoupled from each other. As a matter of fact, if we happen to live in a world with a domain wall, we should perceive this field configuration as our "vacuum state" rather than an excitation, although with respect to the full theory, which includes all sectors (as it might be viewed by God), the domain wall is certainly an excitation.

In supersymmetric gluodynamics the relevant order parameter is the gluino density, $\lambda\lambda$. This parameter is quadratic in the fermion field. This fact alone tells us that we will not be able to treat the wall in a quasiclassical approximation, routinely used in the studies of the wall-like solutions in weakly coupled theories. The literature devoted to the domain walls in field theory is quite rich, but next to nothing is said about the quantitative aspects of the wall configurations in the strongly coupled theories. In supersymmetric gluodynamics the domain wall is a genuinely strong-coupling phenomenon, the realm of non-perturbative physics. Thus, once again, we find ourselves in *terra incognita,* and it is only the power of supersymmetry that will eventually lead us to an exact solution for the wall tension.

Central extension of $N = 1$ superalgebra

The title of this section may sound like a heresy for those who are familiar with the basics of supersymmetry. Indeed, in every respectable text book it is written that in four dimensions only extended supersymmetries ($N = 2$ and higher) admit central extensions, while $N = 1$ theories cannot have central charges. Thus, we are definitely going to violate one of the most sacred theorems of supersymmetry. Some time ago I gave a talk at Imperial College in London. One of the mathematicians in the audience got very excited at this point and said that the existence of the central extensions $N = 1$ theories cannot be true because it can never be true because if "one calculates the Hochschild-Serre spectral sequence for the algebra, the second cohomology should be zero!". Although I did not understand a single word in the above statement, I nevertheless insist that supersymmetric gluodynamics, being $N = 1$ theory, still does have a non-trivial central extension[84].

Since this element is absolutely crucial in the given range of questions, I will explain in detail what a central extension in general means, and how it emerges in supersymmetric gluodynamics.

The defining relation of supersymmetry is the anticommutator of two supercharges,

$$\{Q_\alpha \bar{Q}_{\dot\beta}\} = 2P_{\alpha\dot\beta}\,, \tag{3.2}$$

where $P_{\alpha\dot\beta}$ is the energy-momentum operator. To close the algebra one must consider a few other (anti)commutation relations between Q's, P's and other conserved quantities.

All these relations are well known and are irrelevant for our purposes. Let us concentrate on the only relevant anticommutation relation,

$$\{Q_\alpha\, Q_\beta\} = 2T_{\alpha\beta}\,. \tag{3.3}$$

Note that Eq. (3.2) contains the supercharge Q and its Hermitean conjugate $\bar{Q}$, while Eq. (3.3) contains only Q's. Certainly, one can consider a similar anticommutator of two $\bar{Q}$'s too. Since the supercharge Q is conserved, T must be conserved too. On general grounds one can show[85] that $T_{\alpha\beta}$ must commute with all other conserved operators of the theory. That is why $T_{\alpha\beta}$ is called the central extension. In many instances T reduces to a number (the central charge). Why it was universally believed that $T \equiv 0$ in $N = 1$ theories? The general classification of superalgebras dates back to the classical paper[85]. Should the central extension appear in the anticommutator (3.3), it will clearly belong to (0,1) representation of the Lorentz group. This fact is obvious, since $T_{\alpha\beta}$, by construction, is symmetric with respect to two undotted indices, while the only Lorentz-invariant combination would be proportional to $\epsilon_{\alpha\beta}$. The existence of an extra conserved quantity that is not a Lorentz scalar, in addition to four-momentum, is forbidden by the famous Coleman–Mandula theorem[86] for all theories with non-trivial S matrix. The essence of this theorem is very simple: if we have too many conserved operators that are not Lorentz scalars, the S matrix is constrained too strongly. The energy-momentum conservation still allows two-by-two scattering amplitudes to continuously depend on the scattering angle θ. If we want this property to persist, the only other conserved charges we can introduce in the theory are "external" Lorentz-scalar charges, e.g. the electromagnetic charge, the baryon charge and so on. If we add a conserved operator transforming as (0,1) with respect to the Lorentz group, we will kill any possibility of non-trivial scattering. Note that in two-dimensional theories there is no scattering angle θ, the Coleman-Mandula theorem is not applicable, and the bookkeeping works differently: central extensions are perfectly possible in the minimal $N = 1$ superalgebras. And indeed, centrally extended two-dimensional $N = 1$ theories are very well known in the literature[87].

In order to have a central extension in four-dimensional $N = 1$ theories we must violate one or more assumptions of the Coleman–Mandula theorem. The central assumption is the Lorentz-invariance. Let us look at situations when the Lorentz symmetry is spontaneously broken. This is exactly what happens in the presence of the domain wall. The original theory where the domain wall develops (say, supersymmetric gluodynamics) is perfectly Lorentz-invariant. However, after the wall is formed, in the sector with the domain wall, the translation invariance in the direction perpendicular to the wall is spontaneously broken. The wall has infinite energy – it is impossible to boost it.

It is not difficult to see that in this case a non-vanishing central charge transforming as (0,1) with respect to the Lorentz group does not forbid a non-trivial S matrix. In this way we bypass the Coleman-Mandula theorem and open the possibility for central extensions of $N = 1$ superalgebras. Field configurations of the wall type that interpolate between distinct vacua at spatial infinities are extended objects which are not invariant under the action of the Poincaré group. If we choose such a field configuration as our "vacuum", then it may well happen that $T_{\alpha\beta} \neq 0$. The central charge, however, must vanish in the sector with the Lorentz-invariant vacuum.

The central extension in SUSY gluodynamics: a new old anomaly

The argument presented above does not necessarily mean that $T_{\alpha\beta} \neq 0$ develops in any $N = 1$ theory. Moreover, evaluating the anticommutator (3.3) in supersymmetric gluodynamics using the standard canonic commutation relations in a straightforward manner gives zero. Perhaps, this was the reason why this basic property was not discovered[84] until 1996, 22 years after the discovery of SUSY Yang-Mills theory. The central charge does not appear at the tree level. $T_{\alpha\beta} \neq 0$ is a *quantum anomaly.*

At the classical level supersymmetric gluodynamics is conformally invariant, there are no dimensional parameters in the Lagrangian of this theory. Correspondingly, the trace of the energy-momentum tensor vanishes. Supersymmetry entails then that $\partial^\mu J_\mu = 0$ and $\gamma^\mu J_{\mu\alpha} = 0$ where J_μ is the chiral current (see the section on supersymmetric gluodynamics) and $J_{\mu\alpha}$ is the supercurrent[§]. As a matter of fact, all three "geometric" conserved operators form a unified supermultiplet of currents[88]

$$J_{\alpha\dot\alpha} = R^0_{\alpha\dot\alpha} - \frac{1}{2}\left\{i\theta^\beta(J_{\beta\alpha\dot\alpha} - \frac{2}{3}\epsilon_{\beta\alpha}\epsilon^{\gamma\delta}J_{\delta\gamma\dot\alpha}) + \text{H.c.}\right\} -$$

$$\theta^\beta\bar\theta^{\dot\beta}\left(J_{\alpha\dot\alpha\beta\dot\beta} - \frac{1}{3}\epsilon_{\alpha\beta}\epsilon_{\dot\alpha\dot\beta}\epsilon^{\gamma\delta}\epsilon^{\dot\gamma\dot\delta}J_{\gamma\dot\gamma\delta\dot\delta}\right) + \ldots \tag{3.4}$$

where $J_{\delta\gamma\dot\alpha}$ is the supercurrent and $J_{\alpha\dot\alpha\beta\dot\beta}$ is the energy-momentum tensor (see the Appendix, Eq. (A.28)). The statement of the classical conformal invariance can be written as $\bar D^{\dot\alpha} J_{\alpha\dot\alpha} = 0$. The good old anomalies in the trace of the energy-momentum tensor and the divergence of the chiral current imply that, at the quantum level,[15, 36]

$$\bar D^{\dot\alpha} J_{\alpha\dot\alpha} = -\frac{T(G)}{16\pi^2} D_\alpha\left\{\text{Tr}W^2\right\}. \tag{3.5}$$

All three conventional anomalies reside in Eq. (3.5). The anomaly in the trace of the energy-momentum tensor, in particular, is responsible for the generation of the scale parameter Λ.

Equation (3.5) presents the superanomaly relation of SUSY gluodynamics. It is written in the operator form. In this form the coefficient of the anomaly is exhausted by one loop[36]; the expression on the right-hand side is exact, there are no corrections.

Let us return now to the issue of the central extension $T_{\alpha\beta}$. As was mentioned, the non-vanishing anticommutator $\{Q_\alpha Q_\beta\}$ is not seen at the classical level. Does it mean that in the problem of $T_{\alpha\beta}$ we deal with a new anomaly? Taking into account the geometric nature of the supercharges, the occurrence of a new geometric anomaly seems unlikely. And indeed, one can show[89] that the "old" anomaly (3.5) automatically entails $T_{\alpha\beta} \neq 0$. In view of the importance of the issue, let us sketch the corresponding derivation below, even though it is a little bit more technical than I would like.

The lowest component of the current supermultiplet $J_{\alpha\dot\alpha}$ is the R_0 current, while the θ and $\bar\theta$ components of the supermultiplet are related to the supercurrent $J_{\beta\alpha\dot\alpha}$, see Eq. (3.4). Now, let us investigate the anticommutator of the supercharge $\bar Q_{\dot\alpha}$ with the supercurrent,

$$\{\bar Q_{\dot\alpha}\bar Q_{\dot\beta}\} = \frac{1}{2}(\bar\sigma^0)^{\gamma\dot\gamma}\int d^3x\{\bar Q_{\dot\alpha}\bar J_{\dot\beta\dot\gamma\gamma}\}. \tag{3.6}$$

[§]The precise definition of the supercurrent and many useful relations are given in the Appendix. There the reader will find, in particular, the component form of $J_{\alpha\dot\alpha}$. The matrix $(\bar\sigma)_{\dot\alpha\beta}$ that will appear in our master formula (3.12) for the central extension is also defined there.

By inspecting Eq. (3.4) we readily observe that the supercurrent can be expressed in terms of the $\bar\theta$ component of the current supermultiplet,

$$\frac{-i}{2}\bar J_{\dot\beta\dot\gamma\gamma} = 2\{J_{\gamma\dot\beta}\}_{\bar\theta^{\dot\gamma}} - \{J_{\gamma\dot\gamma}\}_{\bar\theta^{\dot\beta}}\,. \tag{3.7}$$

Then

$$\frac{1}{2}\{\bar Q_{\dot\alpha}\bar J_{\dot\beta\dot\gamma\gamma}\} = 2i\{\bar Q_{\dot\alpha}J_{\gamma\dot\beta}\}_{\bar\theta^{\dot\gamma}} - i\{\bar Q_{\dot\alpha}J_{\gamma\dot\gamma}\}_{\bar\theta^{\dot\beta}}. \tag{3.8}$$

The $\bar\theta$ component of the anticommutators on the right-hand side reduces to the lowest component of the superderivative of the current,

$$2i\{\bar Q_{\dot\alpha}J_{\gamma\dot\beta}\}_{\bar\theta^{\dot\gamma}} - i\{\bar Q_{\dot\alpha}J_{\gamma\dot\gamma}\}_{\bar\theta^{\dot\beta}} = \left\{\epsilon_{\dot\gamma\dot\alpha}\bar D^2 J_{\gamma\dot\beta} - \frac{1}{2}\epsilon_{\dot\beta\dot\alpha}\bar D^2 J_{\gamma\dot\gamma}\right\}_{\theta=0}\,. \tag{3.9}$$

The last term, being antisymmetric with respect to the indices $\dot\alpha$, $\dot\beta$, does not contribute to the anticommutator of the supercharges and I drop it, while for the first term it is easy to obtain,

$$\left\{\epsilon_{\dot\gamma\dot\alpha}\bar D^2 J_{\gamma\dot\beta}\right\}_{\theta=0} = 2\left\{\epsilon_{\dot\gamma\dot\alpha}\bar D_{\dot\beta}\bar D^{\dot\delta} J_{\gamma\dot\delta}\right\}_{\theta=0}. \tag{3.10}$$

Assembling everything together we arrive at

$$\{\bar Q_{\dot\alpha}\bar Q_{\dot\beta}\} = -2i(\sigma^0)^{\gamma}_{\dot\alpha}\int d^3x\left\{\bar D_{\dot\beta}\bar D^{\dot\delta} J_{\gamma\dot\delta}\right\}_{\theta=0}. \tag{3.11}$$

The last step is substituting the anomaly equation (3.5) into the superderivative of the current,

$$\{\bar Q_{\dot\alpha}\bar Q_{\dot\beta}\} = -4(\vec\sigma)_{\dot\alpha\dot\beta}\int d^3x\vec\nabla\left\{\frac{T(G)}{16\pi^2}\,\mathrm{Tr}W^2\right\}_{\substack{\bar\theta=0\\ \theta=0}} = \frac{N}{4\pi^2}(\vec\sigma)_{\dot\alpha\dot\beta}\int d^3x\vec\nabla(\mathrm{Tr}\,\lambda\lambda)\,. \tag{3.12}$$

It is evident that the right-hand side generally speaking is non-vanishing. In deriving Eq. (3.12) we took advantage of the fact that the term with the time derivative ∂_0 is proportional to $\epsilon_{\dot\alpha\dot\beta}$; it cancels out after symmetrization over the indices $\dot\alpha$, $\dot\beta$, while those with the spatial derivatives are proportional to the matrix $(\vec\sigma)_{\dot\alpha\dot\beta}$

$$(\sigma^0)^{\gamma}_{\dot\alpha}(\sigma^i)_{\gamma\dot\beta}\partial^i = (\vec\sigma)_{\dot\alpha\dot\beta}\vec\nabla\,.$$

This rather long technical digression is intended for a single purpose – to demonstrate that the anomalous central charge in supersymmetric gluodynamics is an overlooked consequence of the familiar anomalies. The right-hand side of Eq. (3.12) is the spatial integral of a total derivative. This feature is welcome. The integral of the total derivative vanishes for all field configurations satisfying $\lambda\lambda \to$ const. at all spatial infinities, in full accord with the Coleman-Mandula theorem and common wisdom. If, however, the value of the gluino condensate is different at $z \to \infty$ and $z \to -\infty$, i.e. for a field configuration interpolating between distinct vacua, T_{α_β} is proportional to a jump in the value of $\langle\lambda\lambda\rangle$.

I would like to emphasize that Eq. (3.12) is exact, much in the same way as the calculation of the gluino condensate discussed before was exact. There are no perturbative or non-perturbative corrections. As we will see shortly, this result implies an exact prediction for the wall tension.

SUSY preserving walls

Now we are finally prepared to do the calculation of the wall tension. Assume that the wall lies in the *xy* plane, and the vacua between which it interpolates are characterized by real values of the gluino condensate,

$$\langle\lambda\lambda\rangle_{z\to-\infty} \neq \langle\lambda\lambda\rangle_{z\to\infty}\,.$$

The first assumption is a matter of choice of the reference frame, the second can be readily lifted. We will get rid of it shortly. The gluino condensate is real in the Kovner-Shifman vacuum (where it vanishes), and in one or two chirally asymmetric vacua (provided that the vacuum angle $\vartheta = 0$). For SU(N) gauge groups with even N the gluino condensate is real if k in Eq. (3.1) is chosen 0 or $N/2$, while for $SU(N)$ groups with odd N the chirally asymmetric vacuum with the real value of the gluino condensate corresponds to $k = 0$.

P_z, the generator of translations in the z direction, is spontaneously broken in the sector of the theory with the given domain wall. Since the supercharges are related to the energy-momentum operator, see Eq. (3.2), generically one might expect that the breaking of P_z implies that all SUSY generators are spontaneously broken too. In other words, building a wall in supersymmetric theory, one eliminates all standard consequences of supersymmetry, such as the Fermi-Bose degeneracy.

Is it possible to salvage at least a part of supersymmetry?

Before answering this question, let us pose another one: "can the wall tension be arbitrarily small?" Since the only dimensional parameter of SUSY gluodynamics is Λ, it is natural to expect that the wall width is proportional to Λ^{-1}. Since in the transitional layer the vacuum energy density is $\sim \Lambda^4$, the wall tension ε must be proportional to Λ^3. Can it be $\Lambda^3/10$? Or $\Lambda^3/100$?

It turns out that both questions are interrelated. The following argument answers them simultaneously. Consider a (Hermitean) linear combination of supercharges

$$K = \beta^\alpha Q_\alpha + \bar\beta^{\dot\alpha}\bar Q_{\dot\alpha} \tag{3.13}$$

where β is an arbitrary complex parameter, with two components, $\beta = \{\beta^1, \beta^2\}$ that are treated as c-numbers rather than the Grassmann numbers. Denote the state of the world with a wall in the *xy* plane, centered at $z = 0$, by $|w\rangle$. Since the operator K is Hermitean

$$\langle w|K^2|w\rangle \geq 0\,, \tag{3.14}$$

in accordance with the general rules of quantum mechanics. On the other hand

$$K^2 = \beta^\alpha\bar\beta^{\dot\alpha}\{Q_\alpha \bar Q_{\dot\alpha}\} + \frac{1}{2}[\beta^\alpha\beta^\gamma\{Q_\alpha Q_\gamma\} + \text{H.c.}] =$$

$$2\beta^\alpha\bar\beta^{\dot\alpha}\,P_{\alpha\dot\beta} + [\beta^\alpha\beta^\gamma T_{\alpha\gamma} + \text{H.c.}]\,. \tag{3.15}$$

The second line is a direct consequence of the general relations (3.2) and (3.3). We need to examine it in more detail in our specific circumstances. First, the wall is at rest; therefore, only the time component of the energy-momentum operator contributes to $P_{\alpha\dot\beta}$,

$$\beta^\alpha\bar\beta^{\dot\alpha}\,\langle w|P_{\alpha\dot\beta}|w\rangle \to \beta^\alpha\bar\beta^{\dot\alpha}\,(\sigma_0)_{\alpha\dot\beta}\,\langle w|P^0|w\rangle = \mathcal{E}(|\beta^1|^2 + |\beta^2|^2)\,, \tag{3.16}$$

where $\mathcal{E} = \varepsilon A$ is the total wall energy, A is its area (the matrix σ_0 is defined in the Appendix).

Second, since the wall lies in the xy plane, the central extension $T_{\alpha\beta}$ takes the form

$$\beta^\alpha\beta^\gamma T_{\alpha\gamma} = \frac{NA}{8\pi^2}\beta^\alpha\beta^\gamma(\tau^1)_{\alpha\gamma}\left[\langle\lambda\lambda\rangle_{z\to\infty} - \langle\lambda\lambda\rangle_{z\to-\infty}\right] =$$

$$\frac{NA}{4\pi^2}\beta^1\beta^2\left[\langle\lambda\lambda\rangle_{z\to\infty} - \langle\lambda\lambda\rangle_{z\to-\infty}\right], \tag{3.17}$$

where the master formula (3.12) is used ($(\vec{\sigma})_{\alpha\beta}$ is again defined in the Appendix).

We are already very close to our goal. One last effort: start from the positivity condition (3.14), substitute there Eqs. (3.16) and (3.17) and arrive at

$$\mathcal{E}(|\beta^1|^2 + |\beta^2|^2) \geq -\frac{NA}{8\pi^2}\,\mathrm{Re}\left\{\beta^1\beta^2\left[\langle\lambda\lambda\rangle_{z\to\infty} - \langle\lambda\lambda\rangle_{z\to-\infty}\right]\right\}. \tag{3.18}$$

This inequality must be valid for all values of the parameters $\beta^{1,2}$. We can choose these parameters wisely in order to make this inequality as informative as possible.

By an appropriate definition of the gluino field one can always achieve that the expression in the square bracket (the jump of the gluino condensate) is positive. Then the optimal choice of β's is

$$\beta^1 = -\beta^2 = 1\,.$$

Thus, the wall tension turns out to be constrained from below

$$\varepsilon \geq \frac{N}{8\pi^2}\left[\langle\lambda\lambda\rangle_\infty - \langle\lambda\lambda\rangle_{-\infty}\right]. \tag{3.19}$$

It is clear that the equality is achieved only provided K annihilates the wall state,

$$(Q_1 - Q_2 + \bar{Q}_1 - \bar{Q}_2)|w\rangle = 0\,. \tag{3.20}$$

In this case the wall tension is minimal. If the wall tension exceeds the minimal possible value indicated in Eq. (3.19), K acts on $|w\rangle$ non-trivially.

In the case of the minimal wall tension a linear combination of the supercharges acts on the wall trivially. In other words, if the wall is treated as a "vacuum state" (which is legitimate in the given sector of the theory), a linear combination of the supercharges annihilates it. This means that a part of supersymmetry remains unbroken. Such walls are called Bogomolny-Prasad-Sommerfield-saturated walls, or just BPS walls[90], for historical reasons that I do not have time to explain (the above gentlemen had nothing to do with the wall solutions, neither did they consider supersymmetry; nevertheless, the name became common). What part of supersymmetry is unbroken? This is easy to find out on general grounds. Since P_0, P_x and P_y remain unbroken, effectively we deal with a minimal supersymmetry in three dimensions. If in four dimensions the minimal $N = 1$ supersymmetry requires four complex supercharges altogether, two Q's and two $\bar{Q}$'s, in three dimensions the minimal supersymmetry can be built with four real supercharges, or two complex. Thus, building a BPS wall eliminates 1/2 of the original supersymmetry and preserves the other half. Non-BPS wall destroys all supersymmetry.

Let us return to the beginning of this section and ask ourselves what happens if the values of the gluino condensate in the vacua between which the wall interpolates are not real. The consideration above changes in an insignificant way. We will not go into details here, leaving this straightforward exercise to the reader. In the general case Eq. (3.19) becomes

$$\varepsilon \geq \frac{N}{8\pi^2}\left|\langle\lambda\lambda\rangle_\infty - \langle\lambda\lambda\rangle_{-\infty}\right|. \tag{3.21}$$

A linear combination of the supercharges annihilating the BPS wall is different, but the very fact that the BPS wall preserves 1/2 of supersymmetry remains intact, as well as the exact prediction for the wall tension,

$$\varepsilon = \frac{N}{8\pi^2}\left|\langle\lambda\lambda\rangle_\infty - \langle\lambda\lambda\rangle_{-\infty}\right| . \tag{3.22}$$

The line of reasoning outlined above brings us to the conclusion that SUSY preserving walls are possible. Whether or not they actually exist is a dynamical question which must be addressed separately in every given $N = 1$ theory. In weakly coupled theories, where quasiclassical methods for finding the wall solutions can be exploited, this question can be easily answered, see e.g.[91, 92]. A plethora of various dynamical regimes was observed in these works, with extremes being the models with all walls that are BPS, or all walls non-BPS. In strongly coupled theories, the prime subject of this lecture course, one has to resort to more subtle and sophisticated analysis, however.

In SUSY gluodynamics two arguments make us believe that the walls interpolating between distinct vacua do preserve 1/2 of supersymmetry, which would automatically imply the exact formula (3.22). First, the wall can be explicitly constructed[93] within framework of the (amended) Veneziano-Yankielowicz effective Lagrangian[12, 11]. I mentioned this Lagrangian in passing in the first part of these lectures. The advantage of this approach is that it provides us with an explicit dynamical model for the order parameter $\lambda\lambda$. A disadvantage is obvious too: the Veneziano-Yankielowicz Lagrangian is *not* a genuinely Wilsonian construction; therefore, theoretical derivations based on it are somewhat shaky.

The second argument comes from a totally different direction. Glimpses of the desired walls are seemingly seen by D-braners, from higher dimensions[94]. Both arguments go well beyond the scope of the present lectures, and I will leave this topic here, with a hope that further exciting developments will be reported at the next school.

CONCLUSIONS

This lecture course summarizes advances in theoretical understanding of nonperturbative phenomena in the strong coupling regime. If before the SUSY era, the number of exact nonperturbative results in four-dimensional field theory could be counted on one hand, with the advent of supersymmetry a wide spectrum of problems relevant to the most intimate aspects of strong gauge dynamics found exact solutions. Mysteries unravel. Our understanding of gauge theories is dramatically deeper now than it was a decade ago. When preparing these lectures, I intended to share with you, all the excitement and joys associated with the continuous advances in this field spanning over 15 years. Hopefully, the message I tried to convey will be appreciated in full.

Supersymmetric gauge dynamics is very rich, but life is richer, still. The world surrounding us is not supersymmetric. It remains to be seen whether the remarkable discoveries and elegant, powerful methods developed in supersymmetric gauge theories will prove to be helpful in solving the messy problems of real-life particle physics. So far, not much has been done in this direction. In today's climate it is rare that the question of practical applications is even posed. I hope that we reached a turning point: high-energy theory will return to its empirical roots. The command we obtained of supersymmetric gauge theories will be a key which will open to us Pandora's box of problems of Quantum Chromodynamics, *the* theory of our world. Pandora opened the jar that contained all human blessings, and they were gone. Will the achievements obtained in supersymmetric gauge theories be lost in the Planckean nebula?

Acknowledgments

I am grateful to Pierre van Baal for his kind invitation to lecture at the NATO Advanced Study Institute "Confinement, Duality and Non-Perturbative Aspects of QCD", Isaac Newton Institute for Mathematical Sciences, University of Cambridge, June 26 – 28. I am grateful to colleagues and staff of the Isaac Newton Institute for Mathematical Sciences for hospitality and financial support. This work was supported in part by DOE under the grant number DE-FG02-94ER40823.

APPENDIX: Notation, Conventions, Useful Formulae

In this Appendix the key elements of the formalism used in supersymmetric gauge theories are outlined. Basic formulae are collected for convenience.

The notation we follow is close to that of the canonic text book of Bagger and Wess[21]. There are some distinctions, though. The most important of them is the choice of the metric. Unlike Bagger and Wess, we use the standard metric $g_{\mu\nu} = (+ - - -)$. There are also distinctions in normalization, see Eq. (A.19).

The left-handed spinor is denoted by undotted indices, e.g. η_β. The right-handed spinor is denoted by dotted indices, e.g. $\bar\xi^{\dot\beta}$. (This convention is standard in supersymmetry but is opposite to one accepted in the text-book[95]). The Dirac spinor Ψ then takes the form

$$\Psi = \begin{pmatrix} \bar\xi^{\dot\beta} \\ \eta_\beta \end{pmatrix} . \tag{A.1}$$

Lowering and raising of the spinorial indices is done by applying the Levi-Civita tensor from the left,

$$\chi^\alpha = \epsilon^{\alpha\beta}\chi_\beta \,, \quad \chi_\alpha = \epsilon_{\alpha\beta}\chi^\beta , \tag{A.2}$$

and the same for the dotted indices, where

$$\epsilon^{\alpha\beta} = -\epsilon^{\beta\alpha} \,, \quad \epsilon^{12} = -\epsilon_{12} = 1 \,. \tag{A.3}$$

The products of the undotted and dotted spinors are defined as follows:

$$\eta\chi = \eta^\alpha\chi_\alpha = -\eta_\alpha\chi^\alpha \,, \quad \bar\eta\bar\chi = \bar\eta_{\dot\alpha}\bar\chi^{\dot\alpha} \,. \tag{A.4}$$

Under this convention $(\eta\chi)^+ = \bar\chi\bar\eta$. Moreover,

$$\theta^\alpha\theta^\beta = -\frac{1}{2}\epsilon^{\alpha\beta}\theta^2 , \quad \theta_\alpha\theta_\beta = \frac{1}{2}\epsilon_{\alpha\beta}\theta^2 , \quad \bar\theta^{\dot\alpha}\bar\theta^{\dot\beta} = \frac{1}{2}\epsilon^{\dot\alpha\dot\beta}\bar\theta^2 , \quad \bar\theta_{\dot\alpha}\bar\theta_{\dot\beta} = -\frac{1}{2}\epsilon_{\dot\alpha\dot\beta}\bar\theta^2 \,. \tag{A.5}$$

The vector quantities (representation $(\frac{1}{2}\,,\frac{1}{2})$) are obtained in the spinorial formalism by multiplication by

$$(\,\sigma^\mu\,)_{\alpha\dot\beta} = \{\,1\,,\vec\tau\,\}_{\alpha\dot\beta} \tag{A.6}$$

where $\vec\tau$ stands for the Pauli matrices, for instance,

$$A_{\alpha\dot\beta} = A_\mu(\sigma^\mu)_{\alpha\dot\beta}. \tag{A.7}$$

Note that

$$A_\mu B^\mu = \frac{1}{2}A_{\alpha\dot\beta}B^{\alpha\dot\beta} \,, \quad A_{\alpha\dot\beta}A^{\gamma\dot\beta} = \delta^\gamma_\alpha\, A_\mu A^\mu. \tag{A.8}$$

The square of the four-vector is understood as

$$A^2 \equiv A_\mu A^\mu = \frac{1}{2}A_{\alpha\dot\beta}A^{\alpha\dot\beta} \,. \tag{A.9}$$

If the matrix $(\sigma^\mu)_{\alpha\dot\beta}$ is "right-handed" it is convenient to introduce its "left-handed" counterpart,

$$(\bar\sigma^\mu)^{\dot\beta\alpha} = \{\, 1\,, -\vec\tau\,\}_{\dot\beta\alpha}\,. \tag{A.10}$$

The matrices that appear in dealing with representations (1,0) and (0,1) are

$$(\vec\sigma)^\alpha_\beta = \vec\tau_{\alpha\beta}\,, \quad (\vec\sigma)^{\alpha\beta} = \epsilon^{\beta\delta}\vec\sigma^\alpha_\delta\,, \tag{A.11}$$

and the same for the dotted indices. The matrices $(\vec\sigma)^{\alpha\beta}$ are symmetric, $(\vec\sigma)^{\alpha\beta} = (\vec\sigma)^{\beta\alpha}$. In the explicit form

$$(\vec\sigma)^{\alpha\beta} = \{\, \tau^3\,, -i\mathbf{1}\,, -\tau^1\,\}_{\alpha\beta}\,. \tag{A.12}$$

Note that with our definitions

$$(\vec\sigma)_{\alpha\beta} = \{\, -\tau^3\,, -i\mathbf{1}\,, \tau^1\,\}_{\alpha\beta}\,.$$

The left (right) coordinates $x_{L,R}$ and covariant derivatives are

$$\begin{aligned} &(x_L)_{\alpha\dot\alpha} = x_{\alpha\dot\alpha} - 2\,i\,\theta_\alpha\bar\theta_{\dot\alpha}\,, \quad (x_R)_{\alpha\dot\alpha} = x_{\alpha\dot\alpha} + 2\,i\,\theta_\alpha\bar\theta_{\dot\alpha}\,, \\ &D_\alpha = \frac{\partial}{\partial\theta^\alpha} - i\,\partial_{\alpha\dot\alpha}\bar\theta^{\dot\alpha}\,, \quad \bar D_{\dot\alpha} = -\frac{\partial}{\partial\bar\theta^{\dot\alpha}} + i\,\theta^\alpha\partial_{\alpha\dot\alpha}, \end{aligned} \tag{A.13}$$

so that

$$\begin{aligned} \{D_\alpha\, \bar D_{\dot\alpha}\} &= 2i\partial_{\alpha\dot\alpha}\,, \\ \bar D_{\dot\beta}(x_L)_{\alpha\dot\alpha} &= 0\,, \qquad D_\beta(x_R)_{\alpha\dot\alpha} = 0\,, \\ \bar D_{\dot\beta}(x_R)_{\alpha\dot\alpha} &= -4\,i\,\theta_\alpha\epsilon_{\dot\beta\dot\alpha}\,, \quad D_\beta(x_L)_{\alpha\dot\alpha} = 4\,i\,\bar\theta_{\dot\alpha}\epsilon_{\beta\alpha}\,. \end{aligned} \tag{A.14}$$

The law of the supertranslation is

$$\begin{aligned} &\theta \to \theta + \epsilon\,, \quad \bar\theta \to \bar\theta + \bar\epsilon\,, \quad x_{\alpha\dot\beta} \to x_{\alpha\dot\beta} - 2\,i\,\theta_\alpha\bar\epsilon_{\dot\beta} + 2\,i\,\epsilon_\alpha\bar\theta_{\dot\beta}\,, \\ &(x_L)_{\alpha\dot\beta} \to (x_L)_{\alpha\dot\beta} - 4\,i\,\theta_\alpha\bar\epsilon_{\dot\beta}\,, \quad (x_R)_{\alpha\dot\beta} \to (x_R)_{\alpha\dot\beta} + 4\,i\,\epsilon_\alpha\bar\theta_{\dot\beta}\,. \end{aligned} \tag{A.15}$$

It corresponds to the infinitesimal transformation of the superfield in the form

$$\delta\Phi = i\left(\epsilon Q + \bar\epsilon\bar Q\right)\Phi \tag{A.16}$$

where

$$Q_\alpha = -i\frac{\partial}{\partial\theta^\alpha} - \partial_{\alpha\dot\alpha}\bar\theta^{\dot\alpha}\,, \qquad \bar Q_{\dot\alpha} = i\frac{\partial}{\partial\bar\theta^{\dot\alpha}} + \theta^\alpha\partial_{\alpha\dot\alpha}\,. \tag{A.17}$$

The integrals over the Grassmann variable are normalized as follows

$$\int d^2\theta\,\theta^2 = 2\,, \quad \int d^4\theta\,\theta^2\,\bar\theta^2 = 4\,, \tag{A.18}$$

and we define

$$\{\ldots\}_F = \frac{1}{2}\int d^2\theta\{\ldots\}\,, \quad \{\ldots\}_D = \frac{1}{4}\int d^4\theta\{\ldots\}. \tag{A.19}$$

A generic non-Abelian SUSY gauge theory has the Lagrangian

$$\mathcal{L} = \left\{\frac{1}{4g_0^2}\mathrm{Tr}\int d^2\theta\, W^2 + \mathrm{H.c.}\right\} + \frac{1}{4}\sum_i\int d^4\theta\bar Q_i e^V Q_i + \left\{\frac{1}{2}\int d^2\theta \mathcal{W}(\{Q_i\}) + \mathrm{H.c.}\right\}, \tag{A.20}$$

where

$$\frac{1}{g_0^2}=\frac{1}{g^2}-\frac{i\vartheta}{8\pi^2}\,,$$

is the (complexified) gauge coupling constant, the sum in Eq. (A.20) runs over all matter superfields Q_i present in the theory, and $\mathcal{W}(\{Q_i\})$ is a generic superpotential. Most commonly one deals with the superpotential corresponding to the mass term of the matter fields. In many models, cubic terms are gauge invariant; then they are allowed too (and do not spoil renormalizability of the theory).

Furthermore, the superfield W_α, which includes the gluon strength tensor, is defined as follows:

$$W_\alpha=\frac{1}{8}\,\bar{D}^2\left(\mathrm{e}^{-V}\,D_\alpha\,\mathrm{e}^{V}\right) \tag{A.21}$$

where V is the vector superfield. In the Wess-Zumino gauge

$$V=-2\theta^\alpha\bar{\theta}^{\dot{\alpha}}A_{\alpha\dot{\alpha}}-2i\bar{\theta}^2(\theta\lambda)+2i\theta^2(\bar{\theta}\bar{\lambda})+\theta^2\bar{\theta}^2D\,, \tag{A.22}$$

$V=V^aT^a$ and T^a stands for the generators of the gauge group G. In the fundamental representation of $SU(N)$, a case of most practical interest,

$$\mathrm{Tr}\left(T^aT^b\right)=\frac{1}{2}\,\delta^{ab}\,.$$

The supergauge transformation has the form

$$Q_i\to e^{i\Lambda}\,Q_i\,,\quad e^V\to e^{i\Lambda}e^V\,e^{-i\Lambda}\,,\quad W_\alpha\to e^{i\Lambda}W_\alpha e^{-i\Lambda}\,, \tag{A.23}$$

where Λ is an arbitrary chiral superfield ($\bar{\Lambda}$ is antichiral). In components

$$W_\alpha=i\left(\lambda_\alpha+i\theta_\alpha D-\theta^\beta\,G_{\alpha\beta}-i\theta^2\mathcal{D}_{\alpha\dot{\alpha}}\bar{\lambda}^{\dot{\alpha}}\right) \tag{A.24}$$

where λ_α is the gluino (Weyl) field, $\mathcal{D}_{\alpha\dot{\alpha}}$ is the covariant derivative, and $G_{\alpha\beta}$ is the gluon field strength tensor in the spinorial notation.

The standard gluon field strength tensor transforms as $(1,0)+(0,1)$ with respect to the Lorentz group. Projecting out pure (1,0) is achieved by virtue of the $(\sigma)_{\alpha\dot{\beta}}$ matrices,

$$G_{\alpha\beta}\;=\;-\frac{1}{2}G_{\mu\nu}(\sigma^\mu)_{\alpha\dot{\beta}}(\sigma^\nu)_{\beta\dot{\delta}}\epsilon^{\dot{\beta}\dot{\delta}}=(\vec{E}-i\,\vec{B})(\vec{\sigma})_{\alpha\beta}\,. \tag{A.25}$$

Then

$$G^{\alpha\beta}G_{\alpha\beta}\;=\;2(\vec{B}^2-\vec{E}^2+2\,i\,\vec{E}\vec{B})=G_{\mu\nu}G_{\mu\nu}-i\,G_{\mu\nu}\tilde{G}_{\mu\nu}$$

where

$$\tilde{G}_{\mu\nu}=\frac{1}{2}\epsilon_{\mu\nu\alpha\beta}G^{\alpha\beta}\,,\quad(\epsilon_{0123}=-1)\,. \tag{A.26}$$

The supercurrent supermultiplet has the following general form

$$J_{\alpha\dot{\alpha}}=R^0_{\alpha\dot{\alpha}}-\frac{1}{2}\left\{i\theta^\beta(J_{\beta\alpha\dot{\alpha}}-\frac{2}{3}\epsilon_{\beta\alpha}\epsilon^{\gamma\delta}J_{\delta\gamma\dot{\alpha}})+\mathrm{H.c.}\right\}-$$

$$\theta^\beta\bar{\theta}^{\dot{\beta}}\left(J_{\alpha\dot{\alpha}\beta\dot{\beta}}-\frac{1}{3}\epsilon_{\alpha\beta}\epsilon_{\dot{\alpha}\dot{\beta}}\epsilon^{\gamma\delta}\epsilon^{\dot{\gamma}\dot{\delta}}J_{\gamma\dot{\gamma}\delta\dot{\delta}}\right)+\dots \tag{A.27}$$

where $R^0_{\alpha\dot{\alpha}}$ is the R_0 current, $J_{\beta\alpha\dot{\alpha}}$ is the supercurrent, and $J_{\alpha\dot{\alpha}\beta\dot{\beta}}$ is related to the energy-momentum tensor, $\theta_{\mu\nu}$, in the following way

$$J_{\alpha\dot{\alpha}\beta\dot{\beta}}=-(\sigma^i)_{\alpha\beta}(\sigma^j)_{\dot{\alpha}\dot{\beta}}\left\{\theta^{ij}+\theta^{00}g^{ij}-\epsilon^{ijk}\theta^{0k}\right\}+\frac{1}{2}\epsilon_{\alpha\beta}\epsilon_{\dot{\alpha}\dot{\beta}}\theta^\mu_\mu\,; \tag{A.28}$$

here $i, j, k = 1, 2, 3$, $g_{\mu\nu}$ is the metric tensor and the matrices $(\sigma^i)_{\alpha\beta}$ are defined in Eq. (A.11).

The general anomaly relation (three "geometric" anomalies) is

$$\bar{D}^{\dot{\alpha}} J_{\alpha\dot{\alpha}} = \frac{1}{3} D_\alpha \left\{ \left[3\mathcal{W} - \sum_i Q_i \frac{\partial \mathcal{W}}{\partial Q_i} \right] - \left[\frac{3T(G) - \sum_i T(R_i)}{16\pi^2} \mathrm{Tr} W^2 + \frac{1}{8} \sum_i \gamma_i \bar{D}^2 (\bar{Q}_i e^V Q_i) \right] \right\}, \tag{A.29}$$

where γ_i are the anomalous dimensions of the matter fields Q_i. The general Konishi anomaly has the form

$$\frac{1}{8} \bar{D}^2 (\bar{Q}_i^+ e^V Q_i) = \frac{1}{2} Q_i \frac{\partial \mathcal{W}}{\partial Q_i} + \frac{\sum_i T(R_i)}{16\pi^2} \mathrm{Tr} W^2 . \tag{A.30}$$

In conclusion let us present the full component expression for the simplest $SU(2)$ model with one flavor (two subflavors), assuming that the superpotential in the case at hand reduces to the mass term of the quark (squark) fields. This model was discussed in detail in the first part of these lectures. If the index f denotes the subflavors, $f = 1, 2$

$$\begin{aligned}
\mathcal{L} = {} & \frac{1}{g^2} \left\{ -\frac{1}{4} G^a_{\mu\nu} G^a_{\mu\nu} + \lambda^{\alpha,a} i \mathcal{D}_{\alpha\dot{\alpha}} \bar{\lambda}^{\dot{\alpha},a} + \frac{1}{2} D^a D^a \right\} \\
& + \psi^{f\alpha} i \mathcal{D}_{\alpha\dot{\alpha}} \bar{\psi}^{f\dot{\alpha}} + (\mathcal{D}_\mu \phi^{+f})(\mathcal{D}_\mu \phi^f) + F^{+f} F^f \\
& + i\sqrt{2} (\phi_1^+ \lambda \psi_1 + \phi_1 \bar{\lambda} \bar{\psi}_1 + \phi_2^+ \lambda \psi_2 + \phi_2 \bar{\lambda} \bar{\psi}_2) \\
& + \frac{1}{2} D^a \, (\phi_1^+ T^a \phi_1 + \phi_2^+ T^a \phi_2) \\
& + m \, (\phi_1 F_2 + \phi_1^+ F_2^+ + \phi_2 F_1 + \phi_2^+ F_1^+ + \psi_1 \psi_2 + \bar{\psi}_1 \bar{\psi}_2) .
\end{aligned} \tag{A.31}$$

In this model the component expression for the supercurrent is

$$J_{\alpha\beta\dot{\beta}} = 2 \left\{ \frac{1}{g^2} \left[i G^a_{\beta\alpha} \bar{\lambda}^a_{\dot{\beta}} - 3\epsilon_{\beta\alpha} D^a \bar{\lambda}^a_{\dot{\beta}} \right] + \right.$$

$$\left. \sqrt{2} \left[(\partial_{\alpha\dot{\beta}} \phi^+) \psi_\beta - i \, \epsilon_{\beta\alpha} F \bar{\psi}_{\dot{\beta}} \right] - \frac{\sqrt{2}}{6} \left[\partial_{\alpha\dot{\beta}} (\psi_\beta \phi^+) + \partial_{\beta\dot{\beta}} (\psi_\alpha \phi^+) - 3\epsilon_{\beta\alpha} \partial^\gamma_{\dot{\beta}} (\psi_\gamma \phi^+) \right] \right\} . \tag{A.32}$$

RECOMMENDED LITERATURE

It is assumed that the reader is familiar with the text-books on supersymmetry:

- J. Bagger and J. Wess, *Supersymmetry and Supergravity,* (Princeton University Press, 1983).
- P. West, *Introduction to Supersymmetry and Supergravity* (World Scientific, Singapore, 1986).
- S.J. Gates, M.T. Grisaru, M. Roček and W. Siegel, *Superspace or one Thousand and one Lessons in Supersymmetry* (Benjamin/ Cummings, 1983).
- D. Bailin and A. Love, *Supersymmetric Gauge Field Theory and String Theory* (IOP Publishing, Bristol, 1994).

A solid introduction to supersymmetric instanton calculus is given in:

- *Instantons in Gauge Theories,* ed. M. Shifman, (World Scientific, Singapore, 1994), Chapter VII.

A brief survey of those aspects of supersymmetry which are most relevant to the recent developments can be found in:

- J. Lykken, *Introduction to Supersymmetry,* hep-th/9612114.

Reviews on Exact Results in SUSY Gauge Theories and Related Issues

- N. Seiberg, *The Power of Holomorphy - Exact Results in 4D SUSY Field Theories,* in *Proc. VI International Symposium on Particles, Strings, and Cosmology (PASCOS 94),* Ed. K. C. Wali, (World Scientific, Singapore, 1995) [hep-th/9408013].
- K. Intriligator and N. Seiberg, *Lectures on Supersymmetric Gauge Theories and Electric - Magnetic Duality, Nucl. Phys. Proc. Suppl.* **45BC** (1996) 1 [hep-th/9509066].
- K. Intriligator and N. Seiberg, *Phases of **N=1** Supersymmetric Gauge Theories and Electric - Magnetic Triality,* in Proc. Conf. *Future Perspectives in String Theory (Strings '95)* Eds. I. Bars, P. Bouwknegt, J. Minahan, D. Nemeschansky, K. Pilch, H. Saleur, and. N. Warner (World Scientific, Singapore, 1996) [hep-th/9506084].
- D. Olive, *Exact Electromagnetic Duality, Nucl. Phys. Proc. Suppl.* **45A** (1996) 88 [hep-th/9508089].
- P. Di Vecchia, *Duality in Supersymmetric Gauge Theories, Surveys High Energ. Phys.* 10 (1997) 119 [hep-th/9608090].
- L. Alvarez-Gaumé and S.F. Hassan, *Introduction to S-Duality in **N=2** Supersymmetric Gauge Theories, Fortsch. Phys.* **45** (1997) 159 [hep-th/9701069].
- W. Lerche, *Notes on **N=2** Supersymmetric Yang-Mills Theory, Nucl. Phys. Proc. Suppl.* **55B** (1997) 83 [hep-th/9611190].
- A. Bilal, *Duality in **N=2** SUSY Yang-Mills Theory: A Pedagogical Introduction to the Work of Seiberg and Witten,* hep-th/9601007.
- S. Ketov, *Solitons, Monopoles, and Duality: From Sine-Gordon to Seiberg-Witten, Fortsch. Phys.* **45** (1997) 237 [hep-th/9611209].
- M. Peskin, *Duality in Supersymmetric Yang-Mills Theory,* hep-th/9702094.

REFERENCES

1. A. Polyakov, *Nucl. Phys.* **B120** (1977) 429.
2. S. Mandelstam, *Phys. Reports* **23** (1976) 245.
3. G. 't Hooft, in *1981 Cargése Summer School Lecture Notes on Fundamental Interactions, NATO Adv. Study Inst. Series B: Phys.,* Vol. 85, ed. M. Lévy *et al.* (Plenum Press, New York, 1982) [reprinted in G. 't Hooft, *Under the Spell of the Gauge Principle* (World Scientific, Singapore, 1994), page 514]; *Nucl. Phys.* **B190** (1981) 455.
4. Yu. Golfand and E. Likhtman, *Pis'ma ZhETF,* **13** (1971) 452 *[JETP Lett.* **13** (1971) 323]; see also in *Problems of Theoretical Physics,* I.E. Tamm Memorial Volume (Nauka, Moscow, 1972),

page 37; E.P. Likhtman, *Irreducible representations, extensions of the algebra of Poincaré generators by bispinor generators, Kratk. Soob. Fiz. – Short Comm. Phys. FIAN,* 1971, No. 5, page 197.
5. J. Wess and B. Zumino, *Nucl. Phys.* **B70** (1974) 39; *Phys. Lett.* **B49** (1974) 52.
6. W. Pauli, *Pauli Lectures on Physics,* Vol. 6, *Selected Topics in Field Quantization* (MIT Press, Cambridge, 1973), page 33.
7. A. Salam and J. Strathdee, *Nucl. Phys.* **B76** (1974) 477; *Phys. Rev.* **D11** (1975) 1521.
8. S. Ferrara and B. Zumino, *Nucl. Phys.* **B79** (1974) 413.
9. V. Novikov, M. Shifman, A. Vainshtein and V. Zakharov, *Nucl. Phys.* **B229** (1983) 407.
10. M. Shifman and A. Vainshtein, *Nucl. Phys.* **B296** (1988) 445.
11. A. Kovner and M. Shifman, *Phys. Rev.* **D56** (1997) 2396.
12. G. Veneziano and S. Yankielowicz, *Phys. Lett.* **B113** (1982) 321; T. R. Taylor, G. Veneziano and S. Yankielowicz, *Nucl. Phys.* **B218** (1983) 493.
13. V. Novikov, M. Shifman, A. Vainshtein and V. Zakharov, *Nucl. Phys.* **B229** (1983) 381; *Phys. Lett.* **B166** (1986) 334.
14. A. Salam and J. Strathdee, *Nucl. Phys.* **B87** (1975) 85; P. Fayet, *Nucl. Phys.* **B90** (1975) 104.
15. M. Grisaru, in *Recent Developments in Gravitation* (Cargése Lectures, 1978), Eds. M. Levy and S. Deser (Plenum Press, New York, 1979), page 577, and references therein.
16. E. Witten, *Nucl. Phys.* **B202** (1982) 253.
17. F. Buccella, J.-P. Derendinger, C. Savoy and S. Ferrara, *Phys. Lett.* **B115** (1982) 375, and in Proc. Europhys. Study Conf. *Unification of the Fundamental Particle Interactions. II,* Eds. J. Ellis and S. Ferrara (Plenum Press, New York, 1983), page 349.
18. I. Affleck, M. Dine and N. Seiberg, *Nucl. Phys.* **B241** (1984) 493.
19. I. Affleck, M. Dine and N. Seiberg, *Nucl. Phys.* **B256** (1985) 557.
20. M. Luty and W. Taylor, *Phys. Rev.* **D53** (1996) 3399.
21. J. Bagger and J. Wess, *Supersymmetry and Supergravity,* 2nd Edition (Princeton University Press, Princeton, 1992).
22. E. Witten, *Nucl. Phys.* **B403** (1993) 159.
23. A. Vainshtein, V. Zakharov and M. Shifman, *Usp. Fiz. Nauk,* **146** (1985) 683 [*Sov. Phys. - Usp.* **28** (1985) 709].
24. I. Kogan, A. Morozov, M. Olshanetsky and M. Shifman, *Yad. Fiz.* **43** (1986) 1587 [*Sov. J. Nucl. Phys.* **43** (1986) 1022].
25. N. Seiberg, *Phys. Rev.* **D49** (1994) 6857; *Nucl. Phys.* **B435** (1995) 129.
26. K. Intriligator and N. Seiberg, *Nucl. Phys.* **B431** (1994) 551; K. Intriligator, R. Leigh, and N. Seiberg, *Phys. Rev.* **D50** (1994) 1092.
27. K. Intriligator, R. Leigh, and M. Strassler, *Nucl. Phys.* **B456** (1995) 567.
28. E. Poppitz and L. Randall, *Phys. Lett.* **B336** (1994) 402; J. Bagger, E. Poppitz and L. Randall, *Nucl. Phys.* **B426** (1994) 3.
29. S. Giddings and J.M. Pierre, *Phys. Rev.* **D52** (1995) 6065.
30. T. Gherghetta, C. Kolda and S. Martin, *Nucl. Phys.* **B468** (1996) 37.
31. *3.* Wess and B. Zumino, *Phys. Lett.* **B49** (1974) 52; J. Iliopoulos and B. Zumino, *Nucl. Phys.* **B76** (1974) 310; P. West, *Nucl. Phys.* **B106** (1976) 219; M. Grisaru, M. **Roček,** and W. Siegel, *Nucl. Phys.* **B159** (1979) 429.
32. Y. Meurice and G. Veneziano, *Phys. Lett.* **B141** (1984) 69; I. Affleck, M. Dine and N. Seiberg, *Phys. Lett.* **B137** (1984) 187.
33. D. Amati, K. Konishi, Y. Meurice, G. Rossi and G. Veneziano, *Phys. Rep.* **162** (1988) 557.
34. I. Affleck, M. Dine and N. Seiberg, *Phys. Rev. Lett.* **52** (1984) 1677; Y. Meurice and G. Veneziano, *Phys. Lett.* **B141** (1984) 69.
35. S. Ferrara and B. Zumino, *Nucl. Phys.* **B87** (1975) 207.
36. M. Shifman and A. Vainshtein, *Nucl. Phys.* **B277** (1986) 456.
37. I. Kogan, M. Shifman, and A. Vainshtein, *Phys. Rev.* **D53** (1996) 4526.
38. J. Wess and B. Zumino, *Phys. Lett.* **B49** (1974) 52.
39. M. Shifman and A. Vainshtein, *Nucl. Phys.* **B359** (1991) 571.
40. I. Jack, D.R.T. Jones, and P. West, *Phys. Lett.* **B258** (1991) 382.
41. E. Poppitz and L. Randall, *Phys. Lett.* **B389** (1996) 280.
42. T.E. Clark, O. Piguet, and K. Sibold, *Nucl. Phys.* **B159** (1979) 1; K. Konishi, *Phys. Lett.* **B135** (1984) 439; K. Konishi and K. Shizuya, *Nuov. Cim.* **A90** (1985) 111.
43. V. Novikov, M. Shifman, A. Vainshtein and V. Zakharov, *Nucl. Phys.* **B260** (1985) 157; for a review see[23].

44. A. Vainshtein, V. Zakharov and M. Shifman, *Yad. Fiz.* **43** (1986) 1596 *[Sov. J. Nucl. Phys.* **43** (1986) 1028], Sect. 3; see also *Pis'ma ZhETF* **42** (1985) 182 *[JETP Lett.* **42** (1985) 224]; for a recent discussion see M. Shifman, *Int. J. Mod. Phys.* **A11** (1996) 5761.
45. I. Jack, D.R.T. Jones and C.G. North, *Nucl. Phys.* **B486** (1997) 479.
46. N. Seiberg, *Phys. Lett.* **206B** (1988) 75.
47. N. Seiberg, *Phys. Lett.* **B318** (1993) 469.
48. I. Affleck, M. Dine and N. Seiberg, *Phys. Rev. Lett.* **51** (1983) 1026.
49. I. Affleck, M. Dine and N. Seiberg, *Phys. Lett.* **B137** (1984) 187; *Phys. Rev. Lett.* **52** (1984) 493; *Phys. Lett.* **B140** (1984) 59.
50. L. Dixon, V. Kaplunovsky and J. Louis, *Nucl. Phys.* **B355** (1991) 649.
51. H. Li and K. Mahanthappa, *Phys. Lett.* **B319** (1993) 152; *Phys. Rev.* **D49** (1994) 5532.
52. M. Shifman (Ed.), *Instantons in Gauge Theories* (World Scientific, Singapore, 1994), Chapter VII.
53. B. Zumino, *Phys. Lett.* **B69** (1977) 369.
54. A. D'Adda and P. Di Vecchia, *Phys. Lett.* **B73** (1978) 162.
55. G.C. Rossi and G. Veneziano, *Phys. Lett.* **B138** (1984) 195.
56. N. Seiberg and E. Witten, *Nucl. Phys.* **B426** (1994) 19; (E) B430 (1994) 485; *Nucl. Phys.* **B431** (1994) 484.
57. N. Dorey, V. Khoze, and M. Mattis, *Phys. Rev,* **D54** (1996) 2921; *Phys. Rev.* **D54** (1996) 7832; *Phys. Lett.* **B388** (1996) 324; *Phys. Lett.* **B390** (1997) 205; H. Aoyama, T. Harano, M. Sato, and S. Wada, *Phys. Lett.* **B388** (1996) 331; K. Ito and N. Sasakura, *Nucl. Phys.* **B484** (1997) 141; *Mod. Phys. Lett.* **A12** (1997) 205; F. Fucito and T. Travaglini, *Phys. Rev.* **D55** (1997) 1099; T. Harano and M. Sato, *Nucl. Phys.* **B484** (1997) 167; Y. Yoshida, hep-th/9610211; M. Slater, *Phys. Lett.* **B403** (1997) 57.
58. A. Yung, *Nucl. Phys.* **B485** (1997) 38.
59. S. Dimopoulos, *Nucl. Phys.* **B168** (1980) 69; M. Peskin, *Nucl. Phys.* **B175** (1980) 197.
60. M. Vysotsky, I. Kogan, and M. Shifman, *Yad. Fiz.* **42** (1985) 504 *[Sov. J. Nucl. Phys.* **42** (1985) 318].
61. A.C. Davis, M. Dine and N. Seiberg, *Phys. Lett.* **B125** (1983) 487.
62. G. 't Hooft, in *Recent Developments in Gauge Theories,* Eds. G. 't Hooft *et at.,* (Plenum Press, New York, 1980).
63. A. Dolgov and V. Zakharov, *Nucl. Phys.* **B12** (1971) 68.
64. M. Shifman, *Phys. Rep.* **209** (1991) 161.
65. C. Vafa and E. Witten, *Nucl. Phys.* **B234** (1984) 173.
66. S. Coleman and E. Witten, *Phys. Rev. Lett.* **45** (1980) 100.
67. E. Poppitz and S. Trivedi, *Phys. Lett.* **B365** (1996) 125; P. Pouliot, *Phys. Lett.* **B367** (1996) 151.
68. K. Intriligator and P. Pouliot, *Phys. Lett.* **B353** (1996) 471.
69. P. Cho and P. Kraus, *Phys. Rev.* **D54** (1996) 7640.
70. C. Csáki, W. Skiba and M. Schmaltz, *Nucl. Phys.* **B487** (1997) 128.
71. K. Intriligator and N. Seiberg, *Nucl. Phys.* **B444** (1995) 125.
72. P. Pouliot, *Phys. Lett.* **B359** (1995) 108; P. Pouliot and M. Strassler, *Phys. Lett.* **B370** (1996) 76; *Phys. Lett.* **B375** (1996) 175.
73. I. Pesando, *Mod. Phys. Lett.* **A10** (1995) 1871; S. Giddings and J. Pierre, *Phys. Rev.* **D52** (1995) 6065.
74. C. Csáki, M. Schmaltz, and W. Skiba, *Phys. Rev. Lett.* **78** (1997) 799; *Phys. Rev.* **D55** (1997) 7840.
75. For a concise review and relevant references see e.g. I. Hinchliffe, *Phys. Rev.* **D54** (1996) 77.
76. A. Belavin and A. Migdal, *Pis'ma ZhETF* **19** (1974) 317 *[JETP Lett.* **19** (1974) 181]; *Scale Invariance and Bootstrap in the Non-Abelian Gauge Theories,* Landau Institute Preprint-74-0894, 1974 (unpublished).
77. T. Banks and Zaks, *Nucl. Phys.* **B196** (1982) 189.
78. N. Seiberg, *Nucl. Phys.* **B435** (1995) 129.
79. D. Kutasov, A. Schwimmer, and N. Seiberg, *Nucl. Phys.* **B459** (1996) 455.
80. G. 't Hooft, *Phys. Rev.* **D14** (1976) 3432; (E) **D18** (1978) 2199.
81. P. Pouliot, *Phys. Lett.* **B359** (1995) 108; D. Kutasov, *Phys. Lett.* **B351** (1995) 230; D. Kutasov and A. Schwimmer, *Phys. Lett.* **B354** (1995) 315; P. Pouliot and M. Strassler, *Phys. Lett.* **B370** (1996) 76; **B375** (1996) 175; K. Intriligator, *Nucl. Phys.* **B448** (1995) 187; K. Intriligator, R. Leigh and M. Strassler, *Nucl. Phys.* **B456** (1995) 567.
82. J. Distler and A. Karch, hep-th/9611088.

83. J. Brodie and M. Strassler, hep-th/9611197; P. Cho, *Phys. Rev.* **D56** (1997) 5260; and references therein.
84. G. Dvali and M. Shifman, *Phys. Lett.* **B396** (1997) 64; (E) **B407** (1997) 452.
85. R. Haag, J. Lopuszanski and M. Sohnius, *Nucl. Phys.* **B88** (1975) 257.
86. S. Coleman and J. Mandula, *Phys. Rev.* **159** (1967) 1251.
87. E. Witten and D. Olive, *Phys. Lett.* **B78** (1978) 97.
88. S. Ferrara and B. Zumino, *Nucl. Phys.* **B87** (1975) 207.
89. B. Chibisov and M. Shifman, hep-th/9706089 [Phys. Rev. D, to appear).
90. E. Bogomol'nyi, *Sov. J. Nucl. Phys.* **24** (1976) 449, M.K. Prasad and C.H. Sommerfield, *Phys. Rev. Lett.* **35** (1976) 760.
91. G. Dvali and M. Shifman, hep-th/9611213 [Nucl. Phys. B, to appear].
92. M. Shifman, hep-th/9708060 [Phys. Rev. D, submitted]; M. Shifman and M. Voloshin, hep-th/9709137 [Phys. Rev. D, submitted].
93. A. Kovner, M. Shifman and A. Smilga, hep-th/9706089 [Phys. Rev. D, submitted]; A. Smilga and A. Veselov, hep-th/9710123.
94. E. Witten, hep-th/9706109.
95. V.B. Berestetskii, E.M, Lifshits and L.P. Pitaevskii, Quantum *Electrodynamics,* (Pergamon Press, New York, 1982).

PHASES OF SUPERSYMMETRIC GAUGE THEORIES

Adam Schwimmer

Department of Physics, Weizmann Institute
Israel

SUMMARY

During the last years remarkable progress was made in deriving exact results for the dynamics of supersymmetric gauge theories. In particular the knowledge of the low energy effective action characterizes unequivocally the phases which appear in these theories.

In the lectures presented at the School we reviewed the field trying to emphasize this aspect, i.e. the variety and characterstics of the phases gauge theories can be in. Though some of the features appearing are undoubtedly a consequence of the high supersymmetry these models posess, we believe that the lessons learned could be relevant for QCD.

All the interesting phases can be realized in the $N = 2$ supersymmetric gauge theories with gauge group $SU(2)$ studied by Seiberg and Witten. We list the non-perturbative information about these systems.

- In the $N = 2$ $SU(2)$ gauge theory without matter[1] the effective action has two singularities corresponding to points where magnetic monopoles and dyons become massless, respectively. When a small perturbation breaking the $N = 2$ supersymmetry to $N = 1$ is added the massless particles condense. The two phases obtained this way give a first exact realization in the continuum of the confining phase proposed by 't Hooft and Mandelstam and of the oblique confinement phase proposed by 't Hooft.

- Considering $N = 2$ $SU(2)$ gauge theory with e.g. three hypermultiplets in the fundamental representation[2] there are points where monopoles carrying nontrivial representation of the global symmetry group become massless. When such monopoles condense the global symmetry is spontaneously broken. This mechanism could be relevant for the general problem of breaking chiral symmetries in gauge theories.

- Particularly interesting and novel are the Argyres-Douglas[3] points where mutually nonlocal objects (i.e. objects which cannot be simultaneously described in the same "picture" of the electromagnetic field) condense. These points can be realized in the framework of $SU(2)$ theories[4] by finetuning the mass parameters of the hypermultiplets in such a way that two singularities collide. Such a point necessarilly produces a superconformal theory and some information about the anomalous dimensions of the primary fields can be extracted.

The lectures followed closely the original articles[1, 2, 3, 4] and used extensively the excellent reviews available[5, 6, 7, 8].

REFERENCES

1. N. Seiberg, E. Witten, Monopole Condensation and Confinement in $\boldsymbol{N} = \mathbf{2}$ Supersymmetric Yang-Mills Theory, hep-th/9407087, *Nucl.Phys.* B426:19 (1994).
2. N. Seiberg, E. Witten, Monopoles, Duality and Chiral Symmetry Breaking in $\boldsymbol{N} = \mathbf{2}$ Supersymmetric QCD, hep-th/9408099, *Nucl. Phyg.* B431:484 (1994).
3. P.C. Argyres, M.R. Douglas, New Phenomena in $SU(3)$ Supersymmetric Gauge Theory, hep-th/9505062, *Nucl. Phys.* B448:93 (1995).
4. P.C. Argyres, M.R. Plesser, N. Seiberg, E. Witten, New $\boldsymbol{N} = \mathbf{2}$ Superconformal Field Theories in Four Dimensions, hep-th/9511154, *Nucl. Phys.* B461:71 (1996).
5. A. Bilal, Duality in $\boldsymbol{N} = \mathbf{2}$ SUSY $SU(2)$ Yang-Mills Theory, hep-th/9601077.
6. L. Alvarez-Gaume, S.F. Hassa, Introduction to S-Duality in $\boldsymbol{N} = \mathbf{2}$ Supersymmetric Gauge Theory, hep-th/9701069.
7. W. Lerche, Introduction to Seiberg-Witten Theory and its Stringy Origin, hep-th/9611190.
8. M. Peskin, Duality in Supersymmetric Yang-Mills Theory, hep-th/9702094.

INDEX

Kolibri 360 GmbH
Ferdinand-Jühlke-Straße 7
99095 Erfurt, Deutschland
produktsicherheit@kolibri360.de